Radiative Processes
in Discharge Plasmas

NATO ASI Series

Advanced Science Institutes Series

A series presenting the results of activities sponsored by the NATO Science Committee, which aims at the dissemination of advanced scientific and technological knowledge, with a view to strengthening links between scientific communities.

The series is published by an international board of publishers in conjunction with the NATO Scientific Affairs Division

A	Life Sciences	Plenum Publishing Corporation
B	Physics	New York and London
C	Mathematical and Physical Sciences	D. Reidel Publishing Company Dordrecht, Boston, and Lancaster
D	Behavioral and Social Sciences	Martinus Nijhoff Publishers
E	Engineering and Materials Sciences	The Hague, Boston, Dordrecht, and Lancaster
F	Computer and Systems Sciences	Springer-Verlag
G	Ecological Sciences	Berlin, Heidelberg, New York, London,
H	Cell Biology	Paris, and Tokyo

Recent Volumes in this Series

Volume 142—New Vistas in Electro-Nuclear Physics
edited by E. L. Tomusiak, H. S. Caplan, and E. T. Dressler

Volume 143—Atoms in Unusual Situations
edited by Jean Pierre Briand

Volume 144—Fundamental Aspects of Quantum Theory
edited by Vittorio Gorini and Alberto Frigerio

Volume 145—Atomic Processes in Electron-Ion and Ion-Ion Collisions
edited by F. Brouillard

Volume 146—Geophysics of Sea Ice
edited by Norbert Untersteiner

Volume 147—Defects in Solids: Modern Techniques
edited by A. V. Chadwick and M. Terenzi

Volume 148—Intercalation in Layered Materials
edited by M. S. Dresselhaus

Volume 149—Radiative Processes in Discharge Plasmas
edited by Joseph M. Proud and Lawrence H. Luessen

Series B: Physics

Radiative Processes in Discharge Plasmas

Edited by
Joseph M. Proud
GTE Laboratories Incorporated
Waltham, Massachusetts

and
Lawrence H. Luessen
Naval Surface Weapons Center
Dahlgren, Virginia

Springer Science+Business Media, LLC

Proceedings of a NATO Advanced Study Institute on
Radiative Processes in Discharge Plasmas,
held June 23–July 5, 1985,
in Pitlochry, Perthshire, Scotland

Library of Congress Cataloging in Publication Data

NATO Advanced Study Institute on Radiative Processes in Discharge Plasmas
(1985: Pitlochry, Perthshire)
   Radiative processes in discharge plasmas.

  (NATO ASI series. Series B, Physics; vol. 149)
  "Proceedings of a NATO Advanced Study Institute on Radiative Processes in Discharge Plasmas, held June 23–July 5, 1985, in Pitlochry, Perthshire, Scotland"—T.p. verso.
  "Published in cooperation with NATO Scientific Affairs Division."
  Includes bibliographies and index.
  1. Plasma radiation—Congresses. 2. Electric discharges through gases—Congresses. 3. Plasma diagnostic—Congresses. I. Proud, Joseph M. II. Luessen, Lawrence H. III. North Atlantic Treaty Organization. Scientific Affairs Division. IV. Title. V. Series: NATO ASI series. Series B, Physics; v. 149.
QC718.5.R3N38   1985   530.4′4   86-30563

ISBN 978-1-4684-5307-2   ISBN 978-1-4684-5305-8 (eBook)
DOI 10.1007/ 978-1-4684-5305-8

© 1986 Springer Science+Business Media New York
Originally published by Plenum Press, New York in 1986
Softcover reorint of the hardcover 1st edition 1986

All rights reserved. No part of this book may be reproduced, stored in a retrieval system,
or transmitted in any form or by any means, electronic, mechanical, photocopying,
microfilming, recording, or otherwise, without written permission from the Publisher

PREFACE

An Advanced Study Institute on Radiative Processes in Discharge Plasmas was held at the Atholl Palace Hotel, Pitlochry, Perthshire, Scotland, June 23 through July 5, 1985. This publication is the Proceedings from that Institute.

The Institute was attended by eighty-five Participants and Lecturers representing the United States, Canada, France, West Germany, Greece, The Netherlands, Portugal, Turkey, the United Kingdom, and Switzerland.

A distinguished faculty of eighteen Lecturers was assembled and the topical program organized with the assistance of an Advisory Committee composed of: Dr. John Waymouth, USA; Dr. Timm Teich, Switzerland; Dr. Arthur Phelps, USA; Dr. Nicol Peacock, England; Professor Erich Kunhardt, USA; Dr. Anthony Hyder, USA; and Dr. Arthur Guenther, USA.

The underlying theme and objective of the Institute was the enhancement of scientific communication and exchange among academic, industrial, and national laboratory groups having a common concern for radiative processes in discharge plasmas. The program was organized into four major sessions sequentially treating: the fundamental science of visible and near-visible radiation in plasmas; the technology of discharge light sources; recent and novel methods for the generation of plasmas; and an update on advances in laser-based diagnostics. Each major session culminated in a panel discussion comprised of the Lecturers for that session.

The initial pace of the Institute allowed ample time for informal discussion sessions to be organized and scheduled by the Participants and Lecturers with the encouragement and assistance of the Directors. As the Institute progressed, the interest and demand for these additional sessions grew to consume virtually all of the unscheduled time.

An innovation which greatly enhanced this Institute was the utilization of Poster Sessions to provide a forum for contributed work. A total of thirty-five poster papers were presented in two sessions, which were placed early in the program to allow continued discussions over the duration of the Institute. An important benefit of the posters was the catalytic effect upon the technical interaction among the Participants. A list of these posters and their authors is presented in Appendix A.

We are grateful to a number of organizations for providing the financial assistance that made the Institute possible. Foremost is the NATO Scientific Affairs Division, which provided the single most financial contribution for the Institute. In addition, the following US sources made contributions: Naval Surface Weapons Center, GTE Laboratories, Air Force Office of Scientific Research, and Office of Naval Research.

We would also like to thank the management and staff of The Atholl Palace Hotel for a truly enjoyable and memorable two weeks in the Scottish Highlands. The accomodations, food, service, and meeting facilities were superb. And to the Tenth Duke of Atholl, our thanks for the use of Blair Castle for a magnificent banquet.

Our appreciation to Josephine Antonellis for her invaluable administrative assistance to the Institute, to the EG&G Washington Analytical Services Center in Dahlgren, Virginia, which had the task of centrally retyping every Lecturer's manuscript and producing a camera-ready document for delivery to Plenum, and to Susie M. Anderson of EG&G for once again producing an exceptional product.

And finally to the people of Perthshire, Scotland, who certainly displayed, in every way, 'Ceud Mile Failte' (Gaelic for 'A Hundred Thousand Welcomes').

Joseph M. Proud
GTE Laboratories
Waltham, Massachusetts

Lawrence H. Luessen
Naval Surface Weapons Center
Dahlgren, Virginia

August 1986

CONTENTS

INTRODUCTION

Overview 1
    J. F. Waymouth

Tutorial Lecture - Radiative and Collisional Processes
    in Plasmas 7
    H. R. Griem

PLASMA RADIATION FUNDAMENTALS

Radiative Processes in Discharge Plasmas:
    Questions and Issues 27
    J. H. Ingold

Light Scattering 39
    H. -J. Kunze

Emission Spectroscopy 55
    H. -J. Kunze

Pulsed Optical Pumping in Low Pressure
    Mercury Discharges 65
    P. van de Weijer and R. M. M. Cremers

Transient Phenomena in High Pressure Discharges 95
    E. Marode, F. Bastien and G. Hartmann

Molecular Spectral Intensities in LTE Plasmas 147
    D. O. Wharmby

Self-Reversed Emission Lines in Inhomogeneous
    Plasmas 171
    D. Karabourniotis

Radiation Transport in High Pressure Discharge Lamps 249
    M. A. Cayless

DISCHARGE LIGHT SOURCES

Discharge Light Sources 277
    J. F. Waymouth

Low-Pressure Mercury and Sodium Lamps 309
    A. G. Jack

High-Pressure Sodium (HPS) Arcs 327
    D. O. Wharmby

Metal Halide Sources 347
    J. H. Ingold

PLASMA GENERATION

Laser-Produced Plasmas 363
    C. Grey Morgan

Plasmas Sustained by Surface Waves at Microwave and
    RF Frequencies: Experimental Investigation
    and Applications 381
    M. Moisan and Z. Zakrzewski

Plasmas Sustained by Surface Waves at Radio and
    Microwave Frequencies: Basic Processes
    and Modeling 431
    C. M. Ferreira

RECENT DIAGNOSTIC ADVANCES

Gas Discharge Laser Diagnostics Update 467
    A. Garscadden

Laser (and Other) Diagnostics of RF Discharges 495
    C. E. Gaebe and R. A. Gottscho

Optogalvanic Effects in the Cathode Fall 525
    J. E. Lawler, D. K. Doughty, E. A. Den Hartog
    and S. Salih

Rydberg States: Properties and Applications to
    Electrical Discharge Measurements 547
    A. Garscadden

Resonance Ionization Spectroscopy 569
    C. Grey Morgan

Appendix A: Poster Papers 577

Appendix B: Organizing Committee, Lecturers,
    and Participants 581

Index 589

OVERVIEW

John F. Waymouth

GTE Lighting Products
Sylvania Lighting Center
Danvers, MA, USA

INTRODUCTION

> In The Beginning
> All Was Darkness and Chaos
> and God Said,
> "Let There Be Light"
> And There was Light, and the Light was Good

So wrote the Hebrew philosophers-prophets-scholars in about 1000 BC or earlier, in a description of Creation that is so prescient as to give great credence to the belief of both the Jewish and Christian religions that it was divinely inspired. For, in the beginning, **Radiation** was all there was: Pure energy, in the "Big Bang", which gradually congealed into matter.

Even today, electromagnetic radiation is the most pervasive feature of the universe, far more common than matter. We happen to live and move and have our being on one of the few isolated blobs of the universe that is cold enough not be be plasma, where solid, liquid, and gaseous matter locally outweighs radiation. Yet still we are illuminated by radiation day and night: primordial 3 K radiation, peaking in the microwave region of the spectrum, left over from the "Big Bang"; solar radiation during the day, stellar radiation (the same thing, only from further away) and reflected solar radiation at night; and increasingly, radiation of our own making.

Man first made radiation when he discovered fire, and created incandescent particles of carbonaceous soot at temperatures ca 1000 K. From the dawn of human history until just a few hundred years ago, man-made radiation was entirely from flames. In 1809, Sir Humphrey Davy touched two charcoal sticks together that were connected to "Volta's Crown of Cups", and made the first carbon arc in air, and the science of plasma physics and the technology of plasma radiation sources were born at the same instant.

Of course, Man had been studying radiation and light as an entity in itself long before that, showing the colors of sunlight and the presence therein of invisible ultraviolet and infrared radiation, without knowing how to make it. So the technology of making radiation, and the science of understanding it, proceeded along separate but parallel paths throughout the Nineteenth Century: the former motivated by Man's crying need for light he could control, the latter by his crying need to satisfy his curiosity.

A third thread in Man's interaction with radiation occurred in the Nineteenth Century when Fraunhofer spotted the absorption lines in the spectrum of the Sun. When Bunsen and Kirkhoff created the science of spectroscopy, it was recognized that all were looking at the same thing! Radiation was providing **Information** about inaccessible distant sources. Later, similar absorption lines were seen in the spectrum of nearby stars, which Doppler's principle showed to be moving away from us.

Shortly thereafter, Maxwell conceived the equations that bear his name, showing that electromagnetic radiation needed no material medium for its propagation, and Max Planck applied the quantum hypothesis to the interaction of radiation with matter, deducing the equation for blackbody radiation. With Bohr, Schrodinger, Sommerfeld, and Heisenberg providing the quantum mechanical explanation of atomic and molecular energy levels and the origin of the discrete nature of their radiations, the use of radiation as a diagnostic probe, foreshadowed by Fraunhofer, became a science.

In the hands of the astrophysicists, the tools developed by Planck and Schrodinger were incorporated into the development of radiation-transport theory and line-broadening theory to enhance further the utility of emitted radiation to provide information about the farthest reaches of the Universe. Wood (1902, 1903), Mitchell and Zemansky (1934, 1961), and Kenty (1932) laid the foundations for the understanding of the unique behavior of "Resonance Radiation", that radiation emitted by excited states of atoms but simultaneously absorbed by the much more numerous ground states.

The year 1939 marked the commercial introduction of the fluorescent lamp, a plasma light source, perhaps the most important device ever invented in Man's empirical quest for better and cheaper sources of optical radiation, **Light**. And every aspect of the physical phenomena in the plasma of this light source is dominated by the behavior of resonance radiation, the one so closely identified with R. W. Wood that French publications of the time referred to it as "La Lumiere d'Wood", the mercury 253.7-nm resonance line.

With the contributions of Holstein (1951) to the theory of resonance radiation transport, the use of optical absorption of mercury lines by Kenty (1950) to measure metastable concentrations in a fluorescent lamp plasma, and the development of the theory of LTE arcs and its application to high-pressure-mercury arc light sources by Elenbaas (1951), the three threads of this thumbnail history of Man's involvement with radiation come together:

1) The generation of radiation, especially light, for which plasmas offer some unique advantages, but encompass great complexity requiring in-depth knowledge of radiation interactions to understand;

2) The use of various features and characteristics of emitted radiation for non-intrusive diagnostics of physical and chemical processes in radiation sources, usually plasmas, which may be extemely remote, or otherwise inaccessible;

3) The use of radiation from external sources to interact with plasmas to provide information about plasma processes.

These three, of course, form the basis for this ASI.

You will learn, from experts in the field, a great deal about the generation of radiation by plasmas, and about the processes involved therein. You will learn, again from experts, about the unique "signatures" observable in the spectral characteristics of emitted radiation, and what you can deduce about the source therefrom. And you will learn, once more from experts, about the interaction of external radiation with plasmas and the unique information that it provides.

Plasma light sources (otherwise known as "Electric Discharge Lamps") will figure heavily in the program, for several reasons. First, they are the most pervasive radiation generators on earth, being the only work of man observable from outer space. Second, the study of radiation physics, emission, absorption, transport, and diagnostics has become an integral part of discharge light source R&D, so that they provide many illustrative examples of the fundamental concepts to be presented. And finally, the community of researchers in this field has at last emerged from empiricism into making significant contributions to the science of these processes.

It may be of some interest to review briefly what the apparent impact of man-made discharge sources might be to an observer viewing the Earth from outer space. I have estimated, in very round numbers, that there are approximately 2 billion fluorescent lamps in service in the world, not counting the USSR and mainland China, at an average wattage of about 30 watts. And there are approximately 0.2 billion high-pressure (HID) lamps in service around the world, with an average wattage of about 200 watts. Of the total electrical power they consume, 100 billion watts if they were all burning at once, approximately two-thirds is radiated, and about one half of that is in the visible part of the spectrum, say 33 billion watts, (3.3×10^{10} watts emitted in a 300-nanometer band centered around 550 nanometers). Perhaps one-third of them are in operation at any one time, say 1.1×10^{10} watts/300 nanometers, or 0.37×10^{8} watts per nanometer.

The total surface area of the earth is about 2×10^{8} square miles, of which one-seventh is land, making about 3×10^{7} square miles of land area, with 2.6×10^{10} cm^2/sq mi, a total of about 7.2×10^{17} square centimeters of land area. The spectral exitance of the land area of the earth becomes about 0.5×10^{-10} watts/cm^2-nm.

The temperature of a blackbody that would have a spectral exitance of 0.5×10^{-10} watts/cm^2-nm around 550 nanometers is approximately 825 K, or 550 C.

It can certainly be argued that only a fraction of the total emitted light escapes to outer space; much of it is radiated indoors, and of that which is radiated out of doors, only a fraction is reflected upwards. On the other hand, the surface area employed in the above calculation is the entire land area of the world, including Russia and China, as well as Antarctica and the underdeveloped world. It is probably safe to say that 90% of the above radiation is emitted by 10% of the land area, so at least a fraction of the Earth's land area must look to an observer out there as if it has a surface temperature approaching 1000 K! It would be recognized, however, as a decidedly non-thermal radiator, becuase the color temperature, in the visible, of the emitted radiation is 2000-3000 K, and the Earth as a whole radiates in the IR as a 300-K blackbody, or thereabouts. This would cause no problems in interpretation, however, to any civilization capable of measuring the radiations from the dark side of the Earth in the face of the enormously greater spectral exitance of the Sun. Such a civilization would itself be already well equipped with artificial illumination and would clearly recognize such a radiation pattern as a signal of the existence of intelligent life. So, if anybody is watching, it is unnecessary for us to send coded radio signals into outer space to advertise our presence. We are already providing unmistakable evidence of it.

RADIATIVE PHENOMENA IN DISCHARGES

Radiative phenomena in plasma discharges span an enormous range, from those in which every photon created within the plasma escapes and the rate-limiting step is the creation of excited states, to those in which radiation is so thoroughly imprisoned that essentially all forward processes are balanced by reverse processes, the system is in thermal equilibrium, and the rate limiting process is radiation escape. We shall find examples spanning this entire range in the science and technology of Discharge Light Sources, and they will illustrate many of the concepts that will be presented in the portion of the ASI dealing with fundamental processes. Since radiation is the raison d' etre of Discharge Light Sources, and radiation plays the dominant role in their energy balances, incorporation of detailed understanding of the generation and transport of radiation is essential to the calculation of those energy balances. The latter, in turn, is essential to the determination of the limits to performance, and the proper directions of parameter modification for improved performance. The fundamental physics of radiation emission, absorption, and radiation transport in plasmas will be reviewed for us by Professor Griem.

The application of detailed modeling calculations to Discharge Light Sources is a relatively recent development that had its inception in the pioneering work of Kenty (1950) in calculating rate processes in the plasma of the mercury-rare-gas discharges (used in the fluorescent lamp). This work was extended in the form of a model to explore the influence of adjustable parameters by Bitter and Waymouth (1956), and still further extended and improved by Cayless (1963), expanded in detail by Vriens and co-workers (1973, 1978), and more recently by Lagushenko (1984). Steadily improving models of the key factor - the escape of mercury resonance radiation - have contributed in part to the improved accuracy of the successive discharge models.

It is, of course, the richness and complexity of such radiation sources that has made the discharge model such an important tool in light source development, and has forced the Discharge Light Source Industry to at least supplement its prior empirical mode of investigation (parenthetically, it was the Lighting Industry that gave the world the term "Edisonian Research"). There are five major external parameters involved in a fluorescent lamp, which control about a dozen internal variables, which in turn determine the values of a dozen or so output variables. Since changing any one of the external parameters changes all the internal and output variables, the power of a reasonably accurate arc model to speed up the testing of individual and combined effects of parameter variations is evident. More importantly, the ability of computer models to rapidly present graphic multidimensional results of different sections through a parameter space is an enormous aid to intuition (Smarr, 1985).

Needless to say, the understanding of the transport of resonance radiation in such a plasma (as well as in the low-pressure-sodium discharge plasma) is a major factor in such model calculations, and we shall hear more about this in the presentations of Dr. Jack and Dr. van de Weijer.

Other complexities are evident in the role of radiation generation and transport in Discharge Light Sources involving plasmas at higher pressures (one atomsphere and up), such as the High-Pressure-Sodium (HPS) and Metal Halide (MH) lamp. Both optically thick (strongly-absorbed) and optically thin (non- or weakly-absorbed) radiation transport is a major factor in such devices. Local thermal equilibrium is usually present in the main body of the plasma, but non-LTE effects in radiation transport also exist. Stark, van der Waals, collisional, and resonance broadening play a significant role in the radiation escape mechanisms. Study of such lamps, moreover, has contributed significantly to the understanding of far-wing broadening mechanisms and transient quasi-molecular radiation processes. Useful arc model calculations embodying some of these effects are in existence for HPS lamps, but are much less well established for MH lamps. The presentation of Messrs Cayless, Wharmby, and Ingold will address these considerations.

The fundamental science involved in the transient phenomena in high pressure plasmas will be presented by Dr. Marode.

In contrast to such plasma radiation sources, in which electrical energy input maximizes in the center of the plasma, while radiation is emitted from the surface, we shall also hear about the unique family of plasmas created by surface-wave excitation, in which the electrical energy input maximizes at or near the surface as does the radiation energy loss. I look forward to learning how this reversal of the usual situation affects the radiative properties of the plasma from the presentations of Professors Ferreira and Moisan.

Another unique form of interaction between radiation and a plasma is the creation of plasma by radiation itself at the focus of a high-intensity laser beam, which will be described by Professor Morgan.

INFORMATION FROM EMITTED RADIATION

As already mentioned, there is a wealth of information about the source contained in the radiation emergent from an emitting medium, information that can be obtained from a careful analysis of the absolute and relative intensities as a function of wavelength. Emission and absorption wavelengths permit deduction as to species present; absolute exitances or spectral radiances provide information as to temperatures. Weekly or non-radiating species may give a clue as to their presence by their perturbation of radiating species in the form of the various line-broadening mechanisms which may be present.

Virtually all the information about radial temperature profile in High-Pressure Discharge Light Sources comes from the measurement and analysis of spectral radiance in selected emission lines. Information about plasma chemistry in such sources also comes from spectral measurements identifying concentrations of various species as a function of radial position. Information about molecular and free-radical, or excimer, species comes from their spectral signatures, either isolated or superposed as anomalous broadening on emission lines of principal radiators.

Professors Griem and Kunze will present the fundamentals of these processes, and applications will be evident in the presentations by Professor Karabourniotis, and by Messrs Wharmby, Cayless, Ingold, and Jack.

INFORMATION FROM PROBING RADIATION

A relatively old technique for studying radiation interactions with plasmas, that of studying their absorption and scattering, has been revolutionized in the last decade or so by the availability of a wide selection of high-spectral-radiance, narrow-line-width, tunable lasers, creating a whole new area of science: Laser Plasma Diagnostics. Information available is the wavelength dependence and strength of the absorption or scattering itself, which provides both qualitative and quantitative information about the species present. Entirely new information is available from a determination of the impact of the resultant perturbation on the plasma itself ("optogalvanic effect" and "laser-induced fluorescence"), as well as the rate of relaxation of those effects after cessation of the perturbation, both of which may provide information about otherwise inaccessible rate constants in the plasma. The ease with which laser radiation may be focussed onto a small spot size permits most impressive spatial resolution for all of these techniques.

Professor Kunze and Dr. Garscadden will describe the fundamentals, and applications will be apparent in presentations by Gottscho, Lawler, Garscadden, and Morgan.

POPULATION INVERSIONS IN PLASMAS

Finally, in non-equilibrium plasmas, a fortuitous combination of rate processes may lead to inverted state populations, offering the possibility for laser oscillation in a cavity with optical feedback. This then completes the marriage of the several threads talked about earlier, of radiation generation, and obtaining information from radiation, both emitted and probing. In passing to the limit, the probed plasma may someday provide the primary laser pump to excite the dye laser probing it, and the "side light" radiation provide additional information.

Dr. Peacock will describe the processes by which population inversions may be created in plasmas in his special seminar on this topic.

REFERENCES

Cayless, M. A., 1963, Brit. J. Appl. Phys., 14:863.
Elenbaas, W., 1951, "The High Pressure Mercury Vapor Discharge", North Holland Publishing Co., Amsterdam.
Holstein, T., 1951, Phys. Rev. 83:1159.
Kenty, C., 1932, Phys. Rev. 42:823 (1932).
Kenty, C., 1950, J. Appl. Phys. 21:1309.
Lagushenko, R., 1984, J. IES 14:306.
Mitchell, A. G. and Zemansky, M. W., 1934 and 1961, "Resonance Radiation and Excited Atoms", Cambridge University Press, London.
Smarr, L. L., 1965, Science 228:403.
Vriens, L., 1973, J. Appl. Phys. 44:3980; 1978, J. Appl. Phys. 49:3814
Vriens, L., Keijser, R. A. J. and Ligthart, F. A. S., 1978, J. Appl. Phys. 49:3807.
Waymouth, J. F. and Bitter, F., 1955, J. Appl. Phys. 27:122.
Wood, R. W., 1902, Phil. Mag. 3:128; 1903, Phil. Mag. 6:362.

TUTORIAL LECTURE -

RADIATIVE AND COLLISIONAL PROCESSES IN PLASMAS

H. R. Griem

Laboratory for Plasma and Fusion Energy Studies
University of Maryland
College Park, MD, USA

INTRODUCTION

Plasmas of moderate temperatures, say less than 10^5 K (~ 9 eV), and particle densities, say less than 10^{19} cm^{-3}, occur in a fair number of devices. They are therefore of considerable interest, notwithstanding the fact that higher temperature plasmas have attracted more attention because of their importance for magnetically confined fusion plasmas. As a matter of fact, plasma conditions as desired for inertially confined fusion plasmas seem even more interesting because of their much higher densities and the importance of electromagnetic radiation for their behavior and diagnostics.

Returning to more moderate plasma conditions, it is always useful to remember the similarities among plasmas in stellar atmospheres, and also the differences. While temperatures are quite similar, chemical composition and densities are likely to be very different. Of course, there are indeed devices operating mostly with a hydrogen filling, but more often than not the emphasis is on much heavier and more complex atoms, e.g., argon or even mercury. Densities tend to be much higher in laboratory devices or experiments, a difference which is in many respects more than made up by the large spatial extent of stellar atmospheres and the presence of a strong radiation field over a broad range of the electromagnetic spectrum. These various factors result in a comparatively simple situation for the radiative-collisional coupling problem in laboratory plasmas.

Many stars and laboratory plasmas have sufficiently high temperatures that thermal dissociation of H_2 and other molecules is essentially complete. For this reason and because of the great complexity of molecular spectra and collisional processes, dissociation will be assumed to be complete. At least in the boundary regions of laboratory plasmas, this assumption is certainly not valid. This fact is not only obvious from the presence of many-line molecular spectra, but also the reason for some interesting increases in atomic line intensities (Kunc, 1985) and greatly enhanced Doppler broadening (Freund et al.,1976; Higo and Kamata, 1982).

The following two sections contain descriptions and basic quantitative relations for radiative and collisional processes. A third section consists of a discussion of methods for self-consistent solutions of radiative transfer and collisional rate equations, including, e.g., photo-excitation by inelastic photon-atom collisions.

RADIATIVE PROCESSES

There are three distinct ways in which radiation and matter interact by first-order processes so as to change the number of photons present in some beam, namely, spontaneous emission, stimulated emission, and absorption. The corresponding Einstein coefficients, A (for spontaneous emission) and B (for stimulated emission or absorption), for transitions between a given pair of states (of degeneracies or statistical weights g_n and g_m) are related to each other through the celebrated Einstein relations

$$A_{nm} = \frac{\hbar \omega^3}{4\pi^3 c^2} \frac{g_n}{g_m} B_{mn} = \frac{\hbar \omega^3}{4\pi^3 c^2} B_{nm}, \qquad (1)$$

which can be derived from the requirement that in the case of thermodynamic equilibrium between matter and radiation the spectral intensity $I(\omega)$ should approach the Planck distribution

$$I_p(\omega) = \frac{\hbar \omega^3}{4\pi^3 c^2} [\exp(\frac{\hbar \omega}{kT}) - 1]^{-1}. \qquad (2)$$

The angular frequency obeys Bohr's condition

$$\hbar \omega = E_m - E_n, \qquad (3)$$

if m and n designate upper and lower states of the transition, while the B-coefficients give the corresponding transition rates, per atom or ion in the appropriate initial state, through

$$dW_{mn} = \frac{1}{4\pi} B_{mn} I_\Omega(\omega) \, d\Omega \qquad (4a)$$

$$dW_{nm} = \frac{1}{4\pi} B_{nm} I_\Omega(\omega) \, d\Omega \qquad (4b)$$

in terms of the actual intensity which generally depends on the direction Ω (solid angle). Also, if this intensity varies substantially over the actual frequency range covered by the transition (see Line Profiles), a corresponding frequency-normalized line shape factor $L(\omega) \, d\omega$ should be inserted. Higher order photon-atom interactions, e.g., scattering, are discussed separately (Kunze, 1985).

Line Emission

To relate the rate of spontaneous emission to (measurable) intensities, one must first consider the emission coefficient (power per unit volume, solid angle, and angular frequency interval)

$$\varepsilon(\omega) = \frac{\hbar \omega}{4\pi} A_{nm} N_m L(\omega) \qquad (5)$$

assuming for the time being that a single line dominates in the frequency range of interest. This formula is self-evident, but relating N_m, the density of atoms or ions in the initial state to the general plasma conditions can be quite difficult (see COLLISIONAL PROCESSES). Also, the line shape function appropriate for emission (or absorption) coefficients, $L(\omega)$, is normally not directly reflected by the observed line shape. This complication not only arises because of changes in conditions along the line of sight, but possibly also from reabsorption (opacity broadening, see Radiative Transport).

For diagnostic applications it is usually preferable to use lines for which absorption and induced emission are negligible. In such cases emerging intensities can simply be calculated by integrating along the path taken by the ray traversing the plasma,

$$I(\omega) = \int \varepsilon(\omega)dx + I_i(\omega) = \frac{\hbar\omega}{4\pi} A_{nm} \int N_m L(\omega)dx + I_i(\omega) \, , \qquad (6)$$

with the incident intensity $I_i(\omega)$ in the proper direction at the far side of the plasma. In dense and pulsed plasmas, observable lines are frequently substantially modified by radiation transport, e.g., in laser produced plasmas (Tondello et al., 1977).

Knowledge of transition probabilities and frequencies (or wave lengths) can not always be taken for granted. Unknown wavelengths must frequently be measured in plasma experiments, with atomic structure theory serving as a guide to level identifications. Measurements of transition probabilities are much more difficult, and most published results (Wiese and Martin, 1980) are from atomic structure calculations (for ~ 5000 lines). Except for atoms or ions with a very small number of electrons, it is quite likely that relatively large errors will occur, especially for relatively weak lines.

A natural measure of the strength of a line is the dimensionless oscillator strength or, rather, its product with the appropriate statistical weight,

$$g_n f_{mn} = g_m f_{nm} = \frac{c}{2r_o \omega^2} g_m A_{nm} \, . \qquad (7)$$

Such gf values for some FeX (or Fe^{9+}) and FeXI (or Fe^{10+}) lines as calculated by Mason (1975) and by Bromage et al. (1977) are shown in Table 1 to illustrate the sensitivity to approximations made in typical atomic structure calculations. The remarks on Table 1 are mostly meant for allowed electric dipole transitions involving only singly-excited levels. Lines arising from various so-called forbidden transitions can also be useful for diagnostics, especially for density measurements. Because of their relatively small transition probabilites such lines can only be seen if their upper levels are substantially over-populated compared to levels giving rise to allowed lines. Such deviatlons from essentially statistical populations in turn suggest that collisional excitation energy transfer is not too important, a situation consistent with relatively low electron density.

A third general type of line radiation is associated with doubly-excited states, which involve instead of the usual single optical electron designated by quantum numbers n and ℓ, two such electrons $n\ell$, $n'\ell'$, outside of some more or less frozen core of inner electrons. The additional $n'\ell'$ electron acts mostly as a spectator, but of course causes some change in the transition energies of the $n\ell$ electron, almost always to lower energies. The corresponding lines appear therefore as dielectronic satellites on the long wavelength side of the parent lines produced by transitions of the $n\ell$ electron in ions without the additional $n'\ell'$ electron. The ratio of the two ion populations is mostly a function of electron temperature so that the relative intensities of satellite and parent lines can be used for temperature determinations.

Line Profiles

Spectral lines are not infinitely sharp nor usually exactly centered at the frequency given by Eq. (3). There are three general causes of spectral line broadening as expressed in terms of the line shape $L(\omega)$ of

Table 1. Calculated gf Values for Some Iron X and XI Lines

	Transition		Mason	Bromage et al.
Fe X	$3p^4(^1D)3d\ ^2S_{1/2}$	$- 3p^5\ ^2P_{1/2}$	0.34	0.45
		$- 3p^5\ ^2P_{3/2}$	2.35	1.51
	$3p^4(^3P)3d\ ^2D_{3/2}$	$- 3p^5\ ^2P_{3/2}$	0.347	0.115
		$- 3p^5\ ^2P_{1/2}$	4.34	3.82
	$3p^4(^3P)3d\ ^2P_{1/2}$	$- 3p^5\ ^2P_{3/2}$	0.11	0.43
		$- 3p^5\ ^2P_{1/2}$	1.76	1.40
Fe XI	$3p^33d\ ^1F_3$	$- 3p^4\ ^1D_2$	7.13	5.64
	$^3D_1$	$- 3p^4\ ^3P_0$	1.54	1.31
	$^1D_2$	$- 3p^4\ ^1D_2$	3.03	3.14
	$^3P_2$	$- 3p^4\ ^3P_1$	0.79	0.60
	$3p^33d\ ^3S_1$	$- 3p^4\ ^3P_2$	1.06	0.16
		$^3P_1$	0.3726	0.47
		$^3P_0$	0.10224	0.24
		$^1D_2$	0.0237	1.13

emission or absorption coefficients. Since at least one of the levels has only a finite life time, there is always some broadening according to Heisenberg's uncertainty principle. This natural line broadening gives rise to Lorentzian profiles whose FWHM (full-width between half-of-maximum-intensity points) in the ω-scale is given by the sum of the decay rates for upper and lower levels, including absorption and induced emission rates and also any nonradiative rates, like Auger rates in the case of inner-shell x-ray transitions. With the exception of the latter transitions, it is normally safe to neglect natural broadening because it typically amounts to only $\sim 10^{-4}$ Å.

The second general line broadening mechanism is also always present. It is due to the Doppler shifts associated with the random velocities of emitting or absorbing atoms or ions. For nonrelativistic systems only the velocity component along the line of sight matters, and for thermal plasmas the corresponding velocity distribution and, therefore, also the line profile are Gaussians. The resulting Doppler width is given by $\omega_D \sim v_{th}\omega/c$, where v_{th} is a characteristic thermal velocity. More precisely, the broadening due to thermal Doppler shifts is

$$\omega_D = 2[2kT(\ell n2)/M]^{1/2}\omega/c \tag{8}$$

in terms of the kinetic temperature T and mass M of the emitting or absorbing species. This FWHM (full width between half of maximum intensity points) width is usually orders of magnitudes larger than the natural width.

The third general line broadening mechanism is associated with the perturbations of the emitting or absorbing atoms and ions caused by other particles in the plasma. If these particles are charged, they produce electric fields which change with time more or less rapidly but vary in space only slightly over the region occupied by a given atom or ion in the appropiate states. Often this spatial variation can be entirely ignored and the actual interaction between radiating atoms or ions and the rest of the plasma be replaced by that of a dipole representing the radiating system and the local electric field. Since all radiators experience different fields and since these are time-dependent, the corresponding Stark effects effectively broaden the lines, and in the case of the quadratic Stark effect, also shift them.

Before discussing the various subsidiary approximations made in Stark broadening calculations, it is worth noting that the usual monopole-monopole interactions which dominate the scattering between plasma particles do not explicitly cause any line broadening. The strength of these interactions is the same for upper and lower levels of the line and the corresponding level shifts therefore cancel. However, since these Coulomb interactions do cause changes in the relative motions of perturbing particles and radiators, there is an important influence of interactions via the time-dependent electric field. Interactions with neutral perturbers, on the other hand, tend to be negligible in plasmas (Griem, 1964), except for very small degrees of ionization.

In principle, the response of a given radiator to the various time-sequences of fields must be calculated from time-dependent quantum-mechancal perturbation theory and the corresponding spectrum then be appropriately averaged. These calculations would not only be formidable but would also require more information on plasma fields than we now have. In particular, one would have to carefully join the more or less stochastic particle-produced fields to the wave fields associated with collective motions of the plasma. Fortunately such great detail is normally not needed because a multi-time-scale analysis offers itself.

Fields produced by electrons colliding with the radiators tend to vary so rapidly that corresponding widths and shifts depend only on net changes in the radiator states caused by such collisions. As in natural broadening, the FWHM width is now given by the appropriate collision rate from inelastic and (non-Coulombic) elastic collisions. However, generally there is also a shift which may essentially be viewed as a dynamical generalization of the quadratic Stark effect. This collision or impact theory of Stark widths and shifts has been quite successful for a large number of spectral lines (Griem, 1974), the most notable exception being the far wings of broad lines, especially of hydrogen. Since line profiles are essentially Fourier transforms of time-dependent wave functions, these far wings correspond to short times for which details of the collision dynamics are important. Inclusion of these details through relaxation or unified theories of line broadening indeed improved the agreement with experiment, without major changes in the structure of the basically Lorentzian profiles (width and shift parameters become functions of the frequency.) In either case, for so-called overlapping lines, impact profiles consist of superpositions of Lorentzians with various widths, and of certain interference terms. These superpositions are not exactly Lorentzian.

Ion-produced fields vary so slowly that their time-dependence can be neglected in some important cases, e.g., on the wings of broad hydrogen lines or lines from one-electron ions. The time interval for which accurate wave functions are needed is then $\sim \Delta\omega^{-1}$, where $\Delta\omega$ is the angular frequency separation from the unperturbed line, and this time may indeed be smaller than a relevant collision time r/v in terms of appropriate radiator-perturber separation r and relative velocity v. In such cases the quasistatic or Holtsmark theory can be used, which involves calculating Stark shifts and intensities for assumed electric (micro-) fields and then averaging the corresponding Stark patterns. For this weighted average, the probability distribution of ion micro-fields is required, which for uncorrelated perturbers is given by the Holtsmark distribution. To include such correlation and shielding effects, Hooper and co-workers have performed a series of calculations for dense plasmas consisting of a variety of ions and atoms (see Griem, 1974 and 1983 for references).

In fact, the components of the Stark patterns are also broadened and shifted by electron collisions. So-called standard calculations are therefore convolutions of impact electrons with quasistatic ion-produced profiles. For hydrogen and similar lines these two contributions tend to be comparable, whereas for other lines electron broadening usually dominates. The need to examine the validity of the quasistatic approximation is therefore greatest for the former type of lines and it was indeed found experimentally that ion-dynamical corrections are important in the line cores of lower series members. Various theoretical correction procedures (see Griem, 1979; Griem and Tsakiris, 1982; Cauble and Griem, 1983) reproduce these effects. They are also in reasonable agreement with results of the model microfield method (Brissaud and Frisch, 1971), in which the actual time-dependent field is replaced by a sequence of constant fields with magnitudes of fields and frequencies of steps chosen to give agreement in the extreme impact and quasistatic limits. This means that the model field has the same autocorrelation and distribution functions as the actual smoothly varying field.

An interesting feature of ion-dynamical effects is their correlation with Doppler broadening. The usual convolution procedure for combining Stark and Doppler broadening is therefore, strictly speaking, not valid in cases where dynamical corrections to the quasistatic approximation are important. Finally, there are situations in which higher-than-dipole multipole interactions are noticeable. Quadrupole interactions caused by spatial inhomogeneities in the field acting on hydrogen atoms are responsible for some asymmetry in the otherwise almost symmetrical profiles of lines subject to linear Stark effect. The corresponding shifts, however, tend to be smaller than shifts caused by electron collisions due to interactions involving states of different principal quantum numbers.

Continuum Radiation

Dense plasmas produce spectra which over substantial frequency ranges are continuous, i.e., have no pronounced maxima or minima, i.e., no strong emission or absorption lines. However, unless the spectrum observed is the result of a number of emission and absorption processes in an optically thick plasma, it is not simply described by the blackbody, or, Planck formula, Eq. (2). For a quantitative description of optically thin continua one must extend the relation for line emission coefficients, Eq. (5), from bound-bound to free-bound and free-free transitions. A very physical way of doing this (Griem, 1964; p. 557) is through the use of a line-shape function given by the inverse of the absolute value of the frequency separation between adjacent lines in a series, and of transition probabilities or f-numbers which are asymptotically valid for large quantum numbers (see also Omidvar, 1982). Thirdly, the density N_m of the upper level of the line is expressed in terms of the electron density and the density in the next higher ionization stage using the appropriate Saha equation (see <u>Ionization and Three-body Recombination</u> for the corresponding ionization equilibrium).

The result of these replacements is an approximate expression for the contribution of, e.g., substantially broadened and therefore merged high series members of the Balmer series to the emission coefficient at frequencies below the Balmer edge, down to frequencies corresponding to the Inglis-Teller (1939) limit. By now replacing the energy of the electron in the upper level, $-E_m/n^2$, by the Bohr relation, Eq. (3), this relation can be extrapolated beyond the usual series limit to calculate free-bound transitions into a given lower level. Summing over these levels gives the total free-bound continuum, and extension of the sum over lower levels through an integral over positive energy states also yields the free-free, or bremsstrahlung continuum.

Except for the Gaunt factors $g(\omega)$, which are often close to unity (Griem, 1964, 1983), the above procedure gives a continuum emission coefficient of

$$\varepsilon(\omega) = \frac{32(\alpha a_o)^3 z E_H}{3^{3/2} \pi^{1/2}} \left(\frac{z^2 E_H}{kT}\right)^{3/2} \left[\frac{g_n(\omega)}{n^3} \exp\left(\frac{z^2 E_H}{n^2 kT}\right) \right.$$
$$\left. + \frac{kT}{2z^2 E_H} g(\omega) \exp\left(\frac{z^2 E_H}{n^2 kT}\right) \right] N_e N_z \exp\left(-\frac{\hbar\omega}{kT}\right) \qquad (9)$$

for recombination into, and bremsstrahlung on, fully stripped ions of density N_z and nuclear charge z. The temperature in this relation is that of the electrons whose density is N_e, and E_H is the ionization energy of hydrogen (13.6 eV). Also, α and a_o are the fine-structure constant (1/137) and the Bohr radius (0.529Å), respectively. The sum over boundfree or recombination continua involves lower state principal quantum numbers such that

$$\hbar\omega \gtrsim z^2 E_H \left(\frac{1}{n^2} - \frac{1}{n_\ell^2}\right), \qquad (10)$$

and n_ℓ must be chosen according to the merging of upper levels.

For continua arising from electron interactions with incompletely stripped ions, Eq. (9) must be modified more or less drastically (Kim and Pratt, 1983), especially for recombination into the ground state. There is not only a change in the degeneracy of the final state and, in the extreme of closed shells, even a strict rule (Pauli principle) against such transitions, but also the possibility of a stronger dependence on frequency, e.g., the Cooper minimum in some recombination or photoionization cross sections.

We should also not overlook the fact that in partially ionized plasmas electron-atom interactions can cause continuum emission as well, both by recombination, i.e., formation of negative ions, and by bremsstrahlung. The most famous example is the role of H^- (negative hydrogen) in the solar atmosphere.

Radiative Transport

Only if absorption is negligible can emission coefficients be directly converted into observable intensities according to Eq. (6). Usually the underlying assumption of negligible optical depth first fails in the center of resonance lines. We must therefore consider the magnitude of the line absorption cross section,

$$\sigma_{mn} = 2\pi^2 r_o c f_{mn} L(\omega), \qquad (11)$$

where f_{mn} is the absorption oscillator strength first introduced in Eq. (7) and r_o the classical electron radius. For strong lines, $f \sim 1$, and $L(\omega)$ may be as large as the inverse of the Doppler width in Eq. (8). A numerical estimate for the maximum cross section is

$$\sigma(\omega_o) \sim 3 \times 10^{-17} \left(\frac{AE_H}{kT}\right)^{1/2} \lambda f_{mn} (cm^{-2}). \tag{12}$$

Here the wavelength λ is in Å units and A is the atomic weight of the absorbing atom or ion. Depending on the densities N_n in the lower state, the corresponding mean free paths $(N_n \sigma)^{-1}$ can be quite short. If Stark broadening is important as well, cross sections are smaller according to the ratio of actual and Doppler line shape functions. Absorption into the continuum, if at all important, would be mostly controlled by the photo-ionization cross section, which for hydrogen and one-electron ions of nuclear charge z in state n obeys

$$\sigma_n = \frac{64\alpha}{3^{3/2}} \pi a_o^2 \left(\frac{E_H}{\hbar\omega}\right)^3 \frac{z^4}{n^5} g_n(\omega), \tag{13}$$

where g_n is again the Gaunt factor ($g \sim 1$, $64\alpha\pi a_o^2/3^{3/2} \sim 0.08 \times 10^{-16} cm^2$). This process is, of course, the inverse of recombination, whereas there is no special term for absorption by inverse bremsstrahlung which is important in laser-matter interactions. The corresponding cross section can be obtained from the free-free emission coefficient using Kirchhoff's law.

In addition to absorption and spontaneous emission, it is necessary to consider induced emission. This is often done via the effective absorption coefficient, namely, in the case of a single line

$$k' = 2\pi^2 r_o c f_{mn} (N_n - \frac{g_n}{g_m} N_m) L(\omega), \tag{14}$$

where Eq. (7) was used to relate emission and absorption oscillator strengths. Also, we assumed that the same line shape could be used for both processes which is usually, but not always, true. The change of intensity due to the three processes considered (neglecting scattering) along the line of sight is in terms of ε and k', given by

$$\frac{dI}{dx} = \varepsilon - k'I \tag{15a}$$

or, introducing the optical depth τ through

$$d\tau = k'dx \tag{16}$$

given by

$$\frac{dI}{d\tau} = \frac{\varepsilon}{k'} - I \equiv S - I, \tag{15b}$$

where S is called the source function.

In general, the source function, i.e., the ratio of Eqs. (5) and (14), requires a self-consistent solution of the integral of the radiative transfer equation, i.e., Eq. (15), and the set of rate equations (see COUPLING OF RADIATION AND COLLISIONS) which control the populations in the upper and lower levels of the line. Such calculations are very demanding; they both require a large atomic data set and efficient numerical methods, in particular if they are coupled with a plasma code which predicts input parameters like electron densities and temperatures. Calculations of this kind are very important for the understanding of the radiation dynamics of non-LTE plasmas, i.e., plasmas for which local and instantaneous thermodynamic equilibrium cannot be assumed in calculations of level populations and charge state distributions. On the other hand, if densities are sufficiently high and spatial and time variations reasonably weak, then we can use Boltzmann factors to relate level populations to each other,

$$\frac{N_m}{N_n} = \frac{g_m}{g_n} \exp\left(-\frac{E_m - E_n}{kT}\right). \tag{17}$$

Criteria for the validity of this relation can be obtained using approximations for the various rate processes (see, e.g., Griem, 1964; p. 145) and are often reasonably met for resonance lines from dense plasmas. In such cases, Eqs. (5), (7), and (14) reveal that the source function S reduces to the Planck function I_p in Eq. (2), as to be expected from Kirchhoff's law.

Returning to the radiative transfer equation, we note that its formal solution is

$$I(\tau) = \int_0^\tau S(\tau') \exp(\tau' - \tau)d\tau' + I(0) \exp(-\tau), \tag{18}$$

if the optical depth is calculated from Eqs. (14) and (16) along the ray entering the plasma with intensity $I(0)$ and leaving with intensity $I(\tau)$. Only if the traversed optical depth is large and if S is constant and equal to I_p does the plasma radiate like a blackbody. In the other extreme, i.e., $\tau \ll 1$, Eq. (18) is easily seen to reduce to Eq. (6), the optically thin solution.

COLLISIONAL PROCESSES

Although transitions in excitation and ionization states are usually not directly observable, they nevertheless play a crucial role in the formation of the spectrum emitted by a plasma because they tend to control populations of radiating states. Except for photo-excitation, with a cross section corresponding to Eq. (14) without the density factor, radiation-induced transitions are usually less frequent than transitions caused by electron-ion collisions which may or may not involve photon emission as in radiative or dielectronic recombination.

Knowledge of collisional rates is required at two levels of accuracy. At the lower level of accuracy it is sufficient to find whether or not collisional rates for some process are significantly larger than radiative rates. For Maxwellian distributions of the colliding electrons the steady state solution for the system of coupled rate equations for level populations in a homogeneous plasma then gives thermodynamic equilibrium populations. Such low accuracy estimates of collisional rates can also be used to estimate permissible rates of plasma parameter variations in time and space to which the atomic states could respond without substantial deviations from steady state populations.

The purpose of the following sections is to describe the physics of the collision processes with enough realism to allow us to make estimates of the various rates sufficient for the above purposes. For more demanding applications, e.g., for use in plasma modelling codes, one would have to use measured cross sections or quantum mechanical scattering theory.

Excitation and De-excitation

To calculate electron-atom (or ion) cross sections it is necessary to consider the full bound electron-free electron Coulomb interaction,

$$H = \frac{e^2}{|\underline{r} - \underline{x}|}, \quad (19)$$

where $\underline{r}$ and $\underline{x}$ represent the coordinates of the free and the active bound electron relative to the nucleus, respectively. (This is in contrast to atom or ion interactions with the radiation field, for which the dipole approximation is usually an extremely accurate description.) Fermi's golden rule gives

$$Q_{fi} = \frac{2\pi}{\hbar} |\langle f_t|H|i_t\rangle|^2 dn_f/dE \quad (20)$$

for the transition rate between degenerate states i_t and f_t of the total system consisting of a free electron and the atom or ion to be excited or de-excited. (Since the free electron gains or loses the atomic transition energy, the states of the total system involved here are indeed degenerate.) Moreover, there is actually a continuum of final states whose number per energy interval is dn_f/dE. The states of the total system are assumed to be well approximated by products of bound state wave functions $|i\rangle$, $|f\rangle$, etc., and free electron wave functions, $|k_i\rangle$, $|k_f\rangle$, etc. In other words, the interaction is assumed to be weak on the average.

If we normalize the free electron wave functions in a volume V, the matrix elements between plane wave states are

$$\langle k_f|H|k_i\rangle = V^{-1} \int \exp[i(\underline{k}_i - \underline{k}_f) \cdot \underline{r}] \frac{e^2}{|r - x|} d\underline{r} \quad (21)$$

$$= \frac{4\pi e^2}{Vq^2} \exp(i\underline{q} \cdot \underline{x})$$

in terms of the momentum transfer

$$\underline{q} = \underline{k}_i - \underline{k}_f. \quad (22)$$

Equation (21) suggests that small q values are most effective in causing transitions and we therefore expand the exponential. The first term contributing to the transition matrix element is the linear (dipole) term,

$$\langle f|\langle k_f|H|k_i\rangle|i\rangle \sim \frac{4\pi e^2}{V^2 q^2} i\underline{q} \cdot \langle f|\underline{x}|i\rangle. \quad (23)$$

The absolute value squared of this matrix element, averaged over relative orientations of $\underline{q}$ and $\underline{x}$ is

$$|\langle f_t|H|i_t\rangle|^2 = \frac{16\pi^2 e^4}{3V^2 q^2} |\langle f|\underline{x}|i\rangle|^2 = \left(\frac{4\pi e^2 a_0}{Vq}\right)^2 \frac{E_H}{\Delta E} f_{fi}, \quad (24)$$

where we have used the definition of the oscillator strength to obtain the second version. (The factor $g_i \hat{=} g_n$ is omitted because we require an average over initial states, while $g_f \simeq g_m$ appears on both sides of the defining equation.) Also, ΔE is the energy transfer or transition energy, corresponding to $\hbar\omega = E_m - E_n$ in the radiative case.

The density of final free electron states is as usual, with $E = \hbar^2 k^2/2m$,

$$\frac{d^2 n}{dE d\Omega} = V \frac{k_f^2 dk_f}{(2\pi)^3 dE\, d\Omega} = V \frac{m k_f}{(2\pi)^3 \hbar^2}, \tag{25}$$

where Ω is the solid angle associated with the directions of $\underset{\sim}{k}_f$. Substitution of Eqs. (24) and (25) into Eq. (20) gives

$$Q_{fi} \sim \frac{4 e^4 a_0^2 m k_f}{\hbar^3 v} \frac{E_H}{\Delta E} f_{fi} \int \frac{d\Omega}{q^2}. \tag{26}$$

To relate solid angle and momentum transfer, one considers

$$q^2 = (\underset{\sim}{k}_i - \underset{\sim}{k}_f)^2 = k_i^2 + k_f^2 - 2 k_i k_f \cos\theta \tag{27}$$

which, with $d\Omega = -2\pi(d\cos\theta)$, leads to

$$d\Omega = \frac{2\pi}{k_i k_f} q\, dq. \tag{28}$$

The integral in Eq. (26) therefore becomes $\ln(q_{max}/q_{min})$ where q_{min} corresponds to $|k_i - k_f|$. However, use of $k_i + k_f$ for q_{max} would often be inconsistent with the replacement of $\exp(i\underset{\sim}{q}\cdot\underset{\sim}{x})$ by $1 + i\underset{\sim}{q}\cdot\underset{\sim}{x}$, and Bethe therefore suggested $q_{max} \sim r_{if}^{-1}$, r_{if} being representative of the range of the bound state wave functions. Using this estimate for the integral and replacing k_i by $mv/\hbar$ and V^{-1} by N_e (remember that there was exactly one free electron in the normalization volume) we obtain for the collisional excitation or de-excitation rate

$$Q_{fi} \sim \frac{8\pi}{v} (\frac{\hbar}{m})^2 N_e \frac{E_H}{\Delta E} f_{fi} \ln|(k_i - k_f) r_{if}|^{-1} \tag{29}$$

and for the corresponding cross section

$$\sigma_{fi} \sim 8\pi (\frac{\hbar}{mv})^2 \frac{E_H}{\Delta E} f_{fi} \ln|(k_i - k_f) r_{if}|^{-1}. \tag{30}$$

The Maxwell-averaged excitation rate coefficient is

$$X_{fi} \sim \frac{16\pi^{1/2}}{\alpha c} (\frac{\hbar}{m})^2 (\frac{E_H}{kT})^{1/2} \exp(-\frac{\Delta E}{kT}) \frac{E_H}{\Delta E} f_{fi} \ln|\ldots|^{-1}, \tag{31}$$

if we neglect the velocity dependence of the logarithm. (In the case of de-excitation, there is no exponential factor.)

To derive a similar formula for electron collisions with ions, the plane wave states used in Eq. (21) must be replaced by free Coulomb states. This results in a replacement of the logarithmic factor by Gaunt factors $g(k_f, k_i)$, multiplied by $(\pi/3)^{1/2}$, in Eq. (31),

$$X_{fi} \sim \frac{16\pi}{\alpha c} (\frac{\hbar}{m})^2 (\frac{\pi E_H}{3 kT})^{1/2} \exp(-\frac{\Delta E}{kT}) \frac{E_H}{\Delta E} f_{fi} \bar{g}. \tag{32}$$

Often the averaged Gaunt factors are close to unity, and Eq. (32) then tends to give a reasonably accurate estimate for dipole transition rates. However, it would be wrong to conclude that other transitions are negligible. Higher order terms in the expansion of $\exp(i\mathbf{q}\cdot\mathbf{x})$ are generally not small enough for this conclusion to be valid. Moreover, close coupling effects that come in if the interaction is not actually weak as assumed here may suppress the dipole transition rate.

Ionization and Three-body Recombination

Estimates of the collisional ionization rates can be made in the same way, but it is more convenient to extrapolate our result for the excitation rate. The first step is to consider excitation into a group of states in some interval E, $E + dE$ and to replace the oscillator strength by $(df/dE)dE$. The derivative is, for hydrogenic spectra, of order E_H^{-1}, and assuming $df/dE = 16\, E_{\infty\, n}^2 /3^{3/2} mE^3$ (Kramers approximation, see Omidvar, 1982) for atoms or ions in the n-th level of ionization energy $E_{\infty\, n}$, this estimate for the ionization rate coefficient is

$$S^z \sim \frac{50\nu}{\alpha c n} \left(\frac{\hbar}{m}\right)^2 \left(\frac{E_H}{kT}\right)^{1/2} E_{\infty\, n}^2 E_H \int_{E_{\infty\, n}}^{\infty} g \exp\left(-\frac{E}{kT}\right) \frac{dE}{E^4}$$

$$\sim \frac{50\nu}{\alpha c n} \left(\frac{\hbar}{m}\right)^2 \frac{n}{z^2} \left(\frac{E_H}{kT}\right)^{1/2} \frac{\exp(-E_{\infty\, n}/kT)}{3 + (E_{\infty\, n}/kT)} g_{eff}. \tag{33}$$

The second version is obtained by assuming the Gaunt factor to be essentially constant and by using an interpolation formula between the low and high temperature limits of the remaining integral. With $g_{eff} = (\sqrt{3}/\pi)[1 + n(1 + kT/E_{\infty n})]$ this estimate is then seen to be consistent (for $n = 1$) with the classical Thomson (1912) cross section if $kT \lesssim E_{\infty n}$. It also comes close to the semi-empirical relation of Lotz (1968), although again only for $n = 1$. (Thomson and Lotz have $S^z \sim n^2$ instead of $S^z \sim n$.) Inner shell ionization is normally small, but can be estimated similarly using appropriate ionization energies, etc.

To estimate the rate coefficient for the inverse process, we invoke the principle of detailed balance for the process in question, namely,

$$e + e + I_z \rightleftarrows e + I_{z-1}, \tag{34}$$

where I_z and I_{z-1} stand for the z and (z-1) charged ion of the element in question. In steady state we have

$$C\, N_z\, N_e^2 = S\, N_{z-1}\, N_e \tag{35}$$

in terms of the various densities, and the ionization rate coefficient S, and writing $C\, N_e^2$ for the three-body recombination rate. In thermodynamic equilibrium, the densities are related by the Saha equation, i.e., by the extension (Griem, 1964; p. 135) of Boltzmann factors (17) to continuum states,

$$\frac{N_z N_e}{N_{z-1}} = \frac{g_z}{4 a_0^3 g_n} \left(\frac{kT}{\pi E_H}\right)^{3/2} \exp\left(-\frac{E_{\infty n}}{kT}\right). \tag{36}$$

Here a_o is the Bohr radius and g_z and $g_n < 2n^2$ are statistical weights of the initial and final ions. From Eqs. (33), (35), and (36) follows

$$C_n \sim \frac{740 \nu n^3 a_o^3}{\alpha g_z z^2} (\frac{\hbar}{m})^2 (\frac{E_H}{kT})^2 \frac{g_{eff}}{1 + (E_{\infty n}/3kT)} \quad (37)$$

for recombination into the (unoccupied) states of principal quantum number n. To obtain an effective rate coefficient for recombination leading to lower states of the recombined ion, the C_n should be multiplied by a suitable branching ratio and then be summed over n. Since for $kT \gg E_{\infty n}$, further collisional excitation is certainly dominant over collisional de-excitation, we will simply assume that for $E_{\infty n} < kT/10$ there is no effective recombination, while recombination into lower levels contributes fully. With $\Sigma n^3 \sim \frac{1}{4} n_{max}^4$ and $n_{max}^2 = z_H^2 E/E_{\infty n} \sim 10 \, z^2 E_H/kT$, this estimate for the effective three-body recombination coefficient is then

$$C_{eff} \sim \frac{3.5 \times 10^4 \nu z^2}{\alpha g_z} (\frac{\hbar}{m})^2 a_o^3 (\frac{E_H}{kT})^4 , \quad (38)$$

which is typically in order of magnitude agreement with more detailed calculations or measurements (see Eletsky and Smirnov, 1983). To apply our estimate to recombination into atoms or ions, gz must be interpreted as a sum over all low-lying states which have significant populations.

Radiative and Dielectronic Recombination

Because in many plasmas the collisional ionization rate is not balanced by its inverse in the sense of the principle of detailed balancing (three-body recombination), but rather by radiative and dielectronic recombination (corona ionization equilibrium), it is important to calculate the corresponding rate coefficients as well. As one might expect, this causes no particular difficulty, the radiative recombination process being an extension of line emission. In the same way as we can obtain the number of atoms or ions undergoing a particular radiative decay per unit time and volume from the corresponding line emission coefficient by integrating it over frequency and solid angle and dividing by the photon energy, we can also find the volume rate of recombinations. Basically, this leads to the formula (9) for the recombination continuum, divided by the photon energy and integrated over the angular frequency. Taking all these steps for the partial recombination rate into hydrogen or hydrogenic level n we obtain

$$\alpha_n N_z N_e \sim \frac{2^7 \pi^{1/2} (\alpha a_0)^3 z^4}{3^{3/2} n^3} \omega_L (\frac{E_H}{kT})^{3/2} \exp(z^2 E_H/n^2 kT)$$

$$\times N_z N_e \int_{z^2 \omega_L/n^2}^{\infty} \exp(-\hbar \omega/kT) d\omega/\omega. \quad (39)$$

Here $\omega_L = E_H/\hbar = 2.1 \times 10^{16} \text{ sec}^{-1}$ is the angular frequency corresponding to the limit of the Lyman series of hydrogen. The corresponding radiative recombination rate coefficient is therefore

$$\alpha_n \sim \frac{\alpha}{6c} \left(\frac{\hbar}{m}\right)^2 z (z^2 E_H/n^2 kT)^{3/2}$$

$$\times \exp(z^2 E_H/n^2 kT) \int_{z^2 \omega_L/n^2}^{\infty} \exp(-\hbar\omega/kT) d\omega/\omega , \quad (40)$$

written in a form suggestive of the z dependence remaining after the required approximate scaling of the temperature with z^2 is accounted for, and combining the various constants as was done for the excitation and ionization rate coefficients in Eqs. (32) and (33).

As in case of recombination continua, Eq. (40) should be multiplied, under the integral, with appropriate Gaunt factors g ~ 1 to obtain exact results for recombination with bare ions. In the case of other ions, further corrections are required. The most important of these are factors <1 that are of order $1 - \nu_n/2n^2$, where ν_n is the number of electrons already present in a given principal quantum number shell. These factors take care of the Pauli principle, and other corrections are required to allow for deviations of dipole matrix elements, etc., from hydrogenic behavior.

We finally observe that the remaining z scaling of collisional excitation and ionization rates, after also allowing for the approximate temperature scaling, is ~ z^{-3}. In other words, the ratio of ionization and radiative recombination rate coefficients goes approximately as z^{-4} which shows that the temperature for maximum abundance of a given ion in corona equilibrium actually increases faster than z^2. These effects are even stronger than is implied here because of the elementary process corresponding to dielectronic recombination which should perhaps more properly be called radiationless capture. The name dielectronic, on the other hand, indicates that the transition requires two electrons in an essential way: while the originally free electron is captured, a bound electron is promoted to a higher excited state and takes up the energy set free in the capture process. Normally the captured electron is in a highly excited state of the resulting ion, while the originally bound electron is more likely than not promoted to an excited state corresponding to the upper level of resonance lines of the recombining ion. These doubly excited states can, of course, auto-ionize, i.e., undergo the microscopic inverse of the capture process. In that case, there is no effective recombination. However, it is also possible that the originally bound electron returns to its ground state through a radiative (stabilizing) transition. The corresponding photon corresponds in frequency almost to the resonance lines in the spectrum of the recombining ions, and the slightly displaced and often unresolved lines are called dielectronic satellites. The other electron, at sufficiently low densities, eventually cascades down to complete the process.

The usual procedure for calculations of effective dielectronic recombination rates begins with an estimate of the population in the doubly excited states in terms of the density of free electrons and the ground state population of the recombining ion through the appropriate Saha equation. This estimate is corrected by multiplication with the ratio of auto-ionization to total decay rate, including radiative decay. Such a factor allows for the fact that radiative processes are not in balance and is sometimes called the Saha decrement. The auto-ionization rate, on the other hand, can be expressed in terms of the elementary capture rate through the principle of detailed balancing. To finally obtain the recombination rate, the doubly excited state populations are multiplied with the spontaneous radiative decay rates, and these products are summed over all doubly excited states.

Because of the large number of states involved and because of the uncertainties in cross sections, etc., it is difficult to estimate the accuracy of such calculations. Moreover, especially at laboratory plasma densities, a number of additional processes affecting the doubly excited state densities may have to be included. For example, although the primary capture process will usually only result in small orbital angular momenta ℓ for the captured electron, both electron and ion collisions may result in a redistribution over ℓ-values and therefore a net reduction of the auto-ionization rate and increase of the effective recombination rate. On the other hand, electron collisions may also cause further excitation and even ionization of the $n\ell$-levels of the captured electron and thereby decrease the effective dielectronic recombination rate.

Rather than attempting a quantitative description of the current theoretical research on these complex processes, we shall now return to the description of the capture process as below-threshold excitation of the recombining ion. We estimate the corresponding rate coefficient by taking the difference of the excitation rate coefficients according to Eq. (32) for a fictitious reduced excitation energy and for the actual excitation energy for the "promoted" electron,

$$d \sim \frac{16\pi}{\alpha c} \left(\frac{\hbar}{m}\right)^2 \left(\frac{E_H}{kT}\right)^{3/2} \exp\left(-\frac{\Delta E}{kT}\right) \frac{E_n}{\Delta E} f\bar{g} . \tag{41}$$

Here $E_n = z^2 E_H/n^2$ is an estimate for the extra energy available for excitation because of the binding energy of the captured electron. (We assumed $E_n \lesssim kT$ and expanded the exponential accordingly.)

Since auto-ionization rates decrease as $1/n^3$, while radiative stabilization rates are nearly independent of n, we impose a lower bound on n corresponding to near equality of these competing processes and obtain as an order of magnitude estimate for the effective dielectronic recombination coefficient

$$d \sim \frac{10^3 \nu}{2\alpha c} \left(\frac{\hbar}{m}\right)^2 \left(\frac{E_H}{kT}\right)^{3/2} [\alpha(z+1)/n_\nu]^{8/3} \tag{42}$$

$$\times (z^2 E_H/\bar{E}) \exp(-\bar{E}/kT),$$

where n_ν is the principal quantum number of the ground state (with ν equivalent electrons) of the recombining ion of charge z. The relative values of radiative and auto-ionization rates were chosen as $\sim 20[\alpha(z+1)/n_\nu]^4$ and $\sim (20 n^3)^{-1}$, and $f\bar{g}$ replaced by 0.2ν to obtain this formula, in which $\bar{E}$ is an average excitation energy for the resonance lines of the recombining ion.

Comparison with Eq. (40) suggests that dielectronic recombination is much more probable than radiative recombination, especially at relatively high temperatures, i.e., kT values close to $\bar{E}$. (Note, however, that there is obviously no dielectronic recombination on fully stripped ions.) On the other hand, at low temperatures, three-body recombination may be most important, not to mention dissociative recombination.

COUPLING OF RADIATION AND COLLISIONS

As indicated in Radiative Transport, the combined effects of all rate processes, including absorption and induced emission are generally important in establishing relative populations of upper and lower levels of a line. The ratio of these populations controls the source function, i.e., the ratio of emission and absorption coefficients. Only if the most significant rate processes are microscopic inverses of each other can these populations be assumed close to their local thermodynamic equilibrium (LTE) ratio as given by Eq. (17). In that case the source function equals the Planck function in Eq. (2) at the relevant temperature and the formal solution of the radiative transfer equation, Eq. (18), can be used to calculate the outgoing intensity, provided the temperature distribution along the ray path is known.

It has long been recognized that this LTE assumption is rarely appropriate for the quantitative description of line formation in stellar atmospheres. However, even in many laboratory plasmas electron densities are often too low for electron-ion collisions to control the excited state-ground state ratio for resonance lines. This can be inferred, e.g., from the LTE criteria in Ch. 6 of Griem (1964). These lines, on the other hand, are most affected by radiative energy transfer and are usually responsible for a large fraction of the radiated energy. A non-LTE approach to the radiation-collision coupling problem is therefore indicated for laboratory plasmas as well.

As already mentioned, in laboratory plasmas the source function is almost always well approximated by the ratio of line emission and (effective) absorption coefficients, i.e., according to Eqs. (5), (7), and (14) by

$$S = \frac{\hbar \omega^3}{4\pi^3 c^2} \left[\frac{g_m}{g_n} \frac{N_n}{N_m} - 1 \right]^{-1}, \tag{43}$$

neglecting scattering, i.e., partial redistribution (Mihalas, 1978).

If lower (N_n) and upper (N_m) state population densities are related to each other by Eq. (17), this indeed reduces to the Planck function given by Eq. (2). In the general case, the populations must be obtained by balancing the rates for all processes populating and depopulating the levels. These rates are nearly equal to each other under almost all circumstances, because even local plasma temperatures or densities can hardly change on the time scale of radiative life times (A_{nm}^{-1}) of resonance levels.

A typical balance equation for ground-state populations in neutral atoms is

$$[(\sum_{n'} X_{n'n} + S_n) N_e + \sum_{n'} \int \frac{\sigma_{n'n}(\omega)}{\hbar \omega} 4\pi \bar{I}(\omega) \, d\omega] N_n$$

$$\approx \sum_{n''} (X_{nn''} N_e + A_{nn''} + \int \frac{\sigma_{nn''}(\omega)}{\hbar \omega} 4\pi \bar{I}(\omega) \, d\omega) N_{n''}$$

$$+ (\alpha_n N_e + C_{eff} N_e^2) N_1 \tag{44}$$

with depopulation rates due to (electron) collisional excitation and ionization and due to photo-excitation on the left hand side. The latter term contains the direction-averaged value $I(\omega)$ of the local intensity in the frequency integral. This quantity is normally called J in the astrophysical literature (Mihalas, 1978). The cross sections are as in Eq. (11), while the rate coefficients $X_{n'n}$ and S_n may be estimated from Eqs. (31) or (32) and from Eq. (33). Local values should of course also be used for these coefficients and for the line shape function L(m) in $\sigma_{n'n}$ and for the various densities. In the present example, this mostly implies local values of electron temperatures and densities and of kinetic (Doppler) temperatures.

Such temperature profiles are rarely known a priori and must therefore either be assumed or be calculated from energy balance considerations, i.e., by balancing local energy sources and sinks against the divergence of heat flux and radiative energy flux,

$$\int F_s(\omega)d\omega = \int \int I(\omega, \Omega)\,\mu d\Omega\, d\omega , \qquad (45)$$

where Ω designates the solid angle associated with the various rays and μ is their direction cosine relative to the element of area dS through which the energy flow is calculated. (In the astrophysical literature, H = F/4π is often used instead, both are vector quantities. J and H are the zeroth and first order moments of I with respect to powers of μ. A corresponding second order moment K gives the pressure associated with the radiation field in the interval $d\omega$, except for a factor 4π/c. For isotropic radiation one has the Eddington factor f = K/J = 1/3.)

Suffice it to note that the first moment of the unknown intensity appears in the energy balance equations and the zeroth order moment in the population balance equations, in which photo-ionization was neglected because of its minor role in laboratory plasmas. Also, on the right-hand side of Eq. (44), the population processes included are electron-collisional and spontaneous and induced radiative de-excitation from levels n'', and radiative and collisional three-body recombination from ions of density N_1. The reader will note here that inclusion of three-body recombination using an effective rate coefficient is rather questionable because the primary process is really recombination into excited states. [See the discussion between Eqs. (37) and (38).] Fortunately, ionization and recombination terms in Eq. (44) tend to be rather small in situations where non-LTE radiative transfer is important. Often the same assumptions are made for the population balance of the upper level m, which may then be written as

$$[\sum_{m'} (X_{m'm}N_e + A_{m'm} + \int \frac{\sigma_{m'm}(\omega)}{\hbar\omega} 4\pi\, \bar{I}(\omega)d\omega\,]N_m$$

$$= \sum_{m''} (X_{mm''}N_e + A_{mm''} \int + \frac{\sigma_{mm''}(\omega)}{\hbar\omega} 4\pi\, \bar{I}(\omega)d\omega)N_{mm''}. \qquad (46)$$

Here the spontaneous transition probabilities are of course zero for m' > m and m'' < m.

The simplest non-trivial case (two-level atom model) corresponds to m' = m'' = n, i.e., to assuming that only rate processes between levels n and m are important. In that case the population ratio required for the source function in Eq. (43) becomes

$$\frac{N_m}{N_n} = \frac{X_{mn}N_e + (4\pi/\hbar\omega)\int \sigma_{mn}(\omega)\,\bar{I}(\omega)d\omega}{X_{nm}N_e + A_{nm} + (4\pi/\hbar\omega)\int \sigma_{nm}(\omega)\,\bar{I}(\omega)d\omega}. \qquad (47)$$

If all radiation terms are small, this relation reduces to the Boltzmann relation, Eq. (17), at the electron temperature, because X_{mn} and X_{nm} are related to each other by the principle of detailed balancing. If A_{nm} is much larger than the quenching rate $X_{nm} N_e$ and the radiation field still weak, we obtain the coronal limit

$$\frac{N_m}{N_n} = \frac{X_{mn} N_e}{A_{nm}} , \qquad (48)$$

in which the line emission coefficient, Eq. (5), is independent of the radiative transition probability. Finally, if the electron collisional rates are small compared to absorption and induced emission rates and if $\bar{I}(\omega)$ is essentially flat over the range of $L(\omega)$, one has

$$\frac{N_m}{N_n} = \frac{g_m}{g_n} \{ \frac{I_p(\omega_{mn})}{\bar{I}(\omega_{mn})} [\exp(\frac{E_m - E_n}{kT}) - 1] + 1\}^{-1} . \qquad (49)$$

which reduces to the Boltzmann relation as $\bar{I}$ approaches the Planck function I_p.

However, in general, neither of these limits is appropriate and the radiative transfer and radiative-collisional population balance equations must be solved self-consistently. Depending on the symmetry of the problem, slab geometry, spherical or cylindrical symmetry, the local intensity fortunately depends only on a reduced number of positional and directional variables. The first two cases are again of considerable astrophysical interest (Mihalas, 1978) and because of the symmetry there is only one spatial coordinate, e.g., the radial distance, r, from the origin and one directional variable, the cosine μ of the angle between ray direction and radius. For cylindrical symmetry (Heasley, 1977) one needs three variables, e.g., either radius and two angles or the impact parameter of the ray, distance from the point of closest approach, and angle between ray direction and the axis of symmetry. Naturally, the optical depth must be calculated accordingly, and one must remember that it mostly depends on the lower state density and on the line shape function.

The absolute value of the lower state density N_n in cases of interest here may often be inferred from other considerations. In a recent paper Anderson et al. (1985) obtained ground state densities of mercury isotopes from vapor pressures and gas temperatures. Also, they replaced the iterative solution of the coupled equations by a Monte Carlo simulation of the complicated history of resonance line photons, with appropriate relative probabilities for all relevant processes occurring between original excitation and escape. This method allows for the excitation energy transfer between various isotopes and exhibits the gain in radiative efficiency achieved by using more isotopes radiating at slightly different frequencies. An interesting problem here is also the optical pumping of lines from rare isotopes by the very optical thick lines from more abundant isotopes. This situation is quite similar to the line pumping of lasers (see, e.g., Apruzese and Davis, 1985) which has received considerable attention. The latter authors used a probabilistic coupling constant technique for most lines (Apruzese, 1981) which yields a matrix of probabilities C_{ij} that if a photon is emitted in the ith spatial cell it is absorbed in the jth cell. This technique treats photon pumping and escape realistically, but the detailed profile of line emission is lost. (This contrasts with the Monte Carlo method.) For the pumping line for the upper laser level, a frequency grid was therefore established for its profile, and a coupling matrix was calculated for each of these frequencies. As always in numerical solutions of such transfer problems, an iterative technique (Apruzese et al., 1984) had to be used to achieve self-consistency between level populations and radiation field. Calculations of such line pumping problems have also been done by Alley et al. (1982) and by Kunasz (1985), who used the moment method and allowed for partial redistribution. This method is particularly efficient and adaptable to the various boundary conditions.

There are important boundary conditions on $I(\omega)$ and its moments also on the axis of cylindrical symmetry or in the origin of a spherically symmetric system to ensure that these quantities are continuous along rays crossing this axis or point.

REFERENCES

Alley, E. W., Chapline, G., Kunasz, P., and Weisheit, J. C., 1982, J. Quant. Spectrosc. Rad. Transfer, 27:257.
Anderson, J. B., Maya, J., Grossman, M. W., Lagushenko, R., and Waymouth, J. F., 1985, Phys. Rev., A31:2968.
Apruzese, J. P., 1981, J. Quant. Spectrosc. Radiat. Transfer 125:419.
Apruzese, J. P., Davis, J., Duston, D., and Clark, R. W., 1984, Phys. Rev. A, 29:246.
Apruzese, J. P. and Davis, J., 1985, Phys. Rev. A, 31:2976.
Brissaud, A. and Frisch, U., 1971, J. Quant. Spectrosc. Radiat. Transfer, 11:1767.
Bromage, G. E., Cowan, R. D., and Fawcett, B. C., 1977, Phys. Scr., 15:177.
Cauble, R. and Griem, H. R., 1983, Phys. Rev. A, 27:3187.
Eletsky, A. V. and Smirnov, B. M. 1983, Ch. 1.2 in: "Handbook of Plasma Physics", Rosenbluth, M. N. and Sagdeev, R. Z., eds., "Volume 1: Basic Plasma Physics I", Galeev, A. A. and Sudan, R. N., eds., North Holland, Amsterdam.
Freund, R. S., Schiavone, J. A., and Brader, D. F., 1976, J. Chem. Physics, 64:1122.
Griem, H. R., 1964, "Plasma Spectroscopy", McGraw-Hill, New York; Univ. Microfilms International 212 00000 7559, Ann Arbor.
Griem, H. R., 1974, "Spectral Line Broadening by Plasmas", Academic Press, New York.
Griem, H. R., 1979, Phys. Rev. A, 20:606.
Griem, H. R., 1983, Ch. 1.3, in: "Handbook of Plasma Physics", Rosenbluth, M. N., and Sagdeev, R. Z., eds., "Volume 1: Basic Plasma Physics I," Galeev, A. A., and Sudan, R. N., eds., North Holland Amsterdam.
Griem, H. R. and Tsakiris, G. D., 1982, Phys. Rev. A, 25:1199.
Heasley, J. N., 1977, J. Quant. Spectrosc. Radiat. Transfer, 18:541.
Higo, M. and Kamata, S., 1982, Chemical Physics, 73:99.
Inglis, D. R. and Teller, E., 1939, Astrophys. J., 90:439.
Kim, Y. S. and Pratt, R. H., 1983, Phys. Rev., A, 27:2913.
Lotz, W., 1968, Physik Z., 216:241.
Kunasz, P. B., 1985, J. Quant. Spectrosc. Radiat. Transfer, 33:155.
Kunc, J. A., 1985, J. Quant. Spectrosc. Radiat. Transfer, 33:1.
Kunze, H. J., 1985 (in these Proceedings).
Mason, H. E., 1975, Monthly Notic. Roy. Astron. Soc., 170:651.
Mihalas, D., 1978, "Stellar Atmospheres", 2nd edition, Freeman, San Francisco.; see also Mihalas, D. and Mihalas, B. W., 1984, "Foundations of Radiation Hydrodynamics", Oxford University Press, Oxford.
Omidvar, K., 1982, Phys. Rev. A26:3053.
Thomson, J. J., 1912, Phil. Mag. 23:419.

Tondello, G., Jannitti, E. and Malvezzi, A. M., 1977, *Phys. Rev. A* 16:1705.
Wiese, W. L. and Martin, G. A., 1980, "Wavelengths and Transition Probabilities for Atoms and Atomic Ions," NSRDS-NBS, US Govt. Printing Office, Washington, D.C.

RADIATIVE PROCESSES IN DISCHARGE PLASMAS:

QUESTIONS AND ISSUES

J. H. Ingold

General Electric Company
Lighting Business Group
Cleveland, OH, USA

INTRODUCTION

When a gas is heated to a sufficiently high temperature, it becomes ionized; when the ion density is high enough to ensure approximate neutrality, the ionized gas is called a plasma. In addition to electrons and ions, a plasma may be composed of atomic and molecular species in various states of excitation. Excited atoms and molecules are continually produced by collisions with other particles and by absorption of photons, and continually destroyed by collisions with other particles and by emission of photons. Depending on the spatial extent of the plasma and the density of photon-absorbing species, the average photon may be emitted and absorbed many times before reaching the edge of the plasma. In some applications, such as a fusion plasma escaping photons represent an energy sink which cools the plasma, perhaps preventing the proper fusion temperature from being reached. In other applications, such as the plasma of a discharge light source, photons escaping in the wavelength range detectable by the human eye, i.e., between 400 and 700 nm, are not only desirable, but are necessary for proper functioning of the light source. In either case, fundamental understanding of photon emission, migration, and absorption processes in plasmas is indispensable to the researcher. By charter, this Advanced Study Institute is devoted to the study of radiation from discharge plasmas. Consequently, subsequent remarks will be focused on discharge plasmas, particularly those used for light sources.

By convention, discharge light sources are categorized as "low pressure" or "high pressure," depending on whether the electron temperature is much higher than, or approximately equal to, the gas temperature. For example, the discharge in a fluorescent lamp takes place in a mixture of argon at a partial pressure of a few torr and mercury vapor at a partial pressure of several millitorr. The electron temperature in this discharge is about one eV, or 12,000 K, and the gas temperature is about 350 K. In contrast, the discharge in a high pressure mercury lamp takes place in mercury vapor at a partial pressure of several atmospheres, and the electron temperature in the core is approximately equal to the gas temperature of about 6000 K.

A striking feature of these discharge plasmas is the high value of radiation efficiency, which is defined as the ratio of the power radiated at all wavelengths to the total power dissipated in all loss processes, i.e., the power input required to maintain the discharge in a steady state. Values representative of various discharges used for lighting purposes are listed in Table 1. In the positive column of the fluorescent lamp discharge, for example, about 75% of the input power is radiated in mercury lines. Even in the high pressure mercury discharge, the radiation efficiency is over 50%. Therefore, it is apparent that radiative processes in discharge plasmas used for light sources have a profound influence on discharge behavior.

The purpose of this paper is to focus attention on unanswered questions and unresolved issues in the area of radiative processes in discharge plasmas. The format of this paper is as follows. First, a plea is made for accurate and extensive input data in the form of rate coefficients, transport coefficients, emission/absorption coefficients, etc. Then the question of local thermodynamic equilibrium is addressed, and an example where this widely used principle is violated in a high pressure discharge is described. Next, the question of photon trapping is addressed, and examples of recent attempts to reduce trapping in a low pressure discharge are described.

DATA FOR MODEL CALCULATIONS

It is straightforward to write down the equations which govern the behavior of discharge plasmas. However, it is not so simple to solve them. Furthermore, much of the input data must be estimated. Consequently, calculated results are sometimes questionable. The purpose of this section is to enumerate the kinds of input data that are necessary for model calculations, and to encourage each speaker to make clear which parameters are known with confidence, and which parameters are estimated.

Input data required for model calculations include rate coefficients for particle production and destruction processes, transport coefficients for flow processes, thermodynamic data for chemical processes, and emission/absorption coefficients for radiation transport processes. The equations in which these coefficients appear are briefly described below.

In principle, there are three macroscopic (measurable) variables for each species present in the case of one dimensional cylindrical geometry. These variables, which may vary in both time and space, are density, average velocity, and temperature. The three equations which determine these three variables are the conservation equations, or moment equations, which guarantee conservation of particle density, particle momentum, and particle energy (cf. Hirschfelder, et al., 1954). For low pressure discharges, it is usually assumed that all heavy particles have the same temperature, so that two energy balance equations are required: one for heavy particles and one for electrons. For high pressure discharges, it is usually assumed that all particles have the same temperature, so that only one energy balance equation is needed. The

Table 1. Radiation Efficiency of Discharge Light Sources

Discharge Type	Radiation Efficiency (%)
Low Pressure Ar-Hg	75
High Pressure Mercury	52
High Pressure Sodium	50
High Pressure Metal-Halide	60-70

requirements for the latter simplification to be valid are discussed in a later section. In each of the three types of conservation equations, certain coefficients appear. These coefficients depend on particle-particle interactions of one kind or another. For example, in the continuity equations, which express conservation of particles, terms describing the rates at which new particles are created by ionization, excitation, dissociation, etc., and destroyed by the respective reverse processes appear. The rates at which these processes take place depend on cross-sections which are peculiar to each species involved. Likewise, in the energy balance equations, transport coefficients such as electrical and thermal conductivity must be specified. These coefficients also depend on cross-sections for the various particle- particle interactions. Many of these cross-sections can be found in the literature, but by far the majority of the required cross-sections are not known, and must be estimated to obtain numerical solutions for the macroscopic variables.

When the assumption of local thermodynamic equilibrium (LTE) is valid, then all particles have the same temperature, so that only one energy balance equation is required. Furthermore, the particle densities are related to the common temperature by the equations of equilibrium thermodynamics, so that the rate equations are redundant. However, high temperature thermochemical data is incomplete, making it necessary to extrapolate low temperature data, or to use some other method of estimation. Even with the approximation of LTE, it is necessary to solve the momentum balance equations, which are equivalent to diffusion equations, to get species distributions. Consequently, input data must include diffusion coefficients for each species diffusing relative to other species. Once again, the list of available binary collision integrals, or cross sections, is incomplete, and it is common to use the hard sphere approximation. The momentum balance equations also contain thermal diffusion coefficients (Chapman and Cowling, 1970) about which even less information is available.

The importance of having accurate broadening parameters for radiation transport calculations cannot be overemphasized. For resonance lines, the dominant broadening mechanism is pressure broadening, either by atoms of the same kind, or by foreign atoms. For non-resonance lines, Stark broadening by plasma electrons is usually very important. Output information that depends on line broadening input data includes diagnostic information as well as energy balance of the discharge. For example, the shape of the a resonance line can be used to estimate the partial pressure of the parent species, because the difference in wavelength between the wings of a self-reversed line depends on pressure. The partial pressure of a foreign gas can also be estimated from the shape of a resonance line for the same reason. A good example of both is found in the high pressure sodium discharge, which nominally has 50-100 torr of sodium vapor and 100-400 torr of mercury vapor. It has been shown that the displacement of the short wavelength (blue) wing of the sodium resonance line varies only as the sodium partial pressure, whereas the displacement of the long wavelength (red) wing varies with the partial pressures of both sodium and mercury (Reiser and Wyner, 1985).

LTE OR NOT LTE

The assumption of local thermodynamic equilibrium (LTE) is almost always invoked in the analysis of high pressure discharge plasmas. The LTE assumption implies that all species, including electrons, have the same temperature, and that all species concentrations, or densities, are related to this common temperature in the same way as they are in true thermodynamic equilibrium. It is often asked, How can LTE exist in a discharge plasma when fifty percent of the power input is radiated? The explanation is that species densities can be in LTE, even though the radiation field is not, when the rate of destruction of the species by collisions is large compared with that by photon emission.

In a discharge, the electric field "heats" the electrons, then electron energy is transferred to other species by collisions, both elastic and inelastic. For LTE to obtain, the collision rates must be large compared with other processes. Consequently, the assumption of LTE is not valid for low pressure discharges. For high pressure discharges, the assumption of LTE is valid provided the following conditions are satisfied in each unit volume of plasma:

1. For equal electron and heavy particle temperatures, the forward and backward rates of elastic energy transfer between electrons and heavy particles must, be much larger than the difference between these rates. The difference between these rates accounts for the energy delivered to the wall by gaseous thermal conduction, which is partially supplied by Joule heating of the electrons.

2. For excited states to be in LTE at the local temperature, the rates of excitation and deexcitation by electron impact must be much larger than the difference between these rates. The difference between these rates accounts for the energy radiated by the discharge, which is supplied by the remainder of the Joule heat.

3. For ionic, atomic, and molecular species to be in LTE at the local temperature, the rates of dissociation and recombination of each species must be much larger than the difference between these rates. The difference between these rates accounts for diffusion losses.

Criteria which various collision frequencies must satisfy in order to meet these conditions are straightforward to apply to a given situation. When these criteria are examined, it is usually found that the conditions for LTE to obtain are met in the hot core of the high pressure discharge where the electron density is high, but generally are not met in the cooler region near the wall. For example, it has been shown by calculation that LTE does not obtain at the wall of the high pressure sodium discharge (Ingold, 1974; Waszink, 1974). The departure from equilibrium is manifested by a higher population of the first excited state of sodium than predicted on the basis of LTE at the gas temperature. The higher population of excited atoms near the wall is caused by absorption of pressure-broadened resonance radiation emitted from the hot core of the discharge. Waszink (1974) concluded that this overpopulation resulted in an electron temperature higher than the gas temperature, due to superelastic collisions between excited sodium atoms and electrons. The calculated departure from LTE near the wall of the high pressure sodium discharge is in agreement with measurements of spectral intensity at the center of the resonance line, as shown in Fig. 1 (de Groot, 1974). The non-LTE calculation shown in Fig. 1 corresponds to an electron temperature of 2300 K and a gas temperature of 1600 K at the wall, with a core temperature of 4200 K.

In spite of the large difference between electron temperature and gas temperature near the wall, it was concluded that little error in calculated values of core temperature and electric field resulted from the LTE assumption (de Groot, 1974). However, calculated values of radiation output in the D-line based on the LTE assumption were found to be about 5% smaller than those found for the non-LTE calculation (de Groot, 1974; Ingold, 1974). Therefore, it appears that non-LTE calculations must be made when better than 5% accuracy in calculated radiation output is required.

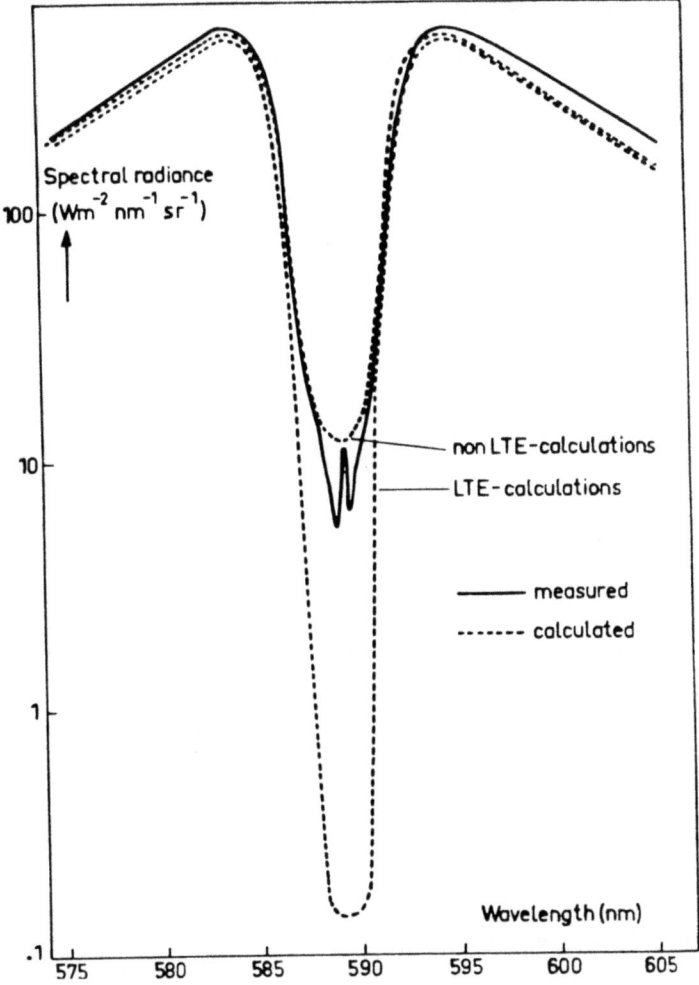

Fig. 1. Relative intensity of the sodium D-line in a high pressure sodium discharge, showing measurement (solid curve) and calculation (dashed curves). Curve for LTE based on equal electron and gas temperature; curve for non-LTE based on electron temperature higher than gas temperature from model calculations (de Groot, 1974).

PHOTON TRAPPING

Concept of Imprisonment Time

In the standard fluorescent lamp, the partial pressure of mercury ranges between 1 and 10 millitorr. The corresponding density of mercury atoms ranges between 3×10^{13} and 3×10^{14} atoms per cubic centimeter. Therefore, when a resonance photon is emitted somewhere in the lamp, it is very likely to be absorbed by by another atom, which then becomes one of the excited atom population. The result of many successive emissions and absorptions is equivalent to the transfer of excitation energy from atom to atom, and the eventual escape of the radiation across the discharge boundary (into the phosphor) requires a large number of such transfers. It is customary to speak of this radiation which undergoes many emissions and absorptions before reaching the wall of the discharge as being "imprisoned." It is also customary to speak of the decay time of the resonance radiation as the "imprisonment time." In this paper, the imprisonment time is denoted by the symbol T. The physical meaning of the imprisonment time T can be demonstrated by monitoring the decay of resonance radiation emitted from a discharge tube. When the discharge is turned off, the radiation begins to decay with time. After a certain time, depending on the details of the discharge tube and the gas, the decay will be exponential with time. The time constant for this exponential decay of radiation is simply the imprisonment time T. In mathematical form, the decay of intensity is given by the relation

$$I = I_o \exp(-t/T) \tag{1}$$

where I is intensity and t is time. The imprisonment time is analogous to the ambipolar diffusion time in an afterglow. The analogy extends to the steady state, i.e., the imprisonment time is also known as the effective lifetime. In the optically thin limit (no absorption), the effective lifetime T is equal to the natural lifetime of the excited atom, which is about 100 ns for the 254 nm resonance line of mercury. In the optically thick limit (copious absorption), the effective lifetime is many times longer than the natural lifetime.

Continuity Equation for Excited States

In 1947, a new approach to radiation transfer in discharge plasmas was developed (Holstein, 1947; Biberman, 1947). The radiation transfer equation was solved for the intensity of radiation and substituted into the continuity equation for the excited state particles. When this continuity equation is integrated over the volume of the discharge, the following approximate equation is obtained for the dependence of excited state density N^* on the electron density N and imprisonment time T:

$$N^* = T A N / (1 + T B N) \tag{2}$$

where A is electron impact excitation frequency and BN^* is electron impact deexcitation frequency. The quantity TBN is a measure of the importance of superelastic collisions in determining the steady state density of excited states. For large values of TBN, N^* approaches the value A/B, which is equal to the value N* would have in thermodynamic equilibrium at the electron temperature. In the fluorescent lamp, however, the value of TBN is about 0.2 for the 254 nm resonance line of mercury (Kenty, 1950). The power radiated in the 254 nm line is proportional to the density of excited atoms divided by their effective lifetime, which from the equation above is

$$N^*/T = A N / (1 + T B N) \tag{3}$$

From this equation, it is apparent that the effect of photon trapping, i.e., imprisonment, is to limit the radiation output. In other words, if there were less trapping, then the radiation output would increase with electron density, or current, without limit. Therefore, it is natural to ask, How can photon trapping be reduced in the low pressure Ar-Hg discharge?

How Can Photon Trapping be Reduced?

To answer this question, it is necessary to understand the factors which determine the magnitude of the effective lifetime, or imprisonment time, T. In general, the imprisonment time is larger than the natural lifetime of the excited atom by a factor which is roughly equivalent to the number of times the average photon is absorbed and reemitted before it escapes the discharge. This factor is called the imprisonment factor, and will be denoted by the symbol g in this paper. The imprisonment factor g is inversely proportional to the volume-averaged probability that a photon emitted somewhere in the discharge will escape to the wall. Holstein (1947) has shown that the imprisonment factor g for a Doppler-broadened resonance line generated in a cylindrical discharge of radius R is given approximately by the expression

$$g = 0.625 \, kR \, (\pi \ln(kR))^{1/2} \tag{4}$$

where k, the absorption coefficient at line center, is proportional to the density of absorbers. According to this relation, the question of how to reduce photon trapping is equivalent to the question of how to reduce the absorption coefficient k without reducing the number of emitters.

It is known that there are ten hyperfine components of the 254 nm resonance line of natural mercury (Mitchell and Zemansky, 1971). The intensity of the Hg196 component is very small compared with the intensities of the other hyperfine components, because the fraction of Hg196 in natural mercury is small. In addition, some of the lines lie so close to each other, they are indistinguishable. Consequently, it is commonly assumed that there are five hyperfine components of the 254 nm resonance line of natural mercury. The relative intensities and calculated Doppler-broadened line shapes of the hyperfine components for mercury atom density corresponding to a vapor pressure at room temperature are shown in Fig. 2. Except for the Hg196 line, the relative intensities are not greatly different, so it is common practice to assume that the 254 nm resonance line of natural mercury is composed of five hyperfine lines of equal intensity. This assumption has been invoked in several published analyses of the low pressure Ar-Hg positive column, such as that found in conventional fluorescent lamps (Kenty, 1950; Waymouth and Bitter, 1956). In other words, the absorption coefficient at the center of the 254 nm resonance line of natural mercury is assumed to be one-fifth of the value obtained on the assumption that the 254 nm resonance line is composed of one line. This approach gives a g-factor of about 100, which has led to fair agreement between theory and measurement of the low pressure Ar-Hg positive column.

Calculated Doppler-broadened line shapes for natural mercury at 6 millitorr vapor pressure are shown in Fig. 3. In the calculation, the dependence of excited atoms on discharge radius is approximately parabolic, and it is assumed that absorption coefficient k and emission profile p have the same shape in frequency. This assumption is valid when the frequency of the photon emitted by an excited atom is uncorrelated with the frequency of the absorbed photon which produced the excited atom. The self-reversal is due to this parabolic distribution of emitters, which can be seen from the radiative transfer equation for intensity I along a line of sight s:

$$I(s,\nu)' = - k(\nu)I(s,\nu) + Ap(\nu)N^*(s) \tag{5}$$

where ν is frequency, the prime means differentiation with respect to s, and the constant A, which has units of energy per unit time, depends on the energy of the transition and the transition probability. The self-reversed line shape shown in Fig. 3 is due to the fact that photons at line center where the optical depth is large come from a thin region next to the edge of the discharge, whereas photons in the wings where the optical depth is of order unity, come from the center of the discharge. For the special case of constant N^*, there is no self-reversal, i.e., the lines are flat across the top. The two lines with asymmetric wing intensities are composed of two or more overlapping hyperfine lines. The line shapes shown in Fig. 3 are similar to those calculated recently by monte carlo methods (Anderson et al., 1985).

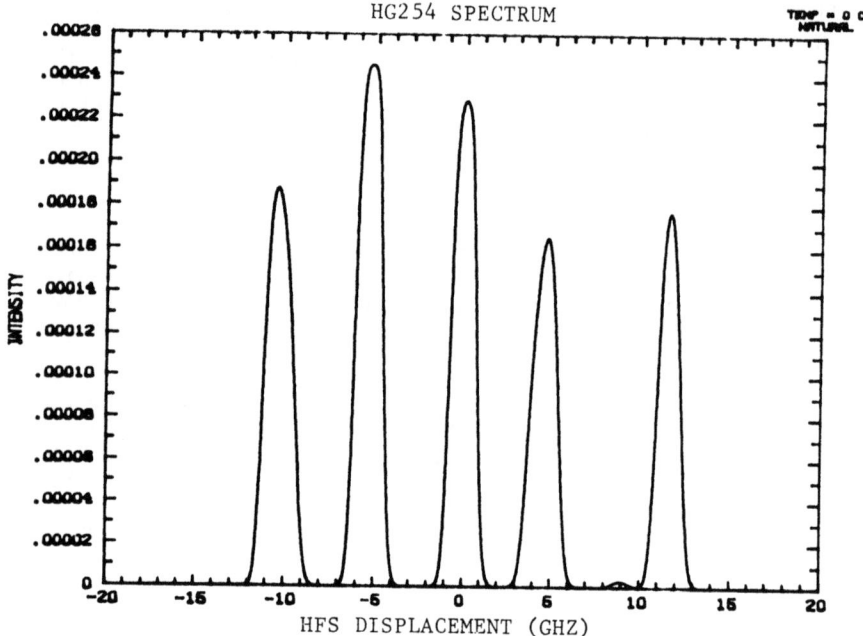

Fig. 2. Calculated Doppler-broadened hyperfine emission lines from excited mercury vapor at 0 C, showing how ten lines appear as five due to overlap and relative absence of Hg196.

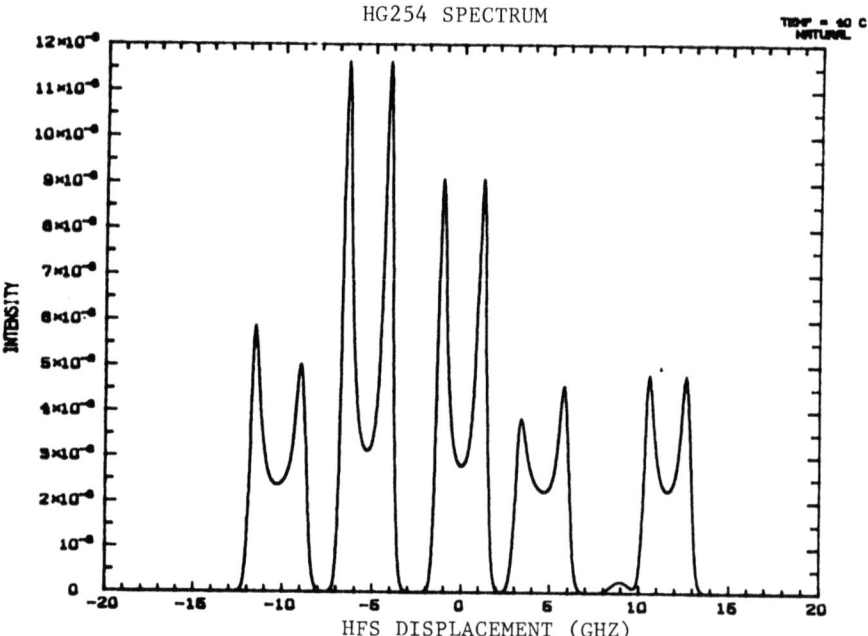

Fig. 3. Calculated Doppler-broadened hyperfine emission lines from excited mercury vapor at 40 C, showing self-reversal due to parabolic distribution of excited atoms.

Once it is realized that the efficiency of 254 nm output from the low pressure Ar-Hg positive column is dependent on the imprisonment time T, it is natural to ask, What improvement can be achieved by changing from five equal hyperfine lines to six, which could be accomplished by artificially adding Hg196 to natural mercury? This question has been answered, at least in part, by Maya et al., 1984, who found that a few percent increase in efficiency could be achieved by the addition of a few percent Hg196 to natural mercury. In a theoretical analysis, the same authors make the point that it is necessary to take into account energy transfer, overlap, and mixing effects, in order to get good agreement between measurement and calculation. Presumably, it is necessary to account for these effects because a simpler model which neglects them gives only 3-4% improvement in 254 nm efficiency for six lines of equal intensity, which requires about 17% Hg196, and little or no improvement for 4% Hg196 added to natural mercury.

It is also interesting to ask: What is the increase in 254 nm efficiency that can be achieved with no trapping? The answer to this question should give the upper limit of improvement that can be expected by decreasing the imprisonment time. Fig. 4 shows calculated results for 254 nm radiation efficiency versus current in a typical low pressure Ar-Hg discharge used for fluorescent lighting. The lower curve in Fig. 4 shows the calculated result for natural mercury with a g-factor of approximately 100, while the upper curve shows that for a g-factor of unity. The standard lamp runs at a current of about 0.4 A, where Fig. 4 shows the maximum gain to be about 10%. The main reason 254 nm radiation efficiency falls with rising current is well-known; it is because the effect of photon trapping is enhanced at higher current where more excited atoms are destroyed in superelastic collisions with electrons.

Another way that imprisonment time can be reduced is by the Zeeman effect. This was pointed out by Hollister and Berman (1983). When an excited atom of an even isotope, which has one emission line, is in a magnetic field, the single emission line of the unperturbed atom is split into three lines of equal strength. The lines are separated in frequency by an amount given by

$eB/4\pi m$

$= 1.4 \times 10^6$ B /sec (B expressed in units of gauss)

The corresponding separation in wavelength is given by

3×10^{-7} B nm (B expressed in units of gauss)

= 3 milli-angstroms for B = 1000 gauss

Intuitively, one would expect a monotonic decrease in imprisonment time with increasing magnetic field, hence a monotonic increase in radiation efficiency, until the emission lines are completely separated. For Hg202, for example, calculation shows no change in imprisonment time occurring beyond 3000 gauss, where the line separation reaches about 10 milli-angstroms. However, an optimum magnetic field of about 700 gauss has been reported for a discharge composed of Ar and Hg202 (Hollister and Berman, 1983). Here, then, is an unresolved issue.

This issue is addressed in a poster paper at this ASI by my colleague, V. Roberts. He has measured the increase in efficiency of a low pressure Ar-Hg discharge placed in an axial magnetic field. He finds a monotonic increase in efficiency up to 110% of that for zero magnetic field, as magnetic field is increased to 1000 gauss. Here is another unresolved issue---model calculations predict a decrease in imprisonment time of a factor of about two for natural mercury in a magnetic field of

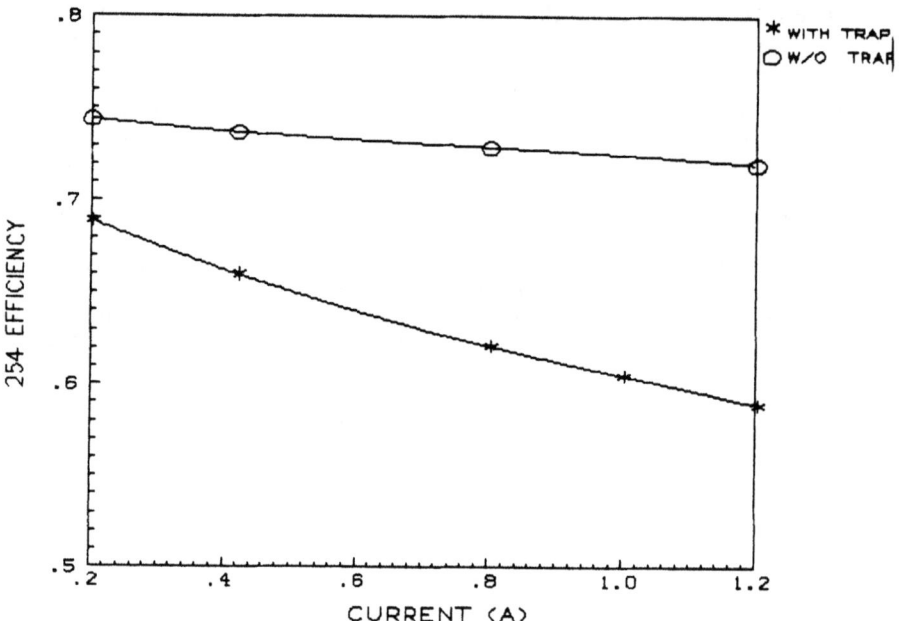

Fig. 4. Calculated variation of 254 nm radiation efficiency with current for 3.6 cm diameter discharge in low pressure Ar-Hg mixture. Lower curve is for normal trapping; upper curve is for same conditions without trapping.

1000 gauss, and a corresponding efficiency increase of a few percent, whereas the measurement results in an efficiency increase of 10%.

SUMMARY

Several unresolved issues concerning radiative processes in discharge plasmas have been discussed in this paper. First, it was emphasized that accurate and extensive input data in the form of rate coefficients, transport coefficients, and emission/absorption coefficients are needed to model discharge plasmas with confidence. Second, the widely used assumption of local thermal equilibrium was discussed, and an example of the failure of this assumption in the high pressure sodium discharge was described. Finally, the subject of photon trapping in a low pressure discharge was discussed, and novel suggestions dealing with isotopic concentration and Zeeman splitting were explored. The aim of this paper has been to pose questions, not answer them. It is anticipated that other questions and issues will be raised throughout subsequent discussions, so that the panel discussion which takes place later on this week will be lively and informative.

ACKNOWLEDGMENTS

The author gratefully acknowledges the help of colleague M. E. Duffy, who provided the graphical portrayal (shown in Figs. 2 and 3) of his calculated shape of Doppler-broadened hyper-fine emission lines of mercury in the low pressure Ar-Hg discharge. In addition, the author acknowledges helpful discussions with colleague J. T. Dakin and V. D. Roberts on the subjects of input data needed for modeling and Zeeman splittings, respectively. Finally, the author wishes to acknowledge the support and encouragement of the management of the Lighting Research and Technical Services Operation of the General Electric Lighting Business Group during preparation of this work.

REFERENCES

Anderson, J. B., Maya, J., Grossman, M. W., Lagushenko, R., and Waymouth, J. F., 1985, Phys. Rev. A,31:2968.
Biberman, L. M., 1947, JETP (Russian), 17:416.
Chapman, S. and Cowling, T. G., 1970, "The Mathematical Theory of Non-Uniform Gases," University Press, Cambridge, pf 268.
de Groot, J. J., 1974, "PhD Thesis," Technische Hogeschool, Eindhoven.
Hirschfelder, R. O., Curtiss, C. F., and Bird, R. B., 1954, "Molecular Theory of Gases and Liquids," Wiley, New York.
Hollister, D. and Berman, S. J., 1983, "Third International Symposium on the Science and Technology of Light Sources," Toulouse, France.
Holstein, T., 1947, Phys. Rev., 72:1212.
Ingold, J. H., 1974, Bull. Am. Phys. Soc., 19:162.
Kenty, C., 1950, J. Appl. Phys., 41:94.
Maya, J., Grossman, M. W., Lagushenko, R., and Waymouth, J. F., 1984, Science, 226:435.
Mitchell, A. C. G. and Zemansky, M. W., 1934, "Resonance Radiation and Excited Atoms," University Press, Cambridge.
Reiser, P. A. and Wyner, E. F., 1985, J. Appl. Phys. 57:1623.
Waszink, J. H., 1974, Bull. Am. Phys. Soc., 19:162; also, 1973, J. Phys. D,6:1000.
Waymouth, J. F. and Bitter, F., 1956, J. Appl. Phys., 27:122.

LIGHT SCATTERING

H.-J. Kunze

Institut für Experimentalphysik
Ruhr-Universität
Bochum, FR Germany

INTRODUCTION

Light incident on a laboratory plasma interacts with all constituents of the plasma (electrons, ions, atoms and molecules), and the fraction re-radiated into all directions usually is referred to as scattered radiation. Various scattering processes are possible, and by selecting suitably the wavelength of the incident radiation, one may achieve one process to dominate. The analysis of the scattered radiation yields a wealth of information on plasma parameters, plasma composition and plasma phenomena, and these diagnostic possibilities have been the prime motivation for the strong interest in these processes during the past two decades. Quite a number of different experimental methods have evolved: the differences arise from different incident radiation, from different interaction processes being effective or from different experimental techniques used specifically with regard to the detection and analysis of the radiation. Figure 1 reveals one great advantage of most of these methods: as the incident radiation traverses the plasma, the "scattering volume" may be selected by the detection optics. Scattered radiation thus is received only from this volume although the plasma radiation is collected along the total line of sight. Light scattering methods allow, therefore, spatially resolved measurements.

In the following we will discuss some of the fundamental aspects, specific applications in diagnostic techniques being the subject of separate lectures. When studying radiative transfer in laboratory plasmas, scattering may become important at high densities for wavelengths coinciding with resonance transitions of atoms and ions.

RADIATION SOURCES

Lasers and their frequency converted radiation cover today the spectral region from about 0.12 µm to 500 µm, and due to the unique properties of this radiation, discharge sources, which were initially used in some applications, have become obsolete. Depending on the goal, one or even several of the following characteristics of laser radiation are made use of in scattering experiments:

- high directionality
- narrow spectral width
- high power
- continuously working or pulsed lasers down to 1 ps pulse duration

- fixed-wavelength or tunable lasers
- spatial and temporal coherence.

At long wavelengths, scattering of microwaves extends the possibilities.

SCATTERING BY PLASMA ELECTRONS

Thomson scattering by plasma electrons is one of the most widely used plasma diagnostic techniques and has been the subject of several reviews (see, e.g., Kunze, 1968; Evans and Katzenstein 1969; DeSilva and Goldenbaum, 1970; Sheffield, 1975). In discharge plasmas, particle velocities are small compared to the speed of light, and we need to consider only the non-relativistic limit.

A polarized monochromatic plane wave may be propagating into plasma:

$$\vec{E}(\vec{r},t) = \vec{E}_o e^{i(\vec{k}_o \vec{r} - \omega_o t)}. \tag{1}$$

All charged particles are accelerated by the oscillating electric field and emit radiation as a consequence: they are nothing else but radiating dipoles, and the emitted intensity and the scattering cross section, respectively, may readily be calculated. Because of the high particle mass, the acceleration of the ions is much lower than that of the electrons, and the contribution from the ions to the scattering thus is negligible. The differential cross section for scattering by an individual electron initially at rest is

$$\frac{d\sigma}{d\Omega} = r_e^2 \sin^2\phi \tag{2}$$

where ϕ is the angle between the electric vector of the incident wave and the direction of observation. $r_e = 2.82 \times 10^{-15}$ m is the classical electron radius. Integration over the solid angle yields the total cross section

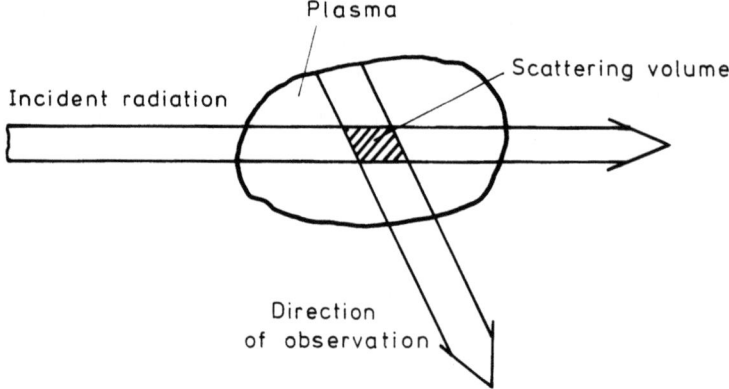

Fig. 1. Scattering Geometry.

$$\sigma_{Th} = \frac{8}{3}\pi r_e^2 = 0.665 \times 10^{-24} \text{cm}^2 \tag{3}$$

which is called the Thomson cross section. This value is extremely small, and it becomes obvious why corresponding experiments need powerful lasers.

Due to its motion, an electron will experience a Doppler shifted frequency in its frame of reference, and the stationary observer will receive radiation Doppler shifted a second time by the moving radiator. The total Doppler shift $\Delta\omega$ becomes

$$\Delta\omega = \omega_s - \omega_o = (\vec{k}_s - \vec{k}_o) \cdot \vec{v}. \tag{4}$$

ω_s and $\vec{k}_s$ are frequency and wave vector of the scattered radiation in the direction of observation, and $\vec{v}$ is the velocity of the electron.

$$\vec{k} = \vec{k}_s - \vec{k}_o \tag{5}$$

is called scattering vector. Equations (4) and (5) are in fact nothing else but statements of the conservation of energy and momentum. In the non-relativistic limit ($v/c \ll 1$) it follows $|\vec{k}_s| \sim |\vec{k}_o|$, and from the wavevector diagram of Fig. 2 we obtain for the scattering vector

$$|\vec{k}| = k \sim 2k_o \sin(\theta/2). \tag{6}$$

θ is the scattering angle.

At any point of observation, the total scattered electric field from a scattering volume containing N electrons is simply the vector sum of the scattered fields $\vec{E}_{sj}$ of all N electrons,

$$E_s = \sum_{j=1}^{N} E_{sj} \tag{7}$$

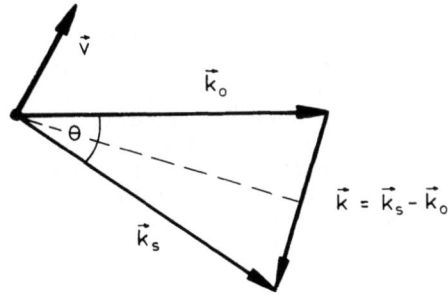

Fig. 2. Wavevector diagram for the scattering.

and the scattered power becomes by taking the time average:

$$P_s \sim \overline{\sum_{j=1}^{N} \vec{E}_{sj} \cdot \sum_{i=1}^{N} \vec{E}_{si}^{*}} \sim \overline{\sum_{j=1}^{N} |E_{sj}|^2} + \overline{\sum_{\substack{j,i=1 \\ j \neq i}} \vec{E}_{sj} \cdot \vec{E}_{si}^{*}}. \tag{8}$$

If the electrons are randomly distributed and scatter independently of each other, the second term is zero since all phases of the electric field will be present in a large ensemble. The scattered power is determined only by the first term, which represents nothing else but the sum of the contributions from all N electrons:

$$P_s = \sum_{j=1}^{N} P_{sj}. \tag{9}$$

We refer to this case as <u>incoherent</u> scattering, and the scattering cross section of the plasma becomes simply

$$\frac{d\sigma}{d\Omega} = N_e \, r_e^2 \, \sin^2\phi \tag{10}$$

where N_e is the electron density.

In order to obtain the frequency distribution of the scattered radiation, one proceeds similarly and adds the contributions of all electrons, which give equal Doppler shifts of the scattered radiation, i.e. which have the same velocity component in the direction of the scattering vector $\vec{k}$. The differential cross section becomes

$$\frac{d\sigma}{d\Omega \, d\omega} = r_e^2 \, \sin\phi \, \frac{1}{k} \, f\left(\frac{\omega_s - \omega_0}{k}\right). \tag{11}$$

$f\left(\frac{\omega_s - \omega_0}{k}\right) = f(v_k)$ is the velocity distribution function of the electrons in the direction of the scattering vector k.

From the analysis of the scattered radiation one can derive, therefore, the electron density and the velocity distribution function, i.e. the temperature in the case of a Maxwellian distribution. The spectrum is a Gaussian profile, the full half-width of which is given by

$$\Delta\lambda_{1/2} = 4\lambda_0 \sin(\theta/2) \sqrt{\frac{2\kappa T}{m} \ln 2}. \tag{12}$$

(Example: $\lambda = 694.3$ nm; $\theta = 90°$; $\kappa T = 1$ eV; $\Delta\lambda_{1/2} = 3.24$ nm.)

We return to Eq. (8). The second term becomes important, if the scattered electric field vectors from the individual plasma electrons are correlated. Such correlations should generally be expected in plasmas, since the electrons not only interact with each other but also with the ions via the Coulomb force. This gives rise to many types of plasma waves, which may be excited, as well as to the well-known Debye shielding effect. The correlations even dominate scattering, if the scattering scale length 1/k is larger than the plasma Debye length λ_D.

$$\frac{1}{k} > \lambda_D, \text{ resp. } \frac{1}{k\lambda_D} > 1. \tag{13}$$

The magnitude of these effects is governed by the scattering parameter α, which is defined by

$$\alpha = \frac{1}{k\lambda_D} = \frac{\lambda_o}{4\pi\lambda_D \sin(\theta/2)}. \tag{14}$$

Equation (8) may be evaluated by calculating the autocorrelation function of the electric field. The Fourier transform of this function yields directly the spectrum of the scattered radiation (theorem of Wiener Khinchine). The differential cross section becomes ($\omega_s - \omega_o = \omega$)

$$\frac{d^2\sigma}{d\Omega\, d\omega} = r_e^2 \sin^2\phi\, S(\vec{k},\omega). \tag{15}$$

One recognizes that the first two factors describe the scattering properties of the single particle, the factor $S(\vec{k},\omega)$ contains the scattering properties of the whole system of interacting particles. $S(\vec{k},\omega)$ is known as the dynamic form factor or spectral density function. It is the time - and space - Fourier-transform of the electron density pair correlation function. Integration over the frequency spectrum results in

$$\frac{d\sigma}{d\Omega} = r_e^2 \sin^2\phi\, N_e\, S(\vec{k}) \tag{16}$$

where

$$S(\vec{k}) = \frac{1}{N_e V} \int_{-\infty}^{\infty} S(\vec{k},\omega)\, d\omega. \tag{17}$$

The dynamic form factor has been calculated by many authors for various plasma conditions, and we refer to the review articles which give account of the plasma theory, (e.g. Sheffield 1975). In general, the form factor separates into two components,

$$S(\vec{k},\omega) = S_e(\vec{k},\omega) + S_i(\vec{k},\omega). \tag{18}$$

The first term contains the contributions from the uncorrelated motion of the electrons as well as from the correlation of the electrons as well as from the electrons with each other. The ion term $S_i(\vec{k},\omega)$ reflects the contributions from electrons correlated with the motion of the ions.

The dynamic form factor takes a relatively simple form in the Salpeter approximation (Salpeter, 1960), which holds for Maxwellian velocity distributions with $T_e \sim T_i$. In the case of one ion species of charge Z, both the electron and the ion component of $S(\vec{k},\omega)$ are approximated by an analytically similar term:

$$S(\vec{k},\omega)d\omega = \frac{N_e}{\sqrt{\pi}} \left[\Gamma_\alpha(x_e)dx_e + \frac{Z\alpha^4}{(1-\alpha^2)^2} \Gamma_\beta(x_i)dx_i \right] \tag{19}$$

and

$$S(\vec{k}) = \frac{1}{1+\alpha^2} + \frac{Z\alpha^4}{(1+\alpha^2)[1+\alpha^2+Z\alpha^2(T_e/T_i)]}. \tag{20}$$

The parameters, variables and functions are:

$$\beta^2 = Z \frac{\alpha^2}{1+\alpha^2} \frac{T_e}{T_i} < 3.5;\quad x_e = \frac{\omega}{k}\sqrt{\frac{m}{2\kappa T_e}};\quad x_i = \frac{\omega}{k}\sqrt{\frac{M}{2\kappa T_i}} \tag{21}$$

M and m are the ion and electron masses, respectively.

$$\Gamma_\alpha(x) = \frac{\exp(-x^2)}{|1+\alpha^2 W(x)|^2}.\qquad(22)$$

$W(x)$ is the plasma dispersion function:

$$W(x) = 1 - 2x \exp(-x^2) \int_0^x \exp(t^2)dt - i\sqrt{\pi}\, x \exp(-x)^2.\qquad(23)$$

This allows convenient calculation of the scattering cross sections for many plasmas of interest.

In the limit $\alpha \to 0$, the ion term $S_i(\vec{k},\omega)$ vanishes and Eq. (19) and (20) reduce to the result of completely incoherent scattering (Eq. 11): one obtains a Gaussian spectral profile of the scattered radiation. For increasing α collective effects not only modify the spectral shape of the electron component but the ion component becomes strong and dominant. Fig. 3 shows a typical spectrum for large α.

Most of the scattered energy is in a narrow high-intensity "ion feature" described by $S_i(\vec{k},\omega)$, the spectral width primarily reflecting the thermal ion motion. The small maximum is observed at the position of highly damped ion acoustic waves. As the temperature ratio T_e/T_i increases, the ion wave resonance becomes sharper. Also electron drifts modify the shape of the resonance, and the heights of the two peaks located symmetrically with respect to $\omega=0$ become different. Finally, this collective ion spectrum will be influenced by collisions if the mean free path of the ions becomes smaller than the fluctuation wavelength: ion-neutral collisions will damp the ion acoustic resonance in partially ionized plasmas, and ion-ion collisions, which become important only in dense cold plasmas, will enhance it.

$S_e(\vec{k})$ decreases rapidly with α (see first term of Eq.(20)), and the spectrum of the electrocomponent $S_e(\vec{k},\omega)$ is characterized by a narrow resonance at the position of electron plasma waves. This position is given by the Bohm-Gross dispersion relation:

$$\omega_{GB}^2 = \omega_p^2 + \frac{3\kappa T}{m} k^2.\qquad(24)$$

For large scattering parameters α, i.e., small scattering vectors $\vec{k}$, the resonance occurs close to the plasma frequency, $\omega \sim \omega_p$, and the electron density can readily be derived simply from the position.

In the presence of magnetic fields the orbits of the charged particles take on a helical form, and it can be surmised that the modulation of the scattered electric field leads to a corresponding fine structure of the spectra, i.e. modulation of the spectra with the electron cyclotron frequency or the ion cyclotron frequency, respectively. Although this was observed experimentally for high magnetic fields, it is extremely difficult to resolve this fine structure in most cases, and such measurements just are not feasible as standard diagnostic technique.

In discharge plasmas, plasma composition is of great interest, and the theory has been extended to multi component systems (Evans, 1970). It was recently verified in a magnetically stabilized arc discharge (Kasparek and Holzhauer, 1983). The effects show up in the central ion feature, the electron component of the scattered light spectrum being only weakly affected.

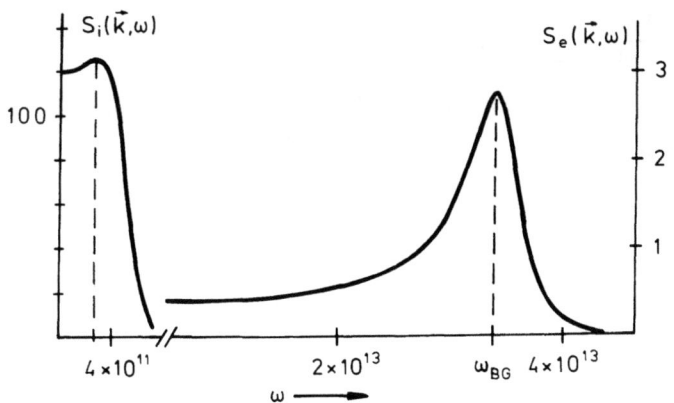

Fig. 3. Dynamic form factor in arbitrary units for k = 1 x $10^5 m^{-1}$ and α = 40.

Figure 4 shows the ion feature of a hydrogen-argon plasma, where the temperatures of both ion species are equal to the electron temperature. The narrow central component reflects the thermal motion of the argon ions, the broader part that of the protons, although the spectrum is not a simple linear superposition of the spectra which correspond to plasmas consisting of one ion species. $S_e(k)$ is modified, too. In principle, ion densities, ion temperatures and the effective charge number may be obtained in this way.

In summary, scattering of electromagnetic radiation by plasma electrons can yield a wealth of information, the scattering by unstable and turbulent plasmas even has not been discussed yet in this context. The realization of proper experiments, however, is not straightforward and many difficulties have to be surmounted. They are discussed in the reviews cited previously.

RESONANT SCATTERING BY ATOMS AND IONS

In the case the wavelength of the incident radiation coincides with transitions in atoms or ions, the effective cross section for scattering become many orders of magnitude larger than the Thomson cross section. In standard resonance experiments, the spectral width of the incident radiation $\delta\omega$ is larger than the width $\Delta\omega_{1/2}$ of the line profile of the transition, and scattering corresponds to independent absorption and emission. The well-known Einstein rate equations may be used to describe the process in this case. Although a two-level approximation is sufficient for many applications, the three-level model describes most experiments extremely well. In a plasma, collisional processes have to be considered, too, and Fig. 5 illustrates the model and the processes. The wavelength of the incident radiation coincides with the transition from level 1 to level 2. Level 3 is a metastable level in many cases, but in general, it may represent all levels of the system to which collisional and radiative transitions from level 2 occur. Collisional transitions are indicated by the rate coefficients X for electron collisions, which usually dominate in discharge plasmas. The "scattered" radiation is observed as emission from the level 2, with the emission coefficients given by

$$\varepsilon_{21} = \frac{H\nu_{21}}{4\pi} A_{21} N_2 \text{ and } \varepsilon_{23} = \frac{h\nu_{23}}{4\pi} A_{23} N_2. \tag{25}$$

In most applications lasers are used as light sources and the scattering process is commonly referred to as "laser induced fluorescence (LIF)" in the literature today.

Finally, all losses from the three-level system are representatively indicated by collisional ionization $N_e S$ from the upper level. Although negligible in most cases, if this ionization or ionization after further collisional excitation become considerable, a change in the electrical properties of the discharge results which is known as the optogalvanic effect.

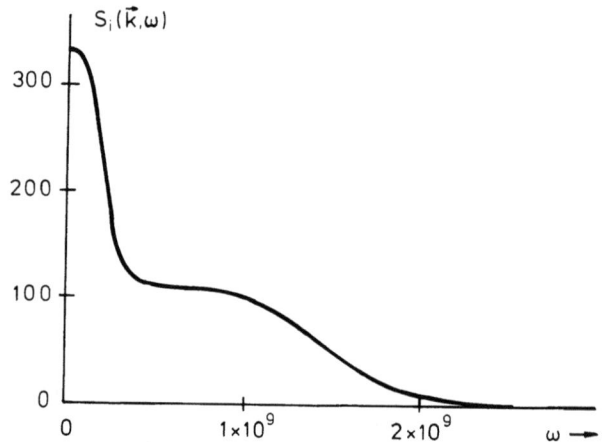

Fig. 4. Dyanmic form factor in arbitrary units for $k = 1 \times 10^5 \text{m}^{-1}$ and $\alpha = 40$.

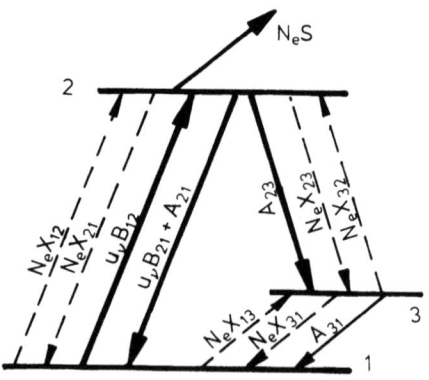

Fig. 5. Energy level diagram for a three-level system; radiative and collisional processes are indicated.

The set of coupled rate equations can be solved analytically for negligible losses ($N_1 + N_2 + N_3 = N_{10}$ = const) and a step function of the incident radiation (Meng and Kunze, 1979). Initially, all atoms were assumed to be in the lower level. The solution is illustrated in Fig. 6. For strong optical pumping ($B_{12} u_\nu \gg A_{21} + A_{23}$), the population density N_2 and hence the "scattered" radiation increase rapidly, the pumping time constant being simply $\tau_p^{-1} = (1 + g_1/g_2) B_{12} u_\nu$, where u_ν is the spectral energy density of the incident radiation and B_{12} is the Einstein coefficient for absorption. The population then decreases through transitions to level 3, where all atoms accumulate or return via collisional or radiative transitions to level 1. The transitions through this "bottleneck" determine the atoms available for pumping and therefore the steady-state density N_2. With increasing power of the incident radiation, the peak density approaches its maximum possible value,

$$N_{2max} = \frac{g_2}{g_1 + g_2} N \qquad (26)$$

which is independent of the radiation power, of atomic as well as of plasma parameters. The peak scattered intensity "saturates", upper and lower state populations are equalized ($N_2/g_2 = N_1/g_1$). This offers the unique possibility of determining the initial population density N_{10} of the lower level, and this is widely exploited in many diagnostic applications (e.g. Hintz, 1982; Gottscho, 1985).

By selecting a narrow bandwidth tunable laser, $\delta\omega < \Delta\omega_{1/2}$, the wavelength may be scanned across the profile of Doppler broadened lines. In this case, only those atoms will be pumped, which have the proper velocity component in the direction of the laser beam, and the saturated "scattered" intensity reflects exactly the velocity distribution function and hence the temperature of atoms or ions in the gas discharge.

Laser pulses of very short duration (~10 ns) may be used as well if the power is sufficiently high that the pumping time constant τ_p shorter than the pulse duration Δt_L of the laser,

$$\tau_p < \Delta t_L \text{ or } \Delta t_L > \frac{g_2}{(g_1+g_2)B_{12}u_\nu} \text{ with } B_{12} u_\nu > A_{21}+A_{23}. \qquad (27)$$

Short radiation pulses offer additional diagnostic possibilities independently whether saturation occurred or not. The decay of the upper state population gives the "life time" of this level which thus can be measured directly. In the case of free atoms, this yields the "radiative life time"

$$\tau = \frac{1}{A_{21} + A_{23}}. \qquad (28)$$

In plasmas, the decay is modified by collisions (Burrell and Kunze, 1972):

$$\tau = \frac{1}{A_{21} + A_{23} + \Sigma\Omega'} \qquad (29)$$

where $\Sigma\Omega$ not only represents collisional transition rates from the pumped level but also collisional transitions leading into this level. Depending on the plasma conditions, electron collisions or atomic collisions

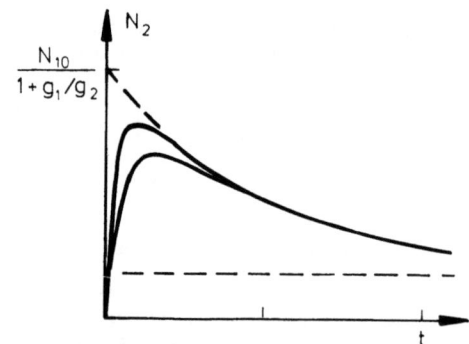

Fig. 6. Population density of level 2 for the case of a step-function laser pulse, when initially all atoms are in the lower level.

may dominate, and under favourable conditions they may be derived from fluorescence signals (e.g. Burgess et al. 1978, Burrell and Kunze, 1978; Dubreuil and Catherinot, 1980). The measurements are not straightforward: in order to take proper account of collisional transitions back into the pumped level, one has to monitor also the population densities of those levels to which collisional transitions from the pumped level occur. The coupled set of rate equations has to include all these levels, the three level system is no longer appropriate anymore to describe the time dependence of the population densities. On the other hand, once collision rates are known, the collision dependent signals may be used to measure the respective densities, (Tsuchida et al., 1983). Finally, trapping of fluorescence signals will take place at high atomic concentrations and should be checked during measurements.

Of relatively little consequence in the work on plasmas have been quite a number of phenomena or elegant methods which were observed or developed, respectively, when investigating the interaction of highly monochromatic intense laser radiation with atomic systems; we refer to the extensive literature only (e.g. Demtroder, 1981).

NEAR-RESONANT, RAYLEIGH AND RAMAN SCATTERING

We return to scattering of monochromatic radiation by atomic systems, if the wavelength of the incident radiation does not coincide with that of an atomic transition. The combined effect of radiative and collisional interaction upon an atom or ion allows the following scattering processes, (see Fig. 7), where the collisional transitions between the three levels are neglected in this approximation.

Light is scattered exactly at the frequency of the incident radiation, $\omega_{Ray} = \omega_L$, and this is nothing else but Rayleigh scattering. In order to distinguish against common cases, where scattering occurs far from any resonance, the name near-resonant Rayleigh scattering has come to use. Scattered light is also observed at a frequency $\omega_{Ram} = \omega_L - \omega_{13}$, and this is due to electronic Raman scattering. In addition to this Stokes component, anti-Stokes components may be seen as well from suitable atoms (Vriens and Adriaansz, 1975). Last but not least, collisions may provide the energy needed to transfer population from the laser-induced virtual level to the real level 2, and collision-induced fluorescence occurs at the frequencies ω_{21} and ω_{23}.

The differential cross section $\sigma(\omega_1,\omega_2)$ for absorbing a photon with frequency ω_1 and emitting one with frequency ω_2 (integrated over outgoing direction and polarization) is written most generally as (Omont et al., 1972):

$$\frac{d}{d\omega_2}\sigma(\omega_1,\omega_2) = \sigma_{tot} L(\omega_1-\omega_0)\, P(\omega_1,\omega_2) \qquad (30)$$

where $\sigma_{tot} = 2\pi^2 r_e c f_{12}$. f_{12} is the absorption oscillator strength and $L(\omega_1 - \omega_0)$ is the general line shape function for a line centered at $\omega_0 = \omega_{12}$. $P(\omega_1, \omega_2)$ is the redistribution function describing the probability that after absorption of a photon of frequency ω_1 one will be emitted at ω_2. $P(\omega_1,\omega_2)$ contains the Rayleigh-, Raman- and fluorescence terms, and it has been the subject of intense theoretical studies (e.g. Burnett et al., 1980 (I, II, III); Cooper, 1979; Voslamber and Yelnik, 1978; Yelnik et al., 1981). Experimental investigations have been limited to specific systems (e.g. Carlsten and Szoke, 1976), but the diagnostic possibilities of near-resonant scattering have already been recognized and taken advantage of (e.g. Chan and Daily; 1979, Vriens and Adriaansz, 1976; Watanabe et al., 1984).

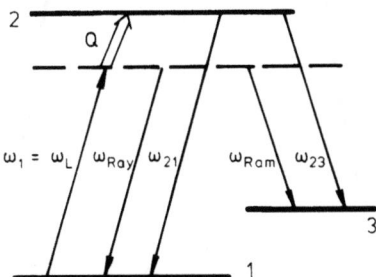

Fig. 7. Energy level diagram and transitions in the case of near-resonant scattering.

The Rayleigh component is not trapped in a plasma even at high atomic concentrations and thus allows unambiguous measurements of the atomic densities in these cases. Its spectral width has the shape of the absorption line profile, and atomic or ion temperatures may be derived in this way (e.g. Dobele and Hirsch, 1975; Stern and Johnson, 1975).

Novel possibilities were explored by Himmel and Sowa (1983) when performing scattering experiments on the wing of Stark-broadened line profiles. The results indicate that the relaxation time of the electric microfield established by the ions may be investigated.

Finally, we should not suppress the fact, that Rayleigh scattering may also be completely undesired. This is the case, for example, when carrying our Thomson scattering experiments on high-density low-temperatures plasmas. In such plasmas, Rayleigh scattering even from excited levels may become very strong. Fig. 8 shows the ion feature $S_i(k,\omega)$ obtained from a pure hydrogen plasma (Maurmann and Kunze, 1983) where the superposed Rayleigh component is shown by the dot-dashed line. Without taking proper account of this contribution, large errors may occur in the interpretation of scattered spectra.

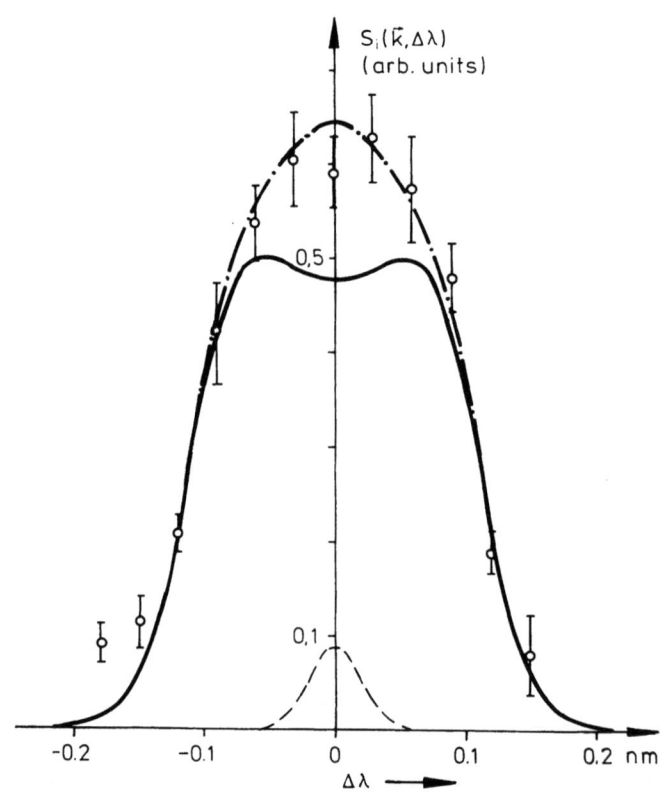

Fig. 8. Spectrum of light scattered from a high-density, low-temperature plasma. $N_e = 1.2 \times 10^{18}$ cm^{-3}; $T_e = T_i = 2.5$ eV; $\theta = 90°$.

MULTI-PHOTON TRANSITIONS

Fluorescence excited by two-photon absorption is of great interest for two reasons: first, neither the primary laser beam nor the fluorescence radiation are trapped in the case of high atomic concentrations, and second, higher levels may be pumped directly, which is important, if the first excited level is already too high for single-photon excitation by the available laser radiation. The feasibility has been demonstrated and the method has been applied by several groups (e.g. Das et at., 1983, DiMauro et al., 1984).

One considers again a three-level system, where the transitions 1 → 3 and 1 → 2 are dipole transitions (Fig. 9). Initially, one may picture a photon being absorbed and the atom reaches a virtual level 3'. The life time of this level is extremely short, but in the case of very high laser powers a second photon may indeed be absorbed during that short time, and the system goes to level 2. The probability of this two-photon absorption not only depends on the second power of the laser intensity but also inversely on the square of the energy difference $\Delta E = E_3 - E_{3'}$.

The impact of such possibilities were tremendous in atomic physics, since by using counter-propagating beams the Doppler effect could be eliminated (e.g. Demtroder, 1981).

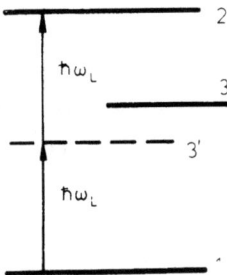

Fig. 9. Energy level diagram for two-photon absorption.

A variant of these techniques allows the local measurement of high-frequency electric fields in plasmas (Kunze, 1981) and has been successfully used to study such fields in a pulsed hollow cathode discharge (Hildebrandt, 1983). Fig. 10 indicates the energy level diagram: the level 3 now is close to level 2, the transition 2 → 1 still forbidden because $\Delta L = 0, \pm 2$. The laser frequency is tuned till a two-quantum transition via the virtual level 3' occurs: a photon from the laser beam and a quantum from the electric wave field in the plasma are absorbed. A second process is possible via the virtual level 3": it corresponds to the absorption of a laser photon and simultaneous emission of a quantum to the electric field. The transition rate is given for low laser intensities P_L by:

$$R \sim N_1 \cdot P_L \cdot \frac{\langle E^2(\Omega) \rangle}{(\omega_L - \omega_{13})^2} \text{ for } \omega_L = \omega_{12} \pm \Omega \ . \tag{31}$$

This reveals that the fluorescence radiation from the level 2 reflects the spectrum $\langle E^2(\Omega)\rangle$ of the electric wave field in the plasma. At higher laser powers saturation will occur and has to be taken into account. In the case of strong wave fields Eq.(31) no longer holds since it was derived within the limits of second-order perturbation theory: higher order transitions become possible, i.e. in addition to absorption of $\hbar\omega_L$ simultaneous absorption or emission of 2, 4, 6 ... quanta $\hbar\Omega$ leading to level 3, or of 3, 5, 7 ... quanta leading to level 2. Theoretical calculations have been confirmed through experiments on lithium atoms in microwave fields (Rebhan, 1983).

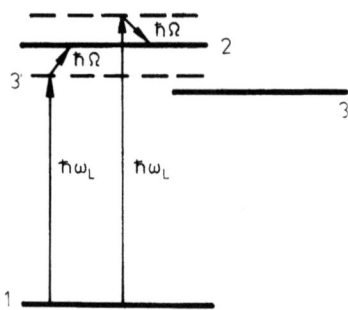

Fig. 10. Energy level diagram pertinent to absorption of laser photons by atoms in high-frequency field Ω.

REFERENCES

Burgess, D. D., Kolbe, G., and Ward, J. M., 1978, J. Phys. B, 11:2765.
Burnett, K., Cooper, J., Ballagh, R. J., and Smith, E. W., 1980 (I), Phys. Rev. A, 22:2005.
Burnett, K., Cooper, J., 1980 (II), Phys. Rev. A, 22:2027.
Burnett, K. and Cooper, J., 1980 (III), Phys. Rev. A, 22:2044.
Burrell, C. F. and Kunze, H. J., 1972, Phys. Rev. Lett., 28:1.
Carlsten, J. L. and Szoke, A., 1976, J. Phys. B, 9:1231.
Chan, C. and Daily, J. W., 1979, J. Quant. Spectrosc. Radiat. Transfer, 21:527.
Cooper, J., 1979, Astrophys. J., 228:339.
Das, P., Ondrey, G., van Veen, N., and Bersohn, R., 1983, J. Chem. Phys. 79:721.
Demtroder, W., 1981, "Laser Spectroscopy". Springer, Berlin.
DeSilva, A. W. and Goldenbaum, G., 1970, Chapter 3, in: "Methods of Experimental Physics", Vol. 9A, Griem, R. H., and Lovberg, eds., Academic Press, New York.
DiMauro, L. F., Gottscho, R. A., and Miller, T. A., 1984, J. Appl. Phys., 56:2007.
Dobele, H. F. and Hirsch, K., 1975, Phys. Lett., 54A:267.
Dubreuil, B. and Catherinot, A., 1980, Phys. Rev. A, 21:188.
Evans, D. E., 1970, Plasma Phys., 12:573.
Evans, D. E. and Katzenstein, J., 1969, Rep. Prog. Phys., 32:207.
Gottscho, R., 1985, this Proceedings.
Hildebrandt, J., 1983, J. Phys. B, 16:149.
Himmel, G. and Sowa, L., 1983, J. Phys. B., 16:4117.
Hintz, E., 1982, Physica Scripta, T2/2:454.
Kasparek, W. and Holzhauer, E., 1983, Phys. Rev. A, 27:1737.

Kunze, H. -J., 1981, in: "Spectral Line Shapes," Wende, B., ed., Walter de Gruyter, Berlin.
Kunze, H. -J., 1968, Chapter 9, in: "Plasma Diagnostics," W. Lochte-Holtgreven, North Holland, Amsterdam.
Maurmann, S. and Kunze, H. -J., 1983, Phys. Fluids, 26:1630.
Meng, H. C. and Kunze, H. -J., 1979, Phys. Fluids, 22:1082.
Omont, A., Smith, E. W., and Cooper, J., 1972, Astrophys. J., 175:185.
Rebhan, U., 1983, Ph. Thesis, Ruhr-University Bochum, to be published in J. Phys. B.
Salpeter, E. E., 1960, Phys. Rev., 120:1528.
Sheffield, J., 1975, "Plasma Scattering of Electromagnetic Radiation," Academic Press, New york.
Stern, R. A. and Johnson, III, J. A., 1975, Phys. Rev. Lett., 34:1548.
Tsuchida, K., Miyake, S., Kadota, K., and Fujita, J., 1983, Plasma Phys., 25:991.
Vriens, L. and Adriaansz, M., 1975, J. Appl. Phys., 46:3146.
Vriens, L. and Adriaansz, M., 1976, Appl. Phys., 11:253.
Voslamber, D. and Yelnik, J. -B., 1978, Phys. Rev. Lett., 41:1233.
Watanabe, Y., Ikegami, T., Nishiyama, T. and Arkazaki, M., 1984, Jap. J. Appl. Phys., 23:904.
Yelnik, J. B., Burnett, K., Cooper, J., Ballagh, R. H., and Voslamber, D., 1981, Astrophys. J. 248:705.

EMISSION SPECTROSCOPY

H.-J. Kunze

Institut für Experimentalphysik
Ruhr-Universität
Bochum, FR Germany

INTRODUCTION

Although light scattering may be considered already a subfield of emission spectroscopy, if this is defined in the broadest sense, orthodox emission spectroscopy commonly is understood to be concerned with the electromagnetic radiation emitted from plasmas without external influence on the radiators. The main objective usually is to obtain information on the plasma state, i.e. for example, on temperatures, densities, plasma composition, magnetic and electric fields, or waves excited in the plasma. This is certainly only possible to that extent that these parameters influence sufficiently the emitted electromagnetic radiation. For applications, on the other hand, one may simply be interested in the emitted spectrum of a specific discharge.

As far as instrumentation goes, emission spectroscopy is probably the simplest among the powerful diagnostic techniques, but we should not conceal the classic drawback that all measurements are integrated along the line of sight. It could certainly be desirable to know the total spectrum, which covers the region from the x-rays to the far-infrared, but it is clear, that this will not be attainable in most cases, be the only reason the fact, that the necessary equipment is not available, which is rather different for the various spectral regions. One naturally will focus first onto the spectral region where the strongest emission occurs. Bremsstrahlung shows a maximum at

$$\lambda_m \cdot (kT) = 620 \text{ nm eV}. \tag{1}$$

For a plasma of $kT = 5$ eV, for example, this maximum is at $\lambda_m = 124$ nm.

The emission decreases slowly to longer wavelengths, reaching 1% of the maximum at about 3.3 µm; the decrease to shorter wavelengths is faster, the 1% level is at about 25 nm. The emission of strong spectral lines will also occur in this spectral region between the two wavelengths.

Measurements in the visible and near ultraviolet are certainly the easiest ones; a sufficient selection of spectrographs and detectors is available, independent whether low or high spectral resolution is desired, high time resolution is necessary or steady-state discharges are investigated. Below 200 nm, air absorbs the radiation and it is necessary to evacuate the entire optical system. It may be kept separated from the discharge vessel by using windows, but there are no materials known suitable for windows below about 110 nm, the short-wavelength transmission limit of LiF. This is also the limit of presently available intensity standards (calibrated arc discharges). Studies below require differentially pumped systems in which the entrance slit is quite narrow and placed in the side wall of the discharge device. Naturally also no lenses are available and mirrors must be used instead. The absolute sensitivity calibration is difficult, and can be accomplished with reasonable means and effort only at single wavelengths using the "branching ratio method" (Fig. 1): two optically thin lines from the same upper level emitted from a suitable source are measured, the wavelengths of which are in the vacuum-uv and the visible spectral region. If the transition probabilities are known, an absolute measurement of the visible line also yields the absolute intensity of the other transition,

$$\varepsilon_{21} = \varepsilon_{23} \frac{\lambda_{23}}{\lambda_{21}} \frac{A_{21}}{A_{23}} . \tag{2}$$

At long wavelengths, quartz transmits to about 4 μm and detectors having high sensitivity to about 3 μm without cooling are available (e.g. the PbS photo-conductive detector).

INHOMOGENEOUS, OPTICALLY THIN PLASMAS

As pointed out already, spectroscopic measurements integrate along the line of sight, and the question arises, whether it is possible at all to derive local emission coefficients and hence local plasma parameters or not. The problem is certainly aggravated by absorption and re-emission, and we shall restrict ourselves, therefore, to the optically thin case.

We consider a cylindrically symmetric plasma column which implies that the emission coefficient is a function of the radius only, i.e. $\varepsilon(\omega) = \varepsilon(\omega,r)$. This column is viewed perpendicular to the axis, and the spectral intensity is obtained by integrating along a chord. Fig. 2 illustrates the geometry. The intensity may be written as

$$I(\omega,y) = 2 \int_0^{\sqrt{R^2-y^2}} \varepsilon(\omega,r) dx = 2 \int_0^R \frac{\varepsilon(\omega,r) r \, dr}{\sqrt{r^2-y^2}} . \tag{3}$$

This integral equation is of the Abel type and can be inverted, the transformation being called Abel inversion:

$$\varepsilon(\omega,r) = \frac{1}{\pi} \int_r^R \frac{I'(\omega,r)}{\sqrt{y^2-r^2}} dy . \tag{4}$$

It can readily be accomplished by means of a computer.

In the case of no symmetry, tomographic methods can be used in principle. The emitted intensity $I(\omega,y)$ is observed not only in one direction (here x-direction) but in many, and the local emission structure is synthesized by a computer. Such rather elaborate methods have so far been applied to fusion plasmas in large devices.

INTERPRETATION OF SPECTRA

The many aspects and diagnostic possibilities of emission spectroscopy as well as techniques and instrumentation are treated in a number of books (e.g. Bekefi, 1976; Griem, 1974; Huddlestone and Leonard, 1965,

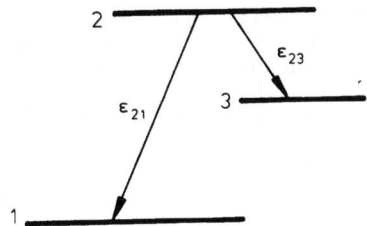

Fig. 1. Principle of branching ratio calibration.

Lochte-Holtgreven, 1968), the monograph "Plasma Spectroscopy" by Griem (1964) still remaining the standard reference.

Line Profiles

The measurement of line profiles is rather straightforward since it does not require the absolute sensitivity calibration of the spectrograph. As discussed in the tutorial lecture by Griem (1985), Doppler shifts associated with the random motion of the radiating atom or ion and Stark broadening and shifts caused by the particle-produced fields in the plasma determine the shape of a spectral line. The profile of the line actually emanating from the plasma will be modified by radiative transfer, and there can be self-absorption and even self-reversal in colder boundary layers (see Karabourniotis, 1985). Zeeman shifts are of the order $\Delta\lambda \sim 10^{-7} \lambda^2 B$) when wavelengths are in nanometer and magnetic fields in Tesla, and they are usually negligible.

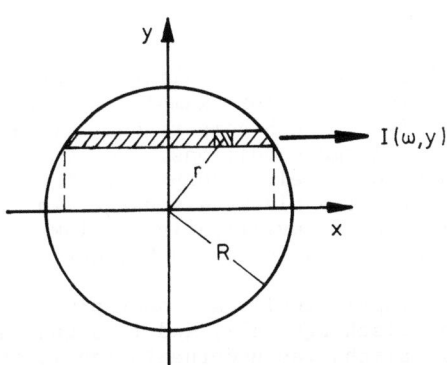

Fig. 2. Schematic view of cylindrical plasma column.

We consider now the optically thin case only; one certainly will attempt to find lines, where one broadening mechanism dominates. From Doppler broadened profiles the atom or ion temperature may be derived, from Stark broadened profiles the electron density, since the shape depends only rather weakly on the temperature. The whole subject of line broadening is treated in the book "Spectral Line Broadening by Plasmas" by Griem (1974), and up-to-date critical reviews and tabulations of experimental results have been published by Konjevic et al. (1984).

The diagnostician primarily is interested in the accuracy of the available data, and typical uncertainties quoted in the reviews are between 30% and 50%, through uncertainties of a factor of two still exist. The standard procedure is to derive the electron density from the half-width $\Delta\lambda_{1/2}$ only. High-precision data are available for example, for H, He I and He II, and hydrogen or helium may be added for density determination to discharges if not already present.

Hydrogen lines, especially those of the Balmer series, exhibit linear Stark effect, are easy to measure and are the most strongly broadened lines under comparable conditions. The most frequently measured Balmer line is H_β at 486.13 nm; it is far less sensitive to radiative transfer effects than H_α, and a comparison of electron densities from independent diagnostic methods (Griem, 1974) indicates an accuracy to within about 5%. A similar accuracy exists for H_γ at 434.05 nm.

In the temperature range 1 eV - 4 eV and for densities between $10^{20} m^{-3}$ and $10^{24} m^{-3}$ we can deduce a simple relation from the tabulated data of Griem (1974) for the H_β line,

$$N_e = 1.09 \times 10^{22} (\Delta\lambda_{1/2})^{1.458} \ m^{-3} \tag{5}$$

where the full halfwidth $\Delta\lambda_{1/2}$ is in nm.

In hydrogenic helium, the P_α and the P_β transitions are at convenient wavelengths, i.e. at 468.6 nm and 320.3 nm, respectively, and they have been studied extensively, too. Pittmann and Fleurier (1983), propose an empirical relation for the P_α line,

$$N_e = 2.04 \times 10^{22} (\Delta\lambda_{1/2})^{1.21} \ m^{-3} \tag{6}$$

and Ackermann et al. (1985) for the P_β transition,

$$N_e = 4.1 \times 10^{22} (\Delta\lambda_{1/2})^{1.35} \ m^{-3}. \tag{7}$$

The broadening of other lines may be calibrated against that of these transitions, and this was done, for example, also by Ackermann et al. (1985) for some C IV lines. Such comparisons certainly have to be performed under excellent plasma conditions, i.e. emission of the lines from a homogeneous plasma with no self-absorption. The authors used the gas-linear pinch, a new type of a plasma discharge, which offers specific advantages for spectroscopic investigations (Finken and Ackermann, 1981). Fig. 3 shows the schematic drawing of such a device.

The discharge is topologically a linear pinch. A fast valve injects a gas shell along the discharge wall, and this gas is preionized by a circular array of pin discharges underneath the lower electrode E2. The magnetic field of the main discharge then accelerates the plasma cylinder towards the axis and, after collapse on this axis, a dense plasma column

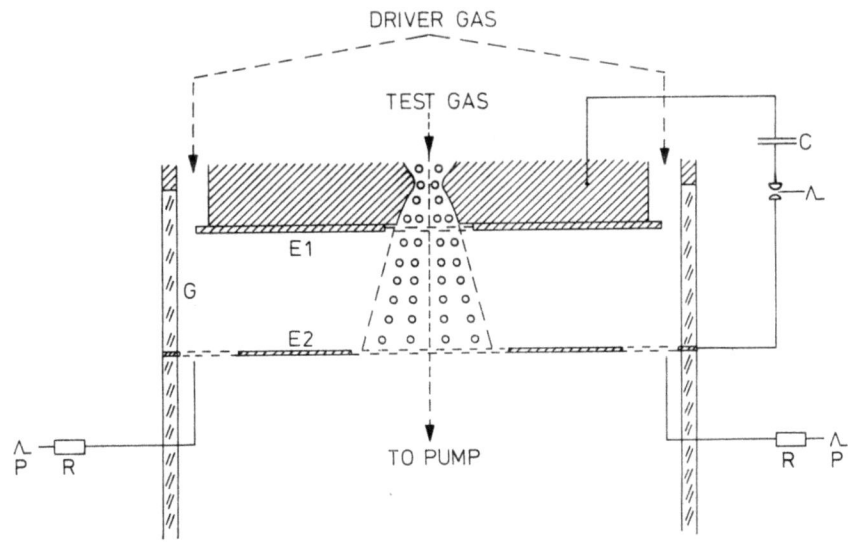

Fig. 3. Schematic drawing of the gas-linear pinch.

is formed which exists for about a sound transit time, i.e. typically for about a few hundred nanoseconds. These dynamically confined plasmas are very reproducible and show steep density gradients at the boundary while being flat at the center. The critical point, however, is the injection of a second gas stream ("test gas") along the axis prior to the compression. If properly operated, the test gas is essentially confined to the center of the discharge tube, and the spectroscopic analyses indeed confirms that the respective atoms or ions emit only from the homogeneous central part of the plasma column; the amount of test gas is easily adjusted to optically thin line emission, and no self-absorption in cold boundary layers occurs.

On the other hand, the concentration of the test gas can be increased without strongly affecting the plasma conditions till resonance lines become optically thick. Figure 4 shows the resonance doublet $2^2P \rightarrow 2^2S$ of C IV for such a condition. The optical depth at the center of the stronger component was $\tau = 17$ (Bottcher, 1985), and the absolute intensity in the line center is simply given by the Planck function.

Finally, also HeI lines are recommended for plasma diagnostic measurements, the accuracy of the electron densities deduced from their Stark-widths being estimated to about $\pm 10\%$ (Konjevic' et al., 1984). Many HeI lines have nearby forbidden transitions $\Delta \ell \pm 1$, and Figure 5 displays the theoretical profile of the line at 492.19 nm for a density of 3×10^{20} m^{-3} (full line) as example. The forbidden line occurs as a result of the breakdown of the parity selection rules induced by the particle-produced fields. As the electron density increases, the line not only broadens but the ratio of the intensity of the forbidden line to the allowed line grows and the separation between the two maxima increases. All three phenomena together lead to the rather accurate determination of the electron density.

In rapidly pulsed discharges, high-frequency oscillations may be excited far above their thermal level, and two satellites displaced

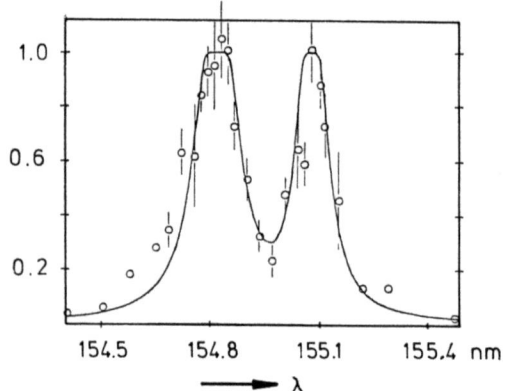

Fig. 4. Measured and computed profile of the C IV resonance doublet for an optical depth $\tau = 17$ at the line center.

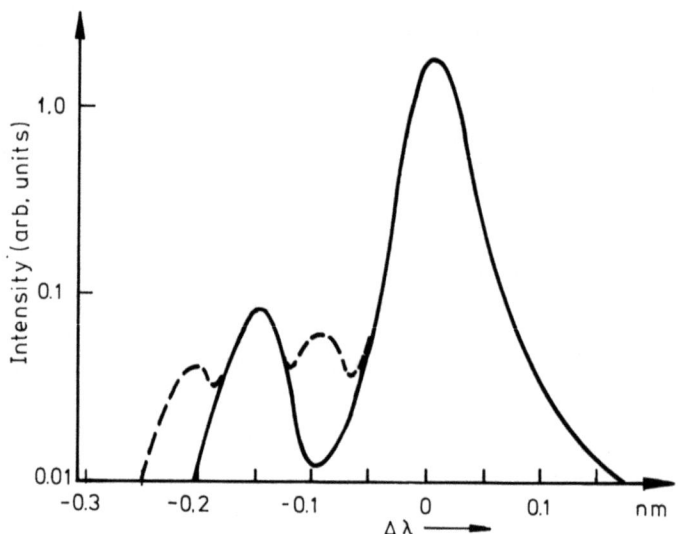

Fig. 5. Theoretical spectrum (full line) of the HeI transition at 492.19 nm for a density of 3×10^{20} m^{-3}.

symmetrically with respect to the forbidden component will appear (indicated by the dashed lines in Fig. 5.). They are caused by the high-frequency Stark effect of the oscillating field (Griem, 1974), their separation from the forbidden component corresponds to the frequency Ω of the oscillations, their intensity is determined by the time average of the oscillating electric field squared, $\langle E^2(\Omega)\rangle$. Second-order perturbation theory yields for the ratio S of the integrated satellite intensity to that of the allowed line (Baranger and Mozer, 1961)

$$S_{\pm} = \frac{4\pi\varepsilon_0 h^2}{6m^2 e^2} R_{\ell\ell'} \frac{\langle E^2(\Omega)\rangle}{(\omega_{ij} \pm \Omega)^2} . \tag{8}$$

m, e, ε_0 and h have their usual meaning, $R_{\ell\ell'}$ is a dimensionless radial matrix element, and ω_{ij} is the frequency separation between the two upper levels.

Fig. 6 shows the time evolution of such a profile in a pulsed linear discharge (from Wiegart, 1983). This illustrates, if higher order effects become important, the spectra become rather complicated in detail.

<u>Line Intensities</u>

The emission coefficient of a transition is given by

$$\varepsilon = \frac{\hbar\omega}{4\pi} A_{mn} N_m \tag{9}$$

where N_m is the density of the atoms or ions in the upper level. As discussed by Griem (1985), relating N_m to the general plasma conditions, however, can be an extremely difficult problem, in transient plasmas N_m may even depend on the time history of the plasma. The problem simplifies in the limit that collisions between the levels dominate all radiative transitions (LTE); the level populations obey a Boltzmann distribution in this case,

$$\frac{N_m}{N} = \frac{g_m}{Z(T)} \exp(-E_m/kT) \tag{10}$$

where Z(T) is the partition function of the system and N is the total number density.

The intensity of a transition thus depends only on the temperature and the total density of the atomic species, but the partition function has to be known,

$$Z(T) = \Sigma\, g_n \exp(-E_n/kT) \tag{11}$$

which involves a summation over all existing levels. The actual evaluation is difficult and has to take into account the lowering of the ionization potential in the plasma. Tabulations are found by Drawin and Felenbok (1965) for many elements.

Temperature measurements from relative line intensities are much easier, since N and Z(T) cancel in the ratio,

$$\frac{\varepsilon_{mk}}{\varepsilon_{n1}} = \frac{\lambda_{n1} A_{mk}}{\lambda_{mk} A_{n1}} \exp\left(-\frac{E_m - E_n}{kT}\right). \tag{12}$$

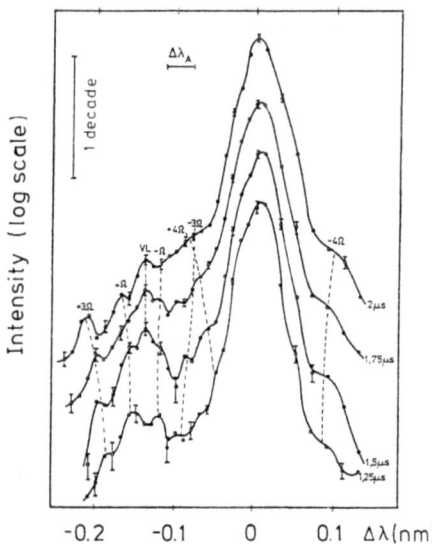

Fig. 6. Time evolution of a HeI line profile in a pulsed discharge (from Wiegart, 1983).

Accurate measurements require

$$kT < E_m - E_n \qquad (13)$$

and for higher temperatures one has to compare transitions in subsequent ionization stages, since now the difference $E_m - E_n$ is augmented by the respective ionization energy of the lower ionization stage.

The population density of ions of charge z-1 in the energy state E_n^{z-1} is related to the ground state ions of charge z by the following Saha equation

$$\frac{N_e N_1^z}{N_m^{z-1}} = \frac{2g_1^z}{g_m^{z-1}} \left(\frac{mkT}{2\pi\hbar^2}\right)^{3/2} \exp\left(\frac{E_\infty^{z-1} - E_m^{z-1}}{kT}\right). \qquad (14)$$

$E_\infty^{z-1} - E_m^{z-1}$ is the ionization energy of the level m.

The existence of LTE must certainly be checked, and general validity criteria are discussed by Griem (1964). With decreasing electron density, collisions will cease to dominate between widely spaced levels, the populations of the other levels, i.e. usually the high-lying ones, still being balanced by collisions; they remain populated according to their respective Boltzmann factors (Partial local thermodynamic equilibrium, PLTE).

At low densities, practically all processes are not balanced in detail, and we reach the coronal limit (Griem, 1985; eq., 48). Radiative transitions dominate, and all atoms and ions are essentially in their respective ground states or in metastable states. Excitation occurs by electron collisions, decay by radiative transitions.

The emission coefficient of a line is given by

$$\varepsilon_{mn} = \frac{\hbar\omega}{4\pi} \frac{A_{mn}}{\Sigma_k A_{mk}} X(1 \to m) N_1^Z N_e \tag{15}$$

if $X(1 \to m)$ is the collisional rate coefficient and excitations from metastable levels are negligible. X depends on the electron temperature, and the intensity ratio of two suitable lines allows again the determination though now the respective rate coefficients must be known. Metastable levels introduce an additional density dependence and thus a new diagnostic possibility (Gabriel and Jordan, 1972).

The steady-state distribution between consecutive ionization stages is determined by the famous corona equilibrium relation,

$$\frac{S_z}{\alpha_{z+1}} = \frac{N^{z+1}}{N^z} . \tag{16}$$

Ionization is balanced by recombination.

Continuum Radiation

The continuum radiation emitted by a plasma is a superposition of bremsstrahlung (free-free transitions) and recombination radiation (free-bound transitions). It is a function of the wavelength, of the electron and ion densities and of the electron temperature.

At short wavelengths ($\hbar\omega \gg kT_e$) both bremsstrahlung and recombination radiation fall off as $\exp(-\hbar\omega/kT_e)$, and the intensity ratio at two wavelengths is extremely sensitive to the temperature.

Recombination continua exhibit discontinuities at series limits, and the ratio of the intensities on either side is only a function of the electron temperature.

Pure plasmas, finally, where calculations of the continuum are precise, allow the determination of the temperature simply from the ratio of a line intensity and continuum intensity. At longer wavelengths, the continuum is only a weak function of the temperature, and densities may be derived from the absolute intensity with reasonable accuracy.

The optical thickness of the bremsstrahlung radiation increases rapidly with wavelength (Griem, 1964)

$$\tau \sim \lambda^2 N_e^2 / T_e^{3/2} \tag{17}$$

and approaches the Planck function. For high plasma densities this may already occur in the visible spectral region and the electron temperature can readily be determined (Finken et al., 1978).

REFERENCES

Ackermann, U., Finken, K.-H., and Musielok, J., 1985, Phys. Rev. A, 31:2597.
Baranger, M. and Mozer, B., 1981, Phys. Rev., 123:25.
Bekefi, G., Deutsch, C., and Yaakobi, B., 1976, Ch. 13, in: "Principles of Laser Plasmas," Bekefi, G. ed., John Wiley & Sons, New York.
Bottcher, F., 1985, private communication.
Darwin, H. W. and Felenbok, P., 1965, "Data for Plasmas in Local Thermodynamic Equilibrium," Gauthiers-Villars, Paris.
Finken, K.-H. and Ackermann, U., 1981, Phys. Lett., 85A:278.
Gabriel, A. H. and Jordan, C., 1972, in: "Case Studies in Atomic Collision Physics," Vol. 2, McDaniel and McDowell, ed., North Holland, Amsterdam.
Griem, H. R., 1964, "Plasma Spectroscopy," McGraw-Hill, New York.
Griem, H. R., 1974, "Spectral Line Broadening by Plasmas," Academic Press, New York.
Griem, H. R., 1985, in this Proceedings.
Huddlestone, R. H., and Leonard, S. L., eds., 1965, "Plasma Diagnostic Techniques," Academic Press, New York.
Karabourniotis, D., 1985, in this Proceedings.
Konjevic, N., Dimitrijevic, N. S., and Wiese, W. L., 1984, J. Phys. Chem. Ref. Data, 13:619; 13:649.
Lochte-Holtgreven, W., ed., 1968, "Plasma Diagnostics," North Holland, Amsterdam.
Pittmann, T. L. and Fleurier, C., 1983, in: "Spectral Line Shapes," Vol. II, Burnett, K. ed., de Gruyter, New York.
Wharmby, D., 1985, in this Proceedings.
Wiegart, N. J., 1983, Phys. Rev. A, 27:2114.

PULSED OPTICAL PUMPING IN LOW-PRESSURE MERCURY DISCHARGES

P. van de Weijer and R.M.M. Cremers

Philips Research Laboratories
Eindhoven, The Netherlands

INTRODUCTION

A good understanding of the fundamental processes occurring in the positive column of low-pressure mercury-noble-gas discharges is of practical interest because these discharges are used in fluorescent lamps (Elenbaas, 1959; Waymouth, 1971). In these lamps the light production is mainly due to mercury UV resonance radiation originating from the $6\ ^3P_1$ and $6\ ^1P_1$ levels at 254 and 185 nm, respectively (Fig. 1). This UV radiation is converted into visible radiation by a coating of fluorescent powder on the inside of the discharge tube. Once a mercury UV resonance photon has been produced in the volume of the discharge it is absorbed and re-emitted many times before it escapes from the discharge. Due to this process, called trapping or imprisonment, the effective radiative lifetime of the $6\ ^3P_1$ and $6\ ^1P_1$ level can be considerably longer than the natural radiative lifetime. The longer the effective radiative lifetime, the greater the possibility that the energy will be released via a non-radiative process. Thus, a detailed knowledge of the effective radiative lifetime of the $6\ ^3P_1$ and $6\ ^1P_1$ levels as a function of the discharge parameters may be of help in the development of more efficient fluorescent lamps.

For discharge conditions relevant to the conventional fluorescent lamp the UV output is dominated by the 254 nm contribution (Barnes, 1960; Koedam et al., 1963). Due to this, the radiation trapping on the 254 nm line for cylindrical discharge geometries has been studied in the past by numerous investigators, both experimentally (Thomas and Gwinn, 1948; Alpert et al., 1949; Helps and McCoubrey, 1960; Yang, 1966; Nussbaum and Pipkin, 1967; Micheal and Yeh, 1970; Hammond and Gallo, 1971; Halstead and Reeves, 1982) and theoretically (Biberman, 1947; 1949; Holstein, 1947, 1951; Kenty, 1950; Phelps, 1958; Walsh, 1959; Payne and Cooke, 1970; van Irigt, 1976). The experiments, however, are almost always limited to pure mercury vapour, whereas we are also interested in mercury-noble-gas mixtures because of their application in fluorescent lamps. Recently, therefore, we performed measurements on the effective radiative lifetime of the $6\ ^3P_1$ level in mercury, mercury-argon, and mercury-krypton discharges (van de Weijer and Cremers, 1985a). We used an experimental method different from the earlier experiments. It is

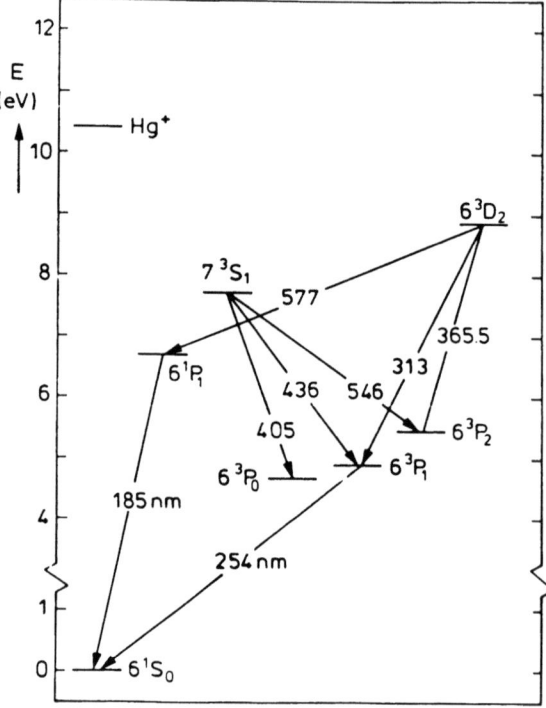

Fig. 1. Energy Level Diagram of Mercury.

based on optical pumping with a dye laser pulse at 405 nm, transferring mercury atoms from the $6\ ^3P_0$ level to the $6\ ^3P_1$.

Until recently the trapping on the 185 nm line had not been studied. In order to calculate the 185 nm output of fluorescent lamps in model calculations, a theory similar to the one applied for the radiation trapping on the 254 nm line is generally used. However, this theory is based on the assumption of complete spectral redistribution. Complete spectral redistribution means that there is no correlation between the frequency of the absorbed and the re-emitted photon, i.e. absorption and (local) emission profile are the same. In most cases of practical interest this assumption is valid for the 254 nm line, since there are many decorrelating collisions with the excited mercury atoms within the natural lifetime of the $6\ ^3P_1$ level. For the 185 nm line the assumption of complete spectral redistribution is not valid, since the natural radiative lifetime of the $6\ ^1P_1$ level (1.3 ns (Lurio, 1965)) is much shorter than that of the $6\ ^3P_1$ level (120 ns, see next section).

For conventional fluorescent lamp conditions the incorrect assumption of complete spectral redistribution does not lead to large errors in the calculation of the UV output, since the contribution of the 185 nm line to the total UV output is only small. Moreover, this assumption does not lead to dramatic errors in the results of the calculation of the 185 nm output itself. This is due to the fact that radiative loss is the dominant loss mechanism for the $6\ ^1P_1$ level. As a consequence a large error in the calculation of the effective radiative lifetime of the $6\ ^1P_1$

level leads to a large error in the calculated density in the $6\ ^1P_1$ level but not to a large error in the 185 nm output.

In recently developed compact fluorescent lamps, mercury discharges are used with a much higher current density than in the conventional fluorescent lamps. As a consequence electron deexcitation processes may become of importance for the population density in the $6\ ^1P_1$ level. Moreover, the contribution of the 185 nm line to the total UV output increases with current density (Barnes, 1960; Koedam et al., 1963). For these two reasons one might expect an increasing error in the calculated UV output if the effective radiative lifetime of the $6\ ^1P_1$ is calculated with the assumption of complete spectral redistribution.

In view of these considerations, Post (1984) performed afterglow experiments on low-pressure mercury discharges. From these experiments the effective radiative lifetime of the $6\ ^1P_1$ level could be derived if the discharge current is low enough to ensure that radiative decay is the dominant loss mechanism for the $6\ ^1P_1$ density. Moreover, Post (1985) applied a theory which accounts for incomplete spectral redistribution. This theory is based on the work of Payne et al. (1974).

However, the effective radiative lifetime of the $6\ ^1P_1$ level could only be derived from the afterglow experiments if direct electron-impact excitation from the ground state to the $6\ ^1P_1$ level contributes significantly to the population density in the $6\ ^1P_1$ level in the steady state of the mercury discharge. This is generally true for low-pressure mercury discharges (especially at low discharge currents), but not for mercury-noble-gas discharges with a few hundred Pascal of noble-gas, as used in fluorescent lamps. Therefore we applied an optical pumping experiment to determine the effective radiative lifetime of the $6\ ^1P_1$ level. This method is similar to the one we used to determine the effective radiative lifetime of the $6\ ^3P_1$ level. In this case the dye laser pulse transfers mercury atoms from the $6\ ^3P_2$ level to the $6\ ^1P_1$ level (Post et al., 1985).

Performing the optical pumping experiments to determine the effective radiative lifetime of the $6\ ^3P_1$ and the $6\ ^1P_1$ level we chance the population density in the 6 P levels in the discharge. For discharge conditions where these 6 P levels are involved in the ionization processes this results in a change of the discharge voltage, i.e. an optogalvanic effect. Though the effect was observed more than 50 years ago (Penning, 1928), the potential of the phenomenon as a method of discharge diagnostics has been recognized only since the development of tunable dye lasers (Green et al., 1976). The optogalvanic effect experiment can be performed with a CW or a pulsed laser. When the discharge is irradiated with a CW laser, a steady state is created in the discharge. With a pulsed laser, however, the dynamic behavior of the discharge can be studied.

During the last few years it has become popular to study this dynamic optogalvanic effect, i.e. the optogalvanic effect induced by a light pulse which is short compared to the time scale of the ionization processes in the discharge (Erez et al., 1979; Miron et al., 1979; Rosenfeld et al., 1979; Kravis and Haydon, 1981; Ben-Amar et al., 1981;

Nippoldt and Green, 1981; Delsart et al., 1981; Shuker et al., 1982, 1984; Nestor, 1982; Ceasar and Heuilly, 1983; Fujimoto et al., 1983; Burakov et al., 1983; Uetani and Fujimoto, 1984; Suri et al., 1984, 1985; Pfaff et al., 1984; Doughty et al., 1984). The pulsed optogalvanic effect in low-pressure mercury discharges had not yet been studied. We discovered some interesting phenomena when inducing this optogalvanic effect in the mercury discharges by optical pumping:

- the time-behavior of the pulsed optogalvanic effect depends on the value of the ballast resistance in the discharge circuit (van de Weijer and Cremers, 1985b),

- two-step photoionization causes a non-linear effect in the magnitude of the optogalvanic effect induced by optical pumping on the 408 nm line (van de Weijer and Cremers, 1985c), and

- amplified spontaneous emission in the direction of the laser beam may occur when inducing the optogalvanic effect by the optical pumping process (van de Weijer and Cremers, 1985c, 1985d).

DETERMINATION OF THE EFFECTIVE RADIATIVE LIFETIME OF THE $6\ ^3P_1$ LEVEL

Principle of the method

We measured the effective radiative lifetime of the $6\ ^3P_1$ level in low pressure mercury (noble-gas) discharges by irradiating these discharges with a 10 ns dye laser pulse at 405 nm (see Fig. 1). As a consequence mercury atoms are excited from the $6\ ^3P_0$ level to the $7\ ^3S_1$ level. The $7\ ^3S_1$ level decays radiatively to the $6\ ^3P$ levels. Then the decay of the temporary overpopulation in the $6\ ^3P_1$ level is recorded by monitoring the 254 nm fluorescence signal. In most experiments described thus far in the literature a population density in the $6\ ^3P_1$ level was created by irradiating a mercury vapour cell with the 254 nm output of a low-pressure mercury discharge. Then the 254 nm signal was recorded.

An advantage of our method is that excitation and fluorescence are at different wavelengths, thus facilitating the suppression of stray light. Moreover, by using the laser pulse at 405 nm as the excitation source we are more flexible in creating different excitation profiles (see Results and Discussion). A disadvantage of our method is that the experiments have to be made in a discharge instead of a cell. In view of this, we performed the fluorescence measurements at very low discharge currents for two important reasons: (a) In order to measure the effective radiative lifetime of the $6\ ^3P_1$ level, the decay of the 254 nm fluorescence signal must be dominated by radiative decay. Other excitation or deexcitation processes of the $6\ ^3P_1$ level must, therefore, be negligible. Electron-impact processes can be eliminated if the electron density is low, i.e., if the discharge current is low; (b) At most 75% of the population in the $6\ ^3P_0$ level can be transferred to the $7\ ^3S_1$ level by the 405 nm laser pulse. Then the $7\ ^3P$ level decays radiatively to the $6\ ^3P$ levels with a time constant of 8 ns (Borisov and Osherovich, 1981; Mohamed, 1983). About one third of the temporary overpopulation in the 7 $^3S_1$ level ends in the 6 $^3P_1$ level (Mosburq and Wilke, 1978; van de Weijer and Cremers, 1983a).

Suppose that the densities in the 6 $^3P_0$ and the 6 $^3P_1$ level are equal for the steady-state condition, which is roughly true for the conventional fluorescent lamp conditions (diameter 36 mm, current 400 mA, argon pressure 400 Pa, wall temperature 42°C) (Koedam and Kruithof, 1962). This means that the temporary overpopulation in the 6 $^3P_1$ level can be, at most, 25% of the steady- state population. As a consequence the 254 nm fluorescence signal is small compared to the 254 nm steady-state emission. If the electron impact excitation rate from the ground state to the 6 $^3P_1$ level is disturbed by the optical pumping process, which may occur if the 6 ^3P atoms are involved in the ionization processes in the discharge the decay of the fluorescence signal cannot be described by a single exponent (see THE OPTOGALVANIC EFFECT).

If the discharge current is decreased, the density in the metastable 6 ^3P levels can be considerably larger than the density in the radiative 6 $^3P_1$ level (Koedam and Kruithof, 1962). This is due to the fact that at low electron densities, the dominant loss process for the metastable 6 ^3P atoms is diffusion to the wall. This process is much slower (Kryukov et al., 1981) than radiative decay of the 6 $^3P_1$ level, even if considerable radiation trapping occurs. For discharge currents in the μA regime we obtained 254 nm fluorescence signals, which are more intense than the 254 nm steady state emission by a factor of 100 or more. Any disturbance of the steady state emission by the optical pumping process will not be noticed then in the 254 nm fluorescence signal.

Experimental arrangement

A schematic diagram of the experimental arrangement is given in Fig. 2. The light source is a dye laser pumped with a pulsed nitrogen laser (Molectron DL II 14 and UV 14, respectively). The dye used to create the 10 ns laser pulse at 405 nm is diphenylstilbene. The spectral width of the pulse is 0.01 nm and its energy is 100 μJ.

The tubes containing the mercury (noble-gas) discharges are centered in wider outer tubes. Water or an ethylene glycol/water mixture, flowing between inner and outer tube, is used to control the temperature of the inner tube. The temperature of the inner tube determines the mercury vapour pressure (Hultgren et al., 1973). An aqueous solution of $CoSO_4$/$NiSO_4$ was used as a filter to separate the 254 nm radiation from other radiation, like stray light of the dye laser pulse.

The fluorescence signal is detected with a photomultiplier, which is connected with an oscilloscope for visual inspection of the fluorescence signal when tuning the dye laser to the correct wavelength. Further a boxcar integrator and an X-Y recorder are used to record the fluorescence signals.

Results and discussion

The natural radiative lifetime of the 6 $^3P_1$ level has been measured by numerous investigators (see Table I). To test our method, we therefore measured this lifetime first. In a 25 mm diameter mercury argon discharge (argon pressure 400 Pa) a coldest spot was introduced (T=77 K). The lifetime measured in this discharge is independent of the current for

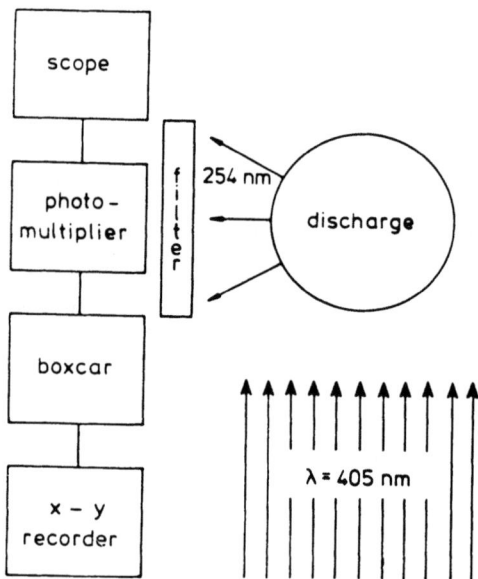

Fig. 2. Schematic diagram of the experimental arrangement for the measurement of the effective radiative lifetime of the 6^3P_1 atomic mercury level.

Table 1. The natural radiative lifetime of the 6^3P_1 atomic mercury level.

Investigators	Lifetime (ns)
Barrat, 1959	118 ± 2
Kaul, 1966	118
Nussbaum and Pipkin, 1967	114 ± 14
Deech and Baylis, 1970	117 ± 1
Popp et al., 1970	120 ± 2
Dodd et al., 1970	117.4 ± 1.0
King and Adams, 1974	120.0 ± 0.7
Osherovich et al., 1975	115 ± 5
Halstead and Reeves, 1982	122 ± 2
Mohamed, 1983	116 ± 5
van de Weijer and Cremers, 1985a	120 ± 2

i=0.1-5 mA. The result found was 120±2 ns, which is in good agreement with the values given thus far in the literature.

For mercury densities and tube radii relevant to fluorescent lamps, the effective radiative lifetime of the $6\ ^3P_1$ level is a unique function of k_0R (Phelps and McCoubrey, 1960). Here R is the tube radius and k_0 is the absorption coefficient on the 254 nm line as defined by Walsh (1959). This absorption coefficient is proportional to the mercury vapour density.

We measured the effective radiative lifetime of the $6\ ^3P_1$ level in a mercury discharge (tube diameter 13 mm, tube length 500 mm) for mercury densities corresponding to a wall temperature between -15 and 84 C. For discharge currents smaller than 1 mA, the decay of the fluorescence signal can be described by a single exponent. The time constant of this decay is independent of the current in this regime. From this we conclude that electron impact processes do not influence the decay rate of the temporary overpopulation in the $6\ ^3P_1$ level for i < 1 mA. Quenching of the $6\ ^3P_1$ level to the $6\ ^3P_0$ level by collisions with mercury ground state atoms is estimated to contribute only 1-2% for the highest mercury densities studied (Waddell and Hurst, 1970; Stock et al., 1977). The results of the measurements are given in Fig. 3. The reproducibility of the measurements is 5 - 10%.

For lifetimes longer than 3 µs, Alpert et al. (1949) measured the $6\ ^3P_1$ lifetime in a mercury vapour cell (diameter 13 mm) using the 254 nm output of a low-pressure mercury discharge in combination with a chopper as the excitation source. Later on, Phelps and McCoubrey (1960) repeated the experiment of Alpert et al. on three mercury vapour cells (diameter 5.3, 13 and 47 mm). Surprisingly, the results of Phelps and McCoubrey for the $6\ ^3P_1$ lifetimes are 20% lower than those of Alpert et al.. No explanation for this discrepancy could be found.

At the mercury vapour density studied, the 254 nm excitation beam will be absorbed over a small distance in the mercury vapour cell, resulting in a rather local population density in the $6\ ^3P_1$ level. With our 405 nm laser pulse we can create a much more uniform excitation profile in the $6\ ^3P_1$ level. This is due to a much higher intensity of our light source, in combination with a much lower density in the $6\ ^3P_0$ level in comparison with the ground state.

Our results are in good agreement with those of Alpert et al. and show a nearly constant difference of 20% with those of Phelps and McCoubrey (see Fig. 3). From this it may be concluded that the results of the measurements are insensitive to the shape of the excitation profile. This conclusion is supported by the fact that a variation of our excitation profile (we irradiated the discharge side-on with laser beam diameters from 2 to 30 mm, and along the axis of the discharge with the same beam diameters) did not affect the resulting lifetimes within the experimental error (5-10%). Though this observation can be well understood from a theoretical point of view (see e.g. van Trigt, 1976), this is, to our knowledge, the first time that it has been proven experimentally.

Both Alpert et al. and Phelps and McCoubrey were unable to measure lifetimes shorter than 2 - 3 µs due to the limitations of the mechanical

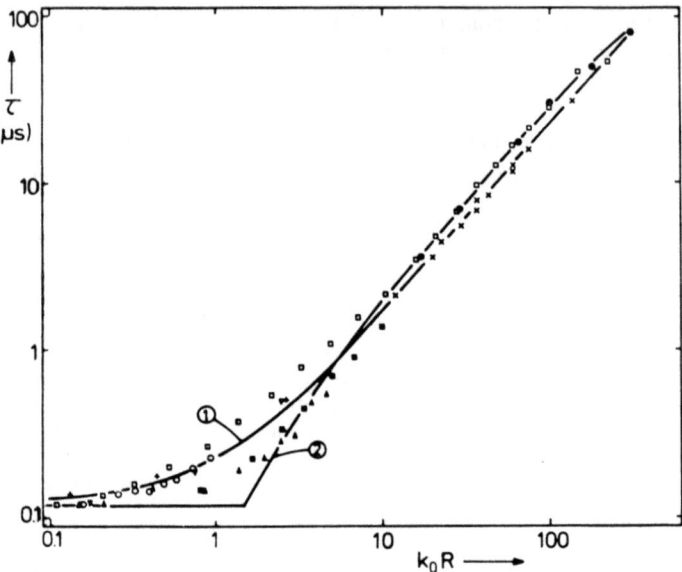

Fig. 3. Effective radiative lifetime of the 6 $^3P_1$ level in mercury as a function of k_0R. Here R is the radius of the mercury vapour cell or the mercury discharge and k_0 is the absorption coefficient as defined by Walsh (1959), which is proportional to the mercury vapour density. ■ = Thomas and Gwinn (1948), ● = Alpert et al. (1949), x = Phelps and McCoubrey (1960), ▲ = Yang (1966), ∇ = Nussbaum and Pipkin (1967), o = Micheal and Yeh (1970), + = Halstead and Reeves (1982), □ = van de Weijer and Cremers (1985a), 1 = Phelps (1958), 2 = Walsh (1959).

chopper. With our excitation source we are able to measure the effective lifetime up to conditions where the 254 nm line is optically thin.

For small k_0R values ($k_0R < 10$), five experiments have been reported in the literature thus far. (In fact there are more, but the cell dimensions are not always specified (Barrat, 1959) or the cell is not a cylinder (Deech and Baylis, 1970; Dodd et al., 1970).) Three of them (Thomas and Gwinn, 1948; Yang, 1966; Micheal and Yeh, 1970) concern quenching experiments of the 6 $^3P_1$ level by hydrogen or ethylene. By extrapolation of the hydrogen/ethylene density to zero, the effective radiative lifetime of the 6 $^3P_1$ level could be determined. However, in these experiments a mercury vapour cell was used whose length was hardly larger than its diameter. For optically thin conditions, the experimental result is independent of the geometry of the cell or discharge tube, but if radiation trapping occurs the geometry and dimensions are important. As the theory applies to an infinite cylinder, and in the other experiments the ratio of length and diameter of the cylinder is large, it is not surprising that the results derived from the quenching experiments are lower than the other results.

Nussbaum and Pipkin (1967) reported a photon-photon time correlation experiment for determining the (effective) radiative lifetime of the $6\,^3P_1$ level. Their results agree with ours within their experimental error (15 - 20%).

Halstead and Reeves (1982) have measured the (effective) radiative lifetime in a 10 mm diameter mercury vapour cell using a laser induced fluorescence technique. Their results are in reasonable agreement with ours.

Our results of the $6\,^3P_1$ lifetimes are 20 - 30% larger than the results of Phelps' calculations (Phelps, 1958). The results of the calculations of Walsh (1959) agree with our experimental data in the k_0R regime where this theory is applicable ($k_0R \gg 1$). However, it must be mentioned that Walsh used an adjustable parameter in his calculations.

Our results for the effective radiative lifetime of the $6\,^3P_1$ level in mercury-argon and mercury-krypton discharges (diameter 36 mm, argon and krypton pressure 400 Pa) are given in Fig. 4. At low k_0R values, the 254 nm emission is determined by the center of the 254 nm line components, whereas at higher k_0R values, the wings of the line components become increasingly important. As collisions between mercury and noble-gas atoms cause a broadening of the wings of the 254 nm line components (Lorentz broadening) the effect of the noble-gas on the $6\,^3P_1$ lifetime is larger if the trapping increases.

For a quantitative description, we used the theory of Walsh (1959). The Lorentz broadening parameter due to argon has been taken from Zemansky (1930). From Fig. 4 it can be seen that the theory gives a reasonable description of the $6\,^3P_1$ lifetime in mercury-argon discharges. The deviations between theory and experiment are always less than 30%. For krypton, no experimental Lorentz broadening parameter is available. Since the $6\,^3P_1$ lifetimes for mercury-krypton mixtures are the same as for mercury-argon mixtures within the experimental error, it can be concluded that the Lorentz broadening parameter of krypton is very close to that of argon (within 10%).

As far as we know, only one experiment has been reported in the literature concerning the determination of the effective radiative lifetime of the $6\,^3P_1$ level in mercury-noble-gas discharges. Hammond and Gallo (1972) measured this lifetime in mercury-argon discharges by observing the 254 nm afterglow output of these discharges. Though the current density in their discharges is rather high (400 mA in a 16 mm diameter discharge tube), they claim that the decay time of the 254 nm output is independent of the discharge current. We cannot compare our results with their experimental data, since considerable gas heating occurs at their high current densities. The gas heating result in a higher collision frequency and a lower mercury density. Under our experimental conditions there is no significant gas heating due to the much lower argon densities and discharge currents ($i < 1$ mA).

The 254 nm output of a low-pressure mercury (noble-gas) discharge, when expressed as the number of photons emitted from the discharge per unit time and per unit volume, can be approximated by the ratio of the density in the $6\,^3P_1$ level, averaged over the cross-section of the

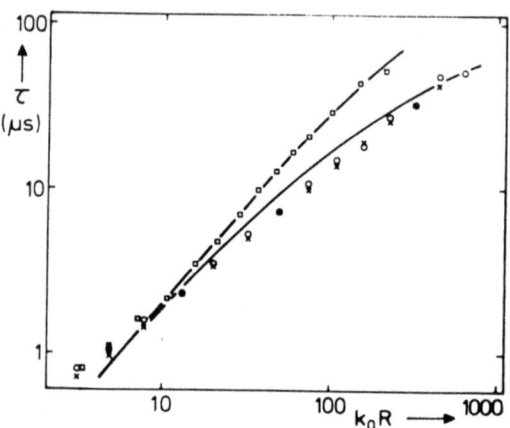

Fig. 4. Effective radiative lifetime of the 6 $^3P_1$ level as a function of k_0R in mercury, mercury-argon (argon pressure 400 Pa), and mercury-krypton (krypton pressure 400 Pa) discharges. □ = mercury, o = mercury-argon, x = mercury-krypton. The solid lines represent calculations of Walsh (1959) (see also text).

discharge, and the effective radiative lifetime of this level. In the past, the density in the 6 $^3P_1$ level has been measured for both mercury and mercury-argon discharges (Koedam and Kruithof, 1962; Uvarov and Fabrikant, 1965a, 1965b, 1965c; van de Weijer and Cremers, 1982, 1983b). We have performed the output calculation for a mercury and a mercury-argon discharge for which the 254 nm output has already been measured directly.

For a mercury discharge (diameter 38 mm, current 450 mA), the 254 nm output calculated with the 6 $^3P_1$ densities given by Uvarov and Fabrikant (1965c) and the effective radiative lifetimes determined in this work, agrees with the 254 nm output measured by Uvarov and Fabrikant (1965a) (see Fig. 5a).

For a mercury-argon discharge (diameter 36 mm, current 400 mA, argon pressure 400 Pa), the 254 nm output has been measured by Koedam et al. (1963). The 6 $^3P_1$ densities have been measured with the hook method (van de Weijer and Cremers, 1982, 1983b). The calculated 254 nm output for this discharge (Fig. 5b) is higher than the measured output for wall temperatures higher than 50°C. The deviation is somewhat larger than the experimental uncertainty in the calculated output which is estimated to be 15%. We could not find an explanation for this slight discrepancy. Despite this, we feel that in our study on the 254 nm radiation trapping, we derived the physically relevant parameter.

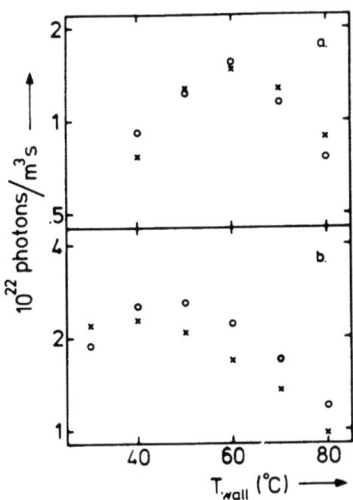

Fig. 5. a. The 254 nm output of a mercury discharge (diameter 38 mm, current 450 mA). x = measured by Uvarov and Fabrikant (1965a), o = calculated with 6 $^3P_1$ densities taken from Uvarov and Fabrikant (1965c).

b. The 254 nm output of a mercury-argon discharge (diameter 36 mm, current 400 mA, argon pressure 400 Pa). x = measured by Koedam et al. (1963), o = calculated with 6 $^3P_1$ densities determined with the hook method (van de Weijer and Cremers, 1982, 1983b).

DETERMINATION OF THE EFFECTIVE RADIATIVE LIFETIME OF THE 6 $^1P_1$ LEVEL

Principle of the Method

We measured the effective radiative lifetime of the 6 $^1P_1$ level in low-pressure mercury (noble-gas) discharges by irradiating these discharges with a 10 ns dye laser pulse at 365.5 nm, thus exciting mercury atoms from the metastable 6 $^3P_2$ level to the 6 $^3D_2$ level (see Fig. 1). The 6 $^3D_2$ level decays radiatively, with a time constant of 9 ns (Borisov et al., 1979), to the 6 $^3P_2$, 6 $^3P_1$ and 6 $^1P_1$ levels. By recording the decay of the temporary overpopulation of the 6 $^1P_1$ level on the 185 nm fluorescence signal we are able to determine the effective radiative lifetime of this level.

In order to measure the effective radiative lifetime of the 6 $^1P_1$ level the decay of the temporary overpopulation in the 6 $^1P_1$ level must be dominated by radiative decay. Electron impact excitation or deexcitation processes must therefore be negligible. This can be achieved by performing the experiments at low discharge currents (0.1 - 10 mA).

The steady state density in the $6\ ^3P_2$ level is much larger than that in the $6\ ^1P_1$ level. As a significant fraction of the $6\ ^3P_2$ density is transferred by the laser pulse to the $6\ ^1P_1$ level, the temporary overpopulation in the $6\ ^1P_1$ level is larger than the steady state density by a few orders of magnitude, as can be seen on the 185 nm emission intensity. This means that any changes in the 185 nm steady state emission, introduced by the pumping process via the optogalvanic effect, can be ignored.

Experimental Arrangement

The experimental arrangement for the determination of the effective radiative lifetime of the $6\ ^1P_1$ level is similar to the one described above for the determination of the $6\ ^3P_1$ lifetimes.

The dye used to create the 10 ns dye laser pulse at 365.5 nm is 2-phenyl- 5-(4-biphenylyl)-1,3,4-oxadiazole (PBD). The spectral width of the pulse is 0.01 nm and its energy is 30 µJ.

The tubes containing the mercury (noble-gas) discharges are centered in wider outer tubes. Water, flowing between inner and outer tube, is used to control the temperature of the inner tube. As water is not transparent to 185 nm radiation, two separate outer tubes are used for each discharge tube. In this way a small part of the discharge tube is not surrounded by the water bath, thus enabling us to detect the 185 nm fluorescence. However, a coldest spot is introduced in this way at water temperatures higher than ambient. In order to prevent this we surrounded the "naked" part of the discharge tube by heating wire. We checked for every data point the mercury vapor density by measuring the effective radiative lifetime of the $6\ ^3P_1$ level which is known as a function of the mercury density. We are able to determine this $6\ ^3P_1$ lifetime on the same pumping wavelength since the 312.6 nm line originating from the $6\ ^3D_2$ level creates a temporary overpopulation in the $6\ ^3P_1$ level. The mercury vapour density was determined in this way with an experimental error, corresponding to an uncertainty in the wall temperature of 1 - 2°C.

The fluorescence signals were detected with a photomultiplier, in combination with two interference filters. This photomultiplier is connected with an oscilloscope for visual inspection of the fluorescence signal when tuning the laser to the correct wavelength. Finally a boxcar integrator and an X-Y recorder were used to record the fluorescence signals.

Results and Discussion

As a test of our method of measuring the effective radiative lifetime of the $6\ ^1P_1$ level, we measured these lifetimes in a 25 mm diameter low-pressure mercury discharge on which Post (1984) performed his afterglow experiments. From Fig. 6 it can be seen that the agreement between the results of our laser induced fluorescence measurements and those of the afterglow experiments of Post is excellent.

The theory for the calculation of the effective radiative lifetime of the $6\ ^1P_1$ level, which accounts for incomplete redistribution (Post, 1985), gives a reasonable description of the experimental results,

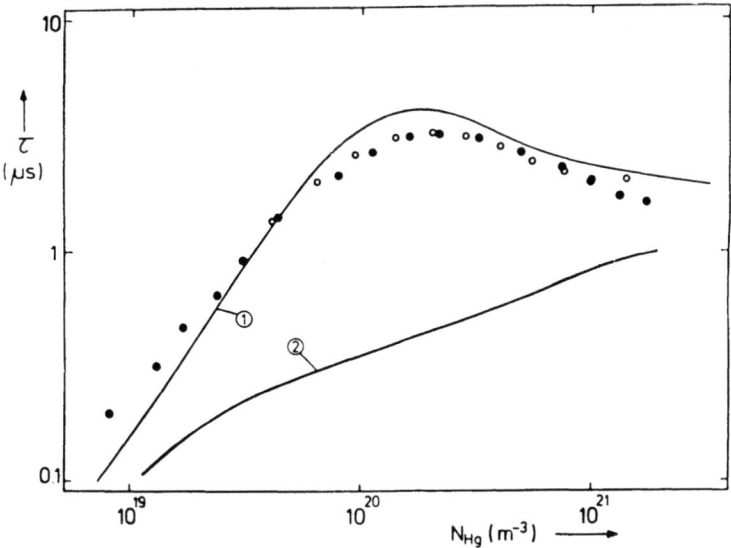

Fig. 6. Effective radiative lifetime of the 6 $^1P_1$ level in a mercury discharge (diameter 25 mm) as a function of the mercury vapor density. o = afterglow measurements (Post, 1985), • = laser induced fluorescence measurements (Post et al., 1985). The solid lines represent calculations by Post (1985), 1 = no complete spectral redistribution, 2 = complete spectral redistribution.

especially when it is compared with the theory which does assume this complete spectral redistribution (Fig. 6).

The effective radiative lifetime of the 6 $^1P_1$ level could only be derived from the afterglow experiments if direct electron impact excitation from the ground state contributed significantly to the population density of the 6 $^1P_1$ in the steady-state discharge. This means that the afterglow experiments are only applicable for mercury discharges, or mercury-noble-gas discharges at very low noble-gas pressures. They are not applicable to mercury- noble gas discharges with 100 Pa noble gas or more. Our laser induced fluorescence measurements could be applied to these mercury-noble-gas discharges. Some results are given in Figs. 7 and 8. Further, we found that the influence of krypton on the effective radiative lifetime of the 6 $^1P_1$ level is the same as that of argon within the experimental error of the measurements (5 - 10%).

If the theory of Post (1985) is applied to calculate the effective radiative lifetime of the 6 $^1P_1$ level in mercury-noble-gas discharges, for instance to the discharge used in the conventional fluorescent lamp (diameter 36 mm, argon pressure 400 pa), a reasonable agreement between theory and experiments is found (Fig. 9). If we assume complete spectral redistribution we would again have a large discrepancy between theory and experiment. However, spectral redistribution is not the only critical question affecting the outcome of the calculations.

Since the 185 nm line is highly trapped, even more than the 254 nm line, special care has to be taken to account for all line broadening

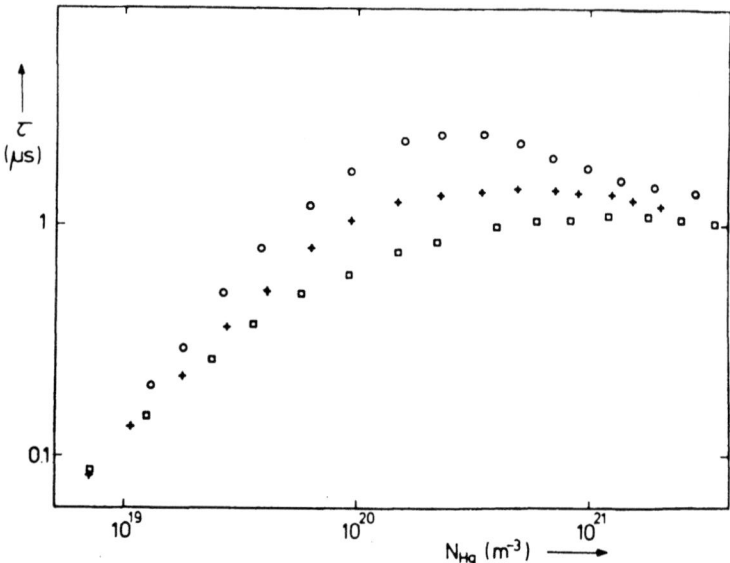

Fig. 7. Effective radiative lifetime of the 6 $^1P_1$ level in mercury-argon discharges (diameter 15 mm) as a function of the mercury vapor density for various argon pressures, o = 0 Pa, + = 133 Pa, □ = 400 Pa.

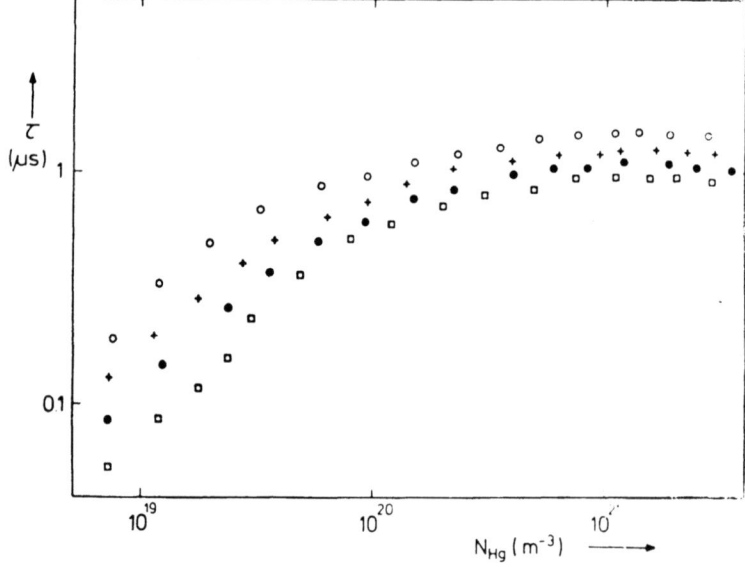

Fig. 8. Effective radiative lifetime of the 6 $^1P_1$ level in mercury-argon discharges (argon pessure 400 Pa) as a function of the mercury vapor density for various tube diameters, o = 36.5 mm, + = 25.0 mm, ● = 15.0 mm, □ = 10.4 mm.

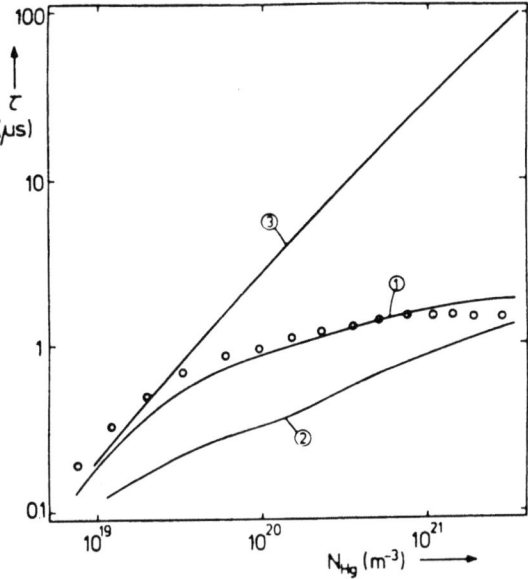

Fig. 9. Effective radiative lifetime of the $6\,^1P_1$ level in a mercury-argon discharge (diameter 36.5 mm, argon pressure 400 Pa) as a function of the mercury vapor density. 1 = theory Post (1985), 2 = theory Post (1985), assuming complete spectral redistribution, 3 = theory Walsh (1959) as applied by Winkler et al. (1983).

mechanisms. If not, severe deviations in the results of the calculations are obtained. This can be illustrated by curve 3 in Fig. 9, which represents the results of Winkler et al. (1983). They applied the theory of Walsh (1959), assuming complete spectral redistribution and only taking Doppler broadening of the 185 nm line into account, thus neglecting natural and collisional broadening. With increasing mercury vapor pressure the deviations increase up to more than an order of magnitude.

THE OPTOGALVANIC EFFECT

The optogalvanic effect (OGE) induced with a laser pulse which is short compared with the time scale of the ionization processes in the discharges was first reported in 1979 by Erez et al. and Miron et al. Since the laser excites atoms to a higher level, thus enhancing electron-impact ionization, these authors assumed that the OGE always starts with a decrease in voltage across the discharge. However, Shuker et al. (1982) observed the opposite effect for some neon transitions: the OGE starts with an increase in voltage. They attributed this phenomenon to a population inversion between upper and lower level of the transition. Later on Ceasar and Heuilly (1983) contradicted this explanation on the basis of a two-photon enhanced optogalvanic signal. Shuker et al. (1984) argued in reply that the experiment of Ceasar and Heuilly cannot be used as a criterion for population inversion. However, Fujimoto et al. (1983) clearly demonstrated the absence of a population inversion, having observed positive fluorescence from the upper level of the transition after laser excitation.

We report on the pulsed optogalvanic effect in a low-pressure mercury discharge for the following transitions: the $6\ ^3P - 7\ ^3S_1$ transitions, the $6\ ^3P_1 - 7\ ^1S_0$ transition, and the $6\ ^3P_2 - 6\ D$ transitions. In all cases we found positive fluorescence from the upper level of the transitions, thus eliminating population inversion in the steady state of the discharge. We believe that the upper level of the transition does not participate in the OGE, as the radiative lifetime of the upper level is too short to alter the ionization frequency of the discharge significantly. The sign of the OGE is determined by the population redistribution created after the decay of the upper level.

The mercury discharge studied (diameter 19 mm, length 500 mm, mercury density $2.5\ 10^{21}\ m^{-3}$, current 10 mA) has electrodes bent sideways (see Fig. 10). This made it possible to illuminate the discharge almost uniformly over its cross-section. The OGE induced across the discharge was detected by means of an oscilloscope. Photomultipliers, in combination with interference filters, were used to detect the fluorescence signals.

The OGE Induced on the $6\ ^3P - 7\ ^3S_1$ transitions

When a mercury discharge is irradiated with the 546 nm dye laser pulse, mercury atoms are excited from the $6\ ^3P_2$ to the $7\ ^3S_1$ level (see Fig. 1). The overpopulation in the $7\ ^3S_1$ level then decays radiatively with a time constant of 8 ns (Borisov and Osherovich, 1981; Mohamed, 1983) to the $6\ ^3P$ levels. As these three lines are optically thin for these discharge conditions (van de Weijer and Cremers, 1982), the population redistribution over the $6\ ^3P$ levels is determined by the transition probabilities (Mosburg and Wilke, 1978; van de Weijer and Cremers, 1983a): 50% decays to the $6\ ^3P_2$ level, 35% to the $6\ ^3P_1$ level and 15% to the $6\ ^3P_0$ level. The resulting OGE is positive (the voltage across the discharge increases) over its total duration (Fig. 11a). If the $7\ ^3S_1$ level is involved in the ionization we would expect a negative OGE initially. This is not observed since the time constant of the detection system is much larger (1 μs) than the radiative lifetime of the $7\ ^3S_1$ level (8 ns) and the length of the dye laser pulse (10 ns). Moreover, due to its short radiative lifetime, only a very small fraction of the temporary overpopulation in the $7\ ^3S_1$ level is ionized.

The positive sign of the OGE on the 546 nm line is due to mercury atoms being transferred from the $6\ ^3P_2$ level (via the $7\ ^3S_1$ level) to the $6\ ^3P_0$ and $6\ ^3P_1$ levels. The cross section for electron impact ionization from these two levels is smaller than that from the $6\ ^3P_2$ level. Moreover, fewer electrons in the discharge have enough energy for ionization. As a consequence the ionization frequency in the discharge is decreased.

If the discharge is irradiated with a 405 nm dye laser pulse, mercury atoms are transferred from the $6\ ^3P_0$ level to the $6\ ^3P_1$ and $6\ ^3P_2$ levels, which are higher in energy. Therefore, the OGE is negative (Fig. 12a). Irradiation of the discharge with a 436 nm dye laser pulse has the

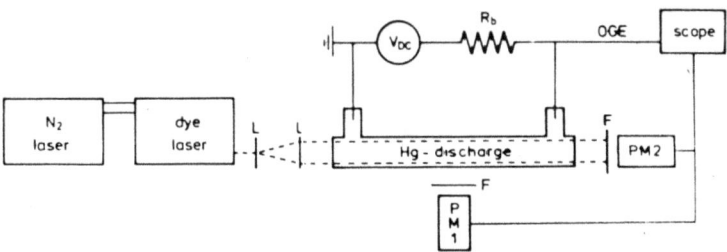

Fig. 10. Schematic diagram of the experimental arrangement for the measurement of the optogalvanic effect (OGE), fluorescence and amplified spontaneous emission. L = lens, F = filter, PM = photomultiplier, R_b = ballast resistor, V_{DC} = DC power supply. PM1 detects spontaneous emission, PM2 detects amplified spontaneous emission.

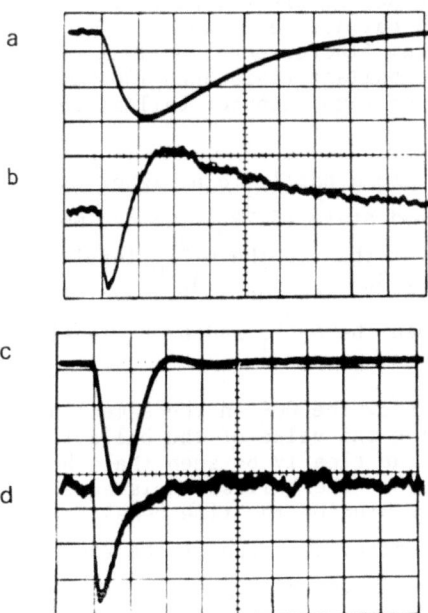

Fig. 11. OGE and 254 nm fluorescence after irradiation of the discharge with the 546 nm dye laser pulse. All time scales are 200 μs/div.
    a. OGE, ballast resistance 2 kΩ.
    b. 254 nm fluorescence, ballast resistance 2 kΩ.
    c. OGE, ballast resistance 100 kΩ.
    d. 254 nm fluorescence, ballast resistance 100 kΩ.

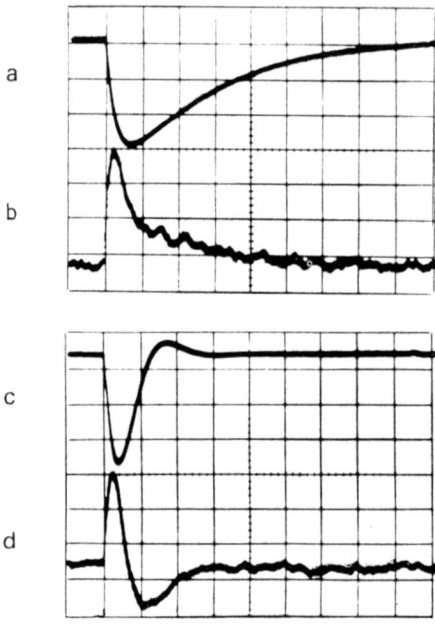

Fig. 12. OGE and 254 nm fluorescence after irradiation of the discharge with the 405 nm dye laser pulse. All time scales are 200 μs/div.
a. OGE, ballast resistance 2 kΩ.
b. 254 nm fluorescence, ballast resistance 2 kΩ.
c. OGE, ballast resistance 100 kΩ.
d. 254 nm fluorescence, ballast resistance 100 kΩ.

effect of transferring atoms partly to a higher excited state ($6\,^3P_2$) and partly to a lower excited state ($6\,^3P_0$). Since more atoms are transferred to the $6\,^3P_2$ level than to the $6\,^3P_0$ level and since the energy difference between the $6\,^3P_2$ and the $6\,^3P_1$ level is larger than the energy difference between the $6\,^3P_1$ and $6\,^3P_0$ level, the OGE is negative (Fig. 13a).

Figures 11c - 13c show the OGE on the same lines as Figs. 11a - 13a, the only difference being the value of the ballast resistance in the discharge circuit. It can clearly be seen that the time behavior of the OGE depends on the value of the ballast resistance. In order to understand this phenomenon we studied the time behavior of the density in the $6\,^3P_1$ level by monitoring the emission of the 254 nm line. The results are given in the b and d curves of Figs. 11 - 13.

Let us consider the 254 nm fluorescence signals following the 546 nm dye laser pulse. For both values of the ballast resistance the initial part of the signal is positive with respect to the steady-state signal because of the transfer of mercury atoms from the $6\,^3P_2$ level to the $6\,^3P_1$ level. The long term fluorescence, which obstructs the measurement of the effective radiative lifetime of the $6\,^3P_1$ level from the initial fluorescence signal, is on the time scale of the OGE itself. However,

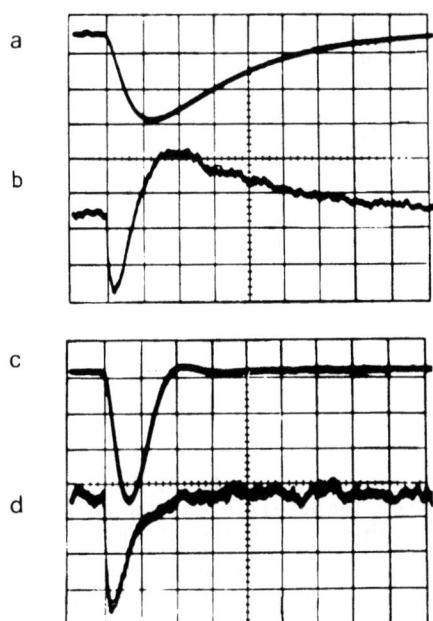

Fig. 13. OGE and 254 nm fluorescence after irradiation of the discharge with the 436 nm dye laser pulse. All time scales are 200 μs/div.
a. OGE, ballast resistance 2 kΩ.
b. 254 nm fluorescence, ballast resistance 2 kΩ.
c. OGE, ballast resistance 100 kΩ.
d. 254 nm fluorescence, ballast resistance 100 kΩ.

for the smaller value of the ballast resistance this long term fluorescence is negative, whereas for the larger value it is positive. The OGE on the 546 nm line is positive, which means that the electric field, and thus the electron temperature, increases. On the other hand the electron density decreases. The increase in electron temperature causes an increase in the excitation frequency from the ground state to the $6\ ^3P_1$ level, whereas the decrease in electron density causes a decrease of this frequency.

Now, following the theory of Doughty and Lawler (1983), it can be understood which alteration has the largest effect on the population of the $6\ ^3P_1$ level. The current through the discharge is a function of electron density n and electric field E:

$$i = F(n,E). \tag{1}$$

The perturbation of the current and the perturbation in electric field are related through the ballast resistance R_b by

$$R_b\ \Delta i + \ell\ \Delta E = 0, \tag{2}$$

where ℓ is the length of the discharge. Neglecting the influence of the change in electric field on the electron density, the perturbed current equation is

$$\frac{\partial F}{\partial n} \Delta n + \frac{\partial F}{\partial E} \Delta E = - \frac{\ell \Delta E}{R_b} \ . \tag{3}$$

This can be rewritten as

$$\frac{\Delta V/V}{\Delta n/n} = - \frac{n \ \partial F/\partial n}{V \ (1/R_b + 1/R_d)} \ , \tag{4}$$

where R_d is the dynamic resistance of the discharge.

The influence of a change in ballast resistance (keeping the discharge conditions constant) can be seen from the above equation. If the ballast resistance decreases the relative change in the discharge voltage decreases with respect to the relative change in electron density. Combining the above equation and the 254 nm fluorescence signals, we conclude that for a 2 kΩ ballast resistance the decrease in electron density has more effect on the density in the $6\ ^3P_1$ level than the increase in electron temperature. On the other hand, for a 100 kΩ ballast resistance the increase in electron temperature has more effect than the decrease in electron density. Though we did not monitor the time behavior of the densities of the metastable $6\ ^3P$ levels, it is very likely that the population of these levels will be influenced in the same way.

The positive OGE on the 546 nm line is due to a decrease in ionization from the $6\ ^3P$ levels. At low values of the ballast resistance, the influence of the OGE is to decrease the population of the $6\ ^3P$ levels. So, the consequence of the OGE intensifies the cause of it. Therefore the OGE has a long lifetime. On the other hand, for large values of the ballast resistance, the influence of the OGE is to increase the population of the $6\ ^3P$ levels, resulting in a shorter lifetime of the OGE. The same arguments can be used to explain the OGE and the fluorescence of the 254 nm line following the 405 nm and the 436 nm dye laser pulse.

For high values of the ballast resistance a sign reversal of the OGE is observed for all three lines. In our opinion this is an oscillation of the electrical circuit.

The OGE induced on the $6\ ^3P_1 - 7\ ^1S_0$ transition

When the mercury discharge is irradiated with the 408 nm dye laser pulse mercury atoms are excited from the $6\ ^3P_1$ level to the $7\ ^1S_0$ level (see Fig. 14). The underpopulation in the $6\ ^3P_1$ level, created by this optical pumping, is a few percent of its steady state population. This can be derived from the time resolved emission signal on the 254 nm line. The steady state density in the $6\ ^3P_1$ level (at the axis of the discharge) as determined with the hook method (van de Weijer and Cremers, 1982, 1983b) is $7 \cdot 10^{17}$ m^{-3}.

Using the same method we tried to measure the density in the $6\ ^1P_1$ level by observing hooks around the 577 nm line ($6\ ^1P_1 - 6\ ^3D_2$, oscillator strength 0.28 (van de Weijer and Cremers, 1983a)). However, for these discharge conditions the density in the $6\ ^1P_1$ level is too low to be detected by this method. At higher discharge currents (100-350 mA) we found that the density in the $6\ ^1P_1$ level is proportional to the

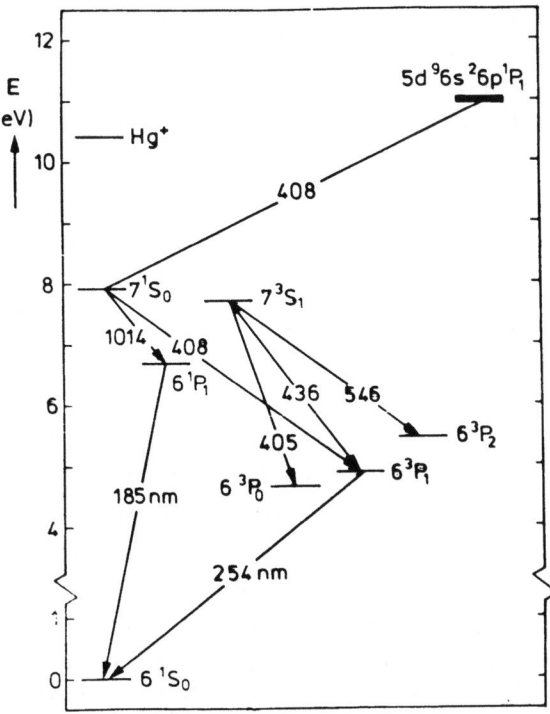

Fig. 14. Partial energy level diagram of mercury, showing the relevant lines in the optical pumping experiment on the 408 nm line.

discharge current, within the experimental error (~ 20%). Extrapolation of this density to lower current provides the density (at the axis of the discharge) for 10 mA: $2 \cdot 10^{15}$ m^{-3}. Though this value for the 6 $^1P_1$ density must be considered as a rough estimate, it indicates that a significant population inversion is created on the 7 $^1S_0$ - 6 $^1P_1$ transition, since a few percent of the 6 $^3P_1$ density is transferred to the 7 $^1S_0$ level by the optical pumping process. This population inversion results in amplified spontaneous emission in the direction of the laser beam on the 7 $^1S_0$ - 6 $^1P_1$ transition at 1014 nm, which has been detected with the photomultiplier.

The radiative lifetime of the 7 $^1S_0$ level is about 30 ns (King et al., 1975; Faisal et al., 1980; Mohamed, 1983). The lifetime of the overpopulation in the 7 $^1S_0$ level is even shorter due to the amplified spontaneous emission process. Due to this short lifetime only a very small fraction of the overpopulation in the 7 $^1S_0$ level is ionized by electron impact. As a consequence the effect of the temporary over population in the 7 $^1S_0$ level will hardly be noticed in the OGE signal.

The OGE signal induced by optical pumping on the 408 nm line is given in Figure 15a. The maximum amplitude of the OGE signal is 0.2% of the steady state discharge voltage. The OGE is negative initially due to the fact that optical pumping on the 408 nm line results in a transfer of mercury atoms from the $6\ ^3P_1$ level, via the $7\ ^1S_0$ level, to the $6\ ^1P_1$ level, which is closer to the ionization limit.

The relaxation time of the temporary overpopulation in the $6\ ^1P_1$ level corresponds to its effective radiative lifetime for these discharge conditions (1.6 μs). We confirmed this value of the relaxation time by measuring the 185 nm fluorescence signal. From the (negative) fluorescence signal on the 254 nm line we determined a relaxation time of 50 ± 5 μs for the underpopulation in the $6\ ^3P_1$ level. The relaxation time is somewhat shorter than the effective radiative lifetime of the $6\ ^3P_1$ level for these conditions (70 ± 7 μs (van de Weijer and Cremers, 1985a)). This indicates that radiative decay is not the only loss term for the population density in the $6\ ^3P_1$ level though it is the largest one. As the relaxation time of the overpopulation in the $6\ ^1P_1$ level is much shorter than the relaxation time of the underpopulation in the $6\ ^3P_1$ level, the $6\ ^3P_1$ level still has an underpopulation, when the density in the $6\ ^1P_1$ level has reached its steady state value. As a consequence a sign reversal in the optogalvanic effect is observed. Similar sign reversals in the pulsed optogalvanic effect were observed for the first time by Miron et al., (1979) in neon and uranium.

After about 130 μs we observe a second sign reversal in the OGE. The time scale of this negative part of the OGE suggests that there is an overpopulation in the metastable $6\ ^3P_0$ and $6\ ^3P_2$ levels. Such an overpopulation is expected to decay by diffusion to the discharge wall, which is a slower process (Kryukov et al., 1981) than radiative decay of the $6\ ^3P_1$ level even if there is considerable radiation trapping.

A possible explanation for an overpopulation in the metastable $6\ ^3P$ levels could be collisional coupling by electrons from the $7\ ^1S_0$ level to the $7\ ^3S_1$ level, followed by radiative decay to the $6\ ^3P$ levels. Indeed we found fluorescence from the $7\ ^3S_1$ level. However, the fluorescence signal was very weak and it increases with discharge current, whereas the relative amplitude of the third part of the OGE decreases with the discharge current.

Collisions between $6\ ^3P$ atoms could be another explanation for an overpopulation in the metastable $6\ ^3P$ levels. The cross sections for $6\ ^3P - 6\ ^3P$ collisional excitation of highly excited mercury levels (Cheron, 1980; van de Weijer and Cremers, 1982) and the cross sections for $6\ ^3P - 6\ ^3P$ associative ionization (Tan and von Engel, 1971; Johnson et al., 1978) seem to be high enough to be of importance for the population density in the $6\ ^3P$ levels for our discharge conditions. Of course collisional excitation or ionization is not a large loss term for the population density in the $6\ ^3P$ levels, but a contribution of a few percent might be sufficient for an overpopulation in the metastable $6\ ^3P$ atoms, large enough to explain the observed OGE. This overpopulation could be created by a decreased excitation and ionization loss of the metastable atoms in collisions with the 6 $^3P_1$ atoms, which are depleted by the optical pumping process. However, we feel that more detailed data must be available, especially on the values of the cross sections for associative ionization, to test this explanation for the third part of the observed OGE quantitatively.

If the intensity of the laser pulse is decreased the magnitude of the OGE decreases due to the fact that fewer mercury atoms are excited from the 6 $^3P_1$ level to the 7 $^1S_0$ level. The anomalous phenomenon, however, is the fact that the amplitude of the initial negative part of the OGE decreases with respect to the amplitude of the rest of the signal (Fig. 15b).

The explanation for this phenomenon is two-step photoionization. After absorption of a first 408 nm photon the mercury atom can be excited from the 7 $^1S_0$ level to an autoionizing level at 11.0 eV (see Fig. 14). Though the 408 nm atomic mercury line is not exactly resonant to this transition (the sum of the energy of the 7 $^1S_0$ level and the energy of the 408 nm photon is 10.96 eV), we still can have significant absorption, as the energy width of the autoionizing level is 0.2 eV (Lincke and Stredele, 1970). This two-step photoionization process is proportional to the laser intensity squared, whereas the processes responsible for the

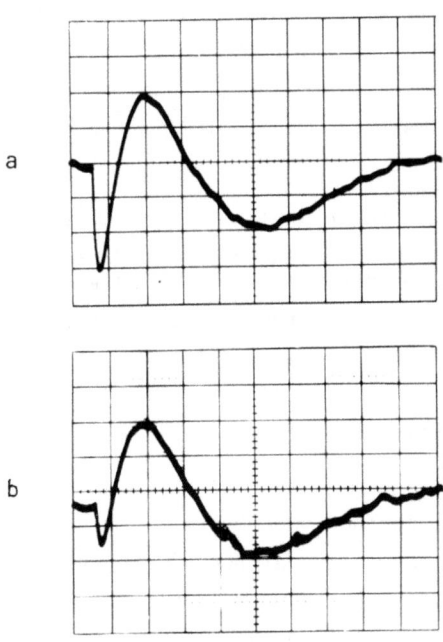

Fig. 15. OGE induced by optical pumping on the 408 nm line. The time scale is 50 μs/div. The value of the ballast resistance is 100 kΩ.
a. laser pulse energy 120 μJ.
b. laser pulse energy 36 μJ.
The sensitivity of the oscilloscope in b is higher than in a.

second and third part are linear with the laser intensity. In our opinion, therefore, the initial negative part in the OGE is not only due to the temporary overpopulation in the $6\,^1P_1$ level but also to two-step photoionization.

An indication that two-step photoionization is responsible for the observed non-linear phenomenon in the OGE is the fact that we could observe an OGE on six mercury lines at 382, 386, 496, 537, 540 and 589 nm, corresponding to mercury transitions from levels in the $5d^9$ configuration to autoionizing levels (Learner and Morris, 1971).

The OGE Induced on the $6\,^3P_2$ - 6 D Transitions

When the mercury discharge is irradiated with the 10 ns dye laser pulse at a wavelength corresponding to one of the $6\,^3P_2$ - 6 D transitions a population inversion is created on a 6D - 6^1P_1 transition (Fig. 16). This can be derived from the 254 nm fluorescence signal, originating from the $6\,^3P_1$ level if it is populated by radiative decay from the 6 D level. For instance, if we pump on the $6\,^3P_2$ - $6\,^3D_2$ transition at 365.5 nm, the amplitude of the 254 nm fluorescence signal is 7% of the steady state emission intensity on the 254 nm line. The steady state density in the $6\,^3P_1$ level (at the axis of the discharge) is $7 \cdot 10^{17}$ m^{-3} (van de Weijer and Cremers, 1982). From this density and the branching ratio of the emission lines from the $6\,^3D_2$ level (Pilz and Seehawer, 1975) it can be derived that the density in the $6\,^3D_2$ level directly after laser excitation is 10^{17} m^{-3}. This is much larger than the density in the $6\,^1P_1$ level, which was estimated to be $2 \cdot 10^{15}$ m^{-3} at the axis of the discharge.

The population inversion on the $6\,^3D_2$ - $6\,^1P_1$ transition results in amplified spontaneous emission (ASE) in the direction of the laser beam at 577 nm, which was so intense that it could be observed by the eye. Due to the ASE process, more mercury atoms in the $6\,^3D_2$ level decay to the $6\,^1P_1$ level than can be expected on basis of the branching ratio of the emission lines from the $6\,^3D_2$ level. As a consequence the density in the $6\,^3D_2$ level directly after laser excitation is even larger than 10^{17} m^{-3}.

A similar ASE phenomenon was observed on the 579 nm line if we pump on the $6\,^3P_2$ - $6\,^1D_2$ transition at 366.33 nm. Pumping on the 366.29 nm line ($6\,^3P_2$ - $6\,^3D_1$) or the 365.0 nm line ($6\,^3P_2$ - $6\,^3D_3$) does not result in ASE to the $6\,^1P_1$ level, since the oscillator strength of the $6\,^3D_1$ - $6\,^1P_1$ transition is very small (Pilz and Seehawer, 1975; van de Weijer and Cremers, 1983a), and the $6\,^3D_3$ - $6\,^1P_1$ transition is optically forbidden.

The optical pumping process on the $6\,^3P_2$ - 6 D transitions induces fluorescence from the $7\,^3S_1$ level. This observation was quite unexpected since the 6 D - $7\,^3S_1$ transitions are optically forbidden and direct

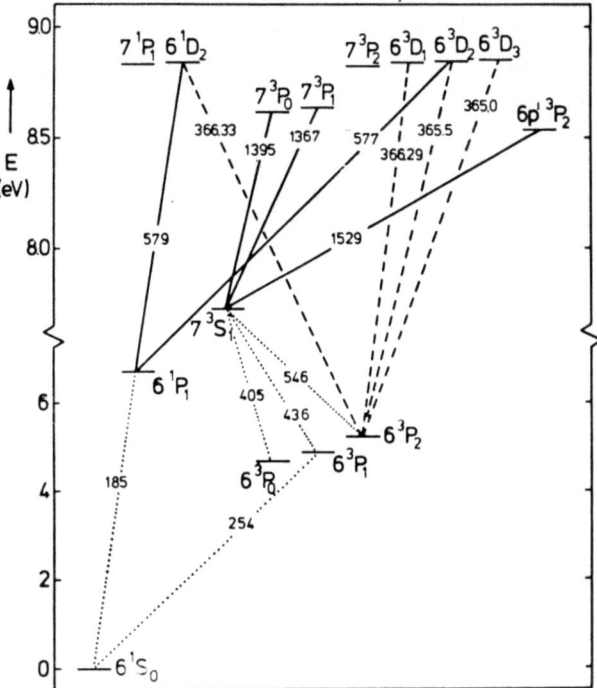

Fig. 16. Partial energy level diagram of mercury, showing the relevant lines in the optical pumping experiment on the 6 $^3P_2$ - 6 D transitions. Note the enlarged scale in the upper part of the level.
---- = pump lines
.... = laser induced fluorescence
——— = laser induced amplified spontaneous emission
All wavelengths are given in nm. The fluorescence signals originating from the 6 D levels are not drawn.

quenching of the 6 D atoms to the 7 $^3S_1$ state by electrons is unlikely due to the large energy gap. A possible explanation could be that mercury atoms in the 6 D states are quenched by electrons to the 7 $^1P_1$ or 7 $^3P_2$ level, which are very close in energy to the 6 D levels. Then radiative decay in the IR from the 7 $^1P_1$ or 7 $^3P_2$ level can create a temporary overpopulation in the 7 $^3S_1$ level. This explanation was supported by the observation of IR amplified spontaneous emission in the direction of the laser beam.

When we spectrally analyzed the IR emission, we found the following results (see also Fig. 16):

- pumping on the 365.0 nm line (6 $^3P_2$ - 6 $^3D_3$) results in ASE on the 6p' $^3P_2$ - 7 $^3S_1$ transition at 1529 nm.

- pumping on the 365.5 nm line (6 $^3P_2$ - 6 $^3D_2$) results in ASE on the 7 $^3P_1$ - 7 $^3S_1$ transition at 1367 nm,

- pumping on the 366.29 nm line ($6\ ^3P_2 - 6\ ^3D_1$) results in ASE on the $7\ ^3P_1 - 7\ ^3S_1$ and $7\ ^3P_0 - 7\ ^3S_1$ transition at 1367 and 1395 nm, respectively, and,

- pumping on the 366.33 nm line ($6\ ^3P_2 - 6\ ^1D_2$) results in ASE on the $7\ ^3P_1 - 7\ ^3S_1$ transition at 1367 nm.

This observation rules out the explanations for the fluorescence from the $7\ ^3S_1$ level by quenching of the 6 D levels by electron impact. First, because no emission was observed from the $7\ ^1P_1$ or $7\ ^3P_2$ level, which are very close in energy to the 6 D levels, and, secondly, because quenching by electron impact is not expected to be so selective. In our opinion, therefore, radiative decay from the 6 D levels is the explanation for the observed phenomena. This radiative decay, probably stimulated, is specifically dependent on the level which is populated in the pump process. For instance, population of the $6\ ^3D_3$ level can only be followed by radiative decay to the $7\ ^3P_2$ or $6\ p'\ ^3P_2$ level (and, of course, back to the $6\ ^3P_2$ level). We observed ASE from the $6\ p'\ ^3P_2$ level but not from the $7\ ^3P_2$ level. This can be explained by the fact that the transition probability of the $6\ ^3D_3 - 7\ ^3P_2$ is very small because of the small energy difference between the two levels. We were unable to observe radiative decay from $6\ ^3D_3$ to $6p'\ ^3P_2$ (at 3.9 μm) for lack of a suitable detector.

The optical pumping process creates a population redistribution in the excited mercury levels. As a consequence an optogalvanic effect (OGE) is introduced. The interpretation of this OGE is not straightforward since the population of several excited states is changed by the optical pumping process and subsequent radiative decay of the 6 D levels. However, we feel that the population redistribution due to radiative decay from 6 D to $6\ ^3P$ and $6\ ^1P_1$ mainly determines the times behavior of the OGE, because radiative decay of 6 D to $6\ ^3P$ and $6\ ^1P_1$ is much more intense than the cascade sequence from 6 D to $7\ ^3S_1$.

So, for instance, pumping on the 365.5 nm line results mainly in a transfer of mercury atoms from the $6\ ^3P_2$ level, via the $6\ ^3D_2$ level (radiative lifetime 9 ns (Borisov et al., 1979)) to the $6\ ^1P_1$ and $6\ ^3P_1$ level. The OGE is negative initially (a decrease in discharge voltage) due to the overpopulation in the $6\ ^1P_1$ level, which has a significantly lower ionization energy than the $6\ ^3P$ levels. However, the amplitude of this negative part is much smaller than the amplitude of the positive part as the relaxation time of the overpopulation in the $6\ ^1P_1$ level, which is dominated by radiative decay, is much faster than the relaxation times of the change in $6\ ^3P$ densities. The second part of the OGE is positive because the underpopulation is in a level ($6\ ^3P_2$) which has a lower ionization potential than the level with the overpopulation ($6\ ^3P_1$). Moreover, the relaxation time of the underpopulation in the $6\ ^3P_2$ level is expected to be longer than the relaxation time of the overpopulation in the $6\ ^3P_1$ level.

The OGE induced by optical pumping on the 366.33 line ($6\ ^3P_2 - 6\ ^1D_2$) is very similar to the one induced by pumping on the 365.5 nm line: it consists of a small negative and a larger positive part, with about the same relative amplitudes and time behavior. The absolute amplitude, however, is somewhat smaller. This is due to a lower oscillator strength of the 366.33 nm line in comparison to that of the 365.5 nm line (Pilz and Seehawer, 1975).

The OGE induced by pumping on the 366.29 nm line ($6\ ^3P_2 - 6\ ^3D_1$) does not show a negative part, as the oscillator strength of the $6\ ^3D_1 - 6\ ^1P_1$ transition is very small.

Though the 365.0 nm line has the largest oscillator strength of the $6\ ^3P_2 - 6\ D$ transitions, the resulting OGE is the smallest. This is due to the fact that radiative decay of the $6\ ^3D_3$ level to the 6 P levels (of the $5d^{10}$ configuration) is limited to radiative decay back to the $6\ ^3P_2$ level, which does not contribute to an OGE. So the observed OGE must be induced by decay of $6\ ^3D_3$ to $6p'\ ^3P_2$. Though the radiative lifetime of the $6p'\ ^3P_2$ level is only 160 ns (Goullet and Pebay-Peyroula, 1974), it may contribute to the OGE as the ionization potential is only 2 eV. As a result the OGE is negative initially. The amplitude of this negative part is comparable to the amplitude of the initial part of the OGE induced by the 365.5 nm laser pulse. The second part of the OGE induced by the 365.0 nm pulse is positive due to the underpopulation in the $6\ ^3P_2$ level.

REFERENCES

Alpert, D., McCoubrey, A. O., and Holstein, T., 1949, Phys. Rev., 76:1257.
Barnes, B. T., 1960, Br. J. Appl. Phys., 31:852.
Barrat, J.-P., 1959, J. Phys. Radium, 19:858.
Ben-Amar, A., Shuker, R., and Erez, G., 1981, Appl. Phys. Lett., 38:763.
Biberman, L. M., 1947, Zh. Eksp. Teor. Fiz., 17:416.
Biberman, L. M., 1949, Zh. Eksp. Teor. Fiz., 19:584.
Borisov, E. N. and Osherovich, A. L., 1981, Opt. Spectrosc. (USSR), 50:346.
Borisov, E. N., Osherovich, A. L., and Yakovlev, V. N. 1979, Opt. Spectrosc. (USSR), 47:109.
Burakov, V. S., Naumenkov, P. A., Razdobarin, G. T., and Tarasenko, N. V., 1983, Sov. Phys. Tech. Phys., 28:1060.
Ceasar, T. and Heuilly, J.-L., 1983, Opt. Commun., 45:258.
Cheron, B., 1980, J. Phys. (Paris), 41:1091.
Deech, J. S., and Baylis, 1970, Can. J. Phys., 40:90.
Delsart, C., Keller, J.-C., and Thomas, C., 1981, J. Phys., B. 14:3355.
Dodd, J. N., Sandle, W. J., and Williams, O. M., 1970 J. Phys. B 3:256.
Doughty, D. K. and Lawler, J. E., 1983, Phys. Rev. A, 28:773.
Doughty, D. K., Salih, S., and Lawler, J. E., 1984, Phys. Lett. A, 103:41.
Elenbaas, W., 1959, "Fluorescent Lamps and Lighting", Philips Technical Library, Eindhoven.
Erez, G., Lavi, S., and Miron, E., 1979, IEEE J. Quant. Electr., QE-15:1328.
Faisal, F.H.M., Wallenstein, R., and Teets, R., 1980, J. Phys. B, 13:2027.
Fujimoto, T., Uetani, Y., Sato, Y., Goto, C., and Fukuda, K., 1983, Opt. Commun., 47:111.
Goullet, G., and Pebay-Peyroula, J. -C., 1974, C. R. Acad. Sci. Paris, 259:93.
Green, R. B., Keller, R. A., Luther, G. G., Schenck, P. K., and Travis, J. C., 1976, Appl. Phys. Lett., 29:727.

Halstead, J. A. and Reeves, R. R., 1982, J. Quant. Spectrosc. Radiat. Transfer, 28:289.
Hammond, T. J. and Gallo, C. F., 1972, Appl. Opt., 11:729.
Holstein, T., 1947, Phys. Rev., 72:1212.
Holstein, T., 1951, Phys. Rev., 83:1159.
Hultgren, R., Desai, P. D., Hawkins, D. T., Gleiser, M., Kelley, K. K., and Wagman, D. D., 1973, "Selected Values of the Thermodynamic Properties of the Elements", American Society for Metals, Metals Park, Ohio.
Johnson, P. C., Cooke, M. J., and Allen, J. E., 1978, J. Phys. D., 11:1877.
Kaul, R. D., 1966, J. Opt. Soc. Am., 56:1261.
Kenty, C., 1950, J. Appl. Phys., 21:1309.
King, G. C. and Adams, A., 1974, J. Phys. B, 7:1712.
King, G. C., Adams, A., and Cvejanovic, D., 1975, J. Phys. B, 8:365.
Koedam, M. and Kruithof, A. A., 1962, Physica (Utrecht), 28:80.
Koedam, M., Kruithof, A. A., and Riemens, J., 1963, Physica (Utrecht), 29:565.
Kravis, S. P. and Haydon, S. C., 1981, J. Phys. D., 14:151.
Kryukov, N. A., Penkin, N. P., and Redko, T. P., 1981, Opt. Spectrosc. (USSR), 51:403.
Learner, R.C.M. and Morris, J., 1971, J. Phys. B, 4:1236.
Lincke, R. and Stredele, B., 1970, Z. Physik, 238:164.
Lurio, A., 1965, Phys. Rev. A, 140:1505.
Micheal, J. V. and Yeh, C., 1970, J. Chem. Phys., 53:59.
Miron, E., Smilanski, I., Liran, J., Lavi, S., and Erez, G., 1979, IEEE J. Quant. Electr. QE-15:194.
Mohamed, K. A., 1983, J. Quant. Spectrosc. Radiat. Transfer, 30:225.
Mosburg, E. R. and Wilke, M. D., 1978, J. Quant. Spectrosc. Radiat. Transfer, 19:69.
Nestor, J. R., 1982, Appl. Opt. 21:4154.
Nippoldt, M. A. and Green, R. B., 1981, Appl. Opt. 20:3206.
Nussbaum, G. H., and Pipkin, F. M., 1967, Phys. Rev. Lett., 19:1089.
Osherovich, A. L., Borisov, E. N., Burshtein, M. L. and Verolainen, Y. F., 1975, Opt. Spectrosc. (USSR), 39:466.
Payne, M. G. and Cooke, J. D., 1970, Phys. Rev. A, 2:1238.
Payne, M. G., Talmage, J. E., Hurst, G. S., and Wagner, E. B., 1974, Phys. Rev. A, 9:1050.
Penning, F. M., 1928, Physica (Utrecht), 8:137.
Pfaff, J., Begemann, M. H., and Saykally, R. J., 1984, Mol. Phys., 52:541.
Pilz, W. and Seehawer, J., 1975, Part I, p. 146, in: "Proceedings of 12th Conference on Phenomena in Ionized Gases", North Holland, Amsterdam.
Phelps, A. V., 1958, Phys. Rev., 110:1362.
Phelps, A. V. and McCoubrey, A. O., 1960, Phys. Rev., 118:1561.
Popp, M., Schafer, G., and Bodenstedt, E., 1970, Z. Physik, 240:71.
Post, H. A., 1984, J. Phys. B, 17:3193.
Post, H. A., 1985, to be published.
Post, H. A., van de Weijer, P., and Cremers, R.M.M., 1985, to be published.
Rosenfeld, A., Mory, S., and Konig, R., 1979, Opt. Commun., 30:394.
Shuker, R., Ben-Amar, A., and Erez, G., 1982, Opt. Commun., 42:29.
Shuker, R., Ben-Amar, A., and Erez, G., 1984, Opt. Commun., 49:263.
Stock, M., Smith, E. W., Drullinger, R. E., and Hessel, M. M., 1977, J. Chem. Phys., 67:2463.
Suri, B. M., Kapoor, R., Saksena, G. D., and Rao, P.R.K., 1984, Opt. Commun., 49:29.
Suri, B. M., Kapoor, R., Saksena, G. D., and Rao, P.R.K., 1985, Opt. Commun., 52:315.
Tan, K. L. and von Engel, A., 1971, Proc. R. Soc. London, Ser. A, 324:183.
Thomas, L. B. and Gwinn, W. D., 1948, J. Am. Chem. Soc., 70:2643.
Uetani, Y. and Fujimoto, T., 1984, Opt. Commun., 49:258.
Uvarov, F. A. and Fabrikant, V. A., 1965a, Opt. Spectrosc. (USSR), 18:323.
Uvarov, F. A. and Fabrikant, V. A., 1965b, Opt. Spectrosc. (USSR), 18:433.
Uvarov, F. A. and Fabrikant, V. A., 1965c, Opt. Spectrosc. (USSR), 18:541.
van de Weijer, P. and Cremers, R.M.M., 1982, J. Appl. Phys., 53:1401.
van de Weijer, P. and Cremers, R.M.M., 1983a, J. Appl. Phys., 54:2835.

van de Weijer, P. and Cremers, R.M.M., 1983b, Appl. Opt. 22:3500.
van de Weijer, P. and Cremers, R.M.M., 1985a, J. Appl. Phys., 57:672.
van de Weijer, P. and Cremers, R.M.M., 1985b, Opt. Commun., 53:109.
van de Weijer, P. and Cremers, R.M.M., 1985c, Opt. Commun., in press.
van de Weijer, P. and Cremers, R.M.M., 1985d, to be published.
van Tright, C., 1976, Phys. Rev. A, 13:726.
Waddell, B. V. and Hurst, G. S., 1970, J. Chem. Phys., 53:3892.
Walsh, P. J., 1959, Phys. Rev., 116:511.
Waymouth, J. F., 1971, "Electric Discharge Lamps", MIT, Cambridge, Mass.
Winkler, R. B., Wilhelm, J., and Winkler, R., 1983, Ann. Phys., 40:90.
Yang, K., 1966, J. Am. Cem. Soc., 88:4575.
Zemansky, M. W., 1930, Phys. Rev., 36:219.

TRANSIENT PHENOMENA IN HIGH PRESSURE DISCHARGES

E. Marode, F. Bastien and G. Hartmann

Laboratoire de Physiques des Décharges (CNRS)
Ecole Supérieure d'Electricité
Gif-sur-Yvette, France

INTRODUCTION

Observation of radiation emitted by transient discharges during the phases in which a gas passes from a non-conducting state to the highly conducting state of the arc phase have been mainly directed toward elucidating the time resolved spatial growth of space charge development. An understanding of the features of the emitted spectrum, however, has become necessary in order to extract information concerning the nature of the various plasma states built up by the discharge phases or to control discharges to achieve some desired gas fluorescence property as in lasers or other transient discharge light sources. An attempt is made here to bring together these two interests; plasma radiation and discharge physics.

Two generic names may be encountered in describing the high pressure plasma transient. Glows: a generic name used for whatever is bright, but not an arc. Corona: a generic name used for whatever happens around a highly stressed electrode, or is driven by a local space charge, and which is not a spark. The two terms have merged gradually during the last two decades with the development of basic studies on time resolved analysis of current growth and discharge development at high pressure (above 100 torr).

Based on earlier work on avalanches (Raether, 1964), space charge growth (Llewellyn-Jones, 1966), and streamer formation (Loeb, 1965), modern high speed measurement techniques have provided new insights into the mechanisms of spark formation. Measurements suggest that the glow discharge is a useful model for all discharges, even those discharges as complex as the corona (Meek and Craggs, 1978; Goldman et al, 1978). The ionized gas generated in these prebreakdown discharges, either cold or warm plasmas are characterized by the presence of an internal electric field. Because of their low mass, electrons are heated at a much more rapid rate than the ions and other heavy species unless fast translational energy transfer to the heavy particles leads to hydrodynamic phenomena and to the spark phase. Roughly speaking, primary reactions, induced by energetic electron collisions, are followed by heavy species reactions controlled by the temperature of the heavy species. Moreover, the plasma will not only be warm in transient discharges, but dynamic changes will take place and phenomena such as "tempering" may occur at this stage. The emitted radiation will obviously be greatly influenced by such processes.

The emitted radiation from transient glows has previously been approached with the assumption of thermodynamic equilibrium. However, the situation outlined above cannot be described by a unique temperature T, and must be considered as one which is clearly not in thermodynamic equilibrium. Highly time resolved spectroscopic studies of transient discharges have been undertaken with the aid of high speed streak cameras. However, only moderate correlation between the development of high pressure discharges and their emission features can be made in view of the statistical nature of the growth of these discharges, their complex spatial and temporal behavior and the fact that, generally, only a single spectral component is recorded.

The approach followed here will be to consider initially the relaxation toward reequilibration of the discharge plasma in order to show how the distribution of excited species may evolve in time by successively different processes. An overview of the main experimental results in transient discharge behavior will then allow the introduction of some of the terms used to designate the various plasma phases. Finally, some aspects will be outlined concerning the discharge parameters such as electric field, electron temperature, and the physico-chemical processes which influence the emission properties during the various phases.

SOME FEATURES OF NON-EQUILIBRIUM

The easiest systems to describe are those for which thermodynamic equilibrium can be assumed because no detailed knowledge is then required about the details of the interaction between the species comprising the system. However, when the distribution of the species concentration does not follow the local thermodynamic equilibrium (LTE) distribution, the microreversibility of each process in the system is not fulfilled. Under these circumstances the actual distribution will be strongly coupled to specific, dominant interactions.

The non-LTE state generally prevails when the system receives energy from an external source which is unbalanced by an equivalent energy loss to the outside of the system, or if balanced, the loss and gain involve different processes. This is the situation in the so-called "corona equilibrium" where stability prevails, but the principle of detailed balance does not hold. Other examples include plasma interactions with laser or nuclear beams and discharges which contain electric and/or magnetic fields which selectively increase the energy of the charged species.

For the moderately ionized gases in a glow discharge, the bulk of the emitted radiation is connected to the distribution of the radiating states and it is necessary to consider individually each process which may influence the production of the radiating states. Simulations of the temporal development of such discharges have not generally dealt with this problem due to the dominant role of space charge and the consequent attention placed on ionizing collisions. In fact, the rate equations for the excited species are rarely written. However, for the purpose of treating the emitted radiation, the master equations governing the populations of the various quantum states are needed.

Let us consider the case where spatial gradients in the excited species densities are small compared to their rates of change, considering, however, that all densities may be spatially dependent. The following kinetic equation may be written, assuming that all heavy species are at the same temperature T:

e-N Processes

$$\frac{dn_r}{dt} = \Sigma_s n_e (n_s q_{sr} - n_r q_{rs})$$

E-E-T Processes

$$+ \Sigma_s \Sigma_j \Sigma_k p_{sr}^{jk} \left\{ n_s n_j - (g_s g_j / g_r g_k) n_r n_k \exp\left[-(\varepsilon_r - \varepsilon_s) + (\varepsilon_k - \varepsilon_j) \right]/kT \right\}$$

E-T Processes

$$+ \Sigma_s \Sigma_r P_{sr} \left\{ n_s n - (g_g/g_r) n_r n \exp\left[-(\varepsilon_r - \varepsilon_s)/kT \right] \right\}$$

Ph Processes

$$+ \Sigma_{s>r} n_s (A_{sr} + u_{sr} B_{sr}) - u_{sr} B_{rs} n_r + \Sigma_{s<r} u_{rs} B_{sr} n_s \qquad (1)$$
$$- (u_{rs} B_{rs} + A_{rs}) n_r$$

with

$$q_{rs} = \int \sigma_{sr}(v) \, vF(\underline{v}) \, dv \qquad (2)$$

Equation (1) gives the rate of change of the density of species in the quantum number r. A similar equation must be written for all different values of r. Each term on the RHS of Eq. (1) corresponds to a specific set of processes as indicated. Thus, e-N represents electron-heavy particle collisions where n_e is the electron density. The density of heavy species in the r and s quantum states is represented by n_r and n_s, respectively. The first term gives the rate of production of species in state r due to electron collisions with species in state s, while the second term holds for the inverse reaction, q_{rs} being the reaction rate from Eq. (2), where $F(\underline{v})$ is the normalized electron velocity distribution. According to the principle of detailed balance, the cross-sections σ_{rs} and σ_{sr} are related as follows:

$$g_r \sigma_{rs}(v) v^2 = g_s \sigma_{sr}(v') v'^2 \qquad (3)$$

with

$$1/2 \, m \, v^2 = 1/2 \, m \, v'^2 + (\varepsilon_s - \varepsilon_r) \qquad (4)$$

The energy level of the quantum state q is ε_q and the g_q terms are the weight factors of the quantum level q.

E-E-T processes include collisions between heavy species involving internal energy transfer. p_{sr}^{jk} is the rate of collisions between species s and j yielding the species r and k after collision. The exponential factor expresses the inverse reaction rate averaged over the heavy species temperature.

The E-T processes are collisions between heavy particles involving translational or kinetic energy transfer only. P_{sr} is the rate of collisions between species r with any heavy species; n is the sum of all n_s.

Ph represents interactions with photons, the first two lines referring to states above r and the second to states below r. The A terms are the usual transition probabilities, while the B terms are either the absorption probability of induced photon transition with photon energy density u.

This set of equations is coupled with the Boltzmann equation for the electron velocity distribution function which, if the energy source is a uniform electric field E, may be written as

$$\frac{\partial f}{\partial t} + \underline{v} \cdot \frac{\partial f}{\partial \underline{r}} \frac{eE}{m} \frac{\partial f}{\partial \underline{v}}$$

$$= \left[J_{eh} - \sigma_{el}(v)v \right] f(v) + \left[\Sigma_s \Sigma_r J_{sr} - \sigma_{sr}(v)v \right] f(v)$$

$$+ \left[\Sigma_s J_{rec} - \sigma_i(v)v \right] f(v) \qquad (5)$$

with

$$n_e = \int f(\underline{v}) \, d\underline{v} \qquad (6)$$

and

$$f(\underline{v}) = n_e F(\underline{v}) \qquad (7)$$

The J terms are collision operators expressing the "in" terms, i.e., the term populating the cell in the phase space, while the $\sigma(v)v$ terms are the "out" terms, depopulating the cell. The subscript "el" represents all elastic collisions with all species including ions and electrons; "sr", "i", and "rec" represent e-N collisions, ionization collisions and electron recombination, respectively. This set of equations may be closed with the neutrality condition.

$$n_+ = n_e \qquad (8)$$

It can be seen in the above equations that each term referring to a direct process, say A, is paired with the corresponding term A' for the inverse process. At LTE, each term A' is exactly balanced by the corresponding A' so that each difference A-A' cancels. If now an external source of energy is coupled to the ionized gas, for example, by an electric field or by laser excitation of a specific state, some of the q_{ij} or the n_j may increase drastically with a resulting unbalancing in one or several of the A-A' pairs, say A_j-A_j'. These unbalanced terms will then dominate the RHS of Eq. (1). Eventually, due to collisional coupling of all terms, all A-A' terms will again become balanced.

As a result of the increase of the rate of production expressed by A_j, the terms A_j' will increase until they approach, after some time interval, the values on the order of A_j. The reequilibration depends upon the ratio of direct to inverse collision cross sections involved in the reactions included in A_j, which, in turn, are functions of the relative kinetic velocity of the colliding species, i.e., roughly, their mean energy or temperature.

Actually, the internal energy gain after the collision by one colliding species is not equal to the loss of internal energy by the other (except in some cases such as those involving quasi-harmonic states). A mechanism to equilibrate the collision energetically is needed which can provide energy exchange of any value, as might be achieved, for example, by photons or translational energy exchange. The final distribution of species in the quantum states, for which the differences A-A' cancel, will exhibit the mean energy characterizing the species which act to equilibrate the process. For example, in Eq. (1), the coupled A-A' terms labeled E-T can cancel only when the distribution of n_r follows the Boltzmann distribution at the temperature T of the heavy species. Couple terms labeled e-N cancel when the distribution of n_r is such that it compensates the ratio of production-to-destruction rates which depend on the electron distribution function. If the electron energy distribution is maxwellian, then the distribution of the n_r must, at equilibrium, be a Boltzmann distribution at the electron temperature. The case of coupled terms labeled E-E-T is not as straightforward and will be commented upon later. In that case, a distribution called the Treanor distribution may exist, for which the coupled terms cancel.

A partial equilibrium exists at the equilibration temperature when $A_j - A_j$ approaches the value of the next larger difference in Eq. (1), since A'_j almost balances A_j. In addition, new groups together with the former one become predominant in the RHS of Eq. (1) so that a new effective driving term exists for the relaxation of the system. With elapsing time several partial equilibria may exist sequentially having time constants which are closely linked to the value of the dominant collision frequency during each step. Comparison of collision frequencies entering into the various portions of Eq. (1) will determine the principal process driving the distribution of the excited n_r species.

If the external source of energy is quasi-permanent when compared to the characteristic collision times, terms written on one line in Eq. (1) may be offset by terms of another line. For instance, excitation rates by electron collision (e-N) may be balanced by radiation losses (Ph), the so-called corona equilibrium, or by diffusion losses of excited species which are not taken into account in Eq. (1).

Although obtained at low pressure, the work of Dubreuil and co-workers (Harnafi and Dubreuil, 1985, Harnafi, 1984) presents a clear experimental illustration of the features just discussed. Their work is based on laser induced fluorescence techniques in which the optical electrons of lithium atoms are pumped to selected n and L energy levels. The technique permits study of the changes in the fluorescence signals from these levels induced by collisions with rare gas atoms. This process is termed "n-mixing" and "L-mixing", n and L being the principal quantum number, and the total kinetic momentum quantum number, respectively.

Using a heat pipe to produce the lithium vapor, they were able to modify the temporal features of the fluorescence of the lithium atoms which follows selective excitation of the Rydberg states n^2D, for n = 3,4...13, by modifying the rare gas pressure. Figure 1 shows the lithium energy level diagram and the indicated two step laser excitation of the ground state to the n^2D levels.

The time resolved fluorescence of the 5^2D state ($5^2D - 2^2P$) is shown in Fig. 2 for different helium pressures, as an example of this technique. The gas temperature of T = 620 K leads to a Li density of the

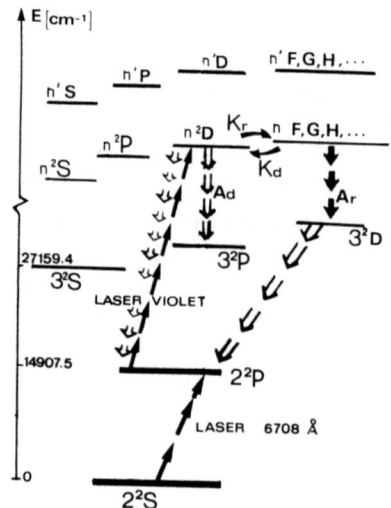

Fig. 1. Lithium energy level diagram (Harnafi, 1984).

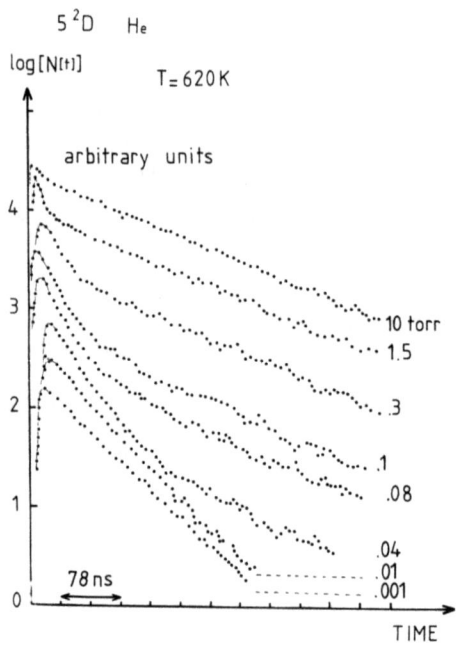

Fig. 2. Time resolved fluorescence of the 5^2D state ($5^2D - 2^2P$) for different helium pressures (Harnafi, 1984).

order of 10^{11} cm^{-3} (Harnafi, 1984). Solving Eq. (1) with only the assumed relevant terms on the RHS leads to the following interpretation.

At 10^{-3} torr, the exponential decay is due only to the radiative depopulation of the 5^2D state, and the radiative decay rate A_d is derived from the measured decay rate. As the helium pressure increases, the decay rate begins to increase due to the depopulation of the 5^2D level toward all 5^2L>2 levels (levels with n = 5 and L > 2) induced by collisions with helium atoms (the sub levels F,G,H...being only about 1 cm^{-1} distant may be considered as a unique level owing to their easy mixing through thermal collisions of some 420 cm^{-1}, or even through black boy radiation of the vessel). This decay rate increases with pressure according to a A_d + K[He] law from which the collision rate k may be inferred. However, the first decay slope is followed by a less steep second decay which is due to the rise in density of the 5^2L > 2 levels that repopulate the 5^2D level through the reversed k' superelastic collision of 5^2L > 2 states with helium. This is exactly the process of balancing a term A_k by the increase in the corresponding term A'_k.

At pressures above about 8 torr the rate of population of the 5^2L > 2 level is so high, owing to the rise of the collision frequency with pressure, that there is an almost simultaneous equilibrium with the reserved process. Now, only the second rate of decrease of the 5^2D level will be effective, maintaining a negligibly small difference $A_k - A'_k$, that is keeping the whole group 5^2L > 2 in equilibrium. Thus, this second depopulation rate corresponds to an average radiation depopulation of the 5^2L > 2 level acting now as a unique level. This behavior is a good example of a succession of relaxation times corresponding to different processes which are controlled by different collision frequencies.

Cross sections, integrated over thermal velocities, could be deduced from these studies for sublevel mixing of lithium by collisions with rare gases. Radiative lifetimes of the various quantum states could also be found. The cross sections (proportional to collision frequency, since $y = \sigma <v>$) for $n^2L \to n^2L > 2$ and the radiative lifetimes increase with the principal quantum number n as seen in Figs. 3 and 4. It was also found that there is an increase with n in the collisional induced transfer from $n^2L > 2$ to n^2P, and $n^2L > 1$ to n^2S. Actually, hydrogenic behavior should be expected for all atoms in high Rydberg states and, in lithium, even for n > 4, since the quantum defect is very small. For hydrogen and hydrogenic atoms, the collision frequencies for the electron-collisionally induced transition n → n' is proportional to n^4, according to Griem (1964), while the radiative transition lifetime increases as $n^{9/2}$.

In both cases, the collision cross sections are increasing with the energy of the excited level while the radiative transition probabilities decrease with this energy. This seems to be a general tendency for all species. It follows that with increasing pressure, or increasing electron density, the distribution of the upper states will be under the combined influence of collision and radiative depopulation. Thus the most probable distribution of the densities of quantum states will be such that

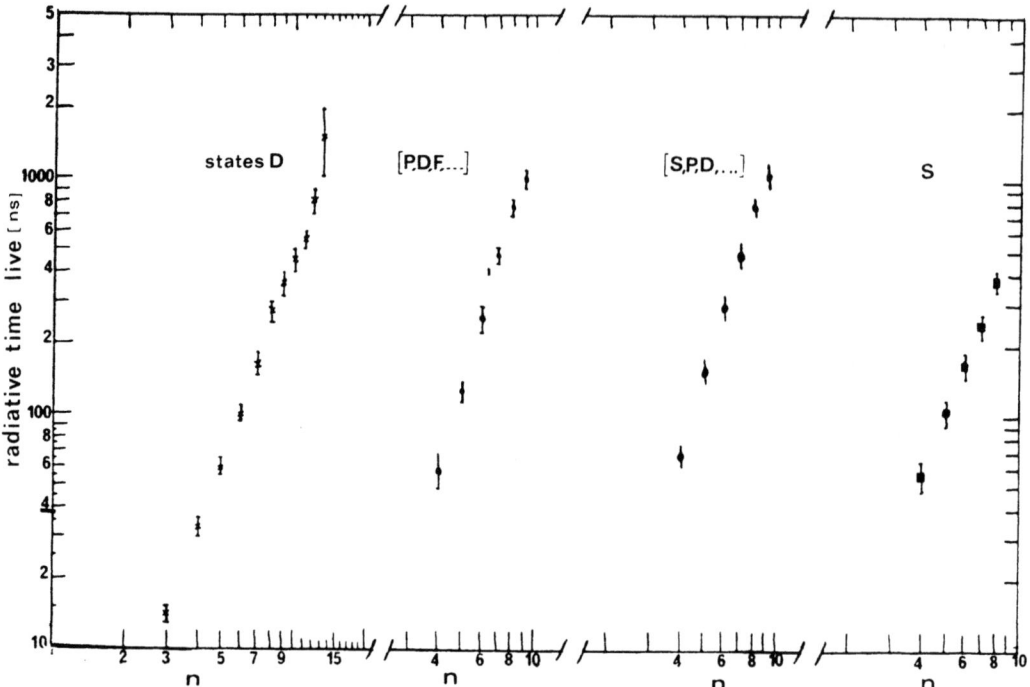

Fig. 3. Radiative lifetimes with the principal quantum number n (Harnafi, 1984).

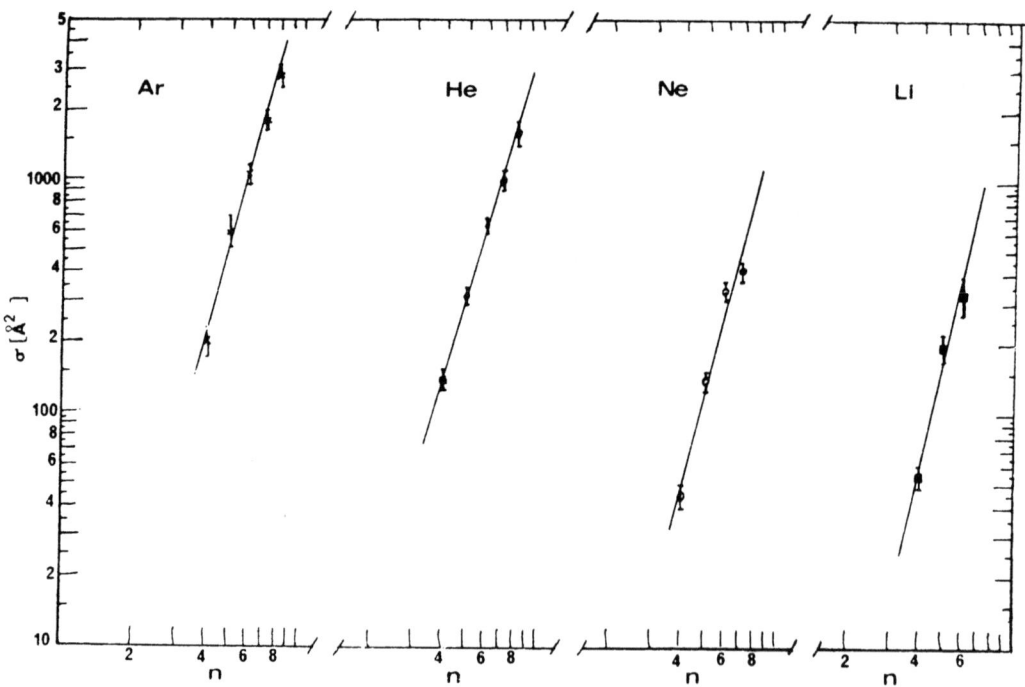

Fig. 4. Cross sections for $n^2D - n^2L > 2$ of lithium atoms with the principal quantum number n induced by Ar, He, Ne, and Li (Harnafi, 1984).

the upper states will exhibit a Boltzmann distribution at the electron or heavy species temperatures, or a distribution between these two temperatures. The lower states will reflect a coronal equilibrium where collisional and radiative mechanisms compete. At the intermediate levels a situation may exist where thermal equilibrium sets up between sublevels having the same n, while coronal equilibrium governs the population of different n.

Obviously, this general description holds mainly for radiative states, while metastable levels, whose lifetimes are much higher, may influence largely the gas chemistry, the distribution of upper states, and the increased role of collisional processes (see, for example, Ricard, 1981). Vibrational levels are an important class of metastable levels in homonuclear molecular gases and may exhibit an intermediate distribution which will be discussed later.

Beyond a particular pressure, the collision frequency becomes greater than the radiative transition probabilities for all levels. Actually, each level has a transition pressure. The "quenching pressure" for a given quantum level is defined as the pressure for which the actual depopulation rate is double that which is expected from radiative transitions alone. Two consequences result from an increasing pressure: a) reduced light intensity, since collisions take away upper states of radiative transitions, and b) reduced time lag between the emitted radiation intensity and the density fluctuations in upper state density. This may be followed in time when the processes which populate the upper states are abruptly stopped. The emitted radiation intensity for a transition $1 \rightarrow j$ being always given by $\phi = A_{ij} n_i$, the relation between the intensity ϕ_o in a pure coronal situation and the intensity ϕ_i at pressure p is

$$\phi_i = \phi_o/(1 + p/p_i) \tag{9}$$

p_i being the quenching pressure of state i. The decrease of $\phi_i(t)$ after the cessation of the upper state production is

$$\frac{1}{\phi_i} \frac{d\phi_i}{dt} = A_{ij}(1 + p/p_i) \tag{10}$$

This concept provides some insights into the effects of collisions without the need for an exact knowledge of the product states (Mitchell, 1970; Hirsh et al., 1970; and Albuques et al., 1974). Quenching pressures are usually in the range of a few torr. Thus, at atmospheric pressure, the distribution of excited states is completely dominated by collisions, and gas fluorescence after excitation will last only for a fraction of a nanosecond due to the fast quenching of the upper state radiative levels. This phenomenon has useful consequences in that it is possible to have a clear spatio-temporal sketch of a developing discharge at high pressure and it results in the short pulse emission observed in a particular high pressure lamp (Hartmann, 1970).

If an electric field is the source of non-equilibrium, it follows that the field also perturbs the populations of the quantum energy levels. While the mechanism of Stark broadening in hydrogen plasmas (Griem, 1964) is mainly related to the local microfield resulting from the interaction of neighboring ions and the emitting hydrogen atom, it will be shown later that the macrofield may combine with the microfield to produce spatial anisotropy in the emitted profile of the hydrogen lines.

The given set of equations serves mainly as a framework for discussing the mechanism of the temporal behavior of a discharge. In practical cases, however, the situation is extremely complex. First of all, the densities n_s should have another subscript, say $n_{s\ell}$, since different chemical species "ℓ" may be present in the gas, and ions may also be considered as reactants. In addition, three body reactions, particularly important in laser physics, should be considered. However, the question here is restricted to glows generated during the transition from a non-conducting gas state to the complete spark formation, and is considered in the context of studies of gaseous dielectric strength and physico-chemical discharge sources. In these studies, three body reactions were considered primarily in connection with physico-chemical yields.

This analysis could obviously not be applied with all the mentioned details to analyze the radiation emitted during the glow to arc transition. However, it gives a basic route for an approach to that question.

TRANSIENT BEHAVIOR OF GLOWS

Observations

Time resolved streak camera and photomultiplier techniques have been extensively used to record the spatial properties of transient gas fluorescence resulting from the discharge build up and its transition towards the spark phase. Generally, these measurements have been performed at pressures of some hundreds of torr (Meek, 1978; Goldman, 1978; Kunhardt, 1983; Renardières, 1972, 1977, 1978). At these pressures the fluorescence relaxation time lies in the nanosecond range so that the discharge brightness closely follows the spatio-temporal discharge activity.

The transient discharge behavior in air exemplifies many features which may be encountered in a large range of gas compositions and pressures. For non-uniform field discharges in air at atmospheric pressure, the spatio-temporal spectral emission, correlated with current and voltage measurements, enables several discharge states to be distinguished with typical space and time dimensions as indicated in Fig. 5.

The streamer is one of the mechanisms which first controls the expansion of the discharge. It is a high field region where electron avalanches lead to a build-up of a growing filamentary plasma. The high field region typically has dimensions of 50 μm and, owing to the high velocity of its advance ($\sim 10^8$ cm/s), exposes each cell of gas that it visits for some 10^{-10} seconds at atmospheric pressure.

The growing filamentary channel built up by the streamer (a high pressure transient positive column) corresponds to a cold plasma state and may reach a spatial extension of several centimeters lasting for some hundreds of nanoseconds. This filament stage may either disappear due to attachment, or transform into an arc phase if the gap is bridged by the discharge, or it may enter into a third typical stage, a leader. This phase corresponds to one or several "aged" filaments merging together and transforming gradually into a medium-hot (some 10^3 degrees, quasi-thermalized plasma which is controlled by the discharge activity evolving at its tip. The leader tends toward a complete bridge of the discharge gap and its lifetime may be as large as $10^2 - 10^3$ microseconds. Leaders of greater spatial and temporal extent may exist in lightning discharges. Fig. 5 illustrates the basic time scales for each process which can be expected to play a major role in the formation and maintenance of each of these discharge states.

Now, if the electrode gap is limited to several centimeters, no leader channel stage will be observed. Fig. 6 summarizes, schematically, the typical sequences which would then be recognized in a large range of

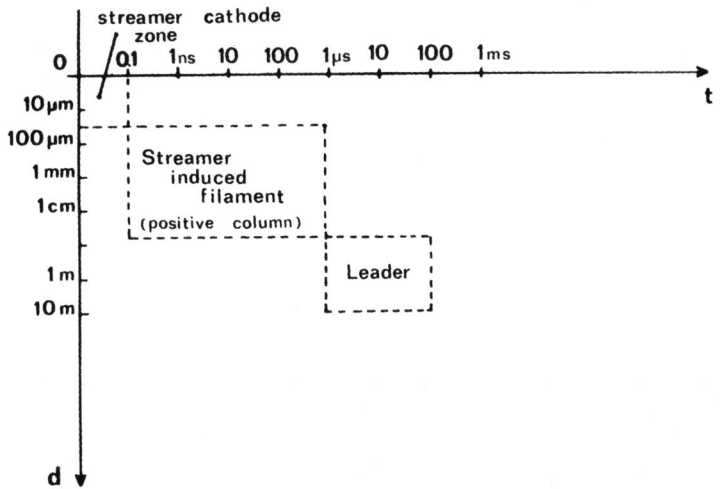

Fig. 5. Typical space dimensions and time duration.

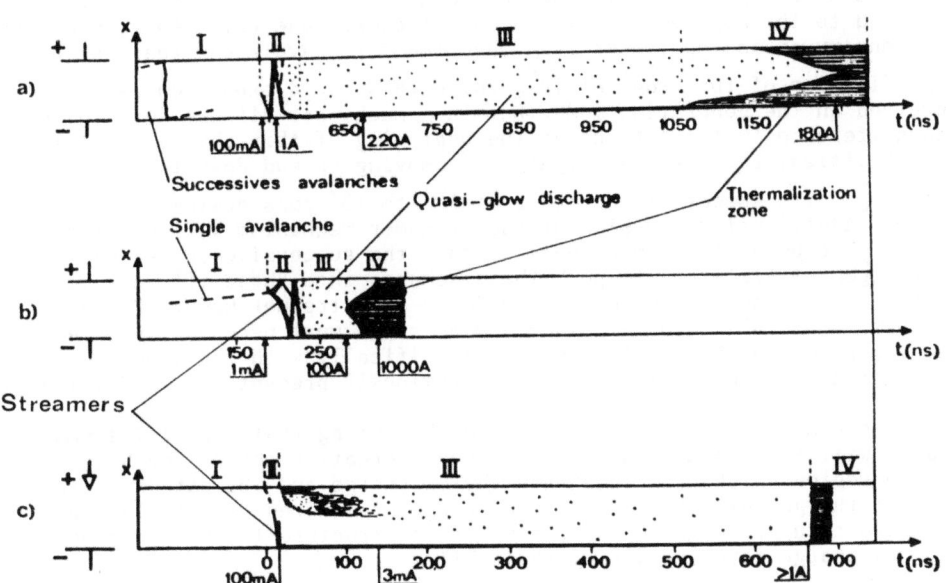

Fig. 6. Comparison between typical streak photographs of spark formation in different cases: a) generation mechanism in uniform field, pulsed gap, nitrogen with small over voltage (7.5%), p = 300 torr, d = 2 cm (Doran, 1968); b) streamer mechanism in uniform field, pulsed gap, nitrogen with high over voltage (35%), p = 300 torr, d = 2 cm (Koppitz, 1971; Chalmers et al., 1972); c) non-uniform field gap, dc potential, air with point radius of 100 μm, p = 760 torr, d = 1 cm (Marode, 1975).

cases. This figure shows the time resolved spatial fluorescence of discharge growth with time elapsing from left to right and space indicated in the ordinate direction. Here the values of the gap parameters are quite different for different situations:

- **gap geometries**: (a,b) uniform field, plane parallel electrodes and (c) non-uniform field, point-to-plane gap.

- **gas species**: (a,b) non-attaching gas, N_2 and (c) attaching gas, air

- **applied potential**: (a) small overvoltage impulse, (b) high overvoltage impulse and (c) DC potential

However, even with these differences, the spark formation exhibits similar sequences in all cases which can be put into four phases. During Phase I, one or more avalanches develop under the applied field, but the space charge field resulting from the charged species does not yet change the local field appreciably. However, with the self-sustained condition already satisfied, each avalanche will statistically be followed by at least one other. In case (1), the so-called "Townsend generation mode", the avalanche amplification factor is small (small overvoltage) compared to that of the so called "streamer mode" development, cases (b) and (c). Phase I will consequently last much longer in case (1) before Phase II is reached, a phase dominated by space charge effects.

Phase II begins when the space charge field becomes non-negligible compared to the applied field. In the streamer mode (b,c) an accelerated filamentary expansion is observed when the first avalanche reaches an amplification of the order of 10^8, while this rapid expansion is also observed in the generation mode, but in a much more diffuse form (note the difference in the current at the beginning of this phase). Axial instabilities, revealed by bright waves moving up and down through the discharge with speeds of the order of 10^7 to 10^9 cm/s dominate this phase. Additional discussion of the streamer mechanism will be given later. The point to be made here is that the successive space charge rearrangement will tend to push the discharge into Phase III, the glow phase, from either the generation mode or the streamer mode. For a discussion of the reason why such a glow stage must necessarily develop, one may refer to the chapter dealing specifically with the glow-to-arc transition and where this figure was previously presented (Marode, 1983).

These mechanisms may be summarized by noting that a positive space charge will tend to move near the cathode, creating there a cathode sheath region which will interface the discharge and the cathode. The rest of the gap will be filled by a bridge of ionized gas similar to a positive column. A glow phase, similar in structure to the low pressure glow discharge develops during Phase III.

The glow phase duration is larger in case (a) than in case (b). In the successive avalanche generation mode (a) a <u>diffuse</u> glow develops, while in the streamer mode (b) a <u>filamentary glow</u> is formed. Since the spark itself is filamentary, a constriction of the discharge must occur in case (a) before it forms, while the already filamentary structure in case (b) leads to much faster glow evolution.

The filamentary non-uniform case (c) in air with the positive polarity applied to the stressed electrode is more puzzling. Because of the small value of the mean electric field, the electrons disappear by electron attachment so that the glow stage tends to be self-quenched, as indicated by the reported current decrease during Phase III. Nevertheless, the final spark formation still takes place. This is because the small amount of gas heating through electron and ion collisions with the heavy species within the filamentary channel is enough to induce a density decrease of the heavy species in the discharge core (Marode et al., 1979; Marode, 1983). The associated increase in the field to density ratio E/N leads the electrons to ionize instead of being attached and, consequently, a spark is formed in the electronegative gas.

In all cases the filamentary channel is formed and preheated during Phases II and III and the final spark development, Phase III, takes place through dissociation of the gas molecules a and thermalization of the channel through electron-ion coulomb interactions from midgap (case b) or from the electrodes (cases a and c). In any of these cases, however, the sequences I, II, III, and IV are recognized despite the large diversity and range of the imposed parameters.

Studies on discharge growth in gaps lying in the meter range have been conducted in air at atmospheric pressure. Frames of pictures of the spatial appearance of the "positive" discharge extension (the highly stressed electrode is positive are shown in Fig. 7. A steak photograph of a similar discharge extension is shown in Fig. 8. A large number of filamentary channels, each being formed by a streamer as illustrated in Fig. 6c forms the so called "leader corona" and creates another discharge phase termed "leader". The leader has a very faint luminosity during its elongation (Fig. 8) and sometimes shows a sudden increase in its core activity called "reillumination". Thus, one area may be visited, first by the leader corona, then by the faint leader channel, then be submitted to reillumination and, finally, when the discharge reaches the anode, the "return stroke", which is the initial part of the final spark formation.

This clearly illustrates that a portion of a gas may be submitted in time to enormous changes in the discharge activity, i.e., electron current and temperature, so that a whole discharge history will determine the physico-chemical and thermodynamic state.

An example of a "negative" discharge is given in Fig. 9 (frames) and Fig. 10 (streak). This example, however, is not in free space (a negative discharge in free space will be seen later) but along a dielectric surface previously charged positively by corona charge deposition. These "gliding" discharges are studied in the general area of lightning interaction with aerospatial vessels. Here again, one sees a typical leading glow forming a thermalized channel. Evidence of its thermalized nature is given by its spectroscopic investigation (Bordage et al., 1982).

Build-up Mechanisms of the Transient Phases

An attempt to summarize the various discharge states and the mechanism of discharge growth is sketched in Fig. 11, where typical spatial distributions of electron densities, electric fields and potential are indicated.

The build-up of a filamentary discharge channel requires the existence of a region where processes are in effect which are able to induce a step increase of the background charged species density n_o to a new value n_1. If it is not in motion but attached to the cathode, it simply becomes the usual cathode region of a glow discharge. In the latter case, if the discharge bridges the whole gap, a complete glow discharge develops whose positive column stage is diffusion controlled (case c). At high pressure, the charge density changes from n_o to n_1 in a thickness on the order of one micron, the cathode sheath thickness.

Obviously, the growth of the channel requires a moving build-up region, the streamer (b). At this stage, the image of a glow discharge

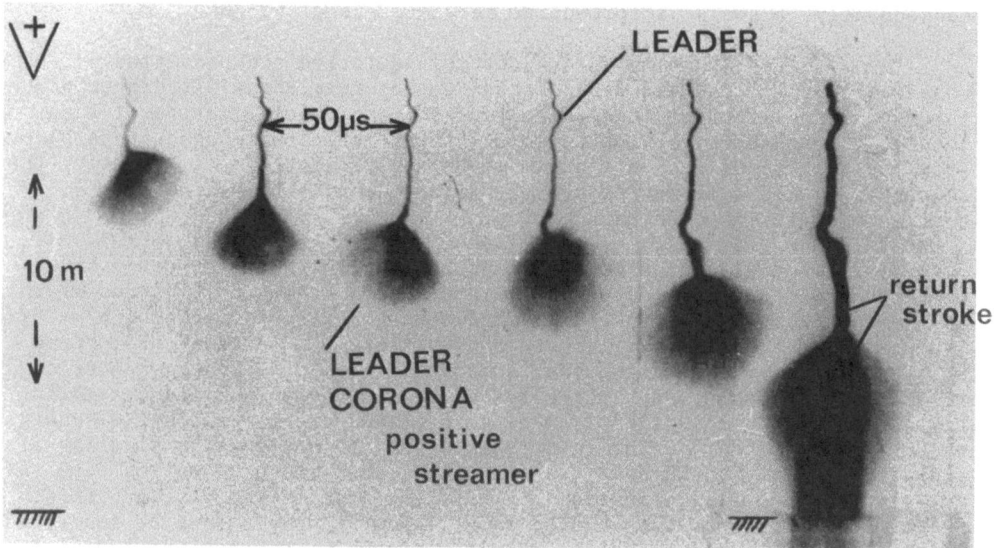

Fig. 7. Frame showing the spatial appearance of the "positive" extension (highly stressed electrode positive), 10 m gap in air. Photo from: "Groupe des Renardières", (Les Renardières, 1972-74).

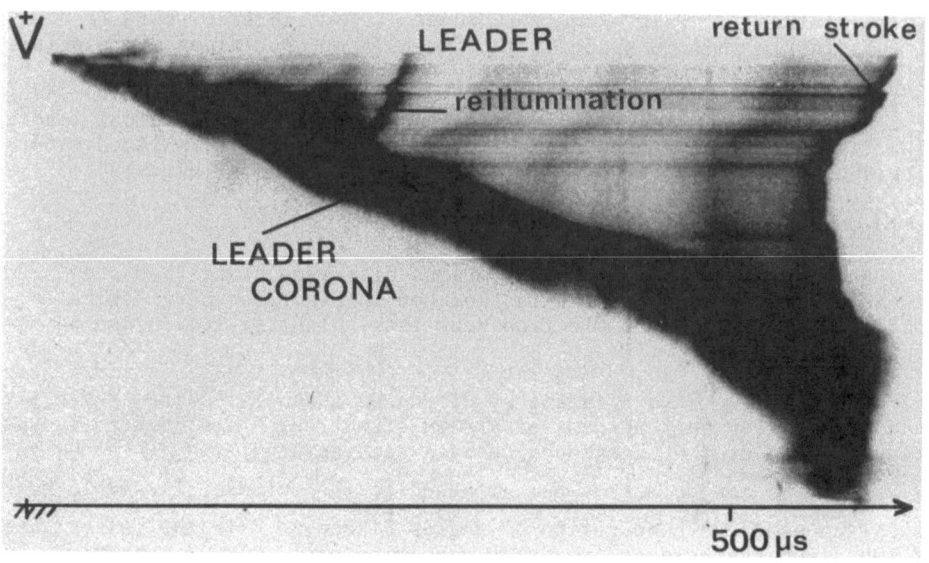

Fig. 8. Streak photograph of "positive" discharge. Photo from: "Groupe des Renardières", (Les Renardières, 1972-74).

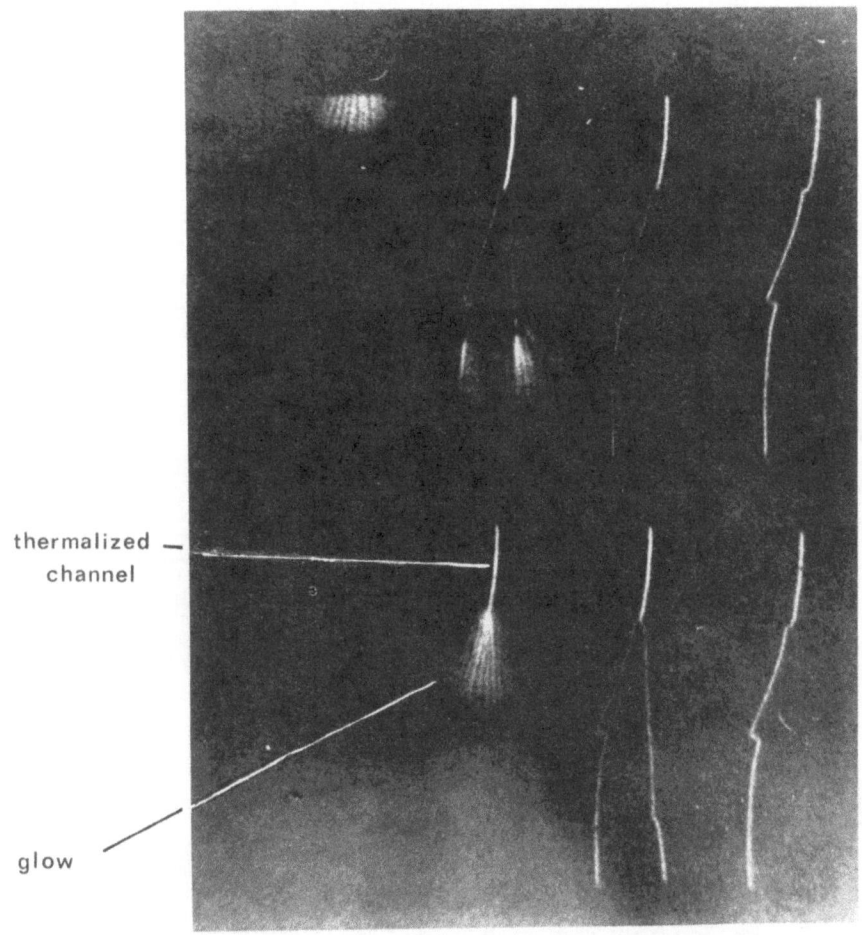

SURFACE DISCHARGE

(frame 10 ns $\Delta t = 50$ ns)

Fig. 9. Frame photo of negative discharge on a dielectric surface. Photo from: CNRS (G. Hartmann) and ONERA (A. Brunet).

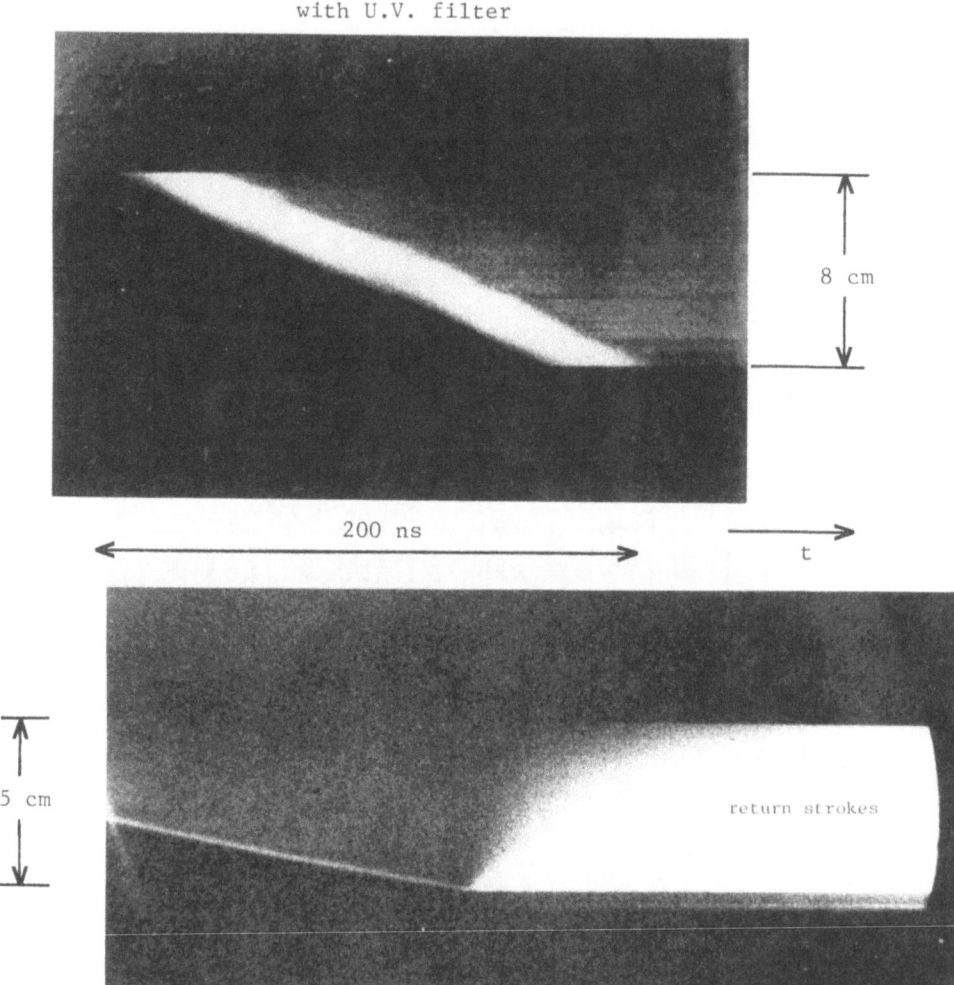

Fig. 10. Streak photo of negative discharge on a dielectric surface. The upper photograph shows part of the filament visible in the lower photograph. Photo from: CNRS (G. Hartmann) and ONERA (A. Brunet).

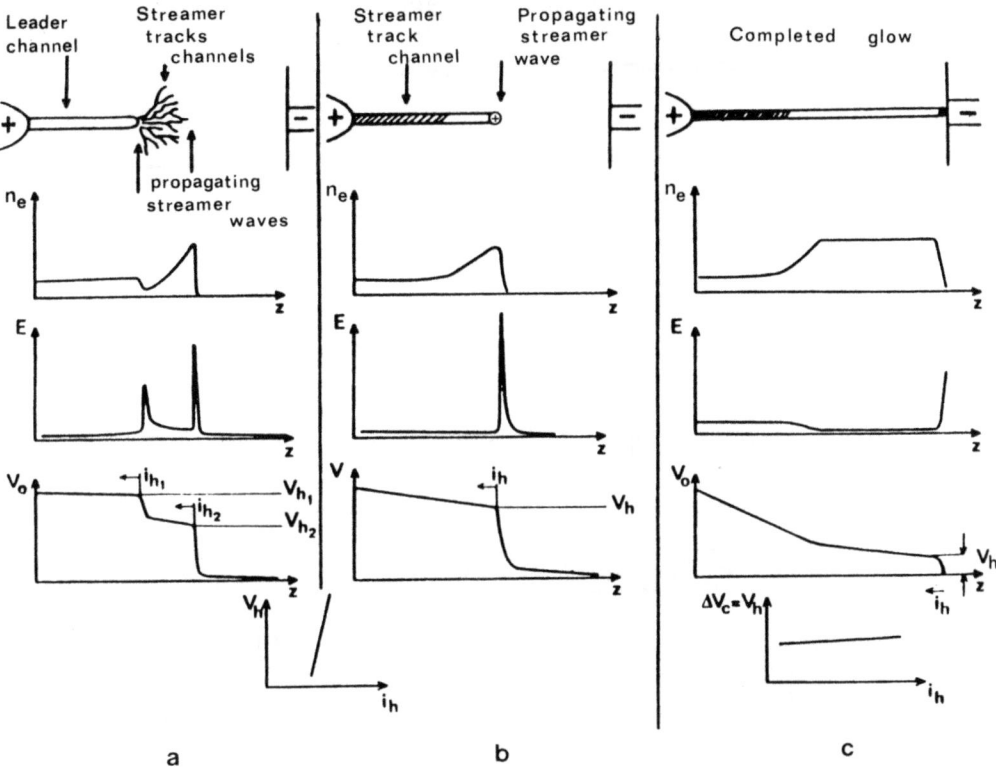

Fig. 11. Various modes of discharge evolution toward spark. V_h and I_h are potential and current, respectively, at the tip of a filamentary channel, behind the active region.

still holds. The filamentary channel is the positive column, while the streamer region is a "moving cathode region" with the gas as the cathode.

With increasing gap length ionizing waves of a second kind exist which act within an already formed channel, leading to the filament-to-leader formation (a). In this case, a localized field enhancement is first encountered with streamers building channels as in case b. A background ionized gas state is created within which another ionizing mechanism, discussed later, develops and leads to the second ionizing wave and growth of the leader channel.

Indicated in Fig. 11 are potentials V_h and current i_h, where the subscript h represents the "head". These are the current i_h produced by an ionizing wave for an existing driving potential V_h. Any increase of V_h at the cathode will lead to a huge increase in i_h, since the cathode field will follow the V_h increase. This not the case in free space, since there the change in the potential V_h will only slightly affect the effective electric field within the ionizing wave, the reference cathode potential being far away. It follows that the discharge may evolve via two basic modes, a channel controlled situation or a channel-head controlled situation. These are now discussed with reference to Fig. 12.

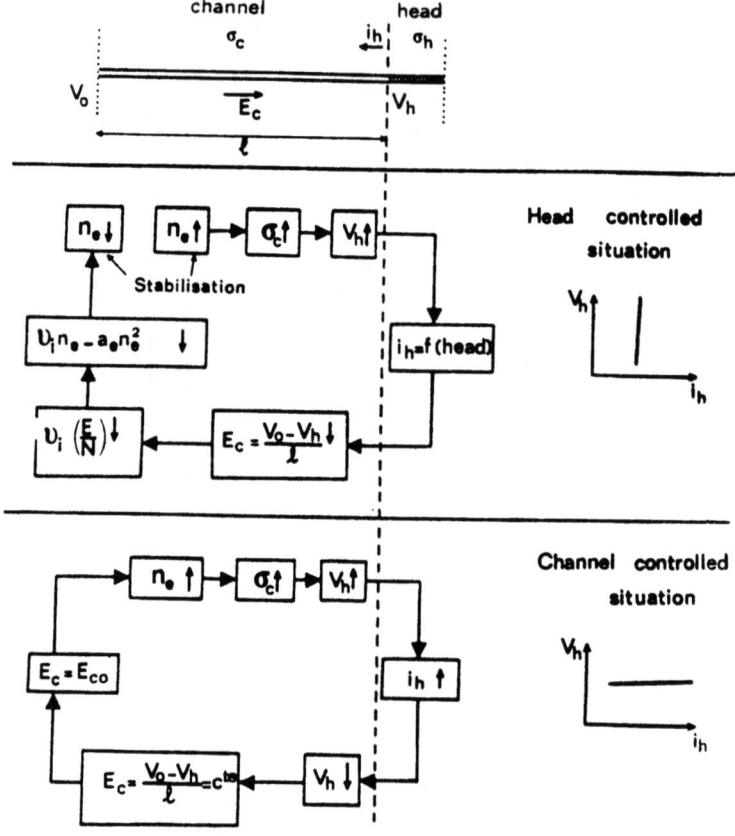

Fig. 12. Diagrams of interactions between a discharge channel and its active tip.

Let us now consider the head controlled case. The channel and the channel-head regions are characterized here by their conductivity σ_c and σ_h. If some change in the ionization rate occurs which tends to increase the electron density within the channel, then σ_c increases and the potential V_h, determined by the conductivities σ_c and σ_h, also increases. However, because i_h is mainly a function of the head space charge and not of V_h, σ_h will not change, V_h will rise, and the field E_c within the channel tends to collapse. The ionization frequency ν_i (E/N) decreases and the source term $\nu_i n_e - a_e n_e^2$, where a_e is the recombination coefficient, becomes negative. An electron density decrease follows which compensates the initial n_e increase. Stabilization occurs which ends with $\nu_i n_e$, the ionization rate, balanced by $a_e n_e^2$, the recombination rate. The electron density is then

$$n_e = \nu_i/a_e \qquad (11)$$

Thus, if s is the channel cross section, μ_e the electron mobility, the relation between i_h and n_e being

$$i_h \simeq e \, s \, (\nu_i/a_e) \, \mu_e \, N \, (E_o/N), \tag{12}$$

it follows that the electron density is entirely determined, given i_h and ν_i/a_e, and therefore the channel is head controlled.

At this point another fundamental process must be considered. The energy released within the channel modifies the channel with a longer time constant than the processes considered above. New chemical species are formed and an overall temperature increase in the channel occurs with time. Hence, the value of $\nu_i(E/N)$ does not change simply because E changes, but also because a heavy species temperature increase leads to changes in N. The function ν_i itself depends on the heavy species and composition. Thus, a quasi-thermalized positive column, similar to a spark positive column phase, may be reached without any discharge bridge of the gap. This is the leader phase.

For a channel controlled situation, the initial increase of V_h leads to an associated increase of the current i_h at the channel head. Consequently, σ_h increases and the relative change between σ_h and σ_c will not allow V_h to change and E_o will remain constant. If for this specific value of E_o, the ionization rate increases, this increase will not be perturbed by a head reaction and, thus, a channel controlled situation exists. It follows that as soon as the corona glow reaches the cathode, the current applied by the cathode is no longer limited. This is readily seen in the "return stroke" phase with a channel quickly reaching the arc stage.

Characterization of the Various Phases

High speed measurements, spectroscopic investigations, and modeling of all discharge phases permits an estimate of the parameters defining each discharge region (Table 1). Let us briefly give some physical reasons which form the background for these value. Avalanches do not greatly modify the applied field, and the field-to-density ratio E/N as well as the electron density n_e may be easily increased.

In the cathode region one must bear in mind that the electron energies do not follow the local value of E/N. It is the so-called "non-hydrodynamic" regime (Marode et al., 1983) and, as such, no direct link between the transport coefficients and the local field is possible. Discussions and measurements of properties of this region are given by A. Garscadden and J. Lawler elsewhere in of this Proceedings.

The streamer phase develops when the electron density amplification factor at the head is about 10^8. Estimates of n_e and j_e are based on the electron diffusion coefficient while the E/N ratio and an estimate of the electron T_e are derived from spectroscopic measurements (Hartmann, 1977). However, as will be seen later, there is strong evidence that the streamer region is in a highly non-equilibrium or non-hydrodynamic state. For the streamer induced filament, the estimates are based on simulations, spectroscopic measures (Hartmann et al., 1975), as well as hydrogen line broadening (Bastien et al., 1977, 1978).

The leader stage is a channel very near LTE and several techniques have been applied to study this phase (Renardières, 1975, 1977, 1978). The final jump corresponds to a large current increase typical of a

Table 1. Some Estimates for Atmospheric Air
References: a) Reather, 1964, b) Boeuf et al., 1982,
c) Marode 1985; Hartmann 1974, 1977 d) Marode 1975;
Bastien et al., 1977, 1978, e) Gallimberti, 1979,
Renardières 1972, 1974, 1977, 1981, f) Renardières, 1977

SOME ESTIMATIONS IN AIR AT ATMOSPHERIQUE PRESSURE

	n_e	J_e	$E/N (Vcm^2)$	T_e	T_N	Caracteristic time
Avalanche (a)	$< 10^{14} cm^3$ $Q < 10^8 e$	Not fixed	$\frac{E}{N} > 10^{-15}$	Not fixed	300°K	
Cathode region (b)		non hydrodynamic regime		within the negative glow Energie from 0 to some keV	1000	Continuous but not at equilibrium
Streamer (c)	$\simeq 10^{15} cm^{-3}$	up to $5 \cdot 10^3 A/cm^2$	$\sim 8.5 \cdot 10^{-15}$	$E \simeq 8$ eV to 12 eV	300	$\simeq 0.1$ ns
Streamer induced filament (positive column) (d)	$10^{13}/10^{15} cm^{-3}$	60 A/cm² ↓ 6 10³	$\sim 5 \cdot 10^{-16}$	1.4 eV	1000	100 ns ↓ 1 µs
Leader (e)	$10^{12}/10^{14} cm^{-3}$	500 A/cm²	$\sim 4 \cdot 10^{-16}$	2 eV	2000 ↓ 8000	10 to 500 µs
Final jump (f)	max $10^{18} cm^{-3}$	j=7000 A/cm²	E/N collapsing	max 3eV → 2.5 eV	$T_n \to T_e$ $\simeq 2.5$ eV	some µs

channel controlled growth of the current (Renardières, 1975). All of these data show that, apart from the non-equilibrium streamer region, one has

$$10^{12} < n_e < 10^{15} \text{ cm}^{-3} \tag{13}$$

$$1 < 3/2 \, k \, T_e < 3 \text{ ev}$$

where k is the usual Boltzmann constant. Let us now compare the various electron collision frequencies in that range of values.

Taking a typical cross section, $\sigma \sim 10^{-15}$ cm², a mean velocity equal to $\sim (8 k T_e / \pi m_e)^{1/2}$ (where m_e is the electron mass), the collision frequency for momentum transfer for electron-neutral collisions is then given by

$$\nu_{en}(s^{-1}) \sim <\sigma v> N_0 \sim 1.75 \cdot 10^{12} \, T_e^{1/2} \text{ (eV)} \tag{14}$$

where N_0, the neutral density at atmospheric pressure, is $2.6 \cdot 10^{19}$ cm^{-3}. The corresponding collision frequency for energy transfer ν'_{en} is given by

$$\nu'_{en} = (2 \, \delta m_e/m_n) \, \nu_{en} \tag{15}$$

m_n being the heavy species mass and δ the collision loss factor taking into account inelastic collisions (Mitchner and Kruger, 1973).

The collision frequency for momentum transfer due to electron-electron collisions, the long range coulomb interactions, can be discussed with the use of a radius cut-off based on the Debye length h (Holt and Haskell, 1965; Delcroix, 1966).

$$\nu_{ee}(s^{-1}) = (2/m_e)^{1/2} \pi e^4/(kT_e)^{3/2} n_e \ln(\Lambda)$$

$$= 1.25 \cdot 10^{-5} n_e/(T_e)^{3/2} \quad (cm^{-3} - eV) \qquad (16)$$

where $\Lambda = h/P_c$, and P_c is the impact parameter for 90° scattering due to coulomb collisions ($\ln \Lambda \sim 6$), and e is the electronic charge.

The corresponding collision frequency for energy exchange is

$$\nu'_{ee} = 1/2 \; \nu_{ee} ; \qquad (17)$$

ν_{en} is generally larger than the other collision frequencies. The most relevant comparison is between ν'_{en} and ν'_{ee}. If $\nu'_{ee} < \nu'_{en} < \nu_{en}$, then the electron distribution function is mainly determined by electron collisions with neutral species.

If, however, $\nu'_{en} < \nu'_{ee} < \nu_{en}$, then, while the electron mean energy is still determined by electron-neutral collisions, and electron thermalization through electron-electron collisions will lead to a Maxwellian distribution. If follows that the electron density n_{ec} for which $\nu'_{ee} = \nu'_{en}$ is very relevant to that transition and, using Eqs. 15 and 17, one finds

$$n_{ec}(cm^{-3}) = (2 \; \delta m_e/m_n) \cdot 6.2 \cdot 10^{16} \; T_e^2 \; (eV) \qquad (18)$$

An order of magnitude estimate for n_{ec} can be derived by recalling that $\delta \sim 50$ in air and ~ 300 in nitrogen (Mitchner and Kruger, 1973). The energy losses in nitrogen are more important due to large cross sections in vibrational excitation. For an electron temperature of 2 eV, n_{ec} would be $3 \cdot 10^{-15}$ cm^{-3} in nitrogen and $5 \cdot 10^{-14}$ cm^{-3} in air at atmospheric pressure.

This means that well before the channel enters the situation where it is heated by the electron-ion coulomb interaction, the electron distribution function is already a Maxwelliam one, and care should be taken in modeling the discharge using rate coefficients based on measurements at E/N for lower n_e values, where the distribution function may be far from Maxwellian.

Another relevant quantity which should be estimated is the optical depth which is descriptive of the radiative emission properties of the discharge. The brightest part of the discharge is the streamer and, in a nitrogen discharge, the dominant transition is the second positive group. Let us estimate the optical thickness of this transition in the streamer. Let us call ℓ a typical streamer dimension. The absorption dI of a beam of intensity I is given by

$$dI/I = (1/\Delta\nu) \cdot (\pi e^2/mc^2) \cdot n_B f_{CB} \cdot \ell \qquad (19)$$

where f_{CB} is the oscillator strength of the transition C → B which is considered, n_B is the concentration of the lower state, and $\Delta\nu$ is the width of the line. The most intense component emitted by the streamer at 300 K is the rotational J = 7 or 8 in the v = 0, v' = 0 transition of the $C^3\pi_u \to B^3\pi_g$ band. The concentration n_B may be estimated from

$$n_B = N_e(4/3\pi \ell^3) \cdot <\sigma_B V_{th}> \cdot (\ell/V_s) \cdot \alpha(J) \tag{20}$$

where N_e is the total number of electrons in the streamer sphere of radius ℓ, σ_B is the electron excitation cross section for the production of $N_2(B^3\pi_g,$ v = 0) from the ground state, V_{th} is the thermal velocity of the electron in the streamer, $\alpha(J)$ being the fraction of molecules in the $B^3\pi_g$, v = 0 state having J as rotational number. $\alpha(J)$ is derived from

$$\alpha(J) = (hc/kT) \ B_v(2J + 1) \exp[- B_v J(J + 1) \ hc/kT] \tag{21}$$

with B_v = 1.628 cm^{-1} and T = 300 k, one finds $\alpha(7) = \alpha(8) = 7.6 \cdot 10^{-2}$. Values of other parameters in Eqs. (19) and (20) are: $\sigma_B \sim 2 \cdot 10^{-18}$ cm^2, $V_{th} \sim 2 \cdot 10^8$ cm/s, $N_e \sim 10^8$ cm^{-3}, and $\pi e^2/mc \sim 2 \cdot 10^{-2}$. f_{CB} for the $(C^3\pi_u,$ v = 0) → $(B^2\pi_g,$ v = 0) transition is $1.9 \cdot 10^{-2}$, and $\Delta\nu$ may be taken as the Doppler profile of width $\sim 2 \cdot 10^9$ Hz. The use of Eqs. (19) and (21) leads to

$$dI/I = (1/\ell) \cdot 3.5 \cdot 10^{-5} \quad (\ell \text{ in cm}) \tag{22}$$

which, for $\ell \sim 20$ μm, gives $dI/I \sim 2 \cdot 10^{-2}$

As seen, the optical depth is negligibly small for this transition, and hence for all of the second positive group of nitrogen. For a resonance transition, having the ground states as the lower state, the situation would be different and the discharge could no longer be considered as optically thin.

Pressure Dependence

Pressure similarity laws are well known (see e.g. Llewellyn-Jones, 1966). Briefly, these laws state that discharges with similar pd and E/p should exhibit similar properties. However, in applying the cathode fall region, Long (1979) adds the following remark. If one assume, as a first approach in a steady state situation, that the energy gain from the electric field should balance the energy loss due to thermal diffusion, the following equation may be written in a one dimensional model

$$J E + \frac{d}{dx}\left[k(T) \frac{dT}{dx}\right] = 0 \tag{23}$$

where k(T) is the coefficient of thermal conductivity. The point to be made is that k(T) is temperature dependent only and is independent of density. If Eq. (23) is divided by N_o^3, to illustrate the scaling parameters, one gets

$$(J/N_o^2) \left(E/N_o\right) + \frac{d}{d(xN)}\left[\frac{k(T)}{N_o} \frac{dT}{d(xN)}\right] = 0 \tag{24}$$

If J/N_o^2, E/N_o, and the normalized dimension xN are kept constant, this equation does not predict a similar discharge since $k(T)/N_o$ is decreasing with increasing density N_o. It follows that with increasing pressure, the heat will be more and more imprisoned, and for similar discharges a greater increase in temperature will be observed at high pressure that at low pressure. A rough simulation of the whole cathode region with a set of simplified equations, including Eq. (24) gave the result shown in Fig. 13 (Long, 1979). Moving from the cathode toward the discharge, a huge increase in temperature is predicted when the pressure is increased. Some experimental results on the temperature increase in the cathode region may be seen in the presentation of J. Lawler elsewhere in this book. This general result is true whatever discharge region is considered: the gas medium is heated more by the discharge at higher pressures.

TRANSIENT EMISSION FEATURES

Going from one discharge state to another, one observes large changes in the electrical and chemical parameters defining each state, and this will greatly influence the discharge radiation features. It is impossible to make a complete and exhaustive list of all the changing

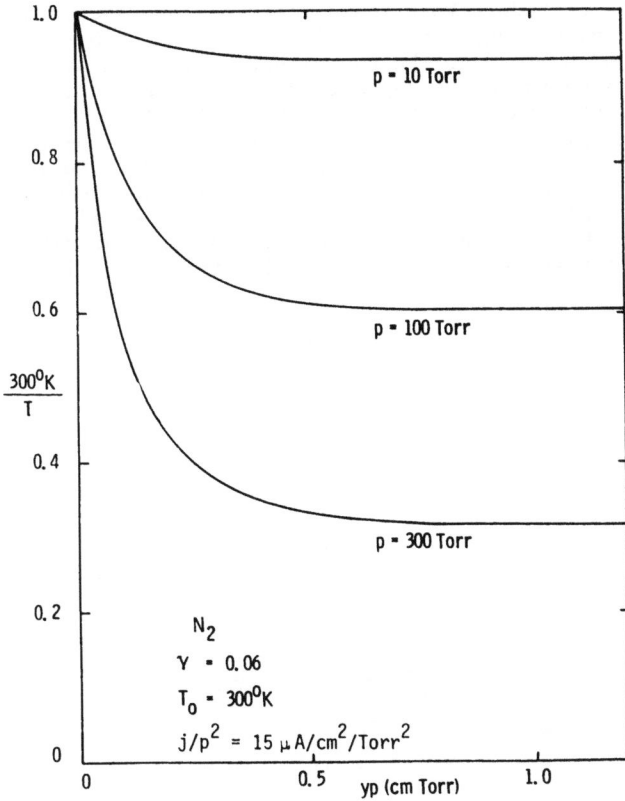

Fig. 13. Temperature variations in the cathode sheath (Long, 1979).

parameters, but one can try to outline the main factors which should influence the emitted radiation. The factors which will be considered now are; the effect of the electrical field, the effect of vibrational excitation of molecular species, and the effect of strong inelastic excitation.

Effects Due to the Electric Field

The value of the electric field acts either as defining the electron mean energy and, thus, the electron collision properties, or on the position in energy of the quantum levels of the heavy species. The second part results in anisotropic radiation emission.

Cathode Region, Streamer and Positive Column. Within the cathode region of a glow discharge the electric field falls quasi-linearly from the cathode toward the anode until it reaches a very small value. This will define the cathode fall section. It is followed by a plasma, the negative glow, where the field is very low and where the charged species are produced by collisions with the energetic electrons which have been accelerated in the cathode fall section and now lose their energy in the low field portion of the negative glow. Clearly, this region is not only in a non-LTE state, but also in a non-hydrodynamic state since electrons are able to ionize in a region where the electric field is negligible (see Marode and Boeuf, 1983 for a discussion of this situation). An example of the variation in space of the electron energy distribution throughout the whole cathode fall and negative glow region as obtained by Monte-Carlo simulation of the electron swarms is given in Fig. 14. The peak near px = 0 corresponds to the initial electron distribution released from the cathode. When these electrons are accelerated in the cathode fall region, they undergo collisions and, as a result, the number of electrons which traverse the gas without collision decreases, and the electrons whose energies are degraded due to inelastic collisions build up the lower energy part of the distribution function. Near the entrance to the negative glow region the general structure of the distribution shows a peak of electrons at very low energy (mainly due to secondary electrons created in ionizing collisions) followed by electrons at medium energy, and a small peak of electrons with energy equal to the full voltage drop (300 volts in this example). The angular distribution function in the cathode fall region is illustrated in Fig. 15. The

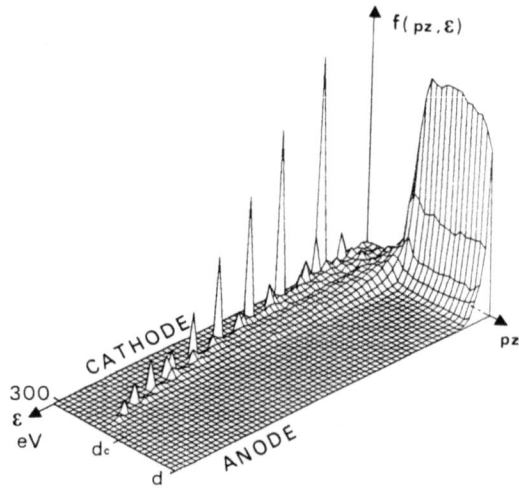

Fig. 14. Variation of the normalized energy distribution function in the cathode region for the abnormal glow discharge case (Boeuf et al., 1982).

electrons at low energies are almost isotropic, while those electrons with energy equal to the voltage drop are entirely forward directed.

Clearly, these distributions are far from Maxwellian and this situation cannot persist on a large scale since the space charge growth does not allow the electric field to remain very high. A condition where

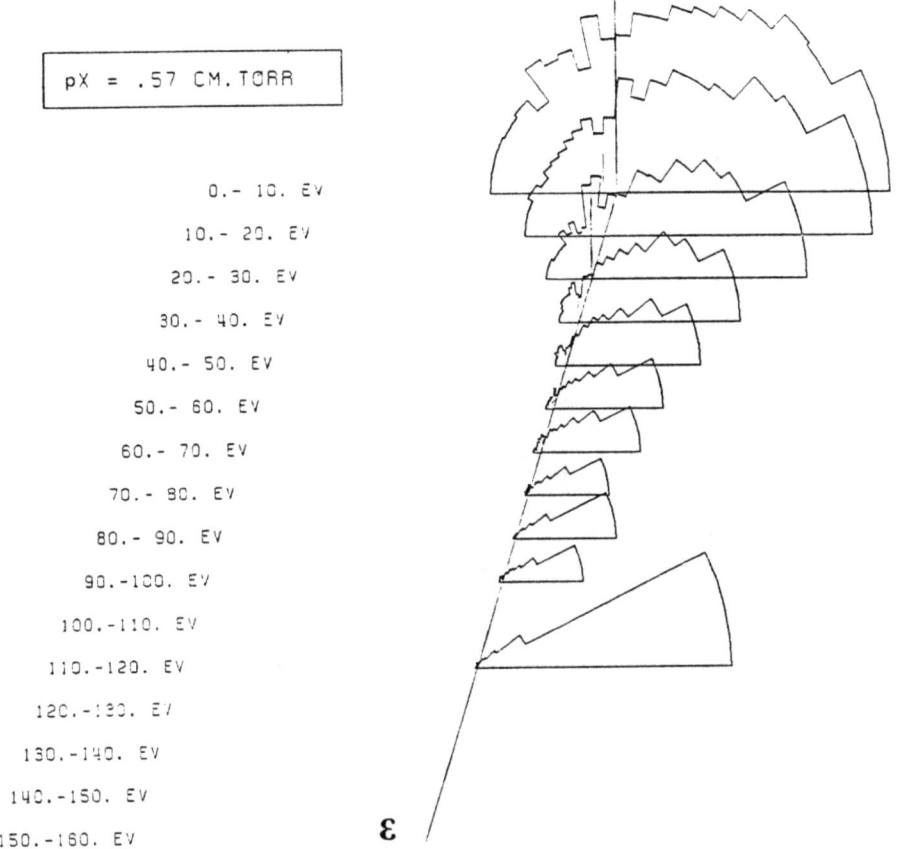

Fig. 15. Angular electron distribution functions for different values of the energy in the cathode region. Monte-Carlo simulation, Px=0.57 torr-cm (Boeuf et al., 1982).

electrons of very high energy exist which are able to pump very high lying states collisionally is confined to a small thickness as illustrated in Fig. 16 (after Badarue and Popescu, 1965). The figure gives the negative glow reduced thickness as a function of applied cathode fall voltage and gas nature. It is seen that this thickness is related to the space interval needed for the primary electrons to lose their energy.

119

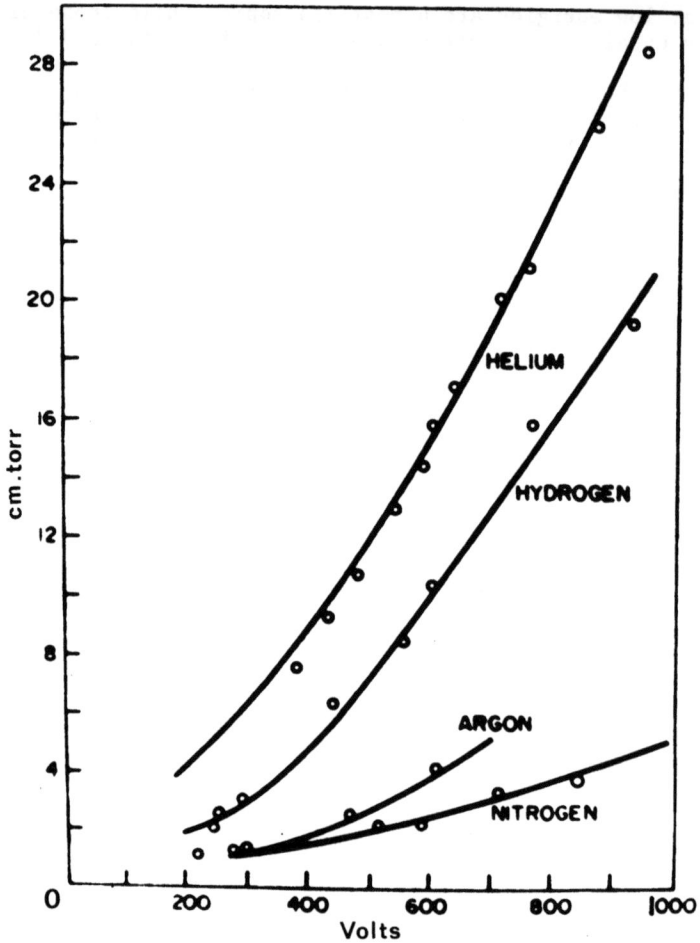

Fig. 16. Comparison between reduced length of negative glow (o are experimental points) and the energy range relaxation distance of the electrons in the gas (in solid line) (Badareu et al., 1965).

Now we discuss the streamer region. At first glance, the region where avalanches build up a streamer, see Fig. 17, has an extent of some 100 μm at atmospheric pressure. An estimate of the number of elastic collisions undergone by an electron in this thickness is about $2.5 \cdot 10^3$ for an electron temperature of 12 eV. For that number of electron collisions, equilibrium between electron excitation rates and the local electric field should be reached. However, this is not only oversimplified, but it obscures an important fact. As seen in the figure, the streamer mechanism is due to avalanches triggered by secondary photoelectrons which are themselves produced by ultraviolet absorption of light emitted by the avalanches (see Teich, 1967 for UV absorption measurements and Dawson et al., 1965; Gallimberti, 1972 for streamer propagation mechanisms). The electrons at the head of the avalanches neutralize the roughly hemispherical positive space charge to which they are attracted. The result is that the hemispherical positive space

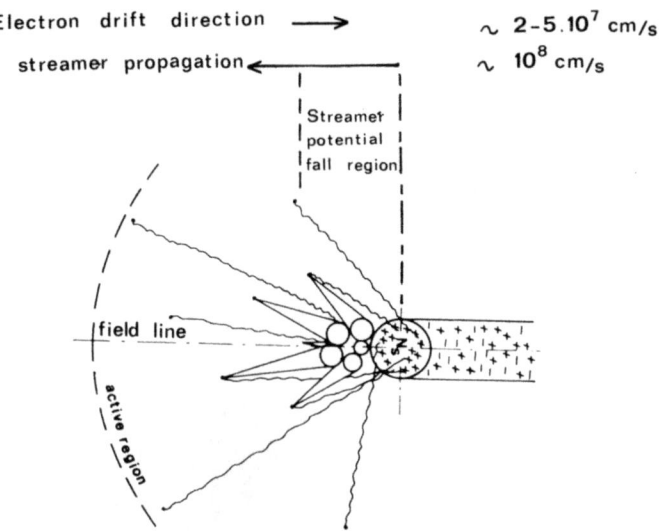

Fig. 17. Streamer mechanism.

charge is moving at the streamer velocity ($V_e \sim 10^8$ cm/s) in the direction opposite to the electrons ($V_e \sim$ 2 to 7 $\cdot 10^7$ cm/s). Consequently, an electron produced in the streamer avalanches "feels" the streamer field only for the brief time that it may be in the streamer region. Taking a typical streamer potential fall thickness of r $\sim$ 100 μm, the maximum time taken by an electron to cross this thickness is $\Delta t = \Delta r/(V_s + V_e) \sim 8 \cdot 10^{-11}$ s. For an electron having a mean energy of 12 eV the thermal speed is about $2 \cdot 10^8$ cm/s, so that it is estimated to undergo $\sigma V_{th} N_o dt \sim 400$ collisions, assuming a typical cross section of 10^{-15} cm^2. This is, however, an overestimate since these electrons are forward directed so that $V_c \sim V_{th}$. With this last assumption, the number of elastic collisions falls to about 40 and the number of inelastic collisions is on the order of unity. The corrected values must be between these numbers so that a relative electric field change $\Delta E/E$ above 10% must exist in the electric field which acts on the electron during two inelastic collisions. No equilibrium should be reached between the excitation rates by electron impact and the local value of the field. Within the streamer region the electron distribution function must be similar to that which exists in a cathode fall region, and this emphasizes further the analogy of a cathode fall and a streamer.

While a mean E/N value of 10^{-14} V cm^2 prevails within the streamer, the situation is very different in the filamentary discharge channel created by the streamer, which is a transient positive column. Recent modeling takes into account neutral density variations within this channel as well as spatial field redistribution due to attachment controlled changes in local conductivity (Marode et al., 1979; Bastien et al., 1985). The resulting field distributions are shown in Figs. 18 and 19. The first curve on the left hand side is the axial field distribution within the filamentary channel created by the advancement of the streamer. The positive point electrode is at z = 0, while the streamer region is located at the very end of this curve. The streamer field

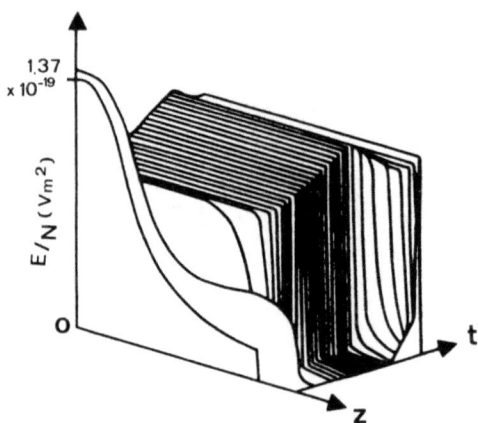

Fig. 18. Spatio-temporal distribution of the axial electric field E. Applied voltage V_a = 17 kV. The point is at position 0. The gap distance is 1 cm and the time interval between curves is 25 ns except for the first curve which illustrates the streamer state 1 ns before it reaches the plane.

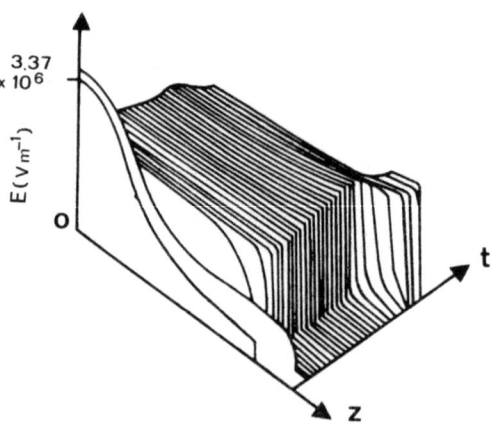

Fig. 19. Spatio-temporal distribution of the E/N ratio (Conditions as in Fig. 18).

itself is not represented here. This explains why the integral of the field from point to plane is not equal to the constant applied potential on the next curve for which the streamer is just about to reach the cathode. The following curves correspond to the complete filamentary glow bridging the entire gap where, again, the field within the cathode is not represented. As predicted (Sigmond et al., 1982) and demonstrated (Bastien et al., 1984, 1985), the E/N rearrangement in two "plateaus" is due to attachment stability conditions. As can be seen, the position which separates the two regions is slowly moving towards the cathode. When this attachment controlled wave reaches the cathode, a constant E/N prevails all along the discharge which leads to the channel controlled spark formation. However, the order of magnitude of E/N is at least one order less than that in the streamer or in the cathode sheath and, thus, the spectrum which will be emitted from the high field plateau side of the discharge will exhibit radiation from heavy species' transitions from much lower energy levels than those emitted by the streamer or the cathode region.

Spectra emitted by the various discharge regions in air at atmospheric pressure are shown in Figs. 20, 21 and 22. These are molecular spectra which exhibit mainly the transitions indicated in Fig. 36. Figure 20 is the spectrum emitted in the cathode region of a continuous discharge near the negative point of a negative point-to-plane discharge (Teich, 1985), while Fig. 21 is the spectrum emitted by the streamer. Both spectra emit about the same radiation. The difference is that neither recombination nor dissociation could have already taken place in the streamer case so that radiation from NO, N(I), and O(I), which are present in the quasi-continuous glow of the negative point, are absent in the positive streamer. (One often adds "positive" or "negative" before streamer to denote the direction of propagation of the ionizing wave relative to that of the electrons.) Obviously, the analog with the cathode region was done with reference to the positive streamer.

Now, if one compares the streamer spectrum in Fig. 21 with that of the filamentary positive column in Fig. 22, two observations can be made. First, one clearly sees that the rotational components in each vibrational band are much increased in the filament compared to that in the streamer, leading to a filament temperature of 1200 K. Second, the first negative band spectrum, clearly apparent in the streamer case (and in the negative glow case), is no longer present in the filament case. Clearly, the upper state production of the first negative system, $e + N_2(^1\Sigma_g^+) \rightarrow N_2^+(B^2\Sigma_u^+) + 2e$, requires some 19 eV and this reaction is much more probable in the streamer region.

This example of air discharge spectra illustrates the difference, well known in laser discharges, between the positive column and the negative glow emission properties. The negative glow properties are often used in hollow cathode devices to enhance spectral lines from the high lying levels. Here, in addition, one sees that the ionizing wave propagating within the gas volume, and not only at the cathode, exhibits features which are very similar to that of the negative glow.

<u>Anisotropic Field Effects</u>. Broadening of the hydrogen Balmer line will now be discussed as an example of anisotropic behavior induced by the presence of a space charge field of the same order of magnitude as the ion microfield.

In a plasma dominated by Coulomb collisions, the high value of the plasma conductivity precludes the existence of a static field (produced externally or by space charge) and the classical line broadening theories apply (Griem, 1964; 1974).

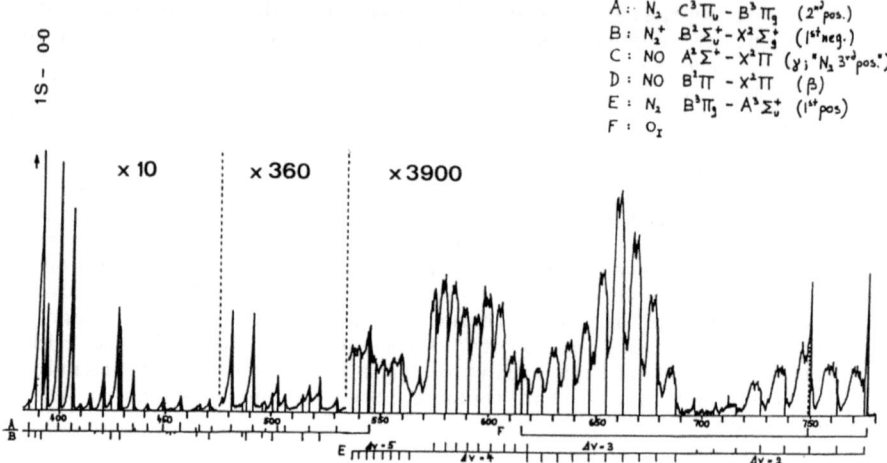

Fig. 20. Continuous discharge near the negative point of a negative point-to-plane discharge. Atmospheric pressure air (20% O_2 - 80% N_2) (Teich et al., 1985).

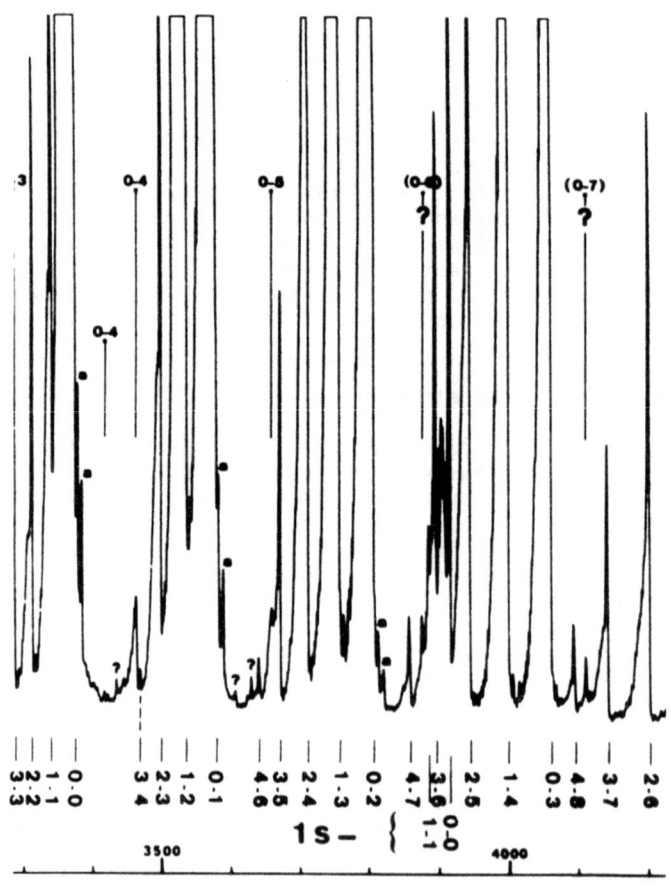

Fig. 21. Streamer spectrum (top) (Hartmann, 1970).

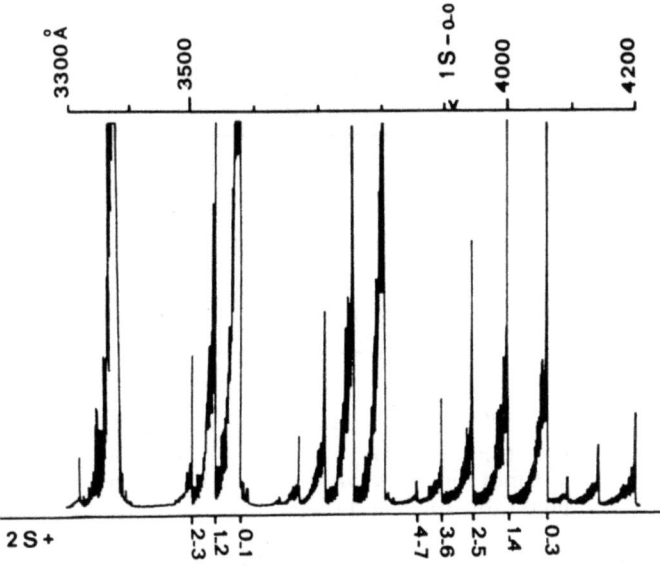

Fig. 22. Transient positive column (Spectrum) (Hartmann, 1970).

As has been shown, a space charge field may develop in the corona glow region and can be on the order of 10 kV/cm at atmospheric pressure. Let us examine this case and recall the principle of the theory of hydrogen Stark broadening. Let us first remark that if a hydrogen atom is in the presence of a uniform field (due to space charge or due to ions), then the emission depends on the field and the direction of observation, so that the components appear as in Fig. 23a in a direction perpendicular to the electric field due to the splitting of lines into components. Before considering the effect of variation in the values of this field, i.e., the quasi-static field distribution, the effect of the broadening due to electron impact must be added. This leads to a final broadening as in Fig. 23b.

With the impact approximation and the Griem-Kolb-Shen (GKS) formalism (Griem et al., 1959) the line profile J_e as a function of $\omega = 2\pi c/\lambda$ and F (the total field) is given by:

$$dJ(\omega, F) = \pi^{-1} \Sigma_{\alpha' \alpha'' m \beta} \text{Re} \langle \beta | \underline{e} \cdot \underline{r} | \alpha' m \rangle \langle \alpha'' m | \underline{e} \cdot \underline{r} | \beta \rangle$$
$$\langle \alpha'' m | [i \Delta\omega_{\alpha'' \beta}(F) - \phi_a]^{-1} | \alpha' m \rangle d\Omega_{obs} \qquad (25)$$

where the radiation is polarized in the $\underline{e}$ direction, $\underline{r}$ is the vector position of the atomic electron, $d\Omega_{obs}$ is the solid angle of observation, and ϕ_a is the collision operator. The transition $|a\rangle \rightarrow |b\rangle$ is split by the Stark effect into substrates $|\alpha\rangle$ and $|\beta\rangle$, respectively. The emitted line ω_{ab} will be shifted toward various $\omega_{\alpha\beta}$ and $\Delta\omega_{\alpha''\beta}(F) = \omega - \omega_{\alpha''\beta}$ is the frequency variable.

We have averaged over fields F, and to do that we need to know the field distribution. When the static field is negligible, it is generally accepted that the field distribution is isotropic. A characteristic

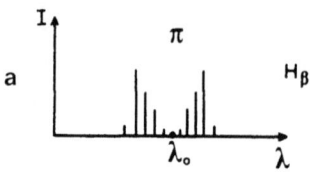

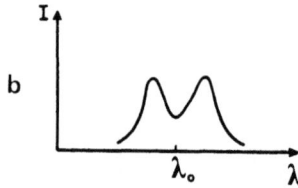

Fig. 23. a) π components of H_β (observation perpendicular to constant field). Scheme of line intensities as a function of wavelength. λ_0 is the initial position of the line.

b) Profile of H_β taking into account the effect of electrons.

field for this distribution is the so-called Holtzmark field $F_0 = 2.6$ $e\, n_e^{2/3}$ (cgs units). Thus, F_0 (statvolt/cm) $= 13 \cdot 10^{-9}\, n_e^{2/3}$ (cm^{-3}), or F_0 (V/cm) $= 3.32 \cdot 10^{-7}\, n_e^{2/3}$ (cm^{-3}).

We must compare this field F_0 with the static field F_c (produced externally or by space charge). If the static field F_c is not negligible compared with F_0, the field distribution is no longer isotropic and the broadening theory must be reconsidered (Bastien et al., 1977). The GKS formalism with a field distribution computed from the composition of the static field and the ionic microfield according to Hopper (1968) has been used.

The direction of observation k and the actual direction of the field F, resulting from the composition of the quasi-static space charge field F_0 and the ionic microfield F_i must be referenced by the angle Θ_{obs}, ϕ_{obs}, Θ and ϕ in a frame where the z axis lies in the static field F_c direction (Fig. 24). In the isotropic case where the microfield results from the ionic microfields which surround each hydrogen atom, the probability of finding the microfield around F_i (if $F_c = 0$) is

$$d^3 P(F) = W(F_i) \cdot \frac{d\Omega}{4\pi} \cdot F_i dF_i \tag{26}$$

where $W(F)$ is the ionic microfield distribution. When the quasi-static field is taken into account ($F_0 \neq 0$), we obtain

$$d^3 P(F) = (1/4\pi) W[(F^2 + F_c^2 + 2FF_c \sin \Theta)^{1/2}].$$
$$[(F^2 \cos \Theta)/(F^2 + F_c^2 - 2FF_c \sin \Theta)]\, d\Theta\, d\phi\, dF \tag{27}$$

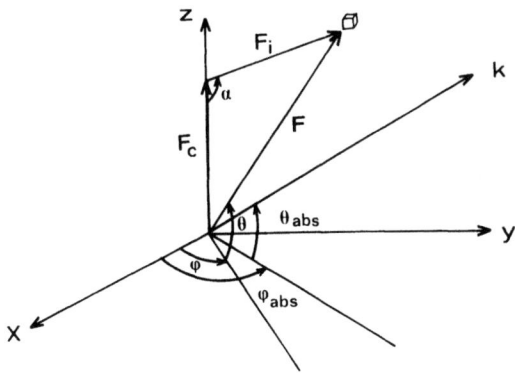

Fig. 24. Coordinate system bound to the static field F_c

The line shape is then obtained by multiplying Eq. (27) with Eq. (25) and integrating over all values of F. It is found that the line shape can be written as

$$L(\omega) = \cos^2 \Theta_{obs} L_\perp(\omega) + \sin^2 \Theta_{obs} L_{//}(\omega) \qquad (28)$$

which shows that the line profile depends on the angle between the static field and the direction of observation.

Before examining some results, we shall point out the possibility of achieving a realistic situation corresponding to the previous computation. In other words, the following assumptions can be fulfilled when F_c and F_o are of the same order of magnitude. The electron distribution is almost isotropic for $E/P < 24$ V cm^{-1} torr^{-1}. Consequently, the collision operator ϕ_a can be computed assuming an isotropic electron distribution. Moreover, the sublevels $|\alpha>$ of the upper level of the hydrogen excited electronic states have almost the same population. This population is essentially controlled by electron-atom collisions (Bastien et al., 1977; Bastien, 1977). Furthermore, $P_c = F_c/F_o$ (anisotropic parameter) can easily reach a value of 3 in the filamentary streamer channel.

Let us now examine some profiles. $L(\omega)$ is a complicated function of n_e, T_c, F_c, and Θ_{obs} (angle between the direction of observation and a direction perpendicular to F_c). As seen in Fig. 25, the difference between these lines and the lines with an isotropic distribution becomes important only when $P_o > 2$. In order to show that this result is not surprising, we have drawn a schematic representation (Fig. 26) of the addition of the constant field F_o, the average length of the microfield vector F_i/F_c is of the order of F_o ($P_c \simeq 1$) in Fig. 26a. In that case, the modulus $|F|$ of the resulting field is between $2|F_o|$ and a small value, so that the mean value remains F_o. Only in the case $P_c > 1$, as in Fig. 26b ($F_c = 3 F_o$), will the quasi-static field really influence the broadening.

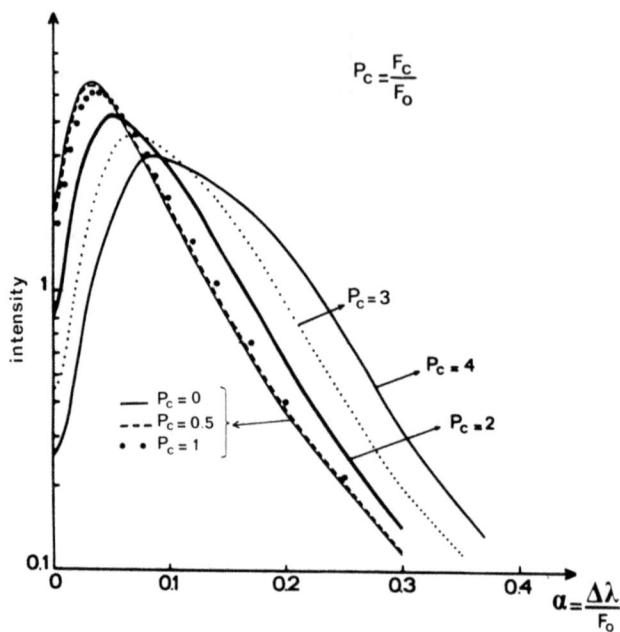

Fig. 25. H_β broadening for $N_e = 10^{15}$ cm^{-3}, $T_e = 20,000$ K, and various values of the anisotropy factor $P_c = F_c/F_0$; $\alpha = \Delta\lambda$ (Å)$/F_0$ (cgs).

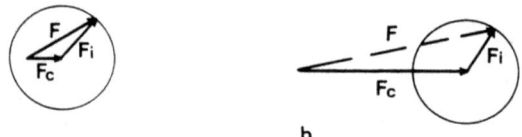

Fig. 26. Schematic representation of the field composition for: a) $P_c < 1$ and b) $P_c > 1$.

In principle, observations in different directions, $L_\perp$ ($\Theta_{obs} = 0$) and $L_{//}$ ($\Theta_{obs} = 90°$), would allow one to obtain information on both n_e and F_c. The difference between $L_\perp$ and $L_{//}$ is presented in Fig. 27. However, even if the line broadening is measured only in the perpendicular direction, it has been demonstrated (Bastien et al., 1977; 1979) that if $1 < P_c < 4$, as a first approximation, the half-maximum intensity width $\Delta\lambda$ may be considered as a function of the product $n_e F_c$ (Fig. 28).

Thus, a measure of $L_\perp$ gives the product $n_e F_c$. The current density j, given by

$$j = e(\mu_e + \mu_i) n_e F_c \simeq e \mu_e n_e F_c \tag{29}$$

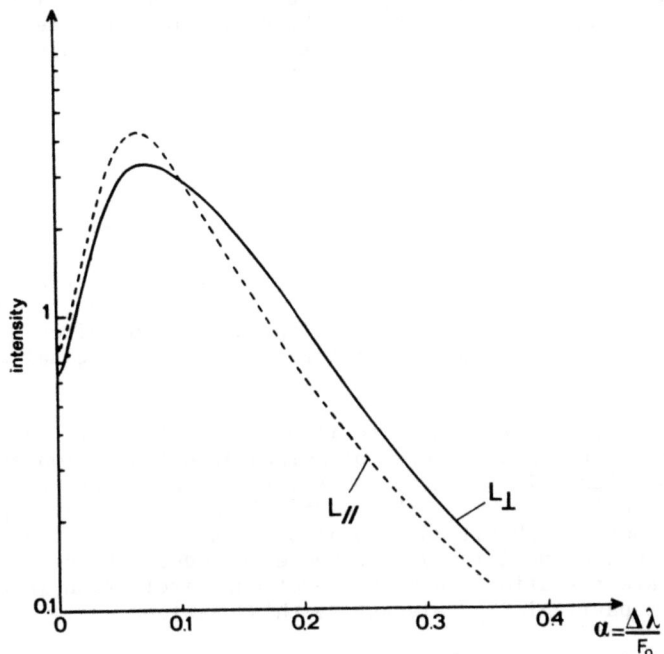

Fig. 27. Comparison of $L_\perp$ and $L_{//}$ profiles for H_β with $N_e = 10^{16}$ cm^{-3}, $T_e = 20{,}000$ K, and $P_c = 3$.

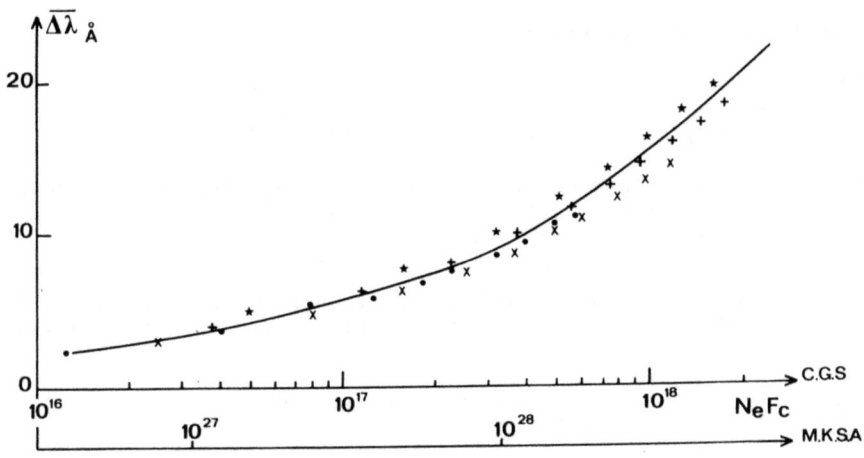

Fig. 28. Half maximum intensity width for H_β as a function of the products $N_e F_c$ and with $P_c = 1(.)$, $2(x)$, $3(+)$, and $4(*)$.

where μ_e and μ_i are the mobilities of electrons and ions, is entirely determined. This method has been applied to the oxygen corona discharge with minute quantities of hydrogen added as a probe in a short gap ($\simeq 1$ cm) to measure the current density j. Further, from j and i (the total current) the average radius of the discharge filament has been obtained at a pressure of ~ 300 torr. This leads to a filament radius of ~ 10 μm at atmospheric pressure for the filamentary streamer channel (Bastien et al., 1979).

Effects Due to Gas Activation

As has been shown, the discharge radiation sequences associated with the spark formation reveal intense physico-chemical activities within the discharge. This activity is triggered by a sequence of electron impact excitation and ionization followed more slowly by transfer reactions between the excited heavy species. The radiation emitted by the discharge after the initial discharge development is thus often generated in a perturbed gas which may be very different from the original gas composition.

For instance, leader channel reillumination is indicative of a gas which has already reached a degree of molecular dissociation corresponding to an equivalent LTE state of several thousand degrees before reillumination. "Positive streamers" evolving in a gas already exposed to "negative streamers" exhibit additional spectral components compared to normal emission from positively stressed electrodes. Processes involving metastable state formation, such as vibrational excitation in homonuclear molecules, are particularly relevant to this question.

Vibrational Excitation Effects. One of the processes which has been recognized as playing an important role in the discharge evolution and, thus, in its emission properties is the excitation and relaxation of the vibrational states in molecules. Due to the relevance in gas laser physics, pulsed power technology, discharge surface interactions, considerable interest has been devoted recently to the question of vibrational distributions (Capitelli, 1981; Cacciatore, 1982; Rich, 1982). Some aspects of this will now be discussed.

Equation (1) may be expressed as

$$\frac{\partial n_r}{\partial t} = \frac{\partial n_r}{\partial t}_{(e-N)} + \frac{\partial n_r}{\partial t}_{(V-V-T)} + \frac{\partial n_r}{\partial t}_{(V-t)} + \frac{\partial n_r}{\partial t}_{(A-V)} \quad (30)$$

where V-V-T and V-T stand for terms equivalent to E-E-T and E-T in Eq. (1). The term A-V will be defined later. The most likely V-V-T exchange implies a transfer of only one quanta, i.e., $r \to r + 1$ and $j \to j + 1$. Since the energy levels ε_r of vibrational states are almost, but not exactly, equidistant from each other (anharmonicity of molecule oscillators), such exchange implies that it is equilibrated by kinetic energy exchange. The term $\frac{\partial n_r}{\partial t}_{(V-V-T)}$ may thus be written as

$$\frac{\partial n}{\partial t}_{(V-V-T)} =$$

$$P^{j,j+1}_{r\pm 1,r} \left\{ n_{r\pm 1} n_j - n_r n_{j\pm 1} \exp\left[-(\varepsilon_{r\pm 1}-\varepsilon_r) - (\varepsilon_{j+1}-\varepsilon_j)\right]/kT \right\} \quad (31)$$

Here the depopulation term A and the population term A' expressing the reversed collision, appear again. Obviously, if the distribution of the n_r follows a Boltzmann distribution, then A' balances A exactly. Let us now try a distribution such as

$$n_r = G(r) \exp(-\varepsilon_r/kT) \tag{32}$$

If the condition A = A' is required, then the following relation must be satisfied

$$[G(r)/G(r\pm 1)] \cdot [(G(j\pm 1)/G(j)] = \tag{33}$$

This condition leads to $G(r) = e^{-r}$ so that the distribution

$$n_r = e^{-r\gamma} \exp(-\varepsilon_r/kT) \tag{34}$$

called the Treanor distribution (Treanor, 1967) is an equilibrium solution of Eq. (31).

Since the rate coefficients for V-V-T are much larger than those for V-T collisions, let us assume for the moment that the V-V-T terms are the only terms on the RHS of Eq. (30) and let us start with any arbitrary distribution. Then, since there is a gain of one quanta for one species at each collision and a loss of one quanta for the other, it follows that the initial distribution maintaining the total number of vibrational quanta constant. The constant γ is then obtained by expressing this condition of constant number of quanta knowing the initial number of quanta. If now one adds the smaller V-T term to the RHS of Eq. (30), a succession of distributions close to a Treanor distribution is obtained with decreasing values of γ. At each V-T collision one quanta is lost until the final distribution reaches thermodynamcic equilibrium with a time constant generally referred to as the vibrational V-T relaxation time. A shorter V-V-T time constant initially controls the evolution toward a Treanor distribution.

Let us now consider vibrational excitation. One of the most recent attempts to model vibrational excitation is the work of Boeuf (1985) as well as Boeuf and Kunhardt (1982, 1986) in nitrogen plasma, confirming and extending earlier work (Capitelli et al., 1981; Rich, 1982). The purpose of the recent work was to derive gas heating effects due to vibrational relaxation of nitrogen molecules. Here, eighteen different types of electron-molecule reactions were taken into account, as were six types of reactions between heavy species. Vibrational excitation was also considered along with transfer via the $A^3\Sigma_u^+$ nitrogen electronic state, termed A-V in Eq. (30). Electron collisions were simulated by Monte-Carlo techniques to derive rate constants for the vibrational species production.

A typical example of time dependent evolution of the distribution of vibrational states is shown in Fig. 29b for the steady state condition $n_e = 10^{12}$ cm^{-3}, E/N = 70 Td, p = 100 torr. At the earlier stages a distribution, slightly overpopulated for large values of the quantum number "r", appears and, later, the Treanor aspect becomes more and more accentuated.

A point which must now be emphasized (see also Cacciotore et al., 1982) is that the tail of the electron distribution function clearly increases as seen in Fig. 29a. A fraction of the energy lost by the electrons to produce the vibrational species is now returned through

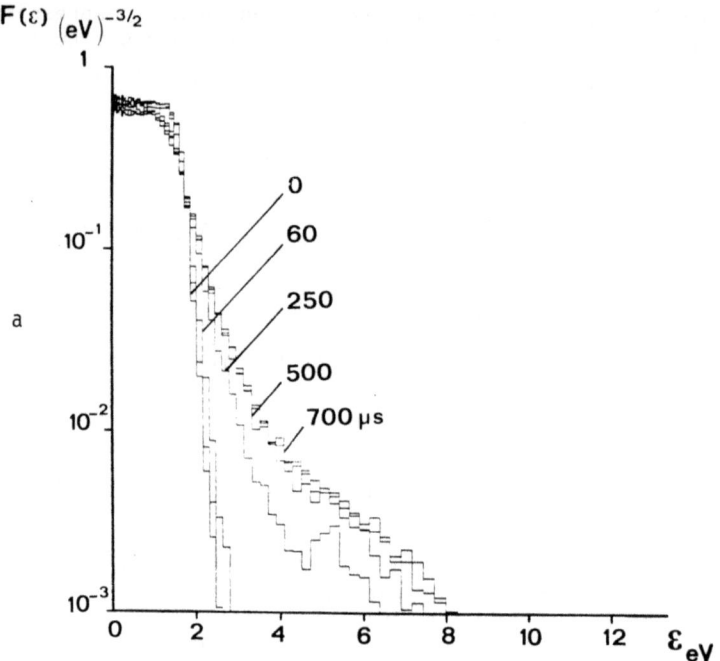

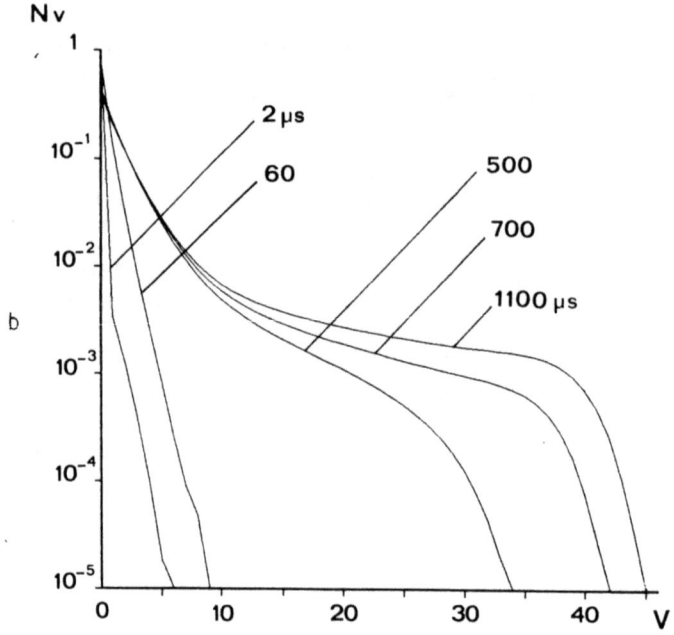

Fig. 29. a) Electron distribution function at several times with E/N = 30 Td, $N_e = 10^{12}$ cm^{-3}, T_{gas} = 500 K, and p = 100 torr.
b) Vibrational level population at several times, E/N = 30 Td (Boeuf, 1985).

superelastic collisions or, in other words, the fraction of energy lost to vibrational excitation decreases in time (Fig. 30) so that the ionization coefficient increases. This may be the basis of an alternative explanation to that suggested by Gallimberti (1974) for the leader formation, i.e., the transition toward a thermalized pre-arc plasma. The process may begin identically with thermal relaxation of vibrationally stored energy leading to gas temperature increases. In one scenario (Gallimberti, 1974) a critical temperature is reached for which electron detachment from O_2^- provides a new electron source. An alternative scenario is suggested here as follows: first, a decrease of the heavy species density occurs at constant pressure due to the increase in gas temperature resulting from vibrational relaxation. This increases the E/N ratio. However, and this is the new aspect to be outlined, the increase of electron energy following the vibrational relaxation also leads to an increase of the electron ionization coefficient. This hypothesis has been recently put forth to explain the formation of the quasi-spark plasma triggered along dielectric surfaces (Larigaldie, 1985).

When the electron source is cut off, as in afterglows, or becomes much less efficient, as in some parts of high pressure glows, a relaxation of the vibrational distribution follows which is similar to that seen in Fig. 31 (Capitelli et al., 1981). Since no more electrons are present, the densities of the first vibrational levels will decrease. Through V-V-T pumping, high "r" levels are populated leaving the gas with a long-lived vibrational population which, as will be seen now, may appear within the spectrum emitted through electron reactivation of the gas.

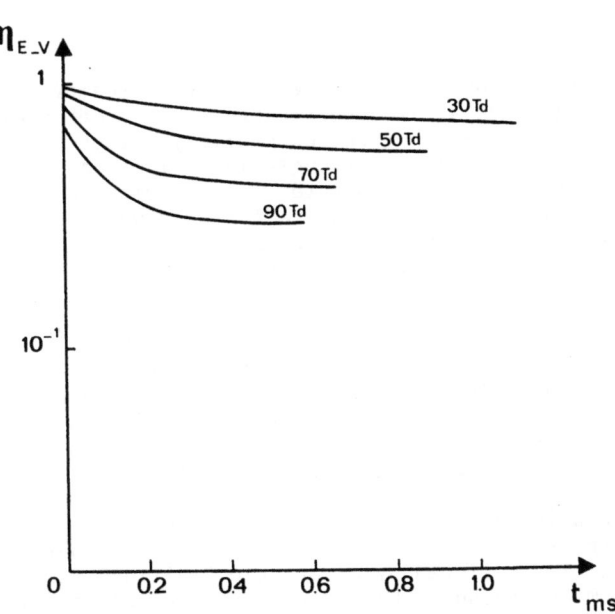

Fig. 30. Temporal variation of the electronic power fraction transferred to molecules as vibrational excitation (Boeuf, 1985).

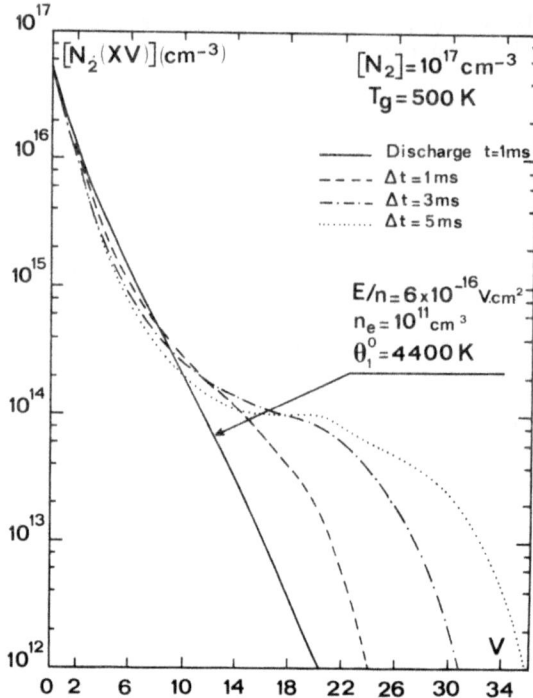

Fig. 31. $N_e(x,v)$ vibrational distribution in a discharge at a residence time $t = 10^{-3}$s and at several times in the post discharge (Capitaelli et al., 1981).

Studies of the corona and the prebreakdown streamer phase in air and in electronegative gases at atmospheric pressures under steady state applied potential have been pursued extensively. These studies are relevant not only to dielectric strength investigations but also to plasma chemistry since corona generates cold plasmas in which hot electrons produce active species such as radicals as well as metastable and vibrationally excited species which are able to initiate specific reaction. The discharge is a high pressure chemical reactor in such a situation which may be used for the production of specific compounds such as ozone, N_xO_y, radicals, and vibrationally excited species for surface interactions. These discharges develop in non-uniform field configurations. A typical gap of 1 cm with a point electrode of about 50 μm radius facing a plane electrode has often been considered. A large number of different plasma states may be generated according to potential value (Sigmond et al., 1983; Goldman et al., 1983) when a positive potential is applied to the highly stressed electrode.

When the applied potential is slightly beyond the critical spark potential, a regular sequence of streamer-induced discharges occurs as sketched in Fig. 17, followed by a variable "waiting" time in the range of hundreds of microseconds. Each streamer discharge and its associated glow will generate a given amount of excited species which will subsequently relax during the ensuing high pressure afterglow. When a new streamer traverses the gap, it encounters a gas not completely free from the remnants of the preceding streamer-glow sequence. The emitted spectrum shown in Fig. 21 will exhibit features connected to the gas states.

Hartmann (1975, 1977, 1978), after identifying almost all the spectrum components, studied the radiation intensity emitted in the band heads of the second positive group of nitrogen ($C^3\pi_u - B^3\pi_g$). This study required the development of a very precise method able to relate the measured intensity through a slit, observing only a small spectral width around different band heads, and the total intensity emitted in a whole band. A so-called slit function (Hartmann et al., 1978) was worked out taking into account the various branches defining a band head.

Let us now analyze the excitation process. As already mentioned, the narrowly viewed position in space will be submitted to the excitation of gas species by the electrons defining the streamer in a time equal to the typical streamer dimension ℓ divided by its velocity V_s. This time interval is between 0.01 to 0.1 ns with ℓ on the order of 10 - 100 μm. Figure 32 shows schematically the energy diagram with the electron induced transition from the ground state to the $C^3\pi_u$ level, the observed transition being the molecular band system generated by $C^3\pi_u - B^3\pi_g$. The excitation is so short that no step-wise excitation is possible and only the direct transitions indicated in Fig. 32 are possible. The population rate for the vibrational levels of the upper state will follow the various Franck-Condon factors and the upper vibrational distribution will be an image of the distribution in the vibrational state of the fundamental state. It follows that the band structure in the $C^3\pi_u - B^3\pi_g$ transition will also be an image of the initial distribution, since the intensity in each v-v' transition is given by the transition probability $q_{vv'}$ times the quenching factor $Q_v = (1 + P/P_v)$. A correction factor has to be applied to the upper vibrational states to take into account the Calo-Axtman v-v' mixing of the $C^3\pi_u$ level through pressure quenching.

The derived first four vibrational levels of the fundamental state show a distribution close to the beginning of a Treanor distribution at a temperature which strongly depends on the time interval between successive streamers as shown in Fig. 33 (Hartmann et al., 1975) and which may be explained by the quenching of the vibrational nitrogen sates due to humidity and CO (Gallimberti, 1979).

Strong Inelastic Excitation. Vibrational excitation is obviously only one of the possible changes in the gas medium. Excitation of electronic states of atoms and molecules, negative ion formation and molecular dissociation open up a whole range of reaction schemes which may further modify the chemical composition and excited state densities of the gas. The change is clearly related to the time during which the electron current flows through the ionized gas. Depending on the duration of this current, greater of lesser changes may be observed in the emitted discharge spectrum. The first level of change involving energies of fractions of electron volts is the vibrational excitation just discussed. Electron excitation and molecular dissociation involve energies of several eV. For times on the order of microseconds and for electron densities in the range of 10^{15} cm^{-3}, the observed spectrum in molecular gases remains a molecular band system, but exhibits new bands indicating the accumulation of metastable species.

Let us give two examples of such observations. If the emitted spectrum in the range 500 - 700 nm of the streamer and the streamer-induced filament are compared (i.e., emission from the region of potential fall in Fig. 17 and region of high field plateau of Figs. 18 - 19), while apparently absent in the streamer, the first positive nitrogen band clearly appears within the filament owing to the continuous current

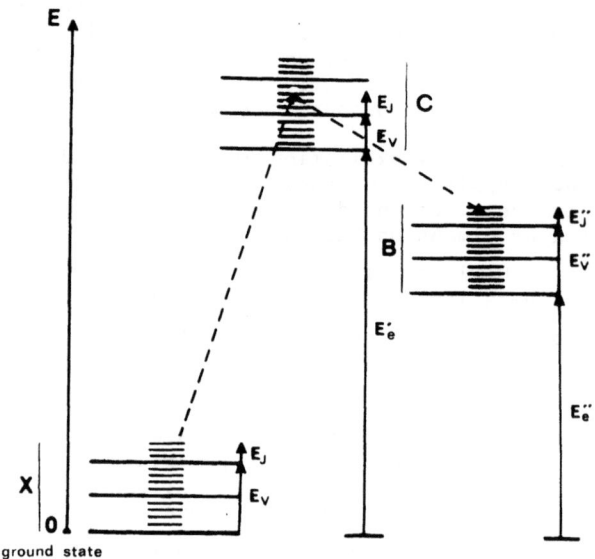

Fig. 32. Molecular energy level diagram.

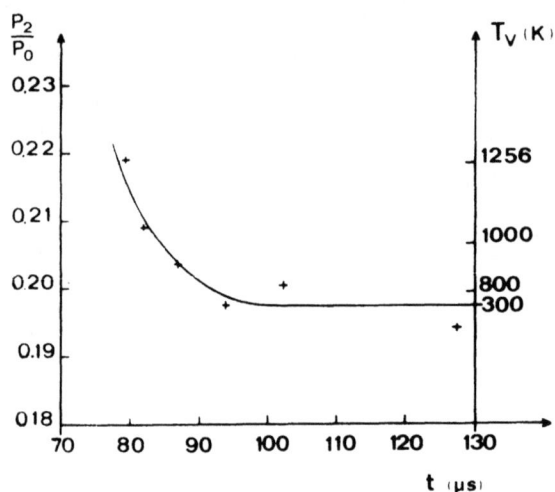

Fig. 33. The population P_2 of the $v = 2$ vibrational level of the N_2 ($C^3\pi_u$) molecules with respect to the population P_0 of the $v = 0$ level as a function of the discharge repetition time (Hartmann et al., 1975).

flow in the filamentary positive column (Fig. 34). (From Table 2, the upper state of this system indicates the presence of a large number of molecules in the $B^3\pi_g$ state.)

An example of streak photography in a long discharge gap between a negative cone-shaped electrode and two positive cone-shaped electrodes is shown in Fig. 35. Here, the anode and cathode-directed streamers are recognized as well as leader formation. As explained earlier, only a portion of the recorded spectrum could be clearly identified due to the statistical nature of the discharge build-up. Those transitions which can be identified without ambiguity are indicated in Table 2. A comparison of the emitted spectrum from various discharge stages as well as the amount of recorded information which could be used are indicated in Fig. 36. As seen, the bands from the highest lying upper levels are no longer present within the leader stage (at least in a first approximation, since only 40% of the leader spectrum would be registered due to discharge fluctuations). But, here again, the point to be stressed is the presence of the first negative system of nitrogen within the positive streamers. This band is not observed in the positive streamer which propagates from positive point electrodes either in large or small gaps. However, in contrast to these two cases, the positive streamer within the negative discharge propagates in an already ionized and perturbed channel and the streamer repetition rate is much higher here ($\sim 10^7$ Hz) than in the positive discharges ($\sim 10^4$ Hz in small gaps).

Confirmation is thus given to the fact that the emitted streamer spectrum reveals the state of the medium it traverses. In the preceding paragraph it was shown that the vibrational streamer band structure was related to the vibrational population. Here, it is suggested that the presence of the first negative system indicates an overpopulation of the nitrogen in the $A^3\Sigma_u^+$ level. An estimate of the increased concentration of $N_2(B^3\pi_g)$ levels due to production of $N_2(A^3\Sigma_u^+)$ levels by the increased repetition rate of streamers in negative discharges has been made by Bordage et al. (1978). Excitation by electron impact (appearing in Table 3 as mean values of cross sections σ, threshold energies ε and collision frequencies ν/N), and collisional de-excitation of upper states by heavy species (appearing in Table 3 as τ with estimated quenching coefficients Q) have been considered according to Fig. 37a. Even if the values given in Table 3 are simple estimates which cannot be taken as absolute values, it follows that a step-wise increase in the concentration of $N_2(A^3\Sigma_u^+)$ is expected as in Fig. 37b. This takes into account that the estimated exposure time due to excitation by the positive streamer-induced filaments is ~ 30 ns and the relaxation time between streamers is ~ 200-300 ns. This leads to narrow pulses of $N_2(B^3\pi_g)$ production at each streamer occurrence from which the first positive system is observed. In Fig. 37b the points are the maximum of each pulse of $N_2(B^3\pi_g)$ production, starting at t=0, from the case where no nitrogen at the $A^3\Sigma_u^+$ level is present.

More dramatic changes are related to the longer discharge "exposure times" associated with leader formation and the return stroke, where substantial emission will arise from atomic compounds. With elaspsing time, the degree of dissociation of gas molecules will increase as will the ion density. Hydrodynamic expansion of the gas results (Marode, 1983). Since the discharge is in a "head" controlled situation, the leader core is not in a highly conductive state. However, evidence that the leader channel is in a state near LTE is obtained. At the 1983

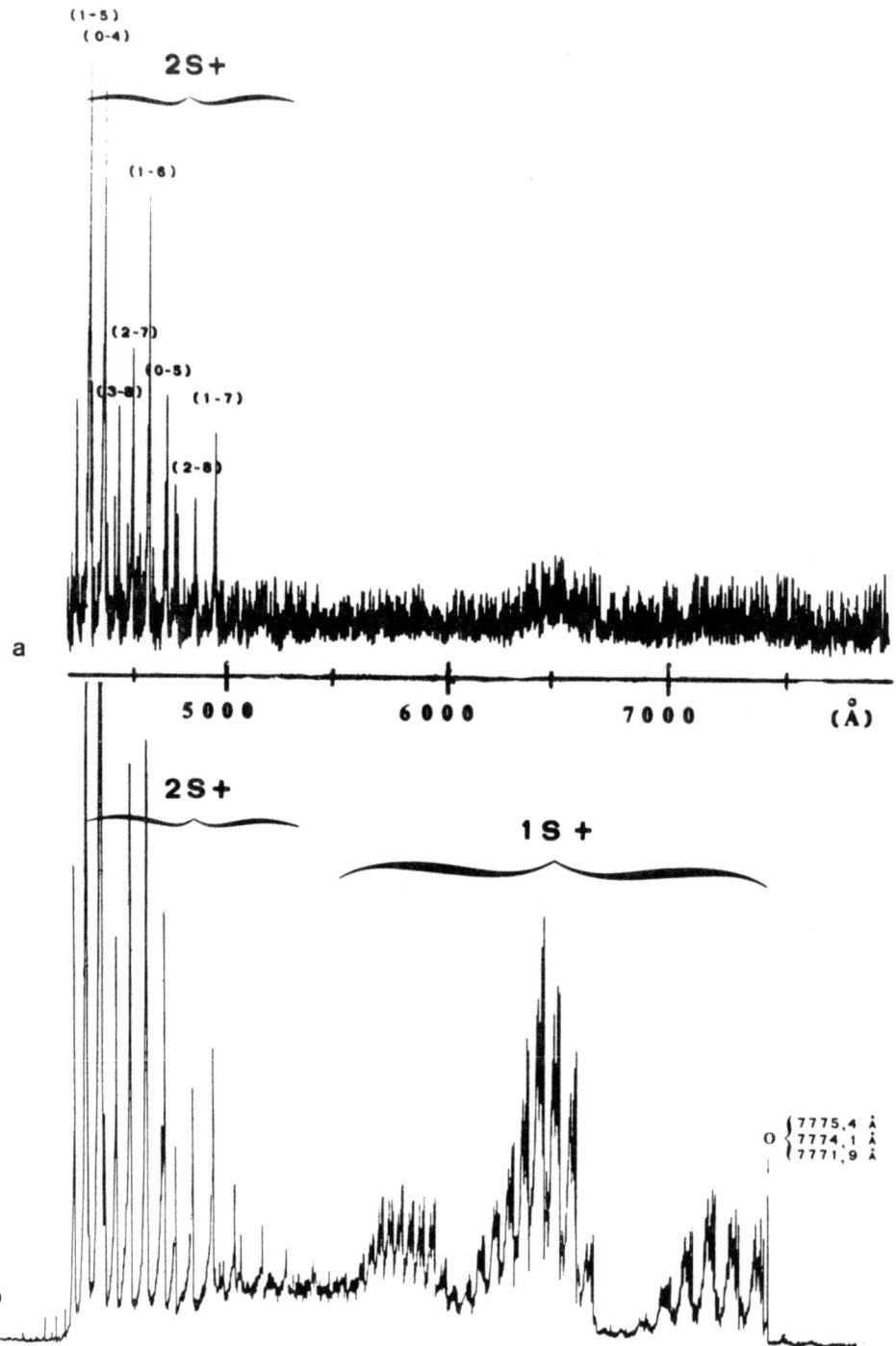

Fig. 34. a) Streamer spectrum and b) streamer-induced filament spectrum (transient positive column)(Hartmann, 1977).

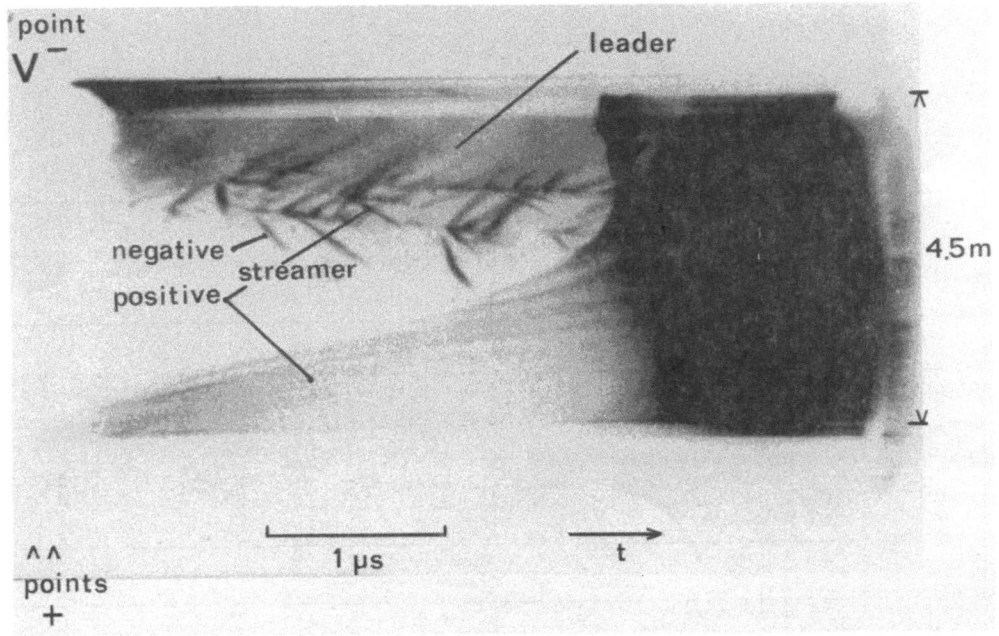

Fig. 35. Streak photograph of negative point-to-point discharge, air gap of 4.5 m at atmospheric pressure. Photo from Cimador (1974).

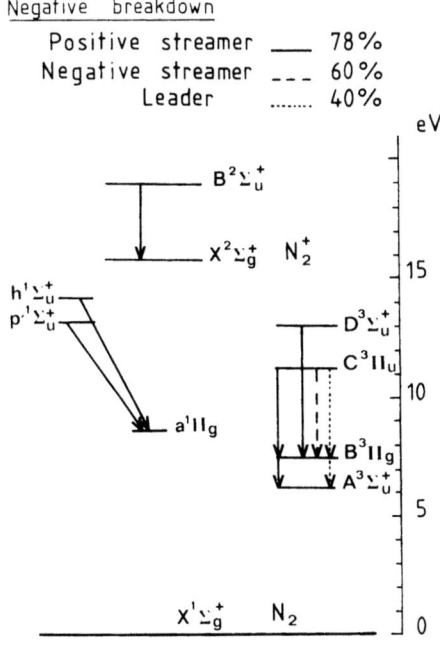

Fig. 36. Radiative states of N_2 for different discharge situations.

Table 2. Molecular band systems effectively identified for the different discharge components. (After Les Renardieres, 1981) (1978 results)

Molécule / Molecule	Système de bandes / Band system	Energie des états supérieurs et inférieurs / Energy of upper and lower states (eV)		Transition	
Streamer positif / Positive streamer					
N_2	$1S^+$: first positive	7.4	6.2	$B^3\Pi_g$	$A^3\Sigma_u^+$
N_2	$2S^+$: second positive	11.0	7.4	$C^3\Pi_u$	$B^3\Pi_g$
N_2	$4S^+$: fourth positive	12.8	7.4	$D^3\Sigma_u^+$	$B^3\Pi_g$
N_2	h : Gaydon-Herman's singlet	12.8	8.5	$h^1\Sigma_u^+$	$h^1\Pi_g$
N_2	p : Gaydon-Herman's singlet	12.9	8.5	$p^1\Sigma_u^+$	$p^1\Pi_g$
O_2^+	$2S^-$: second negative	17.0	12.0	$A^2\Pi_u$	$x^2\Pi_g$
N_2^+	$1S^-$: first negative	18.8	15.6	$B^2\Sigma_u^+$	$x^2\Sigma_g^+$
Streamer négatif / Negative streamer					
N_2	$2S^+$: second positive *	11.0	7.4	$C^3\Pi_u$	$B^3\Pi_g$
N_2^+	$1S^-$: first negative	18.8	15.6	$B^3\Sigma_u^+$	$x^2\Sigma_g^+$
Leader négatif / Negative leader					
OH		4.1	0	$A^2\Sigma_u^+$	$x^2\Pi$
NO	γ	5.5	0	$A^2\Sigma_u^+$	$x^2\Pi$
NO	β	5.7	0	$B^2\Pi$	$x^2\Pi$
N_2	$1S^+$: first positive	7.4	6.2	$B^3\Pi_g$	$A^3\Sigma_u^+$
N_2	B' : Infrared after glow	8.2	7.4	$B'^3\Sigma_u^-$	$B^3\Pi_g$
CO	$3S^+$: third positive	10.4	6.0	$b^3\Sigma^+$	$a^3\Pi_r$
CO	A : Ångström	10.8	8.0	$B^1\Sigma^+$	$A^1\Pi$
CO_2	Fox				
N_2	$2S^+$: second positive	11.0	7.4	$C^3\Pi_u$	$B^3\Pi_g$
CO	TB : triplet bands	7.5	6.0	$d^3\Delta_i$	$a^3\Pi_r$

* à remarquer notamment l'absence du $1S^+$
  to underline namely the absence of $1S^+$

Table 3. Values of the various coefficients entering in the production and destruction processes of the excited species. (after les Renardieres, 1981) (1978 results).

$$X = X^1\Sigma_g^+ \quad A = A^3\Sigma_u^+ \quad B = B^3\Pi_g \quad C = C^3\Pi_u$$

	$X \to C$	$X \to B$	$X \to A$	$A \to C$	$A \to B$	$B \to C$		$\tau_{(s)}$	Q
σ (cm^2)	10^{-17}	$0.7\,10^{-17}$	$3.5\,10^{-17}$	$\sim 10^{-17}$	$\sim 10^{-17}$	$\sim 10^{-17}$	A	1	(10^{-4})?
ϵ_s (eV)	11	7.4	6.2	4.8	1.2	3.6	B	10^{-6}	$\sim 10^{-3}$
v/N ($cm^3 s^{-1}$)	5.10^{-10}	$5.6\,10^{-10}$	$3.2\,10^{-9}$	$5\,10^{-9}$	$1.4\,10^{-9}$	$1.2\,10^{-9}$	C	40.10^{-9}	$1.3\,10^{-2}$

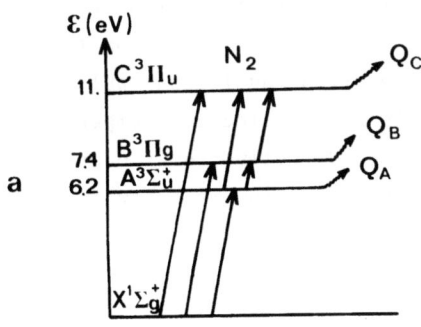

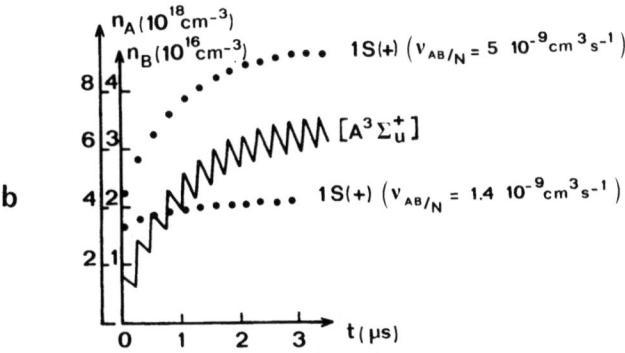

Fig. 37. a) Energy levels and transitions able to occur. The Q_i are the quenching factors of the levels i;
b) temporal evolution of the density n_A (of N_2 ($A^3 \Sigma_u^+$)) and n_B (of N_2 ($B^3 \Pi_g$)). Maxima of the $n_B(t)$ peaks (marked as ·) occurring at times of development of successive streamers ($n_B \simeq 0$ between maxima) (after "Groupe des Renardières", Electra, 1981).

Renardières Campaign, a triggered voltage pulse was added to the main applied pulse some 100 μs after the beginning of the main pulse in a 10 m positive point-to-plane gap in air. Some early unpublished results are shown in Fig. 38, where the additional impulse voltage, having the same polarity as the main pulse, is indicated. At that temporal stage, the leader already extends to mid-gap and, due to the voltage pulse, a strong current pulse is generated associated with the leader reactivation. Spectral components from N^+, N_2, H (or O which follows the same curve) are reported. As seen, the emission from N^+ precedes that from N_2. The upper state for the N^+ transitions are about 22 eV above the N^+ ground state and, if electrons had to excite this level from ground states of either N^+ or N_2, then the emitted spectrum would exhibit emission in the second positive groups of N_2 (referred to here as 2S+) at the same time. A possible explanation of the absence of this N_2 system is that LTE is already reached and that the emission from N^+ is thermal emission. That is, the population of the upper state of N^+ comes from the nearest adjacent levels located only about 2.5 eV from the upper level of N^+. However, with the increase of the additional pulse voltage, the field rises inside the leader, raising the electron temperature. Now the electron excites the N_2 (2S+) band directly from the ground state of N_2 and, since the upper state of this transition is below that of N^+, the emission of N_2 increases drastically ($\sim$ as E^4), while that of N^+ remains controlled by the leader temperature. Further gas dissociation is observed through the delayed emission of H and O. To summarize: plasma at LTE, not yet as conductive as an arc, may characterize the leader properties and its emission features.

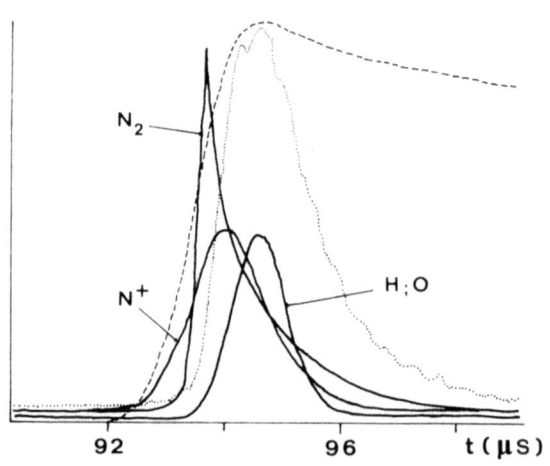

Fig. 38. Time variation of emission of some characteristic lines during induced leader reillumination, point-sphere, gap = 6 m (after "Groupe des Renardières", experiments to be published, 1983).

Further in time, when the return stroke is triggered with the discharge arrival at the other electrode, the whole discharge is reactivated and shows an enhanced emission in the atomic lines (Fig. 39), with a later appearance of the continuum. This picture is in agreement with the idea that the emission features are mainly controlled at this stage by a Boltzmann distribution of the species in excites states at the electron temperature (Gallimberti et al., 1977).

A whole emission sequence starting from molecular bands and showing the rise of atomic line components, including those due to surface interactions, is shown in Fig. 40 for the case of surface discharges on dielectrics. The spectroscope sees a narrow region on the dielectric surface (1.75 mm width) at the center of the dielectric plate.

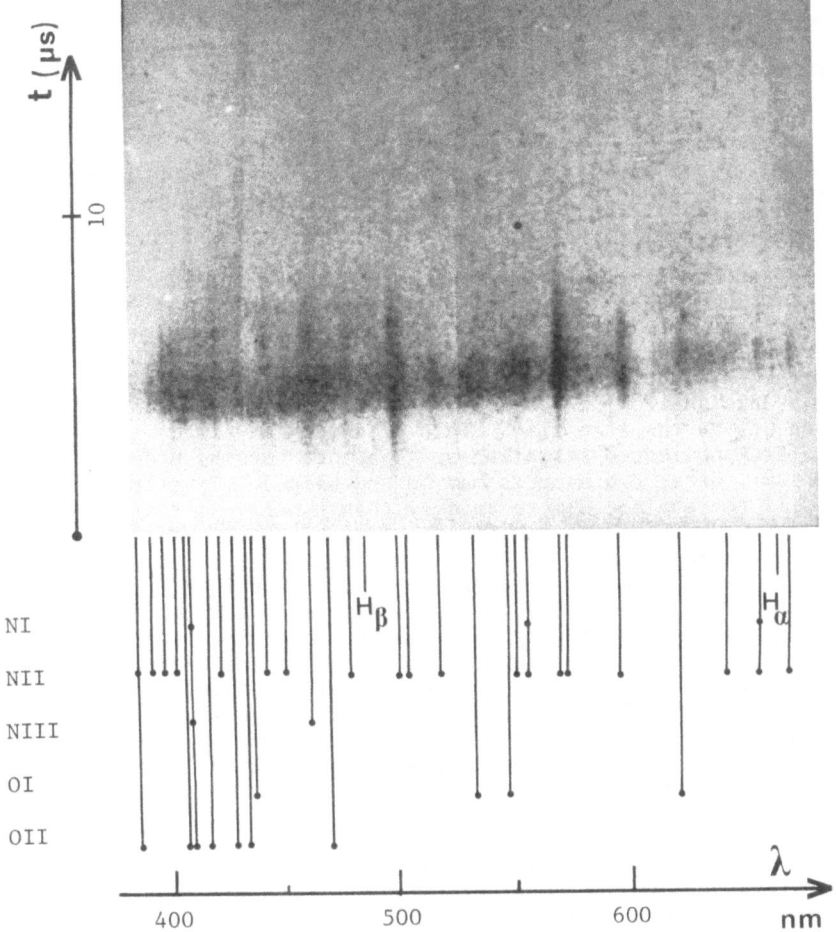

Fig. 39. A typical example of the time resolved spectra of the leader to spark transition. The identification of the main line system is also indicated (after "Groupe des Renardières", Electra, 1977).

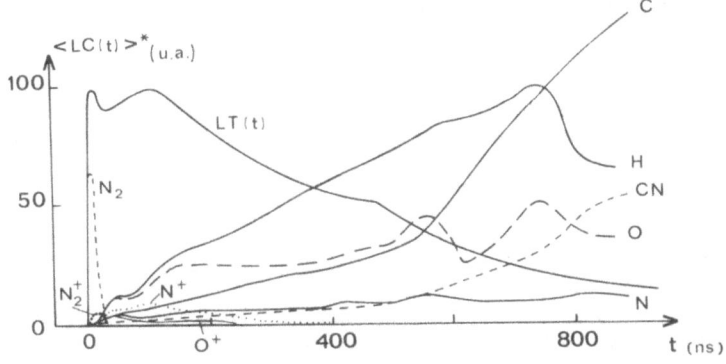

Fig. 40. Characteristic atomic lines and molecular bands of the negative discharge of dielectric surfaces (Bordage, 1981). LT refers to the total light; Vertical axis is arbitrary units.

CONCLUSIONS

As has been shown the emission of transient glows at high pressures offers a variety of features correlated to the numerous parameters defining the state of an emitting discharge.

Let us try to summarize some aspects which have been described here. The emission of the prebreakdown glows do not generally follow thermal equilibrium. Since emission is directly connected to the density distribution of the excited species, the master equation governing the production and destruction processes are relevant to understand what non-equilibrium means. At equilibrium, each process is balanced by the reverse process. However the production rates and destruction rates may be the result of many processes. The presence of an electric field will enhance the electron energy and induce electron collisions which will generally unbalance one or several processes, while rebalancing processes will then act at the same time. This situation may lead either to a steady state unbalanced situation or to a time varying process towards equilibrium. Here, two aspects may be underlined. Firstly, some processes will be rebalanced more quickly than others, so that several time constants may characterize the time evolution. For instance, groups of species at several quantum levels may exchange their energy so quickly that the whole group acts as being composed of a unique type of species at a single level. Secondly, the collision cross section of high lying levels being generally larger than those at lower levels, the part of the distribution of excited species corresponding to the highest levels will reach a Boltzmann distribution much sooner at electron or heavy species temperature than the lower part of the distribution which may be controlled by a so called "corona" distribution, where one process, like collision excitation, is balanced by another process such as de-excitation through photon emission.

It follows that the emitted spectrum will strongly be linked to the way that the levels are "pumped" - to use a laser vocabulary - i.e., deequilibrated. This, in turn, is connected with the local value of the electrical field which depends on the discharge mechanism.

Let us stress the fact that within the variety of plasma discharges, the field change will either be controlled by a "current feeding" ability of one end of the discharge plasma (head controlled) or by the discharge channel itself (channel controlled). In addition a typical discharge structure presents two basic regions showing different emission properties. The positive column region of a discharge, which emits radiations from quantum levels lower than those emitted by the "head" or cathode fall region of a discharge, emitting radiation issuing from high lying quantum levels.

Obviously this is somewhat over-simplified since many additional factors should be considered such as thermal and density diffusion. In that respect, one should recall that the temperature of an "equivalent" discharge increases with increased pressure so that the emitted spectrum may be strongly pressure dependent.

Finally one must stress that what is emitted cannot be inferred from the initial gas composition. The gas nature changes through vibrational, excitation, dissociaton and the purpose of the last figure was actually to demonstrate this point by showing that while being in the same gas, the spectrum changes in time revealing the dynamic of gas chemistry.

REFERENCES

Albuques, F., Birot, A., Blanc, D., Brunet, M., Gally, J., and Millet, P., 1974, J. Chem. Phys., 61:2695.
Badaru, E., and Popescu, I., 1965, "Gas Ionises", Dunod, Meridiane.
Bastien, F. 1977, These d'etat No. 1871, Orsay, France.
Bastien, F. and Marode, E., 1977, J.Q.S.R.T., 17:453.
Bastien, F. and Marode, E., 1977, J. Phys. D., 12:1121.
Bastien, F. and Marode, E., 1978, J. Phys. D., 2:249.
Bastien, F. and Marode, E., 1979, J. Phys. D., 12:249.
Bastien, F. and Marode, E., 1982, in: Gaseons Dielectrics III, L. Crhistophorou, ed., Pergamon, Oxford.
Bastien, F. and Marode, E., 1982, J. Phys. D., 15:2069.
Bastien, F. and Marode, E., 1985, J. Phys. D., 18:377.
Boeuf, J. P. and Kunhardt, E. E., 1982, in: "Proceedings, 35th Gaseous Electronics Conference", Dallas.
Boeuf, J. P., 1985, These, Orsay.
Bondiou, A., 1984, These de docteur-ingenieur, No. 672, Orsay.
Bordage (Prioux), M. C., 1981, These de docteur-ingenieur, No. 503, Orsay.
Bordage, M. C., Gallimberti, I., Hartmann, G., and Marode, E., 1978, Electra, 74:204.
Bordage, M. C., and Hartmann, G., 1982, J. Appl. Phys., 53:8568.
Cacciatore, M., Capitelli, M., and Gorse, C., 1982, Chem. Phys., 66:141.
Calo, J. M. and Axtmann, R. C., 1971, J. Chem. Phys., 54:1332.
Capitelli, M., Gorse, C., and Ricard, A., 1981, J. Physique Let., 42:L185.
Capitelli, M., Dilonardo, M., and Gorse, C., 1981, Chem. Phys., 56:29.
Cimador, A., Rieux, R., and Hutzler, B., 1974 E.D.F. Bulletin - Series B, No. 4, p. 29.
Chalmers, I. D., Duffy, H., and Tedford, D., 1972, Proc. Royal Soc., London A., 239:171.
Dawson, G. A., and Winn, W. P., 1965, Z. Physik, 183:159.
Delcroix, J. L., 1966, "Physique des plasmas T1 and 2", Paris, Dunod.
Doran, A., 1968, Z. Phys., 208:427.
Gallimberti, I. Hepworth, J. K., and Klewe, R. C., 1974, J. Phys. D., 7:880.
Gallimberti, I., Hartmann, G., and Marode, E., 1977 Electra, 53:123.
Gallimberti, I., 1972, J. Phys. D., 5:2179.
Gallimberti, I., 1979, in: "Proceedings, 9th International Conference on Phenomena in Ionized Gases", Grenoble; J. Phys., Sup. 7:40:C7.
Griem, H. R., Kolb, A. C., and Shen, K. Y., 1959, Phys. Rev., 116:4.
Griem, H. R., 1964, "Plasma Spectroscopy", McGraw-Hill, New York.
Griem, H. R., 1974, "Spectral Line Broadening by Plasmas", Academic Press, New York.
Goldman, M. and Goldman, A., 1978, in: "Gaseous Electronics", N. M., Hirsh and H. J. Oskam, eds., Academic Press, New York.
Goldman, A. and Amouroux, J., 1983, in: "Electrical breakdown and Discharges in Gases", E. E., Kunhardt, and L. H. Luessen, eds., Plenum, New York.
Harnafi, M., 1984, These Docteur 3e Cycle, Orleans, France.
Harnafi, M., and Dubreuil, B., 1985, Phys. Rev. A., 31:1375.
Hartmann, G., 1970, C.R.A.S., 270:309.
Hartmann, G., 1970, in: "9th International Congress of Ultra-Speed Photography", Denver.
Hartmann, G., 1974, in: 3rd International Conference on Gas Discharges", IEE.
Hartmann, G. and Gallimberti, I., 1975, J. Phys. D., 8:670.
Hartmann, G., 1977, These No. 1783, Orsay, France.
Hartmann, G. and Johnson, P. C., 1978, J. Phys. B., 11:9.
Hersberg, G., 1950, in: "Molecular Spectra and Molecular Structure", Vol. 1, Nostrand, New York.
Hirsh, M. N., Poss, E., and Eiser, P. N., 1970, Phys. Rev. A., 6:1615.

Holt, E. H. and Haskell, R. E., 1965, "Foundations of Plasma Dynamics", MacMillan Company, New York.
Hooper, C. F., 1968 Phys. Rev., 165:215 and 169:193.
Koppitz, J., 19791, Z. Naturforsch. A., 26:700.
Kunhardt, E. E. and Luessen, L. H., eds., 1983, "Electrical Breakdown an Discharges in Gases", Plenum, New York.
Larigaldie, S., 1985, These, Orsay, France.
Llewellyn-Jones, F., 1966, "Ionization and Breakdown in Gases", Methuen, London.
Loeb, W. H., 1965, in: "Electrical Coronas", University of California, Berkeley.
Long, W. H., 1979, Northrop Res. Tech Cen. Tech. Rep. AFAPL- TR 79-2038.
Marode, E., 1975, J. Appl. Phys, 46:2005 and 46:2016.
Marode, E., Bastien, F. and Bakker, M., 1979, J. Appl. Phys., 50:140.
Marode, E. and Boeuf, J. P., 1983, in: "Proceedings, International Conference on Phenomena in Ionized Gases", Dusseldorf.
Marode, E., 1983, in: 1983, "Electrical Breakdown and Discharges in Gases", E. E. Kunhardt and L. H. Luessen, eds., Plenum, New York.
Meek, J. M. and Craggs, J. D., eds., 1978, "Electrical Breakdown in Gases", Wiley, New York.
Mitchell, K. B., 1970 J. Chem. Phys., 53:1795.
Mitchner, M., and Krugger, H., 1973, in: "Partially Ionized Gases", Wiley-Interscience, New York.
Nicholls, R. N., 1962, J.Q.S.R.T, 2:433.
Raether, N., 1964, "Electron Avalanches and Breakdown in Gases", Butterworth, London.
Les Renardieres Group, 1972, Campagn, 1972, Electra., 23:53.
Les Renardieres Group, 1973, Campagn, 1978, Electra., 35:47.
Les Renardieres Group, 1977, Campagn, 1975, Electra., 53:31.
Les Renardieres Group, 1978, Campagn, 1981, Electra., 74:67.
Ricard, A., 1981, Laboratory Report No. LP 187, Universite Paris-Sud, Orsay Cedex, France.
Rich, W., 1982, in: "Applied Atomic Collision Physics", Vol. 3., Academic Press, New York.
Sigmond, R. S. and Goldman, M., in: 1983, "Electrical Breakdown and Discharges in Gases", E. E. Kunhardt and L. H. Luessen, eds., Plenum, New York.
Treanor, C. E., Rich, J. W., and Rehm, R. G., 1967, J. Chem. Phys. 48:1978.
Teich, T. H. and Braunlich, R., 1984, in: "Proceedings, 4th International Symposium on Gaseous Dielectrics", L. G. Christophorou, and M. O. Pace, eds., Knoxville.
Teich, T. H., 1967, Z. Physik, 199:378 and 395.
Teich, T. H. and Braunlich, R., 1985, in: "Proceedings, 8th International Conference on Gas Discharges and Their Applications", Oxford.

MOLECULAR SPECTRAL INTENSITIES IN LTE PLASMAS

D. O. Wharmby

Thorn EMI Lighting Limited
Research and Engineering Division
Leicester, England

INTRODUCTION

High-pressure arcs which operate in LTE or near LTE have important commercial applications for switchgear, plasma torches, plasma chemistry and analysis, lighting etc. Much modeling has been done on these arcs and it has led to detailed understanding of at least some aspects (Lowke, 1979; Wharmby, 1980; Pfender, 1978). Some arcs contain substantial proportions of molecular species: for example, switchgear arcs in SF_6 (Lowke and Liebermann, 1972) or lighting arcs in tin halides (Fischer, 1974). Molecules have major effects on the transport properties and have been well studied. In theoretical models, radiation from molecules has been treated very incompletely, either by using direct measurements of emission coefficients which apply to very specific conditions (Lowke, 1979; Zollweg et al., 1975), or by simply omitting it (Fischer, 1974). In some arcs with lighting applications up to 75% of the radiation may come from molecules, and so some way of including molecular emission in the theory is very desirable. Even in arcs such as high pressure sodium or mercury, the extreme wing line broadening observed is related to molecular emission and needs correct treatment before realistic spectra can be predicted.

The intensities of molecular spectra are well understood (Tatum, 1967; Armstrong and Nicholls, 1972), but unfortunately the data necessary to calculate all the rotational-vibrational transitions are seldom available, so a simplified treatment would be very useful.

For LTE arcs, the number densities and collision rates are so high that details of rotational and vibrational structure are usually smeared out (e.g., Rehder et al., 1973, Fig. 3). The simple approach to molecular emission (Gallagher 1979) which has been used to understand excimer lasers is then appropriate. There are clear connections between molecular intensities and broadening of spectral lines, so the same approach can be applied to the neutral atom line broadening which is such an important part of the emission from lighting arcs (Wharmby, 1980). The whole subject can be called the 'spectra of colliding atoms', a phrase used by one of the pioneers of the subject (Gallagher, 1975).

My intention here is to take results from the molecular spectroscopy, excimer laser and line broadening literature and to combine it with thermochemical information to give results which are appropriate to calculation of spectra in molecular arcs in LTE. The result is applicable to the emission from strongly bound molecules, such as A C which give emission over wide wavelength regions, or unstable molecules such as NaXe which affect the broadening of Na lines. I shall deal throughout with the spectral absorption coefficient $\kappa(\nu, T)$ since in LTE the radiation transport equation can be used to evaluate emission spectra. MKS units are used throughout.

INTENSITIES OF MOLECULAR SPECTRA

The energy levels of even simple molecules are extremely complicated. For a start, there are far more non-degenerate electronic states than there are in atoms. For example, Fig. 1 shows the recent results of calculations of the lowest molecular levels of Na_2 (Jeung, 1983). The two ^2P levels of the Na atom, which are the upper states of the D lines,

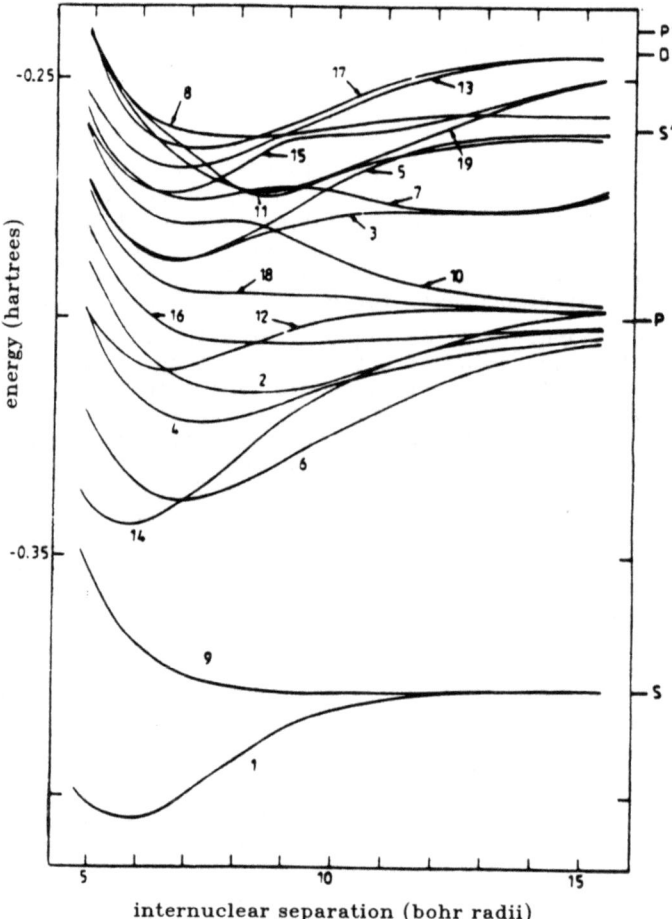

Fig. 1. Recent calculations of the lowest states of Na_2 (Jeung, 1983). Levels are $^1\Sigma_g^+$ (1-3), $^3\Sigma_g^+$ (4-5), $^1\Sigma_u^+$ (6-8), $^3\Sigma_u^+$ (9-11), $^1\Pi_u$, (12,13), $^3\Pi_u$ (14,15), $^1\Pi_g$ (16,17) and $^3\Pi_g$ (18,19). S(3s+3s), P(3s+3p), S'(3s+4s), D(3s+3d) and P'(3s+4p) are dissociation limits.

split into no less than 8 states when the atoms combine into the Na_2 molecule (as shown in Fig. 1). These potential energy diagrams, by the way, are crucial to the simplification of molecular intensities which I shall discuss. They show the potential energy of separate atoms as they are brought together. In some cases there is an attraction (the "bonding case") giving a potential well, and in others a repulsion (the "antibonding case"). In yet others, the potential curve for the electronic states is more complicated, usually as result of avoided curve crossings.

It is important to realize that, although the existence of the large number of states exemplified in Fig. 1 can be inferred from group theory, experimental information about the shape and depth of the potential wells is usually only known for a very limited number of states. In particular, there is virtually no experimental information about the repulsive, 'antibonding' or 'free states'.

The atoms vibrate in the potential well and when the vibrations are quantized, certain fixed transitions are allowed between the vibrational energy levels, now approximately equally spaced in the potential well. Rotations of the molecule are also quantized giving rise to energy levels with a much more finely spaced structure.

Take as an example the ground state $X\ ^1\Sigma_g^+$ and first excited state $A\ ^1\Sigma_u^+$ which are responsible for the well-known A-X transitions in Na_2. There are $\sim 10^5$ rotational-vibrational transitions which make up this band system. Some of these are visible in emission in a discharge in ~100 torr (Whittaker, 1972). (For a concise, simple reminder of molecular state notation see Gaydon, 1968.)

On the face of it, the calculation of 10^5 quantum-mechanical intensities is a daunting task even for cases like this for which molecular constants are known well. Figure 2 (Lam et al., 1974) shows what the result is for Na_2. There is a vast forest of rotational lines even in the small space of $150 cm^{-1}$ (~10nm). In cases of interest to us, the rotational structure is smeared out by line-broadening. After applying a suitable smoothing function to the spectrum in Fig. 2, the result is much more recognizable as the sort of continuous absorption spectrum that we observe (Figs. 3 and 4).

In the course of the calculation of Lam et al., use has been made of the Franck-Condon principle (Herzberg, 1950, p. 194). When the molecule emits a photon, there is a rearrangement of the electron cloud around the nuclei. It is a good approximation that this occurs before the nuclei move substantially. The transitions therefore take place very nearly vertically on the potential energy diagram between pairs of vibrational and rotational states (Fig. 5). The quantum-mechanical FC principle depends on the overlap between the upper and lower state vibrational wave functions. The probability of a transition depends on the amount of overlap through the Franck-Condon factors which are the matrix elements between the vibrational wave functions (Herzberg, 1950, p. 199). This in turn depends on the separability of the electronic from the vibrational and rotational parts of the total wavefunction.

There is a further level of approximation which provides the major simplification necessary to calculate molecular spectra straight-forwardly. This has been called the classical Franck-Condon principle (CFCP) (e.g. Gallagher, 1975). Since the vibrating atoms spend most of their time at the positions of extreme amplitude (as does a child on a swing), it is assumed that transitions only take place at those separations (see Fig. 5). No information is required concerning the vibrational or rotational structure of the levels when calculating the

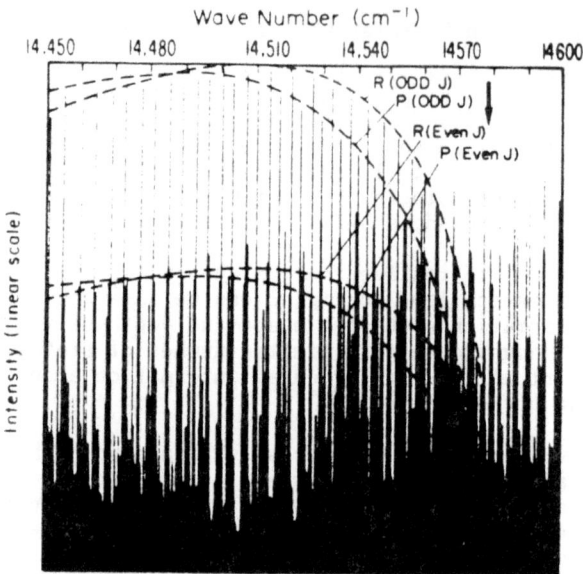

Fig. 2. From Lam, Gallagher and Hessel (1974). Showing 150 cm^{-1} (~ 10nm) in the red region of the Na A-X band. Each vertical line is the emission between two rotational levels (J', J") in the region of the band (v',v") = (2,2). R and P refer to branches of the spectrum. The arrow shows the measured position of the (2,2) band head. After wavelength averaging of this data this feature shows up as a typical band head.

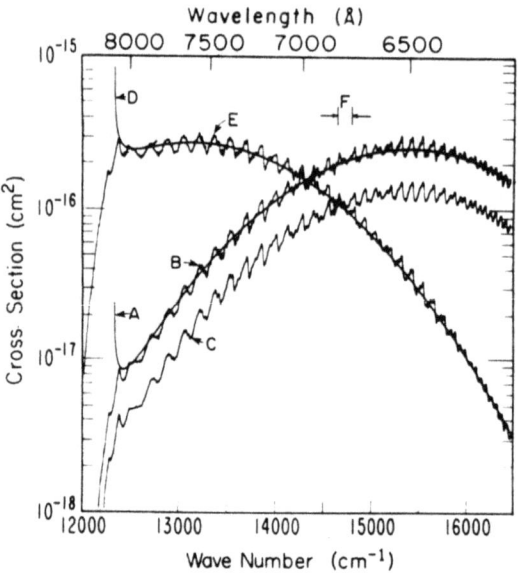

Fig. 3. From Lam, Gallagher and Hessel (1974). Curve B is the absorption cross section for Na$_2$ calculated by averaging rotational structure like that shown in Fig. 2 over a band of 5 cm^{-1} (~ 0.25nm at this wavelength). The region shown in Fig. 2 is denoted by F. C is the same as curve B, but displaced by a factor of 2. The CFCP absorption calculation is curve A. Stimulated emission cross sections are curves E(QM) and D(CFCP).

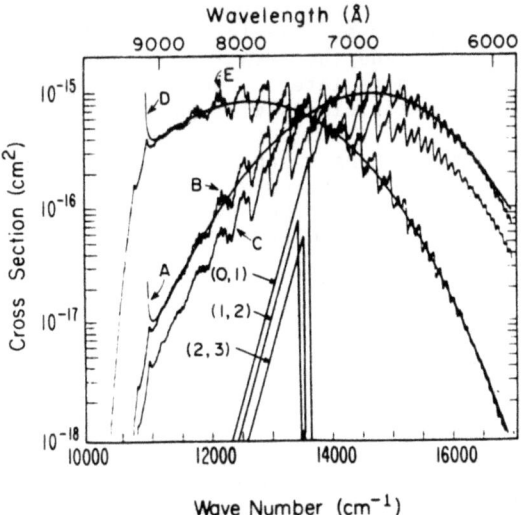

Fig. 4. From Lam, Gallagher and Hessel (1974), but for Li_2 with same caption as Fig. 4. With a lighter molecule the difference between CFCP and QM calculations is greater but still satisfactory for our purposes. The way in which the various bands $(v',v'') = (0, 1), (1, 2)$ etc. contribute to the total is also shown.

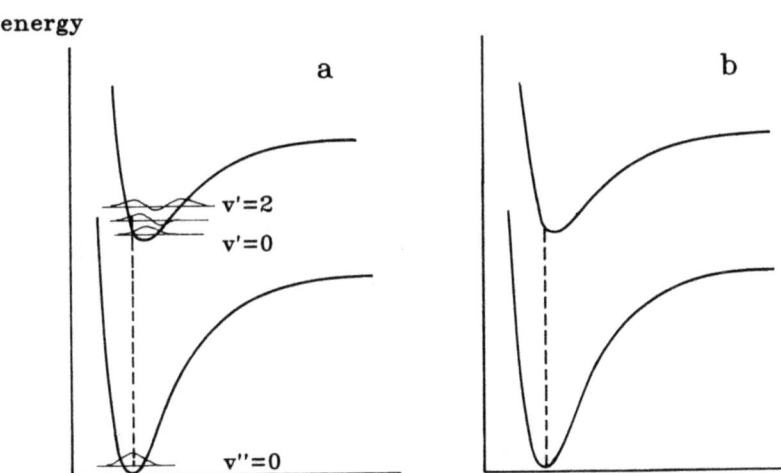

Fig. 5. The Franck-Condon principle (a) and the classical Franck-Condon principle (b). For the FC principle the most probable frequency of emission is that for which the overlap between upper and lower state vibrational wavefunctions is greatest: $(v',v'') = (2,0)$ in this case. The CFCP frequency is given by vertical transitions between the electronic states.

absorption or emission spectrum. All that is required is the shape of the electronic potentials, and the variation of transition probability with separation, of which more later. The appropriate formulae for absorption and emission coefficients will be derived below.

The validity of the CFCP has been tested by a few authors, the paper of Lam et al. (1977) being of particular interest. They made a full quantum-mechanical calculation of the absorption cross sections of the A-X bands in Na_2 and Li_2 with results typified by Fig. 2. The cross sections can be simply related to the absorption and emission coefficients. They also calculated the cross sections on the basis of the CFCP (Figs. 3 and 4). After smoothing to simulate the effect of line-broadening, the QM calculation shows jagged structure from the summation over the vibrational bands: similar results for Li_2 show how typical vibrational bands contribute to the sum (Fig. 4).

These calculations show that for the bound states (that is, those states giving rise to normal molecular band spectra) of even quite light molecules, the CFCP gives excellent results for intensities provided that high resolution spectra are not needed. Line-broadening is so great in high pressure lamps that rotational and vibrational structure is hardly ever seen. The CFCP approximation, in which vibrational and rotational structure is ignored, is therefore an appropriate one to apply in cases where line broadening removes most of the traces of the structure anyway.

TRANSITIONS BETWEEN BOUND AND FREE STATES

Temperatures are high (~6000K) in the center of a high-pressure arc, so high indeed that all molecules are substantially dissociated. As an example, Fig. 6 gives recent calculations of the dissociation of the very stable molecule SnCl into Sn and Cl (Mucklejohn, 1985). Measurements (Rehder et al., 1973) have shown conclusively that radiation in the cool part of tin halide arcs can be attributed to bound SnCl molecules, some details of the band spectra becoming evident at low pressures. Figure 7 gives the continuum radiation from an arc containing Sn and Cl and observation shows that much of this originates from the arc centre: that part cannot be radiation from bound SnCl molecules. On the other hand detailed examination (Mück, 1973) on $AlCl_3$ arcs shows that the extensive continuous radiation near the arc center does not come from atom-ion recombination continua. There is ample evidence of radiation at high temperatures from relatively unstable Na_2 and NaHg, and completely unstable Na - rare gas molecules in high-pressure arcs. In fact, molecular radiation from high temperature regions is likely to be a widespread phenomenon.

How can radiation which is molecular in origin be emitted from temperature regions for which the molecule is almost completely dissociated? The answer lies in the formation of temporary or quasi-molecules. As atoms fly past each other in the gas, they perturb each other's energy levels. If, during the collision time, the atom pair radiates, the frequency of radiation is proportional to the difference potential at the internuclear separation R, at which the photon is emitted. In the CFCP approximation, there is a vertical transition between the free electronic states, and the nuclear kinetic energies adjust to conserve energy (Fig. 8). This figure shows a central collision in which the perturber is scattered back in the same direction as the radiator. Non-central collisions, in which the perturber moves past the radiator with deflection depending on the closeness of collision, are much more common and can be treated by the same methods. The main difference is that only in the case of the central collision is there a

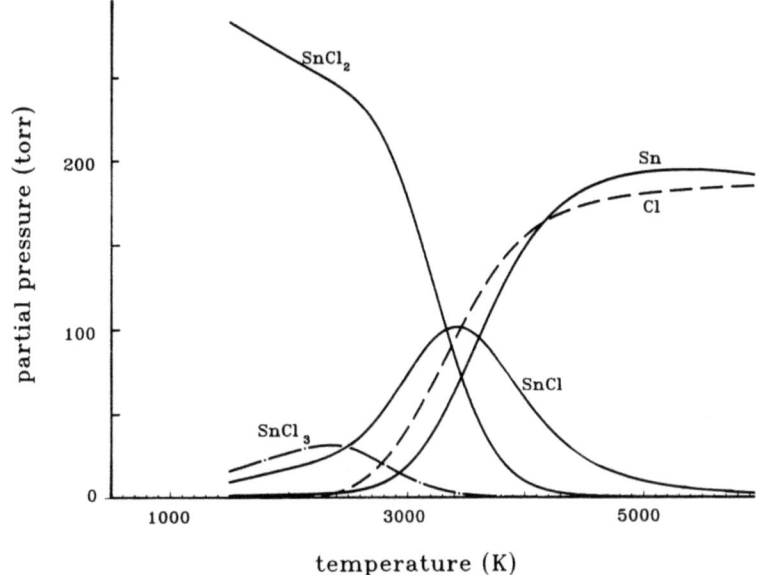

Fig. 6. Calculation of the Sn containing species on a Sn/Na/Hg/Cl LTE plasma based on the recent thermodynamic measurements on tin halides (Mucklejohn, 1985). The composition calculation includes the effects of diffusion. Note that even as stable a molecule as SnCl dissociates below 4000K.

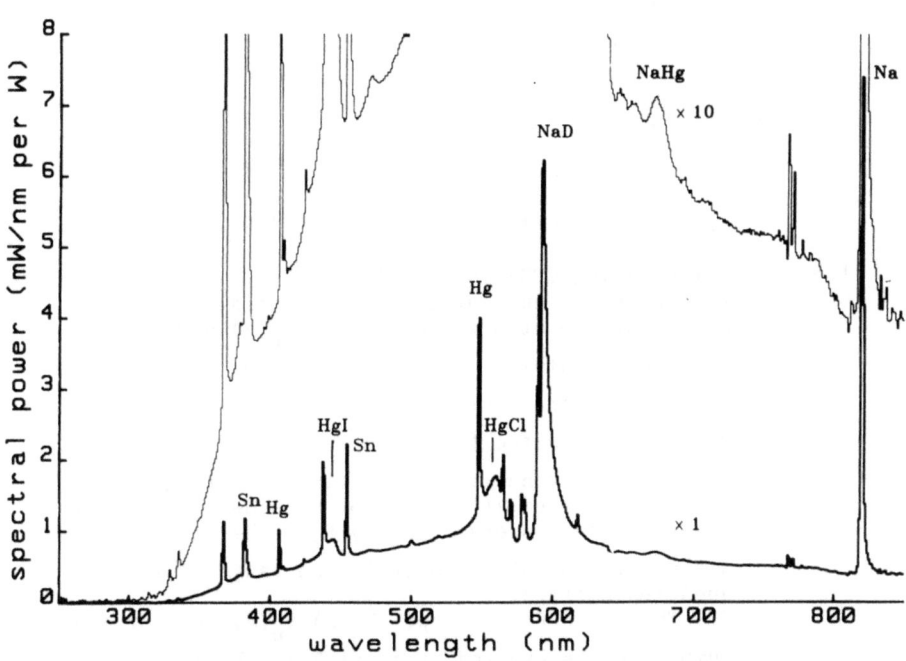

Fig. 7. Spectra of a tin sodium halide discharge having a plasma composition similar to that in Fig. 6. Note how much continuous radiation is emitted; spectroscopic observations indicate that most of this comes from hottest regions (~5500-6000K).

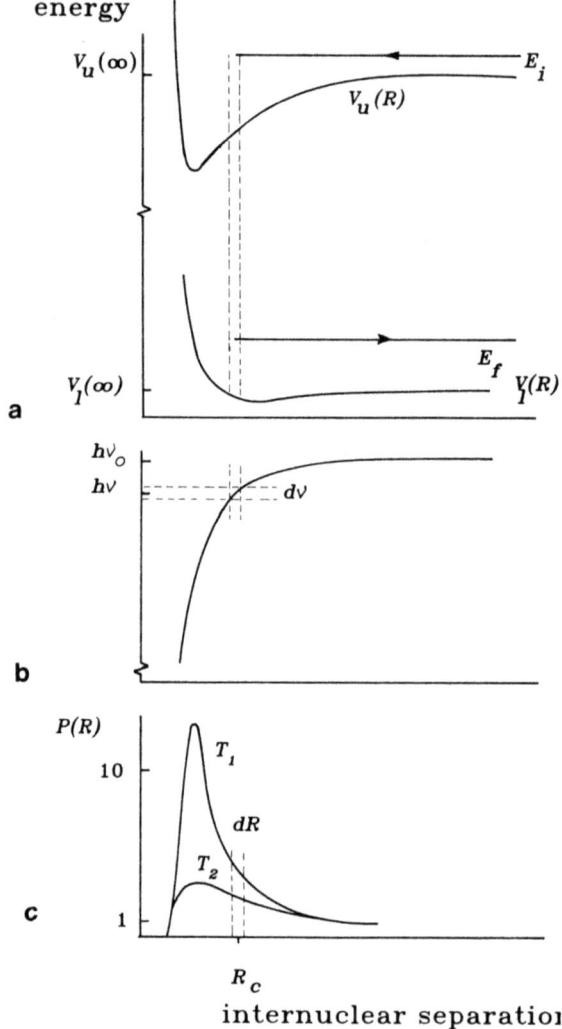

Fig. 8. Illustrating CFCP in emission.
(a) The upper and lower state potentials which dissociate to atom energy levels $V_u(\infty)$ and $V_\ell(\infty)$. An incoming atom at energy E_i perturbs the energy levels, and the system radiates at R_c with energy $h\nu$ differing from the free atom energy $h\nu_o$.
(b) The difference potential gives the range of energies emitted at a range of R values. The atom departs with conserved energy E_f. The slope of the difference potential controls the range of frequencies emitted.
(c) The ratio $P(R) = \exp(-(V_u(R)-V_u(\infty))/kT)$ of the number in the excited state at R to the free excited atom population is also shown, for high (T_2) and low (T_1) temperatures: this could also be interpreted as low and high values of $V_u(R)-V_u(\infty)$ at the same T. (From Phelps, 1972).

turning point to the motion and this leads complications in the quantum mechanical approximation.

Free-free molecular transitions of the type shown in Fig. 8 were first investigated in a series of papers by Jablonski between 1931 and 1968 (see Allard and Kielkopf, 1983 for references). The Franck-Condon factors for these transitions have been evaluated using JWKB wave functions (Allard and Kielkopf, 1983), the result being that the most probable transition occurs when the wave numbers of the initial and final states are equal:

$$k_i(R_c) = k_f(R_c).$$

R_c is the internuclear separation for which this is true. In terms of total energy E and potential energy V of the upper u and lower ℓ levels, this condition becomes

$$E_i - V_u(R_c) = E_f - V_\ell(R_c)$$

which is the classical Franck-Condon principle applied to free-free transitions as illustrated in Fig. 8.

Note that the internuclear separation R_c (the Condon point) is only the most probable: in other words the CFCP is an approximation, albeit a useful one. As Hedges et al. (1972) point out, this principle could equally well have been derived using JWKB wave functions for bound states. The CFCP can therefore be applied to transitions between free and bound, bound and bound, and free and free states, provided that the discrete quantum effects of rotation and vibration can be ignored. As Lam et al. (1974) have shown, this is a viable approximation even for quite light molecules.

The transitions between free-bound, bound-free, and free-free molecular states are therefore simply an extension of the normal rules for transitions between bound molecular states. The principal difference is in the populations of excited molecules available to radiate. If atoms combine to form a stable physical unit in the gas through the reaction

$$A + B \rightarrow AB$$

with an equilibrium constant favouring the RHS, there will be many more AB's available to radiate than if the equilibrium constant favours the dissociation of AB. In this latter case the molecules only have an existence and can only radiate as the atoms fly past each other.

There is a problem here of conservation of energy. If two atoms collide to form a bound state where does their excess energy go? At high pressures, our normal case, three body collisions stabilize the bound states leading to a Boltzmann distribution over bound and free states. At low pressures another distribution, deficient in bound states, pertains (Hedges et al., 1972).

In the high temperature plasmas near the center of our arcs, the stable molecular species are almost completely dissociated. We can therefore expect emission to be mainly between bound states in the cooler regions, and between unbound states in the hotter regions. Since the emission increases exponentially with temperature, there may well be more emission from quasi-molecules in the hot regions than bound molecules, despite the much larger populations of the bound states at the lower temperatures. The quantitative aspects of this are explored later. Bound molecules, even polyatomics are likely to dominate the absorption in cooler parts of the arc and so their effects must be included in radiation transport calculations.

SOURCES OF DATA

We shall assume that the potential energy curves are known for the formation of species AB from A and B. Information of this type is available from conventional molecular spectroscopy at very high accuracy. Although it usually covers a small range of internuclear separations, it can be extrapolated by standard methods (Gaydon, 1968). Unfortunately, the information is usually only known for typically two or three low-lying bound states, although laser spectroscopy is starting to provide information about higher lying states.

Large scale molecular structure calculations are the other main source of data. The potential curves in Figs. 1 and 10 for example, are calculated (see Konawalow et al., 1980 and Laskowski et al., 1981). Such results have the advantage of giving much more complete information than experiments can give. In particular they can give details of the unbound states. The accuracy of calculations is difficult to assess, but it is improving all the time as larger computers and improved techniques are brought to bear. The role of experiments is increasingly to correct, support and verify these theoretical calculations. A review of relevant molecular structure calculations is included in Allard and Kielkopf (1982).

There is one area, where for simple molecules at least, calculations may be as good as or even better than experiments (Konowalow et al., 1983). This is the calculation of transition dipole moments between electronic states: this leads to values of the Einstein A coefficient (the transition probability) as a function of the separation of the atoms.

Recently we have investigated the possibility of deriving absorption coefficients from spectral intensity measurements of discharges as described below. For this, qualitative knowledge of the form of the potentials is required, and this can come from approximate calculations.

CALCULATIONS OF MOLECULAR INTENSITIES USING THE CFCP

The aim of the section is to derive a values of the absorption coefficient $\kappa(\nu,T)$ for all types of transitions between bound and free molecular states in an LTE plasma. These values can then be used in calculations of spectra by employing the radiation transport equation. The derivation presented here is a simple thermodynamic version easily understood by analogy with the LTE thermochemical and atomic radiation expressions usually employed in high-pressure discharge theory. It leads to the same result as more rigorous derivations (Allard and Kielkopf, 1982).

The calculation has several parts. Firstly, we have to calculate from the atom densities in LTE how many molecules, bound and unbound, there are in a given electronic state. This requires enumeration of the states available. We then calculate the absorption coefficient for these molecules and quasi-molecules using the CFCP. An example of the calculation of absorption coefficients is given in a following section. The intensity can be calculated using the radiation transport equation.

The Number of Molecules in the Lower State of the Transition

Figure 9 shows typical bound-bound, bound-free and free-free absorptions from a lower state q to an upper state r of a molecule. To keep it

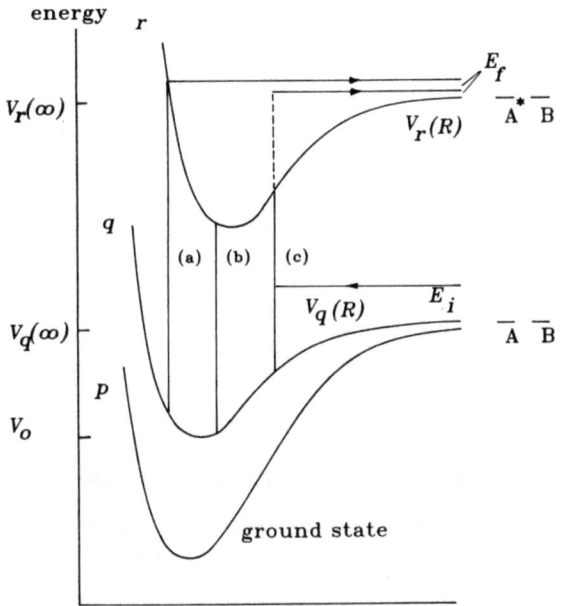

Fig. 9. Showing different types of absorption (between molecular states q and r of AB) that are included in the theory.
(a) is a bound-free absorption occurring as a result of a central collision. The atoms fly apart after absorption, with energy E_f.
(b) is bound-bound absorption.
(c) is a free-free absorption. The atoms move past each other (a non-central collision) with kinetic energy E_i, make a transition at separation R and depart with energy E_f.

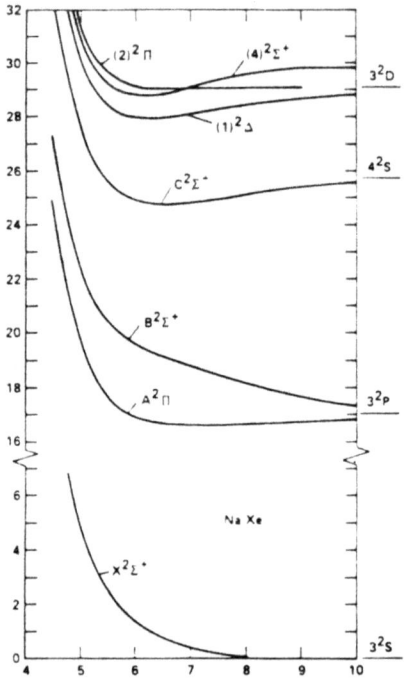

Fig. 10. Low lying states of NaXe from Laskowski et al., (1981). Sodium atom states shown in the dissociation limits.

general we shall assume that q is not the ground state of the molecule. In the CFCP approximation we need not be concerned that the potentials are filled with a great number of vibrational and rotational states. In Fig. 9, I have supposed the lower group of states to dissociate to A and B atoms. The upper state dissociates to excited A and ground state B atoms. This is what happens in Na_2, NaHg and NaXe, for example.

The dissociation constant for the reaction

AB→A+B

is given by the statistical-mechanical expression

$$K_d = \frac{n_A n_B}{n_{AB}} = \frac{Q_A Q_B}{Q_{AB}} s_{AB} \exp\left\{-\left[V_q(\infty) - V_o\right]/kT\right\} \quad (1)$$

where the n's are concentrations or number densities, the Q's are the partition functions, and s_{AB} is a symmetry number (1 for heteronuclear and 2 for homonuclear dimers). The quantity $V_q(\infty) - V_o = D_o$, is the dissociation energy for a molecule in state q. From the chemical point of view the energy in the exponent must be identified with the heat of dissociation, whilst the partition function term is equivalent to the entropy term in the Gibb's free energy of dissociation (Davidson, 1962).

The number in state q of the molecule, for an internuclear separation of R is therefore given by the Boltzmann distribution

$$n_{ABq}(R) = n_{AB} \frac{g_q}{Q_{AB}} \exp\left\{-\left[V_q(R) - V_o\right]/kT\right\} \qquad (2)$$

where g_q is the statistical weight of the qth molecular state. Making substitutions, we find

$$n_{ABq}(R) = n_A n_B \frac{g_q}{Q_A Q_B} \frac{1}{s_{AB}} \exp\left\{-\left[V_q(R) - V_q(\infty)\right]/kT\right\} \qquad (3)$$

This is the Boltzmann distribution over bound and free states. The exponential term contains the temperature dependence and is shown as P(R) in Fig. 8. The atom partition functions at low temperatures can often be replaced by the ground state degeneracies (for example at 5000K for Na, $g_{Na}=2$, $Q_{Na}=2.01$). We have the option of the two equivalent expressions (2) and (3). That is, we can express the lower state density in terms of the bound AB density n_{AB}, or in terms of the free atom densities n_A and n_B. The latter is more convenient for our needs because it automatically includes bound and free state densities, but does not require explicit knowledge of molecular partition functions. Knowledge of the equilibrium constant for dimer dissociation does imply some knowledge of the partition function, but this need not be explicit. It is easy to derive similar expressions to (3) in terms of n_{AB} if needed.

Enumeration of States

Both atomic and molecular states may be degenerate. Proper counting of degenerate sub-levels is essential in order to establish g_q in equation (3). The degeneracies of the molecular states depend on the coupling scheme. For many molecules of interest Hund's case (b) applies, giving

g = (2S+1) for Σ states

= 2(2S+1) for all other states

where S is the multiplicity. For the atom g = 2J+1. Since an atom A in any one of its degenerate sublevels may combine with an atom B in any one of its sublevels

$$\sum_r g_i = g_A \cdot g_B$$

where r is the total number of degenerate sublevels for the molecule. For Na_2 (Fig. 1) in an excited state $Na(^2S) + Na(^2P)$, $g_A \cdot g_B = 24$. For the molecular $^3\Pi_u$ state g = 6, so in the molecule one quarter of the levels are in this state. it is easy to confirm that the total g for all excited Π and Σ states dissociating to the atomic states mentioned is also 24.

Absorption Coefficient

The absorption coefficient is given for upper u and lower states by

$$\kappa(\nu) = \frac{h\nu}{c} n B_{\ell u}(\nu) \qquad (4)$$

where ν is the frequency and n the number density (m^{-3}) of absorbing species. This equation is similar to equation (74) in Richter (1968), but I have used a different version $B_{\ell u}(\nu)(J^{-1}m^3s^{-1})$ of the normal Einstein coefficient $B_{\ell u}$, where

$$B_{\ell u}(\nu) = B_{\ell u} P_{a\nu}$$

Richter uses the version with $P_{a\nu}$, the line profile function for the absorption line. For a narrow line it matters little which formulation is used, although Richter's version is more conventional. For broad molecular bands $B_{\ell u}(\nu)$ is much more appropriate. The B coefficient can always be written in terms of the spontaneous transition probability or Einstein A coefficient, in this case by

$$B_{\ell u}(\nu) = \frac{c^2}{8\pi h\nu^3} \frac{g_u}{g_\ell} A_{u\ell\nu} \qquad (5)$$

where $A_{u\ell\nu}$ is now a spectral quantity with units $s^{-1} \Delta\nu^{-1}$, and g_u and g_ℓ are degeneracies of the upper and lower states involved.

The number of molecules $N_{ABq}(R)$ in state q with internuclear separations between R and R+dR is given by the product of the number density $n_{ABq}(R)$ and the volume available at separation R ($4\pi R^2 dR$). Thus

$$N_{ABq}(R) = 4\pi R^2 dR \, n_A n_B \frac{g_q}{Q_A Q_B} \frac{1}{s_{AB}} \exp\left\{-\left[V_q(R) - V_q(\infty)\right]/kT\right\} \qquad (6)$$

The absorption coefficient is therefore

$$\kappa(\nu,T) = \frac{h\nu}{c} \frac{c^2}{8\pi h\nu^3} 4\pi R^2 dR \, A_{rq\nu} n_A n_B \frac{g_r}{Q_A Q_A s_{AB}} \exp\left\{-\left[V_q(R) - V_q(\infty)\right]/kT\right\} \qquad (7)$$

where g_r is the degeneracy of the upper state of the molecular transition (the result of writing this equation in terms of the Einstein A coefficient). Figure 8(b) shows that dR can be replaced by $|dR/d\nu|d\nu$: this is usually written $d\nu/|d\nu/dR|$.

Note that this equation includes both the free and bound atom pairs (quasi and bound molecules) in a consistent way through the equilibrium constant. If $V_q(R) > V_q(\infty)$, so that the state is repulsive, then the concentration of AB pairs at R is less than the equilibrium number of separated atoms by the exponential factor. Conversely, if $V_q(\infty) > V_q(R)$, then the atoms tend to congregate together giving a concentration of AB pairs at R (i.e., bound molecules), greater than the equilibrium number of separated atom pairs. The ideas used extend easily to the formation of triple and higher atom groupings (West et al., 1978), but only in a formal way because the relevant data are usually unavailable.

There is one further complication. At each internuclear separation there is a corresponding frequency of radiation. Figure 11 shows the difference potentials for two pairs of molecular states of Na_2. A particular internuclear separation is responsible for only one frequency in the case of two transitions. In the other two, the difference potential is multi-valued and there can be more than one separation giving contributions to one frequency. For a particular pair of states with more than one Condon point, there are therefore different values of $V_q(R_i) - V_q(\infty)$, $|d\nu/dR|_i$ and $A_{rq\nu}(R_i)$ for the various separations R_i leading to frequency ν. The final result for the absorption between molecular state q and r is

$$K(\nu,T) = \sum_i \frac{1}{2} \frac{c^2}{\nu^2} \frac{R_i^2 A_{rq\nu}(R_i) d\nu}{|d\nu/dR|_i} n_A n_B \frac{g_r}{Q_A Q_B S_{AB}} \exp\left\{-\left[V_q(R_i) - V_q(\infty)\right]/kT\right\} \quad (8)$$

Allowing for differences of notation this is Gallagher's (1979) equation 5.13b. This is the result for one pair of states. If several pairs of states contribute to the absorption at ν, then the individual absorptions (Eq. (8)) are added together.

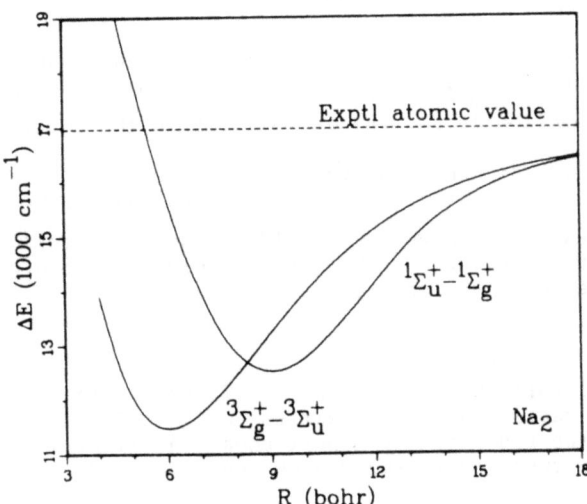

Fig. 11. Difference potentials for states of Na_2 from Konowalow et al., (1983). There is often more than one molecular separation leading to the same frequency as shown here. The A-X transition is between the singlet states.

If the value of $|d\nu/dR|$ becomes zero, as it does at an extremum of the difference potential, the value calculated from Eq. (8) diverges. This is a failure of the CFCP, which can be overcome by more complete theory. Multiple atom effects and the relative motion of the atoms must be included. In classical spectroscopy, the features which appear near extrema in the difference potential are called 'heads of heads'. In line broadening theory they are referred to as satellites. In the later case a unified approach to line broadening (Szudy and Baylis, 1975) predicts the correct form for the satellites, although there is still some doubt about their temperature dependence (Jongerius et al., 1981).

Example of Absorption Coefficient Calculation

An example of the use of this equation has been given by Woerdmann and de Groot (1982) for various transitions in the Na_2 molecule. In this case the Na atom density is $n_{Na} = n_A = n_B$. They expressed the Einstein A coefficient in terms of the oscillator strength $f_{\ell u}$ for the separated atoms and assumed that it was independent of ν and R (justified by Woerdmann, 1981).

$$A_{u\ell\nu} d\nu = \frac{1}{4\pi\varepsilon_o} \frac{8\pi^2 e^2}{mc^3} \frac{g_\ell}{g_u} \nu \; f_{\ell u} \tag{9}$$

Finally they set $Q_{Na} = 2$ for the atom, which is a perfectly reasonable low temperature (<5000 K) approximation. After making these substitutions into Eq. (8), we find

$$\kappa(\nu,T) = \frac{1}{4\pi\varepsilon_o} \frac{\pi^2 e^2}{2mc} f_{\ell u} \frac{R^2}{|d\nu/dR|} g_\ell \; n_{Na}^2 \; \exp\left\{-\left[V_\ell(R) - V_\ell(\infty)\right]/kT\right\} \tag{10}$$

This is identical to Woerdmann and de Groot's (1982) equation (1) with the exception of $1/4\pi\varepsilon_o$ which is the result of my using MKS units in this paper. The summation in Eq. (8) is implicit in the Eq. (10).

Using Eq. 10, de Groot and Woerdmann calculated the absorption coefficient of Na_2 vapour from molecular potentials of Konowalow et al. (1980). The result is shown in Fig. 12. Also given in the figure is the emission spectrum showing how the various molecular absorption features account for the emission features.

Calculation of Emission Spectra

Once the absorption coefficient $\kappa(\nu,T)$ or $\kappa(\lambda,T)$ is known, spectral intensities or radiation flux densities can be calculated by standard methods using the radiation transport equation (Richter 1968 p. 53). In LTE, the source function is the Planck function $B_\nu(T)$ and Kirchoff's law can be used so that

$$I_\nu = \int_{-L}^{L} \kappa(\nu,T) \; B_\nu(T) \; \exp\left[-\int_x^L \kappa(\nu,T) dx'\right] dx \tag{11}$$

for a ray emerging from a plasma of depth 2L. The solution of this equation can be achieved by a variety numerical methods. It is also feasible to do complete radiation flux density calculations with spectra as complicated as this using approximate methods (Jones and Mottram, 1981; Stormberg and Schafer, 1983).

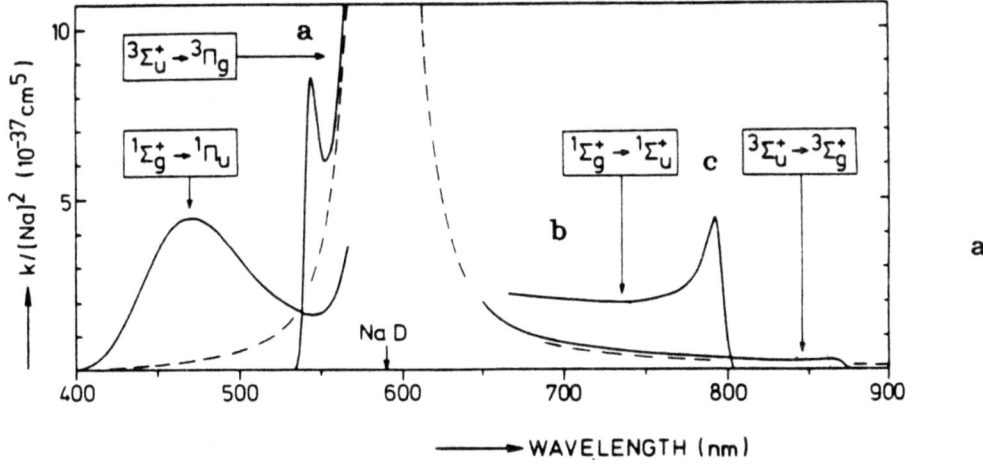

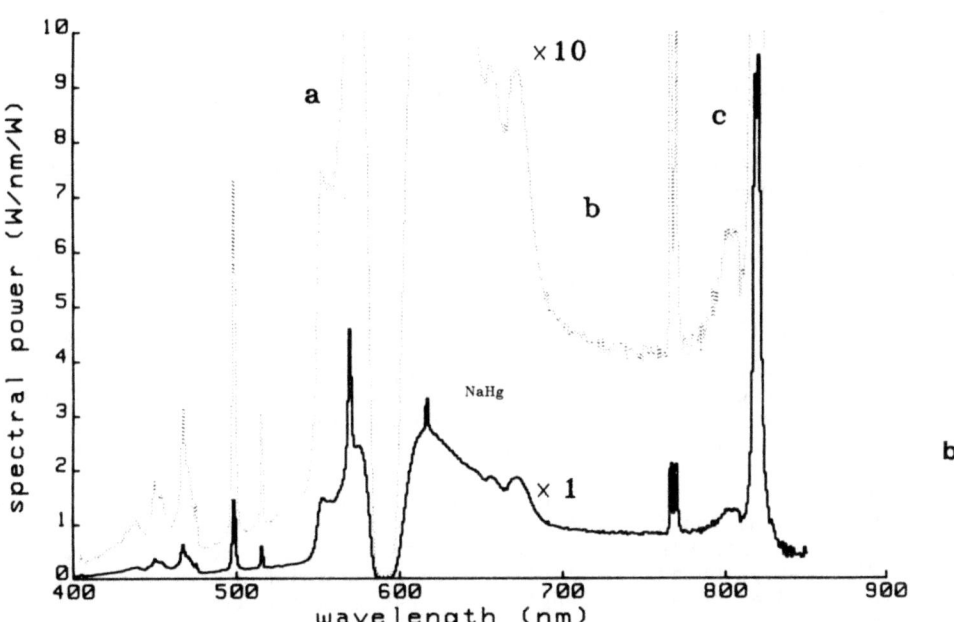

Fig. 12. (a) The absorption coefficient for Na_2 vapour derived by Woerdmann and de Groot (1982) from the data of Konowalow et al. (1982). The total absorption is the sum of the absorption for each pair of states. The dotted line is the absorption coefficient from resonance broadening theory. The experimental high pressure sodium emission spectrum (b) indicates how the Na_2 absorption contributes to the light output and color. The NaHg satellites are also shown.

MOLECULAR TRANSITION PROBABILITIES

The subject of molecular transition probabilities is extremely complicated, as the reviews by Tatum (1967) and Armstrong and Nicholls (1972) show. The latter give an expression for the Einstein A coefficient showing how it depends on internuclear separation R. In the CFCP approximation, Armstrong and Nicholl's equation 4.1.21 can be written

$$A_{u\ell\nu} d\nu \equiv A_{u\ell\nu}(R)d\nu \equiv A_{u\ell}(R) = \frac{1}{4\pi\varepsilon_o} \frac{64\pi^4\nu^3}{3hc^3} \frac{1}{g_u} D^2_{u\ell}(R) \qquad (12)$$

where $D_{u\ell}(R)$ is the dipole matrix element (Herzberg 1950 p. 387).

$$D^2_{u\ell}(R) = \left| \int \Psi^*_u(\underline{r},R) M_e \Psi_\ell(\underline{r},R) d\underline{r} \right|^2 \qquad (13)$$

M_e is the dipole moment operator, as evaluated for example by Konowalow et al. (1983), and the Ψ are the molecular wavefunctions for upper u and lower ℓ states. The variable $\underline{r}$ indicates integration over the configuration space for electrons. The vibrational and rotational contributions to Armstrong and Nicholls' equation have been equated to unity and $2J'+1$ using the sum rules (Armstrong and Nicholls, 1972, equations 4.1.16). This means that only the upper state electronic degeneracy appears in Eq. (12).

Figure 13 shows some examples of D(R) taken from a recent calculation on Na_2 (Konowalow et al., 1983). They can be seen to approach the experimental atomic values asymptotically, in the cases in which the atomic transitions are allowed. So far the effect of this variation on the absorption coefficient has not been checked.

Woerdmann (1981) found evidence in the alkali dimers that the product of $D^2(R)\nu(R)$ was constant, which is equivalent to saying that the oscillator strength f(R), which one can define from Eq. (9) is independent of R. In more detailed work, Konowalow et al. (1983) came up with examples counter to this supposition.

The variation of D(R) shown in Fig. 13 means that it is important to have this information for good calculations of the absorption coefficient. It is particularly so in cases with very great self-absorption (e.g. Fig. 12(b)) because most of the radiation which actually escapes from the arc does so at wavelengths very far from the line centre. These photons are emitted in the course of very rare but very close collisions where details of the potentials and the variation of D(R) are particularly significant. Radiation emitted during weaker collisions, in which D(R) is approximately the separated atom value, is largely self-absorbed. Fortunately molecular structure calculations are really making progress, and this is one area where calculations are starting to yield more complete, and possibly more accurate data than experiments.

LINE BROADENING

Line broadening by neutral particles is also concerned with the effects which result from colliding atoms. What is the connection between line broadening and the molecular picture so far described?

Complete line broadening theory is usually based on a calculation of the Fourier transform of the autocorrelation function of the wave train emitted by a radiator affected by collisions (Hindmarsh and Farr, 1972). It is only in the last decade that this complete line shape theory has been evaluated, and it still is a massive computational problem (Allard

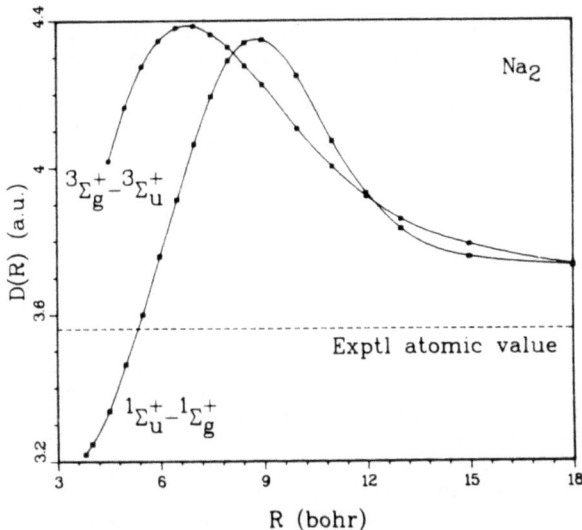

Fig. 13. Values of the dipole moment function D(R) for some states of Na_2 (Konowalow et al., 1983), showing that the variation of D(R) can be large.

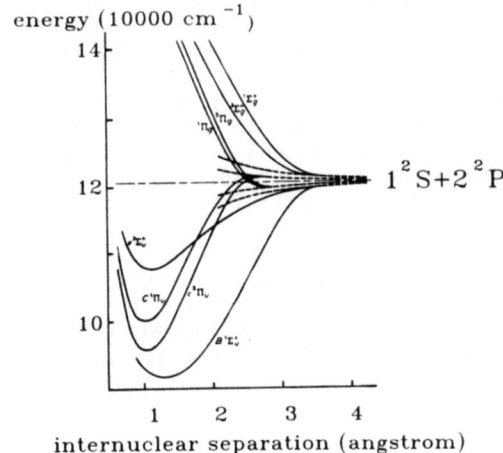

Fig. 14. Illustrating the difference between potentials used in 'far wing' and 'extreme wing' broadening. The dotted lines are calculations of the resonance interaction between H atoms in their 1^2S and 2^2P states. The close range interactions for the H_2 molecule are the solid lines and would be used in the 'extreme wing' case. At long range they approach the resonance curves asymptotically (King and van Vleck, 1939, as quoted by Herzberg, 1950, p. 380).

and Kielkopf, 1982). Historically two approximations have been used, both of which can be derived as limiting cases of this Fourier transform.

In the 'impact' approximation, an excited atom radiates a wave. During this time instantaneous collisions with perturbers occur which change the phase of the wave. The Fourier transform of a randomly interrupted sinusoid is the Lorentzian line shape function

$$L(\lambda) = \frac{1}{\Pi \Delta \nu} \frac{1}{1 + (\nu-\nu_0)^2/\Delta\nu^2} \qquad (14)$$

where $\Delta\nu$ is the semi-half width and ν_0 the line center frequency.

The semi-half width depends on the number density of perturbers and on the strength of the perturbation. It is therefore dependent on the force law describing the interaction between radiator and perturber. I shall not consider the case of charged perturbers. Impact broadening by electrons is important in high-pressure arcs (de Groot, 1974) and the theory has been extensively covered by Griem (1974). The important neutral interactions are dipole-dipole forces between like particles (resonance interaction) and unlike particles (van de Waals interaction) (Hindmarsh and Farr, 1972). The impact approximation is valid if negligible radiation occurs during the collision, that is if the collision time, during which the perturber alters the phase of the wave, is much shorter than the time between collisions. For van der Waals interactions between perturbers the impact region extends only a few wave numbers from the line centre (~ 0.1 nm in the visible). For resonance broadening the region is even smaller because of the much stronger interaction.

In high-pressure arcs, neutral impact broadening is not very important for the following reasons. The power radiated in optically thin lines can be calculated without reference to line broadening. The power radiated in lines with appreciable optical depth can only be calculated if the line broadening is known. Number densities are so high that details of the line broadening close to the line centre are fairly unimportant, because most of the photons at these frequencies do not escape. What must be known is the broadening of the line wings and this is usually well outside the (neutral) impact region.

The other main approximation in the Fourier transform approach to the line-shape is the quasistatic approximation. In this, the radiator is assumed to be at rest relative to the perturber. It emits at a frequency given by the CFCP applied the perturbed radiator states. In other words the radiator radiates at a definite point in the collision, and the motion of the perturber is neglected. The number of perturbers in a shell between R and R+dR is then

$$n \cdot 4\Pi R^2 dR \qquad (15)$$

assuming constant perturber density n. The range dR maps into a frequency interval $d\nu$ through the difference potential. In this simple form of quasistatic theory which applies to the 'far wings' of the line (Gallagher 1975), the potentials representing the perturbed states of the radiator are assumed to be van der Waals C_6/R^6 or resonance C_3/R^3 (Hindmarsh and Farr, 1972). These potentials are purely attractive or repulsive and are therefore not appropriate for very close collisions. Incorporating these potentials into Eq. 15 gives a line shape depending on $(\nu-\nu_0)^{-2}$ for resonance (but with a different line width from that predicted by the impact approximation) and $(\nu-\nu_0)^{-3/2}$ for van der Waals broadening.

'Far wing' broadening is therefore identical in concept to the molecular approach discussed previously. The differences are that the molecular calculation uses the real interatomic potentials and the perturber density is allowed to be affected by the potential well of the radiator. Figure 12 shows that different states contribute to the red and blue wings of the line. Satellites are also explained as resulting from extrema in the difference potentials, and the line profile is temperature dependent. This has been called the 'extreme wing' case. The connection between 'extreme wing' and 'far wing' broadening is most clearly seen in Fig. 14. The real potentials are the solid lines and the dotted lines are a C_3/R^3 resonance potentials. For an accurate description of the 'extreme wing' in which very close collisions are considered, resonance or van der Waals potentials cannot be expected to be correct. We have found extreme wing broadening in the high-pressure sodium arc to be extremely important in the performance of very high pressure versions of the lamps.

It is now possible to calculate complete (unified) line-shapes from real potentials without appeal to impact or quasistatic approximations (Allard and Kielkopf, 1982). Such calculations involve heavy computation, and it is not yet clear how useful they will be in practical cases such as these. There are still important approximations, and the main theoretical problem seems to be how we should incorporate non-isotropic perturber interactions, particularly when more than one perturber is involved.

A practical problem in using 'extreme wing' line broadening (or the molecular approach) is illustrated in Fig. 14. The absorption coefficients calculated from the molecular approach apply only to a limited wavelength range because of the availability of the data. Equally the absorption coefficient calculated from quasi-static resonance broadening theory is satisfactory near the line center but cannot be correct in the extreme wings. Both sets of absorption coefficients can be expected to be correct in their appropriate wavelength regions, but how can they be joined up satisfactorily? This is where unified line shape calculations would be helpful. At present, however, the best approach seems to be to adjust the points at which the two curves overlap so as to secure agreement with experiment. Satellites are explained naturally as positions in which the slope of the difference potentials $|d\nu/dR|$ becomes small.

All types of broadening are important in the sodium lamp. Line core and far wing broadening are important near the center of the emission line, where strong self-absorption affects temperature profiles. Far wing broadening is important in determining the actual spectrum emitted in commercial lamps. Extreme wing broadening is crucial at high pressures, and can explain the satellite features which give rise to important bumps in the emission lines of lamps. Extreme wing broadening and the CFCP approach to molecular band emission are synonymous, so the same approach can be applied to any molecular band provided that rotational and vibrational transitions are sufficiently broadened.

EXPERIMENTAL APPROACH

Many of the difficulties in direct calculation of extreme wing absorption coefficients, in particular the difficulty of matching up to the far wing data, could be overcome by measuring $\kappa(\nu,T)$. However the temperature and wavelength ranges pose considerable experimental problems. The whole line profile can be explored by fluorescence (Hedges et al., 1972) but it is difficult to see how this could be done at the high temperatures encountered in arcs. One could use an arc to produce the high temperature, and measure absorption or emission along its axis.

This approach has worked well in some cases (Mück and Popp, 1973) but the practical problems of doing it in lighting arcs seem insuperable.

Can anything be deduced from relatively straightforward radial intensity measurements? The intensity of an LTE arc in the radial direction can be calculated from the radiation transport Eq. (11) if $\kappa(\nu,T)$ and the radial temperature profile $T(r)$ are known. If $T(r)$ and I_ν are measured, can we invert Eq. 11 to obtain $\kappa(\nu,T)$? The answer is yes, if a simple model for $\kappa(\nu,T)$ is available. If we consider the simplest case in which only one pair of states and one internuclear separation R contribute at a frequency ν, then Eq. 8 can be expressed as

$$\kappa(\nu,T) = n_A n_B \, a(\nu) \, \exp(b(\nu)/T)$$

where n_A and n_B are number densities and the constants a and b incorporate the various factors in Eq. 8. Two measurements $I_{\nu 1}$ and $I_{\nu 2}$ for two well-chosen temperature distributions $T_1(r)$ and $T_2(r)$ are then sufficient to recover these two unknown constants from the radiation transport equation. In practice it is better to make measurements for ~10 different temperature distributions and to obtain $a(\nu)$ and $b(\nu)$ by at least squares procedure. With suitable assumptions about the way in which $a(\nu)$ and $b(\nu)$ are made up, it should be possible to recover molecular potential curves in the manner of Hedges et al. (1972). The method just described has the advantage that very high temperature regions can be explored: the data will thus emphasize very close collisions.

This approach is being actively pursued at present. It has obvious applications for molecules with simple electronic structure such as NaHg. Extensive computer modelling shows that even when more than one atomic separation and more than one pair of states contributes to the absorption at ν, the temperature dependence of absorption can be subsumed into a single exponential dependence. This is still useful for discharge modelling but gives insufficient information to sort out potentials. The method is also applicable to obtaining the free-free absorption coefficients for molecules such as SnCl. However, in such cases the dependence of n_A and n_B on the temperature must also be known from the chemical thermodynamics of the system.

CONCLUSIONS

The calculation of extreme wing broadening of spectral lines, or the application of the classical Franck-Condon principle to molecular bands, techniques which were developed to understand excimer lasers, yield a simple means for calculation of the absorption coefficients for discharge lamps. The data required are interatomic potentials for the various electronic states involved, and the variation of the transition probability with internuclear separation. Some of these data can be obtained from classical molecular spectroscopy, but we need to encourage large scale molecular structure calculations of relevant species. Meanwhile an experimental method for obtaining absorption coefficients from discharge lamp intensities is being investigated. This work shows that free-free molecular emission can play an important role in the high temperature regions of discharge lamps.

ACKNOWLEDGMENTS

I am particularly grateful to Drs. A. Gallagher, E. L. Lewis and B. F. Jones, and M. J. Davis for helpful discussions.

REFERENCES

Alkemade, C. Th. J., Hollander, Th., Snelleman, W. and Zeegers, P. J. Th, 1982, "Metal Vapors in Flames", Pergamon, Oxford.
Allard, N. and Kielkopf, J., 1982, Revs. Mod. Phys., 54:1103.
Armstrong, B. H. and Nicholls, R. W., 1972, "Emission, Absorption and Radiation from Heated Atmospheres", Pergamon, Oxford.
Davidson, N., 1962, "Statistical Mechanics", McGraw-Hill, New York.
Fischer, E., 1974, J. Appl. Phys., 45:3365.
Gallagher, A., 1975, in: "Atomic Physics IV", G.zu Pulitz, E. W. Weber, and A. Winnacker, eds., Plenum, New York.
Gallagher, A., 1979, in: "Excimer Lasers: Topics in Applied Physics", Vol. 30, Ch. K. Rhodes, ed., Springer, Berlin.
Gaydon, A. G., 1968, "Dissociation Energies and Spectra of Diatomic Molecules", 3rd edition, Chapman and Hall, London.
Griem, H. R., 1974, "Spectral Line Broadening in Plasmas", Academic Press, New York.
de Groot, J. J., 1974, "Investigation of High-Pressure Sodium and Mercury Tin Iodide Arc", Ph.D. Thesis, Technical University, Eindhoven.
Hedges, R. E. M., Drummond, D. L. and Gallagher, A., 1972, Phys. Rev. A., 6:1519.
Herzberg, G., 1950, "Molecular Structure and Molecular Spectra," Van Nostrand, Princeton.
Hindmarsh, W. R. and Farr, J. M., 1972, Prog. Quantum Elect., 2:143.
Jeung, G., 1983, J. Phys. B: At Mol Phys., 16:4289.
Jones, B. F. and Mottram D. A. J., 1981, J. Phys. D: Appl. Phys., 14:1183.
Jongerius, M. J. Hollander, Tj and Alkemade, C. Th. J., 1981, J. Quant. Spectrosc. Radiat. Transfer, 26:285.
King, G. W. and van Vleck, J. H., 1939, Phys. Rev., 55:1165.
Konowalow, D. D., Rosenkrantz, M. E. and Olson, M. L., 1980, J. Chem. Phys., 72:2612.
Konowalow, D. D., Rosenkrantz, M. E. and Hochhauser, D. S., 1983, J. Mol. Spect., 99:321.
Lam, L. K., Gallagher, A. and Hessel, M. M., 1977, J. Chem. Phys., 66:3550.
Laskowski, B. C., Langhoff, S. R. and Stallcop, J. R., 1981, J. Chem. Phys., 75:815.
Lowke, J. J., 1979, J. Appl. Phys., 50:147.
Mück, G., 1973, in: Proceedings, 11th Int. Conf. Phen. Ionized Gases, Prague.
Mück, G. and Popp, H-P., 1973, in: "Proceedings, 11th Int. Conf. Phen. Ionized Gases," Prague.
Mucklejohn, S. A., 1985, Private Communication.
Pfender, E., 1978, in: "Gaseous Electronics", Vol. 1., M.N. Hirsh and H. J. Oskam, eds., Academic Press, New York.
Phelps, A. V., 1972, "Tunable Gas Lasers Utilizing Ground State Disslocation", Joint Institute for Laboratory Astrophysics Report 110.
Rehder, L., Fischer, E. and Lorenz, R., 1973, in: "Proceedings, 11th Int. Conf. Phen. Ionized Gases," Prague.
Richter, J., 1968, in: "Plasma Diagnostics", W. Lochte-Holtgreven, ed., North Holland, Amsterdam.
Stromberg, H. P. and Schafer, R., 1983, J. Appl. Phys., 54:4338.
Szudy, J. and Baylis, W., 1975, J. Quant. Spectrosc. Radiat. Transfer, 15:641.
Tatum, J. B., 1967, Astrophysical Journal Supplement Series, 14:21.
West, W. P., Shuker, P. and Gallagher, A., 1978, J. Chem. Phys., 68:3864.

Whittaker, F. L., 1972, "The Spectrum of the High-pressure Sodium Lamp", Thorn Lighting Limited, Technical Report, LRD2005.
Woerdmann, J. P., 1981, J. Chem. Phys., 75:5577.
Woerdmann, J. P. and de Groot, J. J., 1982, in: "Metal Bonding and Interaction in High Temperature Systems", J. L. Gole and W. C. Stwalley, eds., American Chemical Society, Washington.
Zollweg, R., Lowke, J. J. and Liebermann, R. W., 1975, J. Appl. Phys., 46:3828.

SELF-REVERSED EMISSION LINES IN INHOMOGENEOUS PLASMAS

D. Karabourniotis

Department of Physics
University of Crete
Iraklion, Crete, Greece

INTRODUCTION

Information about a plasma can be obtained using emission spectroscopy methods, i.e., from the measurement of the radiation generated within the plasma and emitted by it. The line profile emitted by a volume element of a plasma depends on the properties of both the isolated radiating atom and the plasma in the immediate environment of the radiator. The observed line contour outside the luminous source is the superposition of different line profiles corresponding to the radiation emitted in volume elements along a given optical path.

In an inhomogeneous discharge plasma with cylindrical symmetry, the radial variation of the temperature, the radiating atom density, and the charged particle density causes large changes in both the broadening and the shift of the line profile of a thin layer along the radial direction and in the parameters of the resulting contour.

Furthermore, a part of the line radiation is trapped in the plasma by atoms of the same kind that cause the emission. This phenomenon of absorption is called self-absorption. A line for which the self-absorption of radiation is important is said to be optically thick. If the self-absorption can be neglected, the line is said to be optically thin; this is generally the case for spectral lines arising from transitions between highly excited levels. Assuming that the outer plasma layers are cooler than the interior, it may happen that the absorption at the center of the line is so much stronger than in the wings that the emerging line is darker in the center than in the wings. When this happens we speak of self-reversal which is therefore only an extreme case of self-absorption (Cowan and Dieke, 1948).

In an inhomogeneous high-pressure discharge plasma the most important spectral lines undergo considerable self-reversal. The contour of a self-reversed line may contain valuable information about the radial variation of the physical conditions in the plasma from which the line originates and the nature of the initial line profile (i.e., undistorted by reabsorption). Thus, the study of these lines opens up the possibility of assessing the plasma structure (temperature, concentration, and spatial distribution of emitting and absorbing atoms) and of ascertaining the broadening constants and the transition probabilities of the lines. However, an analysis of these lines which allows one to take into account the inhomogeneity of the plasma and the radial variation of the line broadening and line shift is a rather complicated problem and suitable for application only in particular cases. In the following we will first try to describe the formation of the line contour under reabsorption conditions by considering the plasma properties and the line broadening mechanisms. Furthermore, we will show how it is possible in the case of self-reversed line contours to obtain experimental evaluation of the broadening constants and of the transition probabilities by comparison of experimental and theoretical results.

Secondly, a survey of Bartels' theory for the treatment of self-reversed line contours will be presented and the approximation conditions necessary for the application of Bartels' model to plasma diagnostics will be given.

Special emphasis will be finally placed on the presentation of Cowan and Dieke's model for the self-absorption of spectral lines. Recent extensions to this model and its applicability to the analysis of self-reversed line contours emitted by inhomogeneous discharge plasmas will be reviewed. Most of the application examples given below concern high pressure mercury arc discharges.

FORMATION OF THE LINE PROFILE

Radiation Intensity of a Spectral Line

The spectroscopically measured line profile is related to the radiation intensity $I(\nu, x)$ at frequency ν within the line in a direction x at point x (measured along the line of sight), defined as the radiant energy flowing per unit surface perpendicular to the x direction, within a unit solid angle in the x direction, per unit time and frequency interval

$$I(\nu, x) = \frac{dE}{dt\, d\nu\, d\Omega\, dS} . \qquad (1)$$

The variation $dI(\nu,x)/dx$ of the intensity along the x direction at point x per unit length (Fig. 1) depends on the emission and the reabsorption of the radiation.

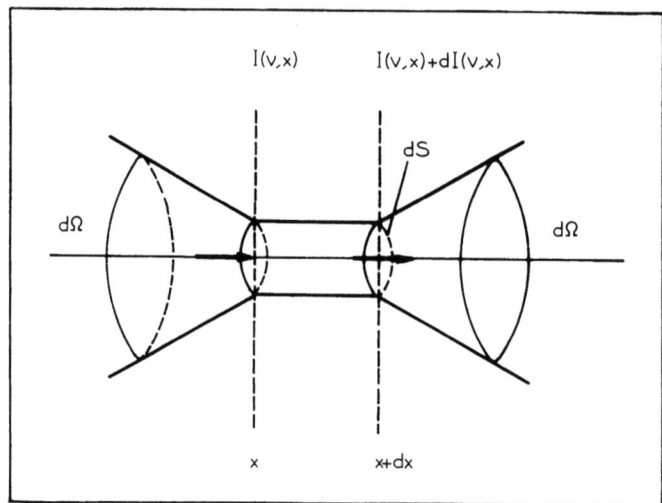

Fig. 1. Geometric volume-element considered in the derivation of the variation of intensity per unit length.

If $(dI(\nu,x)/dx)_{em}$ is the increase of the intensity in the x direction per unit length and $(dI(\nu, x)/dx)_{ab}$ is the diminution of the intensity due to absorption of the radiation traversing the unit length in the x direction, then the variation of the radiation intensity $I(\nu,x)$ per unit length is:

$$\frac{dI(\nu, x)}{dx} = \left[\frac{dI(\nu, x)}{dx}\right]_{em} + \left[\frac{dI(\nu, x)}{dx}\right]_{ab} \quad (2)$$

The interpretation of the measured line intensity and line shape at the plasma surface depends on the degree of plasma equilibrium.

Local Thermodynamic Equilibrium

Thermodynamic equilibrium is said to prevail in a plasma when the collisional processes dominate the population and depopulation of the excited levels. Thus a unique value of temperature is sufficient to describe the distribution of population among energy levels within a given species by Boltzmann's equation, and the ionization distribution by Saha's equation (Cooper, 1968).

The Boltzmann equation can be written in the form

$$\frac{n_e}{g_e} = \frac{n_a}{g_a} \exp\left(-\frac{E_e - E_a}{kT}\right) = \frac{n_0}{Z(T)} \exp\left(-\frac{E_e - E_0}{kT}\right) \quad (3)$$

where n_0 refers to the total density of (neutral) atoms of the particular species and the partition function $Z(T)$ is given by:

$$Z(T) = g_0 \left[1 + \sum_j \frac{g_j}{g_0} \exp\left(-\frac{E_j - E_0}{kT}\right)\right] \quad (4)$$

The Saha equation is written

$$\frac{n_i n_{e\ell}}{n_0} = \frac{(2\pi mkT)^{3/2}}{3} \frac{2Z_i}{Z} \exp\left(-\frac{E_i - \Delta E_i}{kT}\right), \quad (5)$$

where subscripts i and eℓ refers to ions and electrons, E_i is the ionization energy and ΔE_i is the lowering of the ionization energy.

In situations where the thermodynamic equilibrium is established within each small volume element, although plasma temperature is a slowly varying function of the coordinates, the plasma is said to be in local thermodynamic equilibrium (LTE). The dimension of the small volume element must be chosen large in comparison with the mean free path but small enough so that the temperature can be assumed to be constant within it.

For example, in a steady-state discharge, the mean free path of the electrons is determined (Lochte-Holtgreven, 1968) by the formula:

$$\frac{T_e - T_g}{T_e} = \frac{(deE)^2}{(\frac{3}{2} kT_e)^2} \frac{m}{4m_e}, \quad (6)$$

where d is the mean free path between two successive collisions of an electron of mass m_e with particles of mass m, E is the field strength, e is the electron charge, and T_e and T_g are the electron and the gas temperatures. Thus, for a high pressure mercury discharge with $T \approx 6000K$, $E \approx 15V/cm$, and $T_e - T_g \approx 100K$ (Gurevich and Podmoshenskii, 1963), one obtains $d \approx 1.5 \times 10^{-5}$ cm.

The population density of an excited level is determined by the local value of the electron density and temperature. In the case of an optically thin line this is valid (LTE) when the probability of de-excitation of the upper transition level by inelastic collision is much larger than by spontaneous emission. Neglecting absorption and stimulated emission this equilibrium condition is expressed (Karabourniotis, 1977) by the inequality:

$$n_{e\ell} \gg n_e^* = \frac{\Sigma_a A_{ea}}{\Sigma_a Z_{ea}} \tag{7}$$

where A_{ea} is the transition probability for emission and Z_{ea} is the average number of inelastic collisions per incident particle per unit time, which lead an atom from state e to state a (excitation rate). Thus, for the mercury $e=7^3S_1$ level one has:

$$\sum_a A_{ea} = 1/\tau_{7^3S_1} = 9.26 \times 10^7 \text{ s}^{-1}$$

where $\tau_{7^3S_1} = 1.05$ ns (Verolainen and Oscherovich, 1966) is the mean life time of this level and $a=6^3P_2$, 6^3P_1 and 6^3P_0. The values of Z_{ea} have been obtained from the curves $Z_{ae} = f(T_e)$ (Fig. 2) given by Cayless (1959) and known as the Klein-Rosseland principle:

$$Z_{ea} = Z_{ae} \frac{g_a}{g_e} \exp\left(\frac{E_e - E_a}{kT}\right) . \tag{8}$$

Table 1 gives the data used for T=6500K. The obtained value of $n_{e\ell}^*$ is:

$$n_{e\ell}^* = 1.4 \times 10^{15} \text{ cm}^{-3} .$$

Drawin (1969) gives the following general condition for the validity of LTE for optically thin plasmas:

$$n_{e\ell}^* = 6.5 \times 10^{16} \frac{g_e}{g_\alpha} \left(\frac{E_e - E_\alpha}{E_H}\right)^3 \left(\frac{kT}{E_H}\right)^{1/2} \phi(x) \tag{9}$$

where E_H (= 13.58 eV) is the ionization energy for the hydrogen atom and $x = (E_e - E_\alpha)/kT$. Table 2 gives the data used and the critical values of $n_{e\ell}^*$ obtained following the Drawin criterion for the mercury 4916, 5770, and 4047Å lines.

Griem's criterion (Griem, 1963) for the validity of the Boltzmann statistics in the case of hydrogen-like atoms for levels with critical number n and upwards is given by

$$n_{e\ell}^* = 7 \times 10^{18} \left(\frac{kT}{E_H}\right)^{1/2} n^{-8.5} . \tag{10}$$

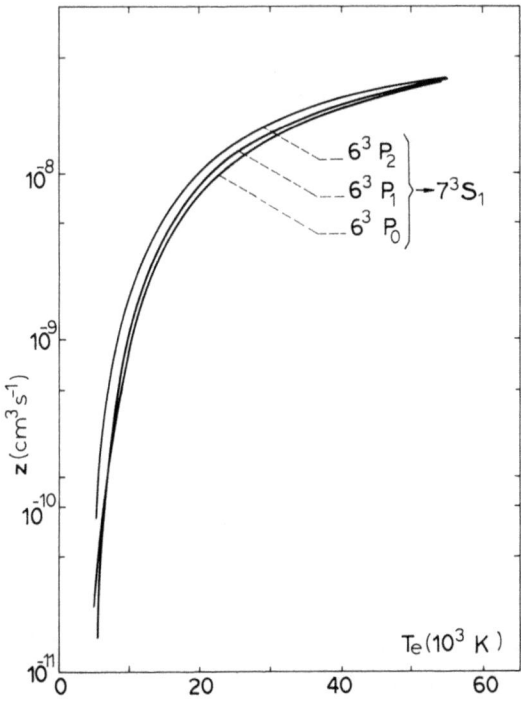

Fig. 2. Excitation rates for the transitions 6^3P_2, 6^3P_1, $6^3P_0 \rightarrow 7^3S_1$ versus electron temperature (Cayless, 1959).

Table. 1. Data used for the calculation of the critical electron density n_e (Eq. (7)) for T=6500K, concerning the mercury level 7^3S_1.

Transition	$6^3P_2 \rightarrow 7^3S_1$	$6^3P_1 \rightarrow 7^3S_1$	$6^3P_0 \rightarrow 7^3S_1$
E_{ea} (eV)	2.27	2.84	3.063
g_a	5	3	1
Z_{ae}	3.86×10^{-10}	1.45×10^{-10}	9.4×10^{-11}
Z_{ea}	3.72×10^{-8}	2.32×10^{-8}	7.46×10^{-9}
$\Sigma_a Z_{ea} = 6.78 \times 10^{-8}$ cm^3 s^{-1}			

Table. 2. Data used and critical electron densities for the mercury lines 4047, 5770 and 4916 Å, following Drawin's criterion (Eq. 9)

Line (Å)	4916	5770	4047
Transition	$8^1S_0 \to 6^1P_1$	$6^3P_2 \to 6^1P_1$	$7^3S_1 \to 6^3P_0^0$
E_{ea} (eV)	2.522	2.149	3.063
X	4.50	3.84	5.47
$\phi(X)$	5.1	4.5	6.15
g_e	1	5	3
g_a	3	3	1
n_{e1} (cm^{-3})	1.4×10^{14}	3.9×10^{14}	2.8×10^{15}

For atoms not having hydrogen-like characteristics, this criterion can be expected to be valid when the quantum number n is identified with an effective quantum number n* of the level under consideration (Szymanski, 1979). For $n^*[E_H/(E_i-E_n)]^{1/2}$ where E_n is the energy of the level n, one obtains for the mercury level 7^3S_1 with T=6500K

$$n^*_{e\ell} = 1.5 \times 10^{15} \text{ cm}^{-3}$$

Further experiments (Assous, 1980) in a mercury-lead iodide arc lamp have shown that the populations of the two upper 6^1D_2 and 6^1P_2 levels of the 3906 and 5790 Å Hg lines have a Boltzmann distribution when the electron density is above 4×10^{15} cm^{-3}. Departures from this distribution occur when the temperature falls below 0.25 eV (~ 2900K) with an electron density around 2×10^{15} cm^{-3}.

For equilibrium between the ground state and the first excited level in an optically thin plasma, Griem's calculation leads to:

$$n^*_{e\ell} = 9 \times 10^{17} \left(\frac{kT}{E_H}\right)^{1/2} \left(\frac{E_e}{E_H}\right)^3 \qquad (11)$$

Thus one has for the 6^3P_1 level of Hg at 6500 K:

$$n^*_{e\ell} = 8.3 \times 10^{15} \text{ cm}^{-3},$$

and for the $2P_{3/2}$ level of Na at 4000 K:

$$n^*_{e\ell} = 5 \times 10^{14} \text{cm}^{-3}.$$

In the case of optically thick plasma where the optical depth is so large that almost all of the radiation is trapped in the plasma, the mean free path of the photons is small, the radiative processes as well as the collisional ones occur in detailed balance, and the population densities of the excited levels depend solely on the local values of temperature and density. Criterion (11) may then be relaxed by about an order of magnitude.

For the mercury resonance line at 2537 Å, the mean free path of the resonance photons (Damelincourt et al., 1977) in the core of the line is very short (on the order of 2×10^{-7} cm) and the value of the absorption oscillator strength (Barrat, 1959) is relatively low (f=0.0028), so that we have to expect a fairly strong reversal, no large broadening, and high a degree of self-absorption.

For the sodium resonance line at 5890 Å, emitted by a high pressure sodium arc (Ozaki, 1971a), the high value of the oscillator strength (f = 0.624) implies a large broadening of the line. In the core of the self-reversed resonance line the optical path is very large, about 10^5, and in both lateral peaks the optical depth is on the order of unity (Ozaki, 1971b). This allows the resonance photons at the lateral peaks of the line, whose mean free path is comparable to the discharge radius, to escape from the arc. Therefore, as far as the region where the optical thickness is of order unity in the resonance line, the LTE approximation does not hold in the entire wavelength range of the resonance line.

In a high pressure arc discharge, the decrease of the temperature and of the electron density as a function of the radius implies deviations from LTE. Figure 3 shows the calculated temperatures as a function of ρ, where $\rho=r/R$ in a Na-Xe discharge (Waszink, 1975). It results in a significant deviation from LTE for $\rho > 0.6$ in terms of the difference between electron and gas temperatures. Figure 4 shows the radial and temporal variation of the electron density in a high pressure (~3atm) mercury discharge operated at 100 W/cm (50Hz). The electron density has been obtained from the measured temperature profile using the Saha equation, and from the measured intensity of the continuum radiation (Damelincourt et al., 1978b). The difference in the electron densities estimated from the two kinds of measurements increases with a decrease in the electron density (i.e., is greatest at the outer region of the discharge and the time zero of the electric field).

Emission Coefficient and Absorption Coefficient

The radiation is described in terms of the Einstein A and B coefficients. For this discussion we will use the representation of Fig. 5. In terms of isotropic intensity, the emission coefficient $\varepsilon(\nu)$ is defined as the power radiated per unit volume per unit solid angle and per unit frequency interval:

$$\varepsilon(\nu) = \frac{h\nu}{4\pi} A_{ea} n_e P_e(\nu) \tag{12}$$

Here A_{ea} is the transition probability for spontaneous emission and n_e is the population density of the emitting level.

The emission line profile $P_e(\nu)$ characterizes the probability of emission in the interval ν, $\nu+d\nu$ within the line and satisfies the relations:

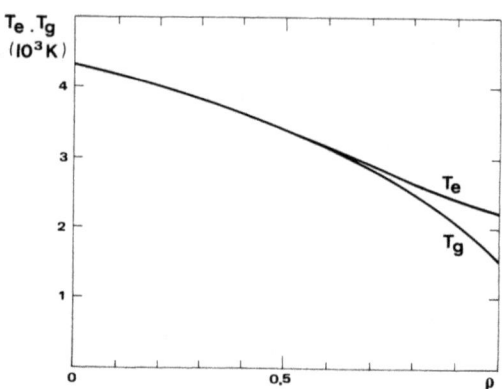

Fig. 3. Calculated electron and gas temperatures as a function of $\rho=r/R$, for $T(0)=4225K$, $T(R)=1500$ K, $P_{Na}=0.36\times10^5$ Nm^{-2}, $P_{Xe}=0.23\times10^5$ Nm^{-2}, R=3.75mm (Waszink, 1975)

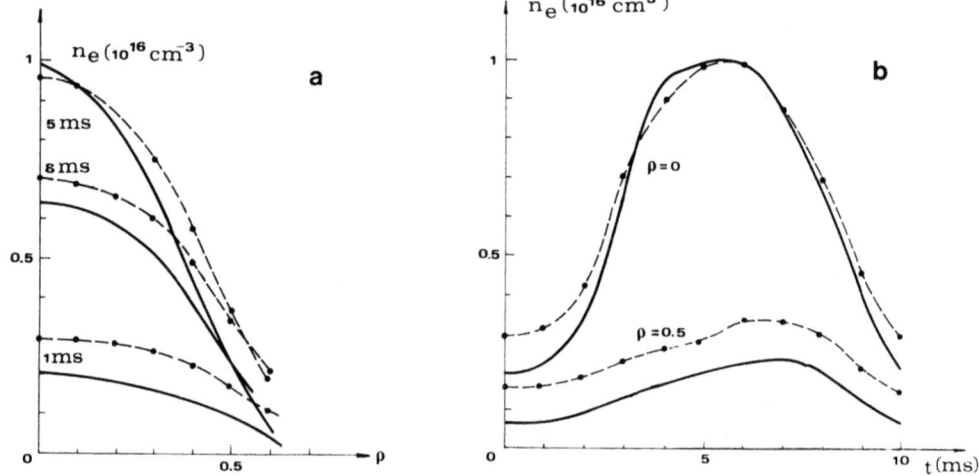

Fig. 4. Radial (a) and temporal (b) variation of the electron density in a mercury discharge operated in ac (50Hz) at 100 W/cm, with P_g=3atm, R=1cm, T(R) = 1000 K, and T(0)=6410K at 5ms and 5500K at 0 ms (Damelincourt et al., 1978b)

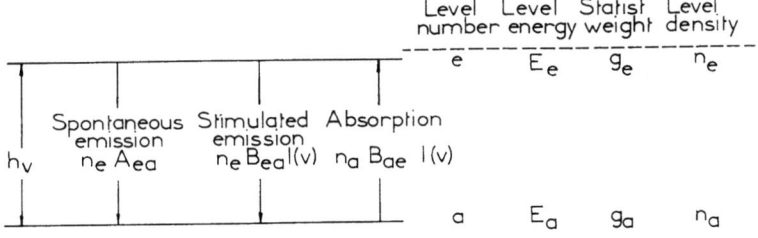

Fig. 5. Notation scheme.

$$\int_{line} P_e(\nu)d\nu = 1 \tag{13}$$

and

$$dn_{e;\nu} = n_e P_e(\nu)d\nu \tag{14}$$

where $dn_{e;\nu}$ is the density of levels which emit the frequency ν within the line. The absorption and the stimulated emission depend on the intensity of the radiation and on the coefficients B_{ae} and B_{ea}. They are defined such that an atom in state a absorbs a photon $h\nu(=h\nu_{ea})$ and $B_{ea} \cdot I(\nu)$ is the probability per unit time that an atom in state e is induced to emit a photon $h\nu$.

For an isotropic intensity distribution, the photon flux per unit frequency range is $4\pi I(\nu)/h\nu$. Therefore, the intensity decrease is expressed by:

$$k'(\nu) I(\nu) = \frac{h\nu}{4\pi} \left[n_a B_{ae} P_a(\nu) - n_e B_{ea} P_e(\nu) \right] I(\nu) \tag{15}$$

where n_a is the population density of the absorbing level. The absorption line profile $P_a(\nu)$ characterizes the probability of absorption in the interval ν, $\nu+d\nu$ within the line and satisfies the relations

$$\int_{line} P_a(\nu)d\nu = 1 \tag{16}$$

and

$$dn_{a;\nu} = n_a P_a(\nu)d\nu. \tag{17}$$

where $dn_{a;\nu}$ is the density of levels which absorb the frequency ν within the line.

Thus, the function $k'(\nu)$, defined as the absorption coefficient, expresses the percentage of the intensity decrease per unit volume and can be written in the form:

$$k'(\nu) = \frac{h\nu}{4\pi} n_a B_{ae} P_a(\nu) \left[1 - \frac{n_e}{n_a} \frac{B_{ea}}{B_{ae}} \frac{P_e(\nu)}{P_a(\nu)} \right]. \tag{18}$$

In the above equation we have considered the profile for stimulated emission to be identical to that for spontaneous emission (Oxenius, 1967), and the resonance fluorescence to be negligible (Heitler, 1954).

Close to LTE, where Boltzmann statistics are valid, one has:

$$P_e(\nu) = P_a(\nu) = P(\nu) \tag{19}$$

$$g_a B_{ae} = g_e B_{ea} \tag{20}$$

and:

$$\frac{A_{ea}}{B_{ae}} = \frac{g_a}{g_e} \frac{2h\nu^3}{c^2}. \tag{21}$$

Thus, for $n_e \ll n_a$, it is possible to neglect stimulated emission, and the absorption coefficient is written in the often used form:

$$k(\nu) = (h\nu/c) B n_a P(\nu) \tag{22}$$

where B is the Einstein transition probability for absorption, related to the Milne transition probability B_{ae} by $B = (c/4\pi) B_{ae}$.

Line Profiles

The line profile is described by the distribution function $P(\nu)$. It depends, apart from the temperature and the atom and electron concentrations, upon the constants that characterize the interaction between the radiating atom and the neighboring particles, and upon the broadening mechanisms. The line profile broadening and shift can arise from a number of causes. We shall consider here the most important forms of $P(\nu)$ which can take place in a high pressure discharge.

Doppler broadening. Because radiating atoms in a gas discharge have various velocities, the observed frequency depends upon the particular velocity of the radiating atom. Assuming thermodynamic equilibrium, the distribution of the emitted frequency is given by:

$$P(\nu) = \left(\frac{\ln 2}{\pi\delta}\right)^{1/2} \exp\left[-\frac{\ln 2}{\delta^2} (\nu-\nu_0)^2 \right] \tag{23}$$

where δ is the half-width at half the maximum amplitude:

$$\delta = \left[\ln 2 (2kT\nu_0^2)/mc^2) \right]^{1/2}. \tag{24}$$

Pressure broadening. The major source of line broadening in high pressure discharges is the interaction of the radiating atom with surrounding particles (atoms, electrons, ions).

Impact approximation. The essential idea of impact broadening is very simple. The collisions which cause broadening are supposed to occur successively (binary approximation) and to occupy a typical time small compared to the mean time between collisions. This condition leads to the relation (Sobelman et al., 1981):

$$h \equiv 2\pi\rho_0^3 n \ll 1 \qquad (25)$$

where:

$$\rho_0 = \left(\alpha_p \frac{C_p}{\bar{v}}\right)^{1/(p-1)} \qquad (26)$$

is the effective interaction radius (Weisskopf radius), $\bar{v}$ is the mean velocity of the perturbers, n is their density, and p, α_p and C_p are the characteristic parameters of the interaction with $\alpha_4 = (1/2)\pi$ and $\alpha_6 = (3/8)\pi$.

Table 3 gives the values of h for the electrons (h_{4e}), the ions (h_{4i}) and the neutral atoms (h_6) of mercury for two temperatures and with atom and electron densities approximately equal to 2×10^{18} cm^{-3} and 10^{16} cm^{-3}, respectively. For the mercury triplet lines, the values $C_4 = 2.73 \times 10^{-14}$ rad cm^{4}s^{-1} and $C_6 = 2.32 \times 10^{-30}$ rad cm^{6}s^{-1} have been used (Karabourniotis, 1977), where C_4 is the Stark constant and C_6 the van der Waals constant. The results show that the inequality (25) is fulfilled and therefore, at the conditions given above, corresponding to a mercury arc discharge at a pressure about 2 atm, impact broadening plays a decisive role. The frequency distribution is described in the impact theory by a dispersion (Lorentzian) profile:

$$P(\nu) = (\delta/\pi) / \left[(\nu-\nu_0-\Delta)^2 + \delta^2\right] \qquad (27)$$

where Δ is the line shift

<u>Electron impact broadening</u>. The line broadening and shift due to collisions with electrons, according to the Lindholm-Foley collision theory, (Traving, 1968) is given (in rads^{-1}) by:

$$2\delta_4 = 11.37 \; C_4^{2/3} \; \bar{v}_{e\ell}^{1/3} n_{e\ell} \qquad (28)$$

and:

$$\eta_4 = \Delta_4/\delta_4 = -1.73 \qquad (29)$$

Table 3. Values of h from Eq. (26) for the electrons (h_{4e}), the ions (h_{4i}) and the neutral atoms (h_6)

h \ T(K)	4000	6000
h_{4e}	3.45×10^{-5}	2.8×10^{-5}
h_{4i}	1.47×10^{-2}	1.2×10^{-2}
h_6	1.2×10^{-2}	1.1×10^{-2}

It is possible to calculate the broadening and shift of isolated lines due to electrons and ions in a plasma using the Sahal-Brechot (1969a, 1969b, 1974) quantum-mechanical impact approximation. A short survey of the Sahal-Brechot model and its application to the mercury visible triple lines has been given by Karabourniotis (1977). The results obtained (Fig. 6) are practically the same for the three lines and can be expressed in analytical form (Damelincourt et al., 1983a) by:

$$2\delta_{ei}/n_e = C_e T^{\gamma_e} + C_i T^{\gamma_i} \tag{30}$$

$$\eta_e = \Delta_{ei}/n_e = (\beta_e T^{\beta_e} + B_i T^{\beta_i}) \tag{31}$$

where $2\delta_{ei}$ and Δ_{ei} are expressed in rads^{-1}, T in K, and n_{ei} in cm^{-3}.

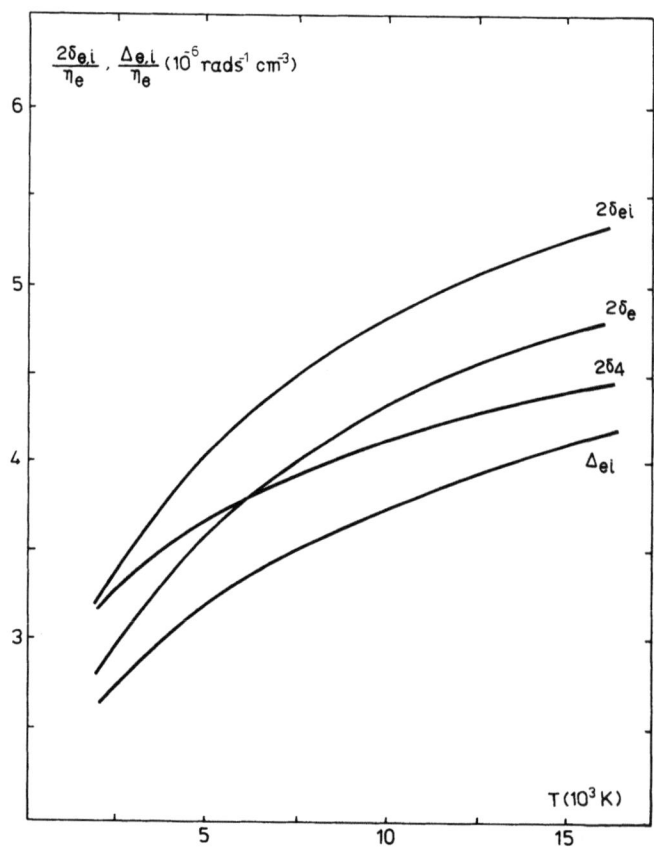

Fig. 6. Variation of the line broadening and shift due to electrons and ions as a function of the temperature. δ_{ei}, δ_e, Δ_{ei} line broadening and shift following the Sahal-Brechot model; δ_4 line broadening due to electrons following the Lindholm Foley collision theory (Karabourniotis, 1977)

Atom-impact broadening. The Lennard-Jones interaction potential between neutral atoms is given by (Hindmarsh and Farr, 1972):

$$V(r) = \hbar(C_{12}r^{-12} - C_6 r^{-6}) \tag{32}$$

where C_6 and C_{12} are constants. Using this model, the impact approximation gives for the total half-width (in rad s^{-1}):

$$2\delta_n = 8\pi \left(\frac{3\pi}{8}\right)^{2/5} n_0 \bar{v}^{-3/5} C_6^{2/5} B(\alpha) \tag{33}$$

and for the shift/width ratio:

$$\eta_n = \frac{S(\alpha)}{2B(\alpha)} \tag{34}$$

in which:

$$\alpha = \frac{63\pi}{256} \left(\frac{8}{3\pi}\right)^{11/5} \bar{v}^{6/5} C_{12}^{11/5}/C_6 \tag{35}$$

where n_0 is the density of the neutral atoms, and $\bar{v}$ is the mean relative velocity of the atoms. The functions $B(\alpha)$ and $S(\alpha)$ are given in Fig. 7.

Omitting the repulsive part of the potential curve, i.e., taking a van der Waals potential, $V(r) = \hbar C_6 r^{-6}$, one has:

$$2\delta_n = 8.08 C_6^{2/5} \bar{v}^{3/5} n_0 \tag{36}$$

and:

$$\eta_n = -0.7 . \tag{37}$$

Quasistatic approximation. The profile of a line according to the quasistatic approximation is given by

$$P(\omega) = \frac{3}{p} \left(\frac{\overline{\Delta\omega}}{\omega-\omega_0}\right)^{(p+3)/p} \exp\left[-\left(\frac{\overline{\Delta\omega}}{\omega-\omega_0}\right)^{3/p}\right] \frac{1}{\overline{\Delta\omega}} . \tag{38}$$

with:

$$\overline{\Delta\omega} = C_p r_0^{-p} \tag{39}$$

and:

$$r_0 = \left(\frac{4}{3}\pi n\right)^{-1/3} \tag{40}$$

r_0 being the mean distance between the particles and n their density.

For a line broadened by the quadratic Stark effect, p= 4; for a line broadened by van der Waals interaction p= 6. In the latter case the line width is given by:

$$2\delta = 0.822 \ \pi^3 C_6 n^2 \tag{41}$$

and the shift by:

$$\Delta = (8/27) \ \pi^3 C_6 n^2 . \tag{42}$$

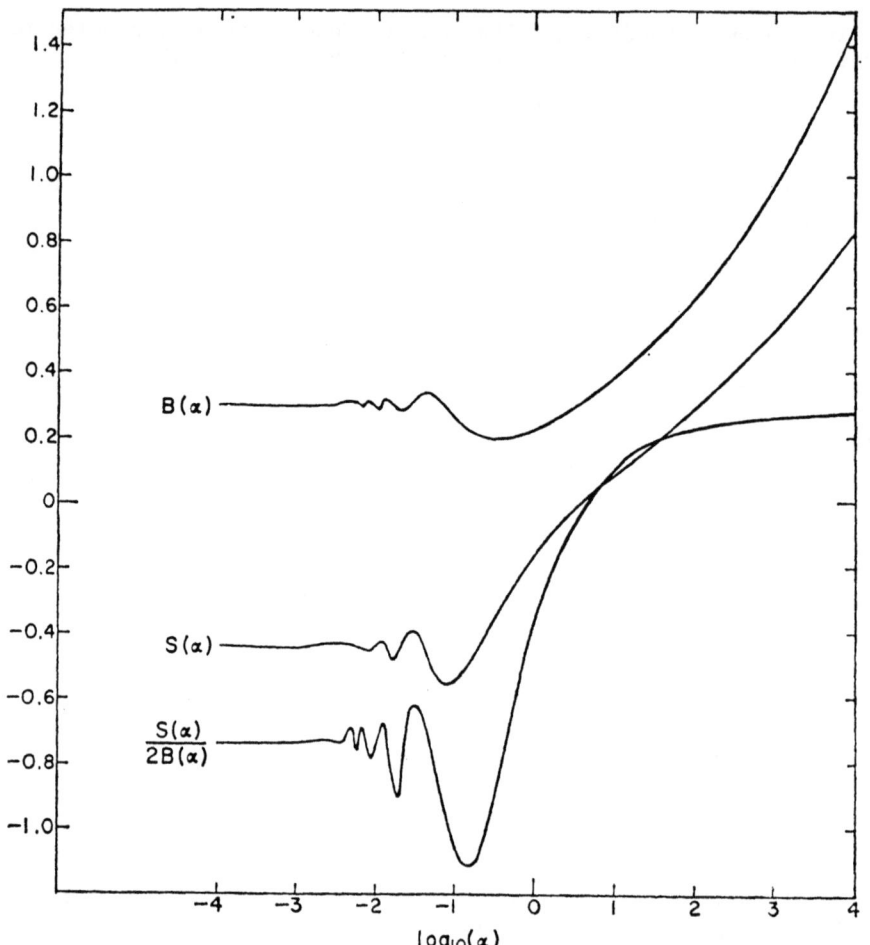

Fig. 7. The function $S(\alpha)$, and $B(\alpha)$ and $S(\alpha)/2B(\alpha)$ computed for a Lennard-Jones potential (Hindmarsh and Farr, 1972)

At high pressures and low velocities, when the inequality (25) is broken, i.e. when

$$h = 2\pi\rho_0^3 n \simeq 1 \qquad (43)$$

the impact approximation is inapplicable.

In fact relation (43) means that the Weisskopf radius ρ_0 is approximately equal to the mean distance r_0 between pertubers, i.e.,

$$\rho_0 = r_0 \qquad (44)$$

Obviously the expression (38) is applicable to the wing of a line and the quasistatic contribution becomes significant for:

$$\omega - \omega_0 > \Delta\bar{\omega} = \frac{2\pi c}{\lambda^2} \overline{\Delta\lambda} \qquad (45)$$

where from equations (26) and (38):

$$\Delta\bar{\omega} = (\bar{v})^{p/(p-1)} / (\alpha_p C_p)^{1/(p-1)}. \qquad (46)$$

Table 4 gives the values of $\overline{\Delta\lambda}$ in the case of the mercury visible triplet lines for T = 4000 and 6000 K. Table 4 also shows the limits of applicability of the impact approximation. Taking a Lennard-Jones interaction the impact boundary on the high-wavelength side is not affected by the presence of the repulsive term, and that on the low-wavelength side is the order of 10 Å (Hindmarsh and Farr, 1972).

__Resonance broadening.__ During the collision of two identical atoms, one of which is excited, a resonance transfer of the excitation energy is possible. Spectral lines beginning (or ending) on the resonance level must be broadened as a result of this. The line profile is then described in the impact approximation by an unshifted dispersion function and the line width (in rad s^{-1}) is given by

$$2\delta_r = K_r \frac{\pi e^2}{m_e \ell \omega_0} f n_0 , \qquad (47)$$

where ω_0 and f are the frequency and the absorption oscillator strength of the resonance line.

Table 4. Limits of applicability of the impact approximation for collisions with electrons ($\overline{\Delta\lambda_e}$), ions ($\overline{\Delta\lambda_i}$) and atoms ($\overline{\Delta\lambda_n}$).

T(K)	$\overline{\Delta\lambda}$(nm)	546.1nm	435.8nm	404.7nm
4000	$\overline{\Delta\lambda_e}$	60.4	38.5	33.2
4000	$\overline{\Delta\lambda_i}$	0.019	0.012	0.01
4000	$\overline{\Delta\lambda_n}$	0.125	0.081	0.070
6000	$\overline{\Delta\lambda_e}$	79.15	50.4	43.5
6000	$\overline{\Delta\lambda_i}$	0.024	0.016	0.013
6000	$\overline{\Delta\lambda_n}$	0.159	0.104	0.090

Theoretical and experimental results concerning the value of the constant K_r have been collected by Perrin-Lagarde and Lennuier (1975). The experimentally obtained value from the same authors is:

$$K_r = 2.1 \pm 0.2.$$

<u>Comparison between broadening mechanisms</u>. For the sodium D-line doublet, the half widths due to different broadening mechanisms are of the following orders of magnitude (Ozaki, 1971b):

natural broadening	10^{-4} Å
Doppler broadening	10^{-3} Å
Stark broadening	$10^{-18} \times n_{e\ell}$ Å
resonance broadening	$10^{-18} \times n_0$ Å

where $n_{e\ell}$ is the electron density in cm^{-3} and n_0 the neutral atom density in cm^{-3}. In a high pressure sodium discharge with $n_{e\ell} \sim 10^{16}$ cm^{-3} and $n_0 \sim 10^{18}$ cm^{-3} the main broadening mechanism is resonance broadening.

Table 4 shows that in a high pressure mercury discharge, the impact approximation may describe the central part of the mercury triplet lines; as far as the total intensity is concerned the impact approximation gives correct values. Figure 8 compares the contribution of the different broadening mechanisms for the 5461 Å line in the central region of a mercury discharge ($\sim$ 3 atm) with 2 cm internal diameter, under ac operation (50 Hz) at $\sim$ 100 W/cm, at the moments of minimum (0 ms) and maximum (5 ms) emission corresponding to the minimum and maximum electron density, versus $\Delta\lambda(=\lambda-\lambda_0)$ (Damelincourt et al., 1983a).

The broadening due to charged particles is negligible at 0 ms. On the other hand, this kind of broadening at 5 ms is of the same order of magnitude as that due to neutral atoms. In all these figures one remarks that quasistatic neutral atom broadening becomes predominant compared to impact neutral atom broadening for $\Delta\lambda$ greater than 1Å. This result is in agreement with the limits of applicability for the impact and quasistatic approximation (set forth in Table 4).

Figure 9 shows the radial variation of the different contributions to the width of the 5461 Å line for the $\sim$3-atm mercury discharge at 0 and 5 ms (Damelincourt et al., 1983b). The corresponding temperature distributions are given in Fig. 10.

<u>Combination of the broadening mechanisms</u>. The combination of the different broadening effects is accomplished by convolution of the corresponding profiles. Let us introduce the notation:

L_n = Lorentzian profile due to neutral atoms;

S_n = quasistatic profile due to neutral atoms;

L_{ei} = Lorentzian profile due to charged particles (electrons and ions);

D = Doppler profile.

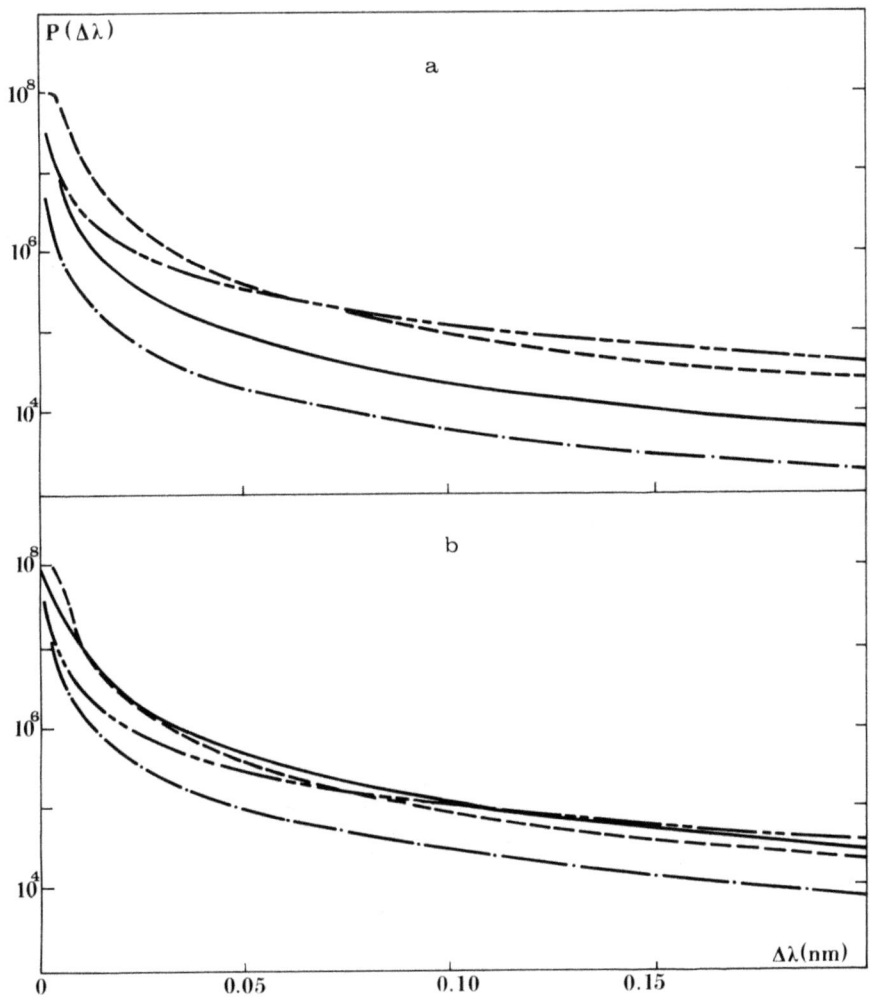

Fig. 8. Evolution of the red wing of the line Hg-5461 Å at the central region of a discharge at 0 (a) and 5 ms (b). Discharge parameters as for Fig. 4 Solid curves: electron impact approximation; dashed curves: neutral atom impact approximation; dot-dashed curves: ion quasistatic approximation; double dot-dashed curves: neutral atom quasistatic approximation. (Damelincourt et al., 1983a).

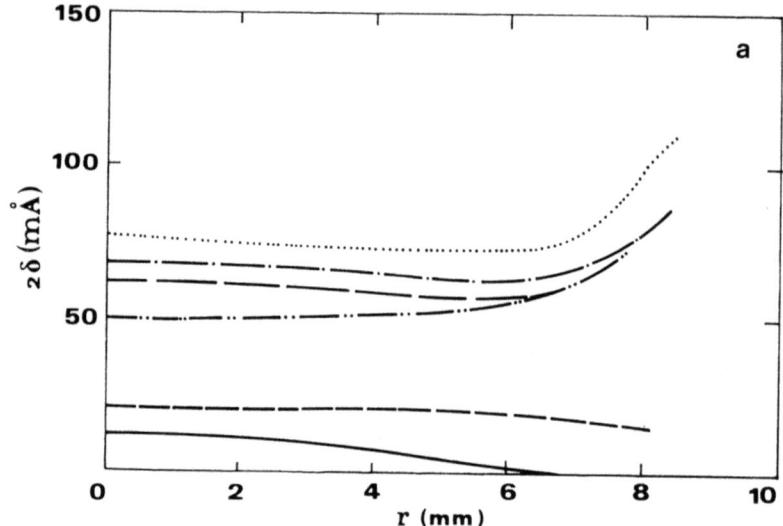

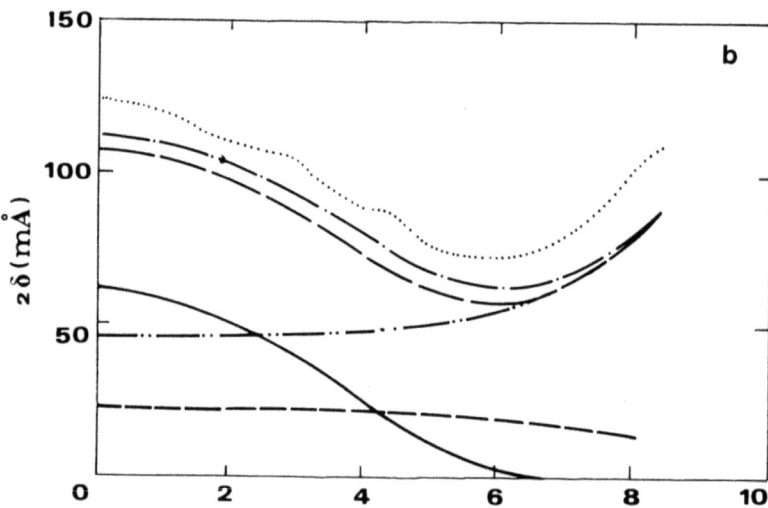

Fig. 9. Variation of the line width for the Hg-5461Å line as a function of the radius at 0 (a) and 5 ms (b). Discharge parameters as for Fig. 4.
——————: broadening due to charged particles; —————: Doppler broadening; —..—..—..—: broadening due to atoms in impact approximation; — — —: broadening due to electrons, ions and atoms in impact approximation; —.—.—.—: broadening of the Voigt profile (convolution of the profiles due to atoms, electrons and ions in impact approximation with Doppler profile); ..... total broadening (convolution of a Lorentzian, Doppler and quasistatic profile) (Damelincourt et al., 1983b).

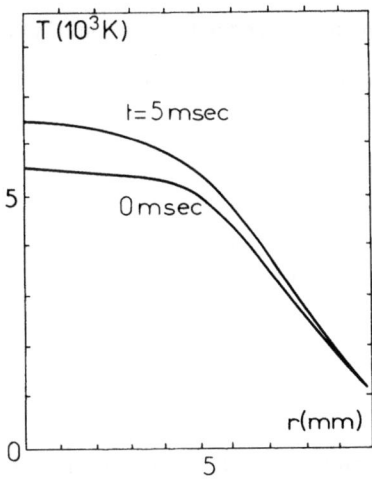

Fig. 10. Radial variation of the temperature at 0 and 5 ms for the discharge as in Fig. 4 (Damelincourt et al., 1983a).

Thus we will write the resulting profile function in the form

$$P = [(L_n * S_n) * L_{ei}] * D \tag{48}$$

where the asterisk (*) indicates the operation of convolution.

Stormberg (1980) has proved that the convolution of a Lorentzian profile (L) with a quasistatic profile (S) can be expressed in an analytical form:

$$L * S = L + C \tag{49}$$

where C is a corrective term. Therefore, Eq. (48) is expressed by:

$$P = (D * L) + (D * C) \tag{50}$$

with:

$$L = L_{ei} * L_n. \tag{51}$$

The convolution of a Doppler with a Lorentzian profile leads to the well-known Voigt profile:

$$V = D * L \tag{52}$$

We designate by C' the term $D * C$, and so equation (50) takes the form:

$$P = V + C' \tag{53}$$

This profile is normalized to unit area. The term C' does not have, however, an analytical expression. On the other hand, in comparison with C, D may be considered to be a Dirac distribution function. Thus one may write:

$$C' = D * C \approx C . \tag{54}$$

In this way Eq. (53) can be expressed in an analytical form.

Figure 11 gives the relative error E within the line profile caused by approximation (54) defined by:

$$E = (V+C) / (V+C') \qquad (55)$$

and the variation of the maximum relative error E_{max} (maximum value of Eq. (55)) as a function of the temperature. The radial variation of the broadening corresponding to the Voigt profile and that of the resulting profile (taken into account the quasistatic effect) for the 5461 Å line is given in Fig. 9.

Figure 12 gives the evolution of the line profile along a diameter of the 3-atm mercury discharge at the minimum (0 ms) and the maximum (5 ms) of the electron density. Around the arc axis and at 5 ms, the normalization condition and the relatively large broadening leads to a lowering of the profile intensity.

RADIATIVE TRANSFER AND SPECTRAL LINE CONTOUR

Radiative Transfer

From the definition of the emission and absorption coefficients in terms of an isotropic intensity distribution, it follows that:

$$\varepsilon(\nu,x) = \left(\frac{dI(\nu,x)}{dx}\right)_{em} \qquad (56)$$

and

$$k'(\nu,x) = -\frac{1}{I(\nu,x)}\left(\frac{dI(\nu,x)}{dx}\right)_{ab}. \qquad (57)$$

Thus, Eq. (2) takes the form:

$$\frac{dI(\nu,x)}{dx} = \varepsilon(\nu,x) - k'(\nu,x)I(\nu,x) \qquad (58)$$

which is known as the equation of radiative transfer.

Referring to Fig. 13, the solution of Eq. (58), with boundary condition $I(\nu,-x_0)=0$, gives:

$$I(\nu) = \int_{-x_0}^{+x_0} \varepsilon(\nu,x) \exp\left\{-\int_{x}^{+x_0} k'(\nu,x')\,dx'\right\} dx \qquad (59)$$

where $I(\nu)$ is the intensity radiated from a plasma column of unit cross-section contained between the coordinates $-x_0$ and $+x_0$.

The ratio:

$$\varepsilon(\nu,x)/k'(\nu,x) = S(\nu,x) \qquad (60)$$

is known as the source function.

For LTE conditions and taking into account the relations (12) and (18) - (22) one obtains:

$$S_\nu(x) = \frac{2h\nu^3}{c^2} \exp-\left(\frac{h\nu}{kT}\right) = W_\nu(T) \qquad (61)$$

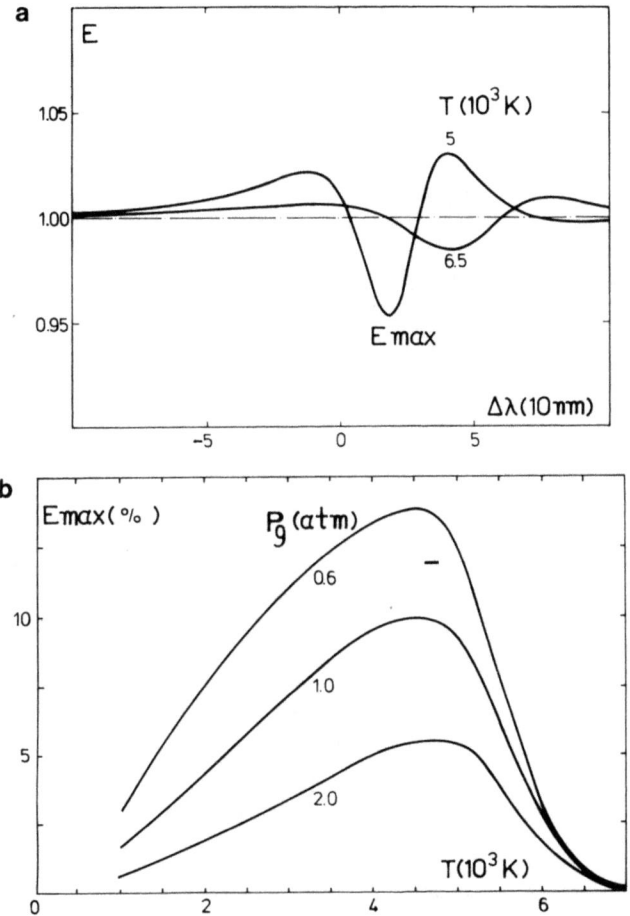

Fig. 11. (a) Relative error E caused by approximation (54) within the line profile for $n_o = 3 \times 10^{18}$ cm^{-3} at T=5000 and 6500K. (b) Variation of the maximum relative error E_{max} as a function of the temperature for three discharge pressures (Fragnac, 1981).

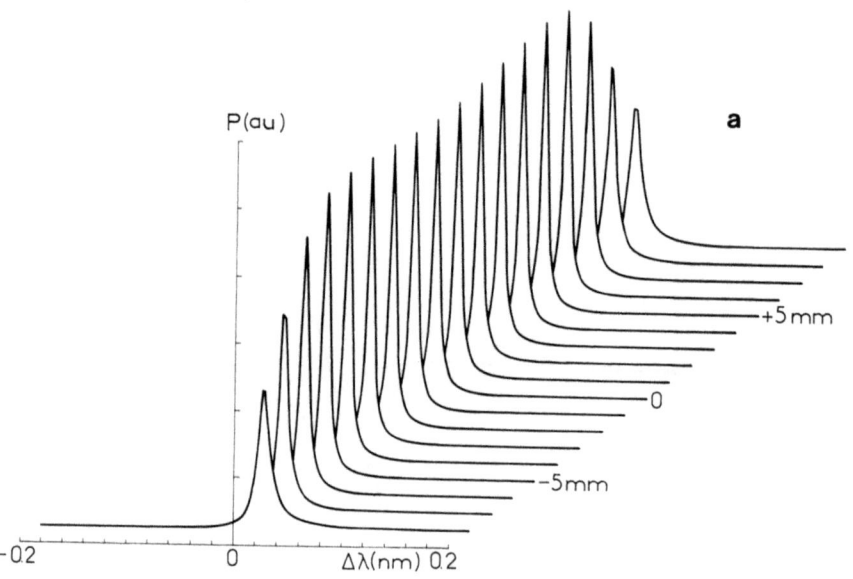

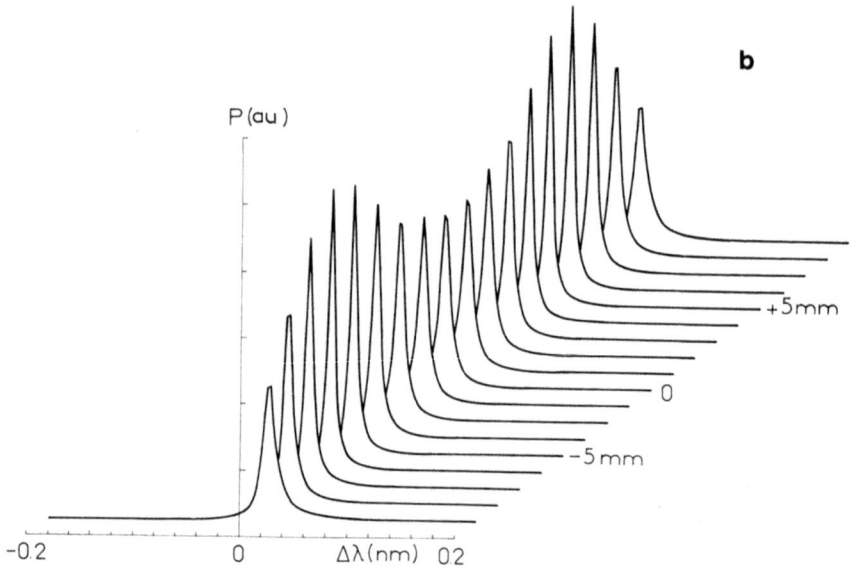

Fig. 12. Evolution of the profile of the mercury 5461Å line at 0 ms (a) and 5 ms (b). Discharge parameters as for Fig. 4. (Fragnac, 1981).

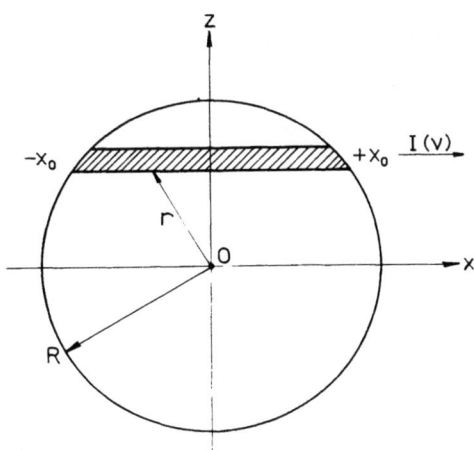

Fig. 13. Illustration of the coordinate system.

i.e., the absorption and emission coefficients are related by Kirchhoff's law, with the Wien's function $W_\nu(T)$ being evaluated at the local temperature T (a function of the position only). Following Hearn (1964), the error of $S_\nu(x)$ is lower than 10% for radiation emitted normally to the plasma boundary, even in non-LTE conditions.

We introduce the optical thickness $\tau(\nu)$ defined by:

$$\tau(\nu) = \int_0^x k'(\nu,x)dx \qquad (62)$$

In the optically thin approximation it is assumed that $\tau \ll 1$. In this case Eq. (59) may be written in the approximate form:

$$I(\nu) = \int_{-x_0}^{+x_0} \varepsilon(\nu,x)dx \qquad (63)$$

The total line intensity is then

$$I_t = \int_{-x_0}^{+x_0} J(x)dx \qquad (64)$$

where J is the line emissivity defined by

$$J = \int_{\text{line}} \varepsilon(\nu)d\nu. \qquad (65)$$

$I(\nu)$ is experimentally measured and $\varepsilon(\nu)$ may be obtained by the solution of Eq. (63). We will assume that the arc discharge, viewed side-on, possesses cylindrical symmetry (Fig. 13).

The emission coefficient is then a function of radius:

$$I(z) = 2 \int_z^R \varepsilon(r) \frac{r\,dr}{(z^2-r^2)^{1/2}} \tag{66}$$

from which $\varepsilon(\nu)$ is found by the well-known Abel inversion (Lochte-Holtgreven, 1968)

$$\varepsilon(r) = -\frac{1}{\pi} \int_r^R \frac{[dI(z)/dz]\,dr}{(r^2-z^2)^{1/2}}. \tag{67}$$

For lines which are not optically thin, additional absorption measurements are required which leads to a more complex set of integral equations (Elder et al., 1965; Birkeland and Oss, 1968; Usher and Campell, 1972; Luizova, 1975; Engelsht and Larkina, 1977).

Calculation of the Spectral Line Contour

Knowledge of the plasma structure (temperatures and densities) and of the line properties (broadening constants and transition probabilities) allows the integration of the equation of radiative transfer and the calculation of the self reversed contour of a spectral line.

Figure 14 illustrates the growth and the formation of the contour of the self reversed 5461Å line along a diameter within a 3-atm mercury discharge at 0 and 5ms. In the same figure, the evolution of the total intensity along a line of sight is given. It is easily seen that:

a) For $x > \frac{7}{5}(R)$, where x is the distance from the origin of the optical path and R the discharge radius, the total intensity remains constant; this means that for $x > \frac{7}{5}(R)$ the radiation trapping is equal to the emission.

b) The self-reversal begins to appear at a distance of about 6 mm from the origin of the optical path.

c) As we move toward the wing of the line, the optical thickness becomes smaller. As a result, somewhere in the wing, the output intensity will exceed that at the center of the line. In this frequency range the plasma is sufficiently opaque to radiate significantly and at the same time sufficiently transparent to let us "see" the central hot region of the discharge.

Experimental Results

In obtaining experimental line contours we have obtained resolution in time and space by using time scanning and sharp focusing of the image of the plasma on the entrance slit of a THR (Jobin-Yvon) monochromator with a bandpass of about 20 mÅ. Under these conditions the error caused by the instrument function is lower than 0.5% at the line maxima. Figure 15 gives, as an example, the experimental contours of the mercury 4047Å line emitted by a 2 atm mercury discharge operated under AC conditions (50Hz), at 100 W/cm at the moments of minimum (0 ms) and maximum (5 ms) emission.

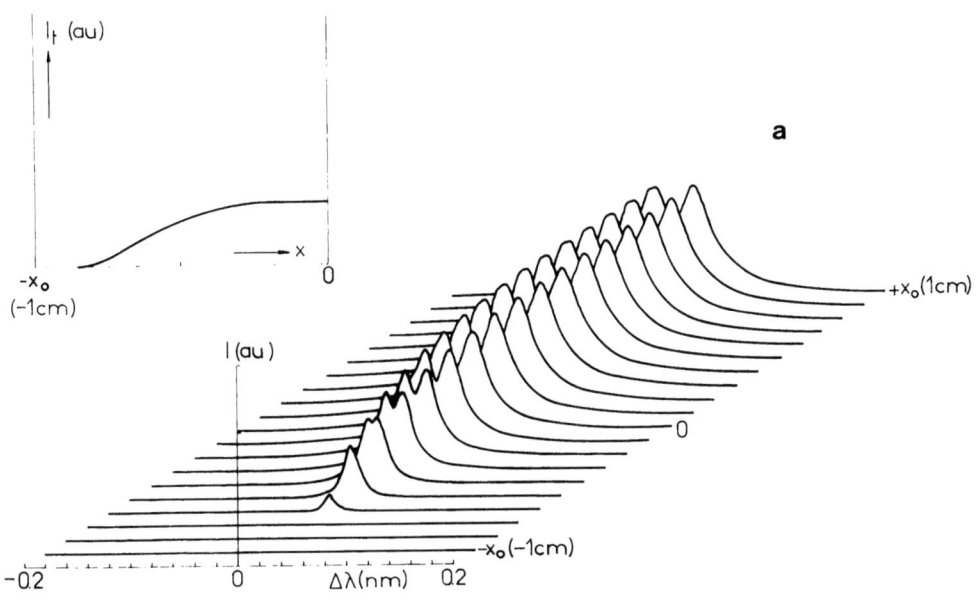

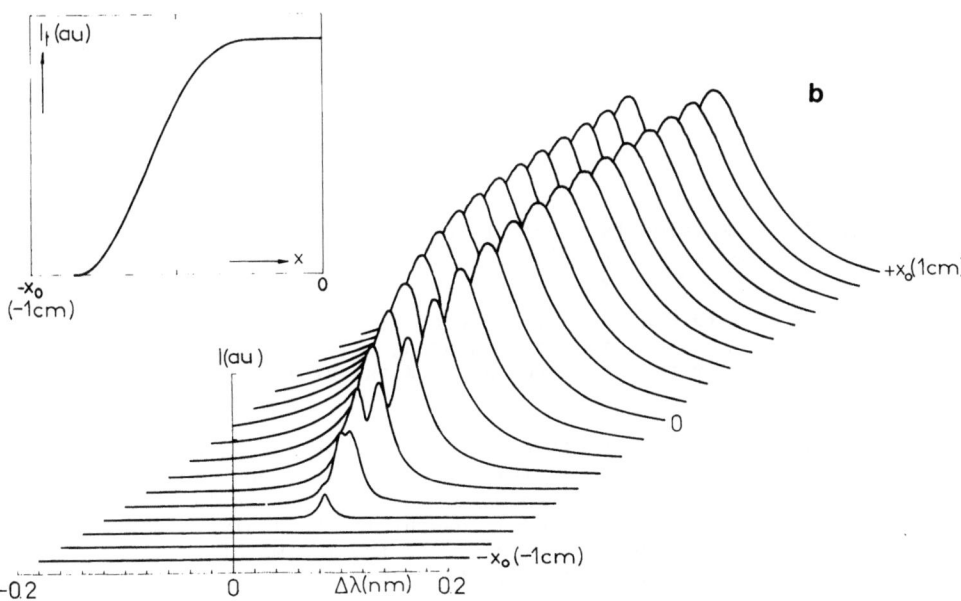

Fig. 14. Spectral formation and total intensity growth of the Hg -5461Å line along a diameter of the discharge at 0 ms (a) and 5 ms (b). au: arbitrary units; I: intensity distribution; I_t: total intensity (Fragnac, 1981).

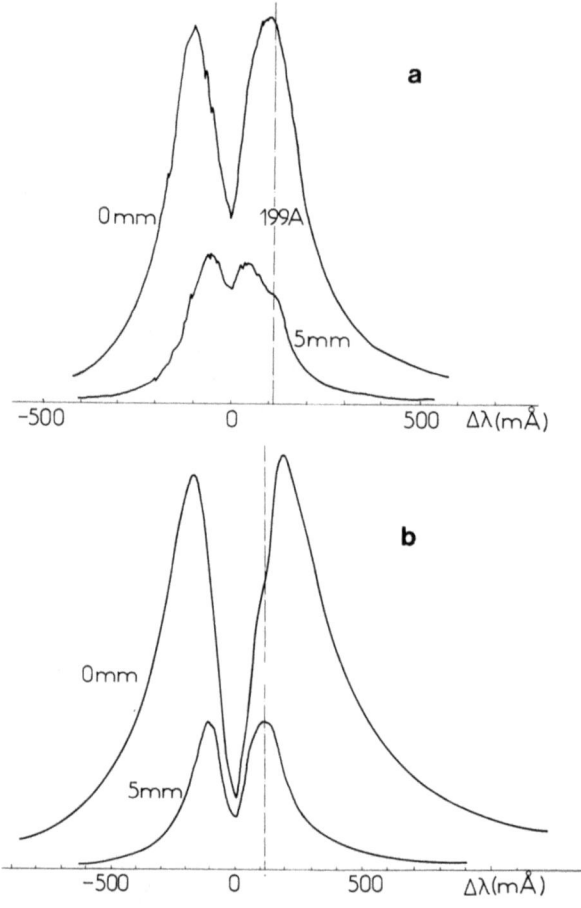

Fig. 15. Experimental contours of the Hg 4047 Å line emitted by a mercury discharge operated in ac (50 Hz), at 100 W/cm, with P_g=2atm, R=1cm, T(R)= 1000 K, T(0) = 5715 K at 0 ms (a) and 6550K at 5 ms (b), along a diameter (z=0 mm) and along a line of sight at 5mm from the discharge axis (Karabourniotis, 1977).

The investigated mercury lines show self-reversal at relatively high concentrations with nearly equal maximum intensities on both sides of the minimum. The existence of the isotopic lines (in particular 199 C and 199 A) on these contours is extremely useful in the analysis of the lines. We also notice the inversion of the two maxima of the lines at 5mm (the blue maximum becomes greater than the red).

For the above lines one can easily obtain contours corresponding to different stages of self absorption as a function of the distance to the arc axis and as a function of the time. The definition of the relevant contour parameters is given in Fig. 16, where $2t = t_r + t_b$ is the width of the observed line contour at a height $1/2\ I_{max}$, $2s = s_r + s_b$ is the distance between the two side maxima of the line, and $2k = k_r + k_b$ is the sum of the distances from line center of the points in the red and blue wings which are at a height of $0.2\ I_{max}$ and $k' = k_r - k_b$; I_{min} is the intensity of the central minimum. Figure 17 shows the temporal evolution of $2t$, $2s$, k' and I_{max}/I_{min} for the 4047 and 5461Å lines emitted from the 3-atm discharge along a diameter. Figure 18 gives the variation of t_r, s_r, k_r, t_b, s_b and k_b for the line 5461Å as a function of the radius for the 3 atm discharge at 0 ms and for the 2 atm discharge at 5 ms.

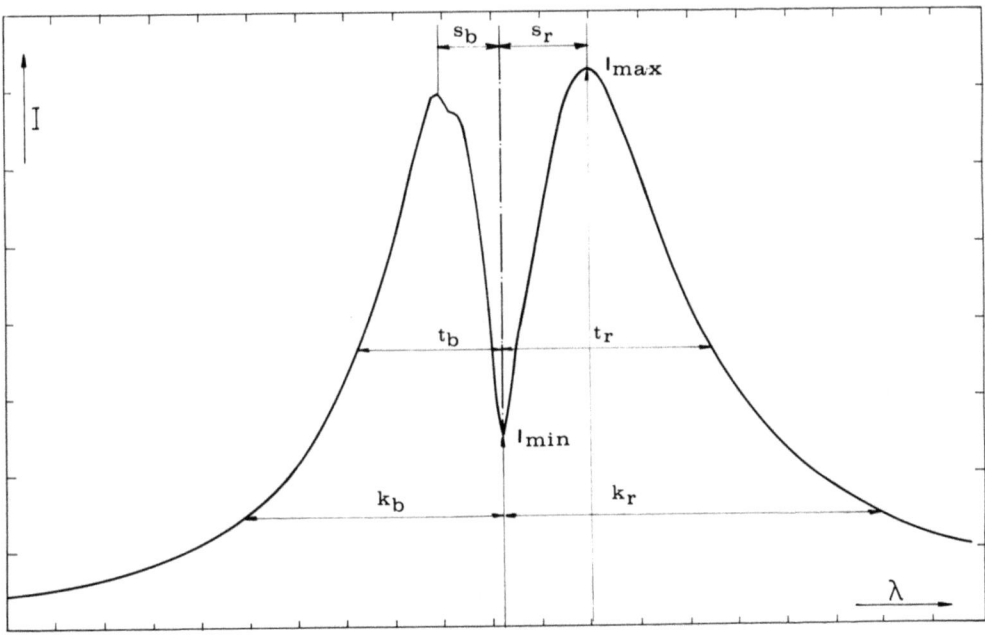

Fig. 16. Designation of the characteristic experimental line contour parameters.

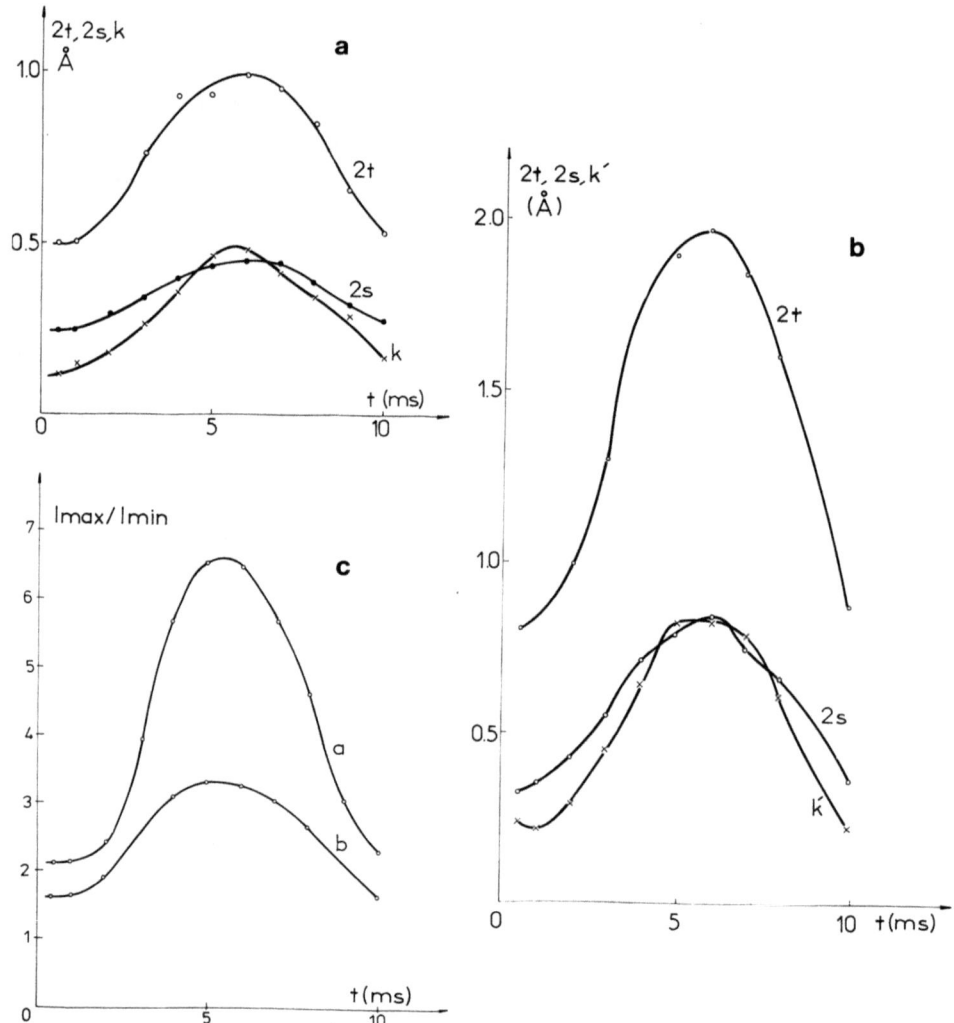

Fig. 17. Temporal evolution of the characteristic contour parameters 2t, 2s, K' and I_{max_o}/I_{min} (defined in Fig. 16) for the Hg 4047 and 5461 Å lines
(a): 4047 Å; (b) 5461 Å; (c) a: 4047 Å, b: 5461 Å
(Karabourniotis, 1977).

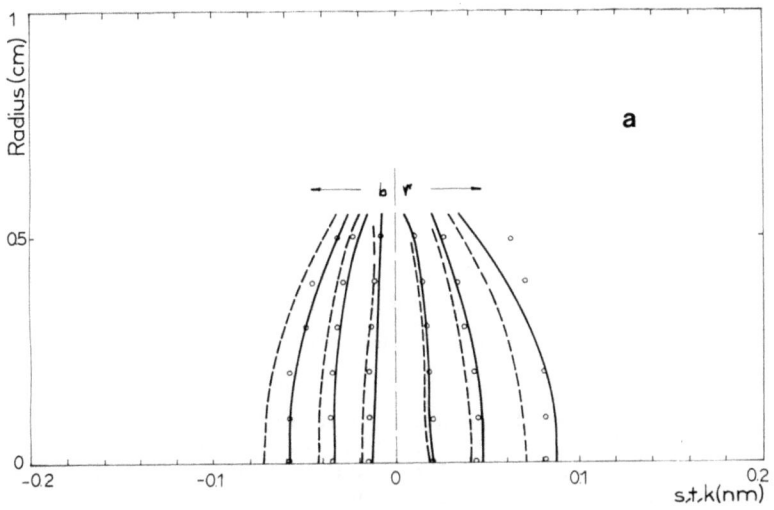
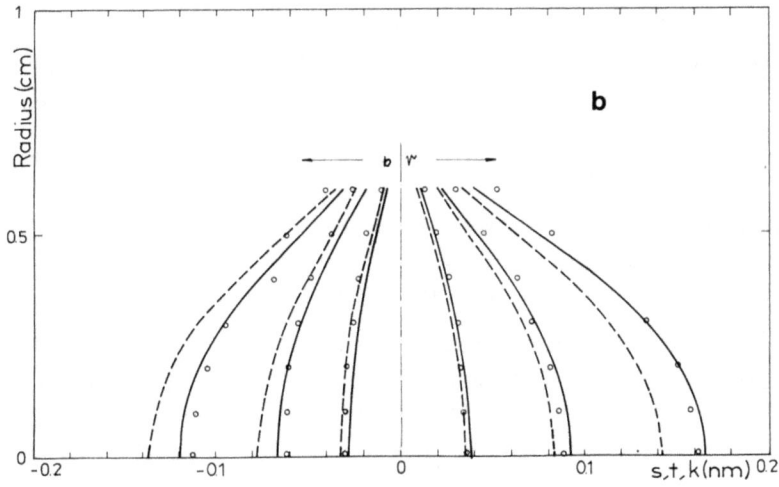

Fig. 18. s_r, t_r, k_r, s_b, t_b and k_s (definitions are given in Fig. 16) for the Hg 5461 Å line against discharge radius. (a): 3 atm discharge, 0 ms (discharge parameters as for Fig. 4); (b): 2 atm discharge, 5 ms (discharge parameters as for Fig. 15). Solid curves: calculated by taking into account the quasistatic effect (Eq. (53)); dashed curves: calculated neglecting quasistatic effect (Eq. (52)). o: experimental points (Fragnac, 1981).

Comparison of Experimental and Theoretical Results

The usefulness of the results obtained from an alternating discharge lies in the strong variation of the electron density between 5 and 0 ms (Fig. 4).

We describe below an iterative method proposed by Karabourniotis (1977) and Damelincourt et al. (1983a) for the determination of the broadening constants and the transition probabilities of self-reversed lines.

We will first consider the 3-atm mercury discharge at the moment of minimum emission (0 ms), i.e., when the electron density attains its minimum value, whereas the atom density becomes maximum. The broadening effect of the electrons is then negligible.

From the position of the experimentally measured points (I_{max}/I_{min}, 2t) for the lines emitted along a diameter of the cross section of the plasma column (Fig. 19) we evaluate the transition probabilities (A) and the van der Waals constants (C_6) for the visible mercury triplet lines. In the calculations, a van der Waals potential has been assumed for the line broadening in the atom-impact approximation (Eqs. (36) and (37)). For the 2-atm discharge at 5 ms, the contribution of the charged particles and of the neutral atoms to the line broadening for the lines traversing the central region of the arc are of the same order of magnitude.

Figure 20 gives the dependence of I_{max}/I_{min} upon 2t for the 4358 and 5461 Å lines as a function of the parameter A and for two values of the width δ_{ei} due to charged particles. The position of the experimental point (I_{max}/I_{min}, 2t) gives a new value for A and the values of C_e and C_i which are defined by Eq. (30).

It is now possible to correct the values of A and C_6 by taking into account the influence of the electrons at 0 ms. The corrected values obtained by this iterative method are given in Table 5.

Thus, with the determined set of parameters (A, C_6, C_e and C_i) we have finally calculated the line contours as a function of the distance to the arc axis for different experimental conditions. Figure 18 gives an example of the results obtained for the 5461 Å line. The theoretical results are in satisfactory agreement with the experimental ones.

Figure 9 shows the radial variation of the line widths caused by different effects as well as the total broadening for the 3-atm discharge at 0 and 5 ms (Damelincourt et al., 1983b). Thus, by comparing calculated and experimental results, it is possible to estimate a set of constants (broadening constants and transition probabilities) which lead to a satisfactory representation of the observed line contours.

BARTELS' THEORY OF RADIATION

Bartels (1949a, 1949b) has developed a theoretical model to describe the contour of a self absorbed line emitted from an inhomogeneous plasma which allows for a radial variation of the width of the initial profile. However, the method of calculation which he has proposed is cumbersome and contains a number of simplifying assumptions.

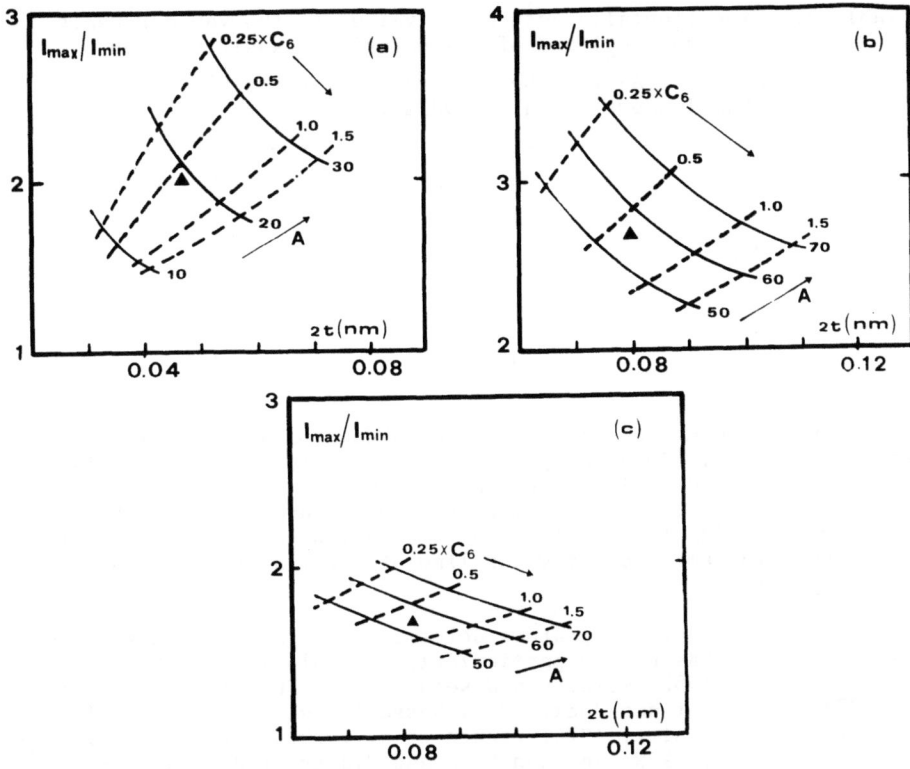

Fig. 19. Dependence of I_{max}/I_{min} upon 2t for the 3 atm discharge at 0 ms. (a) 4047 Å; (b) 4358 Å; (c) 5461 Å; ▲ experimental point. Initial value of $C_6 = 2.32 \times 10^{-30}$ rad $cm^6 s^{-1}$; A in $10^6 s^{-1}$ (Damelincourt et al., 1983 a).

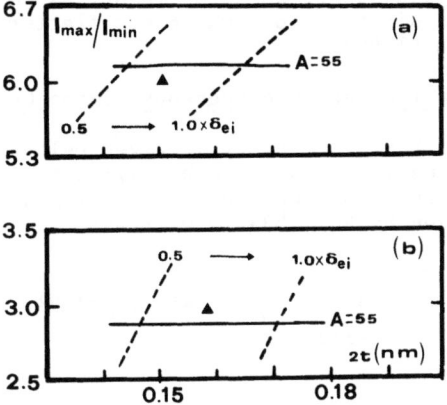

Fig. 20. Dependence of I_{max}/I_{min} upon 2t for the 2 atm discharge at 5 ms as a function of A for two values of δ_{ei}. (a) 4358 Å line; (b) 5461 Å line; ▲: experimental point (Damelimcourt et al., 1983a).

Table 5. Experimentally determined values of the van der Waals constant (C_6) and of the emission transition probability (A) for the mercury triplet visible lines (Damelincourt et al., 1983a).

Line (nm)	404.7	435.8	546.1
C_6 (10^{-30} cm^6 rad sec^{-1})	1.4	1.5	1.6
A (10^6 sec^{-1})	18	55	55

When the temperature profile is the unknown parameter, Bartels (1950a, 1950b, 1953, 1957) has proposed a method to solve the radiative transfer equation under simplified conditions. He has shown that for a plasma in LTE with cylindrical symmetry, the temperature distribution can be found by measuring the intensity at the maximum of a non-resonant self-reversed line emitted along the line of sight at various positions across the plasma.

Bartels' method of temperature determination is fairly easy to use in practice and it has proved particularly valuable in high pressure discharges (Going, 1952; Meiners and Weiss, 1970; Funk et al., 1970; Ozaki, 1971a; de Groot and Jack, 1973; Wesselink et al., 1973; Kolobova, 1973; Albrecht and Schiff, 1976; Finken et al., 1978; Fromm et al., 1978; Aleksandrove et al., 1978; van den Hoek and Visser, 1980; Davenport and Lapworth, 1980; Salakhov and Fishman, 1981; El Hasnaoui, 1983).

The Bartels theory has also been used for the determination of the plasma pressure (Funk and Kloss, 1973 and 1977; Karabourniotis et al., 1986b), and of the total intensity of a self-reversed line (El Hasnaoui, 1983; Couris, 1983; Damelincourt et al., 1983b).

Abridged accounts of Bartels' theory in English (Zwicker, 1968; Neumann, 1975; Wharmby, 1976) merely quote its theoretical results and the conditions that are necessary for the theory to be valid. A rather complete English presentation of Bartels' theory as well as an extension to include temperature distributions which are flatter than parabolic has been given by Lapworth (1980).

Spectral Line Emission from an Inhomogeneous Plasma Column

A short survey of Bartels' theory is given below. The plasma is assumed to be in LTE and it is taken to be cylindrically symmetric. Referring to Fig. 13 and using the quantities defined in Eqs. (60) and (62) the intensity given by Eq. (59) is written:

$$I(\nu) = \tau_0 \exp\left(-\frac{\tau_0}{2}\right) \int_{-1/2}^{+1/2} S_\nu(\xi) \exp(\tau_0 \xi) \, d\xi, \qquad (68)$$

where:

$$\tau_0 = \int_{-x_0}^{+x_0} k(\nu, x) \, dx \qquad (69)$$

is the total optical thickness and:

$$\xi = \tau/\tau_0 \tag{70}$$

is the relative optical thickness.

Using the relative source function defined by:

$$\sigma(\xi) = \frac{S_\nu(\xi)}{\int_{-1/2}^{+1/2} S_\nu(\xi) d\xi} . \tag{71}$$

Eq. (68) becomes:

$$I(\nu) = S\nu_{max} \cdot \int_{-1/2}^{+1/2} \frac{S_\nu(\xi)}{S_{\nu max}} d\xi \cdot \tau_0 \exp\left(-\frac{\tau_0}{2}\right) \int_{-1/2}^{+1/2} \sigma(\xi) \exp(\tau_0 \xi) d\xi \tag{72}$$

where $S\nu_{max}$ denotes the maximum value of $S_\nu(\xi)$ along the line of sight. Eq. (72) may now be expressed in the form:

$$I(\nu) = S\nu_{max} \cdot M \cdot Y(\tau_0) \tag{73}$$

with:

$$M = \int_{-1/2}^{+1/2} \frac{S_\nu(\xi)}{S_{\nu max}} d\xi \tag{74}$$

and:

$$Y(\tau_0) = \tau_0 \exp\left(-\frac{\tau_0}{2}\right) \int_{-1/2}^{+1/2} \sigma(\xi) \exp(\tau_0 \xi) d\xi . \tag{75}$$

If it is assumed that local thermodynamic equilibrium is valid, then the source function may be replaced by the Wien function given by Eq. (61). Wien's function has its maximum value $W_\nu(T_m)$ at the axis of the plasma column, where the temperature attains its maximum value T_m. Consequently Eq. (73) becomes:

$$I(\nu) = W_\nu(T_m) \, M \, Y(\tau_0) \tag{76}$$

If we write Eq. (74) as:

$$M \cdot W_\nu(T_m) \cdot \tau_0 = \int_{-1/2}^{+1/2} W_\nu(x) \, k(\nu, x) dx \tag{77}$$

it is evident that M is the factor by which the total emission along the line of sight is smaller than the emission from an homogeneous plasma of temperature T_m with the same optical thickness τ_0. On the other hand, the function $Y(\tau_0)$ given by equation (75) represents the influence of the optical thickness τ_0 upon the measured intensity.

Behavior of the Function $Y(\tau_0)$

The value of $Y(\tau_0)$ for a given τ_0 depends on the distribution of the relative source function $\sigma(\xi)$. Therefore $Y(\tau_0)$ represents a two-fold infinite diversity of curves corresponding to the infinite diversity of distributions $\sigma(\xi)$.

Two extreme cases will now be considered. First, a uniformly excited source in which excitation is homogeneous (isothermic plasma) and secondly a completely inhomogeneous source, where all of the emitting atoms are concentrated at the arc axis, while all of the absorbing atoms are distributed outside this region.

In the first case, the source function is independent of the coordinate x (constant along the line of sight) and the relative source function, defined by Eq. (71), takes the value:

$$\sigma(\xi) = 1 \tag{78}$$

Therefore,

$$\int_{-1/2}^{+1/2} \sigma(\xi) e^{\tau_0 \xi} d\xi = e^{\tau_0/2} - e^{-\tau_0/2} \tag{79}$$

and Eq. (75) becomes:

$$Y(\tau_0) = 1 - e^{-\tau_0} \tag{80}$$

In the second extreme case $\sigma(\xi)$ can be considered as a δ-function around the arc axis, where $\xi=0$. Thus:

$$\int_{-1/2}^{+1/2} \sigma(\xi) e^{\tau_0 \xi} d\xi = 1 \tag{81}$$

and Eq. (75) reduces to:

$$Y(\tau_0) = \tau_0 e^{-\tau_0/2} \tag{82}$$

From Eqs. (80) and (82) it follows that:

$$\tau_0 e^{-\tau_0/2} < Y(\tau_0) < 1 - e^{-\tau_0} \tag{83}$$

The function $Y(\tau_0)$ for these two extreme cases are shown in Fig. 21.

In the general case, the function $Y(\tau_0)$ is shown by Bartels (1949a) to be expressed by a family of curves in which each curve depends on a parameter p, given by:

$$p = 12 \int_{-1/2}^{+1/2} \xi^2 \sigma(\xi) d\xi \tag{84}$$

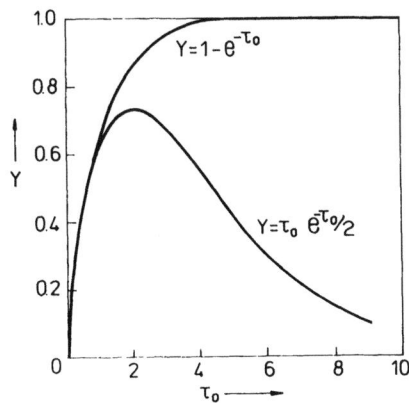

Fig. 21. The two extreme forms of the function $Y(\tau_0)$ (Bartels, 1949a)

The parameter p lies in the range:

$$0 \leq p \leq 1 \qquad (85)$$

and indicates the inhomogeneity of the source; p=1 corresponds to a homogeneous plasma column and p=0 to a completely inhomogeneous source.

Each value of the parameter p defines a subset of the whole family of curves lying between two limiting curves

$$Y_1(\tau_0,p) = \frac{2}{\sqrt{p}} e^{-\tau_0/2} \sinh\left(\frac{\tau_0 \sqrt{p}}{2}\right) \qquad (86)$$

and:

$$Y_2(\tau_0,p) = 2e^{-\tau_0/2}\left[\frac{\tau_0}{2}(1-p) + p \sinh\left(\frac{\tau_0}{2}\right)\right] \qquad (87)$$

The curves Y_1 and Y_2 are not very divergent for small values of τ_0 (generally for $\tau_0 \leq 10$). They can then be replaced by a single curve which is the average of the curves Y_1 and Y_2. The family of such average curves represented by $\bar{Y}(\tau_0,p)$ is given by:

$$\bar{Y}(\tau_0,p) = e^{-\tau_0/2}\left[\frac{1}{\sqrt{p}} \sinh\left(\frac{\tau_0 \sqrt{p}}{2}\right) + \frac{\tau_0(1-p)}{2} + p \sinh\left(\frac{\tau_0}{2}\right)\right] \qquad (88)$$

and the function has a maximum for all values of p less than unity. This maximum describes conditions at the intensity maximum of a self-reversed line. The error of this approximation is smaller than 10% for $\tau_0 \leq 7$ and smaller than 2% for $\tau_0 \leq 4$.

Figure 22 gives the functions Y_1/Y_{1max}, Y_2/Y_{2max} and $\bar{Y}/\bar{Y}_{max}$ for p=0.8 where $\bar{Y}_{max}$ denotes the maximum value of the function $\bar{Y}(\tau_0,p)$. For $\bar{Y}/\bar{Y}_{max} = 0.95$ the uncertainty of τ_0 is about 3%. The function $Y(\tau_0,p)$ is plotted in Fig. 23 for different values of the parameter p.

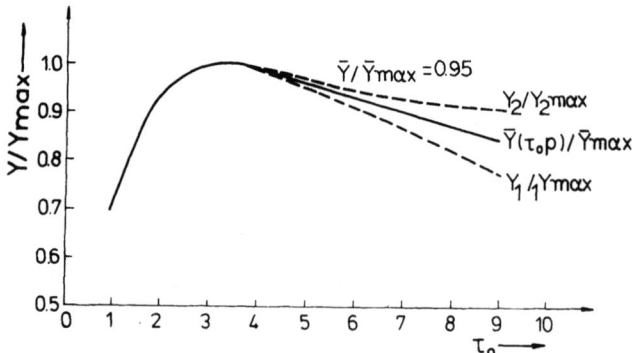

Fig. 22. Total optical thickness τ_0 as a function of Y/Y_{max} for p=0.8 (Bartels and Zwicker, 1962).

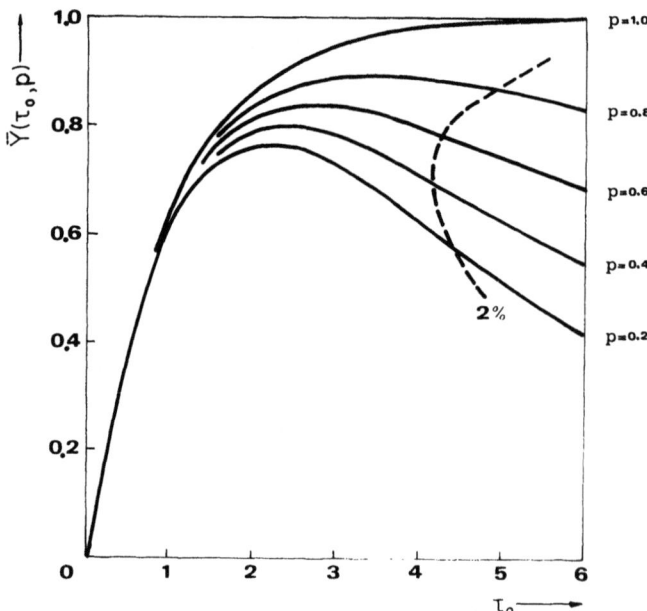

Fig. 23. Variation of $\bar{Y}(\tau_0, p)$ as a function of optical thickness τ_0. The two percent error line is also shown (Bartels, 1949a).

The two percent error line is also shown indicating the error incurred by replacing the pair of limiting curves Y_1 and Y_2 corresponding to each value of p by the average curve $\bar{Y}$. If the value of p is known then $\bar{Y}_{max}$ may be found from Fig. 24 where $\bar{Y}_{max}$ is plotted against p.

The greatest value that $\bar{Y}_{max}$ can have occurs when p=1 and $\tau_0 \to \infty$ (practically when $\tau_0 > 4$) and is equal to 1, and the smallest value occurs at p = 0 and $\tau_0 = 2$, and is $2e^{-1} = 0.736$. Therefore:

$$0.736 < \bar{Y}_{max} < 1 \tag{89}$$

The maximum $\bar{Y}_{max}(p) = \bar{Y}(\tau_0, p)$ of Eq. (88) can be calculated either roughly by an expression originally given by Bartels:

$$\bar{Y}_{max}(p) \approx 2 e^{-1} + (1 - 2 e^{-1})p^2 = 0.736 + 0.264p^2, \tag{90}$$

or with 0.4% accuracy by:

$$\bar{Y}_{max}(p) \approx \frac{2}{1-p} \left(\frac{1-\sqrt{p}}{1+\sqrt{p}}\right)^{\frac{1}{2\sqrt{p}}} \tag{91}$$

The corresponding optical thickness then becomes:

$$\hat{\tau}_0 \approx \frac{1}{\sqrt{p}} \ln \frac{1+\sqrt{p}}{1-\sqrt{p}} \tag{92}$$

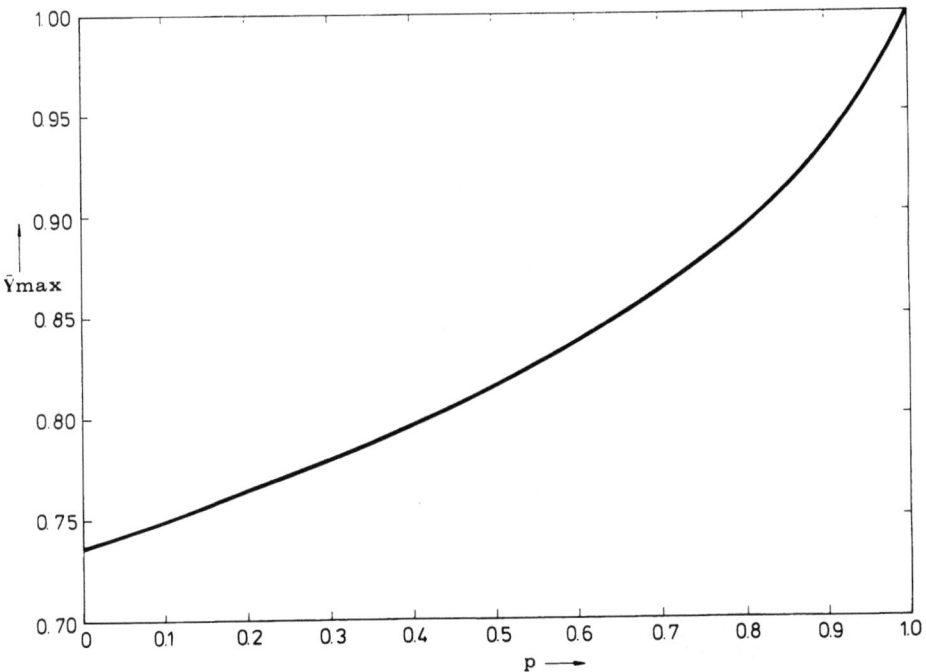

Fig. 24. Variation of $\bar{Y}_{max}$ with p (Lapworth, 1980).

Expressions for the Factor M

From the definitions of M (Eq. (74)), of the optical thickness (Eqs. (62), (69) and (70)), and of the absorption coefficient (Eq. (22)), it is readily shown that M is given by

$$M = \frac{1}{S_{\nu max}} \frac{\int_{-x_0}^{+x_0} S_\nu(x) \, k(\nu,x) dx}{\int_{-x_0}^{+x_0} k(\nu,x) dx} =$$

$$= \frac{1}{S_{\nu max}} \frac{\int_{-x_0}^{+x_0} S_\nu(x) \, n_a(x) P(\nu,x) dx}{\int_{-x_0}^{+x_0} n_a(x) P(\nu,x) dx} \, . \quad (93)$$

We now make the following assumptions:

(i) The line profile is Lorentzian (Eq. (27)).

(ii) There is no shift in the line center frequency.

(iii) The source function is given by Wien's function (Eq. (61)).

(iv) The number density is given by Boltzmann's formula (Eq. (3)).

(v) The equation of state for ideal gases is well fulfilled.

(vi) Depletion of the ground state population due to excitation and ionization can be neglected.

For $|\nu-\nu_0| \gg \delta$, i.e., for frequencies in the wings of the line:

$$P(\nu,x) \simeq P_\infty(x) = \delta(x)/(\nu-\nu_0)^2, \quad (94a)$$

and for $|\nu-\nu_0| \ll \delta$, i.e., for frequencies around the center of the line profile:

$$P(\nu,x) \simeq P_0(x) = 1/\delta(x), \quad (94b)$$

the expression for M becomes independent of frequency, and the asymptotic expressions of M (M_∞ and M_0) are then written in the form

$$M_{\infty,0} = \exp\left(\frac{E_e - E_a}{kT_m}\right) \frac{\int_{-x_0}^{+x_0} f_{\infty,0}(x) \exp\left[-\frac{E_e}{kT(x)}\right] dx}{\int_{-x_0}^{+x_0} f_{\infty,0}(x) \exp\left[-\frac{E_a}{kT(x)}\right] dx} \quad (95)$$

in which:

$$f_\infty(x) = \frac{\delta(x)}{T(x)} \frac{P_g(x)}{Z(T)} \quad (96a)$$

From Eqs. (100) and (101) it is seen that as $\theta_e, \theta_a \to 0$, then:

$$M \to \beta \equiv \left(\frac{\theta_e}{\theta_a}\right)^{1/2} = \begin{cases} \left(\dfrac{E_a}{E_e}\right)^{1/2} & \text{for atom broadening} \\ \\ \left(\dfrac{E_a + \frac{1}{2}E_i}{E_e + \frac{1}{2}E_i}\right)^{1/2} & \text{for electron broadening} \end{cases} \quad (104)$$

with:

$$M_0 > \beta > M_\infty. \quad (105)$$

The inequality shows that as the frequency goes from the center of the line profile to the wings, M decreases from its maximum value M_0 to its minimum value M_∞.

From Eq. (101) M_∞ may be approximated by β if:

$$\frac{1}{2}\theta_e \left(\ell - \frac{1}{2}\right) \ll 1 \quad (106)$$

and from equation (101) M_0 may be approximated by β if:

$$\frac{1}{2}\theta_e \left(\ell + \frac{1}{2}\right) \ll 1. \quad (107)$$

In the case of atom collision broadening for the "worst" case (i.e., $\ell=1$), and from the definition (102), the condition (106) leads to:

$$\frac{kT_m}{E_e} \ll 4 \quad (108)$$

and the condition (107) leads to:

$$\frac{kT_m}{E_e} \ll \frac{4}{3}. \quad (109)$$

Thus, the condition (relation (109)) that M_0 be approximated by β is more stringent than the condition (relation (108)) that M_∞ by approximated by β.

In the case of electron collision broadening, θ_e and θ_a (definition (103)) have values generally smaller than those appropriate to neutral line broadening. Thus, for electron collision broadening the approximation $M_\infty = \beta$ is valuable for the condition:

$$\frac{kT_m}{E_e + \frac{1}{2}E_i} \ll 2. \quad (110)$$

The relative difference between M_∞ and β is given by:

$$\Delta M_\infty^* = (M_\infty - \beta)/\beta \quad (111)$$

while the relative difference between M_0 and β is given by:

$$\Delta M_0^* = (M_0 - \beta)/\beta. \quad (112)$$

and:

$$f_0(x) = \frac{P_g(x)}{T(x)\delta(x)Z(T)}, \tag{96b}$$

where $P_g(x)$ is the partial pressure of the emitting element, and $Z(T)$ is the partition function.

The half-width for collision broadening by neutral particles may be expressed by:

$$\delta(x) = A/T^\ell(x) \tag{97}$$

where A is a constant and $0 \leq \ell \leq 1$.

In elementary collision broadening theory, the half-width is given by

$$\delta(x) = \frac{B}{T^\ell} \exp\left[-\frac{E_i}{2kT(x)}\right] \tag{98}$$

where B is a constant, $\ell = -5/12$ and E_i is the ionization energy. We further assume that:

(vii) The partition function $Z(T)$ is constant.

(viii) The partial pressure of the emitting element is constant throughout the plasma.

(ix) The temperature profile along a diameter is parabolic:

$$T(x) = (T_m - T_w)\left[1 - (x/x_0)^2\right] + T_w. \tag{99}$$

where T_w is the temperature at the boundary x_0. The asymptotic expressions for M then become:

$$M_\infty = \left(\frac{\theta_e}{\theta_a}\right)^{1/2} \frac{1 + \frac{1}{2}\theta_e(\ell - \frac{1}{2})}{1 + \frac{1}{2}\theta_a(\ell - \frac{1}{2})} \tag{100}$$

and:

$$M_0 = \left(\frac{\theta_e}{\theta_a}\right)^{1/2} \frac{1 + \frac{1}{2}\theta_e(\ell + \frac{1}{2})}{1 + \frac{1}{2}\theta_a(\ell + \frac{1}{2})} \tag{101}$$

where:

$$\theta_{e,a} = \frac{kT_m}{E_{e,a}} \tag{102}$$

in the case of atom collision broadening, and

$$\theta_{e,a} = \frac{kT_m}{E_{e,a} + \frac{1}{2}E_i} \tag{103}$$

for the case of the electron collision broadening.

Following Bartels (1950b) Fig. 25 gives ΔM_∞^* as a function of Θ_a.

For the mercury 4358 Å line emitted by a high pressure mercury arc with $T_m=5700K$, ΔM_∞^* is of the order of 0.9×10^{-2} and ΔM_0^* is of the order of 3×10^{-2} (Lapworth, 1980).

Therefore, for this line, the total change in M from line center to the far wing does not exceed 4% and the maximum deviation from the approximation $M=\beta$ does not exceed 3%.

When the temperature profile is of the form:

$$T(x) = (T_m - T_w)\left[1 - (x/x_0)^{2n}\right] + T_w \qquad (113)$$

with $n = 1, 2, \ldots$, Lapworth (1980) has proved that M may be approximated by

$$M = \beta^{1/n} . \qquad (114)$$

Expressions for the Factor p

In order to calculate $Y(\tau_0, p)$ given by Eq. (88), it is necessary to derive an approximation for p.

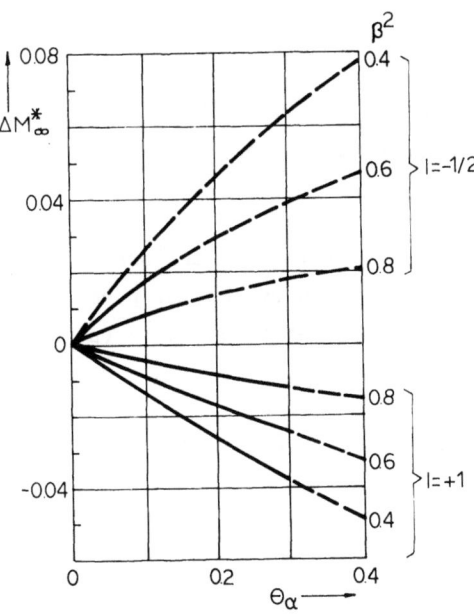

Fig. 25. ΔM_∞^* (Eq. 111) as a function of Θ_a for a parabolic temperature distribution and different values of β^2 (Bartels, 1950b).

From the definitions of p (Eq. (84)), of the relative optical thickness (Eq. (70)), and of the relative source function (Eq. (71)) it is readily shown that p is given by:

$$p = 12 \frac{\int_{-\tau_0/2}^{+\tau_0/2} (\frac{\tau}{\tau_0})^2 S_\nu(\tau) d\tau}{\int_{-\tau_0/2}^{+\tau_0/2} S_\nu(\tau) d\tau} . \quad (115)$$

From Eqs. (69) and (74) it follows that:

$$\int_{-\tau_0/2}^{+\tau_0/2} S_\nu(\tau) d\tau = \tau_0 \int_{-1/2}^{+1/2} S_\nu(\xi) d\xi = \tau_0 \, MS_{\nu max} . \quad (116)$$

Therefore, Eq. (115) is written:

$$p = \frac{12}{M} \int_{-\tau_0/2}^{+\tau_0/2} (\frac{\tau}{\tau_0})^2 \frac{S_\nu(\tau)}{S_{\nu max}} \frac{d\tau}{\tau_0} . \quad (117)$$

From the definition of the absorption coefficient (Eq. (22)), and the expression for M (Eq. 93), p is written in the form:

$$p = \frac{3}{\left[\int_{-x_0}^{+x_0} n_a(x) P(\nu,x) dx\right]^2}$$

$$\frac{\int_{-x_0}^{+x_0} \left[\int_{-x}^{+x} n_a(x) P(\nu,x) dx\right]^2 S_\nu(x) n_a(x) P(\nu,x) dx}{\int_{-x_0}^{+x_0} S_\nu(x) n_a(x) P(\nu,x) dz} . \quad (118)$$

Using the consequences of the assumptions (i)-(vi), and for $|\nu-\nu_0| \gg \delta$ and $|\nu-\nu_0| \ll \delta$, the expression for p becomes independent of frequency and the asymptotic expression for p (p_∞ and p_0) takes the form:

$$P_{\infty,0} = \frac{3}{\left\{\int_{-x_0}^{+x_0} f_{\infty,0}(x) \exp\left[-\frac{E_a}{kT(x)}\right] dx\right\}^2}$$

$$\frac{\int_{-x_0}^{+x_0} \left\{\int_{-x}^{+x} f_{\infty,0}(x') \exp\left[-\frac{E_a}{kT(x')}\right]\right\}^2 f_{\infty,0}(x) \exp\left[-\frac{E_e}{kT(x)}\right] dx}{\int_{-x_0}^{+x_0} f_{\infty,0}(x) \exp\left[-\frac{E_e}{kT(x)}\right] dx} \quad (119)$$

where $f_\infty(x)$ and $f_0(x)$ are given by Eq. (96).

If we consider $\delta(x)$ to be given by Eq. (97) or (98) and using the consequences of the assumptions (vii)-(ix), it has been proven (Bartels, 1950b; Lapworth, 1980) that for $\theta_a \to 0$ we may write:

$$p \to \frac{6}{\pi} \arctan \frac{\beta^2}{\sqrt{1+2\beta^2}} = \gamma \qquad (120)$$

with:

$$p_0 > \gamma > p_\infty \qquad (121)$$

Figure 26 shows γ plotted against β. The relative difference between P_∞ and γ is given by:

$$\Delta p_\infty^* = (p_\infty - \gamma)/\gamma \qquad (122)$$

while the relative difference between P_0 and γ is given by:

$$\Delta p_0^* = (p_0 - \gamma)/\gamma \qquad (123)$$

Figure 27 gives (Bartels, 1950b) Δp_∞^* as a function of θ_a for different values of β^2.

For the mercury 4358Å line emitted by a high pressure mercury discharge with $T_m = 5700K$, Δp_∞^* is of the order of 1.2×10^{-2} and Δp_0^* is of the order of 3.6×10^{-2} (Lapworth, 1980).

Therefore, for this line, the total change in p from line center to the far wings does not exceed 4.8% and the maximum deviation from the approximation $p=\gamma$ does not exceed 3.6%. As seen from Fig. 24, the corresponding change in Y_{max}, is of the order of 1.3%.

It should be pointed out that for resonance lines the conditions (108)-(110) are not fulfilled, and, consequently, the approximations (104) and (120) for M and p are not valid.

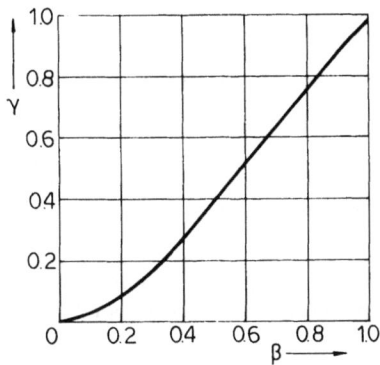

Fig. 26. γ (Eq. (120) plotted against β (Eq. (104)). (Bartels, 1950b).

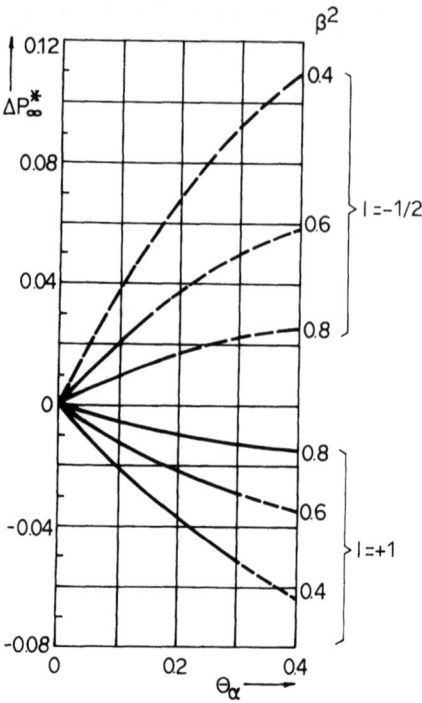

Fig. 27. Δp_∞^* (Eq. 121) as a function of Θ_a for a parabolic temperature distribution and different values of β^2 (Bartels, 1950b).

Description of the Spectral Line Contour

We now consider the variation of the intensity within the contour of a self-reversed line emitted by an inhomogeneous plasma column.

Assuming LTE, the expression (73) for the line contour is written:

$$I(\nu) = W_\nu(T_m) \cdot M \cdot \bar{Y}(\tau_0, p) . \qquad (124)$$

Over the frequency range of the line contour, $W_\nu(T_m)$ remains sensibly constant.

Expressions (104) yield values for M which are constant within the line contour. This is correct in principle only when the line profile is constant throughout the inhomogeneous plasma (position-independent line profile).

Under the conditions established above, M and p are constant within the line to a good approximation, and $\bar{Y}$ depends on the frequency only through τ_0.

Consequently, Eq. (124) shows that the shape of the line contour is determined only by the function $\bar{Y}[\tau_0(u), p^*]$, where $u = \nu - \nu_0$, and p^* is the special value of p, following from Eqs. (104) and (120). In the case of a position independent line profile the total optical thickness may be written in the form:

$$\tau(u) = 2fN_a P(u) \tag{125}$$

where f is the oscillator strength for absorption, N_a is given by:

$$N_a = \int_0^R n_a(r)dr \tag{126}$$

and P(u) is given by Eq. (27), neglecting the line shift. From Eq. (125) it is seen that the variation of $\tau_0(u)$ follows the variation of P(u).

The maximum value of τ_0 (τ_{0max}) will occur at the center of the line profile (u = 0):

$$\tau_{0max} = 2fN_a (\pi\delta)^{-1}. \tag{127}$$

For frequencies remote from the line center, τ_0 is small and as the line center is approached, τ_0 increases to its maximum value. Further change in frequency in the same direction, beyond the line center, leads to a decrease in τ_0. In the $\tilde{Y}$-τ_0 plane (Fig. 23), as the frequency approaches the line center, τ_0 increases, and as the frequency passes the line center τ_0 decreases.

For $\tau_{0max} > \hat{\tau}_0$, where $\hat{\tau}_0$ corresponds to the maximum value of $\tilde{Y}$, the function $\tilde{Y}(u)$ will exhibit a dip, i.e., it will be self-reversed with the two line peaks corresponding to the maximum value of Y on the appropriate Y-τ_0 plane.

Figure 28 shows the relation between the optical thickness $\tau(u)$ (part a), the function $\tilde{Y}$ [$\tau_0(u), p^*$] (part b), and the resulting (relative) line contours (part c) for five different cases.

In part b of this figure the dashed line gives the linear relation $\tilde{Y}(\tau_0, p^*)$ which is correct for $\tau_0 \ll 1$; the dotted line gives the curve for the homogeneous plasma (M = 1, p = 1, Y = $1-e^{-\tau_0}$). The dotted curve in part c of the figure shows the shape of the line contour emitted from the homogeneous plasma.

Bartels' theory has been expended to include stimulated emission and the influence of the continuum (Bartels, 1953; Bartels and Beuchelt, 1957)

Diagnostic Methods Based on Bartels' Theory of Radiation

Temperature determination. Since the self-reversal maxima, I_{max}, always occur when $Y(\hat{\tau}_0, p) = Y_{max}(p)$ (as Fig. 28 shows), expression (124) is written

$$I_{max} = \frac{h\nu_0^3}{c^2} \exp\left(-\frac{h\nu_0}{kT_m}\right) M \tilde{Y}_{max}(p), \tag{128}$$

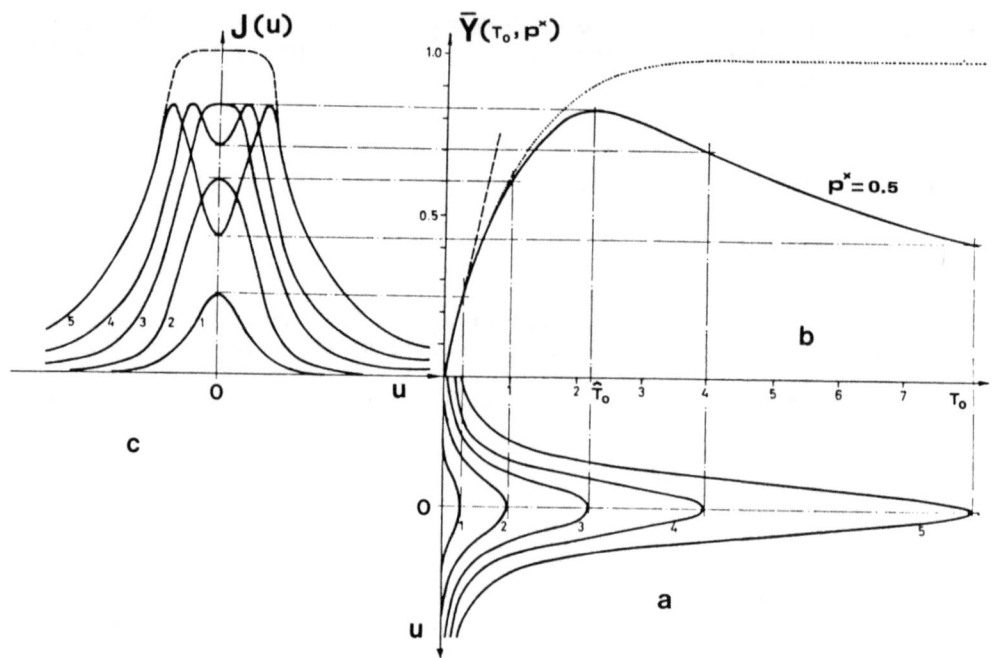

Fig. 28. Relation among the optical thickness $\tau_0(u)$, the function $\bar{Y}(\tau_0,p)$ and the relative intensity distributions $J(u)$ for five different cases.
(a): Distribution $\tau_0(u)$ within the line.
(b): The function $Y(\tau_0, p)$ for $p=0.5$; dotted line: homogeneous case. (c): Resulting relative intensity distributions (Zwicker, 1968).

where ν_0 is the frequency at the center of the self-reversed line.

Therefore, for a plasma in LTE with cylindrical symmetry, the temperature distribution can be found by measuring the absolute intensity at the maximum of a non-resonant self-reversed line emitted along the line of sight at various positions across the plasma.

Wesselink et al. (1973) have proposed a two line method which requires only relative intensity measurements. The temperature is obtained from

$$\frac{I_{1max}}{I_{2max}} = \frac{\nu_1^3}{\nu_2^3} \exp\left[-\frac{h}{kT_m}(\nu_1 - \nu_2)\right]\frac{(M\,Y_{max})_1}{(M\,Y_{max})_2} \qquad (129)$$

where 1 and 2 refer to the two lines.

The temperature distribution evaluated in this way is only a first approximation because Eqs. (104) and (120) are only first approximations. If necessary, this result can be further improved by iterative methods.

Assuming that the reversal peaks occur in the far wings of the line profile, M and p are independent of the frequency and may be approximated by M_∞ and p_∞ (Eqs. (95) and (115)), using the approximately determined temperature distribution.

This is referred to in Bartels' publications as the "third reversal state". By introducing the values of M_∞ and p_∞ in Eq. (128) we find a better approximation for the temperature distribution (Going, 1952; Meiners and Weiss, 1970; Kolobova, 1973).

De Groot and Jack (1973) have studied the effect of the temperature profile and of the line profile in the calculation of M and p. They have proved that finite values of M and p may also be found using the integral expressions (95) and (119) for resonance lines.

The main problem in the application of Bartels' method is connected with asymmetrically self-reversed lines where the two reversal maxima differ in height, causing an uncertainty in the temperature determination. This is a consequence of an essential weakness of Bartels' formalism in the conventional form, since it does not account for the influence of line shifts and of asymmetric broadening of the line profile with increasing pressure of the vapor.

Determination of partial pressures. On the basis of Bartels' theory, Funk and Kloss (1973 and 1977) have proposed a method for the determination of relative partial pressures in metal halide discharges from the separation of the maxima of self-reversed spectral lines. This method has been developed by Couris (1983), Karabourniotis et al. (1985), and Karabourniotis et al. (1986b).

For the determination of the relative partial pressure of the emitting gas, the following relation is used (Karabourniotis et al, 1986b):

$$2s = c P_g^{\phi_g} P_p^{\phi_p} U^\phi , \qquad (130)$$

where 2s is the wavelength distance between the two side maxima of the self-reversed line, c is a constant, P_g is the partial pressure of the emitting gas, and P_p is the partial pressure of the perturbers; P_g and P_p are considered to be constant throughout the plasma, ϕ_g and ϕ_p depend on the form of the line profile with:

$$1/2 < \phi_{g,p} < 2/3, \qquad (131)$$

and:

$$U = \int_{-R}^{+R} [Z(T)]^{-1} [T(r)]^{-2} \exp[-E_a/kT(r)] dr, \qquad (132)$$

where T(r) is the temperature profile, Z(T) is the partition function of the emitting atoms and E_a is the lower energy level of the transition which generates the self-reversed line.

For example, in the determination of the partial pressure of neutral thallium in a H_g-TlI discharge using the Tl-5350 Å line, $P_g = P_{Tl}$ and $P_p = P_{Hg}$. Figure 29 gives the relation between 2s for the Tl-5350Å line and the partial pressure P_{Tl} for two different discharges operated at

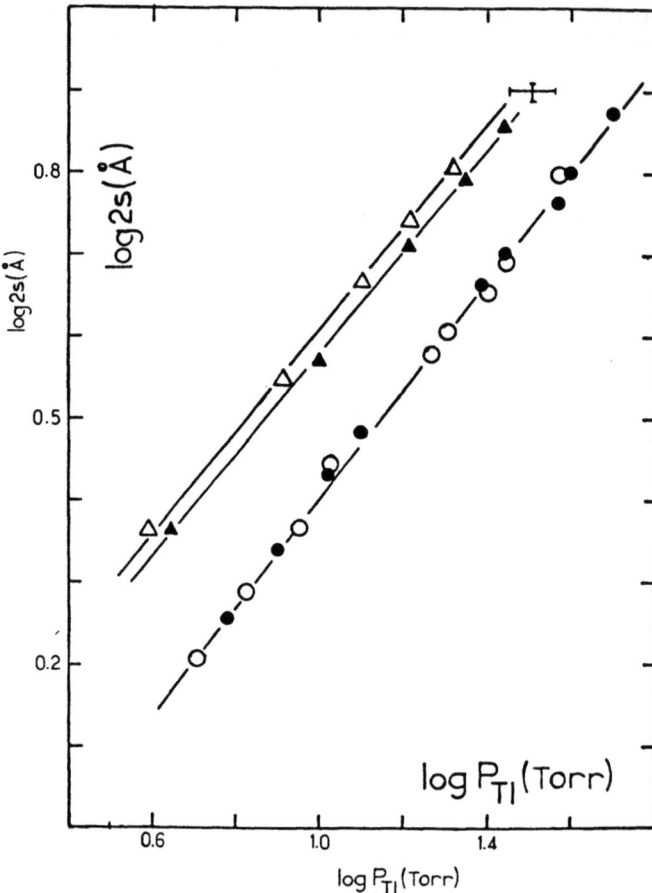

Fig. 29. Experimental relation between the maxima distance 2s of the self-reversed line 5350 Å of thallium and the partial pressure of the (neutral) atoms of thallium for two Hg-TlI discharges with R=1cm, operated at 100 W/cm (50 Hz), at the moments of minimum (0 ms) and maximum emission (5 ms). ●: P_{Hg} = 1.5 atm, 5 ms; o: P_{Hg} = 1.5 atm, 0 ms; ▲: P_{Hg} = 4.6 atm, 5 ms; Δ: P_{Hg} = 4.6 atm, 0 ms.

50Hz, at the moments of minimum (0 ms) and maximum (5 ms) emission. The partial pressure of thallium has been measured from the intensity of the optically thin Tl-6650Å line and controlled by the minimum temperature of the discharge wall. From this figure, we have found:

$$\phi_g \simeq 0.65 \tag{133a}$$

and since U is approximately constant with $\phi \simeq 1/2$, we obtain

$$\phi_p \simeq \begin{cases} 0.58 \text{ at 5 ms} \\ 0.64 \text{ at 0 ms} \end{cases} \tag{133b}$$

COWAN AND DIEKE'S THEORY OF LINE SELF-ABSORPTION

Cowan and Dieke (1948) in an article entitled "Self-absorption of spectrum lines" proposed a model which made it possible to relate the spectroscopic characteristics of an optically thick inhomogeneous discharge plasma to the contour of the spectral line emitted by such a plasma. Their approach is based on the radiative transfer equation. The formulation is one-dimensional and does not take into account multiple absorption and re-emission in the surrounding medium. They have considered that the line profile is constant throughout the plasma column and the function which characterizes the inhomogeneity of the light source has a concrete form.

Ilin and Fishman (1966) have proposed a diagnostic scheme (the multi-parameter method) for the determination of the parameters which characterize a symmetrically self-seversed line following the Cowan-Dieke model. This method has been extended to include asymmetric self-reversed contours (Ilin et al., 1969; Karabourniotis and Karras, 1985).

The Cowan-Dieke model has been used to determine plasma temperature (Karabourniotis et al., 1982; Karabourniotis, 1983; Karabourniotis and Karras, 1985), plasma pressure (Karabourniotis and Damelincourt, 1982; Karabourniotis et al., 1985), transition probabilities (Karabourniotis et al., 1986), and deviations from equilibrium (Karabourniotis, 1986; Karabourniotis et al., 1986c).

General Expression for the Contour of a Self-Absorbed Line

The plasma is taken to be axially symmetric. The function which describes the contour of the line, taking into account the self-absorption and the inhomogeneity along the line of sight, is given (neglecting the stimulation emission) by the integral form of the radiative transfer Eq. (59).

Let N_e denote the total number of emitting atoms along the line of sight (Fig. 13):

$$N_e = \int_{-x_0}^{+x_0} n_e(x) dx \tag{134}$$

$$\bar{n}_e(x) = n_e(x)/N_e \tag{135}$$

and I_0 the total intensity of the line in the absence of self-absorption:

$$I_0 = (h\nu_0/4\pi) A N_e$$

where $A(=A_{ea})$ is the transition probability for emission. The emission coefficient then becomes (Eq. 12):

$$\varepsilon(\nu,x) = I_0 \bar{n}_e(x) P_e(\nu,x) . \tag{137}$$

Let $2N_a$, the total number of absorbing atoms that the light has to traverse, be defined as follows:

$$N_a = (1/2) \int_{-x_0}^{+x_0} n_a(x) dx \tag{138}$$

and

$$\bar{n}_a(x) = n_a(x)/N_a .$$ (139)

Then the absorption coefficient (Eq. 22) is written

$$k(\nu,x) = (h\nu_0/c)BN_a\bar{n}_a(x) P_a(\nu,x) .$$ (140)

Combining Eqs. (136) and (139), the Eq. (59) of radiative transfer may be written in the form:

$$I(\nu) = I_0 \int_{-x_0}^{+x_0} \bar{n}_e(x) P_e(\nu,x) \exp\left[-\frac{h\nu_0}{c} BN_\alpha \int_{-x_0}^{+x_0} \bar{n}_a(x') P_a(\nu,x') dx'\right] dx .$$ (141)

We now assume that the absorption and emission line profiles are the same:

$$P_a(\nu,x) = P_e(\nu,x) = P(\nu,x)$$ (142)

We further let:

$$y = \int_x^{+x_0} \bar{n}_a(x') dx' ,$$ (143)

i.e., y is the relative number of absorbing atoms present per unit cross section between the point under consideration and the outside of the source with:

$$y = \begin{cases} 2 & \text{for } x = -x_0 \\ 1 & \text{for } x = 0 \\ 0 & \text{for } x = +x_0 \end{cases}$$ (144)

$$dy = -\bar{n}_a(x) dx$$

It follows that:

$$\int_{-x_0}^{+x_0} P(\nu,x') \bar{n}_a(x') dx' = \int_0^y P(\nu,y') dy' .$$ (146)

We define a relative excitation function E(y) by:

$$E(y) = \bar{n}_e(x)/\bar{n}_a(x)$$ (147)

Therefore, Eq. (141) becomes:

$$I(\nu) = I_0 \int_0^2 E(y) P(\nu,y) \exp\left[-\frac{h\nu_0}{c} BN_a \int_0^y P(\nu,y') dy'\right] dy.$$ (148)

Effect of Self-Absorption on the Shape of the Line Contour

In order to investigate the effect of the self-absorption in some simple special cases, Cowan and Dieke have assumed a position independent line profile:

$$P(\nu,x) = P(\nu) . \tag{149}$$

This is a remarkable first approximation in ordinary light sources where there are no extreme variations in pressure or Doppler broadening. Defining:

$$u = \nu - \nu_0, \tag{150}$$

with:

$$P(\nu) \to P(u) \tag{151}$$

and by introducing the abbreviation:

$$\mu(u) = p[P(u)/P(0)] \tag{152}$$

where:

$$p = (h\nu_0/c) \, BN_a P(0) \tag{153}$$

is the absorption parameter, Eq. (148) takes the simplified form that was used by Cowan and Dieke for the study of self-absorption:

$$I(u) = I_0 P(u) \int_0^2 E(y) \exp[-\mu(u)y] dy . \tag{154}$$

As Cowan and Dieke have pointed out, the shape and the intensity of a spectral line depends on only three factors:

(i)   the line profile $P(u)$,

(ii) the absorption parameter p, and

(iii) The excitation function $E(y)$.

The function $E(y)$ measures the relative excitation at a particular point and so specifies the type of the light source. It can take values from zero (when there is no excitation) to infinity (when all the atoms are excited).

Let us now assume that the atoms are excited symmetrically with respect to the axis of a cylindrical discharge and that the radiation is emitted in a direction perpendicular to the discharge axis.

If self-reversal is present, differentiation of Eq. (154) at frequencies other than that for which $dP(u)/du=0$ gives as a condition for reversal:

$$\int_0^2 (1-qy) E(y) \exp(-qy) dy = 0 \tag{155}$$

where q is the value of μ at frequencies $u = \pm s$ at which the line reversal occurs.

From the definition of q and Eq. (152), we obtain for an unshifted Lorentzian profile:

$$s = \pm \delta \left[(p/q) - 1 \right]^{1/2}. \qquad (156)$$

Since the intensity maxima of a self-reversed line lie at frequencies satisfying Eqs. (155) and (156), the maximum value of I(u) from Eqs. (154) and (136) is

$$I_{max} = \frac{c}{4\pi} \frac{AN_e}{BN_a} q \int_0^2 E(y) \exp(-qy) \, dy. \qquad (157)$$

Furthermore, from the definitions of $E(y)$ (Eq. (147)), $\varepsilon(\nu)$ (Eq. 12)), and $k(\nu)$ (Eq. (22)), and assuming that the source function at the arc axis is given by Wien's function, Eq. (157) is written:

$$I_{max} = W_\nu(T_m) q \int_0^2 \frac{E(y)}{E(1)} \exp(-qy) dy. \qquad (158)$$

We now consider two limiting cases:

i) the uniformly excited source in which the excitation is homogeneous; and

ii) the source having emitting and absorbing atoms spatially separated, i.e.; all the emitting atoms are concentrated in the arc center while all the absorbing atoms are distributed outside this region.

In the first case where E(y) is constant, Eq. (155) is reduced to:

$$\exp(-2\mu) = \exp\left[-p \frac{P(u)}{P(0)}\right] = 0 \qquad (159)$$

This expression has no solution for finite values of p. Therefore, reversal can never occur in a plasma where E(y) is constant. The constancy of E(y) means that $n_e(r)$ and $n_a(r)$ are not necessarily constant but that the ratio $n_e(r)/n_a(r)$ must be constant. This is the case for a plasma in which the temperature is constant throughout the region occupied by the radiating atoms.

The intensity distribution is found from Eq. (154) to be:

$$i(u) = I_0 \frac{P(0)}{P} \frac{\bar{n}_e}{\bar{n}_a} \left\{ 1 - \exp\left[-2p \frac{P(u)}{P(0)}\right] \right\}$$

$$= I_0 \frac{P(0)}{2p} \left\{ 1 - \exp\left[-2p \frac{P(u)}{P(0)}\right] \right\}. \qquad (160)$$

Uniformly excited sources are easily recognizable in that a high degree of self-absorption does not result in reversal but only in flat-topped broadened spectral lines.

In the case of a source having emitting and absorbing atoms spatially separated, E(y) is non-zero only at y=1 (i.e., r=0) and Eq. (154) is then reduced to:

$$I(u) = I_0 P(u) \exp\left[-p \frac{P(u)}{P(0)}\right]. \tag{161}$$

This expression is independent of n_e and n_a, and it is also applicable to the case in which the arc consists of a central core that contains all the emitting atoms surrounded by a layer which contains all the absorbing atoms.

For a Lorentzian line profile $I(u)$ is written:

$$I(u) = \frac{I_0}{\pi\delta} \frac{\delta^2}{u^2+\delta^2} \exp\left(-p \frac{\delta^2}{u^2+\delta^2}\right). \tag{162}$$

In the line center (u=0):

$$I(0) = \frac{I_0}{\pi\delta} \exp(-p) \tag{163}$$

Differentiation of Eq. (162) gives as a condition for reversal:

$$q = 1 \tag{164}$$

and with the aid of Eq. (156), we obtain the frequencies at which line reversal occurs:

$$s = \pm\, \delta(p-1)^{1/2}. \tag{165}$$

It is then obvious that reversal occurs if p>1. The intensity at the side maxima, specified by equation (165), is

$$I_{max} = \frac{I_0}{\pi\delta} \cdot \exp(-1)/p \tag{166}$$

and the ratio of maximum to minimum is:

$$I_{max}/I(0) = [\exp(p-1)]/p \qquad p \geq 1. \tag{167}$$

Description of the Line Contour

In an actual source, an intermediate state between the two limiting cases discussed above will be the case: the excitation decreases gradually from the center source to the edge of the source. For the description of the line contour emitted by such an inhomogeneous light source it is obvious from Eq. (148) that the functions $E(y)$ and $P(\nu,y)$ must be known.

The excitation function. Very little is known about the exact form of the excitation function. For an axially symmetric source, Cowan and Dieke have proposed the following relation between $n_e(x)$ and $n_a(x)$:

$$\bar{n}_e(x) = \frac{n}{2} \bar{n}_a(x) \left[\int_{|x|}^{+x_0} \bar{n}_a(x')dx'\right]^{n-1} \tag{168}$$

where n is a positive number (the inhomogeneity parameter) and the factor n/2 arises from the normalization condition (138). Thus, with the aid of Eqs. (143), and (168), the excitation function (Eq. (147) is written:

$$E(y) = \begin{cases} (n/2)y^{n-1} & \text{for } 0 \leq y \leq 1 \\ (n/2)(2-y)^{n-1} & \text{for } 1 \leq y \leq 2 \end{cases} \tag{169}$$

From Eqs. (12) and (22), the source function (Eq. (60)) is written:

$$S(\nu,x) = (c/4\pi)(A/B) [n_e(x)/n_a(x)], \qquad (170)$$

Therefore, the excitation function becomes:

$$E(y) = \frac{4\pi}{c} \frac{B}{A} \frac{N_a}{N_e} S(\nu,x) \qquad (171)$$

So, the physical interpretation of the excitation function becomes obvious: it is proportional to the source function. On the other hand the inhomogeneity parameter n determines the trend of the source function and indicates the state of excitation decrease from the center of the source to the edge of the source.

For y = 1 (x=0) one has:

$$E(1) = \frac{n}{2} = \frac{n_e(x=0)}{n_a(x=0)} \cdot \frac{N_a}{N_e} . \qquad (172)$$

Hence, the parameter n takes into account the plasma structure without making any particular assumptions about the thermodynamic equilibrium of the plasma, the shape of the temperature profile, or the radial variation of the initial spectral profile.

The inhomogeneity parameter depends on the individual radiator species density and the individual line. The value of n increases with the inhomogeneity of the radiators. n=1 corresponds to uniformly distributed radiators in which excitation is homogeneous. For n=+∞ the radiator density is completely inhomogeneous.

For the case of a source in LTE, where the partial pressure of the radiating specie is constant, and the depletion of the ground state population due to excitation and ionization can be neglected, the validity of Boltzmann's formula (3) leads to the relation:

$$n = \exp(-\frac{E_e - E_a}{kT_m}) \frac{\int_{-x_0}^{+x_0} \frac{1}{T(x)} \exp[-\frac{E_a}{kT(x)}]dx}{\int_{-x_0}^{+x_0} \frac{1}{T(x)} \exp[-\frac{E_e}{kT(x)}]dx} \qquad (173)$$

where T_m is the arc axis temperature.

From this relation it is found that the value of the parameter n depends upon the excitation energies of both the upper and lower levels of the transition, and upon the shape of the temperature profile. The smallest values of n correspond to lines in which the lower level is high. For resonance lines, n is larger.

To obtain the effect of the temperature profile on the parameter n, it is assumed that along the diameter of a cross-section of the plasma column (z=0):

$$T(x) = T_m - (T_m - T_w)(x/x_0)^\ell \qquad (174)$$

where T_w is the temperature at x_0 (boundary of the temperature profile) and ℓ is a parameter which determines the temperature distribution, with $\ell > 0$.

Figure 30 gives the values of ℓ versus n for the mercury lines at 5461 Å and 4047 Å with $T_w = 1 \times 10^3$ K and $T_m = 6.2 \times 10^3$ K. n depends very weakly upon T_m and T_w.

For the 5461 Å line emitted by a high pressure (~ 3 atm) mercury discharge operated at 100W/cm (Table 6), the experimentally found value of n is n ≃ 1.3, which corresponds (Fig. 30) to $\ell \simeq 1.2$, while the measured value of , using the 5770 Å line is $\ell = 2.7$. This difference may be interpreted as being due to deviations from LTE (Karabourniotis, 1986).

So, if we assume that LTE is valid at the discharge axis, and that the 7^3S_1 level of the transition $7^3S_1 \to 6^3P_2$ (5461Å) is in equilibrium with the higher atomic levels from which the 5770 Å line originates, i.e., the 7^3S_1 level is populated according to Eq. (174) with $\ell = 2.7$, then from Eq. (173) calculation gives n = 1.3, if the lower 6^3P_2 level is populated according to Eq. (174) with $\ell = 3.2$. This means an overpopulation of the 6^3P_2 level in the outer region of the discharge.

In the following, the calculation of the line contour is made assuming an excitation function given by equation (169) and a Lorentzian line profile.

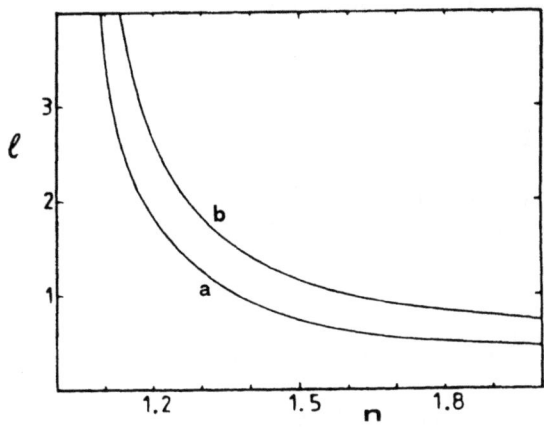

Fig. 30. Theoretical dependence of the parameter ℓ which determines the temperature profile upon the inhomogeneity parameter n for the mercury 5461 Å (curve a) and 4047 Å (curve b) lines with $T_m = 6.2 \times 10^3$ K and $T_w = 1 \times 10^3$ K (Karabourniotis, 1983).

Table 6. Data on Discharge Tubes, Filling, and Operating Conditions

Discharge	Internal Diameter (mm)	Electrode Distance (mm)	Filling	Discharge Current (A)	Tube Voltage (V)
a	20	100	80mg Hg+1mg HgI$_2$+ 8 torr Ar	6.2 rms	162 rms
b	20	150	80mg Hg+10 torr Ar	8.2 rms	198.5 rms
c	20	150	80mg Hg+10 torr Ar	8.2	207

<u>Position - independent line profile</u>. In the case of a position - independent line profile (or a line profile averaged along the direction of observation), Eq. (154) may be written in the form:

$$I(w)=I_0(n/2)P(0)J(w) \tag{175}$$

where $J(w)$ is the relative intensity at the dimensionless frequency:

$$w = (\nu-\nu_0)/\delta_0 . \tag{176}$$

The relative intensity is given by:

$$J(w) = \frac{P(w)}{P(0)} \int_0^1 y^{n-1} \exp[-\mu(w)y]\{1+\exp[2\mu(w)(y-1)]\}dy . \tag{177}$$

In the calculations we have assumed that $P(w)$ is a dispersion profile:

$$P(w) = \left[\pi\delta(w^2 + 1)\right]^{-1} \tag{178}$$

therefore:

$$P(0) = (\pi\delta)^{-1}. \tag{179}$$

Figure 31 shows the influence of the variation in the parameters n and p on the characteristics of the line shapes.

The condition for reversal (Eq. (155)) becomes:

$$\int_0^1 (1-qy) \, y^{n-1} \exp(-qy) \left\{1+\exp\left[2q(y-1)\right]\right\} dy = 0 \tag{180}$$

and the relation (158) for the line maximum is written:

$$I_{max} = W_\nu(T_m) K_0 \tag{181}$$

with

$$K_0 = q \int_0^1 y^{n-1} \exp(-qy)\{1+\exp[2q(y-1)]\}dy. \tag{182}$$

Therefore, it can be seen from equations (180) and (182) that K_0 depends only upon the value of the inhomogeneity parameter n.

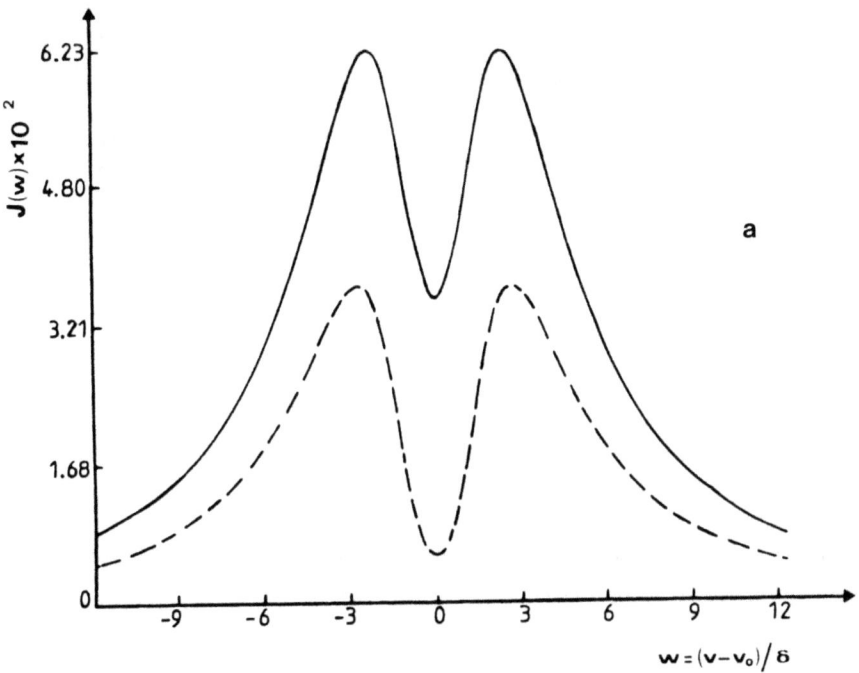

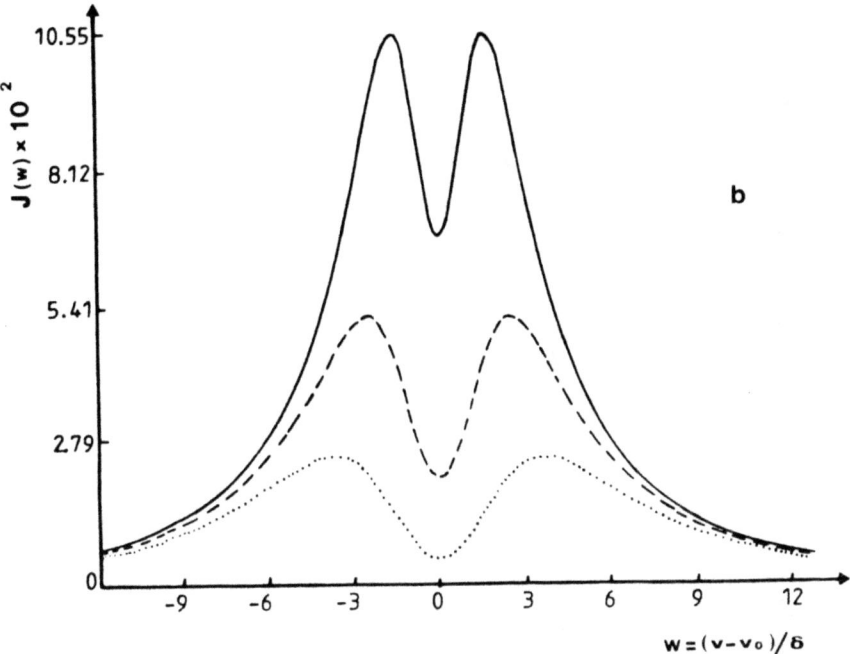

Fig. 31. Influence of the inhomogeneity parameter n and of the absorption parameter p on the relative intensity J(w) of a self-reversed line. (a): p=10; solid curve: n=1.4; dashed curve: n=2.2. (b): n=1.6; solid curve: p=5; dashed curve: p=10; dotted curve: p=20. (Karabourniotis, 1983)

It is therefore possible, using equation (180), to determine the arc axis temperature T_m from the measurement of I_{max} if the value of the inhomogeneity parameter n is known. Figure 32 gives q and K_0 as a function of the inhomogeneity parameter n.

<u>Position - dependent line profile.</u> We now assume that the line profile is described by a dispersion function with a shift given by Eq. (27).

This equation may be written in the form:

$$P(w,y) = (1/\pi\delta_0)P'(w,y), \qquad (183)$$

with:

$$P'(w,y) = \delta'(y) \left\{ [w-\eta\delta'(y)]^2 + [\delta'(y)]^2 \right\}^{-1}, \qquad (184)$$

$$\delta'(y) = \delta(y)/\delta_0, \qquad (185)$$

$$\eta = \Delta(y)/\delta(y), \qquad (186)$$

where δ_0 is the maximum half-width at the discharge axis, $\delta'(y)$ is the relative half-width, and $\Delta(y)$ is the line shift.

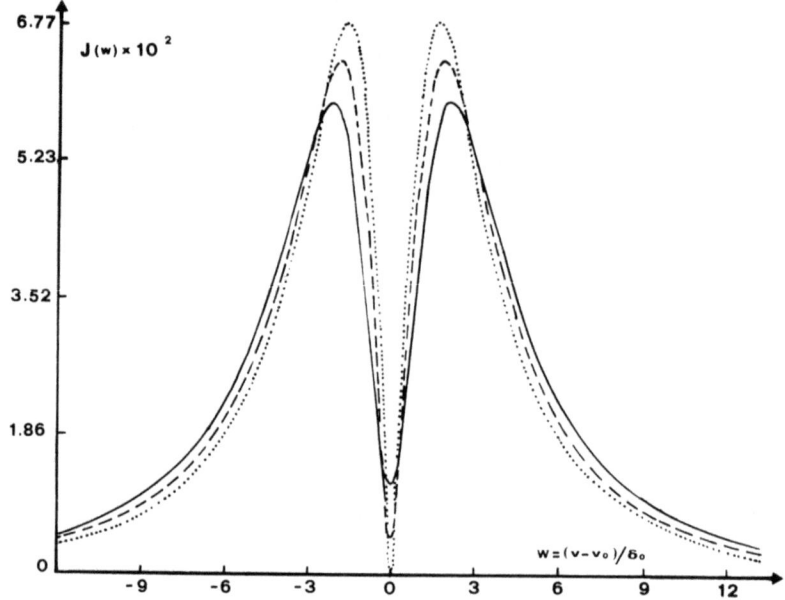

Fig. 32. Theoretical dependence of q, K_0, D and MY upon the inhomogeneity parameter n. (Karabourniotis, 1983).

Therefore, in Eq. (175), the relative line intensity is written:

$$J(w) = \int_0^1 y^{n-1} P'(w,y) \exp\left[-p_0 \int_0^y P'(w,y')dy'\right] \cdot$$

$$\left\{1 + \exp\left[2 p_0 \int_0^y P'(w,y')dy' - \int_0^1 P'(w,y')dy'\right]\right\} dy \quad (187)$$

in which:

$$p_0 = (h\nu_0/c) \, BN\alpha \, (1/\pi\delta_0) . \quad (188)$$

For the relative half-width the approximation:

$$\delta'(y) = \exp\left[-b(1-y)^\ell\right], \quad (0 \leq y \leq 1), \quad (189)$$

is used, where b is constant, and ℓ is defined in Eq. (174). The shift/width ratio is assumed to be constant throughout the discharge plasma.

Figure 33 shows the effect of the variation of the parameters b, ℓ, and η on the shape of the line contours. The lines are symmetrically self-reversed for b=0 and η =0, b≠0 and η =0, and b=0 and $\eta \neq$0. They are shifted and asymmetrically self-reversed if b≠0 and $\eta \neq$0.

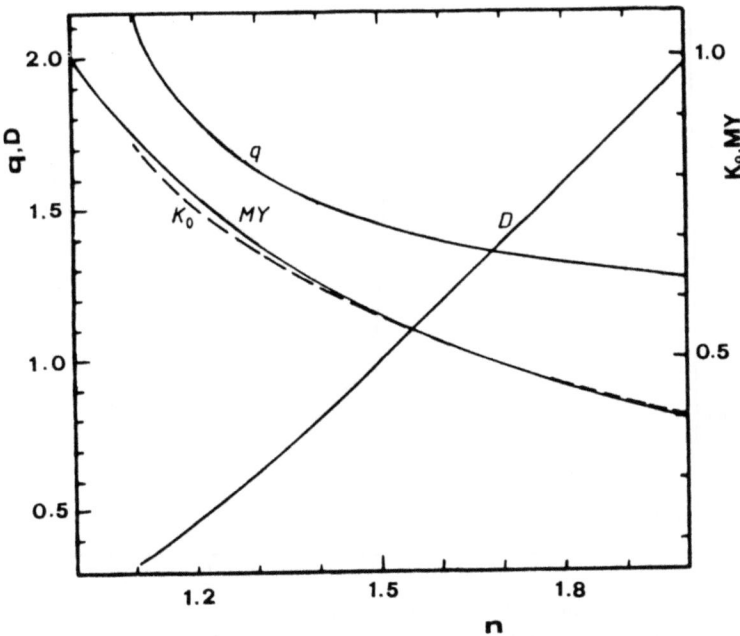

Fig. 33. Influence of the parameter b, ℓ, and η on the contour of a self-reversed line with n = 1.6, and p_0=10.

(a) Influence of the parameter b, for ℓ=2, and η = 0. Solid curve: b=1, dashed curve: b=2; dotted curve; b=3;

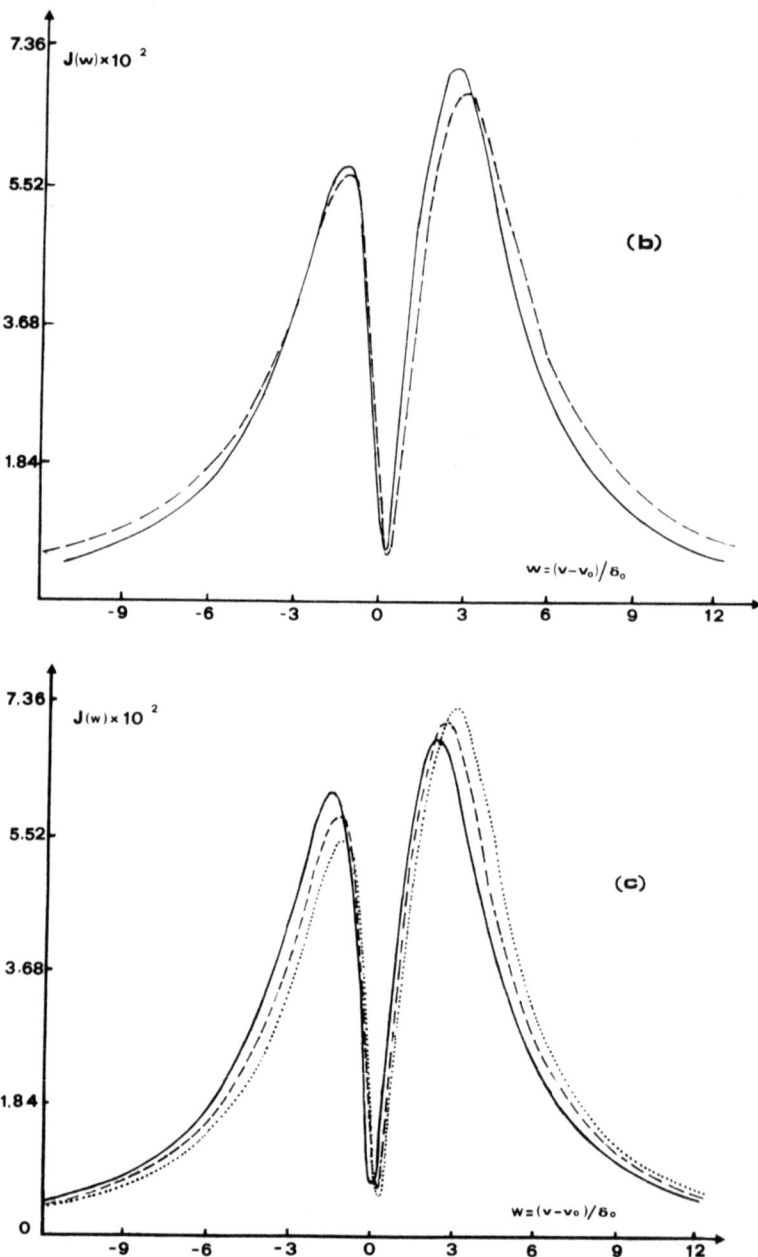

Fig. 33. Influence of the parameter b, ℓ, and η on the contour of a self-reversed line with n=1.6, and $p_0=10$.

(b) Influence of the parameter ℓ, for b=2, and η = 1.2. Solid curve: ℓ=2; dashed curve: ℓ=4;

(c) Influence of the parameter η for b=2, and ℓ=2. Solid curve: η = 0.7, dashed curve: η = 1.2; dotted curve: η = 1.74 (Karabourniotis and Karras, 1985).

In particular, on the frequency scale for $\eta>0$, i.e., for a blue shift of the line profile, the line contour shows a blue shift and an enhancement of the blue wing and for $\eta<0$, i.e., for a red shift the line contour shows, symmetrically to $w=0$, a red shift and an enhancement of the red wing. It is obvious that on the wavelength scale $\eta>0$ corresponds to a red shift of the line profile, and $\eta<0$ to a blue shift.

Relationship between the parameters. From the calculated line contours, the relation between $\log(J_{max;1}/J_{min})$ and $\log 2s_0$ is plotted in Fig. 34 for different values of the parameters n, b, ℓ, and η, by varying the value of p_0. Here $J_{max;1}/J_{min}$ is the ratio of the larger maximum to central minimum of the relative intensity, and $2s_0$ is the distance between the two side maxima of the line in units of δ_0, i.e.

$$2s_0 = 2s/\delta_0 \qquad (190)$$

It appears that, apart from the parameter n, the variation of the other parameters does not appreciably affect the slope of the curves shown in Fig. 34.

Therefore, for a given value of n, the slope of these curves coincides with the slope of the corresponding values calculated for $b=0$ and $\eta=0$ (position - independent profile). Figure 32 gives the slope $D=d\ell \log(J_{max;1}/J_{min})\ell/d(\log 2s_0)$ versus the inhomogeneity parameter n. Therefore, if the slope D of the experimental line profiles is known, one may determine the value of n.

Increasing the value of the parameter b, for given values of n, ℓ, and η, causes a shift of the curves $\log(J_{max;1}/J_{min})=f(\log 2s_0)$ towards the lower values of $\log 2s_0$. This shift is designated by the quantity F which is the distance between the $\log(J_{max;1}/J_{min})=f(\log 2s_0)$ plotted for a given value of b, and the corresponding curve plotted for $b=0$, $\eta=0$ (Fig. 34). Thus we have plotted (Fig. 35) the dependence of the shift F, upon b, for different ℓ and for different values of the inhomogeneity parameter n.

As can be seen from Figs. 34 and 35, an increase of ℓ and/or η causes a decrease of F, but the dependence of F upon η is particularly weak.

From the definition of the quantity F, neglecting the influence of $\eta(\eta=0)$, one may write:

$$F = \log(2s/\delta) - \log(2s/\delta_0) = \log(\delta_0/\delta). \qquad (191)$$

The asymmetry of the line contour is characterized by the ratio $J_{max;1}J_{max;2}$, where $J_{max;2}$ is the smaller maximum. Fig. 36 shows the calculated dependence of the ratio $J_{max;1}/J_{max;2}$, upon η for three sets of the parameters n, b, ℓ, and p_0.

On the frequency scale, for $\eta>0$, $J_{max;1}$ and $J_{max;2}$ are the blue and red maxima, respectively, and for $\eta<0$, $J_{max;1}$ and $J_{max;2}$ are the red and blue maxima.

On the wavelength scale, for $\eta>0$, $J_{max;1}$ and $J_{max;2}$ are the red and blue maxima, respectively, and, for $\eta<0$ $J_{max;1}$ and $J_{max;2}$ are the blue and red maxima.

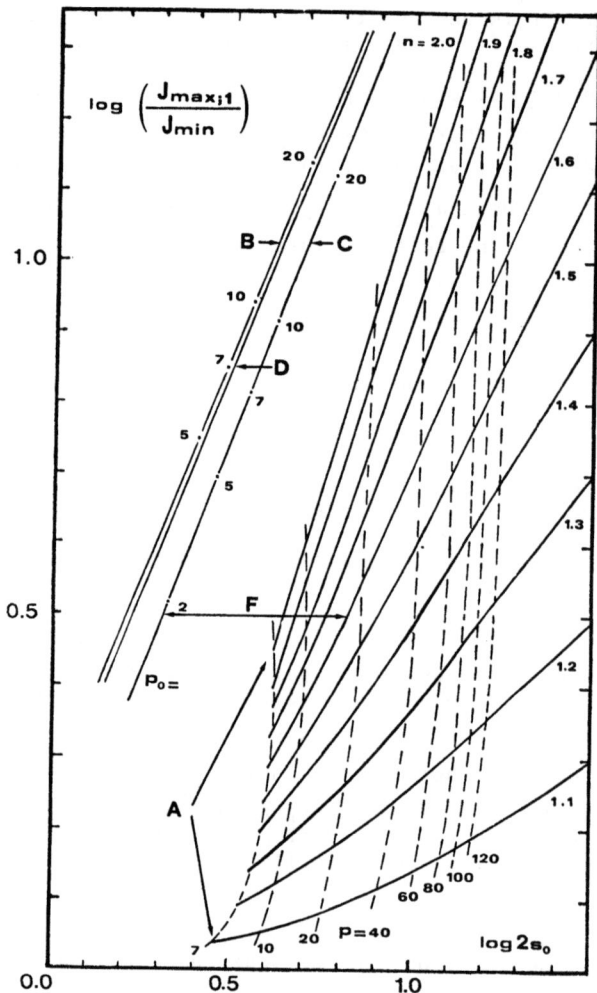

Fig. 34. Theoretical dependence of log ($J_{max;1}/J_{min}$) upon log $2s_o$. $J_{max;1}/J_{min}$: ratio of the larger maximum to the central minimum of the calculated contour; $2s_o$: distance of the two side maxima of the line in units of δ_o (maximum half-width). Curves A: position-independent line profile (b=0, η=0) for different values of n and p. Curves B, C, D: position-dependent line profile with n=1.6 and different values of p_o. B: b=2, η=0, ℓ=2; C: b=2, η=0, ℓ=4; D: b=2, η=1.74, ℓ=2. F is the distance between the curve C and the corresponding curve (n=1.6) with b=0 and η=0 (Karabourniotis and Karras, 1985).

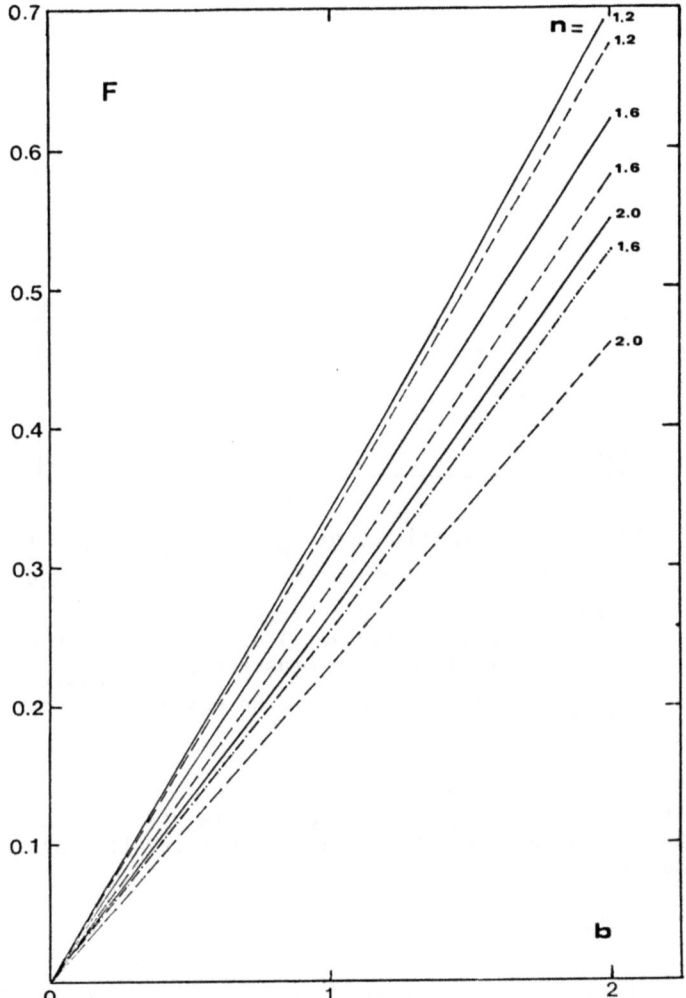

Fig. 35. Calculated dependence of F upon b for different values of n. Solid curves: $\ell=2$; Dashed curves: $\ell=3$; dot-dashed curves; $\ell=4$. (Karabourniotis and Karras, 1985).

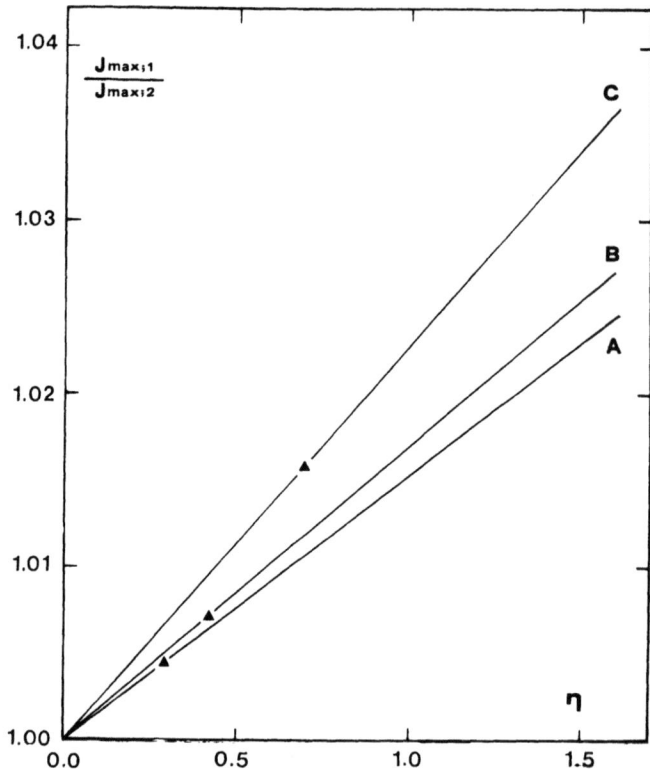

Fig. 36. Theoretical dependence of $J_{max;1}/J_{max;2}$ upon η for three sets of the parameters n, b, ℓ, and p_o, corresponding to the discharges a (curve A), b (curve B), and c (curve C) of Table 1; Δ experimental points obtained from the measurement of $I_{max;1}$ and $I_{max;2}$ are the red and the blue maximum of the line, respectively.

On the other hand from Eq. (175) and assuming thermodynamic equilibrium at the discharge axis, the maximum line intensity may be expressed in the form:

$$I_{max;1} = W_{\nu_0}(T_m) \, p_0 J_{max;1} \tag{192}$$

Calculation has shown that the product $p_0 J_{max;1}$ (=K) is independent of the value of p_0. Thus, for a given value of the inhomogeneity parameter n, K is a function of b, η, and ℓ. Figure 37 gives K plotted against n for different values of b, η, and ℓ.

Therefore, knowing K, it is possible to determine the temperature at the arc axis from the measurement of the absolute intensity $I_{max;1}$, by taking into account the radial variation of the line broadening, the line shift, and the temperature distribution.

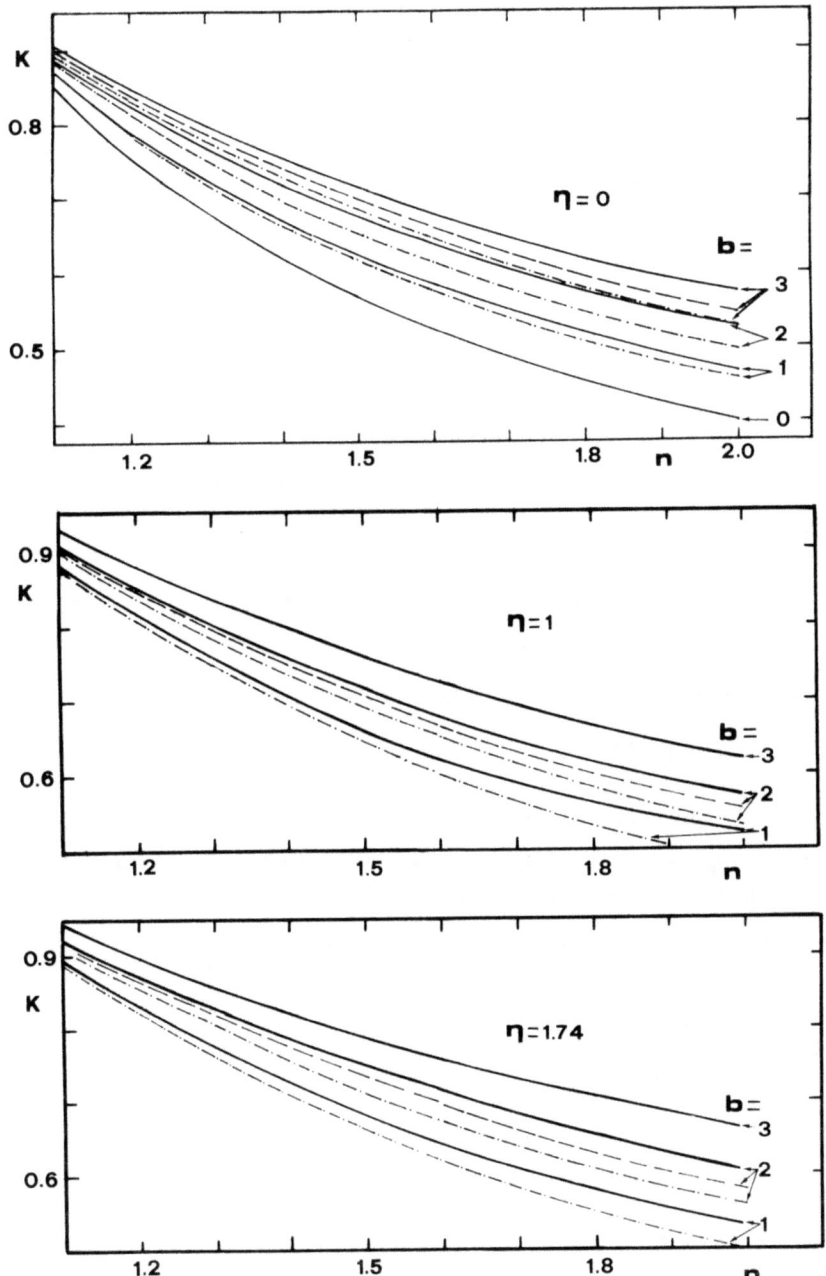

Fig. 37. Theoretical dependence of K upon the inhomogeneity parameter n for different values of b, η, and ℓ. Solid curves: $\ell = 2$; dashed curves: $\ell = 3$; dot-dashed curves: $\ell = 4$.

If the self-reversed maxima lie in the wings of the line profile with $s_0 \gg \delta'$ and $\eta\delta'$, then for $w = \pm s_0$ Eqs. (156) and (184) become:

$$P'(s_0, y) \simeq \delta'(y)/s_0^2 \tag{193}$$

and

$$q \simeq p_0/s_0^2. \tag{194}$$

Therefore K is written:

$$K = q \int_0^1 y^{n-1} \delta'(y) \exp\left[-q \int_0^y \delta'(y')dy'\right] \left\{1 + \exp\left[2q \int_1^y \delta'(y')dy'\right]\right\} dy. \tag{195}$$

For the calculation of K it is now necessary to know n and the variation of $\delta'(y)$.

<u>Radial variation of the line broadening</u>. In order to derive the approximation (189) concerning the radial variation of the relative half-width we assume that the total half-width is given by:

$$2\delta = 2\delta_n + 2\delta_{ei} \tag{196}$$

where δ_n is caused by collisions with neutral atoms and δ_{ei} is due to collisions with charged particles.

In the impact approximation, $2\delta_n$ may be expressed by:

$$2\delta_n(T) = A_n P_g T^{-7/10} \tag{197}$$

where P_g is the pressure of the emitting element, and A_n is a constant. The value of $2\delta_{ei}$ is calculated from Eq. (30) in which the electron density has been found using the Saha Eq. (5).

Thus, knowing the temperature profile given by Eq. (174), one may calculate the radial variation of the relative half-width by:

$$\delta'(\rho) = \frac{2\delta_n(\rho) + 2\delta_{ei}(\rho)}{2\delta_{0;n} + 2\delta_{0;ei}} \tag{198}$$

where $\rho = x/x_0$.

Figure 38 shows the radial variation of δ' for the mercury 5461 Å line emitted by a fictitious mercury discharge with $P_g = 3$ atm, $T_m = 6000K$, $T_w = 1000K$, $A_n = 1.53 \times 10^4$ mA atm^{-1}k$^{0-7}$, and for $\ell = 2, 3,$ and 4.

On the other hand if we assume the validity of the Boltzmann formula and constancy of the pressure, one may calculate from Eqs. (138), (139) and (143), the radial variation of the parameter y by:

$$y(\rho) = V(\rho)/V(0) \tag{199}$$

in which:

$$V(\rho) = \int_\rho^1 [1/T(\rho')] \exp[-E_a/kT(\rho')] \, d\rho \tag{200}$$

Figure 39 shows the radial variation of the function $y(\rho)$ given by Eqs. (199) and (200).

The dependence of δ' on $(1-y)^{\ell}$ for different values of ℓ and A_n is plotted in Fig. 40. It is seen that δ' is very sensitive to A_n (Fig. 40(a), but depends weakly upon ℓ (Fig. 40(b)). Hence, when the value of ℓ changes from $\ell=3$ to $\ell=2$, the variation of δ', for $(1-y)^{\ell} = 0.5$, amounts to less than 2%, and when ℓ changes from $\ell=3$ to $\ell=4$, to less than 4%.

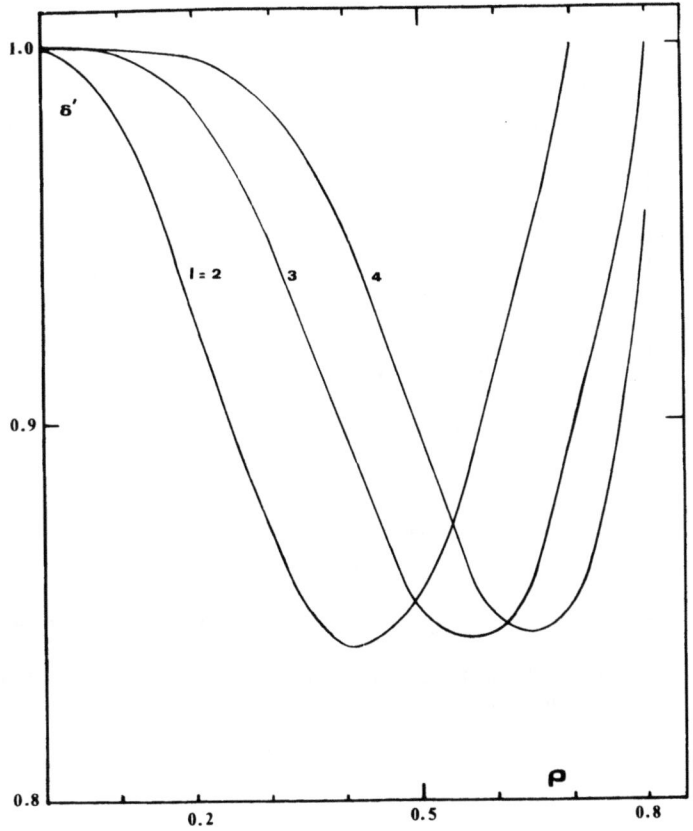

Fig. 38. Radial variation of the relative half-width for different values of the parameter ℓ.

As can be seen from Fig. 40(a), the dependence of δ' upon $(1-y)^{\ell}$ in the range $0 < (1-y)^{\ell} \leq 0.92$ may be expressed by the approximation given in equation (189).

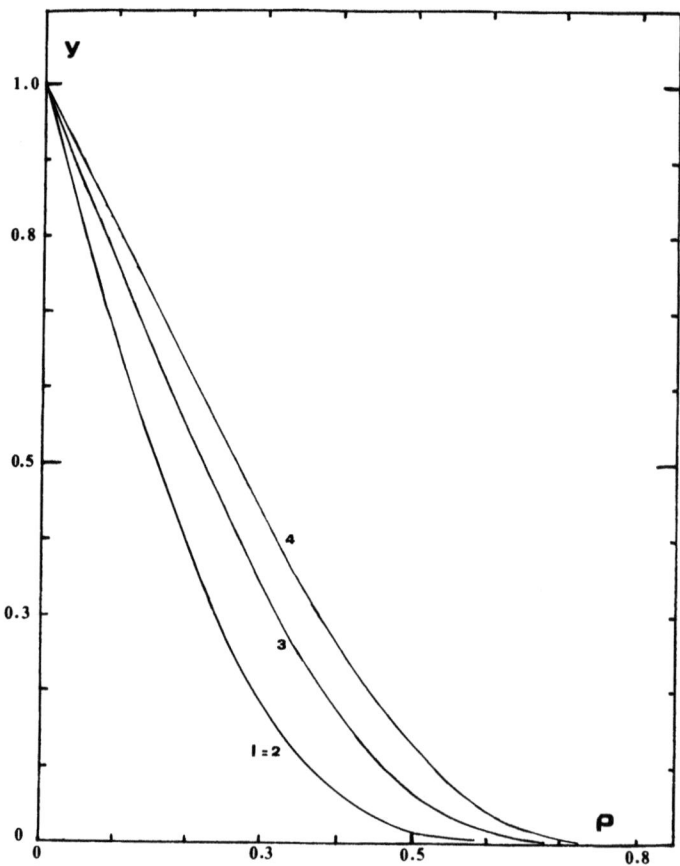

Fig. 39. Radial variation of the parameter y for different values of the parameter ℓ.

In the peripheral region of the discharge, corresponding to $0.92 \leq (1-y)^{\ell} \leq 1$, one obtains $0 \leq y \leq 0.04$, 0.03, and 0.02 for $\ell = 2, 3,$ and 4, respectively. This means that the number of absorbing atoms in this

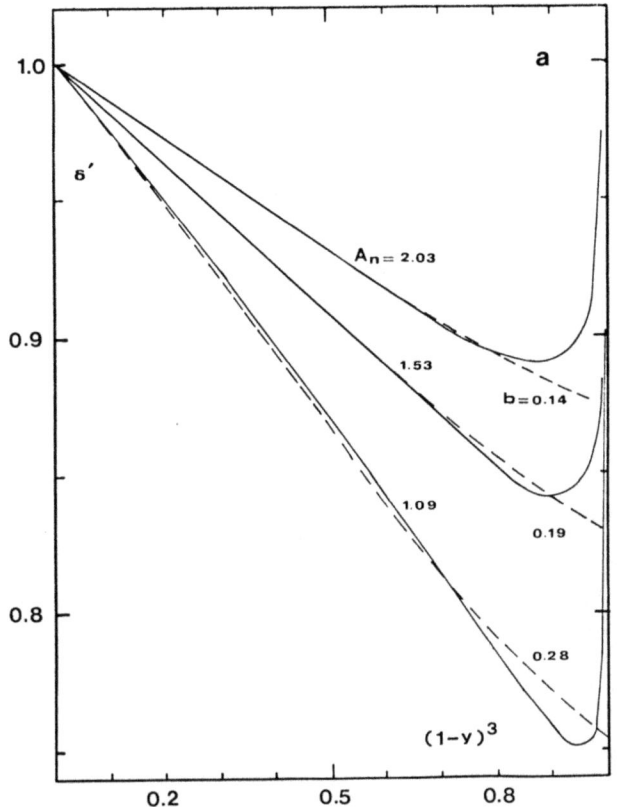

Fig. 40. Dependence of the relative half-width δ upon $(1-y)^{\ell}$ for the Hg-5461 Å line emitted by a fictitious discharge with $T_m = 6000$ K, $T_w = 1000$ K and $P_g = 3$ atm. (a): for different values of A_n (in 10^4 mÅ atm^{-1} K$^{0.7}$), with $\ell = 3$; solid curves: exact relation; dashed curves: approximation given by equation (189).

region is lower than 4, 3 and 2%, respectively, of the total number of absorbing atoms. Therefore, the dependence of the relative half-width δ' upon y can be described with quite good accuracy by the relation (189).

Determination of the parameters. Figure 41 gives the measured dependence of $I_{max;1}/I_{min}$ upon $2s$ for the 5461Å line emitted along different distances from the discharge axis for different mercury discharges (Table 6). $I_{max;1}$ is the red maximum of the line.

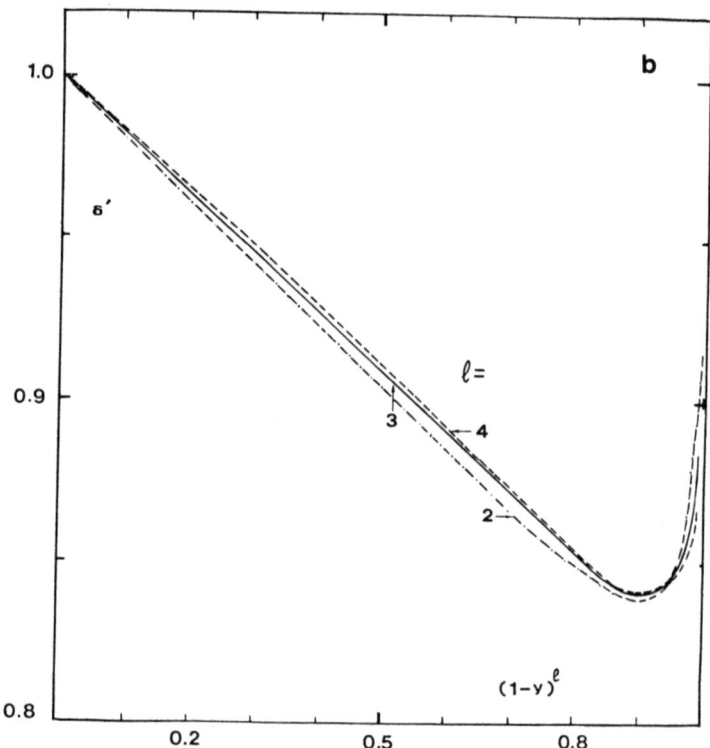

Fig. 40. Dependence of the relative half-width δ upon $(1-\gamma)^\ell$ for the Hg-5461 Å line emitted by a fictitious discharge with $T_m=6000$ K, $T_w=1000$ K and $P_g=3$ atm.

(b): for different values of ℓ with $A_n=1.53.19^4$ mA atm^{-1} K$^{0.7}$.

From Fig. 32 (curve D) the slope of this curve gives the value of n. Knowing n, and from the measurement of $I_{max;1}/I_{min}$ at $z=0$, we find from Fig. 34 the values of $2s_0$ and p for $b=0$, $\eta=0$. So, from Eq. (190) and the measured value of $2s$, we determine the value of δ, where now $\delta=\delta_0$ is the half-width of the averaged line profile. The results are listed in Table 7. Using:

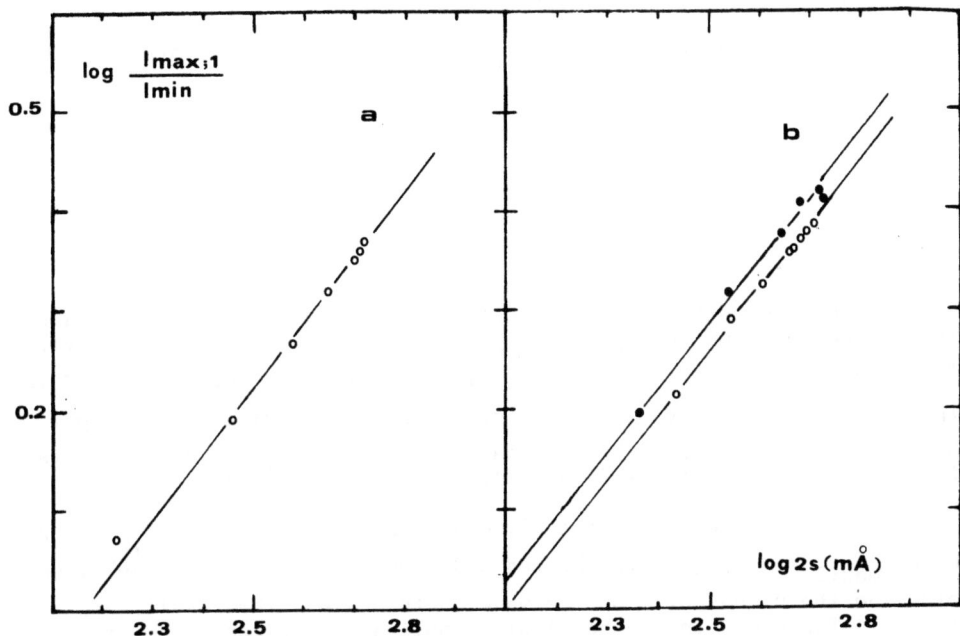

Fig. 41. Experimental dependence of $\log(I_{max;1}/I_{min})$ upon $\log 2s$ for the Hg 5461Å line; $I_{max;1}$ is the red maximum of the line intensity. o: ac operation; ●: dc operation, (a): discharge a (b): discharges b and c.

Table 7. Experimentally determined values of the inhomogeneity parameter n, of the absorption parameter p and of the half-width δ for b=0, η=0.

Discharge	n	I_{max}/I_{min} at z = 0	$2s_o$ at z = 0	2s(mÅ) at z = 0	p	δ (mÅ)
a	1.33	2.35	8.91	524	34.8	57.5
b	1.29	2.43	11.48	513	56.0	44.7
c	1.31	2.63	11.61	525	56.2	45.2

(i) the dependence of A_n upon b (Fig. 40),

(ii) the dependence of F upon b (Fig. 35), and

(iii) Eqs. (191), (196) and (197)

we determine the values of A_n, b, δ_{0n} and δ_0. The results are listed in Table 8.

With an error 6% in the calculation of δ, and maximum error of about 2% in the determination of the axis temperature, the errors in the determination δ_0 and b are 12 and 15%, respectively, while the errors in the determination of $\delta_{0;n}$ and A_n are 8%. The mean value of A_n is then

$$A_n = (1.43 \pm 0.11) \times 10^4 \text{ m\AA atm}^{-1} K^{0.7}.$$

In order to estimate the parameter η (shift/width ratio), we have plotted the relation $\log(J_{max;1}/J_{min}) = f(\log 2s_0)$ using the known values of n, b, and ℓ, for different values of p_0, as for the curves B and C in Fig. 34. Knowing $2s_0$ ($=2s/\delta_0$), we graphically determine value of p_0. Therefore we obtain from the computation the ratio $J_{max;1}/J_{max;2}$ in terms of the unknown parameter η (Fig. 36).

Hence, from the experimentally measured ratio $I_{max;1}/I_{max;2}$, at z=0, where $I_{max;1}$ and $I_{max;2}$ are the red and blue maximum, respectively, we determine the values of η on the wavelength scale. The results are listed in Table 8. In the particular applications given above, an error of up to 30% is possible in the determination of the ratio η, due to the nearly symmetric shape of the lines and the experimental uncertainties.

Diagnostic Methods Based on Cowan-Dieke's Model

Temperature determination. In the case of a position - independent Lorentzian line profile, Eq. (181) is written:

$$I_{max} = \frac{2h\nu_0^3}{c^2} \exp\left(-\frac{h\nu_0}{kT_m}\right) \cdot K_0 \tag{201}$$

where K_0 depends on the inhomogeneity parameter n (Fig. 32) thus, the measurement of I_{max} and n permits the determination of the temperature T_m if LTE is valid at the discharge axis (Karabourniotis, 1983).

It is also possible to determine the temperature from the relation (Karabourniotis et al., 1982):

$$I_t = \frac{2h\nu_0^3}{c^2} \exp\left(-\frac{h\nu_0}{kT_m}\right) \cdot pJ_t \tag{202}$$

where I_t is the total intensity of the self-reversed line and J_t is the total relative intensity:

$$J_t = \int_{-\infty}^{+\infty} J(u)du \tag{203}$$

Table 8. Determined values of A_n, b, $2\delta_{0;n}$, $2\delta_0$, p_0 and η (on the wavelength scale).

Discharge	A_n (10^4 mÅ atm^{-1}K$^{0.7}$)	b	$2\delta_{0;n}$ (mÅ)	$2\delta_0$ (mÅ)	p_0	η
a	1.41	0.21	95.0	135	28	0.29
b	1.51	0.32	69.6	113	42	0.42
c	1.38	0.42	61.7	128	37	0.69

In the case of a position-dependent Lorentzian line profile, (Eq. (192)) it is possible to determine the temperature from the relation:

$$I_{max;1} = \frac{2h\nu_0^3}{c^2} \exp\left(-\frac{h\nu_0}{kT_m}\right) \cdot K \tag{204}$$

where K depends on n, b, η and ℓ (Fig. 37).

Pressure Determination. On the basis of the Cowan-Dieke simplified model (Eqs. (175)-(177)), Karabourniotis and Damelincourt (1982) have proposed the following relation for the determination of the pressure of the emitting element from self-reversed lines:

$$P_g = \frac{kc}{h\nu_0 B} \frac{Z}{g_a} \frac{p}{V} \frac{1}{P(0)} \tag{205}$$

where Z is the partition function and:

$$V = \int_0^{+x_0} [1/T(x)] \exp[-E_a/kT(x)]dx \tag{206}$$

The pressure P_g is considered to be constant throughout the plasma.

A relation of the same kind has been proposed by Karabourniotis et al., (1985b) in order to determine the averaged partial pressure of thallium in a mercury-thallium iodide discharge from the self-reversed thallium 5350Å line.

Relationship Between the Inhomogeneity Parameter and Bartels' Parameters

Preobrazhenskii (1959) has made a comparison between the Cowan-Dieke and Bartels models in the case of a position-independent line profile. He has proved that even in this case the theory of Bartels is broader and more complete than the Cowan-Dieke theory.

The reason for this is the different degree of generality achieved in solving the equation of radiative transfer, which, in the case of Cowan-Dieke's theory, consists of giving concrete form to the excitation function.

The imposition of the condition

$$p_B = 6[(n+1)(n+2)]^{-1}, \qquad (207)$$

where p_B (=p) is given by Eq. (84), makes it possible to obtain the contour of a self reversed line corresponding to the Cowan-Dieke simplified model (Eqs. (175 - 177), from the manifold of Bartels' contours as a particular case.

Comparing Eqs. (95) and (173) we find that in the case of a position-independent line profile:

$$n = 1/M. \qquad (208)$$

An approximate expression of the product $M Y_{max}$ appearing in Eq. (128) is obtained:

$$MY_{max} \simeq \frac{1}{n}\left\{0.736 + 0.264\left[\frac{6}{(n+1)(n+2)}\right]\right\} \qquad (209)$$

For the mercury 5461Å and 4047Å lines from the relation (208) with $M = (E_a/E_e)^{1/2}$ one obtains n=1,190 and 1.287, respectively, which corresponds to ℓ=1.19 (Fig. 32). The corresponding values of K_0 (0.765 for 5461Å and 0.690 for 4047Å) are very close to the values $M Y_{max}$ (0.763 for 5461Å and 0.687Å for 4047Å) obtained from equations (90), (104) and (120).

The difference between K_0 and the constant value $M Y_{max}$ obtained by Bartels' method, is small for plasmas of low inhomogeneity; it becomes large for highly inhomogeneous plasmas. Likewise, the difference between $T_m(K_0)$ and $T_m(M Y_{max})$ increases with plasma inhomogeneity; this difference depends on the individual source and it appears as a systematic error when one uses the limiting values of Bartels' method.

SUMMARY

We have discussed the physical conditions and the mechanisms in discharge plasmas which lead to the creation of a spectral line. We have especially studies the situation in a high pressure mercury discharge. The important simplifying assumption of local thermodynamic equilibrium has been presented and the necessary conditions for its validity have been discussed. The relative importance as well as the simultaneous action of different effects caused by electrons and atoms on the formation of the local line profile have been calculated.

The use of an alternating current discharge (50 Hz) has allowed us to separate the effect of the electrons from that of the atoms on the line broadening. This was accomplished by taking advantage of the strong variation in the electron density between 0 ms (minimum of the electron density) and 5 ms (maximum of the electron density). It was also shown that the asymmetry of the line wings can be partly explained by the quasistatic effect of the atoms.

The calculation of the self-reversed line contours has been carried out using the experimentally determined radial temperature profiles. The atom density and the electron density have been calculated by varying the parameters which characterize the properties of the spectral line. We have determined the parameters to which the dependences in the observed lines are most sensitive on the basis of calculated data and whether there are other parameters which could be determined independently of the values of these parameters. In order to do this we have chosen characteristic line quantities which are not very sensitive to certain parameters but, at the same time, quite sensitive to other parameters.

It has thus been possible to determine broadening constants and transitions probabilities by comparison between theoretical predictions and observed line contours at different instants in the current cycle of the discharge. The obtained set of constants leads to a satisfactory representation of the observed self-reversed line contours. Determination of such constants was, until now, only possible under otpically thin plasma conditions, where special experimental conditions and rigorous allowance for reabsorption and inhomogeneity of the plasma are required. A survey of Bartels' theory for the treatment of self- reversed line contours has been given. The behavior of the function $Y(\tau_0)$ has been analyzed.

In order to derive approximations for the parameters M and p it is necessary to know the form of the line profile and the temperature distribution. We have considered the case of an unshifted Lorentzian profile and approximations have been derived for M and p at the line wing and at the line center. Furthermore, assuming a parabolic temperature distribution, it has been shown that M and p may be approximated by very simple expressions in the case of a nonresonant line.

An evaluation of the uncertainties due to these approximations has been given. From this analysis a simple experimental method has been derived for the determination of the plasma temperature from a nonresonant symmetrically self-reversed spectral line, in the case of a plasma with parabolic temperature distribution.

The Cowan and Dieke theory of the self-absorption of spectral lines has been presented. In the simplified Cowan-Dieke model, the self absorption is completely described by three quantities; the line profile $P(u)$, the absorption parameter p, and the excitation function $E(y)$. A full description of the self-reversed line contours requires the knowledge of the exact form of the excitation function and of the line profile function. A concrete form for the excitation function has been proposed by introducing the inhomogeneity parameter n. As a consequence, the generality of the excitation function is restricted. The dependence of the inhomogeneity parameter upon the temperature profile has been determined and its relation to Bartels' parameters has been given. A Lorentzian distribution has been considered for the description of the local line profile.

We have proposed an approximation for the radial variation of the line broadening, while the shift/width ratio has been assumed to be constant throughout the plasma. The influence of the different parameters on the line contour has been investigated and we have described the interrelations between the parameters in graphical form. By implementing the multi-parameter method using a graphical technique and by comparing the theoretical and experimental results, we have found the parameters of the observed self-reversed contours.

Thus, by taking into account the asymmetry of the line contour, we have determined the line broadening at the discharge axis, its radial variation, and the shift/width ratio. At the same time, from the analysis of the theoretical results, a method for determining the discharge temperature has been derived, by taking into consideration the plasma structure, the variation of the line broadening and the self-reversed line asymmetry.

ACKNOWLEDGMENT

This work was performed as part of a Joint Research Program between our laboratory and the Centre de Physique Atomique of Toulouse University (France) and has been supported by the Greek Ministry of Research and Technology.

REFERENCES

Albrecht, H., and Schiff, A., 1976, Z. Naturforsch., 31a:196.
Aleksandrov, A. F., Galuzo, S., Yu Sanichev, A. T., and Timofeev, I. B., 1978, Sov. Phys. Tech. Phys., 23:42.
Assous, R., 1980, J. Quant. Spectrosc. Radiat. Transfer, 23:435.
Barrat, J. P., 1959, J. Phys. Rad., 20:42.
Bartels, H., 1949a, Z. Phys., 125:597.
Bartels, H., 1949b, Z. Phys., 126:108.
Bartels, H., 1950a, Z. Phys., 127:243.
Bartels, H., 1950b, Z. Phys., 128:546.
Bartels, H., 1953, Z. Phys., 136:411.
Bartels, H. and Beuchelt, R., 1957, Z. Phys., 149:594.
Bartels, H. and Zwicker, H., 1962, Z. Physik, 166:148.
Birkeland, J. W. and Oss, J. P., 1968, Appl. Opt., 7:1635.
Cayless, M. A., 1959, Brit. J. Appl. Phys., 10:186.
Cooper, J., 1968, Reports on Prog. in Phys., 29:36.
Couris, S., 1983, These de Specialite, Univ. P. Sabatier, Toulouse.
Cowan, R. D. and Dieke, G. H., 1948, Rev. Mod. Phys., 20:418.
Damelincourt, J. J., Asselman, A., El Hasnaoui, A., Guilhem, D., and Karabourniotis, D., 1983b, CIE 20th Session, Amsterdam.
Damelincourt, J. J., Aubes, M., Fragnac, P., and Karabourniotis, D., 1983a, J. Appl. Phys., 54:3087.
Damelincourt, J. J., Bordas, M., Karabourniotis, D., Scoarnec, L., and Villain, J., 1977, Rev. Gen. Electr. (Fr), 86:749.
Damelincourt, J. J., El Hasnaoui, A., Guilhem, D., Salon, J., and Karabourniotis, D., 1983b, in: "Proceedings of Third Intern. Symp. Sci. Tech. Light Sources," Toulouse.
Damelincourt, J. J., Karabourniotis, D., Karabourniotis, M., and Scoarnec, L., 1978a, J. Phys. D: Appl. Phys., 11:2207.
Damelincourt, J. J., Karabourniotis, D., Scoarnec, L., and Herbet, P., 1978b, J. Phys., D: Appl. Phys., 11:1029.
Davenport, A. J. and Lapworth, K. C., 1980, Nat. Phys. Lab. (UK), Report Qu 54.
de Groot, J. J., and Jack, A. G., 1973, J. Quant Spectrosc. Radiat. Transfer, 13:615.
Drawin, H. W., 1969, Z Physik, 228:99.
Drawin, H. W. and Felenbonk, P., 1965, in: "Data for Plasma in Local Thermodynamic Equilibrium," edited by Gauthier-Villars, Paris.
Drawin, H. W. and Felenbonk, P., 1965, in: "Data for Plasma in Local Thermodynamic Equilibrium," edited by Gauthier-Villars, Paris.
El Hasnaoui, A., 1983, These de Specialite, Univ. P. Sabatier, Toulouse.
Elder, P., Jerrick, T., and Birkeland, J. W., 1965, Appl. Opt., 4:589.
Engelsht, V. S. and Larkina, L. T., 1979, J. Quant. Spectrosc. Radiat. Transfer, 21:65.
Finken, K. H., Bertschinger, G., Maurmann, S., and Kunze, H. J., 1978, J. Quant. Spectrosc. Radiat. Transfer, 20:467.
Fragnac, P., 1981, These de Specialite, Univ. P. Sabatier, Toulouse.
Fromm, D. C., Seehawer, J., and Wagner, W. J., 1978, in: "Proc. Symp. on High Temp. Metal Halide Chemistry," edited by D. L. Hildenbrand and D. D. Cubiciotti, Princeton.
Funk, W., and Kloss, H. G., 1973, Beitr. Plasmaphysik, 13:101.
Funk, W., and Kloss, H. G., 1977, in: "Proc. 13th Int. Phen. Ion Gases," Berlin.
Funk, W. Kloss, H. G., and Serick, F., 1979, Beitr. Plasmaphysik, 10:487.
Going, W., 1952, Z. Phys., 131:603.

Griem, H. R., 1963, Phys. Rev. 131:1170.
Gurevich, D. B. and Podmoshenskii, I. V., 1963, Opt. Spectrosc., 15:319.
Hearn, A. C., 1964, Proc. Phys. Soc., 84:11.
Heitler, W., 1954, in: "The Quantum Theory of Radiation," Oxford University Press, London.
Hindmarsh, W. R., and Farr, J. M., 1972, in: "Progress in Quantum Electronics," Pergamon Press, Oxford, Vol 2.
Ilin, G. G. and Fishman, I. S., 1966, Opt. Spectrosc., 20:21.
Ilin, G. G. Protasevich, V. I., and Fishman, I. S., 1969, Opt. Spectrosc., 26:21.
Karabourniotis, D., 1977, These de Doctorat d' Etat, Univ. P.. Sabatier, Toulouse.
Karabourniotis, D., 1983, J. Phys. D: Appl. Phys., 16:1267.
Karabourniotis, D., 1986, Opt. Commun. (submitted).
Karabourniotis, D., Couris, S., and Damelincourt, J. J., 1986a, J. Quant. Spectrosc. Radiat. Transfer (to be published).
Karabourniotis, D., Couris, S., Damelincourt, J. J., and Aubes, M., 1986b, IEEE, Trans. Plasmas Sc., August.
Karabourniotis, D., Couris, S., and Karras, C., 1985, J. Appl. Phys. 58:2786.
Karabourniotis, D., Couris, S., Ziggig, G., and Damelincourt, J. J., 1986c, Int. Symp. Light Sources, Karlshrue.
Karabourniotis, D. and Damelincourt, J. J., 1983, J. Appl. Phys., 53:7249.
Karabourniotis, D. and Karras, C., 1985, J. Appl. Phys., (in press).
Karabourniotis, D., Karras, C., Drakakis, M., and Damelincourt, J. J., 1982, J. Appl, Phys., 53:7259.
Kolobova, G. a., 1973, J. Appl. Spectrosc., 14:187.
Lapworth, K. C., 1980, Nat. Phys. Lab. (UK), Report Qu 55.
Lochte-Holtgreven, W., 1968, in: "Plasma Diagnostics," edited by W. Lochte-Holtgreven, North-Holland, Amsterdam.
Luizova, L. A., 1975, Opt. Spectrosc., 38:362.
Meinerl, D., and Weiss, O., 1970, Z. Angew. Phys., 29:35.
Neumann, W., 1975, in: "Progress in Plasma and Gas Electronics," edited Rompe and Steenbeck, Akademie-Verlag, Berlin.
Oxenius, J., 1967, J. Quant. Spectrosc. Radiat. Transfer, 7:837.
Ozaki, N., 1971a, J. Quant. Spectrosc. Radiat. Transfer, 11:1111.
Ozaki, N., 1971b, J. Quant. Spectrosc. Radiat. Transfer, 11:1464.
Perrin-Lagarde, D., and Lennuier, R., 1975, J. Physique, 36:357.
Sahal-Brechot, S., 1969a, Astr. Astrophys., 1:91.
Sahal-Brechot, S., 1969b, Astr. Astrophys., 2:322.
Sahal-Brechot, S., 1974, Astr. Astrophys., 35:319.
Salakhov, M., Kh. and Fishmann, I. S., 1981, High Temp., 18:561.
Sobelman, I. I., Vainshtein, L. A., and Yukov, E. A., 1982, in: "Springer Series in Chemical Physics," Springer-Verlag, Berlin.
Stormberg, H. P., 1980, J. Appl. Phys., 51:1963.
Szymanski, Z., 1979, J. Quant. Spectrosc. Radiat. Transfer, 22:577.
Traving, G., 1968, in: "Plasma Diagnostics," edited by W. Lochte-Holtgreven, North-Holland, Amsterdam.
Usher, J. L., and Campell, H. D., 1972, J. Quant. Spectrosc. Radiat. Transfer, 12:1157.
Van den Hoek, W. J., and Visser, J. A., 1980, J. Appl. Phys., 51:174.
Verolainen, Ya., F., and Oscherovich, Al. L., 1966, Opt. Spectrosc., 20:517.
Waszink, J. H., 1975, J. Appl. Phys., 46:3139.
Wesselink, G., de Mooy, D., and an Gemert, M. E. ·C., 1973, J. Phys. D: Appl. Phys., 6:L27.
Wharmby, D. O., 1976, Thorn Lighting Ltd, Tech. Rep. No. LRR 2010.
Zwicker, H., 1968, in: "Plasma Diagnostics," edited by W. Lochte-Holtgreven, North-Holland, Amsterdam.

RADIATION TRANSPORT IN HIGH PRESSURE DISCHARGE LAMPS

M. A. Cayless

Thorn EMI Lighting Limited
Research and Engineering Division
Leicester, England

INTRODUCTION

In a gas discharge, radiation of various frequencies is emitted, absorbed and scattered in various regions throughout the plasma. In passing through the plasma it acts as a means of conveying energy from one region to another, and participates in associated phenomena such as excitation, ionization and dissociation.

Ultimately, some of this radiation may emerge from the surface of the discharge to become the light perceived by an observer, or detected by some external device. But before reaching the surface, radiation first produced inside the body of the discharge has had to traverse the outer regions, and may be scattered, absorbed and re-emitted many times.

In some cases this may simply cause an attenuation of light produced in an intense region of the discharges in cooler outer regions. In extreme cases this may be very marked and we talk of "imprisonment" or "trapping", particularly of resonance radiation. This can have a marked effect on the characteristics of the discharge, and is a major factor governing the performance of many discharge light sources.

In other circumstances light from the inner regions is augmented by light of a different frequency distribution from the cooler outer regions; or line radiation may be absorbed in surrounding continuum. What is perceived externally is a composite of many processes occurring in different parts of the discharge.

We are now going to study this whole process of radiation transport as it applies to discharge lamps. It plays a crucial role in determining the energy balance of the discharge plasma, and is often one of the main factors governing the temperature distribution, and the distribution of excited or ionized species. Also, of course, it determines its performance as a light source.

For the purpose of this lecture, discussion will be largely confined to high pressure discharge lamps. I make no apologies for this because these are what I know about! They are discharges in a closed vessel, or arc tube, containing a mixture of gases and vapors, which may include the rare gases, volatile metals or compounds such as halides, at operating pressures between about 0.1 and 10 atmospheres. The gas and electron temperatures are usually about equal and vary typically from a maximum of

4000 to 7000 K near the center of the arc tube to below 1500 K near the walls. Local thermodynamic equilibrium, or near LTE, conditions hold throughout, though there may be appreciable "de-mixing" of the various constituents.

Although any treatment of this subject is necessarily mathematical, I shall emphasis throughout the physical interpretation of the various relationships, rather than dwell on specific details. Emphasis will also be given to those aspects of special importance in this particular kind of discharge and, because of this, it will be found to differ in places from some more conventional treatments. Remember, in many discharges, radiation is a nuisance, a loss, to be minimized or got rid of. In discharge lamps it is the be-all and end-all, to be encouraged and maximized by all means possible!

The general equations of radiation transport are readily derived, and formal "solutions" may easily be obtained. The whole process and the results are deceptively simple. And I emphasize deceptively: actual evaluation is often notoriously difficult, and gives rise to some of the most intractable problems in classical mathematical physics.

This is especially so when radiation scattering plays a prominent role, as in many astrophysical situations. Fortunately in most discharge lamp plasmas collisions are frequent and absorption and emission predominate over scattering. We are therefore able to omit scattering from our analysis, with great simplification. (See many standard text for the treatment of scattering, e.g., Chandrasekhar 1960, or Mihalas, 1978.)

We assume, in fact, "complete redistribution": when a photon is absorbed, and the energy later emitted as another photon, this latter process is completely independent of the former: the second photon carries no "memory" of the first. (See e.g. Mihalas, 1978, for a full discussion.)

Similarly, we consider only a static plasma, or one changing only slowly with time. Neither are polarization nor coherence effects taken into account - they are, of course important in plasmas associated with laser action (see e.g. Maitland and Dunn, 1969; Born and Wolf, 1964.) Refractive index is practically constant and dispersion is negligible. We shall incorporate stimulated emission in our analysis, although in these circumstances it makes only a small correction. Its inclusion does, however, ensure correct asymptotic convergence to the black body limit at great optical depths.

The plasma may be inhomogeneous, often very markedly so: that is, the gaseous and electrical species present, their degree of excitation or ionization, and the local temperature, may vary markedly with position, often, though not necessarily, with cylindrical symmetry. We do, however, assume spacial isotropy at every point, so that the emission and absorption coefficients are independent of the direction of the radiation. The radiation field may, of course, be highly anisotropic.

RADIATION TRANSPORT EQUATION

We start with the spectral intensity $I_\nu(\underline{r},\underline{u})$ of radiation of frequency ν passing through a position $\underline{r}$ in the direction defined by unit vector $\underline{u}$ (Fig. 1), defined so that the radiation power at frequencies between ν and $\nu + d\nu$ passing through an element of surface $\underline{d\sigma}$ at position $\underline{r}$ in an element of solid angle $d\Omega$ in direction $\underline{u}$ is

$$dW = I_\nu(\underline{r},\underline{u}) \; \underline{u}\cdot\underline{d\sigma} \; d\Omega \; d\nu \qquad (1)$$

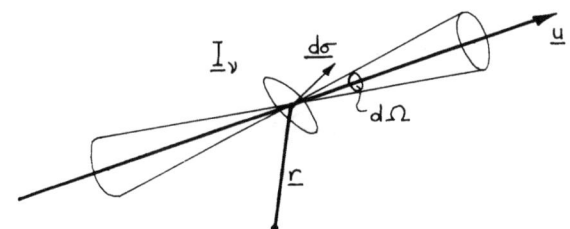

Fig. 1. Spectral intensity of radiation. A spectral intensity $I_\nu(\underline{r},\underline{u})$ passes through a surface element $\underline{d\sigma}$ at a position $\underline{r}$ in a direction of unit vector $\underline{u}$ in a pencil of solid angle $d\Omega$ and with a frequency between ν and $\nu + d\nu$.

Through any point $\underline{r}$ there may be pencils of radiation passing in all directions $\underline{u}$ (Fig. 2).

In general $I_\nu(\underline{r},\underline{u})$ is a function of six independent variables: the three coordinates of $\underline{r}$, the two independent direction cosines of $\underline{u}$, and the frequency ν. This number may be reduced by symmetry, for example in cases of planar, cylindrical or spherical symmetry.

$I_\nu(\underline{r},\underline{u})$ is the fundamental quantity from which all others are derived. When there is no emission or absorption of light in a medium of constant refractive index it possesses the valuable property of invariance along a ray. (Born and Wolf, 1964). We shall have occasion to write it as a vector, thus

$$\underline{I}_\nu(\underline{r},\underline{u}) \equiv I_\nu(\underline{r},\underline{u})\underline{u}. \qquad (W\ m^{-2} sr^{-1} \nu^{-1}) \qquad (2)$$

Note, however, that it does not form a vector field in the same sense as do, e.g., the electric and magnetic vectors $\underline{E}$ and $\underline{H}$. Thus the sum of two intensities at some point $\underline{r}$ in directions $\underline{u}_1$ and $\underline{u}_2$, say (Fig. 2), do not combine to form a single intensity in a resultant direction $\underline{u}_3$, i.e.,

$$\underline{I}_\nu(\underline{r},\underline{u}_1) + \underline{I}_\nu(\underline{r},\underline{u}_2) \neq \underline{I}_\nu(\underline{r},\underline{u}_3). \qquad (3)$$

Note also that this quantity is quite distinct from the "intensity" used in photometry, where it is normally referred to as the "radiance". (CIE, 1970). The photometric spectral radiation intensity $I_{\nu e}$ is, in fact,

$$I_{\nu e}(\underline{u}) = \int I_\nu(\underline{r},\underline{u}) \cdot \underline{dS} \qquad (W\ sr^{-1} \nu) \qquad (4)$$

where $\underline{S}$ is the area of the emitting source.

In ascribing units to radiation quantities it is useful to retain the unit of frequency as a separate entity, $\nu\ (= s^{-1})$, so that the units of I_ν are $W\ m^{-2} sr^{-1} \nu^{-1}$, while those of $I_{\nu e}$ are $W\ sr^{-1} \nu^{-1}$. Frequency is used rather than wavelength to avoid complications in dispersive media, though these are negligible for the plasmas considered here.

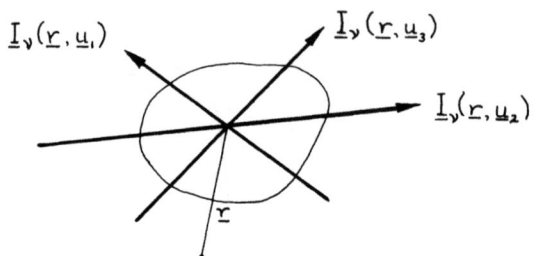

Fig. 2. Intensities in various directions $\underline{u}_1$, $\underline{u}_2$,... do not combine as vectors to form a resultant intensity at any point $\underline{r}$, but form a totality in themselves.

Integration over all frequencies provides the overall intensity

$$I(\underline{r},\underline{u}) = \int I_\nu(\underline{r},\underline{u}) \, d\nu \qquad (W\,m^{-2}sr^{-1}) \qquad (5)$$

Similar relations apply to all spectral quantities characterized by the suffix ν. The notation or the units are usually sufficient to make clear which quantity is being used, but where necessary the adjectives "spectral" and "overall" are used to avoid ambiguity. When frequency dependence does not result in a derivative quantity, the functional notation $a(\nu)$ is used.

Careful adherence to these seemingly trivial precepts avoids much confusion and ambiguity in this field, which has more than its share of such annoyances.

Now consider a discharge plasma, each elementary volume $d\underline{r}$ at position $\underline{r}$ of which is emitting and absorbing radiation in direction $\underline{u}$ with spectral emission coefficient

$$\varepsilon_\nu(\underline{r},\underline{u}) \qquad (W\,m^{-3}sr^{-1}\nu^{-1}) \qquad (6)$$

and effective spectral absorption coefficient

$$\kappa'(\nu,\underline{r},\underline{u}), \qquad (m^{-1}) \qquad (7)$$

where the dash in κ' indicates that stimulated emission is incorporated as a small, but significant, negative contribution to the true absorption coefficient κ.

These are defined so that the spectral power spontaneously emitted from $d\underline{r}$ into solid angle $d\Omega$ about direction $\underline{u}$ is

$$dW_e = \varepsilon_\nu(\underline{r},\underline{u}) \, d\underline{r} \, d\Omega \, d\nu \qquad (W) \qquad (8)$$

in all directions of polarization, and the attenuation by absorption of a ray of intensity $I_\nu(\underline{r},\underline{u})$ in passing through a distance ds in direction $\underline{u}$ is

$$dI_\nu(\underline{r},\underline{u}) = -\kappa(\nu,\underline{r},\underline{u})I_\nu(\underline{r},\underline{u}) \, ds. \qquad (W\,m^{-1}sr^{-1}\nu^{-1}) \qquad (9)$$

If there were a significant amount of scattering, in addition to absorption, κ would be replaced by quantity

$$\chi(\nu,\underline{r},\underline{u}) = \kappa(\nu,\underline{r},\underline{u}) + \sigma(\nu,\underline{r},\underline{u}) \qquad (10)$$

variously called the extinction coefficient, or opacity, where σ is the corresponding scattering term. In our circumstances we simply replace κ in this definition by

$$\kappa'(\nu,\underline{r},\underline{u}) = \kappa(\nu,\underline{r},\underline{u}) - \kappa_s(\nu,\underline{r},\underline{u}) \qquad (11)$$

where κ_s is the corresponding term for emission stimulated by I_ν in the same direction.

Note that whereas we can usefully define a total emission coefficient

$$\varepsilon(\underline{r},\underline{u}) \equiv \int \varepsilon_\nu(\underline{r},\underline{u}) \, d\nu \qquad (W \, m^{-3} sr^{-1}) \qquad (12)$$

the corresponding quantity

$$\kappa(\underline{r},\underline{u}) \equiv \int \kappa(\nu,\underline{r},\underline{u}) \, d\nu \qquad (13)$$

has little physical significance or value. It is always necessary to consider absorption processes as explicit functions of frequency. In the case of line spectra this comprises the detailed line profile, including where necessary the hyperfine structure and the relevant line broadening processes.

Evaluation of these coefficients is not the subject of this lecture, but it may be remarked that they are normally determined from the concentrations of the interacting species and the relevant lifetimes, transition probabilities, Einstein coefficients, or corresponding quantities. (See Wharmby, 1985a, this seminar; also e.g. Griem, 1964; Mihalas, 1978; Mitchell and Zemansky, 1934; Richter, 1968; Penner, 1959.) For example in the case of a simple spectral line involving transitions between a concentration n_1 (m^{-3}) of atoms in a lower state to n_2 in an upper state, then

$$\varepsilon_\nu(\underline{r},\underline{u}) = \frac{h\nu_{12}}{4\pi} A_{21} n_2(\underline{r}) P_{e\nu}(\underline{r}) \qquad (14)$$

where $P_{e\nu}$ is the emission line profile, $\int P_{e\nu} d\nu = 1$, and

$$\kappa'(\nu,\underline{r},\underline{u}) = \frac{h\nu_{12}}{c} \left[n_1(\underline{r})B_{12} - n_2(\underline{r})B_{21} \right] P_{a\nu}(\underline{r}) \qquad (15)$$

where A_{21}, B_{12} and B_{21} are the Einstein coefficients and $P_{e\nu}(\underline{r})$ and $P_{a\nu}(\underline{r})$ the emission and absorption line profiles. In most cases these last are the same, as we shall see.

Expressions of this kind are summed at each frequency for each process. Note that, through the concentrations $n(\underline{r})$, the coefficients are indirectly dependent on the radiant energy density, $u_\nu(r)$, since this in turn influences the population balance at every point in the plasma. Note also that the B coefficients themselves are frequency-dependent, having indeed the dimensions ($m^3 \, sec^{-1} \nu$) in the form in which we have quoted them (Richter, 1968; Penner, 1959). The temptation to write

$\kappa'(\nu,\underline{r}) = \kappa'(\underline{r})P_{a\nu}(\underline{r})$ should therefore be resisted, as giving a false indication $\kappa'(\underline{r})$ is frequency-independent. The exact meaning of $P_{a\nu}(\underline{r})$ therefore needs careful consideration wherever it is used, as will be seen.

This point is emphasized, since it causes considerable confusion in extending treatments originally developed for homogeneous plasmas and relatively narrow lines to the highly inhomogeneous plasmas with very broad lines met in high pressure discharge lamps.

One further quantity has great utility, the net directional spectral emission coefficient

$$\hat{\varepsilon}_\nu(\underline{r},\underline{u}) \equiv \varepsilon_\nu(\underline{r},\underline{u}) - \kappa'(\nu,\underline{r},\underline{u})I_\nu(\underline{r},\underline{u}), \qquad (16)$$

the difference between emission and absorption at each point. Note particularly that, though we have assumed spatial isotropy in the plasma, and can thus legitimately write $\varepsilon_\nu(\underline{r})$ and $\kappa'(\nu,\underline{r})$ omitting the directional dependence, this must be retained for the net coefficient, which is directionally dependent, through I_ν, giving

$$\hat{\varepsilon}_\nu(\underline{r},\underline{u}) \equiv \varepsilon_\nu(\underline{r}) - \kappa'(\nu,\underline{r})I_\nu(\underline{r},\underline{u}). \qquad (17)$$

It is often convenient, however, to retain the longer forms for ε_ν and $\kappa'(\nu)$ as a reminder of their nature.

If we now consider (Fig. 3) a pencil of radiation in direction $\underline{u}$ passing through a surface element dS normal to $\underline{u}$, and measuring distance along the pencil by s, so that $\underline{s} = s\underline{u}$, then the specific intensity $I_\nu(\underline{r},\underline{u})$ may be written $I_\nu(s)$ and clearly changes along the pencil as

$$dI_\nu(s)dSd\Omega = \varepsilon_\nu(s)dsdSd\Omega - \kappa'(\nu,s)ds \cdot I_\nu(s)dSd\Omega \qquad (18)$$

or

$$\frac{dI_\nu(s)}{ds} = \varepsilon_\nu(s) - \kappa'(\nu,s)I_\nu(s), \qquad (19)$$

the equation of radiative transport.

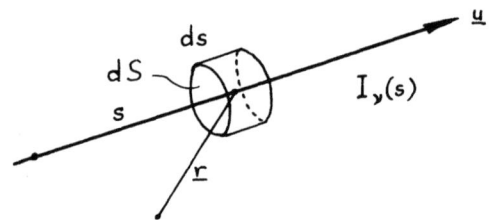

Fig. 3. Radiation transport equation. The intensity $I_\nu(s)$ in the direction $\underline{s} = s\underline{u}$ passes through an elementary cylinder of cross-sectional area dS normal to s and depth ds at a position $\underline{r}$.

Nothing could look more deceptively simple: the physical meaning is quite clear and such apparent simplicity could never be more deceptive!

We can even write down a "solution", almost by inspection. Using the notation in Fig. 4 this is

$$I_\nu(s) = I_\nu(s_0)e^{-\tau(\nu,s_0,s)} + \int_{s_0}^{s} \varepsilon_\nu(s')e^{-\tau(\nu,s',s)} ds' \qquad (20)$$

where

$$\tau(\nu,s_1,s_2) \equiv \int_{s_1}^{s_2} \kappa'(\nu,s) \, ds \qquad (21)$$

is our old friend the optical path between s_1 and s_2.

If you want to convince yourself that this is a solution, just put it in, work it out, and see! I believe the exponential term is what mathematicians call an "integrating factor". I prefer to call it the spectral transmission factor

$$T(\nu,s',s) = e^{-\tau(\nu,s',s)} \qquad (22)$$

the probability that a photon emitted at s' in the direction $\underline{u}$ actually reaches s. Trivial, but integrated forms are more useful, as will be seen.

Physically, it is easy to see that the first term is simply the incoming radiation from outside the plasma reaching the point s_0, attenuated as it passes through the optical path τ. The second term is the emitted radiation from all points s' between s_0 and s, similarly attenuated. In most cases we shall not be considering radiation from outside the plasma, so by choosing s_0 outside the active region, and extending the integration to include the whole emitting region before s, we can write

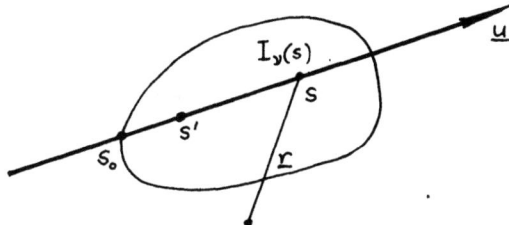

Fig. 4. Intensity $I_\nu(s)$ at point s is the integral of the intensity emitted from each point s' before s, attenuated by absorption along the optical path between s' and s, together with any incident intensity $I_\nu(s_0)$ similarly attenuated.

$$I_\nu(s) = \int_0^S \varepsilon_\nu(s')e^{-\tau(\nu,s',s)}ds'$$

or

$$I_\nu(s) = \int_0^S \varepsilon_\nu(s')T(\nu,s',s)\,ds' \tag{23}$$

Physically, therefore, straightforward, but reflection shows that the mathematical complexity is beginning to emerge. To obtain the radiation intensity in any one direction at any point, there is a triple integration, over s, s' and ν, in which arguments ε_ν and $\kappa'(\nu)$ may vary rapidly over many orders of magnitude, particularly if line radiation is involved, so that numerical integration is far from straightforward. Furthermore, ε_ν and $\kappa'(\nu)$ are themselves indirectly non-linear functions of the intensities in all directions. And this still gives only the intensity in a single direction $\underline{u}$ at a single point s. Further integrations are needed to determine the other quantities if required.

RADIATION QUANTITIES

Several further quantities are needed to provide a complete description of the radiation characteristics of the discharge. First, the spectral radiation flux density at each point (Fig. 5)

$$\underline{F}_\nu(\underline{r}) \equiv \int \underline{I}_\nu(\underline{r},\underline{u})\,d\Omega \qquad (W\,cm^{-2}\nu^{-1}) \tag{24}$$

i.e. the radiation power per unit area through a surface element at $\underline{r}$.

This is a vector quantity, and the integral must be evaluated vectorially over all solid angles about the point $\underline{r}$. This normally involves two further variables, nominally the polar and azimuthal angles θ and ϕ, so evaluation of the total flux density $\underline{F}(\underline{r})$ involves a complex five-fold integration.

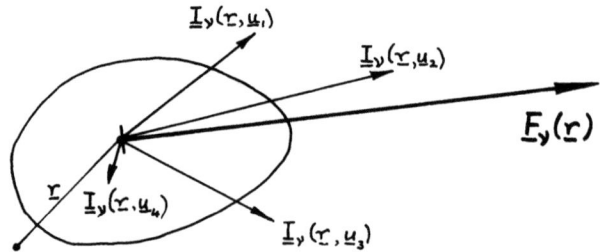

Fig. 5. The spectral radiation flux density $\underline{F}_\nu(\underline{r})$ is the vector integral of the spectral intensity $\underline{I}_\nu(\underline{r},\underline{u})$ over all solid angles at the point $\underline{r}$.

Note that, unlike $\underline{I}_\nu(\underline{r},\underline{u})$, $\underline{F}_\nu(\underline{r})$ does form a normal vector field and can be added, subtracted and manipulated just as other vector fields such as electrical or gravitational fields.

First and foremost $\underline{F}_\nu(\underline{r})$ is needed to determine the radiation output of the discharge. The radiation flux through any surface $\underline{S}$ (Fig. 6) is

$$Q_\nu(S) = \int \underline{F}_\nu(\underline{r}) \cdot \underline{dS} \qquad (W\ \nu^{-1})$$

$$= \iint \underline{I}_\nu(\underline{r},\underline{u}) \cdot \underline{ds}\ d\Omega \qquad (26)$$

and if S is the external surface of the discharge, then this is the light output. For this purpose $\underline{F}_\nu(r)$ need only be evaluated at the outside surface, but is needed also throughout the plasma in connection with the energy balance. This, in turn, also affects the flux at the surface, so there's no escaping it. (I have changed the order of integration in the second expression: the inner integral in this form may be recognized as the photometric radiation intensity mentioned earlier). Before proceeding with this, we need to consider some further quantities. The first is (Fig. 7) the scalar spectral radiation flux density

$$\tilde{F}_\nu(\underline{r}) \equiv \int I_\nu(\underline{r},\underline{u})\ d\Omega \qquad (W\ m^{-2}\nu^{-1}) \qquad (27)$$

i.e. the radiation power passing through a volume element at $\underline{r}$ in all directions.

Although superficially similar to $\underline{F}_\nu(\underline{r})$ this is a quite different quantity. The difference is most obvious in the case of a uniform plasma, or at a center of symmetry (Fig. 8). Here

$$\underline{F}_\nu(\underline{r}) = 0 \qquad (28)$$

but

$$\tilde{F}_\nu(\underline{r}) = 4\pi\ I_\nu(\underline{r},\ \underline{u}) \qquad (29)$$

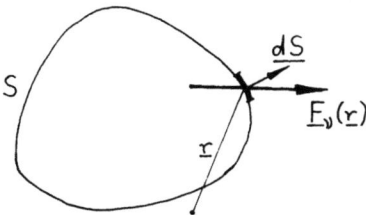

Fig. 6. The spectral radiation flux through any surface S is the surface integral of the flux density $\underline{F}_\nu(\underline{r})$.

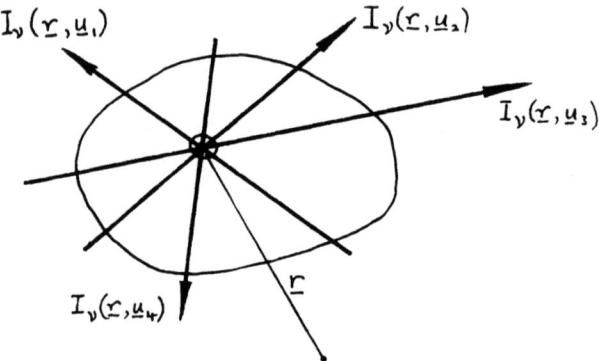

Fig. 7. The scalar spectral radiation flux density $\tilde{F}_\nu(\underline{r})$ is the scalar integral of the spectral intensity $I_\nu(\underline{r},\underline{u})$ over all solid angles at the point $\underline{r}$.

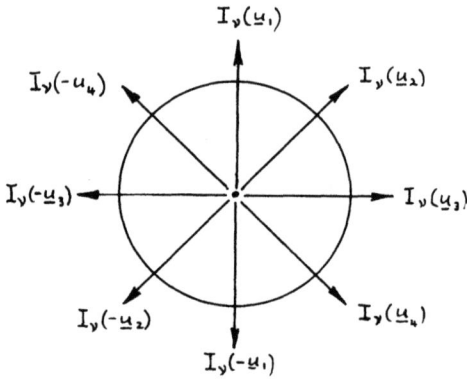

Fig. 8. Illustrating the difference between the vector and scalar flux densities at a center of symmetry. Opposing intensities $I_\nu(\underline{u})$ and $I_\nu(-\underline{u})$ cancel when added vectorially, but produce the sum $2I_\nu(\underline{u})$ when added as scalars.

since in every direction $\underline{u}$ there are equal opposing intensities $I_\nu(\underline{r},\underline{u})$ and $I_\nu(\underline{r},-\underline{u})$: there is flux without flow, a concept familiar in photometry as scalar illuminance. Closely related to this is the spectral radiation energy density

$$u_\nu(\underline{r}) = \frac{1}{c}\tilde{F}_\nu(\underline{r}) = \frac{1}{c}\int I_\nu(\underline{r},\underline{u})\,d\Omega \qquad (\text{J m}^{-3}\nu^{-1}) \qquad (30)$$

and the spectral photon concentration

$$n_\nu(\underline{r}) = \frac{1}{h\nu}u_\nu(\underline{r}) = \frac{1}{ch\nu}\int I_\nu(\underline{r},\underline{u})\,d\Omega \qquad (\text{n m}^{-3}\nu^{-1}) \qquad (31)$$

which is the quantity required to determine the rates of the various interactions between components which involve radiative transitions, which in turn affect the population balance of the various active species in the plasma - and hence, in turn, the emission and absorption coefficients. This emphasizes the non-linearity of the whole radiation transport process, and indicates that practical solutions to the transport equations will involve iterative procedures, with questions of convergence and stability.

Now, a bit of vector analysis.

If we now consider any volume of plasma surrounded by a closed surface S, we may apply Green's (Gauss') theorem to the flux density $\underline{F}_\nu(\underline{r})$ to express the radiation flux through S thus:

$$Q_\nu(S) = \oint_S \underline{F}_\nu(\underline{r}) \cdot \underline{dS}$$

$$= \int_S \text{div } \underline{F}_\nu(\underline{r}) \cdot d\underline{r} \; . \qquad (W \; \nu^{-1}) \qquad (32)$$

Now, noting the vector relation

$$\text{div } a\underline{b} = \underline{b} \cdot \text{grad } a + a \text{ div } \underline{b} \qquad (33)$$

so that

$$\text{div } a\underline{u} = \underline{u} \cdot \text{grad } a \qquad (34)$$

since the divergence of a unit vector in any given direction is zero, we may write

$$\text{div } \underline{I}_\nu(\underline{r},\underline{u}) = \text{div } I_\nu(\underline{r},\underline{u})\underline{u}$$

$$= \underline{u} \cdot \text{grad } I_\nu(\underline{r},\underline{u})$$

$$= \frac{dI(s)}{ds} \; . \qquad (35)$$

So the radiation transport equation can be written in its vector form

$$\text{div } \underline{I}_\nu(\underline{r},\underline{u}) = \varepsilon_\nu(\underline{r}) - \kappa'(\nu,\underline{r})I_\nu(\underline{r},\underline{u}) \; . \qquad (W \; m^{-3} sr^{-1} \; \nu^{-1}) \qquad (36)$$

Integration over all angles $d\Omega$ then gives

$$\text{div } \underline{F}_\nu(\underline{r}) = 4\pi\varepsilon_\nu(\underline{r}) - \kappa'(\nu,\underline{r})F_\nu(\underline{r}) \; . \qquad (W \; m^{-2} \; \nu^{-1}) \qquad (37)$$

Thus, writing the net spectral flux emission coefficient

$$\tilde{\varepsilon}_\nu(\underline{r}) \equiv \varepsilon_\nu(\underline{r}) - \frac{\kappa'(\nu,\underline{r})}{4\pi} \tilde{F}_\nu(\underline{r}) \qquad (38)$$

we have

$$\text{div } \underline{F}_\nu(\underline{r}) = 4\pi\tilde{\varepsilon}_\nu(\underline{r}) \qquad (W \; m^{-3} \; \nu^{-1}) \qquad (39)$$

and

$$Q_\nu(S) = \int \text{div } F_\nu(\underline{r}) \, d\underline{r} = 4\pi \int \tilde{\varepsilon}_\nu(\underline{r}) \, d\underline{r} \qquad (W \; \nu^{-1}) \qquad (40)$$

as the spectral radiation output from the plasma. Integrating over all frequencies gives

$$\text{div } \underline{F}(\underline{r}) = 4\pi \, \tilde{\varepsilon}(\underline{r}) \qquad (W \; cm^{-3}) \qquad (41)$$

where

$$\hat{\varepsilon}(\underline{r}) = \int \hat{\varepsilon}_\nu(\underline{r}) \, d\nu$$

$$= \int \hat{\varepsilon}_\nu(\underline{r}) \, d\nu - \frac{1}{4\pi} \int \kappa'(\nu,\underline{r}) \tilde{F}_\nu(\nu) \, d\nu \tag{42}$$

is the overall net emission coefficient. This term is required to evaluate the energy balance, which usually takes a form such as

$$\underline{E}(\underline{r}) \cdot \underline{j}(\underline{r}) = 4\pi\hat{\varepsilon}(\kappa) + \text{terms representing transport by thermal conduction, diffusion, convection etc.} \tag{43}$$

where $\underline{E}$ and $\underline{j}$ are the electrical field and current density. (Lowke, 1969; Jones and Mottram, 1981).

LINE RADIATION

Before considering the transport of radiation over the whole of a line profile, we first transform some of the preceding expressions into volume integrals. Noting that, if $s = 0$ at $\underline{r}$, (Fig. 9),

$$s' = |\underline{r} - \underline{r}'| \tag{44}$$

$$d\underline{r}' = s'^2 \, ds' \, d\Omega \tag{45}$$

then the scalar flux density

$$\tilde{F}_\nu(\underline{r}) = \iint_0^s \varepsilon_\nu(s') e^{-\tau(\nu,s',s)} ds' \, d\Omega \qquad (\text{W m}^{-3} \nu^{-1}) \tag{46}$$

becomes

$$\tilde{F}_\nu(\underline{r}) = \int \frac{\varepsilon_\nu(\underline{r}') e^{-\tau(\nu,\underline{r}',\underline{r})}}{|\underline{r} - \underline{r}'|^2} d\underline{r}' \tag{47}$$

where the optical path

$$\tau(\nu,\underline{r}',\underline{r}) = \int_{|\underline{r}-\underline{r}'|}^{0} \kappa'(\nu, \underline{r} - \underline{s}) \, ds. \tag{48}$$

The spectral photon concentration becomes

$$n_\nu(\underline{r}) = \frac{1}{ch\nu} \int \frac{\varepsilon_\nu(\underline{r}') e^{-\tau(\nu,\underline{r}',\underline{r})}}{|\underline{r} - \underline{r}'|^2} d\underline{r}' \qquad (\text{n m}^{-3} \nu^{-1}) \tag{49}$$

and

$$\text{div } \underline{F}_\nu(\underline{r}) = 4\pi\varepsilon_\nu(\underline{r}) - \kappa'(\nu,\underline{r}) \int \frac{\varepsilon_\nu(\underline{r}') e^{-\tau(\nu,\underline{r}',\underline{r})}}{|\underline{r} - \underline{r}'|^2} d\underline{r}'$$

$$(\text{W m}^{-3} \nu^{-1}) \tag{50}$$

These are spectral forms of the well-known Biberman-Holstein equations, applicable to an inhomogeneous plasma. (Biberman, 1947, 1949; Holstein, 1947, 1951).

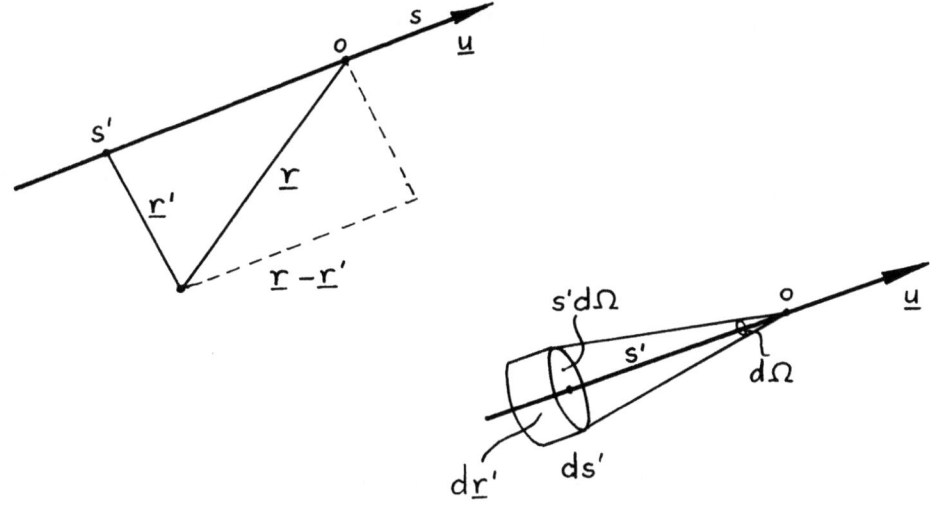

Fig. 9. Line and volume elements along a pencil through a point $\underline{r}$ in direction $\underline{u}$. If the coordinate s along $\underline{u}$ is zero at $\underline{r}$, then $\bar{s}' = |\underline{r} - \underline{r}'|$ and the volume $d\underline{r}' = s'^2 ds' d\Omega$.

In the case of line radiation with an emission line profile $P_{e\nu}(r)$ density and a total emission density $\varepsilon(\underline{r})$

$$\varepsilon_\nu(\underline{r}) = \varepsilon(\underline{r}) P_{e\nu}(\underline{r}) \tag{51}$$

we can integrate over the whole profile to give the scalar line flux density

$$\tilde{F}(\underline{r}) = 4\pi \int \varepsilon(\underline{r}') \int P_{e\nu}(\underline{r}') \frac{e^{-\tau(\nu,\underline{r}',\underline{r})}}{|\underline{r} - \underline{r}'|^2} d\nu d\underline{r}' \quad (W\ m^{-2}) \tag{52}$$

and

$$\text{div}\ \underline{F}(\underline{r}) = 4\pi\varepsilon(\underline{r}) - 4\pi \int \varepsilon(\underline{r}') \int P_{e\nu}(\underline{r}') \kappa'(\nu,\underline{r}) \frac{e^{-\tau(\nu,\underline{r}'\underline{r})}}{|\underline{r} - \underline{r}'|^2} d\nu d\underline{r}'.$$
$$(W\ m^{-3}) \tag{53}$$

These rather complex expressions have a straightforward physical interpretation which can be seen by writing them as

$$\tilde{F}(\underline{r}) = 4\pi \int \varepsilon(\underline{r}') g(\underline{r}',\underline{r})\ d\underline{r}' \tag{54}$$

and

$$\text{div}\ \underline{F}(\underline{r}) = 4\pi\varepsilon(\underline{r}) - 4\pi \int \varepsilon(\underline{r}') G(\underline{r}',\underline{r})\ d\underline{r}' \tag{55}$$

where

$$g(\underline{r}',\underline{r}) \equiv \int P_{e\nu}(\underline{r}') \frac{e^{-\tau(\nu,\underline{r}',\underline{r})}}{4\pi|\underline{r}-\underline{r}'|^2} d\nu \quad (m^{-3}) \tag{56}$$

and

$$G(\underline{r}', \underline{r}) \equiv \int P_{e\nu}(\underline{r}') \kappa'(\nu,\underline{r}) \frac{e^{-\tau(\nu,\underline{r}',\underline{r})}}{4\pi|\underline{r}-\underline{r}'|^2} d\nu \quad (m^{-3}) \tag{57}$$

are the transfer functions from $\underline{r}'$ to $\underline{r}$.

$g(\underline{r}',\underline{r})$ is the probability that a line photon emitted in the volume element $d\underline{r}'$ reaches the volume element $d\underline{r}$ per unit projected area of $d\underline{r}$ normal to $\underline{r} - \underline{r}'$.

$G(\underline{r}',\underline{r}) \, d\underline{r}$ is the probability that a line photon emitted in the volume element $d\underline{r}'$ is absorbed in the volume element $d\underline{r}$.

The net emission coefficient for the line becomes

$$\tilde{\varepsilon}(\underline{r}) = \varepsilon(\underline{r}) - \int \varepsilon(\underline{r}') G(\underline{r}',\underline{r}) d\underline{r}' \quad (W \, m^{-3}) \tag{58}$$

the physical interpretation of which is immediately obvious.

The closely related quantity

$$\Lambda(\underline{r}) \equiv \frac{\tilde{\varepsilon}(\underline{r})}{\varepsilon(\underline{r})} = 1 - \int \frac{\varepsilon(\underline{r}')}{\varepsilon(\underline{r})} G(\underline{r}',\underline{r}) d\underline{r}' \tag{59}$$

is variously referred to as the Biberman-Holstein coefficint, or net radiative bracket (Irons, 1979; Mihalas, 1978). The net radiation emission from the whole plasma is

$$\tilde{\varepsilon} \equiv \int_V \tilde{\varepsilon}(\underline{r})d\underline{r} = \int_V \varepsilon(\underline{r})d\underline{r} - \int_V\int_V \varepsilon(\underline{r}') G(\underline{r}',\underline{r}) d\underline{r}' \, d\underline{r} \quad (W) \tag{60}$$

and the quantity

$$\Lambda \equiv \frac{\int_V \tilde{\varepsilon}(\underline{r})d\underline{r}}{\int_V \varepsilon(\underline{r})d\underline{r}} = 1 - \frac{\int_V\int_V \varepsilon(\underline{r}')G(\underline{r}',\underline{r})d\underline{r}'d\underline{r}}{\int_V \varepsilon(\underline{r})d\underline{r}} \tag{61}$$

is variously referred to as the escape factor, imprisonment factor, or a variety of other names. These quantities, their detailed interpretation and interrelationships are discussed at length by Irons (1979).

Useful as they may be in situations which are optically thin, or at least not too thick, there are obvious difficulties in evaluation, particularly by numerical methods, in optically thick cases, where there is a small difference between two relatively large terms. Indirect methods are therefore necessary. First we consider some further properties of the quantity $G(\underline{r}',\underline{r})$.

We may write

$$G(\underline{r}',\underline{r}) = -\frac{1}{|\underline{r}-\underline{r}'|^2} \frac{dT(\underline{r}',\underline{r})}{d|\underline{r}-\underline{r}'|} = -\frac{1}{4\pi s^2} \frac{dT(\underline{r}',\underline{s})}{ds} \tag{62}$$

where

$$T(\underline{r}',\underline{s}) = \int P_{e\nu}(\underline{r}')e^{-\int_0^s \tau'(\nu,s)ds} d\nu \tag{63}$$

is the transmission coefficient, the probability that a line photon emitted at $\underline{r}'$ travels a distance s in the direction $\underline{u}$. It is an integrated form of the factor $T(\nu,s',s)$ quoted earlier.

Holstein (1947, 1951), in fact, derives his form of these relations starting from the transmission coefficient T and thence deriving $G(\underline{r}',\underline{r})$. This procedure is more appropriate for the case of a uniform plasma density considered by him. In this case also G has the important symmetry property

$$G(\underline{r}',\underline{r}) = G(\underline{r},\underline{r}') \tag{64}$$

which is central to Holstein's treatment, using a variational method which requires this symmetry.

This symmetry does not necessarily apply in the case of a non-uniform plasma, particularly because the line profile is not constant, i.e.

$$P_{e\nu}(\underline{r}) \neq P_{e\nu}(\underline{r}'). \tag{65}$$

Indeed, in many discharge lamp plasmas, the profile may vary very substantially indeed through the plasma, so this powerful method of solution cannot be used.

We can express the total intensity in the spectral line $I_\nu(s)$ in terms of $T(s',s)$ by integrating the first integral of the transport equation over the line-width, thus

$$I(s) = \int I_\nu(s) d\nu = \iint_0^s \varepsilon_\nu(s') e^{-\tau(\nu,s',s)} \, ds' d\nu$$

$$= \int_0^s \varepsilon(s') \, T(s',s) \, ds' \tag{66}$$

since

$$T(s',s) = \int P_{e\nu}(s') \, e^{-\tau(\nu,s',s)} \, d\nu$$

$$= \int P_{e\nu}(s') \, T(\nu,s',s) \, d\nu \tag{67}$$

where $T(\nu,s',s)$ is the spectral transmission factor we defined earlier.

$T(s's)$ is therefore the analogue for the whole line of the exponential absorption at a single frequency. Because of the profile factor $P_{e\nu}$, this may be far from exponential.

In the case of a homogeneous plasma $\kappa'(\nu)$ is independent of position, and we may write

$$T(s) = \int P_{e\nu} \, e^{-\kappa'(\nu)s} \, d\nu, \tag{68}$$

the form used by Holstein (1947).

It may be necessary to re-iterate that in all these relations the frequency dependence of $\kappa'(\nu,\underline{r})$, and its relation to the frequency profile of the radiation intensity at $\underline{r}$ is vital, even to the extent of hyperfine structure, and the temptation to write $\kappa'(\nu,\underline{r}) = \kappa'(\underline{r})P_{av}(\underline{r})$ should be resisted, as leading to endless confusion.

Numerous special procedures have been devised to evaluate the imprisonment of resonance radiation in low pressure positive column discharges, particularly the mercury rare-gas discharge which forms the basis of fluorescent lamps. In many cases these evaluate "effective diffusion coefficients" or "effective radiative lifetimes" for the resonance radiation. Earlier work is described in Cayless (1963) and the references cited there, and more recent work in van de Weijer and Cremers (1985).

EVALUATION OF THE FLUX DENSITY

We have referred to the possibility of approximate treatments applicable to optically thick or optically thin situations. Many of these are treated in various books and publications. However, in the plasmas of many practical light sources, the regions of greatest interest are just those where these approximations are invalid, for example, the spectrum of the light generated by the D-lines in a high pressure sodium lamp (Fig. 10). The frequency regions in the center and in the wings of the self-reversed spectrum can be evaluated by suitable approximations. But the very regions of greatest importance, where most of the light actually comes from, and which transport the most power, are those of optical depth of the order of unity, and approximations of this kind cannot be used (Cayless, 1966). There is no help for it but to carry out a full-blooded numerical evaluation.

In cases of symmetry, planar, cylindrical or spherical, there is some simplification, though the situation remains essentially 3-dimensional, as consideration of Fig. 11 shows: the intensities in various directions are functions of the angles the pencils make with the z-direction, or radius, and the longitudinal axis.

The key calculation is the evaluation of the flux density vector $\underline{F}_\nu(\underline{r})$ at all points through the plasma.

In the general case it is necessary to express the components of the intensity $\underline{I}_\nu(\underline{r},\underline{u})$ in an appropriate coordinate system, and evaluate the components of $\underline{F}_\nu(\underline{r})$ through the several components of the general integral

$$\underline{F}_\nu(\underline{r}) = \int \underline{I}_\nu(\underline{r},\underline{u}) \, d\Omega \tag{69}$$

This is straightforward, though rather complicated.

Fortunately, in cases of planar and cylindrical symmetry, the flux is entirely in the direction of symmetry, and is a function of a single variable only, $F_\nu(z)$ or $F_\nu(r)$. Thus, in a planar geometry, homogeneous in Cartesian coordinates x and y, the transport equation becomes

$$\cos\theta \frac{dI_\nu(z,\theta)}{dz} = \varepsilon_\nu(z) - \kappa'(\nu,z) I_\nu(z,\theta) \tag{70}$$

where θ is the angle of the ray to the normal z to the plane of symmetry, giving (omitting the first term)

$$I_\nu(z,\theta) = \int_0^z \varepsilon_\nu(z') e^{-\tau(\nu,z',z)} \sec\theta \, dz \tag{71}$$

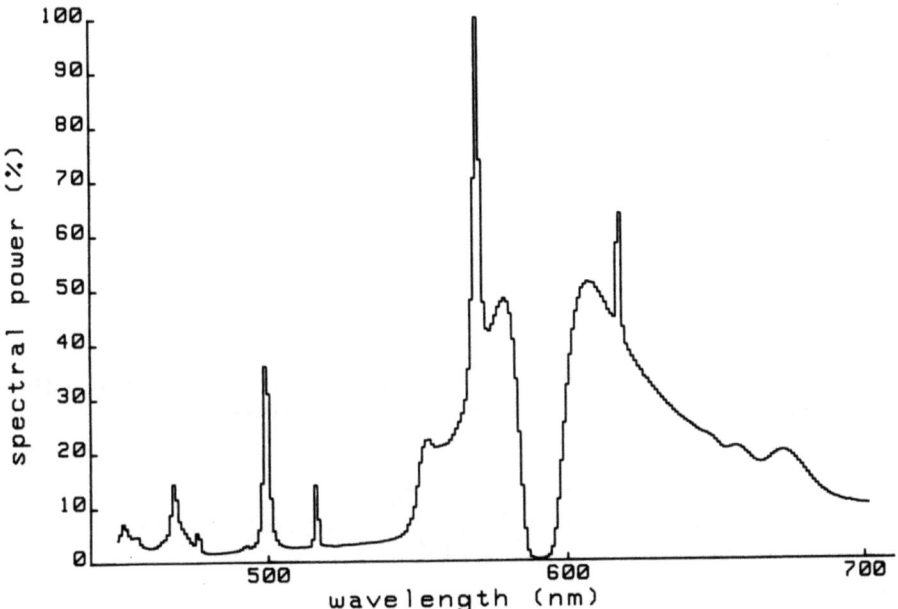

Fig. 10. Spectrum of typical high pressure sodium lamp showing highly broadened self-reversed D-lines, together with other lines and molecular emission features.

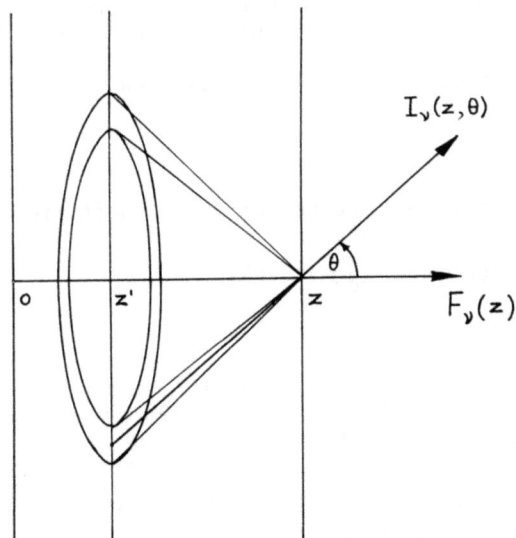

Fig. 11. The spectral flux density from a planar source comprises intensity contributions $I_\nu(z,\theta)$ from cones of all vertex angles θ to the normal coordinate z.

with

$$\tau(\nu,z',z) = \int_{z'}^{z} \kappa'(\nu,z') \sec\theta \, dz' \qquad (72)$$

Now the flux becomes

$$F_\nu(z) = 2\pi \int_{-\frac{\pi}{2}}^{\frac{\pi}{2}} I_\nu(z,\theta) \cos\theta \sin\theta \, d\theta \qquad (73)$$

so that

$$F_\nu(z) = 2\pi \int_{-\frac{\pi}{2}}^{\frac{\pi}{2}} \int_0^z \varepsilon_\nu(z') \, e^{-\tau(\nu,z',z)} \sin\theta \, dz' d\theta . \qquad (74)$$

In the case of cylindrical symmetry, the flux density has only a radial component $F_\nu(r)$ which becomes, after a similar analysis (Church et al., 1966)

$$F_\nu(r) = 4 \int_0^\pi \cos\phi \int_0^{s_n(\phi)} \varepsilon_\nu(s_n') \, G_1\left[\int_0^{s_n'} \kappa'(\nu,s_n'') ds_n'' \right] ds_n' \, d\phi \qquad (75)$$

where

$$G_1(x) \equiv \int_0^{\frac{\pi}{2}} \sin\theta \, \exp\left(-\frac{x}{\sin\theta}\right) d\theta \qquad (76)$$

and s_n is the projection of the direction of a general pencil s on the plane normal to the axis, as shown in Fig. 12. (Note that all the variables s_n, s_n', s_n'' are functions of the azimuthal angle ϕ.)

For energy balance purposes it is the total flux density

$$F(r) = \int F_\nu(r) \, d\nu \qquad (77)$$

that is required, and thence the net emission coefficient

$$\varepsilon(r) = \frac{1}{4\pi} \text{div} \, \underline{F}(r) = \frac{1}{4\pi r} \frac{d}{dr}\left(rF(r)\right) \qquad (78)$$

in this case.

Thus full evaluation requires computation of a five-fold integral throughout the plasma. In this case, the symmetry allows the longitudinal integral over θ to be performed separately as the function $G_1(x)$, but the integrations over s_n and ϕ have to be performed for each position and direction in the normal plane, a matter of some complexity if ε_ν and $\kappa'(\nu)$ vary substantially with the radial coordinate r and the frequency ν, as they usually do. (Lowke, 1974; Jones and Mottram, 1981).

Nevertheless, straightforward finite difference integration using a simple trapezoidal rule with an adequate number of intervals in each coordinate usually suffices. The radius is divided into zones, typically 21, the emission and absorption coefficients for each zone evaluated and

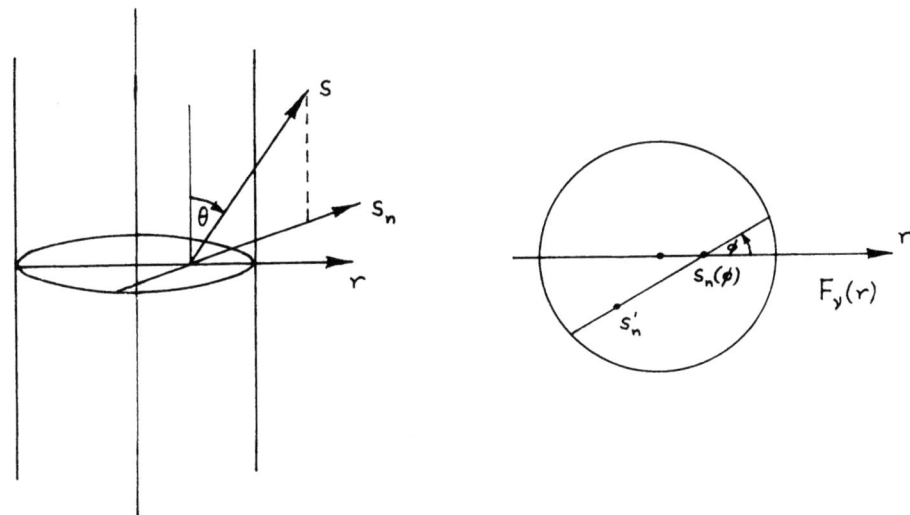

Fig. 12. Coordinate system for a cylindrical discharge. s_n is the projection of s on the plane normal to the axis. Note that variables such s_n are functions of the angle ϕ.

tabulated, and then $G_1(x)$ likewise, ready for interpolation. The integrations over s'_n, ϕ and ν are then performed in turn, the finite difference intervals being adjusted as the computation proceeds, particularly in regions of high optical depth, where some of the summations can be curtailed after a small number of closely-spaced intervals.

The principal constraint is the sheer computer time consumed. Our original program took from 24hr upwards to get adequate accuracy for $F(r)$ in a typical case. It was normally left running overnight, or over the weekend. However, many devices for speeding the computation, and increasing the accuracy in particular regions, have been devised, and these, combined with a faster computer, now enable typical evaluations to be made in about 20 minutes. Another lecture would be needed to describe some of these devices (Jones and Mottram, 1981).

By evaluating the net emission coefficients via the flux in this way, the inaccuracies associated with taking the difference of two nearly equal numbers are largely prevented. Although the terms in the basic transfer equation are also close together, correct formulation can avoid problems arising at this stage. Integration itself acts as a smoothing process and after five integrations the data are pretty well smoothed! If necessary a further numerical smoothing technique can be applied before the final differentiation in taking the divergence. There are many other ways of overcoming these problems, some of which were mentioned yesterday, and others which will be referred to later.

Similar procedures can be used for the evaluation of the scalar flux density $\tilde{F}_\nu(\underline{r})$ and thence the spectral photon concentration $n_\nu(\underline{r})$, developing the basic definition

$$\tilde{F}_\nu(\underline{r}) = \int I_\nu(\underline{r},\underline{u}) \, d\Omega \tag{79}$$

in the appropriate coordinate system.

However, if the net emission coefficient is derived from div $\underline{F}(\underline{r})$, as above, explicit evaluation of $\tilde{F}(\underline{r})$ is not needed. Further, in cases of equilibrium or LTE, $n_\nu(\underline{r})$ is not required to determine the population balance equations for the various species in the plasma, since these are determined from the local temperature by normal chemical equilibrium calculations, supplemented Boltzmann's and Saha's relations for excited and ionized species. Nearly all high pressure light sources fall into this category.

I have referred to the need for correct formulation, particularly to ensure proper asymptotic behavior at great optical depths. This gives rise to the considerations in the next section.

SOURCE FUNCTION

The emission and absorption coefficients $\varepsilon_\nu(\underline{r},\underline{u})$ and $\kappa'(\nu,\underline{r},\underline{u})$ are functions of the concentrations of all the particles at each point in the plasma, including the photons (or the radiation field), and of their various interactions. These are dealt with elsewhere in this institute, but certain general relationships are important in determining the nature of the radiation transport process and its evaluation.

We continue to assume "complete redistribution" on absorption and reemission, and negligible scattering. Also, that emission and absorption are spacially isotropic, so we may write $\varepsilon_\nu(\underline{r})$ and $\kappa'(\nu,\underline{r})$ in place of $\varepsilon_\nu(\underline{r},\underline{u})$ and $\kappa'(\nu,\underline{r},\underline{u})$, though retaining the longer forms for some of the more general relations.

The ratio

$$S_\nu(\underline{r},\underline{u}) \equiv \frac{\varepsilon_\nu(\underline{r},\underline{u})}{\kappa'(\nu,\underline{r},\underline{u})} \tag{80}$$

is defined as the source function at the point $\underline{r}$ in the direction $\underline{u}$. We may therefore write

$$S_\nu(\underline{r}) \equiv \frac{\varepsilon_\nu(\underline{r})}{\kappa'(\nu,\underline{r})} . \tag{81}$$

The transfer equation can then be written

$$\frac{1}{\kappa'(\nu,s)} \frac{dI_\nu(s)}{ds} = S_\nu(s) - I_\nu(s) \tag{82}$$

or,

$$\frac{dI_\nu(\tau)}{d\tau(\nu)} = S_\nu(\tau) - I_\nu(\tau). \tag{83}$$

(This latter form makes for elegant analytical discussions (Chandrasekhar, 1960), but not usually for convenient numerical evaluation, so we will normally stick to the former.)

If the plasma is in thermal equilibrium, at some temperature T, say, then clearly the radiation absorbed at each point must equal that emitted, and we have Kirchhoff's law in its strict form

$$\frac{\varepsilon_\nu(\underline{r},\underline{u})}{\kappa'(\nu,\underline{r},\underline{u})} = B_\nu(T) \tag{84}$$

where

$$B_\nu(T) = \frac{2h\nu^3}{c^2} \frac{1}{\exp\left(\frac{h\nu}{kT}\right) - 1} . \qquad (W\ m^{-2} sr^{-1} \nu^{-1}) \tag{85}$$

In other words, the source function $S_\nu(T)$ is equal to the black-body spectral radiation intensity $B_\nu(T)$ in equilibrium.

If this were valid only in conditions of full thermal equilibrium (TE), it would be of little interest from the point of view of radiation transport. But the emission and absorption coefficients are determined solely by the microscopic properties of the medium, and it is evident that this law is valid for any conditions in which these are the same as in TE. In particular, it is evident that this form of the law is valid in local thermodynamic equilibrium, LTE, conditions, even though the radiation fluxes in the region may be appreciably, or even substantially, different from the equilibrium black body fluxes.

In these circumstances, as I have remarked, the evaluation of radiation transport is enormously simplified, since it is no longer necessary for all the rate processes for the various species to be evaluated in order to determine the radiation properties: these can be evaluated from the equilibrium concentrations alone, which can be determined from the temperature profile. It is simply necessary to link this to the radiation flux via an energy balance relation. Normally the radiation transport and energy balance relations, which determine the temperature profile, are solved in an iterative manner.

Furthermore, only the emission coefficient or the absorption coefficient needs to be known: usually it is the latter. In the case of line radiation, the line profiles are the same in emission and absorption, and the relevant absorption profile is that for absorption of black-body radiation, or slowly varying continuum. The profile may be obtained by measurement, or by computation, taking into account the appropriate line broadening processes: pressure, resonance, Stark, Doppler, etc. (e.g. Griem, 1964; Mitchell and Zemansky, 1934; Breene, 1961; Penner, 1959; and Wharmby, 1985a).

The conditions in which LTE is a valid approximation are considered elsewhere in this ASI. It is valid for most high pressure discharge lamp plasmas. Note that this quite separate from "demixing" of the constituent elements in the plasma: it is quite possible to have substantial "demixing", while still retaining LTE to a very good approximation. In other cases, the assumption of the Planck function as source function is a sufficiently good approximation, even in the absence of LTE. Each case must be examined on its merits - obviously it must be possible to define a local temperature, or effective temperature, at the very least.

There are still other conditions in which Kirchhoff's law, although not valid in the strict form we have considered so far, is valid in a modified form: the emission and absorption coefficients may be proportional to each other, as functions of frequency, though the constant of proportionality is not equal to the Planck function. It may, for example, be independent of frequency over the range of interest, as in the cases treated by Holstein (1947).

Such a relaxed form of Kirchhoff's law is often applicable to collision dominated plasmas in which the populations of excited and ionized states are determined primarily by collisions rather than by radiation processes, but in which LTE is not established. A good example is the fluorescent lamp plasma, in which the excited state populations are determined primarily by collisions, but which are well below their LTE values (at the electron temperature). We heard some detailed consideration of this case yesterday.

Each such case must be examined on its merits and the appropriate constant of proportionality (i.e. constancy of the source function) established. See for example those considered by Holstein (1947).

In dense plasmas, or regions of great optical depth, the intensity $I_\nu(s)$ becomes very close to the source function $S_\nu(s)$ or Planck function $B_\nu(T)$, if near to LTE. This form of the transport equation, with its prominent dependence on the term $S_\nu(s) - I_\nu(s)$, highlights this, together with the problems of evaluation we have referred to.

In this connection, note that the correct approach to equilibrium is only obtained if stimulated emission is included, even though it may make only a small contribution under most conditions. We have incorporated it into the effective absorption coefficient $\kappa'(\nu,\underline{r})$, thus ensuring that the source function S_ν tends towards the Planck function B_ν as optical density increases, e.g. with increase in current, pressure or depth within the plasma, or as the frequency approaches the center of a spectral line. Otherwise the mathematical expression will tend towards Wien's law instead - i.e., without the -1 in the denominator.

Clearly it is desirable that the envelope of continuum emission at high plasma densities, or of maximum line intensities should be the Planck black-body curve, as observed experimentally, and this is only so if stimulated emission is incorporated. (It was, of course, just this sort of inconsistency which led to Planck's original quantum concept, some years before Einstein pursued the logic to deduce the inevitability of stimulated emission.)

We have preferred to incorporate stimulated emission as a modification to the absorption coefficient because it fits more neatly with its directional and coherence properties. But some authors prefer to incorporate it as a modification to the emission coefficient. Either way, as long as the modification is performed correctly and consistently, the final results are the same.

EVALUATION BY NET EMISSION COEFFICIENT

Although the straightforward evaluation methods I have described are sufficient to determine the radiation flux, together with the temperature profile, with which it is linked by the energy balance, a great deal of computation is involved even for a cylindrically symmetrical discharge with LTE, and some means of reducing this is a necessity. The following method is due to Jones and Mottram (1981) and uses an approximate expression for the net emission coefficient in terms of the flux density and temperature profile. Once determined, this is used in subsequent iterations with great saving in computation time. It is particularly appropriate for discharges comprising several self-absorbed spectral lines, such as the high pressure sodium lamps.

Typically (Fig. 13) the radiation flux density $F(r)$ increases rapidly from zero at the center of the discharge because of strong generation at the high temperatures near the axis. It reaches a maximum at about half-way along the radius, and then starts to fall because of the combined effects of increasing absorption, falling emission and increasing surface area in the cooler outer region of the discharge.

The net emission coefficient is a measure of the difference between the radiated power generated and absorbed within a volume element and is given, as we have seen, by the divergence of the flux density. It is therefore positive near the center, where generation prevails and negative near the edge, where absorption prevails.

We therefore postulate that it can be described by an expression in which a term representing the absorption of radiation is subtracted from a term representing the generation of radiation. It seems reasonable to represent the former by a term proportional to the concentration of absorbers and the local radiation flux density, and the latter by a term proportional to the concentration of emitters, weighted by a Boltzmann-type factor. An appropriate empirical expression is therefore

$$\varepsilon\big(T(r)\big) = n\big(T(r)\big)\left[a \exp\left(-\frac{b}{T(r)}\right) - cF(r) \exp\left(-\frac{d}{T(r)}\right)\right] \qquad (86)$$

where $n(T(r))$ is the number density of the emitter at temperature $T(r)$, and a,b,c and d are empirical constants. d is normally zero, or very small, for resonance transitions.

Normally, a set of net emission coefficients at radial intervals is only valid for a particular radial size of discharge, temperature profile and pressure of emitting atoms. But by choosing the expression in this form, once the coefficients have been determined for a typical profile etc., it gives a good approximation to ε for a wider range of parameters.

A typical set of parameters is chosen for a particular discharge, including a typical temperature profile, and the net emission coefficients are calculated at various radial positions by evaluating $F(r)$ and div $F(r)$ in the way described above. Two methods are used to perform the differentiation. The first is a straightforward finite difference technique, though second-order differences are necessary to get sufficient accuracy. In the other method, a set of cubic splines is fitted to the function $rF(r)$, and these are differentiated analytically. Both methods give comparable accuracy, but the former turns out to have advantages for non-resonance lines and the latter for resonance lines.

Having determined ε for this typical case at a case number radial positions (typically 21) the empirical constants a, b, c and determined by least-squares fitting, using a numerical procedure. The expression for ε can then be used in place of the full calculation (the five-fold integral) to evaluate the flux density energy balance and temperature profile for other discharge conditions.

For instance, the flux density is obtained by integrating the basic expression

$$\frac{1}{r}\frac{d}{dr}\big(rF(r)\big) = 4\pi\varepsilon(r) \qquad (87)$$

to give

$$F(r) = \frac{4\pi}{r} \int_0^r r'\varepsilon(r')\, dr'. \qquad (88)$$

Since the local flux density occurs in the approximation to ε it is

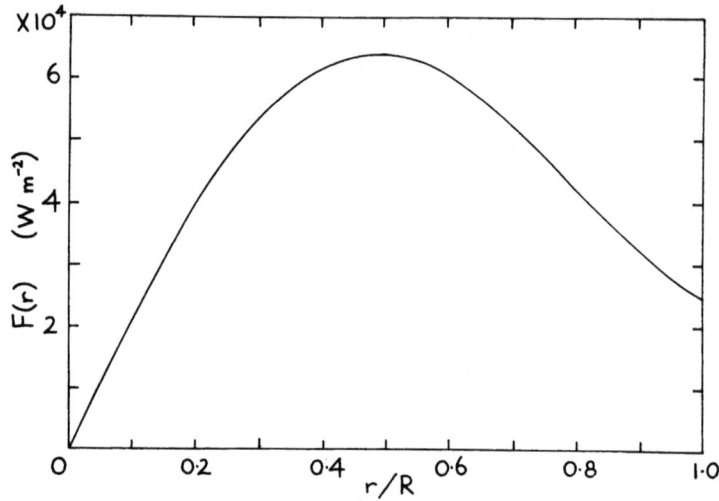

Fig. 13. Typical variation of the radiation flux density F(r) in the positive column of a high pressure sodium lamp of radius R. (After Jones and Mottram, 1981).

necessary to iterate to convergence at each radial position.

$\bar{\epsilon}$ fits equally well into the procedure for computing the temperature profile T(r) from the energy balance. Since the whole procedure for evaluating F(r) and T(r) proceeds iteratively, a very substantial reduction in computation time is achieved. If necessary, a full-scale calculation of F(r) can be carried out from the final profile T(r).

For a typical high pressure sodium lamp, the approximation predicts (Jones and Mottram, 1981) the flux density for the D-lines to within 10% (Fig. 14) for changes of up to 15% in tube diameter, axial temperature and partial pressure from the original conditions. This makes the use of these techniques in lamp design a practical proposition. (Wharmby, 1985b, these Proceedings)

This example of a practically useful technique illustrates some of the devices which can be used to obtain useful evaluations at an adequate accuracy with an acceptable effort. Other authors have used other procedures. For example Stormberg and Schafer (1983) use an approximation to the net emission coefficient in terms of a series of Legendre polynomials. Lowke (1974) uses yet another form. Experimental determinations of net emission coefficients (e.g. Zollweg, 1978) can also be used (Eardley et al., 1979), while a combination of experimental and theoretical coefficients can be useful in complex cases involving several vapors (Zollweg et al., 1981).

Diffusion approximations have been used for some low pressure discharges (Kenty 1932; Cayless 1963) and for some high pressure optically thick discharges (e.g. Lowke and Capriotti, 1969) but are not adequate for cases in which most of the radiation is at intermediate optical depths. Lowke (1974) lists a number of disadvantages.

Methods using "escape factors", the mean probability that an emitted photon reaches the surface, were pioneered by Holstein (1947, 1951), but

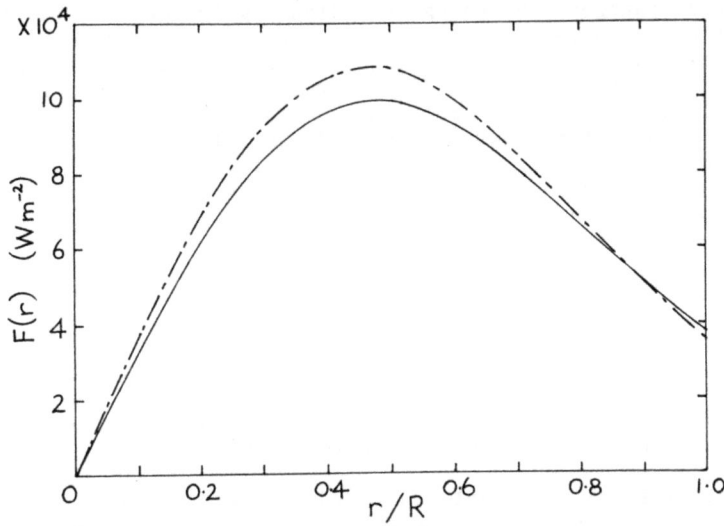

Fig. 14. Variation of radiation flux density with radius in a typical high pressure sodium lamp. Full curve calculated directly; broken curve calculated using the approximate net emission coefficient. (After Jones and Motram, 1981).

his methods are more appropriate to low pressure discharges where the generation of photons is much more uniform than in high pressure discharges. Irons (1979) recently analyzed a variety of "escape factors" procedures, helping considerably to resolve the confusion in this field.

Van Trigt (1970, 1971) developed an analytical technique for solving the transfer equation for optically thick plasmas in planar geometries which circumvents the numerical inaccuracies arising from differences between large numbers, though the resulting complexity restricts application to relatively simple geometries and density distributions. Payne and Cook (1970) extended it to cylindrical geometries and van Trigt and van Laren applied it to evaluating the radiation from optically thick low pressure sodium lamp discharges. Waszink (1973) used this approach to determine an escape factor for a high pressure sodium discharge which is spacially inhomogeneous, though this is valid only for resonance lines.

Methods for use in non-LTE situations are considered by Avrett (1971) and Auer (1971), and a good source for many others is the book by Mihalas (1978).

CONCLUSION

This talk has necessarily dealt with radiation transport in a very restricted set of conditions, namely those pertaining to many forms of high pressure discharge lamp. The extent to which these are valid may be debateable. For example, it is not clear that radiation scattering processes are entirely negligible in all such discharges. The only justification for omitting them is two-fold: the great increase in complexity their inclusion would involve; and the fact that satisfactory evaluations appear to be possible without including them. Some may consider this not too good a justification.

These discharges have enough complicating features of their own, which must be adequately incorporated in any analysis which is going to be practically useful, and this accounts for some of the divergences I have made from some more conventional treatments. The more important of these complicating features are: the considerable spatial inhomogeneity in terms of temperature, composition and excitation; the variety of reacting species present, molecular and atomic, and their interactions; and above all, the high proportion of the input power which is converted to radiation, necessarily produced and transported at optical depths which are neither thick nor thin.

There are many extensions possible, and recent literature deals with a great variety. For example: non-stationary discharges, including ac and high frequency discharges, and discharge plasmas which include convective flow; high energy pulsed discharges; plasmas affected by electric and magnetic fields; and, of course, discharge light sources involving laser action an other coherent radiation processes. I can do no more than mention these in passing in a single lecture.

Nevertheless, I hope that I have shown that, in spite of these complexities, and the drastic assumptions that have to be made in other directions to cope realistically with them, the transport of radiation in these discharges can be analyzed and evaluated in ways which are not only practically useful, at reasonable effort and expense, for lamp design and similar purposes, but which also provide considerable insight into the way in which the various physical processes interact and the intellectual stimulus which leads to future innovation and progress.

REFERENCES

Auer, L. H., 1971, JQSRT, 11:573.
Avrett, E. H., 1971, JQSRT, 11:511.
Biberman, L. M., 1947, Zh. Eksper, i Teor. Fiz., 17:416 (Trans., 1949: Sov. Phys. JETP, 19:584).
Biberman, L. M., 1948, Dokl. Akad, Nausk SSSR, 59:659.
Born, M. and Wolf, E., 1964, "Principles of Optics", 2nd Edn., Pergamon, Oxford.
Breene, R., 1961, "The Shift and Shape of Spectral Lines", Pergamon, Oxford,
Cayless, M. A., 1963, Brit. J. Appl. Phys. 14:863.
Cayless, M. A., 1966, in: "Proceedings, 7th Int. Conf. Phenom. Ionized Gases," Beograd.
Church, C. H., Schlecht, R. G., Liberman, I. and Swanson, B. W., 1966, AIAA J., 4:1947.
Chandrasekhar, S., 1960, "Radiative Transfer", Dover, New York.
CIE, 1970, Publication No. 17 (E-1.1): "International Lighting Vocabulary", Commission Internationale de L'eclairage, Paris.
Eardley, G., Jones, B. F., Mottram, D. A. J. and Wharmby, D. O., 1979, J. Phys. D: Appl. Phys., 12:1101.
Griem, H. R., 1964, "Plasma Spectroscopy", McGraw-Hill, New York.
Holstein, T., 1947, Phys. Rev., 72:1212.
Holstein, T., 1951, Phys. Rev., 83:1159.
Irons, F. E., 1979, JQSRT, 22:1.
Jones, B. F. and Mottram, D. A. J. M., 1981, J. Phys. D: Appl. Phys., 14:1183.
Kenty, C., 1932, Phys. Rev., 42:823.
Lowke, J. J., 1969, JQSRT, 9:839.
Lowke, J. J., 1974, JQSRT, 14:111.
Lowke, J. J. and Capriotti, E. R., 1969, JQSRT, 9:207.
Maitland, A. and Dunn, M. H., 1969, "Laser Physics", North Holland, Amsterdam.
Mihalas, D., 1979, "Stellar Atmospheres", 2nd Edn., Freeman, San Francisco.

Mitchell, A. C. G. and Zemansky, M. W., 1934, "Resonance Radiation and Excited Atoms", University Press, Cambridge.
Payne, M. G. and Cook, J. D., 1970, Phys. Rev. A., 2:1238.
Penner, S. S., 1959, "Quantitative Molecular Spectroscopy and Gas Emmisivities", Pergammon, London.
Richter, J., 1968, in: "Plasma Diagnostics", W. Lochte-Holtgreven, ed., North Holland, Amsterdam.
Stormberg, H. P. and Schafer, R., 1983, J. Appl. Phys., 54:4338.
van Trigt, C., 1969, Phys. Rev., 181:97.
van Trigt, C., 1970, Phys. Rev. A, 1:1298.
van Trigt, C., 1971, Phys. Rev. A, 4:1303.
van Trigt, and van Laren, J. B., 1973, J. Phys. D: Appl., 57:672.
van de Weijer, P. and Cremers, R. M. M., 1985, J. Appl. Phys., 57:672.
Waszink, J., 1973, J. Phys. D: Appl. Phys., 6:1000.
Wharmby, D. O., 1985a, Molecular Spectral Intensities in LTE Plasmas (this ASI).
Wharmby, D. O., 1985b, High Pressure Sodium Arcs (this ASI).
Zollweg, R. J., 1978, J. Appl. Phys., 49:1077.
Zollweg, R. J., Liebermann, R. W. and McLain, D. K., 1981, J. Appl. Phys., 52:3293.

DISCHARGE LIGHT SOURCES

J. F. Waymouth

GTE Lighting Products
Sylvania Lighting Center
Danvers, MA, USA

INTRODUCTION

As already outlined, electric discharge lamps are extremely common, with some two billion fluorescent lamps and 0.2 billion high intensity discharge (HID) lamps in service around the world (not counting USSR and mainland China), not to mention perhaps 5-10 million low-pressure sodium lamps. Taking 2500 lumens as the average light output per fluorescent lamp, and 15,000 lumens as the average light output per HID lamp, if all were lighted at once, about 8×10^{12} lumens would be generated. In turn, if this were distributed over the entire land area of the globe (7.2×10^{17} sq cm), the net illumination would be 10^{-5} lumens/cm^2, 0.1 lumens/m^2 = 0.1 lux (ca 0.01 footcandle). This is about the illumination level of bright moonlight from a full moon.

In addition, the 8×10^{12} lumens from discharge lamps accounts for 60% of the total light output from all kinds of electric light sources, despite the fact that incandescent lamps vastly outnumber (ca five billion in service) all other types combined. Discharge lamps achieve this level of dominance in the world of artificial illumination essentially without penetration of the largest single market area, the home. The reason is simple: they produce light at lower total cost, but the source itself has a higher initial cost.

Typically, considering life-cycle cost as the sum of the cost of the source itself plus the cost of the electric power to operate it over its burning lifetime, 70-80% of that life cycle cost is represented by the cost of electric energy. Discharge lamps typically are three to ten times more efficient in converting electric energy to light energy than incandescent. Said another way, discharge lamps require only one-third to one-tenth the electric power to produce a given amount of light as incandescent, a fact which has major impact upon the cost of producing light. Table 1 shows a comparison of relative lamp costs, energy costs, and life cycle costs for incandescent and the principal discharge lamps. Although low-pressure sodium lamps are not included in this table, they would come out close to high-pressure sodium in life cycle costs. As can be seen, sodium lamps produce light at the least overall cost, despite a high value of source cost per million lumen hours. The dominance of discharge lamps in the world of commerce and industry results entirely from the low life-cycle cost ("TOTAL COST OF LIGHT"). Their failure to penetrate the residential market results from their high initial cost per

Table 1. Comparison of Lamp Costs, Operating costs and Total Life Cycle Costs for Major Lamp Families (1)

Lamp Family	Lamp Cost $/1000 LM	Lamp Cost $/Million Lumen-Hours	Electrical Power Cost $/Million Lumen Hours	Life Cycle cost $/Million Lumen Hours (2)
Incandescent	0.56(3)	0.75	3.16	3.91 (81%)
T-H	1.87	0.94	2.33	3.26 (71%)
Fluorescent	0.90	0.038	0.72	0.76 (95%)
Mercury	1.24	0.052	1.00	1.05 (95%)
M-H	1.91	0.096	0.78	0.88 (89%)
HPS	1.51	0.063	0.44	0.51 (86%)

(1) 1980 Lamp prices and electrical costs; relative rankings not changed appreciably since then. Does not include ballast, fixture, or installation cost.

(2) Figures in parentheses are percentage of life cycle costs represented by electrical energy.

(3) Least expensive highlighted.

lumen, even per lumen hour. That market is extremely sensitive to first costs - the housewife does not think like a banker.

Let us examine the reasons for the high efficiency of conversion of electrical energy to light in discharge lamps. It originates, first of all, in a high efficiency of conversion of electrical energy into radiation and, secondly, in a favorable distribution in wavelength of that radiation, with a high fraction of it in the visible part of the spectrum. One of the reasons for this behavior results from the characteristics of the blackbody radiation law, which essentially sets the upper limit to spectral exitance at every wavelength in thermodynamic equilibrium. Although real light sources deviate in varying degrees from thermodynamic equilibrium, it is generally true that their spectral exitances do not exceed at any wavelengths those of a blackbody at the highest temperature of any species they contain.

Fig. 1 shows the spectral exitance for blackbodies versus wavelength at several different temperatures, illustrating the well-known decrease of the wavelength of maximum emission with increasing temperature. Fig. 2 plots the wavelength of maximum emission, in microns, and the fraction of the total radiation emitted between 0.4 and 0.7 microns ("visible"), both as a function of temperature. Since the luminous efficacy of the visible radiation varies only slowly with temperature, varying only between about 246 lm/visible-watt at 4000 K and 206 lm/visible-watt at 10,000 K, the overall luminous efficacy of the blackbody will vary in a manner similar to the fraction of radiation in the visible. Finally, Fig. 3 plots total and visible radiation per cm^2-vs-temperature.

It can be seen that incandescent lamps suffer from being constrained to too low an operating temperature. Even though tungsten is, in fact, a selective radiator (visible emittance about 0.4, decreasing to about 0.1 at 10.0 microns), too small a fraction of its radiation is in the visible part of the spectrum. A very great advantage enjoyed by discharge lamps

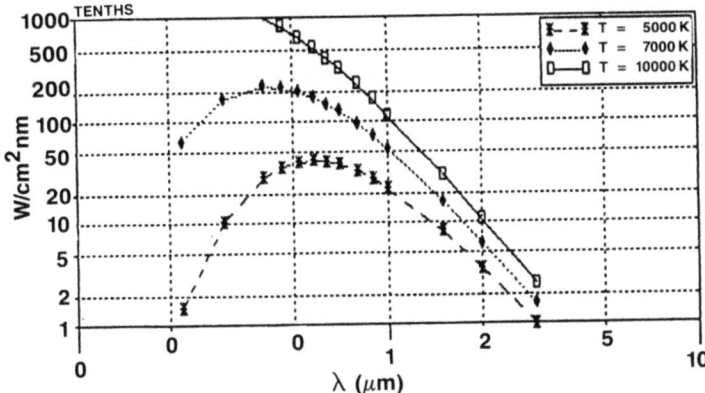

Fig. 1. Spectral exitance (watts/cm^2-nm) of blackbodies at various temperatures.

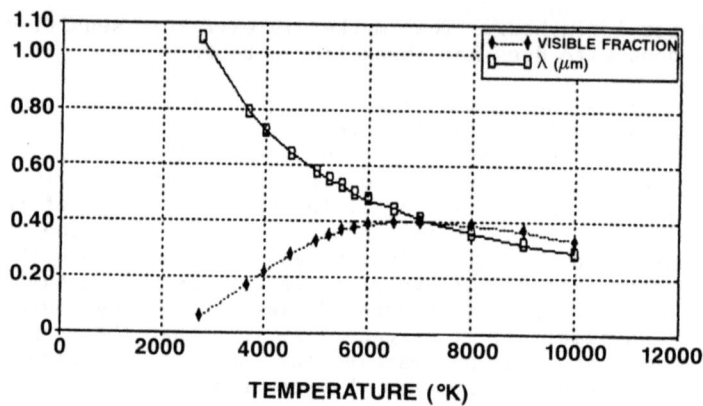

Fig. 2. Wavelength of maximum emission (microns), and fraction of total radiation in the visible, for blackbody-versus-temperature.

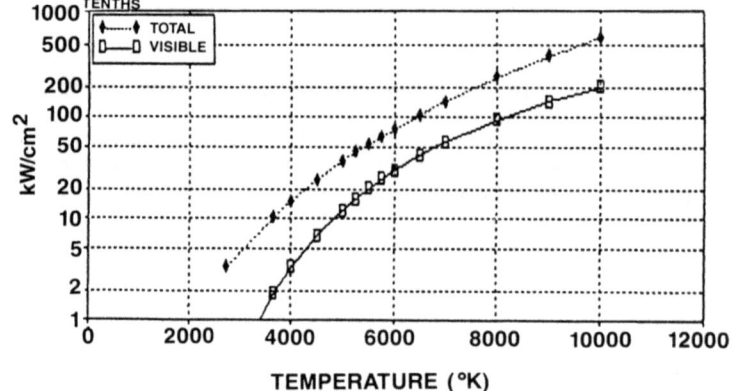

Fig. 3. Total radiant exitance (kw/cm^2) and visible radiant exitance (kw/cm^2) of blackbody-versus-temperature.

derives from the already gaseous state of the radiating medium. Thus, the controlling temperature can be very much higher than that achievable by any solid or liquid material. The resulting shift of the wavelength of maximum emission to a more favorable region then has the effect of permitting a much greater fraction of the radiation to be useful.

A second advantage enjoyed by discharge lamps is a much more favorable dependence of emittance on wavelength. By suitable selection of the radiating gases, the emittance may be made high in the visible, and low elsewhere. Most neutral metals have a number of relatively strong emissions in the visible part of the spectrum, with fewer, weaker emissions in the infrared.

In addition, high temperatures of the radiating media can be achieved in plasmas at such sufficiently low electron density that continuum emission is weak. Recall from the discussions earlier that the free-free continuum absorption coefficient varies as wavelength cubed and electron density squared. Thus, continuum emission is primarily an infrared emission, and increases the emittance in non-useful parts of the spectrum.

Finally, as Fig. 3 shows, at temperatures accessible to electric discharges in gases, the radiant power per unit area can be very large, and increases nearly exponentially with temperature. On the other hand, thermal conduction and convection losses in the unavoidable temperature gradients (dictated by the material properties of transparent containers) increase only linearly with temperature. High enough temperatures can be achieved that radiation loss is the dominant mode of energy dissipation of the discharge plasma.

In summary, light sources based on electric discharges in gases are more efficient because they permit employment of much higher temperatures than could be achieved by any solid or liquid radiator, and they permit a high degree of selectivity in the emittance spectrum. They thus combine a favorable position for the wavelength of maximum emittance of the corresponding blackbody, together with high emittance in the visible and low emittance elsewhere, especially in the infrared.

With this as a preamble, let us now turn to a closer examination of the principal discharge light sources, and study more in detail the role that radiative processes play in their operation, and the level of understanding of those processes that has been incorporated into the overall understanding of their behavior. The discussion will begin with the high-pressure discharge sources, the so-called high intensity discharge (HID) lamps, because the simplifying assumption of local thermodynamic equilibrium (LTE) may be applied to many aspects of their behavior. Subsequently, we will deal with the added complexities of the low-pressure, resonance-radiation discharge family.

HIGH-INTENSITY DISCHARGE (HID) LAMPS

General

HID lamps involve a discharge through relatively high pressure gases or vapors, typically at a total pressure of one atmosphere or more, and produce radiation featuring the strong emission lines of some of the species present. The discharge medium is contained within a relatively small, thick-walled transparent or translucent vessel called the arc tube, contained within a larger glass bulb called the outer jacket. The arc tube is fabricated from transparent fused silica (quartz) in the case of high-pressure mercury (mercury) lamps and metal halide (MH) lamps, but is fabricated of translucent polycrystalline alumina ceramic (PCA) in high-pressure sodium (HPS) lamps. Fig. 4 illustrates a generic diagram of an HID Lamp, illustrating the principal components. The outer jacket serves several functions: first, to protect the metal components of the arc tube seals from oxidation; second, to provide for thermal insulation of the arc tube; and third (in mercury and MH lamps only), to absorb harmful ultraviolet radiation transmitted through the walls of the arc tube.

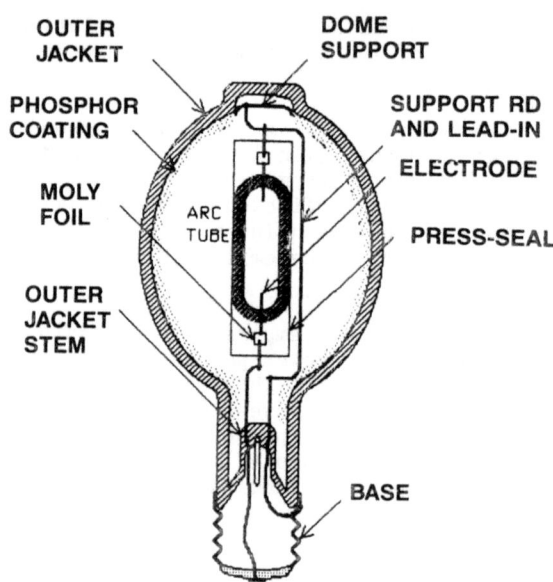

Fig. 4. Generic diagram of a HID lamp illustrating arc tube and outer jacket.

Substantial power input per unit volume and surface area of the arc tube is employed in all three types. Table 2 shows typical values of wall loading, arc loading, power per unit volume of arc tube, and maximum and minimum values of arc tube temperature in the several types. At the indicated levels of power input, electron densities in the center of the plasma are typically into the $10^{15}/cm^3$ range, and momentum-transfer collision frequencies are in the range of $10^{11}/s$. Thus, there is very substantial collisional heating of the gas in the center of the arc tube. The axis temperature of the gas therefore approaches within a few tens of degrees the electron temperature at the axis, but is necessarily constrained to the wall temperature at the outer boundary. As a result, a substantial gradient in gas kinetic temperatures exists between center and wall, which leads to both conductive and convective transfer of heat. The electron temperature also decreases between center and walls because of collisional energy transfer to the cooler gas. Because electrons are strongly coupled to excited states, and through them to the radiation field, the difference between electron and gas-kinetic temperatures in the outer fringes of the plasma are somewhat larger, as much as 100 degrees. It is common in HID lamp models to treat the system as being in LTE at the local electron temperature for calculating radiative and electrical properties, and to accept the small error in calculated heat conduction loss that may result from the disparity between gas-kinetic and electron temperatures.

Considering the energy balance of a section of the column of such a discharge, the power input will be equal to the product of axial electric field times the discharge current. The axis temperature of the discharge will rise to the point that the plasma dissipates the power supplied to it. The two principal energy dissipation modes are radiation and gas conduction/convection. The former typically varies approximately exponentially with temperature, while the latter varies linearly with the temperature difference between axis and wall. Since radiation loss varies steeply with temperature, axis temperature does not vary much with

Table 2. Comparison of Power Input and Wall Temperatures of HID Lamp Families

Lamp	Wall Loading (W/cm^2)	Arc Loading (W/cm)	Power Density (W/cm^3)	Arc Tube Temperature	
				Min. (C)	Max. (C)
MERCURY	7-10	50	16-20	550	750-800
METAL HALIDE	12-17	75	25-40	650	900
HPS	15-20	40	71-100	650	1200

input power, and, therefore, neither does the conductive/convective power loss. To a first approximation, if radiative power per unit length of plasma column is plotted as a function of input power per unit length, the result is a straight line with intercept on the power axis at the heat conduction/convection loss, as in Fig. 5, a result first described by Elenbaas (1951). As a consequence of this behavior, it can be seen that radiation power efficiency is maximized by operating at high values of input power per unit length. HID lamps are typically operated at input powers such that 65% or more of the power dissipation of the plasma column is by radiation.

Typical values of heat conduction losses are in the vicinity of 10 watts/cm in most HID lamps, so that the resulting desired range of power input is 30-80 watts/cm. Necessarily, this results in relatively high values of operating temperature for the arc tube, and dictates the choice of materials of fabrication. Advantage is taken of this high temperature in all HID lamps to provide a heat source for volatilizing the species whose radiations are desired. Thus, the arc tubes can be filled at manufacture with primarily condensed material together with a low pressure of a starting gas (in which a discharge may be easily established at relatively low voltages). The heat from the discharge in the starting gas then vaporizes the condensed additives and establishes the vapor composition desired for the radiating medium.

The radiation process in such lamps is usually calculated on the basis of the LTE model. The absorption coefficient is calculated using densities of lower states given by a Boltzmann population at the local temperature. Frequency-dependent absorption coefficients may be calculated from a knowledge of the line broadening mechanisms which may be present.

Radiation emission and escape from such discharges ranges from optically very thick to optically thin, frequently spanning the entire range in a single emission line. Fig. 6 illustrates the spectral radiance as a function of position along a diameter through a discharge with the indicated temperature profile, for three different values of the optical depth, illustrating that where optical depth is very large, the radiance is everywhere constrained to be the local blackbody value. Where optical depth is intermediate, radiance increases with distance wherever it is less than the blackbody values, and decays wherever it is greater than the blackbody value. However, the distance scale over which such changes can take place is determined by the optical depth so that, in this case, the spectral radiance does not fall to the blackbody level

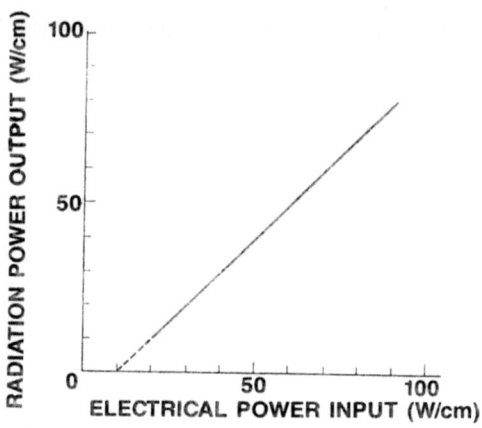

Fig. 5. Radiative power per unit length of HID plasma column vs input power per unit length; extrapolated intercept is approximately equal to heat conduction power per unit length (schematic).

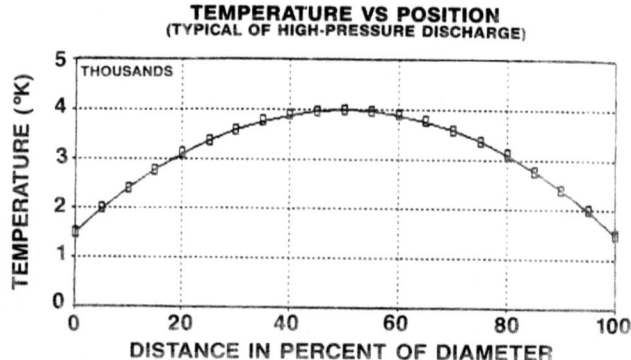

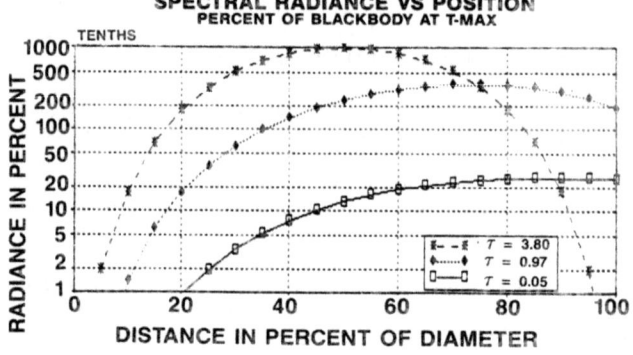

Fig. 6. Positively directed spectral radiance vs position along a diameter through an arc column in LTE with indicated radial temperature profile, for three different values of optical depth. Expressed as percent of blackbody at maximum temperature. Curve for optical depth at 380 is indistinguishable from local black-body value.

at the surface before the radiation escapes the tube. The emitted radiation is a substantial fraction of the blackbody radiance at the arc center. In the optically thin case, there is substantially no absorption of any escaping radiation, and the spectral radiance increases monotonically with distance through the plasma, but never achieves large levels. As Fig. 7 shows, these three conditions, and the entire range in between, can be achieved within a single emission line which has a large optical depth at line center, but has a finite width dictated by the various broadening mechanisms, over which the absorption coefficient and optical depth decrease to negligible values. Such lines are called "self-reversed", for reasonably obvious reasons. The outstanding example of this line shape is found in the emission of the sodium-D resonance radiation from a high-pressure sodium lamp.

The radial temperature profile itself is controlled by an energy balance; the governing equation is an equation of continuity for the heat flux, known as the Elenbaas-Heller equation. Stated in words, the radial divergence of the heat flux (neglecting convection) is equal to the net local production of heat which is, in turn, equal to the difference between electrical power input and net radiation escape. We note in the mathematical expression, Eq. (1) below, that there are two terms to the net radiation escape: radiation emitted and radiation absorbed. Thus, both the emission and absorption of radiation per unit volume have significant effects upon the radial temperature profile:

$$\nabla \cdot \left(- H \frac{dT}{dr} \right) = P_e - \left(P_r - P_a \right) \tag{1}$$

It is plain that the solution to this differential equation must involve as boundary conditions the value of temperature of the wall, as well as the fact that the gradient of temperature at the axis must be zero and the second derivative at the axis must be negative. Inspection of Eq. (1) shows that the second derivative at the axis can only be negative if the electric power input at the axis is greater than the radiation escape term. This condition establishes a minimum axial electric field required for maintaining the discharge. The nature of the balance of the solution depends on the relative temperature dependencies of electrical power input per unit volume and net radiation emission per unit volume. If net radiation emission decreases more rapidly with decreasing temperature than does electrical power input, the second derivative of temperature vs radius will remain negative for all values of radius and temperature, and the radial temperature profile will curve smoothly down toward lower values of temperature with ever-increasing gradient, and a "parabolic" or "fat" type of temperature profile will result.

A unique development that emerged from the investigation of metal halide lamps in the earliest days of their development was the discovery of a "radiative constriction" process that had not ever been previously observed in high-pressure arcs (to the best of my knowledge and belief). This results from the employment of discharge media in which the net radiation loss per unit volume decreases more slowly with decreasing temperature than does electrical power input per unit volume. In this circumstance (although, as before, electrical power input must be greater than net radiation loss at the axis so that the second derivative of T-vs-r is negative at the axis) at some lower temperature the electrical power input per unit volume becomes less than the net radiative loss. The sign of the right-hand side of Eq. (1) reverses, and with it the sign of the second derivative of T-vs-r. Instead of the resulting temperature profile being a smooth parabolic function, it must become a bell-shaped curve, leading to a very constricted arc in the center of the tube.

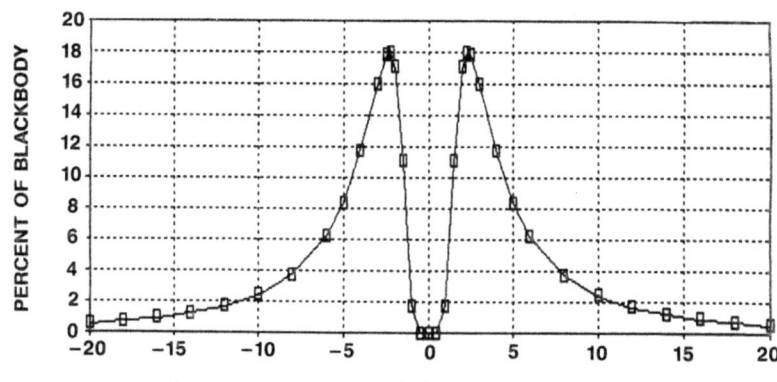

Fig. 7. Spectral radiance of self-reversed line, expressed as percent of blackbody at maximum temperature. Optical depth at line center is 380, at line-wing maximum about unity, and .05 at + 10 nm from line center.

We may gain some insight as to the mechanism that creates this result by incorporating some simple models for radiation loss and electrical power input per unit volume in Eq. (1). Let electrical power input be given by:

$$P_e = J \cdot E = \sigma E^2 \qquad (2)$$

Assume that electrical conductivity is governed by the Saha equation for electron density:

$$\sigma = e\mu_e n_e \sim \exp\left(-eV_i/2kT\right) \text{ and} \qquad (3)$$

$$P_e \propto E^2 \exp\left(-eV_i/2kT\right) . \qquad (4)$$

Further assume that radiation loss is an optically thin process without absorption, is entirely due to line radiation, and may be described in terms of a single exponential function of temperature, in which $\bar{V}$ is the average excitation potential of the radiating atoms, i.e.,

$$P_r \propto \exp\left(-e\bar{V}/kT\right) . \qquad (5)$$

Fig. 8 illustrates the dependencies of P_e and P_r vs. temperature for two different relative values of V_i and $\bar{V}$. If ionization potential (controlling conductivity and electrical power input) is less than twice the average excitation potential (controlling radiation power loss per unit volume), the radiation loss decreases more rapidly than electrical power input with temperature, and the sign of the second derivative of T-vs-r remains unchanged, leading to parabolic-type radial temperature profiles. If, on the other hand, the ionization potential is more than twice the average excitation potential, the net radiation loss will decrease more slowly than, and at some radius become less than, electrical power input

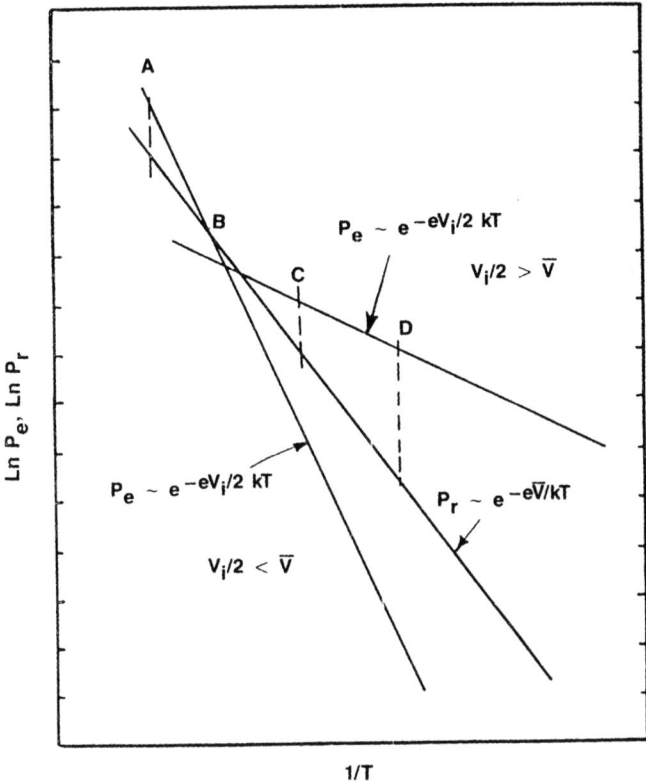

Fig. 8. Dependencies of P_e and P_r on temperature for two different cases: $V_i/2 < \bar{V}$ and $V_i/2 > \bar{V}$.

per unit volume. The sign of the right-hand side of Eq. (1) reverses, and with it the sign of the second derivative of T-vs-r, so that a constricted bell-shaped temperature profile results. More detailed calculations show the critical ratio is at $V_i = 1.71 \bar{V}$, because of temperature-dependent terms in pre-exponential factors of P_e and P_r. Fig. 9 shows the dramatic shift in temperature profile that occurs as assumed ionization potential is varied from 6.5 to 7.5 volts for an assumed average excitation potential of 4.0 volts.

Prior to investigation of metal halide arcs, this phenomenon had not been recognized, because the common monoatomic gases studied in arcs (rare gases and mercury) all exhibited ionization potentials less than twice the average excitation potential. For many of the metals made accessible by iodide-cycle introduction in metal halide lamps, however, ionization potentials are 1.7 or more times average excitation potentials, and these all yield radiatively constricted arcs. Moreover, it is plain from inspection of Eq. (1) that absorption of radiation may locally reduce the net emission. Therefore, when significant absorption is present in the cooler fringes of the discharge, this may prevent the right-hand side of Eq. (1) from changing sign, and may, therefore, prevent the development of a constricted temperature profile. These factors will be discussed more in detail in connection with metal halide lamps presently.

All the information about axis temperatures and radial temperature profiles in HID lamps has come from radiation measurements assuming LTE; the temperatures thereby obtained are the excitation temperatures, which

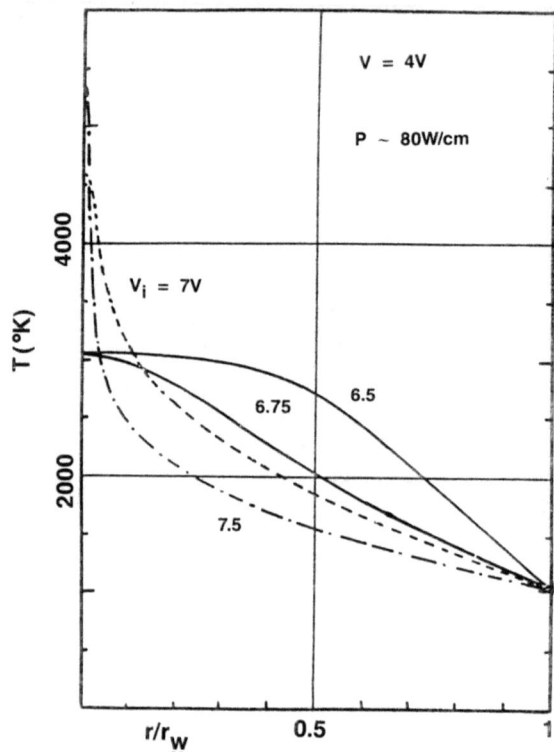

Fig. 9. Calculated radial temperature profiles for a radiation-dominated plasma with assumed average excitation potential V = 4 volts, illustrating transition to a radiatively constricted profile as assumed average ionization potential increases from 6.5 to 7.5 volts.

are substantially equal to the electron temperatures. Methods employed have been the single-line and two-line Bartels, methods (Bartels, 1950; de Groot and Jack, 1973; Wesselink et al., 1973), Abel inversion of absolute radiometric measurements of an optically thin line, and channel-model analysis of total radiant power per unit length in an optically thin line. Continuum-radiation methods (such as have been extensively used in cascade channel arcs) have not been used in discharge lamps because of the relative weakness of the continuum.

High-Pressure-Mercury Lamps

High-pressure-mercury (mercury) lamps are the oldest form of HID lamp, and consist of a discharge through mercury vapor at pressures of 1-10 atmospheres, contained in a fused quartz arc tube. The axis temperature of the discharge is typically about 6000 K, the wall temperature about 1000 K, and the radial temperature profile is of the parabolic "wall-stabilized" type. The radiation emission is predominantly the line spectrum of mercury, but the resonance lines are strongly self absorbed because of the high density of ground state mercury atoms. Fig. 10 shows a typical spectral power distribution of the emitted radiation.

Because of the strong absorption of the resonance lines, the average excitation potential of the radiation which is emitted is about 7.8 electron volts, more than half the ionization potential (10.4V). As

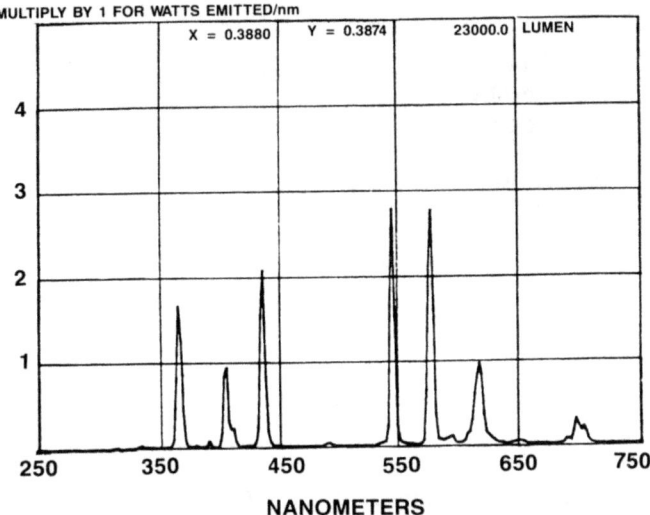

Fig. 10. Spectral power distribution of radiation emitted from a 400-watt high-pressure-mercury lamp. Red emission at 610 nm and 700 nm is from yttrium vanadate:europium phosphor on outer jacket.

noted above, the radial temperature profile is of the parabolic type. There is little or no continuum radiation, and not very much radiation of infra-red lines, which originate at too high an energy level in the atom to be significantly excited at the 6000 K arc temperature.

However, in addition to the visible line emission of mercury, there is considerable non-resonant ultraviolet line emission. Only about 23% of the total radiation is visible, resulting in a net radiant efficiency about 16% of the input power in the visible part of the spectrum, after heat conduction and electrode losses are accounted for. Moreover, of the four visible lines of mercury that account for the vast majority of the visible radiation, two are in the blue and violet, a region of low sensitivity of the eye. The luminous efficacy of the visible radiation is then of the order of 300 lumens per visible watt, and the total luminous efficacy of the lamp is about 50 lumens per electrical watt. The spectrum is deficient in red, leading to poor color rendering qualities of the light output. This is corrected in most lamps in service today by adding a phosphor coating to the inside of the outer jacket, which has the added benefit of converting otherwise wasted ultraviolet light into visible energy. Clear lamps with efficacy of 50 lumens per watt may be increased to 55-60 lumens per watt by phosphor coating, with substantial increase in color quality as well.

The emission lines are broadened to about 0.1-nm width by a combination of Stark, collisional, and resonance broadening. Lines whose lower level is in the 3P manifold are all self-reversed, because the population of those states in the core of the arc is significant. An example is shown in Fig. 11. Thus, these lines are suitable for Bartels' method measurements of axis temperature, or other radiative diagnostics involving self-reversed lines, such as described by Professor Karabourniotis in this Proceedings. The yellow line, terminating on the 1P level, is optically thin in many mercury lamps, and can be used for measurements of radial temperature profile. Also useful for this purpose are some of the infrared lines.

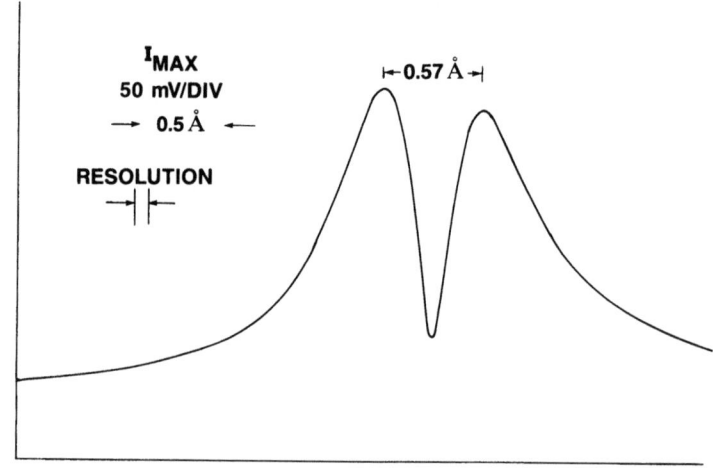

Fig. 11. Spectral radiance of the 365.0-nm non-resonance line of mercury from a high-pressure-mercury lamp, illustrating partial self-reversed character.

A good understanding of the energy balance and radiation transport in mercury lamps has been available for many years, from the pioneering work of Elenbaas, (1951). Appropriate coefficients have been determined for a highly useful two-temperature (channel) model, which can be used for accurate engineering design of lamps of other wattages. Design strategy for maximizing efficiency is based on using the shortest possible arc length (to maximize power per unit length, reducing the relative heat conduction loss). Constraints limiting the extent of this choice are necessary to maintain a given surface area of arc tube to prevent overheating, and the fact that mercury pressure must be increased to maintain the correct value of arc voltage. The extent to which mercury pressure can be increased is limited by the onset of turbulent convection.

Metal Halide Lamps

Metal halide lamps are similar in many aspects of their design and appearance to mercury lamps. The discharge takes place in a gas consisting predominantly of mercury vapor in a quartz arc tube within a glass outer jacket. However, in addition to the mercury, the arc tube also contains a small quantity of metal iodide salts. As indicated in Fig. 12, in operation, these salts evaporate, and the metal iodide molecules diffuse into the high-temperature arc core. In this region they dissociate, freeing the metal atoms to be ionized and excited to contribute their characteristic spectral lines. As metal atoms and iodine atoms diffuse out of the arc core into the cooler gas of the mantle, they recombine to reform the iodide molecules.

The reason for this roundabout method of introduction of metal atoms into a high temperature arc is that it greatly increases the number of accessible species. Only a few metals have vapor pressures suitable for volatilization at temperatures accessible to quartz. Of those that do, many, especially the alkali metals, react vigorously with quartz. By contrast, nearly fifty metals of the periodic table have iodide salts with vapor pressures of one torr or more at temperatures obtainable in a quartz envelope, and very few of these react in any significant way with quartz.

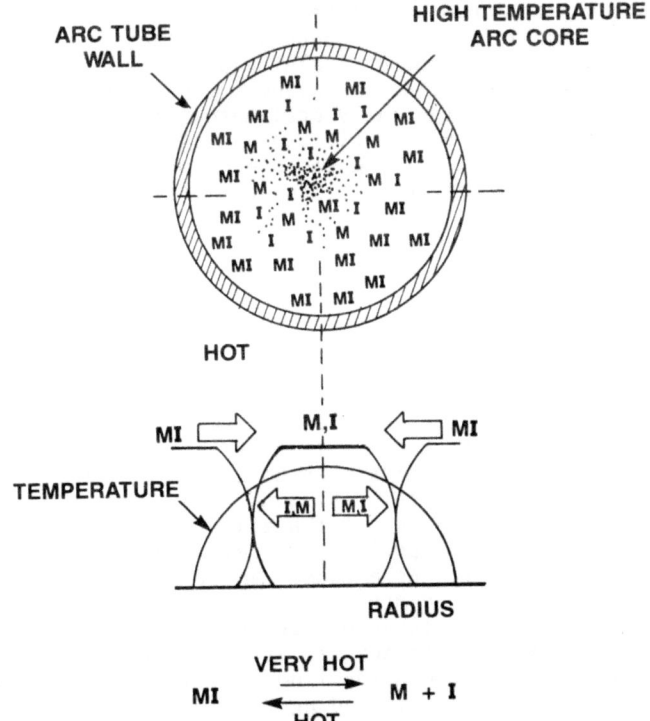

Fig. 12. Illustrating volatilization and subsequent dissociation of metal iodide molecules, liberating metal atoms to be excited to emit characteristic radiations in high-temperature arc core.

Many of the metals of choice have average excitation potentials of 4 eV or less. For such an average excitation potential, in an arc at 6000 K, equal radiation power will be obtained for the added metal as for mercury at 22,000 times higher pressure. Therefore, the addition of only a few torr of added metal to the mercury arc at several atmospheres can add radiation channels which dominate the radiative properties of the discharge.

The very large number of possibilities, especially if combinations of additives are considered, posed quite a problem for the lighting industry in selecting optimum additive mixes when the feasibility of this technology was first demonstrated. Not surprisingly, the choices made were different in different organizations and different marketplaces. In a broad sense, the possible metal radiators may be classified into two categories: "single-color" emitters and "white-light" emitters. The alkali metals, with their strong resonance-line emission, together with indium and thallium, are the outstanding examples of the former category; the rare earths, and metals such as thorium, uranium, and scandium, are the outstanding examples of the latter.

One of the favored early choices was a blend of indium (blue), thallium (green), and sodium (yellow), all basically single-color emitters, with relatively strongly self-reversed resonance-line emissions. Indium and thallium, moreover, also emit a broad continuum-like emission throughout the visible which increases the contribution of red emission from the discharge, improving the color rendering qualities.

The origin of this continuum is not yet known for certain, but its character suggests a transition from a bound single excited state of the monoiodide molecule to an unbound singlet ground state. By suitable adjustment of the mixture ratios of indium to thallium to sodium, a range of apparent source colors (typically designated in terms of "color temperature", the temperature of the blackbody having the closest color match) can be achieved, all with reasonable color-rendering qualities. Lamps embodying this type of formulation are still in wide use today.

The alkali metals, thallium, and indium all have ionization potentials less than 1.7 times the average excitation potential in such discharges (when account is taken of the strong degree of self-absorption of the resonance lines). Moreover, because of the strong absorptions in the resonance lines, there is a great deal of radiative energy transport from arc core to cooler fringes. Thus, as discussed above, this system tends to result in parabolic type, "fat" temperature profiles for the arc, so that the arc is well stabilized by the tube walls. As metal iodide arcs go, they are "well-behaved", and can tolerate significant losses of one or another of the additive species with nothing more damaging than a shift of the color of the emitted radiation. Their greatest failing is color shift, however, because two of the additive species (indium and thallium) are usually fully vaporized, whereas the sodium iodide operates in the saturated-vapor mode. As a consequence, the sodium concentration in the plasma varies with cold-spot temperature and anything that affects it (arc-tube manufacturing tolerances, outer jacket fill pressure, lamp wattage as affected by arc drop and ballast tolerances, as well as line voltage variations). Indium and thallium concentrations are relatively insensitive to these variations, leading to shifts in the ratios of the species concentrations. Moreover, because the three individual colors are widely separated, a large gamut of colors is accessible in the color space they enclose, and significant color shifts are perceptible for relatively small differences in ratios. For all these reasons, this system tends not to be a preferred system for any application in which variability of individual source color is an applications concern, such as interior commercial lighting. The most common application of MH lamps embodying these formulations is for outdoor floodlighting and sports lighting, together with some industrial lighting.

The alternative possibility, that of using "white-light" emitters directly, frequently encounters the problem that most such metals have ionization potentials that are high enough relative to their average excitation potentials to result in radiatively constricted arcs, which are totally unsuitable for light source use. In the vertical position, they result in unstable, swirling discharges, while horizontally, they result in severe local overheating of the upper wall of the quartz arc tube. A major step forward in the technology of MH lamps was the discovery that these arcs could be "tamed" and "fattened" by the addition of lower ionization-potential additives, or by the addition of radiating species with strong self absorptions, or both. These additives changed the relative temperature dependencies of electrical power input and net radiation loss in the Elenbaas-Heller equation, Eq. (1), from the constricting domain to the fat domain, even though substantial radiation contribution from the white-light emitter remained.

Fig. 13 is a photomontage of three arcs in 20-mm ID quartz tubing: 1) pure mercury; 2) mercury plus thorium iodide, showing the strong radiative constriction due to thorium; and 3) mercury plus thorium iodide plus sodium iodide, showing the "fattening" of the thorium-constricted arc by sodium. Similar effects, but to a much lesser degree, can be observed in the fattening of a thorium-mercury arc by the addition of indium or thallium. Indium and thallium provide arc fattening by adding the absorptions of their near-resonant radiations to the net radiation loss term in Eq. (1), but do not significantly affect the average ioniza-

Fig. 13. Photomontage of the negative image of three arcs in 20-mm ID quartz tubes: Bottom, mercury-vapor; Middle, mercury varpor plus thorium iodide; Top, mercury vapor plus thorium iodide plus sodium iodide. Thorium is an outstanding example of a radiative constrictor. (Tube wall is visible only in top image, where tipped-off exhaust tubulation may be seen at left).

tion potential of the gas mixture. Sodium, on the other hand, not only adds absorption, but also reduces the average ionization potential of the mixture. Cesium has been shown to be five times as effective as sodium, so far as arc fattening is concerned, but contributes significant infrared resonance radiation, which reduces lamp efficacy. Sodium resonance radiation, on the other hand, contributes in a region to which the eye is highly sensitive.

The employment of "white-light" emitter formulations has been embodied in the scandium-sodium formulation extensively used in the US and Japan, whereas the dysprosium-holmium-thallium formulation has been used in Europe. A thulium-thallium-sodium combination has also been announced. The first permits a range of color temperatures between about 3000 K and 5000 K to be achieved, dependent on the scandium-to-sodium ratio, whereas the second covers the range 4500-6000 K, and the last achieves about 4200 K.

In formulations containing about ten mole percent scandium, the color temperature of the scandium-sodium system is 4000-4500 K, and the ratio of scandium to sodium in the gas phase is relatively independent of arc-tube cold spot temperature. Volatilization of scandium and sodium from a molten condensate pool of this composition is principally as the complex salt $NaScI_4$, rather than as the individual salts. The color shift of this formulation with lamp wattage and other variables affecting arc tube cold spot temperature is somewhat smaller, and results only from variation in the ratio of mercury radiation to additive radiation. Lower color temperature formulations, which are more sodium rich, result in much more significant volatilization of NaI as an independent species. Since NaI has a lower vapor pressure than the complex salt, color temperature decreases markedly with increasing cold-spot temperature.

Color rendering of the Sc-Na combination is essentially equivalent to the standard fluorescent lamp colors, and the 4000-4500 K color temperature range matches that of the most popular fluorescent color, "cool white". Thus the scandium-sodium lamps have found wide acceptability in the US wherever such fluorescent lamps might be used - in interior commercial and industrial illumination. It has also become the system of choice in the Japanese marketplace. Fig. 14 shows a spectral power distribution of a 400-watt scandium-sodium MH lamps.

The dysprosium-holmium-thallium system has the best color rendition of any (with the possible exception of the thulium-thallium-sodium blend), but it has a relatively high color temperature which has not been favored for interior commercial lighting. Its principal application has been in interior industrial illumination, outdoor floodlighting, and sports lighting. I personally have no current experience with the thulium-thallium-sodium blend, and therefore cannot make any comments about it.

A final formulation type of great interest, both technically and commercially, is a system based on tin iodides, which have high vapor pressures, and emit a strong continuum believed to originate in collisionally broadened band emission from tin monoiodide molecules. Because the continuum peaks in the green portion of the spectrum, the light has an undesirable greenish cast. It is generally used in combination with

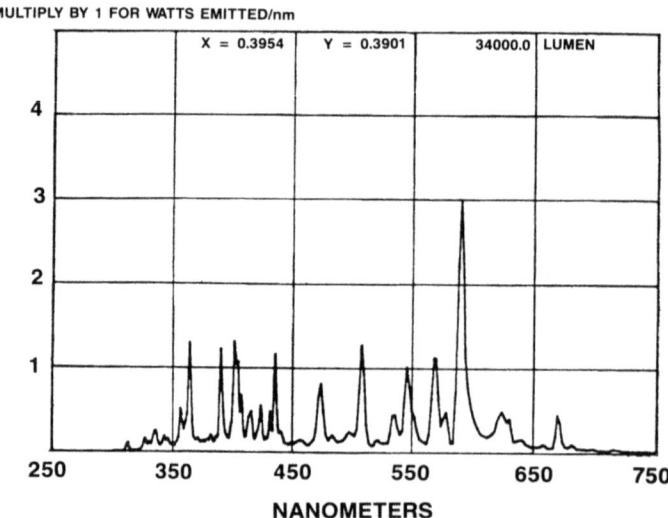

Fig. 14. Spectral power distribution of radiation emitted from 400-watt scandium-sodium MH lamps.

other additives to add red and blue components to the radiation, modifying the color to something found more acceptable.

The efficiency of all of these types of lamps is very high. They are typically operated at higher input power per unit length than mercury lamps, to maximize radiation over conduction/convection losses, and the metal radiators employed emit a greater fraction of their total radiation in the visible than does mercury. Although the operating vapor composition is heavily dominated by mercury, the radiations from the metal additives reduce the arc temperature so that mercury radiation is significantly suppressed. Therefore, the fraction of total radiation emitted into the visible is 50% or more, and up to 30% of the input power is radiated in the visible part of the spectrum. There is still very little infrared, and the ultraviolet component is minimized in comparison to mercury. Accordingly, the luminous efficacy of such lamps is typically 80-100 lumens per electrical watt, 80-100% greater than that of mercury lamps.

The design strategy for maximizing efficiency generally involves operating at as high an electrical power input per unit area of wall surface ("wall loading") as permitted by the quartz material itself, to maximize cold spot temperatures, and hence the volatilization of the additives, together with additive blends that permit operation with the "fattest" arc with the lowest axial temperature possible. The latter minimizes the excitation of the relatively inefficient mercury spectrum, and minimizes the depletion of neutral metal species by ionization in the arc core. Not only does ionization of the metal additives reduce the concentration of neutral radiators, but also, in the case of rare-earth type "white-light" emitters, creates ions which are also excited but radiate predominantly in the ultraviolet.

All of these varieties of combinations of additives, and the lamps based thereon, were developed empirically, with only the sketchiest of qualitative understandings as to the interactions of the several species in the arc. Channel models of the arcs, borrowed from the mercury model, were used to estimate the electrical parameters, but gave little or no guidance to the radiative properties of the plasmas. In addition, there is a strong interaction of the chemistry of the system on the arc temperature profile, and vice versa.

The problem may be seen to originate in the following instability. In a system employing a normal "arc-constrictor" balanced by an "arc-fattener", the energy balance, and consequently the local temperature, in the outer fringes of the arc may be strongly affected by the local concentration of arc-fattening metal atoms. Higher concentrations lead to lower average ionization potentials, and higher absorption of radiation emitted from the core, both of which result in increasing the local temperature. Lower concentrations have, of course, the opposite effect. On the other hand, because this is the temperature zone in which chemical recombination of metal atoms with iodine atoms to reform the iodide molecules is also taking place, the concentration of the arc-fattener metal atoms is strongly dependent on local temperature. Higher temperatures lead to higher concentrations, lower temperatures lead to lower concentrations. Since, as can be seen, the feedback in this coupled system is positive, it is inherently unstable.

The role of chemistry interacting with radiation in this system may be rather forcefully illustrated for the Sc-Na system in Fig. 15, which shows the apparent visual diameter of the arc as a function of the percentage iodine excess over stoichiometric in the formulation. Zollweg et al. (1925) have explained similar constrictions in an arc containing only mercury and iodine as originating from radiations by I_2 molecules, cooling off the outer fringes of the arc. For the data of Fig. 15, the

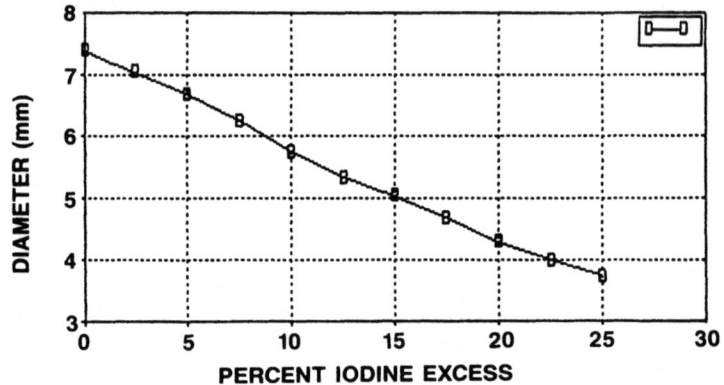

Fig. 15. Apparant visual diameter of arc core in scandium-sodium MH lamp vs percent stoichiometric excess of iodine.

concentration of excess iodine is ten-fold less, and the concentration of I_2 perhaps 100-fold less than that used by Zollweg. I believe that the mechanism of the effect in this case is the reduction in local concentration of free sodium atoms in the arc fringes, as a result of recombination with the higher content of gaseous iodine atoms. Thus, there are fewer sodium atoms in a critical region of the arc temperature profile. Reduction of absorption of sodium radiation emitted from the core, and reduction of the contribution of free electrons from ionized sodium to increase conductivity and local electrical power input, both combine to reduce the local temperature, which further favors recombination and so on.

In terms of the problem of arc modeling and the solution of the energy balance to provide a calculated arc temperature profile, the significance of these effects are that this calculation cannot really be done independently of a calculation of the chemical equilibria to determine the species which are present. In addition, the calculation of the chemical equilibria cannot be done independently of the radiation-transport energy-balance, since that determines the temperature profile. Finally, as a result of ambipolar diffusion and convection, there is a large degree of de-mixing in these arcs. A complete arc model will have to treat all three of these equations at the same time. To date, to the best of my knowledge and belief, this has not yet been done.

Calculations involving radiation transport itself in MH lamps are extraordinarily complex, especially for broad-band or multiline emitters, since significant radiation power may be emitted over nearly the entire optical spectrum, and there may be very large differences in optical absorption coefficients. Since solutions of the Elenbaas-Heller equation involve including absorption of radiation emitted from anywhere in the tube, and involve total radiation, i.e., integration over frequency as well as spatial integration, it is clear that very time-consuming calculations must be involved for each case considered. A common approach to this kind of complexity is to approximate the actual situation by means of a semi-empirical "net-emission-coefficient" for radiation which is a function of temperature only. Empirical coefficients are calculated by a best fit to the complete calculations for one case, and are then used for calculations of energy balances and temperature profiles for different tube radii, and arc currents involving the same species. Investigators employing this approach have included Zollweg (1978), Jones and Mottram (1981), and Stormberg (1983).

Experimental information on radial temperature profiles has been obtained by Abel inversion of absolute spectral radiances on the mercury yellow line, which is optically thin in many of these discharges, (Keeffe, et al., 1985). Axial temperatures have been estimated by two-line Bartels' methods using the green and blue lines of mercury by Waymouth (unpublished); and by channel-model analysis of absolute spectral radiance of the mercury yellow line, by Rothwell et al. (1980). The latter investigators have also used spectral radiance data of selected lines of neutral and ionized scandium and thorium, and of neutral sodium, to estimate concentrations of all these species in the arc, and to observe changes in these concentrations with burning time of the lamps.

A notable lack has been diagnostic techniques able to determine temperature and concentrations of molecular species in the dark region between the arc column and the wall. Since much of the lamp chemistry occurs in this region, this is an area of concern in our Laboratories, and in others as well, I am sure.

In sum, as we shall hear more in detail in the presentation by Ingold, the MH lamp is an extraordinarily complex system, in which radiation emission and absorption play a dominant role in all aspects of lamp behavior, and in which spectroscopic diagnostics has contributed significantly to the current level of understanding.

High-Pressure-Sodium (HPS) Lamps

HPS lamps are the third major category of HID lamps. The discharge in these lamps takes place in a gaseous mixture of mercury vapor plus 10-20% sodium vapor at a total pressure of about one atmosphere. These lamps emit a strong, self-reversed resonance line of sodium. Because of extensive resonance broadening to the blue side, and resonance plus van-der-Waals' and excimer-broadening by mercury to the red side, the absorption coefficient of the sodium-D-related emission has significant values over a range of nearly one hundred nanometers. Strong emission of lines originating in the upper S and D states is also seen. Fig. 16 is a spectral power distribution of an HPS lamp. Approximately 65% of the input energy is radiated, and perhaps 35-40% of the radiation is in the band between 500 and 700 nm. Since this energy peaks at 580 nm, near the maximum of the eye sensitivity curve, the luminous equivalent of this radiation is around 450 lumens per radiated watt, so that the lamp as whole has an efficacy of 120-125 lumens per electrical watt, eight times higher than a typical incandescent lamp.

The color temperature of the emitted radiation is rather low, about 2000 K, and the color rendition is only moderate, largely because of the very small amounts of blue and green radiations emitted. The lamps are widely used for outdoor floodlighting, streetlighting, and indoor industrial applications where color discrimination is not of significant concern. They have been used for interior commercial lighting in combination with decor schemes emphasizing "earth colors" (reds, browns, and oranges), with minimum admixture of blues and greens. Visual appearance of people (probably the most important single determinant of color rendering quality) is acceptable under HPS illumination.

Because of the extreme reactivity of sodium, the only suitable arc tube materials have been some form of alumina or other refractory-stable oxide, such as yttria. Commercial lamps employ sintered polycrystalline

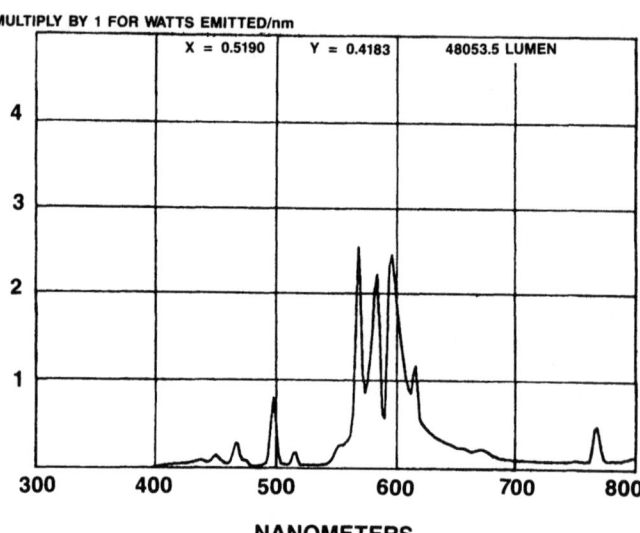

Fig. 16. Spectral power distribution of radiation emitted from 400-watt HPS lamp.

alumina (PCA) which, although an optically scattering material, is translucent with typically 95% diffuse transmittance. Seals are made to the ends of the arc tube for electrodes, employing niobium (columbium) metal components with a crystalline cement, fusible at lower temperatures than the melting points of alumina or niobium. Commercial lamps employ excess sodium-mercury amalgam, and the vapor pressures of sodium and mercury are determined by the amalgam ratio and by the cold-spot temperature. Fig. 17 shows a diagram of a typical HPS arc tube and lamp.

The maximum arc temperature at the axis in an HPS lamp is typically about 4000 K, at which temperature the excitation of mercury is essentially negligible. Data on arc temperature primarily come from single-line Bartels' method measurements using either the 586-nm or 819-nm self-reversed sodium lines. There are very few published measurements of radial temperature profiles. Only those of De Groot (1974) come readily to mind. In part, the problem is that neither PCA nor tubular-drawn sapphire arc tubes are adequate for such measurements. Only core-drilled and polished sapphire has adequate optical quality to permit spectroscopic measurements of radial temperature profiles.

In contrast to the MH lamp system, the HPS arc can be said to be reasonably well understood. The arc temperature profile, dominated by the electrical and optical properties of sodium, is "fat", and does not change very much with variation of parameters. Therefore, approximate models retain their validity over wide ranges of conditions. The dominant feature of the spectrum, the broadened self-reversed D-Band radiation, and the emission and absorption processes which create it, are sufficiently well characterized that the line profiles can be used for accurate in-situ diagnostics of vapor concentrations of both sodium and mercury. The profiles can be used for determination of condensed amalgam ratios and cold-spot temperatures, as recently reported by Reiser and Wyner (1985). Perhaps the only aspects of the radiation processes that are not thoroughly understood are the origins of weak continuua in the infrared. Perhaps we shall hear more about these from Dr. Wharmby at this ASI.

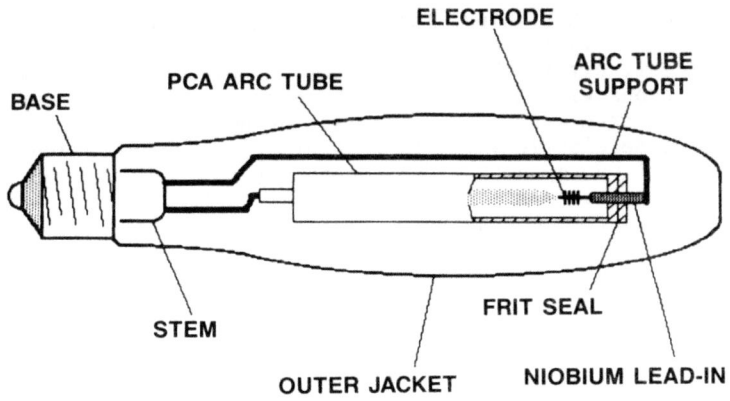

Fig. 17. Schematic diagram of an HPS lamp.

A variety of quite detailed arc models are available in the laboratories of the lamp companies, differing somewhat from one to another in the degree of use of empirically determined constants to represent some of the processes. Some incorporate direct calculation of the radial temperature profile, others employ channel models, but all seem to be quite successful in their capability to permit design of HPS lamps by computer, and come very close to the final optimum result. It is difficult to compare such models, however, since many of them are not published, and, of those that are, what is published is typically only a sketchy description of the features incorporated. There has been no publication of the computer codes themselves, and consequently no opportunity to compare one against the other.

The major thrust of current HPS lamp R&D is in the area of high-temperature chemistry and materials technology rather than radiation physics. Materials are needed which will react less with sodium, and permit the use of higher sodium pressures without penalty in operating life. There is a resurgence of interest, however, in the possible use of alkali metal discharge lamps in a pulsed mode, for possible application for laser pumping.

LOW-PRESSURE DISCHARGE "RESONANCE-RADIATION" LAMPS

General

In complete contrast to the HID type of discharge lamp, the other major type of electric discharge lamp employs a discharge at a relatively low current, through a buffer gas at a few torr pressure, plus a metal vapor at a few millitorr pressure to generate resonance radiations of the metal atoms. Efficiency of such discharges can be extremely high. Up to 55-60% of the electrical input power to the positive column of a rare-gas/mercury-vapor discharge can be emitted as the 254-nm resonance radiation of mercury, and a substantial fraction of the balance is emitted as 185-nm resonance radiation. This discharge forms the basis for the well-known fluorescent lamp, in which the ultraviolet resonance radiation is converted into visible light by a fluorescent powder disposed on the inside of the discharge tube wall. The rare-gas/sodium-vapor discharge of the low-pressure sodium lamp is similarly efficient, converting perhaps 40% of the electrical energy of the plasma into the yellow resonance lines of sodium.

In contrast to HID lamps, these discharges are far from thermal equilibrium. The electrical energy input from the passage of current is given almost entirely to the electrons of the plasma. These undergo elastic collisions with gas atoms, exchanging momentum but very little energy, but nevertheless being scattered through large angles, so that they also collide elastically with each other. Such collisions exchange large amounts of energy as well as momentum because of the equal masses of the colliding partners. The energy input to the electrons is therefore shared among them in the form of an electron energy distribution, which is sufficiently close to a Maxwellian that it can be assigned a "temperature". The energy balance of the discharge and the partition of power dissipation among the several processes is then primarily determined by the electron temperature.

Because of the strong coupling to the electric field, electron temperatures can be quite high, typically 11,000 K, or thereabouts, in the argon-mercury discharge. Because gas density is relatively low, however, only some 10^{17} per cm^3, electron density is only a few times $10^{17}/cm^3$, and collision frequency for momentum transfer is only a few times 10^7 (with energy exchanges of only about 10 parts per million per collision), the kinetic energy imparted to the gas atoms is insufficient to heat the gas more than a few tens of degrees above the wall temperature. Although gas heating is small, the aggregate elastic collision loss is a significant fraction of total energy dissipation by electrons in the plasma, about 25%.

The electrons are more closely coupled to the excited states of the atoms of metal vapor, because a significant fraction of them have energy sufficient to excite metal atoms. A very small fraction of the electrons have energy enough to ionize metal atoms. In fact, in both discharges, much of the ionization is two-stage, via ionization of an already excited atom. The axial electric field for steady-state operation is determined by the necessity to maintain ionization density constant against ambipolar diffusion loss. Increasing the axial field results in more energy imparted to electrons, and higher electron temperatures. Higher electron temperatures result in more excitation and ionization. There is, therefore, for any given set of number density and electron density parameters, a unique value of electric field at which production of ionization and loss by diffusion are in balance. The ballast reduces voltage across discharge terminals as current increases. The steady-state operating current is that for which the discharge voltage establishes this unique axial electric field in the discharge column, and is largely determined by the ballast.

The excited-state population density, unlike that in a HID lamp, is not in Boltzmann equilibrium at the electron temperature, because radiation-escape competes strongly with collisions of the second kind ("quenching collisions") by the electrons in depopulating the excited levels. The calculation of excited-state density then must involve solving coupled rate equations for production and destruction of the individual states, including exciting and quenching collisions, direct and stepwise, as well as depopulation by radiation emission. The calculation of radiation emission involves first a calculation of the population of excited states, followed by calculation of radiation escape. In the case of resonance radiations, these involve repeated emission, absorption, and re-emission, perhaps 100 times in ultimate escape from the tube.

Unlike the situation in HID lamps, in the low-pressure discharge lamps there is an optimum power input for maximum efficiency. This is the product of two factors: the fact that at constant vapor pressure of resonance radiation emitters, the discharge is more efficient at low power density than at high; and the fact that there is an optimum vapor pressure for the resonance-radiation emitter, and that pressure is determined by cold-spot temperature, which varies with power input.

Let us consider the first of these effects, considering a very elementary situation involving two-level atoms in a plasma, being excited by impact of high-energy electrons, quenched to the ground state by impact with slow electrons, and also radiating. In the steady state, the transition rate populating the upper state must be equal to the transition rate depopulating it. Assuming that the excited state population is a negligible fraction of the total, we can equate these two rates as below:

Excitation = De-excitation

$$k_x n_e n_o = k_q n_e n_x + \frac{n_x}{\tau} \tag{6}$$

Solving for n_x, the concentration of excited atoms, gives

$$n_x = \frac{k_x n_e n_o}{1/\tau + k_q n_e} \tag{7}$$

The radiation power is proportional to the concentration of excited atoms divided by their effective radiative lifetime, i.e.:

$$P_r = \frac{n_x}{\tau} = \frac{(k_x n_o) n_e}{1 + k_q (n_e \tau)} \tag{8}$$

We see that at low values of electron density, for which the second term in the denominator is very much less than unity, radiation power increases in proportion to electron density. At high values of $n_e \tau$, for which the second term in the denominator is large in comparison to unity, the radiation output becomes independent of electron density, and approaches:

$$P_r \rightarrow \frac{k_x n_o}{k_q \tau} \tag{9}$$

If the electron energy distribution is truly Maxwellian, then the ratio of k_x/k_q is in fact the Boltzmann ratio, and we see that the dependence calculated for excited-state population and radiation output simply represents the approach of the excited states to LTE with the electron gas.

As we have seen, the electric field in the plasma column is relatively constant and determined by ionization balance. Electron density varies approximately linearly with current density, i.e., power input per unit length of plasma column. Although the resonance radiation loss may saturate, and become independent of electron density (i.e., current), all other losses (elastic collision, non-resonant radiation, etc) continue to increase approximately proportionately with electron density. Thus, the fraction of the total power dissipation by the electron gas that is represented by resonance radiation loss decreases monotonically as electron density and current increase. Therefore, assuming constant vapor pressure, efficiency of generation of resonance radiation decreases with increasing current. These dependencies are illustrated in Fig. 18.

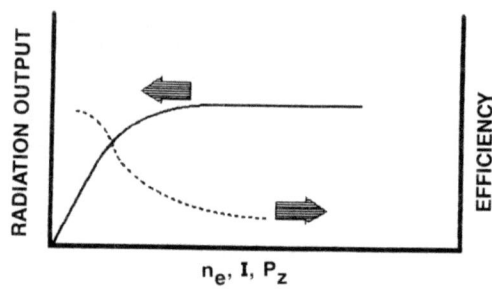

Fig. 18. Resonance radiation power and efficiency vs correlated variables electron density, current density, or power per unit length (schematic).

An examination of Eq. (7) shows that the denominator contains the product of electron density and radiation lifetime. The radiation lifetime, which is important here, is not the natural lifetime of the excited state, but rather the effective lifetime of the ensemble of the excited states, i.e., the "imprisonment time" for the resonance radiation. The radiation emitted by an excited atom in relaxing back to the ground state creates a photon which, in interacting with other ground state atoms, may be absorbed, recreating an excited atom, which subsequently radiates another photon, which is reabsorbed, etc., etc. As many as 100 emissions and absorptions of a resonance photon may be involved in its escape from the tube, and therefore the effective lifetime may well be 100 times longer than the natural lifetime. Intuitively, we can see that the more ground-state atoms there are per unit volume, the more times a photon will be absorbed and re-emitted in travelling a given distance. Therefore, imprisonment time increases with vapor pressure of the resonance-radiation emitter. Similarly, the farther the photons have to travel in escaping from the tube, the more times they will be absorbed and re-emitted. Imprisonment time also increases with tube diameter.

It is plain from Eq. (7) that the longer the imprisonment time the larger the denominator will be for a given electron density, and the smaller will be the resonance radiation output and the lower the efficiency. Thus, we expect efficiency to decrease when the second term in the denominator becomes of importance, i.e., as imprisonment time increases as a result of increasing vapor density of the emitter. On the other hand, at sufficiently low vapor density and imprisonment time, for which the second term in the denominator is small in comparison to unity, resonance radiation output increases with increasing vapor density. Therefore, at low vapor pressures efficiency of generation of resonance radiation increases with increasing vapor pressure. The two combined effects result in an optimum vapor pressure. Vapor pressure of mercury in fluorescent lamps, and of sodium in LPS lamps, is determined by cold-spot temperature, and, therefore, is an increasing function of power input. There is, therefore, an optimum power input for a tube of given size to achieve maximum efficiency.

In addition, since tube diameter is also a determinant of imprisonment time, and of electron density for a given total current, it enters into the determination of efficiency according to Eq. (7) in two different senses. Small diameter reduces imprisonment time, in the direction of increasing efficiency. Small diameter also increases electron density at a given current, in the direction of decreasing efficiency. There is, therefore, in addition to an optimum vapor pressure and power input, an optimum diameter for maximum efficiency.

The optimum vapor pressure for mercury in a typical fluorescent lamp is about 6-10 millitorr, achieved at a cold-spot temperature of 40 C. For LPS lamps, it is about the same, but a cold-spot temperature of about 250 C is required to achieve it. For both lamps, the optimum diameters are in the range of 20-25 mm. Consequently, the low-pressure resonance-radiation lamps are long and slender for a given power input. In fluorescent lamps, the technical problem is usually that the tube wall runs too hot, in enclosed fixtures at least, at power densities that would be efficient otherwise. In low-pressure sodium lamps the technical problem is that of suitable insulation to permit the tube wall to reach 250 C without having to employ power densities that are far beyond the point of declining efficiency.

Fluorescent Lamps

Fluorescent lamps are based upon a discharge in mercury vapor at about 6 millitorr, plus a rare gas, usually argon, at a pressure of 2-3 torr. Approximately 55% of the input power is radiated as the mercury 254-nm, 3P resonance radiation, and 8-9% as the mercury 185-nm, 1P resonance radiation. About 25% of the electrical power input is dissipated in elastic collisions between electrons and gas atoms, 9-10% as electrode loss, and the balance goes into non-resonant line emission and ionization loss. There is essentially no emission of argon lines. The ultraviolet radiations are converted to visible by fluorescent powders applied to the inside surface of the tube wall, at quantum efficiencies of 85-90%, but at energy efficiencies less than 50% by virtue of the smaller energies of visible photons in comparison to ultraviolet. The overall conversion efficiency of electrical energy to visible light is, therefore, about 25%; luminous efficacies are 75-100 lumens per watt, principally dependent on phosphor characteristics.

Optimum tube diameters are 1.0-1.5" (25-35mm), and power loadings are 10-25 watts per foot (ca 0.3-1 W/cm) of arc length. Thus, in comparison to HID lamps, such lamps are relatively bulky for a given light output. For many interior commercial applications, however, this is not a disadvantage, since relatively diffuse illumination from low-brightness sources is desired anyway. The long slender configuration, and source bulk, prove to be great disadvantages in current efforts to develop fluorescent lamps which can replace incandescent lamps, however. Although, in principle, equivalent amounts of light to popular incandescent lamps can be generated for less than one-third the electrical power consumption, similar levels of compactness are difficult to achieve. Engineering solutions to date have employed the topological equivalent of spaghetti; bend long slender straight tubes into twisted clumps that fit into a small volume. Although technically sound, such configurations are costly to fabricate, and the resulting lamps are relatively expensive. While they still retain an excellent payback for the commercial user, the high first cost has, to date, prevented any significant penetration of the residential market.

The understanding of the discharge plasma is reasonably complete. Many arc models exist, with varying degree of sophistication, that account for the main features of the discharge. Unlike the LTE models for HID lamps, however, these models must calculate from first principles the rates of 25 or more individual collision processes to determine the concentration of all species in order to analyze the energy balance. Most of the models deal with electron and excited-state densities averaged across the tube diameter, whereas, in fact, these have strong radial dependencies, maximizing in the center and decreasing toward zero at the walls. Only the model of Cayless (1963) treats the radial dependence of these quantities explicitly.

All models employ an average imprisonment time for the resonance-radiation escape. It is plain, however, from the description of the resonance-radiation escape process, that resonance photons created nearer the outer walls will experience a shorter imprisonment time than those created near the center. The use of an average imprisonment time will therefore underestimate the approach to saturation of the excited states near the center of the tube, and overestimate that near the walls. Moreover, the formula most commonly used to determine the imprisonment time, that of Holstein (1951), determines it as the lowest-mode decay time for an originally arbitrary distribution of excited atoms decaying spontaneously by radiation without any other processes involved. The actual average in any given lamp will depend on the actual distribution of excited states as a function of radius, which is not necessarily the same as the lowest mode distribution in the absence of electron exciting and quenching collisions.

The proper solution is, of course, to solve the radiation transport equations numerically, simultaneously with the continuity equations for excited state populations and for electrons. Computational complexity has hitherto made this impossible, but increasing availability of supercomputers makes it quite feasible. It is doubtful whether the effort is really worthwhile from a practical sense. Existing models employing average imprisonment times agree quite well with experiment, suggesting either that the results are not particularly sensitive to the actual radial distributions, or that the radial distributions themselves are close to those assumed.

Another point of interest with regard to the transport of resonance radiation in this plasma is that there is a significant isotope effect. As shown in Fig. 19, the 254-nm line of mercury exhibits an easily resolvable isotope shift. This has the effect of significantly reducing the imprisonment time since, to a first approximation, each isotope can only absorb radiation emitted by the same isotope. Because there are five components to the line, in effect the imprisonment time is only one-fifth as great as it would be at the same total density if mercury only existed in a single isotope. It has recently been discovered (Work and Johnson, 1983) that enrichment of natural mercury in the 196 isotope (normally present in negligible proportions) opens up a new channel for radiation escape as in Fig. 20. This has the effect of reducing the imprisonment time by 10-15%, thereby increasing the efficiency of generation of resonance radiation by 2-4%. Because of rapid excitation transfer among the mercury isotopes, it is not necessary to enrich the 196 isotope to a level of equal concentration; enrichment to 4% concentration suffices to achieve about 80% of the gain.

Aside from measurements of escaping radiation for diagnostic purposes, radiation absorption of selected emission lines has been used to determine metastable atom concentrations. Earlier work by Koedam and Kruithof (1962) using transmission of mercury lines from an external source has recently been supplemented by measurements of van de Weijer and Kremers using the "hook" method (1982). The relaxation of laser-induced fluorescence (LIF) has been used by van de Weijer (1985) to measure radiation trapping time (imprisonment time) directly, and has also been reported at this ASI.

Low Pressure Sodium Lamps

The last major discharge light source to be discussed in this presentation is the low pressure sodium (LPS) lamp. It is a resonance-radiation emitter closely akin to the mercury/rare-gas discharge of the fluoresecnt lamp, and many of the considerations discussed there apply here as well. The major difference is that the resonance radiation involved, the yellow D-lines of neutral sodium, is already visible, and no phosphor is required. In addition to simplifying construction, this makes the LPS lamp much more efficient in converting electrical energy into visible radiant energy, because no energy loss occurs in quantum

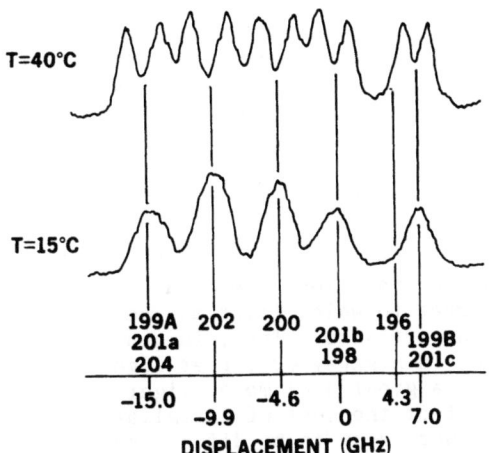

Fig. 19. Hyperfine structure of 254-nm resonance line of natural mercury, displaying isotope shifts. Note self-reversal of individual components at higher vapor density (40C cold spot).

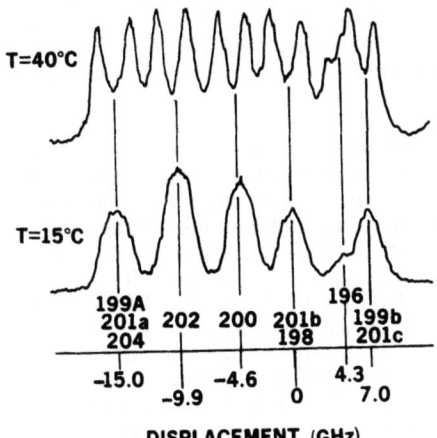

Fig. 20. Hyperfine structure of 254-nm resonance line of mercury, plus 3.9% added Hg196, illustrating additional radiation channel.

conversion. Energy efficiencies are of the order of 40% in the most efficient versions. In addition, the yellow line of sodium is near the maximum of the eye sensitivity curve, with a luminous efficacy of 525 lumens per visible radiant watt, so that the overall efficacy of the source is approximately 200 lumens per electrical watt, twelve or more times the efficacy of an incandescent lamp.

However, the light emission is entirely monochromatic and, therefore, the color rendering quality of the illumination is nil. Vision in pure LPS illumination is essentially in black and white, like newspaper photographs. Applications for such lamps are therefore in those areas where no color discrimination is required, primarily in street and highway illumination, as well as security lighting. By choice of size and operating current, LPS lamps can be produced in power input ratings between 18 and 200 watts. Because of their high efficacy, they require the lowest consumption of electrical energy to produce a given amount of light. Because of their size for a given light level, and the relatively low source brightness, however, they are at a disadvantage in comparison to HID lamps in the degree to which light may be projected a great distance by fixtures of compact size. Thus, in general, LPS lamps must be mounted closer together in highway or street lighting than HPS lamps, for example, and experience a great deal more "light spill" than do HPS lamps. Such factors affect the cost of installation, and therefore the total cost of light. There seem to be differences of opinion on the two sides of the Atlantic as to the resulting total cost of light of an installation designed for equivalent illumination with HPS vs LPS lamps.

Another major feature of difference between LPS and fluorescent lamps is the thermal insulation of the discharge tube. As described earlier, LPS lamps require a coldspot temperature of order 250 C for optimum sodium pressure. This is achieved by operating the lamp at sufficiently high power density that the tube wall achieves that temperature. Because of the decreasing efficiency of resonance-radiation lamps with increasing electron density and current, this means that all LPS lamps are "overdriven" in comparison to their fluorescent counterparts. The major gains in efficacy that have been achieved in LPS lamps over the span of the last twenty years have been primarily due to improving the efficiency of thermal insulation of the discharge tube to permit achievement of optimum vapor pressures at progressively lower and lower power input.

The first line of defense in the thermal insulation war is to enclose the discharge tube itself in a vacuum environment, so that radiation is the principal cooling mechanism. For convenient size, this has dictated that the normally long and slender discharge tube be bent in the form of a "U". The second line of defense is a heat reflecting coating on the inside of the outer jacket to reflect heat back to the discharge tube wall. At about 500 K, the peak of the blackbody radiation from the discharge tube wall would be at 5.5 micron, at which wavelength glass is pretty much a blackbody. It is possible to put a selective reflector on the outer jacket having excellent reflectance at that wavelength and having high transmittance at the 589-nm wavelength of the sodium resonance line. Such a reflector is basically a Lorentz reflector with an electron density such that its plasma frequency corresponds to about 1 micron optical wavelength. Tin oxide, indium oxide, and indium-tin oxides have been used, with ever increasing improvements. Fig. 21 shows how such improved reflectance, and reduced radiative cooling of the discharge tube, are translated into improved efficiency by the steadily reduced power necessary to achieve optimum vapor pressure.

A third point of difference between LPS and fluorescent lamps is the fact that LPS lamps operate at sufficiently high current density as to encounter significant depletion of neutral sodium atom density at the center of the discharge due to ionization. The problem is that ions

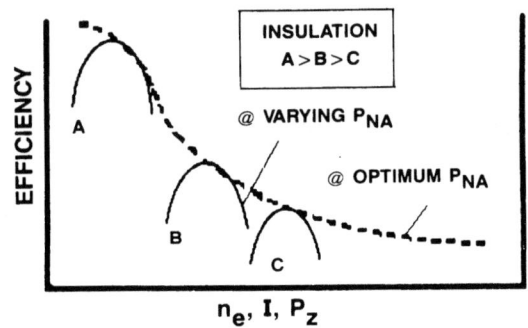

Fig. 21. Efficiency of LPS lamps vs correlated variables electron density, current density, power per unit length. Solid curves A,B,C: sodium vapor pressure varies with power input, for three different degrees of insulation (A>B>C); dashed curve, constant optimum sodium vapor pressure (schematic). Improved insulation permits achieving optimum vapor pressure at lower power, at which intrinsic efficiency is higher.

diffuse radially to the wall by ambipolar diffusion, with diffusion coefficient controlled by the electron temperature, while atoms diffuse back from the walls toward the center as neutrals, with diffusion coefficient controlled by the much lower gas temperature. Thus, there must be a large density gradient between wall and center to provide return flux of neutrals equal to the outward flux of ions, and neutral atom density in the core is correspondingly depleted. The effects of this process on the energy balance and the efficiency of LPS lamps has been treated in model calculations by van Tongeren (1975). It is particularly a problem so far as alternating current operation is involved, since the depletion is dependent on instantaneous current, which at maximum is normally 1.4-1.8 times rms current. It has been found advantageous to operate LPS lamps on ballasts providing nearly a square wave of current, to minimize the peak current and consequently the ionization depletion during the main part of the half cycle.

Most of the considerations already outlined with regard to model calculations, their complexity, and the approximations involved, also apply to LPS lamps, except for the added complexity introduced by the ionization depletion. Resonance radiation transport and imprisonment is treated on the basis of an averaged imprisonment time, computed according to the formalism of van Trigt and van Laren (1973). The same caveats concerning the necessity for simultaneous solution of the radiation transport, excited state, electron, ion, and neutral particle continuity equations also apply, with the same caution that they are not likely to be able to yield enough improvement in comparison to experiment to make the effort worth while.

SUMMARY AND CONCLUSIONS

A cursory survey of the main features of electric discharge lamps has focused on the role that radiation emission and absorption processes dominates their respective energy balances, and the degree to which the understanding of these processes has been developed. Further elaboration of many of these points will be apparent in other presentations at this ASI.

REFERENCES

Bartels, H., 1950, Z. Phys., 127:243.
Bartels, H., 1950, Z. Phys., 128:546.
Cayless, M. A., 1963, Brit. J. Appl. Phys., 14:863.
de Groot, J. J., 1974, Thesis, Tech. Univ. Eindhoven, Chapter III.
de Groot, J. J. and Jack, A. G., 1973 J. Quant. Spect. Rad. Transfer, 13:615.
Elenbaas, W., 1951, "The High Pressure Mercury Vapor Discharge", North Holland Publishing Co., Amsterdam.
Holstein, T., 1951, Phys. Rev., 83:1159.
Jones, B. F. and Mottram, D. A. J., 1981, J. Phys. D. Appl. Phys., 14:134.
Keeffe, W. M., Krasko, Z. K. and Rothwell, H. L., 1985, in: "Proceedings Electrochem. Soc.", 85-2:15.
Koedam, M. and Kruithof, A. A., 1962, Physica, 28:80.
Reiser, P. A., and Wyner, E. F., 1985, J. Appl. Phys., 57:1623.
Rothwell, H. L. and Keeffe, W. M., 1980, J. Illum, Eng. Soc., 10:40.
Stormberg, H. P. and Schafer, R., 1983, J. Appl. Phys., 54:4338.
van de Weijer, P. and Cremers, R. M. M., 1982, J. Appl. Phys., 53:1401.
van de Weijer, P. and Cremers, R. M. M., 1985, J. Appl. Phys., 57:672.
van Tongeren, H., 1975, Philips Res. Rpts. Suppl. 3:1.
van Trigt, C. and van Laren, J. B., 1973, J. Phys. D. Appl. Phys., 6:1247.
Wesselink, G., de Mooy, D. and van Germert, M. J. C., 1973, J. Phys. D. Appl. Phys, 6:L27.
Work, D. E. and Johnson, S. G., 1983, U.S. Patent No. 4,379,252.
Zollweg, R. J., Lowke, J. J. and Liebermann, R. W., 1975, J. Appl. Phys., 46:3828.
Zollweg, R. J., 1978, J. Appl. Phys., 49:1077.

LOW-PRESSURE MERCURY AND SODIUM LAMPS

A. G. Jack

Nederlandse Philips Bedrijven B.V.
Lighting Division, EDW-5
Eindhoven, The Netherlands

INTRODUCTION

The low-pressure discharge forms the basis for two important families of lamps, namely the fluorescent lamp, based on the low-pressure mercury-rare-gas discharge, and the low-pressure sodium-rare-gas discharge lamp. In this paper some of the aspects of the low-pressure discharge, as found in these two families of lamps, will be discussed. The published work on the low-pressure mercury-rare-gas discharge is more extensive and will first be considered. In later sections the similarities and differences as far as the low-pressure sodium-rare-gas discharge is concerned will be examined.

THE LOW-PRESSURE MERCURY-RARE-GAS DISCHARGE

Brief Introductory Description

The type of discharge under discussion is one which is diffusion controlled, i.e., a particle (atom, electron, etc.) undergoes many collisions before it reaches the wall. However, due to the relatively low pressure the number of collisions is not so great that thermal equilibrium among all the particles is approached. One group of particles, namely the electrons, has an average energy of about 1 eV, with an approximately Maxwellian energy distribution, while the ions and neutral atoms have an energy only slightly above the value they would have at room temperature, namely about 0.03 eV. Such is the situation in a mercury-rare-gas discharge where the mercury pressure is in the 1 to 20 millitorr range and the rare gas pressure is in the torr range.

Why is this two-component system necessary? Mercury is the component which, after excitation, radiates mainly via its two resonance lines at 254 and 185 nm. As the mercury density increases the radiation output at first increases, due to the increased number of excited mercury atoms, but then decreases. This latter effect is due to the fact that resonance radiation is absorbed and reemitted many times before it finally escapes from the discharge vessel. Every time a photon is absorbed there is the possibility that it is not reemitted, but that the energy is released via a non-radiative process. Thus, for an efficient discharge, the amount of radiation trapping and, hence, the mercury particle density is limited. In practical situations a mercury vapor pressure in the range 1 to 20 millitorr is necessary.

With such a mercury pressure the electron mean-free-path between collisions is 10^{-3} to 10^{-2} m while the discharge tube radius is typically 10^{-2} m. Thus many of the electrons are lost due to collisions with the wall before they have the possibility of transferring their energy to mercury atoms. In order to limit the loss of electrons and ions to the wall a second gas is added. It is chosen so that electrons only have elastic collisions with this gas, i.e., there is no excitation or ionization of the gas. A suitable choice for this buffer gas is the rare gas argon. As the rare gas pressure is increased, the electron and ion flux to the wall, which is determined by ambipolar diffusion, is diminished. However, as the rare gas pressure is increased the electrons undergo more elastic collisions with the rare gas atoms. This energy transfer from the electrons results in a heating of the rare gas atoms and is known as the volume loss. Because of their opposite pressure dependence there is a rare gas pressure, typically between 0.1 and 2 torr, where the sum of the volume losses and ionization losses is a minimum. In actual lamp situations often a rare gas pressure a little in excess of the optimum value is chosen in order to realize a sufficiently long electrode life.

Experimental Investigations

Measurements of various plasma parameters can give insight into the processes which take place in the discharge and enable a more detailed comparison to be carried out with numerical models. Langmuir probe techniques have been extensively carried out on the low-pressure mercury-rare-gas discharges, for example by Easley (1951) and Verwey (1961), in order to determine the electron concentration and the electron energy distribution. The elastic losses, i.e., volume losses, have been determined by Kenty et al. (1951) by measuring the rare gas pressure rise when the discharge is ignited. Koedam and Kruithof (1962) determined the concentration of the 6^3P states by transmission measurements of the visible mercury triplet. Similar experiments have been carried out by Kagan and Kasmaliev (1967). Uvarov and Fabrikant (1965) measured the concentration of excited atoms in the discharge and the effective probability of photon emission. Kaga et al. (1968, 1971) have measured the electron energy distribution in the positive column of the discharge. Measurement of the 185-nm and 254-nm radiation normal to the surface of the discharge vessel has been carried out by Barnes (1960). Absolute measurements of the total radiation output, both ultraviolet and visible, were done by Koedam et al. (1963). More recently Stern and Schaal (1973) have measured the ultraviolet output for a wide range of conditions, but using the results of Koedam et al. (1963) as their "calibration". The most recent diagnostics of the discharge have been carried out using laser techniques; a recent overview has been given by van den Hoek (1983). Interferometric measurements have been carried out by van de Weijer and Cremers (1982) to determine the dependences of the Hg^3P state densities on the tube temperature and thus on the mercury vapor pressure. Laser techniques have also been used to study pulsed optogalvanic effects (van de Weijer and Cremers (1985a)), and the effective radiative lifetime of the 6^3P level (van de Weijer and Cremers (1985b)).

It may be concluded that for "standard" types of fluorescent lamps there is a substantial amount of experimental data available about the positive column.

Modeling of the Positive Column

The first detailed attempt to carry out a quantitative analysis of the positive column was presented in the classic paper by Kenty (1950). He calculated the rates of excitation, de-excitation, and radiation from measured values of electron density and temperature together with estimates of cross sections. This analysis has formed the basis of much subsequent work. Waymouth and Bitter (1956) constructed a model for the positive column in which they were able to predict the effect of changing parameters such as discharge current and mercury vapor pressure. They averaged the variables over the cross-section and fitted the analysis to certain experimental data. Subsequently, Cayless (1961, 1963) developed a model so that no averaging of species over the cross-section was carried out and no use was made of any arbitrary fitting constants.

More recently the Maxwellian electron energy distribution has been replaced by a two-electron group model (Vriens, 1973), and Vriens and Ligthart (1977)). The importance of excited-atom collisional ionization has been recognized and included in the model by Vriens, Keijser and Ligthart (1978), and Ligthart and Keijser (1980). A comparison between the results given by this model and various plasma diagnostics and radiation measurements has been carried out by Denneman et al. (1980). The most recent modeling has been carried out by Lagushenko and Maya (1984). In this work the electron energy distribution is based on an approximate analytical solution of the Boltzmann equation and an artificial atomic mercury level at 8.6 eV is introduced. This results in a better agreement with experimental data, especially at low mercury vapor densities.

Radiation Transfer

Since a large amount of the input power, typically 50 to 80%, leaves the discharge as radiation it is essential that the radiation transfer in the discharge is adequately described. This topic was briefly discussed in Experimental Investigations and is covered in more detail in the accompanying paper by van de Weijer and Cremers which deals with pulsed optical pumping in low-pressure mercury discharges.

THE FLUORESCENT LAMP

In the above sections it has been shown that there is a large body of knowledge, both theoretical and experimental, dealing with the positive column of the mercury-rare-gas discharge which forms the heart of the fluorescent lamp. Many details relating to fluorescent lamps can be found in review articles and books, e.g., Elenbaas (1959), Waymouth (1971), Polman et al. (1975), Jack and Vrenken (1980), Cayless and Marsden (1983). In this paper the many details of the fluorescent lamp will not be reconsidered. Rather, the power balance for a typical fluorescent lamp will be presented and, using this as a basis, recent advances which have lead to, or can lead to, improvements in lamp performance will be discussed.

POWER BALANCE

Figure 1 shows the power balance of a typical fluorescent lamp as determined by Dorleijn and Jack (1985). The tube internal diameter is about 24 mm while the overall length is about 1.20 m. In this lamp the buffer gas is 75% Kr, 25% Ar with a total pressure of 1.5 torr, while the fluorescent powders are of the narrow emission band type (see section Conversion of Ultraviolet Radiation into Visible Radiation). On a standard ballast the lamp dissipates 36 W when connected to a 220V/50Hz ballast. The main loss processes, i.e., processes which do not lead to the generation directly or indirectly of visible radiation, are:

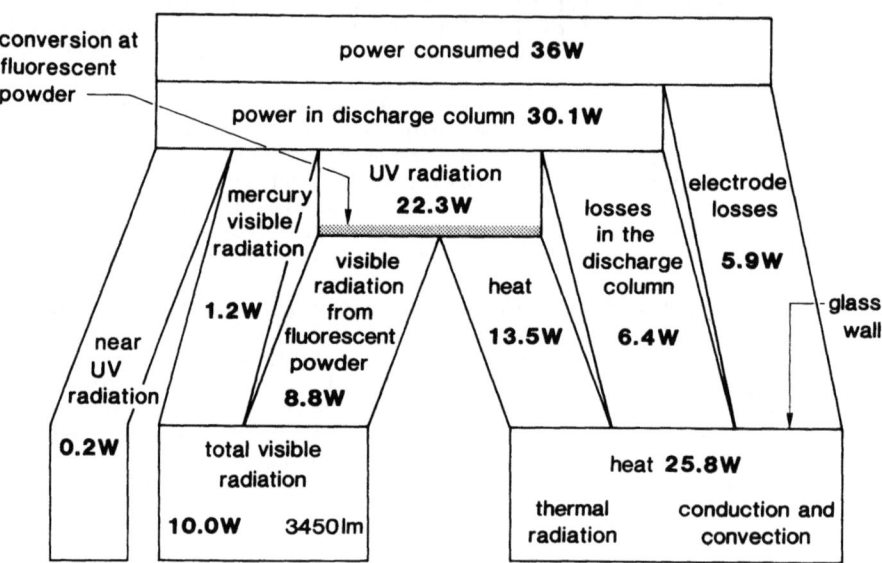

Fig. 1. Power Balance for a 36 W Fluorescent Lamp.

    a. Electrode losses.

    b. Losses in the discharge column (volume and wall losses).

    c. Losses in the fluorescent powder where, at maximum, each UV photon leads to the generation of one visible photon.

Although this is a very efficient lamp, 96 lumen/W compared with the 12 to 14 lumen/W for an incandescent lamp, only 10 W of the 36 W input power is converted into visible radiation. In the following sections some methods of improving this power balance will be considered.

Reduction of Volume and Wall Losses

    Compared with older types of lamps, the buffer gas in this lamp has been optimized to realize a decrease in volume and wall losses. This is discussed in detail by Denneman et al. (1980). It can be shown theoretically and experimentally that krypton, with a pressure in the 0.5 to 2-torr range, leads to the lowest volume and wall losses. However, the lamp designer often has other constraints, e.g., the lamp may have to operate on existing control gear, ignition must be reliable, the gas discharge must be stable (i.e., no visible striations) at low current levels and/or low environment temperatures, losses in electrodes and control gear must be considered, etc. Often the final selection of the buffer gas is influenced by these constraints.

Improved Radiation Transfer in the Discharge

    Each time radiation trapping occurs there is a finite chance that the energy will be lost via non-radiative processes. Thus the efficiency of the radiation generation can be improved if, for a given radiation output, the radiation trapping is decreased. One way of influencing the radiation trapping is to modify the natural mercury isotope distribution. For example, if the natural mercury isotope mix is replaced by a single mercury isotope, then the hyperfine structure of the 254-nm line is reduced and radiation transfer is more difficult. This leads to a lower lamp luminous efficacy as was found by Franck and Schipp (1975). More recently, the natural isotope mix has been modified so that the hyperfine structure is better balanced and thus there is less radiation trapping of the 254-nm radiation. By this means Grossman et al. (1983) have improved the efficacy of a fluorescent lamp by about 3%.

High-Frequency Operation

It has been known for a considerable time that high-frequency operation, as opposed to 50 or 60-Hz operation, of fluorescent lamps leads to improvements in the lamp luminous efficacy (see for example Meijers and Strojny, 1959; Campbell, 1960). In general, the changes in the radiation efficiency of the positive column are not all that great as the supply frequency is increased. Numerical modeling of ac discharges has been carried out by Drop and Polman (1972). Such calculations show a very slightly decreasing column efficiency from dc to about 500Hz, and then an increase to slightly above the dc value at some tens of kHz. The main effect of high frequency operation is the decrease in the anode fall as described by Koedam and Verweij (1965). Above a characteristic frequency, which is typically in the 500-Hz to 5-kHz range, and which is determined by the ambipolar diffusion time constant around the electrode, extra ionization due to an increased anode fall or anode oscillations is no longer necessary.

Although high frequency operation of fluorescent lamps has been applied on a small scale for special applications for some considerable time, it is only very recently that much larger-scale applications have become both technically and economically feasible.

Conversion of Ultraviolet Radiation into Visible Radiation

In order to realize the highest luminous efficacy the fluorescent powder, which converts the ultraviolet radiation into visible radiation, must have the following characteristics:

a. As high as possible a quantum efficiency; ideally each UV photon should be converted to two visible photons if this is energetically possible.

b. An emission spectrum in the visible, which leads to the highest luminous efficacy for the specified chromaticity coordinates and color-rendering index.

c. Properties which do not diminish significantly during lamp life.

Theoretical studies by Koedam and Opstelten (1971) and Thornton (1971) have shown that, by limiting the visible radiation to specific narrow bands, it is possible to obtain a high luminous efficacy together with a high color-rendering index. Further, the correlated color temperature can be chosen over a wide range by varying the balance between the emission bands. Fluorescent powders with the necessary narrow emission bands have been developed, thus transforming the theoretical predictions into working lamps (Verstegen et al., 1975; Vrenken, 1976, 1978; Haft and Thornton, 1972). A typical narrow emission band spectrum is shown in Fig. 2.

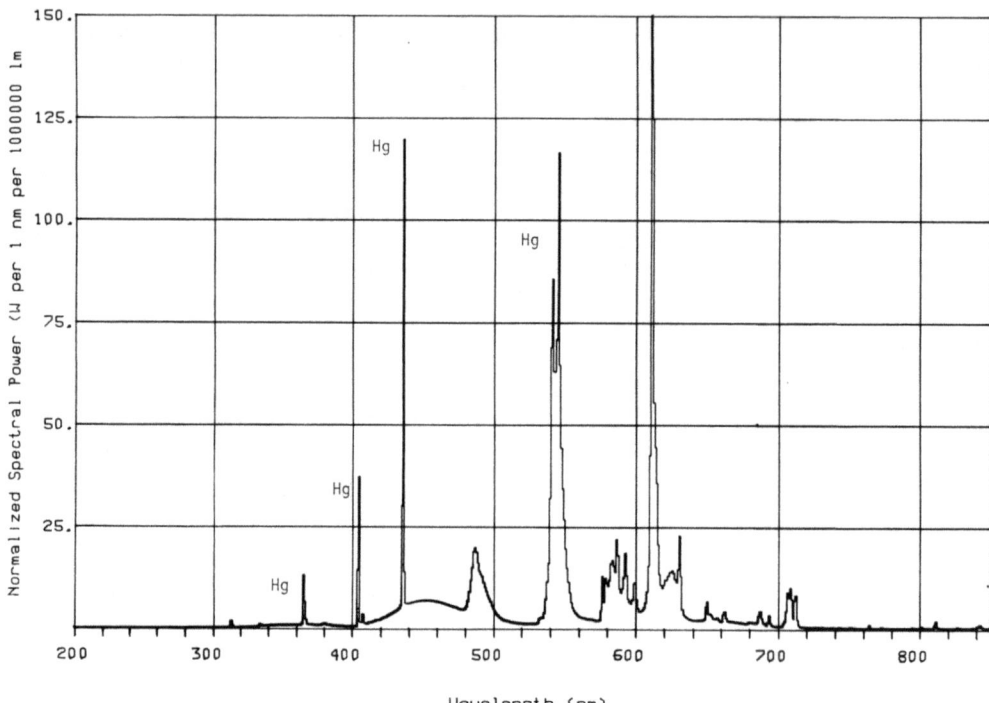

Fig. 2. The emission spectrum of a fluorescent lamp with a mixture of three narrow emission-band fluorescent powders. The intense visible mercury lines are also indicated.

These new fluorescent powders also have another advantage. Compared with the standard halophosphate fluorescent powders these new narrow emission-band fluorescent powders can withstand a higher ultraviolet radiation intensity. This leads to an improved lamp life and makes possible a drastic further miniaturization of fluorescent lamps as described in the following section. If a still better color rendering is required for certain special applications, then this can be realized by using a well balanced five-band spectrum rather than the three-band spectrum (van Kemenade et al., 1983; 1984).

COMPACT FLUORESCENT LAMPS

Fluorescent lamps which are very efficient light sources are very widely used but, until a few years ago, they were little used in the domestic area. During the last decade there has been an increasing interest in energy efficient light sources, but fluorescent lamps were often too large for domestic applications. However, the development of new fluorescent powders, as described above, has made miniaturization of fluorescent lamps possible. First, the miniaturization of the discharge tube will be discussed, and then various ways of realizing compact lamps will be described, together with other aspects, such as the proper control of the mercury vapor pressure.

Miniaturization of the Discharge Tube

In order to realize a compact lamp it is necessary to increase the power dissipation per unit length of the discharge while, at the same time retaining, as much as is possible, the efficiency of the discharge. Increased dissipation can be obtained by either increasing the discharge current or by increasing the discharge electric field.

A viable solution is not normally realized when the discharge current is increased. Due to the discharge characteristics an increase in the current leads to a proportionally smaller increase in the power dissipated and to a decrease in the radiation efficiency. However, the electrode losses are proportional to the discharge current, and the losses in the control gear are often more than proportional to the discharge current. All this leads to a relatively small increase in the light output at the expense of a relatively large increase in the system power.

The other alternative is an increase in the electric field. As described by Bouwknegt (1982) this can be realized via three different ways: (1) a reduction of the discharge tube diameter, (2) the use of a recombination structure, or (3) the use of Ne and/or He as the buffer gas.

When the tube diameter is reduced there is an increase in the loss rate of ions and electrons to the wall. This must be compensated by an increase in the ionization rate, i.e., an increase in the electron temperature and the axial electric field. A variation is to use an elliptical rather than a circular cross-section discharge tube (see also Cayless (1960)). The use of a recombination structure, as proposed by Hasker (1976, 1980), is also a way of increasing the loss rate of ions and electrons. This can be realized by introducing into the whole of the discharge tube a small amount of glass wool which acts as a recombination structure. The third alternative involves replacing the usual argon or krypton buffer gas by neon and/or helium. The electron mobility is decreased and there is a corresponding increase in the electric field.

Naturally, various combinations of the above three options are possible. A comparison of the various alternatives has been given by Bouwknegt (1982) and both the theoretical and experimental results are shown in Fig. 3, where the discharge column efficacy is plotted as a function of the electric field for a given discharge current. It is seen that, for a specified electric field, the discharge efficiency is almost independent of the measure which is used, provided that a proper optimization has been carried out.

In making a final selection other aspects must be taken in consideration. If Ne or He is used then the electrode life may be problematical, while there are technological problems associated with a recombination structure. A reduction of the tube diameter is often the most acceptable solution.

Various Types of Compact Lamps

As can be seen from Fig. 3, the higher the electric field, and thus the shorter the lamp, the lower the luminous efficacy. For a specified luminous flux there are too many disadvantages if the discharge tube is too narrow. For domestic areas where a luminous flux per lamp in the range 600 to 1200 lm is often required, an optimum solution is often a discharge tube typically 300 to 500 mm long and about 10 mm internal tube diameter. In order to realize a still more compact lamp various further size reducing measures can be applied. Some techniques used at the

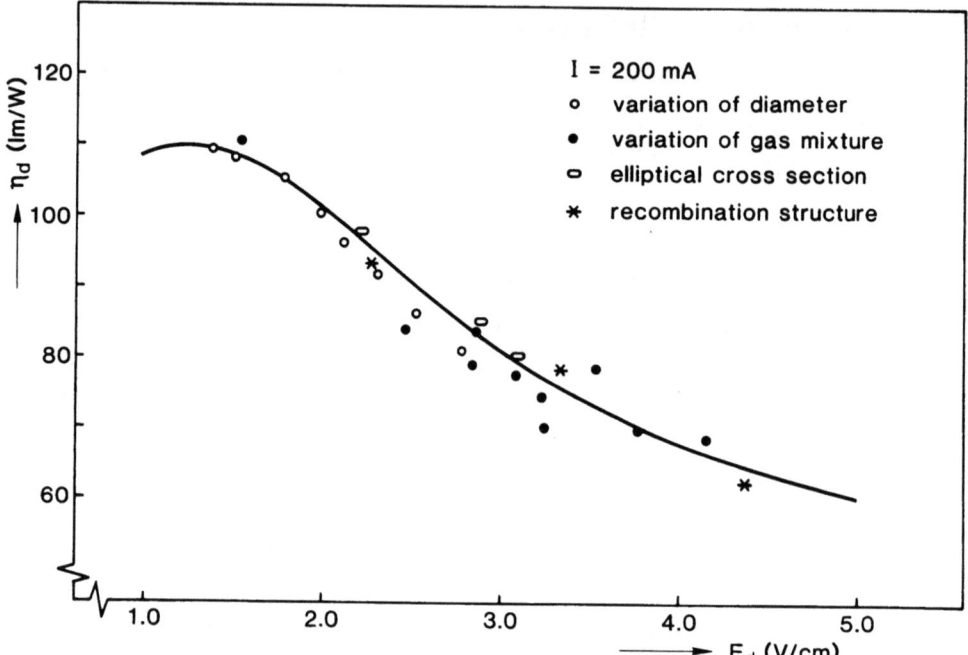

Fig. 3. Efficiency of the discharge column as a function of the electric field. The points are experimental data. The drawn line is the relation found from model calculations which are fitted to the experimental data at E_d = 1.8 V/m (from Bouwknegt, 1982).

present time are: the discharge can be arranged in the form of two or four parallel tubes with external control gear as shown in Fig. 4 (Bouwknegt, 1982; Vrenken and Veenstra, 1983; Verheij, 1985); the discharge can be arranged in a square-shaped form, again with external control gear (Willoughby and Cannell, 1982); or a doubly folded discharge tube with an integrated control gear as shown in Fig. 5 (Bouwknegt, 1982). Other variations have been described by Kamei et al. (1980) and MacDonald et al. (1983).

Various other techniques have also been described in the literature. One example is multiple folding of the discharge tube, this often being realized using new glass technologies (Wesselink, 1980; 1983; Burgmans and van IJzendoorn, 1983). Further, there are various constructions making use of open-ended discharge tubes sealed into an outer bulb (Watanabe and Yamane, 1982; Murayama et al., 1984).

Control of the Mercury Vapor Pressure

For standard fluorescent lamps the temperature of the discharge tube wall is such that the optimum mercury vapor pressure ($\sim 7 \times 10^{-3}$ torr) for efficient radiation generation is realized. In compact, narrow diameter lamps the optimum mercury vapor pressure is slightly higher (typically $\sim 1 \times 10^{-2}$ torr), but the discharge tube is almost always too hot. Thus, extra measures have to be taken to realize the optimum vapor pressure. One alternative is to create a lower temperature region. This has been done for compact fluorescent lamps with two or more parallel tubes (see Fig. 4). The bridge connecting the two discharge tubes is

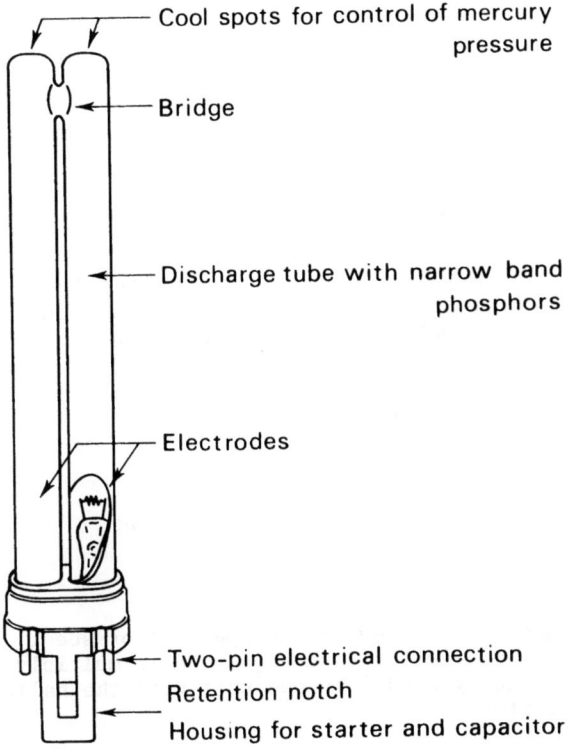

Fig. 4. A compact fluorescent lamp with two narrow diameter discharge tubes connected via a bridge. The starter and capacitor are mounted in the lamp cap.

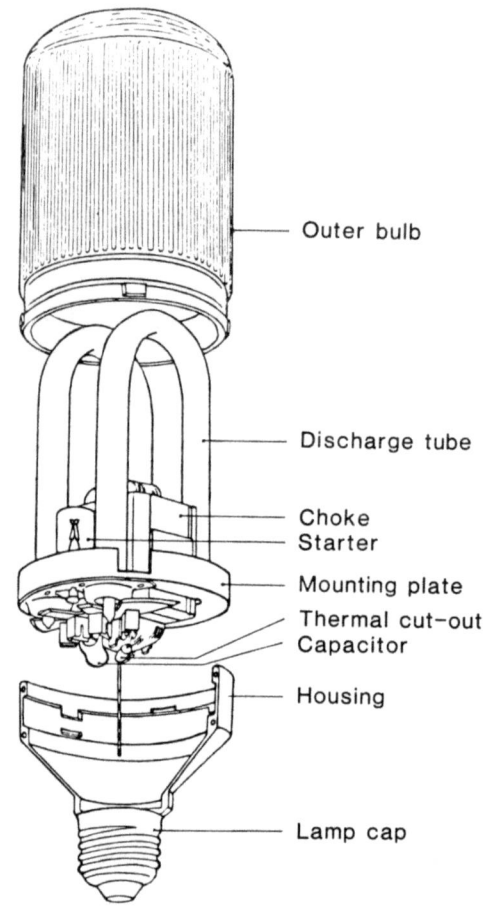

Fig. 5. An exploded view of a compact fluorescent lamp. The narrow diameter discharge tube is multiply folded and integrated into one unit with the ballast and starter.

located a little distance from the ends of the tubes. In this way there is no power dissipation at the tube ends, which then act as a lower temperature appendix.

However, if the complete discharge tube is integrated with the control gear to form one unit, as shown in Fig. 5, then it is extremely difficult, if not impossible, to create the necessary cool area. Another alternative is the use of an amalgam as described by, among others, Eckhardt and Kuhl (1970), Franck (1971), and Bloem et al. (1977). Fig. 6 shows, as a function of temperature, the mercury vapor pressure above mercury and above a typical amalgam, namely In-Bi-Hg. The mercury vapor pressure, for which the lamp luminous flux is at least 90% of its maximum value, is also indicated. From this figure it is seen that with an amalgam not only is a higher wall temperature possible but the temperature range within which the lamp operates close to its optimum is significantly increased. When the lamp is initially switched on at room temperature the presence of an amalgam results in a lower initial mercury

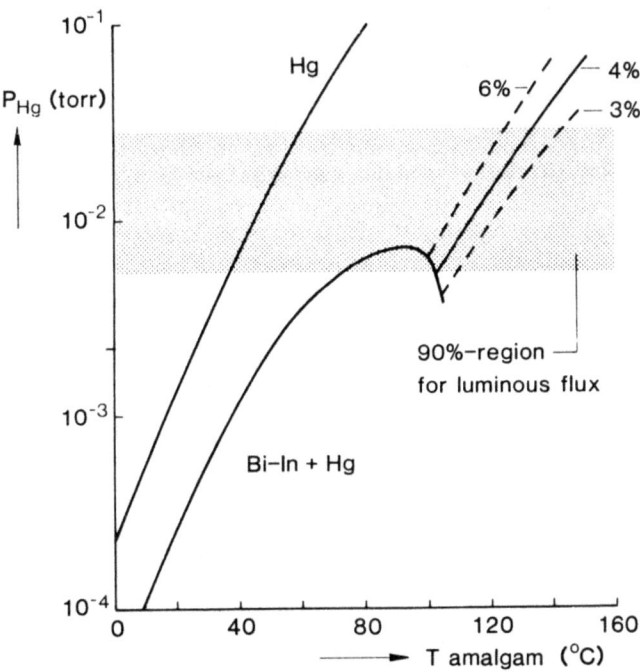

Fig. 6. The mercury vapor pressure above a In-Bi-Hg amalgam compared with the vapour pressure above pure mercury. The various mercury contents are given in atomic percentages (from Bloem et al., 1977).

vapor pressure, and thus a lower light output than would have been the case with only mercury. Thus, an auxiliary amalgam system is often necessary to generate a sufficiently high mercury vapor pressure immediately after ignition. Only with a well controlled mercury vapor pressure is optimum generation of radiation in the discharge possible.

THE LOW-PRESSURE SODIUM-RARE-GAS DISCHARGE

As with the mercury discharge already discussed, this is a metal-vapor-rare-gas discharge where the metal vapor pressure is in the millitorr range and the rare gas pressure is in the torr range. Since the sodium resonance lines are in the visible wavelength range no fluorescent powder is necessary. However, a cold spot temperature of 250 to 260 C is necessary in order to realize the necessary sodium vapor pressure. This results in the need for a good thermal insulation and a glass which is resistant to sodium at the operating temperature. As far as the positive column of the discharge is concerned, there are naturally many similarities with the mercury discharge, but in the sodium discharge depletion of the sodium ground state is a dominant effect. These three differences, namely the need for a sodium-resistant glass, thermal insulation, and the occurrence of depletion, will first be discussed before dealing with other aspects of the discharge and recent improvements in low-pressure sodium lamps.

Sodium Resistant Glass

Before a sodium-rare-gas discharge can be studied and optimized it is necessary to have a discharge vessel which does not react with sodium at temperatures of up to about 300 C. At present a two-ply tube is used which consists of a soda lime tube coated on the inside with a thin layer (typically 50-μm thick) of sodium-resistant borate glass (Adams (1978)).

Thermal Insulation

In order to limit the energy loss from the hot discharge tube there have been a succession of improvements in the thermal insulation. Early constructions are described by Dorgelo and Bouma (1937), while later developments are described by van de Weijer (1961) and Elenbaas et al. (1969). A most significant improvement was the use of a tin-oxide film, as described by Groth and Kauer (1965), and its qualitatively better successor, the indium-oxide film (van Boort and Groth, 1968). Such films are electrically conductive, transparent in the visible range of the spectrum, and have an excellent reflection in the near infrared (Kostlin et al., 1975). Figure 7 shows the construction of a present day low-pressure sodium lamp (SOX lamp). In addition to the indium-oxide film the discharge tube is often U-shaped resulting in a higher dissipation per unit lamp length.

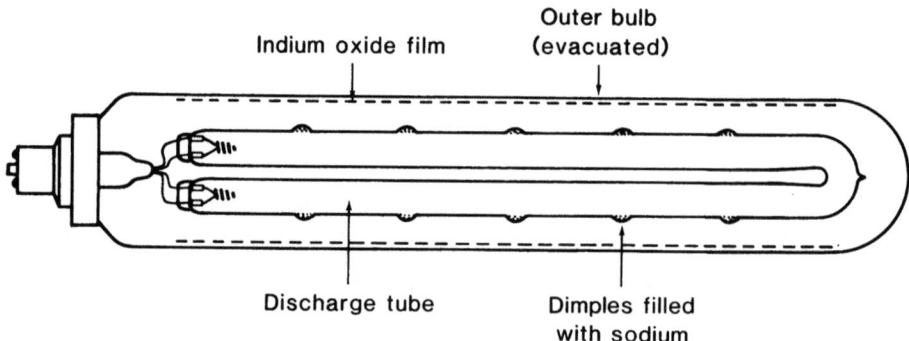

Fig. 7. Sketch showing the main features of a low-pressure sodium discharge lamp (SOX lamp).

Radial Depletion of the Sodium Ground State

In the low-pressure sodium-rare gas-discharge the current density is usually significantly higher than in the mercury discharge, and this often leads to a high degree of ionization of the sodium atoms. Depletion of the ground state arises due to the relatively rapid ambipolar diffusion of sodium ions to the wall where recombination takes place, compared to a slow back diffusion of neutral atoms into the discharge space. Druyvesteyn (1933) measured the very high ion concentration compared with the neutral sodium atom concentration while Uyterhoeven (1938) described early studies of the effect of neutral atom depletion in the sodium discharge.

Insight into depletion has also been gained from studies on cesium-rare-gas discharges which are experimentally somewhat easier to study and very akin to the sodium discharge. Measurements by Bleekrode and van der Laarse (1969) of the radial concentration of the cesium ground state atoms clearly demonstrated radial depletion and its increasing magnitude as the discharge current and, hence, the degree of ionization increase. Depletion can also occur in the low-pressure mercury-rare-gas discharge at higher current densities (Cayless, 1963; Read, 1963), but it is not usually an important effect. Very recently Cornelissen and Merks-Eppingbroek (1985) have determined the radial sodium ground-state density distribution using laser absorption techniques (see Fig. 8).

Experiments on the Positive Column

Early experimental work was carried out by Druyvesteyn and Warmoltz (1934) and Uyterhoeven and Verburg (1939). More recently an extensive theoretical and experimental study has been carried out by van Tongeren (1974, 1975). Figure 9 shows the measured electron temperature as a function of the discharge current for various tube wall temperatures. The abrupt rise in the electron temperature at a given current is due to the fact that the sodium ground state depletion at the axis has reached such a degree that ionization of the rare gas occurs. Detailed studies of this abrupt change in the discharge characteristics have shown that the V-I characteristics at this point are multivalued (van Gemert and Verbeek, 1977). The afterglow decay time at the transition has been investigated (van Gemert, 1975; van Gemert et al., 1977). Measurements on a discharge operated on a 50-Hz supply have also been carried out (Verbeek and van Tongeren, 1977; Jack 1978). The most recent diagnostic work on the sodium-neon discharge has been the determination of local electron densities via Doppler-free two-photon spectroscopy (Cornelissen and Burgmans, 1982).

Modeling of the Positive Column

The first extensive modeling of the sodium-rare-gas positive column was carried out by van Tongeren (1974, 1975), who used a 3-level model of the sodium atom and a simplified version of the radiation transfer theory developed by van Trigt et al. (1973, 1975). Depletion of the sodium ground state was accounted for but no excitation or ionization of the rare gas was included. This 3-level model gave a fairly adequate description of the discharge parameters with the exception of the electric field. More recently Vriens (1978) has considered in detail the ionization processes in these discharges. Multistep electron impact ionization via many higher excited states is found to be a dominant process.

THE LOW-PRESSURE SODIUM LAMP

Recent descriptions of the low-pressure sodium lamp have been given by Pfaue and Schirrwitz (1967), Beijer et al. (1974), and Koedam et al. (1975), and very comprehensively by Denneman (1981). Consider first the power balance of a low-pressure sodium lamp. This is shown in Fig. 10 for a 180-W lamp (Jack and Vrenken, 1980). Here the conversion efficiency of power in the discharge column into radiation is about 43%, which is considerably lower than the 79% realized in a fluorescent lamp (see Fig. 1).

In the sodium lamp the working point of the discharge is so adjusted that the thermal loss in the positive column is that needed to keep the discharge tube at the correct working temperature. For the sodium lamp 35% of the input power is converted to visible radiation, resulting in a lamp luminous efficacy in excess of 180 lm/W. For a typical fluorescent lamp the figures are 28% and 96 lm/W. The significantly lower luminous

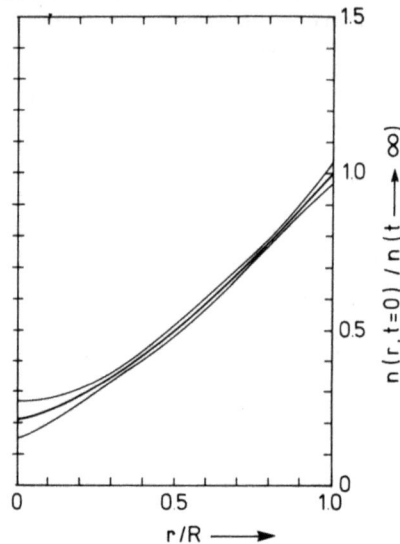

Fig. 8. Measured radial sodium ground-state density distribution. Tube internal diameter is 19 mm, current 0.4 Å, sodium particle density at wall 5.2×10^{19} m^{-3}, and neon particle density 1.8×10^{23} m^{-3} (from Cornelissen and Merks-Eppingbroek, 1985).

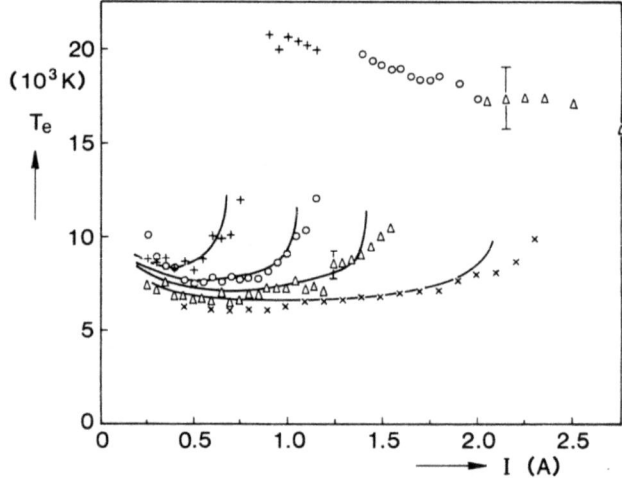

Fig. 9. Measured electron temperature Te in a low-pressure sodium discharge as a function of the dc discharge current at four tube wall temperatures;
+ 520 K, o 527 K, Δ 533 K, x 540 K.
The solid curves are calculated using a three-level model. Na-Ar discharge, P_{argon}=10 Torr, tube diameter=28 mm (from van Tongeren, 1975).

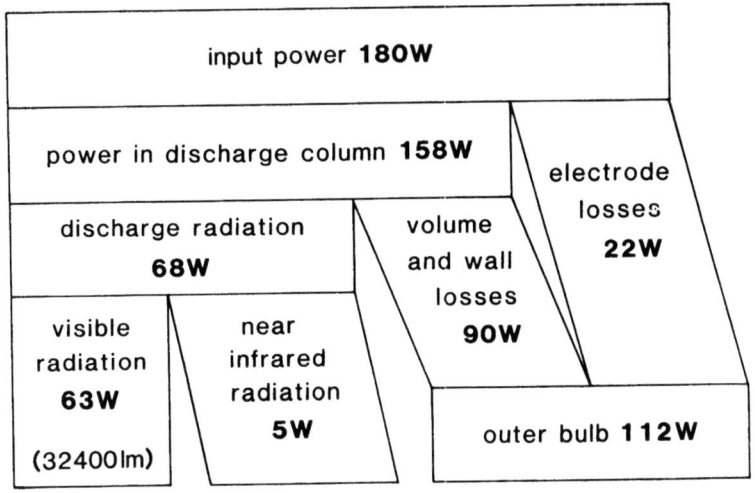

Fig. 10. Power balance for a 180-W low-pressure sodium lamp.

efficacy is due to the fact that white light with a very good color rendering is generated rather than almost monochromatic yellow light.

Further Improvements in the Thermal Insulation

It is essential that the thermal conditions in the lamp remain stable during life and that sodium does not migrate to cooler regions, resulting in an increase in the lamp voltage. Techniques have developed in order to determine the local sodium vapor pressure, either directly, (Vriens, 1977) or indirectly (Jack, 1978). Jacobs et al. (1978) have described measures which can be taken to diminish sodium migration and thus to improve lamp voltage control during life.

More recently the thermal insulation of the lamp has been further increased by improving the quality of the indium oxide layer. This results in an increase in the lamp luminous efficacy to 200 lm/W (Sprengers et al., 1985).

Influence of Current Waveform and Operating Frequency

If the normal sinusoidal 50 or 60-Hz current through the discharge is replaced by a rectangular current waveform then there is an increase in the radiation efficiency of the discharge (Koedam et al., 1975). This is due to the fact that the peak current is decreased and thus there is less depletion of sodium atoms. Another method of increasing the lamp radiation efficiency is by operating it at higher frequencies. As with the fluorescent lamp there is a decrease in the electrode fall of the order of 5 volts but, unlike the fluorescent lamp, there is also a significant effect on the column efficiency. The gain in column efficiency is about 10% but a gain is only realized for frequencies above 100 kHz, and the maximum efficiencies are realized at frequencies above 200 to 400 kHz (see Fig. 11). These effects have been studied by de Groot et al. (1984) but at present there is not yet a full quantitative understanding of the effects. High frequency operation of low-pressure sodium lamps makes possible lamp luminous efficacies of 225 lm/W and system efficacies in excess of 200 lm/W.

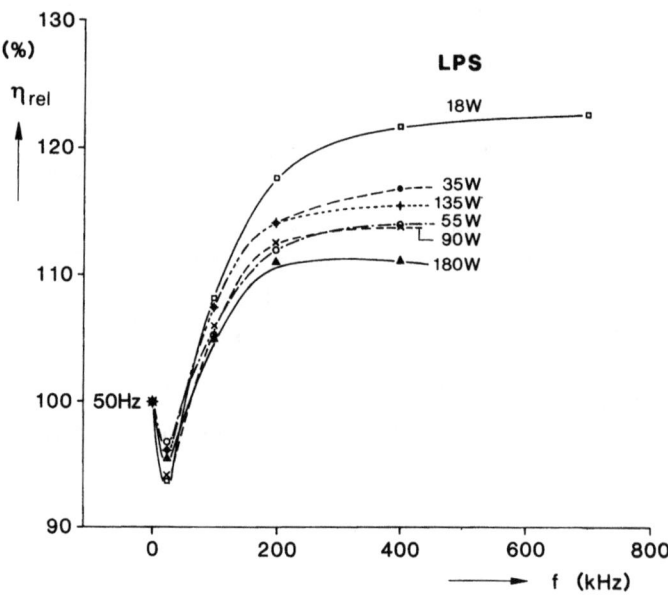

Fig. 11. The relative luminous efficacy η_{rel} (100% at 50-Hz operation) as a function of frequency for low-pressure sodium lamps of various wattages. The current wave form is sinusoidal (from de Groot et al., 1984).

CONCLUDING REMARKS

Although both the fluorescent lamp and the low-pressure sodium lamp have been produced commercially for a number of decades it is seen that progress is still being made. This is true both as far as an understanding of the physics of the devices is concerned and in lamp performance. As for all lamps, there is a continuous effort directed towards the improvement of the radiation efficiency of existing lamp types. For these systems the radiation represents 35 to 80% of the input power and thus radiation transport is a vital factor which determines their performance.

REFERENCES

Adams, O., 1978, Lighting Res. Technology, 10:83.
Barnes, B. T., 1960, J. Appl. Phys., 31:852.
Beijer, L. B., van Boort, H. J. J. and Koedam, M., 1974, Lighting Des. and Applic., 4:15.
Bleedkrode, R. and van de Laarse, J. W., 1969, J. Appl. Phys., 40:2401.
Bloem, J., Bouwknegt, A. and Wesselink, G. A., 1977, J. Illum. Eng. Soc., 6:141.
Bouwknegt, A., 1982, Journal IES, 11:204.
Burgmans, A. L. J. and van IJzendoorn, P. R., 1983, Philips, J. Res., 38:180.
Campbell, J. H., 1960, Illum. Eng., 55:247.
Cayless, M. A., 1960, Brit. J. Appl. Phys., 11:492.
Cayless, M. A., 1961, in: "Proc. 5th Int. Conf. Phenomena Ionized Gases", Vol 1.
Cayless, M. A., 1963, Brit. J. Appl. Phys., 14:863.
Cayless, M. A., and Marsden, A. M., 1983, "Lamps and Lighting", 3rd Edition, Edward Arnold.

Cornelissen, H. J. and Burgmans, A. L. J., 1982, Opt. Comm., 41:187.
Cornelissen, H. J. and Merks-Eppingbroek, H. J. H., 1985, in: "Proceedings, 17th Int. Conf. Phenomena Ionized Gases".
de Groot, J. J., Jack, A. G. and Coenen, H., 1984, Journal IES, 14:188.
Denneman, J. W., de Groot, J. J., Jack, A. G. and Ligthart, F. A. S., 1980, Journal IES, 10:2.
Denneman, J. W., IEE Proc., 128A:397.
Dorgelo, E. G. and Bouma, P. J., 1937, Philips Techn. Review, 2:353.
Dorleijn, J. W. F. and Jack, A. G., 1985, Journal IES, 15:75.
Drop, P. C. and Polman, J., 1972, J. Phys. D: Appl. Phys., 5:562.
Druyvesteyn, M. J., 1933, Physica, 1:14.
Druyvesteyn, M. J. and Warmoltz, N., 1934, Phil. Mag. Ser. 7, 17:1.
Easley, M. A., 1951, J. Appl. Phys., 22:590.
Eckhardt, K. and Kuhl, B., 1970, Lichttechnik, 22:389.
Elenbaas, W., 1959, "Fluorescent Lamps and Lighting", Philips Technical Library.
Elenbaas, W., van Boort, H. J. J. and Spiessen, R., 1969, Illum. Eng., 64:94.
Franck, G., 1971, Z. Naturforsch., 26a:150.
Franck, G. and Schipp, F., 1975, Lighting Res. and Technology, 7:49.
Grossman, M. W., Johnson, S. G. and Maya, J., 1983, Journal IES, 13:89.
Groth, R. and Kauer, E., 1965, Philips Techn. Rev., 26:105.
Haft, H. H. and Thornton, W. A., 1972, J. Illumn. Eng. Soc., 2:29.
Hasker, J., 1976, Illum. Eng. Soc., 6:29.
Hasker, J. and van IJzendoorn, P. R., 1980, Journal IES, 9:134.
Jack, A. G., 1978, Lighting Res. and Technology, 10:150.
Jack, A. G. and Vrenken, L. E., 1980, IEE Proc., 127A:149.
Jacobs, C. A. J., Sprengers, L. and de Vaan, R. L. C., 1978, J. Illum. Eng. Soc., 7:125.
Kagan, Yu.M. and Kasmaliev, B., 1967, Opt. Spectrosc., 22:293.
Kagan, Yu.M. and Kasmaliev, B., 1968, Opt. Spectrosc., 24:356.
Kagan, Yu.M., Kolokolov, N. B., Lyaguschchenko, R. I., Milenin, V. M. and Mirzabekov, A. M., 1971, Sov. Phys. Techn. Phys., 16:561.
Kamei, T., Osada, K., Haysahi, M. and Ikeda, S., 1980, Toshiba Review, 129:5.
Kenty, C., 1950, J. Appl. Phys., 21:1309.
Kenty, C., Easley, M. A. and Barnes, B. T., 1951, J. Appl. Phys., 22:1006.
Koedam, M. and Kruithof, A. A., 1962, Physica, 28:80.
Koedam, M., Kruithof, A. A. and Riemens, J., 1963, Physica, 29:565.
Koedam, M. and Verweij, W., 1965, in: "Proceedings, 7th Int. Conf. Phenomena Ionized Gases".
Koedam, M. and Opstelten, J. J., 1971, Lighting Res. and Technology, 3:205.
Koedam, M., de Vaan, R. L. C. and Verbeek, T. G., 1975, Lighting Des. and Applic., 5:39.
Kostlin, H., Jost, R. and Lems, W., 1975, Phys. Stat. Sol., a29:87.
Lagushenko, R. and Maya, J., 1984, Journal IES, 14:306.
Ligthart, F. A. S. and Keijser, R. A. J., 1980, J. Appl. Phys., 51:5295.
MacDonald, W. T., Johnson, S. G. and Latassa, F. M., 1983, Paper No. 25, IES National Conf., Los Angeles.
Meyer, G. A., and Strojny, F. M. W., 1959, Illum, Eng., 54:65.
Murayana, S., Matsuno, H., Watanabe, Y., Ono, T., Hosoya, K. and Hirota, T., 1984, Journal IES, 14:298.
Pfaue, J. and Schirrwitz, H., 1967, "Technisch - Wissenschaftliche Abhandlungen der Osram-Gessellschaft," Springer Verlag, Berlin.
Polman, J., van Tongeren, H. and Verbeek, T. G., Philips Technical Review, 35:321.
Read, T. B., 1963, Brit. J. Appl. Phys., 14:36.
Sprengers, L., Campbell, R. and Kostlin, H., 1985, Journal IES, 14:607.
Stern, W. and Schaal, G., 1973, Beit. Plasma Phys., 13:27.
Thornton, W. A., 1971, J. Opt. Soc. Am., 61:1155.

Uvarov, F. A. and Fabrikant, V. A., 1965, Opt. Spectrosc., 18:323:433:541.
Uyterhoeven, W., 1983, Philips Technical Review, 3:197.
Uyterhoeven, W. and Verburg, C., 1939, C. R. Acad. Sci. (Paris), 208:503.
van Boort, H. J. J. and Groth, R., 1968, Philips Techn. Rev., 29:17.
van de Weijer, M. H. A., 1961, Philips Techn. Review, 23:246.
van de Weijer, P. and Cremers, R. M. M., 1985, Optical Communications, 53:109.
van de Weijer, P. a nd Cremers, R. M. M., 1985, J. Appl. Phys., 57:672.
van den Hoek, W. J., 1983, Philips J. Res., 38:188.
van Gemert, M. J. C., 1975, J. Appl. Phys., 46:4899.
van Gemert, M. J. C., Lincolne, S. C. and Heuvelmans, J., 1977, Philips Res., 32:8.
van Gemert, M. J. C. and Verbeek, T. G., 1977, Appl. Phys. Letters, 31:500.
van Kemenade, J. T. C., Berns, E. G. and Peters, R. C., 1983, CIE 20th Session, 1:D702.
van Kemenade, J. T. C., van Ooijen, M. H. F. and Dorleijn, J. W. F., 1984, in: "Proceedings, CIBS National Lighting Conference", 208.
van Tongeren, H. and Heuvelmans, J., 1974, J. Appl. Phys., 45:3844.
van Tongeren, H. F. J. J., 1975, Philips Res. Rep. Suppl. No. 3.
van Trigt, C. and Laren, J. B., 1973, J. Phys. D: Appl. Phys., 6:1247.
van Trigt, C. and Blom, N., 1975, J. Qunat. Spectrosc. Radiat. Transfer, 15:905.
Verbeek, T. G. and van Tongeren, H., 1977, J. Appl. Phys., 48:577.
Verjeij, C. M., 1985, in: "Proceedings, IES National Conf.", Detroit.
Verstegen, J. M. P., Radielovic, D. and Vrenken, L. E., 1975, J. Illum. Eng. Soc., 4:90.
Verwey, W., 1961, Philips Res. Repts, Suppl. No. 2.
Vrenken, L. E., 1976, Lighting Res. and Technology, 8:211.
Vrenken, L. E., 1978, Journal IES, 7:154.
Vriens, L., 1973, J. Appl. Phys., 44:3980.
Vriens, L. and Ligthart, F. A. S., 1977, Philips Res. Repts., 32:1.
Vriens, L., 1977, J. Appl. Phys., 48:653.
Vriens, L., Keijser, R. A. J. and Ligthart, F. A. S., 1978, J. Appl. Phys, 49:3807.
Vriens, L., 1978, J. Appl. Phys., 49:3814.
Watanabe, Y. and Yamane, M., 1982, J. Appl. Phys., 53:6724.
Waymouth, J. F. and Bitter, F., 1956, J. Appl. Phys., 27:122.
Waymouth, J. F., 1971, "Electric Discharge Lamps", M.I.T. Press.
Wesselink, G. A., 1980, Philips J. Res., 35:337.
Wesselink, G. A., 1983, Philips J. Res., 38:166.
Willoughby, A. H. and Cannell, B. H., 1982, Thorn Lighting Journal, 24:2.

HIGH-PRESSURE SODIUM (HPS) ARCS

D. O. Wharmby

Thorn EMI Lighting Limited
Research and Engineering Division
Leicester, England

INTRODUCTION

The high-pressure sodium lamp, invented some 25 years ago (Schmidt, 1961), has become one of the most successful commercial lamps. It has a high luminous efficiency (80-140 lm/W depending on input power) and an acceptable color; it is only surpassed in luminous efficiency by the monochromatic low-pressure sodium arc. Its uses are in outdoor lighting, especially street lighting, and in special indoor applications for which good color quality is not required. The increasing cost of electrical power inevitably means that this lamp is being used to replace less efficient sources such as high-pressure mercury (~60 lm/W), despite the apparent longevity of the latter.

The high-pressure sodium arc is contained in a ceramic tube typically 40 to 100 mm long, with a bore in the range 4 to 10 mm for input powers of 50 to 400 W. Electrodes are emitter-coated tungsten. The method of construction has been the subject of intense development but will not be covered here (see reviews by McVey, 1980; van Vliet and de Groot, 1981; Akutsu, 1984).

High-pressure sodium (HPS) arcs are, like the low-pressure mercury discharges used in fluorescent lamps, unusual in that the theoretical modeling of the arc processes has reached a very complete state (for reviews, see Wharmby, 1980; van Vliet and de Groot, 1981). The reasons for this are comparative simplicity. The arc is in near local thermodynamic equilibrium (LTE) and approximates to an infinite cylinder; the increased availability of large scale computing (radiation transport has to be treated satisfactorily); and the high quality data available (widespread academic interest in alkalis in the last decade has helped considerably). It is very likely that the HPS arc models will play an important part in the slow but continuous improvement in the performance of commercial lamps, and they are likely to play a dominant role in any major new development of this light source.

From the scientific point of view, the HPS arc has a number of interesting features which touch on basic problems of physics and chemistry. In this paper, I shall review the important processes which must be included in the arc models, describe the content of various models, and show the extent of the agreement between experiment and theory.

ARC OPERATION

A typical operating HPS arc contains sodium at 60 torr, mercury at 500 torr, and a rare gas at 400 torr. The metals are in excess and exert saturated vapor pressures (Hirayama et al., 1981) over an amalgam at the coldest point; the cool spot temperature is typically 700 C. The mercury (termed the buffer gas) produces a number of effects, including the reduction of thermal conduction losses, the reduction of radial diffusion rates, and the modification of the spectrum. The rare gas which is put in as a cold fill (~40 torr) has similar effects, and in addition is required for starting. All the numbers here are typical for a standard street lamp, but are widely varied for other purposes. The optimization of these quantities has been reviewed by Wharmby (1980).

Observations show that the high-pressure sodium arc has a well developed uniform positive column throughout most of its length, the length being ~10 times the diameter. Within about one tube radius of the electrode tip, the arc contracts diffusely onto the electrode and the spectrum shows evidence of increased electron temperature. There is no satisfactory understanding of the electrode region yet, and I shall concentrate on the positive column which forms the major part of the light-emitting region. Fig. 1 shows typical spectra measured at constant power. Spectra for the region 200-2500 nm have been given by Wharmby (1984). These spectra refer to radiation from the positive column with an allowance made for tube transmission (Wharmby, 1984). The measurements are histograms showing spectral power integrated over the 1 nm bands, which can be seen in the spectra.

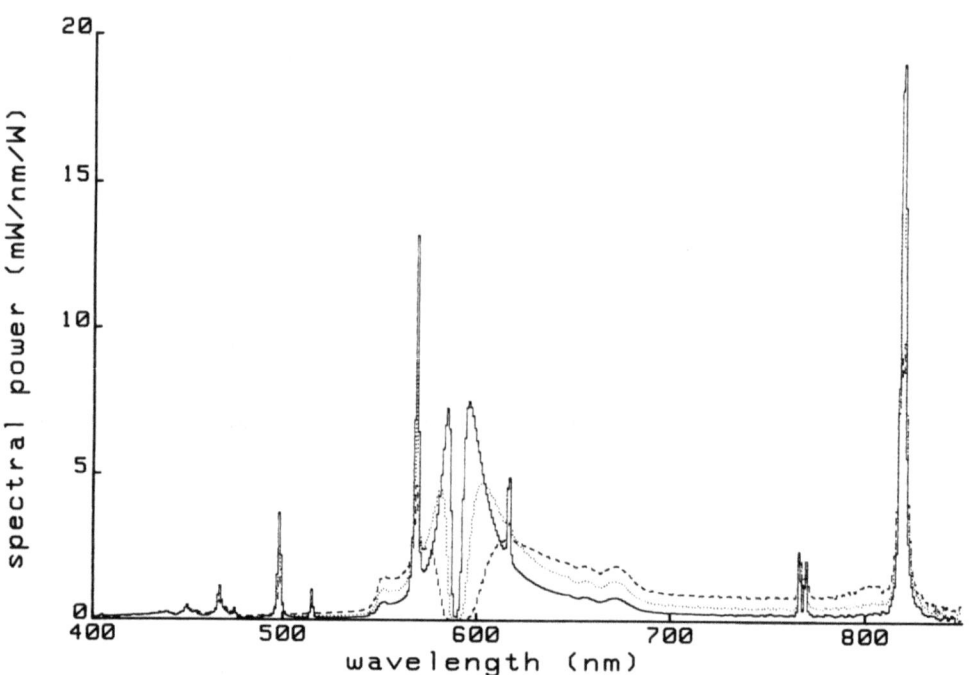

Fig. 1. Positive column spectral power per input W measured in 1 nm bands at approximately constant power input per unit length. Cool-spot temperatures of 700 (——), 750 (....) and 800 C (---) give sodium pressures of 65, 130 and 240 torr, and mercury pressures of 441, 805, 1350 torr, respectively.

Measurements and calculations (e.g., de Groot, 1974; de Groot and van Vliet, 1975) show the characteristic properties of a high-pressure arc. Collision rates are so high that the gas temperature heats to just short of the electron temperature, which is about 4000 K on axis. There is a steep temperature gradient of ~1000 K/mm from axis to wall, with the wall typically at 1500 K. The electrical conduction is mainly in the axis region, and most of the radiation is emitted in this region too.

ENERGY BALANCE

The energy or power balance is concerned with establishing the various power dissipation mechanisms and is a prerequisite for a satisfactory theory. Table 1 shows the energy balance based on spectral measurements such as those shown in Fig. 1 for HPS arcs operated at constant sodium pressure (65 $\pm$ 4 torr), constant power (400 W), and the varying mercury pressures shown. The operating rare gas pressure was chosen to be negligible compared with sodium and mercury pressures, so that the properties shown are those of pure sodium or sodium/mercury arcs. I have also shown the axis temperature measured by Bartels' method (see Arc Temperature Measurement).

Table 1 demonstrates one of the principle effects of adding mercury. It reduces the thermal conduction, with the effect that the axis arc temperature increases slowly with mercury pressure. Because of the exponential dependence of emission on temperature, the radiation processes increase rapidly, restricting the rate of rise of axis temperature. Mercury has nearly ideal properties in this respect. Its excitation potentials are so high that virtually no mercury emission is seen, nor is there significant Hg^+ production to interfere with the arc properties. Its thermal conductivity is very low and it has no effect on starting since it condenses on turn off. It has only one adverse effect: it broadens the self-reversed sodium D lines to the red and so limits the luminous efficiency (lm/W), and, thus, the proportion of mercury that can be used in practice (see Radiation Processes).

The radiation accounts for 50 to 65% of the input power and it is distributed as shown in Table 2. Note that this is arc radiation; thermal radiation from the tube walls, which dissipates the gas conduction losses, has been subtracted. About three-quarters of the radiation is in the NaD (3^2P-2^2S), 568-nm (4^2D-3^2P), 818-nm (3^2D-3^2P), and 1139-nm (4^2S-3^2P) lines. Spectral measurements show these lines to be strongly self-absorbed, and they must be included in calculations of the radiation flux density, describing the radial flow of excitation through the plasma. This is just as important as conduction in determining the arc temperature profile (see Energy Balance Calculations). In only one publication so far has the radiation flux density been calculated for more lines; the 20 used account for about 90% of the radiated power in Table 2 (Dakin and Rautenberg, 1984). Although it is desirable to include as many radiation processes as possible, the computational cost is high. Because the temperature dependence of some of the lines usually omitted is similar to those included, the penalties for omitting some of the weaker lines are not as severe as one might imagine (Wharmby, 1984).

Electron-ion recombination to the 3p ^2P levels of sodium contributes to the spectrum at wavelengths shorter than 408 nm (de Groot, 1974). Recombination to higher levels can only account for ~10% of the IR continuum radiation in Table 2 (Wharmby, 1984).

The large proportion of infra-red radiation suggests the possibility of reflecting some back into the arc. Although it was suggested long ago (Cayless and Clarke, 1963), I know of no practical embodiment of the

Table 1. Percentage of total radiation from positive column which is emitted in the bands shown for an input power of 400W and sodium pressure 65 torr

Mercury Pressure (torr)		0	56	441	1053
NaD lines		47.0	46.9	50.6	52.7
568 nm group		4.3	4.6	4.1	3.7
819 nm group		12.6	13.3	12.0	11.6
1139 nm group		8.5	7.7	7.6	6.8
Strongest lines	Sub total	72.4	72.5	74.3	74.8
UV (<380 nm)		0.6	0.7	0.6	0.7
Continua 380-540 nm		2.0	2.2	1.8	1.6
Lines 380-540 nm		2.9	3.2	3.4	3.4
616 nm group		0.9	0.9	0.8	0.6
K resonance lines		2.5	2.8	1.0	1.6
1847 nm line		4.3	4.2	4.0	3.4
2207 nm line		2.1	2.1	2.1	2.0
2238 nm line		1.5	1.3	1.3	1.1
Other IR lines		5.3	4.9	3.5	2.8
IR continuum		3.9	4.5	5.8	6.7
IR >2500 nm		0.9	0.7	1.4	1.2
Power radiated as a percentage of arc input power		42.6	50.5	56.5	62.5
Arc temperature (K)		4030	4080	4100	4140

Table 2. Experimental energy balance for sodium-arc with 65 torr Na and input power of 400 W (Wharmby, 1984).

Mercury Pressure (torr)	0	56	441	1053
UV <380 nm	0.27	0.36	0.32	0.45
Visible 380-780 nm	24.4	29.1	33.0	37.5
IR >780 nm	18.0	21.1	23.2	24.5
Total arc radiation 200-6000 nm	42.6	50.5	56.5	62.5
Electrode loss	7.9	5.3	3.6	2.7
Balance (thermal conduction, etc)	49.5	44.2	39.9	34.8
Axis temperature (K)	4030	4080	4100	4140

idea. I have not even seen results or calculations which indicate that the effects might be beneficial. Superficially, one might expect that by blocking some channels, radiation might increase through the others, but this ignores the complex self-compensating behaviour found in such arcs.

The insensitivity of HPS properties to orientation shows that convection is not an important process. Recently, Otani (1983) has shown that this cannot be assumed for large values of tube bore (> 10 mm), and 2-dimensional models (Lowke, 1979) might be needed in these cases.

Thermal conduction accounts for the major part of the non-radiative losses. The elastic collisions between neutrals must be included. This is a standard calculation described, for example, by Hirschfelder et al. (1954). At high temperatures it is also necessary to include the electronic thermal conductivity (elastic collisions between electrons and neutrals). Fig. 2 shows some typical calculations.

The other major dissipation process that must be considered is the radial diffusion of species. In sodium vapor electrons and ions are formed in collisions in the high temperature regions. The electrons diffuse rapidly outwards, dragging the ions with them, a process called ambipolar diffusion. They recombine in lower temperature regions, releasing their ionization energy. Similar processes, usually negligible in HPS arcs, but very important in metal halide arcs, occur for molecules. The energy transported per unit area per second by the diffusive flux for species i is

$$\Delta H_i \, J_i \tag{1}$$

where ΔH_i is the enthalpy of formation of i from its constituents (the ionization energy in the case of ions and electrons), and J_i is the net radial flux density of species i. The J_i can be written

$$J_i = - D_{ib} \, n_i \, \frac{dc_i}{dt} \tag{2}$$

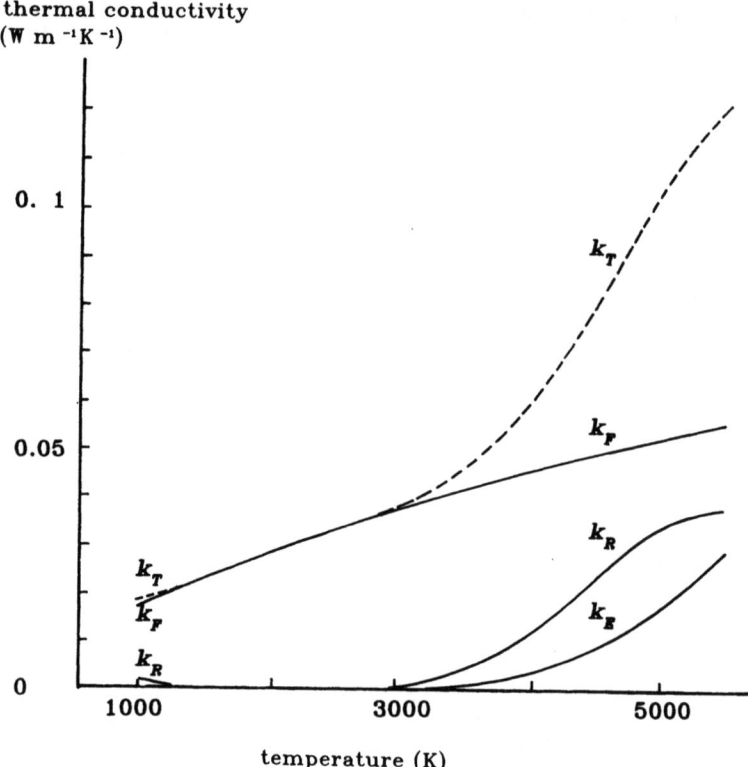

Fig. 2. Effective thermal conductivity for a plasma containing 10 torr Na, 2000 torr Xe, 200 torr Ne, 400 torr Hg. Reaction contributions k_R are calculated using multicomponent diffusion (Mottram, 1980). Note the small contribution from Na_2. Electron conductivity is k_E and normal conductivity k_F. The total effective "thermal conductivity" is k_T.

where D_{ib} is the diffusion coefficient of i against the buffer gas b, n is the total number density, and c_i is the fractional concentration $c_i = n_i/n$ of species i. A more complex treatment includes multicomponent diffusion (Mottram, 1980). For a uniform positive column only radial diffusion need be included. It can then be shown (Cayless unpublished; Mottram, 1980) that dc_i/dr can be expressed as a parametric function of temperature:

$$\frac{dc_i}{dr} = \frac{dc_i}{dT} \cdot \frac{dT}{dr} .$$

This simplifies calculations enormously, because c_i can be calculated as a function of T just once for each sodium partial pressure. The diffusive energy flux then has the same form as conduction,

$$F_c = - k_R \nabla T = - k_r \frac{dT}{dr} \tag{3}$$

where, for species i

$$k_R = -\sum_i \Delta H_i \, n \, D_{ib} \frac{dc_i}{dT}. \tag{4}$$

This value of k_R can then, as a convenience, be added to the normal conduction k_F and the electronic conduction k_E to form a tabulation of the total "thermal conductivity" k_T as a function of temperature, which can then be interpolated during energy balance calculations. Fig. 2 shows the factors making up the total "thermal conductivity"; a particularly complicated example has been chosen for illustrative purposes. The ambipolar contribution is significant in increasing energy loss at high temperatures. The very minor contribution from Na_2 formation can also be seen.

Another important aspect of ambipolar diffusion is its effect on the Na^+ and electron number densities near the axis. In the steady state, there can be no net radial flux of species containing Na. Since the ions, dragged by electrons, diffuse outwards more rapidly than do neutrals inwards, the steady state is achieved by a lowering of the ion and electron densities. Figure 3 shows this demixing effect.

Electrical conductivity is calculated by standard methods from these number densities. Electron-electron, electron-ion and electron-neutral collisions of various types must be included, with appropriate averaging of cross-sections over the Maxwellian distribution. The effects of ambipolar diffusion on electron density are important here.

RADIATION PROCESSES

All the important spectral lines are self-absorbed to some extent, some of them grossly. The radiation transport equation (Cayless, 1985) can be used to calculate the spectral intensity or radiance I_ν (for example, in $W \, m^{-2} \, sr^{-1} \, nm^{-1}$). In LTE the radiation transport equation for a ray crossing a diameter of the arc (radius R) is

$$I_\nu = \int_{-R}^{R} K(\nu,T) \, B_\nu(T) \, \exp\left(-\int_x^R K(\nu,T) \, dx'\right) dx \tag{5}$$

where $B_\nu(T)$ is the full radiator intensity at frequency ν, and the temperature is a function of the distance x across the arc. Incidentally, this accounts for why there is so little UV radiation (Table 1), the value of the full radiator intensity at 250 nm being only ~1/50th of its value at 600 nm.

The most important material function is the absorption coefficient $K(\nu,T)$. If this is known the spectrum can be calculated straightforwardly from Eq. (5). For atomic lines

$$K(\nu,T) = \frac{1}{4\pi\varepsilon_0} \frac{\pi e^2}{mc} n_\ell \, f_{\ell u} \, P_\nu(T), \qquad (m^{-1}) \tag{6}$$

where e and m are the electron charge and mass, c the velocity of light, and ε_0 the permittivity of free space. The oscillator strength $f_{u\ell}$ for the transition with upper state u and lower state ℓ can be obtained from tables (Wiese et al., 1969). n_ℓ is the lower state number density and $P_\nu(T)$ is the line profile function normalized so that

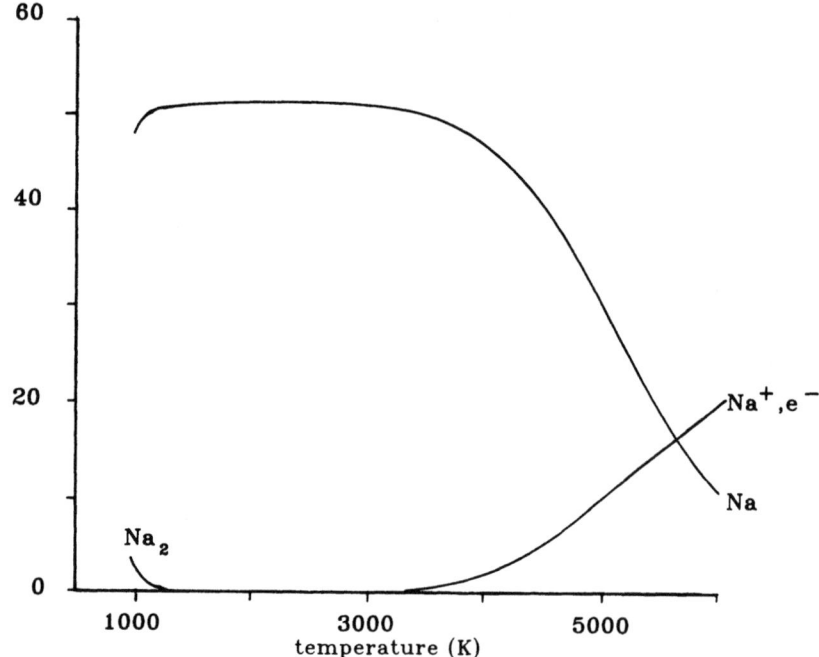

Fig. 3. Species present in a high-pressure sodium plasma. The low concentration of Na+Na$^+$ at 6000 K is the result of the demixing caused by ambipolar diffusion.

$$\int_0^\infty P_\nu(T) \, d\nu = 1 \ . \tag{7}$$

Calculation of $\kappa(\nu,T)$ for sodium lines is straightforward apart from the line broadening implicit in $P_\nu(T)$.

Repeated evaluation of Eq. (5) is required during the calculation of the radiation flux in the energy balance (see below). For this purpose, standard line broadening theory, for at least the major lines shown in Table 2, is adequate; I shall deal with this in the following. Once the energy balance has been computed, so that the temperature profile is known, the emitted spectrum can be calculated. This requires far more information about line broadening, and particularly the extreme wing broadening of the D lines, a matter I have covered in detail elsewhere (Wharmby, 1985).

Resonance broadening dominates the D lines. For virtually all wavelengths of interest it must be considered in the quasi-static approximation (Hindmarsh and Farr, 1972). The quasi-static line profile is Lorentzian:

$$P_\nu = \frac{1}{\pi \delta \nu} \cdot \frac{1}{1+(\nu-\nu_0)^2/\delta\nu^2} \tag{8}$$

where ν_0 is the line center frequency and $\delta\nu$ the semi-half-width (Hz). If n_ℓ is the lower state number density, then

$$\delta \nu = \frac{1}{4\pi\varepsilon_o} \cdot k \; \frac{e^2 f_{\ell u} \; n_\ell}{4\pi \; m\nu_o} \; . \tag{9}$$

There is some uncertainty about the value of k depending on the approximations used in calculating the line broadening. For the sodium D lines, it is best to make minor adjustments to k so that the spectrum has the self-reversal width observed in experiments, with known sodium number densities (Denbigh et al., 1985). The simplicity of resonance line broadening, with its very weak dependence on temperature through n_ℓ and the absence of any line shift, leads to some useful but straightforward diagnostics, for sodium pressure (de Groot, 1973; Zollweg and Kussmaul, 1983; Reiser and Wyner, 1985), and for arc temperature (see section below on Temperature Measurement). At very high pressures none of these methods work satisfactorily because of the extreme wing broadening effects (Wharmby, 1985).

When mercury is included there is extra broadening of the red wing of the D lines as experimental information (Fig. 4) shows. Apart from the region near 760 nm, which results from the extrema of the A-X difference potential for the NaHg molecule (Duren, 1977), the red-wing broadening is well described by a quasistatic van der Waals profile (Sobel'man, 1972):

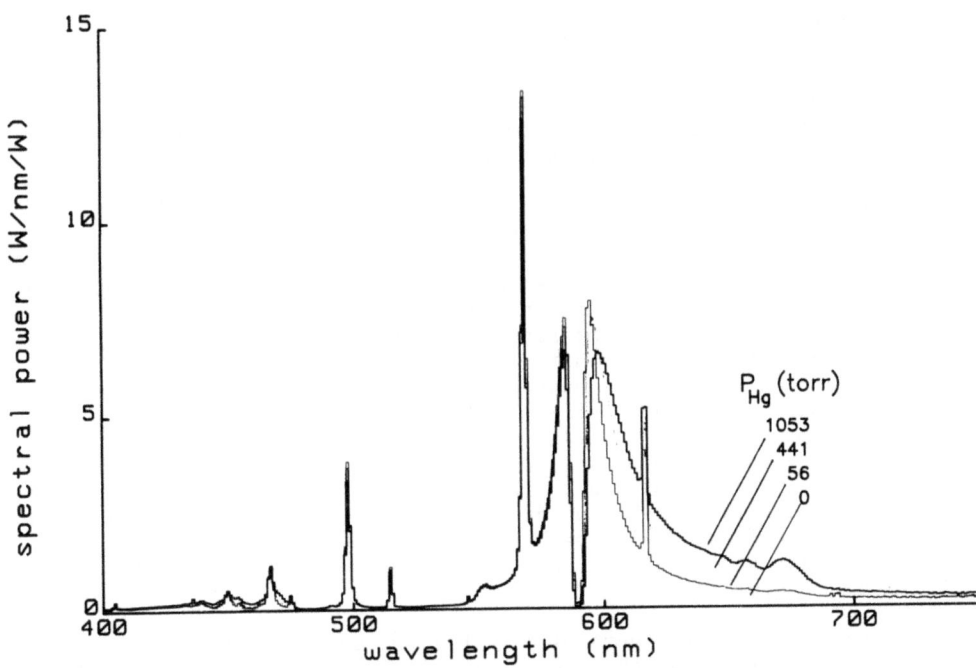

Fig. 4. The effect of mercury at the pressures indicated on the red-wing broadening of the D lines for the plasma shown in Table 1. Minor increases in continuum at short wavelengths are probably the increase in electron-ion recombination as the axis temperature increases.

$$P_\nu = \frac{n_{Hg}}{3(2\pi)^{1/2}} \frac{|C_6|^{1/2}}{(\nu_o-\nu)^{3/2}} \exp\left(-\frac{2}{9} C_6 \frac{n_{Hg}^2}{(\nu_o-\nu)}\right) \quad (\nu_o < \nu) \qquad (10)$$

$$P_\nu = 0 \quad (\nu_o > \nu).$$

Far from the line center, but not in the extreme wing region, this gives the familiar $(\nu_o-\nu)^{3/2}$ behaviour. Comparison with experiment leads to a value of $C_6 \simeq 2 \times 10^{-43}$ m^6 s^{-1}. The magnitudes of blue and red wing self-reversal widths (as seen in Fig. 4) can be used for diagnostics for mercury pressure in a similar manner to that used for sodium pressure.

The combination of the Lorentz and van der Waals profiles can be obtained by convolution in analytical form (Stormberg, 1980). In the line wings (~10 line widths from ν_o), convolution is equivalent to addition, after correction of the normalizing factor (Froment et al., 1983). This is in accord with physical intuition: convolution applies to statistically independent processes happening simultaneously. In the line center, where a radiator is perturbed by distant collisions with many perturbers, convolution seems the correct choice. In the line wings close collisions, one at a time, by individual perturbers, would lead one to expect to add line profiles (with due attention to normalization). Convolution, therefore, covers both these cases naturally.

For the non-resonant spectral lines, Stark broadening becomes important. This is treated in the impact approximation, because of the high velocity of electrons, using formulas and Stark-broadening constants calculated by Griem (1964, 1974). Line shifts are comparable with widths and must be included. The quasistatic ion broadening can also be included (de Groot and van Vliet, 1975). The dependence of width and shift on electron density leads to a strong temperature dependence of the absorption line shape. This results in an emitted line profile which is very temperature dependent and which can lead to uncertainties in Bartels' method of temperature measurement (see below). Despite all these uncertainties, the agreement between emitted and measured profiles for the 568-nm ($4d\ ^2D - 3p\ ^2P$) and 818 nm ($3d\ ^2D - 3p\ ^2P$) transitions is remarkably good (de Groot, 1974; de Groot and van Vliet, 1975).

The broadening of the D lines by rare gases is relatively minor and it does not need to be included in radiation flux density calculations. At high sodium and rare gas pressures, the extreme wing broadening by rare gases can significantly affect the emission from the sodium D lines. For calculations of spectra, however, these extreme wing broadening effects should be included. Fig. 5 (Denbigh, unpublished data) shows normalized spectra for lamps with sodium at ~500 torr and each of the five rare gases at ~1200 torr. Such data contains information about the Na$^-$ rare gas interactions and can be used to check the validity of calculated potentials (Woerdmann and de Groot, 1982). Part of the broadening in Fig. 5 is self-broadening by Na itself and this also ought to be included in calculations of the spectra. Unfortunately, the necessary data are as yet incomplete.

There is always the possibility that another perturber than Hg or a rare-gas may be found which offers possible improvements in this light source. The requirement would be substantial broadening, particularly to the green side of the NaD lines, without any adverse effects. Unfortunately, this is rather against the normal trends, if van de Waals bonding is considered, because upper states are usually more strongly bound than

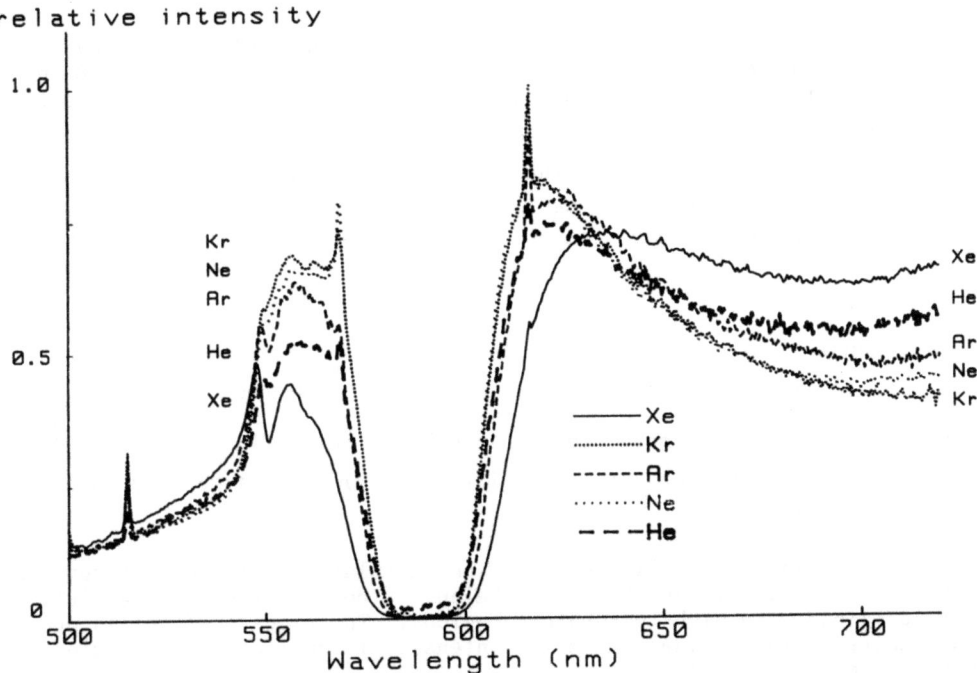

Fig. 5. Experimental relative intensities measured at ~1200 torr rare gas, ~500 torr Na, and at constant input power, showing the different effects of the various Na-rare-gas excimers.

lower states. N_2 would seem to be a possible candidate in that it seems to broaden more strongly than Ar (Jongerius et al., 1981), but it is unfortunately ruled out because of its adverse effect on starting.

ENERGY BALANCE CALCULATIONS

Energy balance calculations have been very helpful in understanding the subtle interplay between conduction, diffusion, and the various radiative processes in the HPS arc. The result of an energy balance calculation is the prediction of the arc temperature profile, and from this all arc properties follow if the arc is in LTE. This requirement is examined in the following section.

Most of the arc models reported on so far are for steady-state dc conditions. They have been compared with average spectral intensities measured in arcs operated from 50 to 60 Hz supplies. The success of the steady-state models results from the fact that for frequencies ~50 Hz, which are not too much less than the inverse of the arc temperature relaxation time, the cycle-averaged values of spectral intensity, arc temperature, etc., are very close to the steady-state values. This has been carefully examined by de Groot (1974). These steady-state models are described overleaf. Finally, increase in computer power has made it practicable to consider time-dependent models and these are dealt with in the last section.

Non-LTE Calculation

One of the most important calculations is a non-LTE one made by Waszink (1973), which established that there are deviations from LTE in the outer parts of the arc. The calculation is based on level-population equations for the ^2S and ^2P Na levels and the $^1S_0$ Na^+ level. Ionization from both ^2S and ^2P levels are included, as are collisional and radiative transitions between ^2P and ^2S Na levels. The radiation is treated using a position-dependent escape factor, which is positive near the arc axis where there is net emission, and negative in the outer parts where there is net absorption. The calculations show that where the escape factor is positive the electron temperature T_e and gas temperature T_g are approximately equal. For negative escape factors, T_e-T_g is substantial, increasing to ~700 K near the wall for an arc with 250 torr Na. Thus T_e is higher than T_g because more electron energy is gained in de-exciting ^2P levels (populated by the outward radiation flux) than is lost by collisional excitation.

This deviation has little effect on calculations of the spectrum, as Waszink has shown, or on the energy balance. Most of the spectral intensity originates in high temperature regions. The biggest effect is near the centre of the D lines where the intensity is emitted in the region very close to the wall.

For sodium pressures of 250 torr or greater, or in cases where there is a substantial mercury pressure, LTE can be expected to be a satisfactory approximation for the purpose of calculating spectra. At lower sodium pressures (~100 torr, sodium only), not explored by Waszink, there is circumstantial evidence that LTE still gives satisfactory spectral intensities from the agreement between calculated and measured intensities for the 568, 818 and Na D lines (de Groot, 1974).

A model for a pulsed HPS arc far from equilibrium has been presented by Shuker et al. (1980). This involves ~20 levels, including Na_2 and NaXe excimer states. The Na states are populated according to a Boltzmann factor at the electron temperature ~3000-6000 K; the gas temperature was ~700 K. This model indicates the level of complexity required in arcs far from LTE.

Steady State LTE Models

In these calculations, a radial temperature distribution is found which is consistent with

$$\sigma \underline{E}^2 = \nabla \cdot (\underline{F}_C + \underline{F}_A + \underline{F}_R), \qquad (11)$$

where $\underline{F}_C$, $\underline{F}_A$, and $\underline{F}_R$ are conduction, diffusion, and radiation flux densities (W m^{-2}). The electric field and conductivity are $\underline{E}$ and σ. As described previously, it is convenient to take $\underline{F}_C$ and $\underline{F}_A$ together to form an effective thermal conduction flux $\underline{F}'_C$.

$$\underline{F}'_C = \underline{F}_C + \underline{F}_A = -k_T \nabla T. \qquad (12)$$

The radiation flux density is the integral (Lowke, 1969; Cayless, 1985)

$$\underline{F}_R = \iint_{4\pi} I_\nu(\underline{n},\underline{r})\, \underline{n}\, d\Omega d\nu \qquad (13)$$

over all solid angles, of rays of intensity $I_\nu(\underline{n},\underline{r})$ crossing unit areas at $\underline{r}$, in the direction defined by the unit vector $\underline{n}$. The intensity is calculated using Eq. (5). Rays coming from all other parts of the plasma contribute to the value of $\underline{F}_R$ at $\underline{r}$. Note that the integration is over all frequencies contributing to the emitted band or line being considered. For reasons of economical computation, it is necessary to reduce the number of frequencies as far as possible. This is the reason why the line broadening, used in the evaluation of Eq. (13), must be considered so carefully (see Radiation Processes). The problem of calculating $\nabla \cdot \underline{F}_R$, or the equivalent net emission coefficient, are covered by Cayless (1985).

In cylindrical coordinates, Eq. (11) can be written

$$\sigma E^2 = \frac{1}{r}\frac{d}{dr}\left(-rk_T \frac{dT}{dr}\right) + \frac{1}{r}\frac{d}{dr}\left(rF_R\right) \qquad (14)$$

where E is now the longitudinal field, and r the radial variable. The behaviour of the radiation term for resonance and non-resonance lines is shown in Fig. 6. Actual calculations show that near the axis where there is net emission $\nabla \cdot \underline{F}_R \sim 10^8$ W m^{-3}, and near the wall where there is net absorption $\nabla \cdot \underline{F}_R \sim -10^7$ W m^{-3}. At about 70% of the tube radius, $\nabla \cdot \underline{F}_R \sim 0$, and arc properties are determined by $\nabla \cdot \underline{F}_C \sim \nabla \cdot \underline{F}_C \sim 10^7$ Wm^{-3}. A number of numerical methods which have been published for the solution of Eq. (14) (Lowke, 1979; de Groot and van Vliet, 1975; Eardley et al., 1979) will not be considered further here. Results from LTE steady-state calculations are considered below.

Time-Dependent LTE Models

Despite the satisfactory results of dc models for predicting average properties of 60-Hz arcs (see Predictive Properties of the Theory), the time-dependent behaviour is of interest. For operation on other types of circuitry (e.g., pulsed), an understanding of the time dependence is very informative. The prototype ac calculation was made for argon and mercury arcs by Lowke et al. (1975), where the energy balance equation was solved simultaneously with the circuit equation. The energy balance is

$$\rho C_p \frac{\partial T}{\partial t} = \sigma E^2 - \frac{1}{r}\frac{\partial}{\partial r}(r F_R) - \frac{1}{r}\frac{\partial}{\partial r}\left(-rk_T \frac{\partial T}{\partial r}\right) - \rho C_p v_r \frac{\partial T}{\partial r} + \frac{\partial p}{\partial t}, \qquad (15)$$

where ρ is density, C_p specific heat at constant pressure, v_r the radial gas velocity, and p the pressure. The LHS of Eq. (15) is the power required to heat unit volume of gas as the temperature is changed. The two extra RHS terms added to Eq. (14) are the radial mass flow resulting from contraction or expansion of the hot column, and the adiabatic heating. The last term was included by Chalek and Kinsinger (1981). In view of the extremely small pressure differences required to drive the flow (Lowke et al., 1975), it is not clear to me how important this term is to the energy balance. As an example of the circuit equation, consider an inductive circuit for which the supply voltage V_S and the lamp voltage V_L are related by

$$V_S = V_L - L\frac{di}{dt} \qquad (16)$$

with

$$V_L = V_E + I / \left(\ell \int_0^R 2\pi r \sigma dr\right) \qquad (17)$$

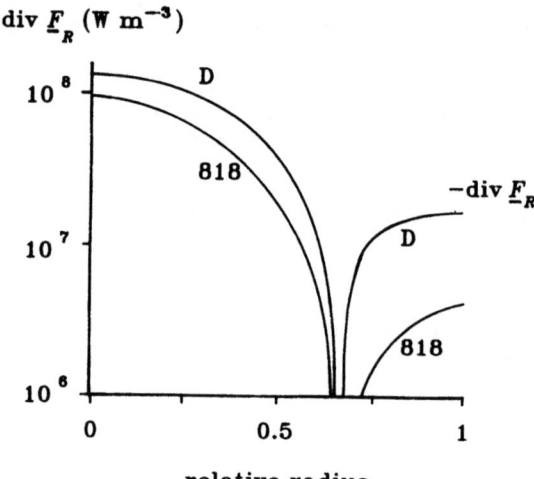

Fig. 6. div $\underline{F}_R$ (i.e., $\nabla \cdot \underline{F}_R$, which is proportional to net emission coefficient) for the D lines and 818-nm lines in a typical HPS arc. Note that div $\underline{F}_R$ becomes negative in the outer parts of the arc because of self-absorption.

where ℓ is the arc length, R the tube radius, and V_E the electrode voltage drop.

Solutions of these equations for HPS arcs, with some or all of the terms in Eq. (15) included, have been presented by van Vliet and de Groot (1981) for 50-Hz operation, and by Chalek and Kinsinger (1981) and Dakin and Rautenberg (1984) for pulsed arcs.

For 50-Hz ac operation, the theory gives satisfactory predictions of electrical characteristics and of the time dependence of axis temperature, which typically fluctuates between 4500 K and 2200 K in a sodium-xenon arc operated with resistive ballast. Because of the comparatively long zero-current period with resistive ballast, the behavior of the temperature profile is very interesting. In the last half of the power dissipation cycle the temperature relaxes uniformly with slight broadening of the temperature profile. As soon as the power starts to increase in the early parts of the cycle, the axis temperature rises very rapidly, the edges catching up much later (van Vliet and de Groot, 1981). Measurement and calculation show discrepancies which may be a result of failure of LTE when heating rates are so high. Behavior on inductive ballast is much smoother, because the power is off for a much shorter time.

Similar rapid heating phenomena occur in pulsed arcs in which the off-to-on ratio is typically ~10:1. The purpose of pulsed operation is to achieve a whiter color appearance, which does indeed happen. The arc must be held off long enough for substantial cooling to occur, the relaxation time being ~10 ms depending on the buffer gas. If the required average power is then applied during a short period (~1 ms), the axis temperature increases to >5000 K so that lines from high-lying levels are excited, giving more blue and green radiation. The otherwise extremely complete calculations by Dakin and Rautenberg (1984) appear to omit electron-ion continua which are an important part of the spectra in these conditions. In order to increase the ratio of on-to-off time, and reduce the peak current required, the gas must have high thermal conductivity. This is the opposite condition for achieving high radiation efficiency and so a compromise has to be found. Pulsed discharges would have many attractions, were it not for their tendency to acoustic oscillations which result from the small pressure fluctuations mentioned earlier. At best these can present themselves as annoying flicker; at worst, the kinked arc may crack the tube.

ARC TEMPERATURE MEASUREMENT

Arc temperatures are the key quantity in an LTE plasma. Conversely, arc models must be capable of predicting the correct temperature profile. Many methods have been used to measure arc temperatures in HPS arcs, and they are generally in agreement (Wharmby, 1980). In HPS arcs there are unique problems associated with poor optical quality, commercial lamps being made from polycrystalline alumina having strong scattering properties. Experimental arcs can be operated in synthetic sapphire tubes, but even the best of these have poor optical quality, suffering from drawlines. Except in circumstances of very high tube quality, no method which requires Abel inversion can be regarded as practical, partly because of the procedure itself, and partly because of the difficulty of deciding if lines are optically thin.

For these reasons Bartels' method, which requires the measurement of the peak spectral intensity I_λ^{sr} of a self-reversed line at wavelength λ, has been extensively used. Bartels showed (Zwicker, 1968; de Groot and Jack, 1973; Lapworth, 1980) that

$$I_\lambda^{sr} = \frac{1}{D_B} B_\lambda(T_m) \qquad (18)$$

where T_m is the maximum temperature along the line of sight for the ray of intensity I_λ^{sr}; $B_\lambda(T_m)$ is the corresponding value of the full radiator spectral intensity. (As in the astrophysical literature, intensity is used synonymously with radiance here: units are W m^{-2} sr^{-1} nm^{-1}). The constant D_B depends on the temperature profile and on the atomic properties of the radiator. Bartels' contribution was two-fold: first, he realized that I_λ^{sr} depends strongly on T_m, only weakly on temperature profile and not at all on radiator number density; second, he gave approximations for D_B. While the former is of lasting value, the latter was so complicated that it only served to cause confusion, although the position has now been clarified by Lapworth (1980). The conditions required to use the approximations are very restrictive.

The difficulties are completely removed, as are most of the restrictions on the use of Bartels' method, if D_B is calculated directly by use of the radiation transport Eq. (5), as was first done by de Groot and Jack (1973). They provided values of D_B^{818} for the Na ^2D-^2P transitions. The values of D_B given by de Groot (1973) have since been slightly revised (de Groot, private communication). The calculation of D_B requires knowledge of the line broadening and is described in the section on Radiation Processes. The broadening of the 818-nm lines is complicated by the combination of Stark and resonance broadening. The Stark shifts, in particular, cause the relative heights of the four 818 nm self-reversals to depend on temperature, which leads to some uncertainty in measurements. A more practical problem is that the 818-nm line is only self-reversed at the highest temperatures, so other methods are needed for lower temperatures.

The difficulties are overcome if Bartels' ideas are applied to the D lines. Bartels' papers specifically forbade the use of resonance lines, but the restriction only applies to the approximations to D_B, as de Groot and Jack have noted. Fig. 7 shows the results of radiation transport calculations of D_B for the sodium D lines for temperature profiles of the form

$$T(r) = T_m - (T_m - T_w)(r/R)^\alpha \qquad (19)$$

where r is the radial coordinate, R the tube radius, and T_w is the wall temperature. The index α characterizes the shape of the profile. The simplicity of the resonance line broadening at sodium pressures of ~250 torr and less, means that the value D_B is established with great certainty. There is a slight dependence on pressure, ignored here, resulting from the changes in Boltzmann factor which occur as the self-reversal wavelength moves out from the line center with increasing number density. The disadvantage of the D-line method is that the value of D_B depends more strongly on temperature profile then it does for lines from higher levels. For example, if for T_m = 4000 K, the profile index α is assumed to be 3, when it is actually 2, the errors in T_m are

D-line method: 270 K

818-line method: 100 K

It requires only very rough measurement of temperature profile, using the same method, to establish the value of α to satisfactory accuracy. Measurements show the 818-nm line and D-line method to give the same results within the experimental accuracy.

The agreement between experimental arc temperatures and the results of theoretical models is generally quite satisfactory, the correct trends being predicted (de Groot, 1973). However, there is some residual disagreement, experimental profiles being broader than theoretical ones, perhaps because of the omission of some of the radiation in the model (Wharmby, 1984), and perhaps because of the great difficulty in measuring temperature profiles.

PREDICTIVE PROPERTIES OF THE THEORY

I have already mentioned many points of comparison between experiment and theory. In particular, the intensities and profiles of self-reversed lines and the arc temperature measurements made by de Groot (de Groot and van Vliet, 1975). I want to draw attention here to two cases in which LTE steady-state theory has successfully shed light on phenomena which are very complicated balances between arc processes. These are the variation of the luminous efficiency (lm W^{-1}) with wall temperature and with bore (luminous flux (lm) is proportional to spectral power weighted by the eye sensitivity). Repeated attempts in these laboratories to do

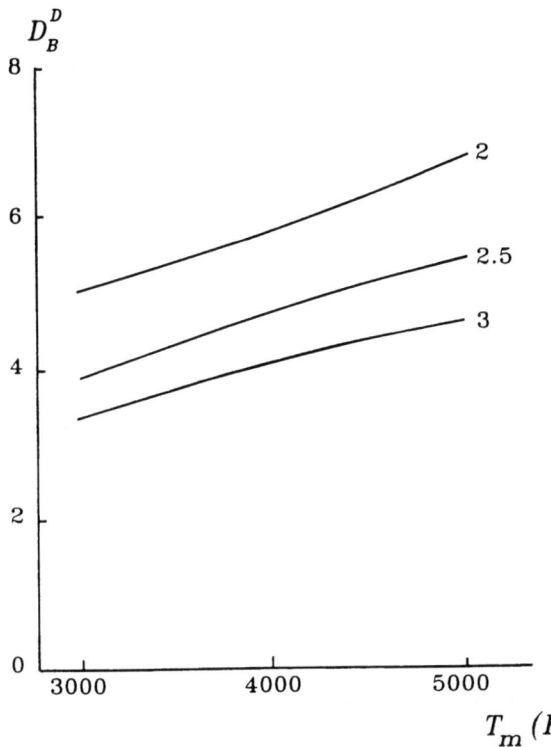

Fig. 7. Bartels' constant for the blue wing self-reversal of the sodium D lines calculated using resonance broadening for three temperature profile indexes α.

convincing experiments produced no satisfactory conclusion, because alteration of one variable changed others which reduced or negated the effect to be observed.

In the case of wall temperature, measurements of luminous efficiency (Denbigh et al., 1983) were made at different powers, with wall temperature being changed by surrounding the arc tube with gases of differing thermal conductivities and with a silica sleeve. The calculations for the same conditions predicted, to within a factor of two, the same dependence on wall temperature and power found in the experiment. Moreover, the dependence had the correct sign, something not achieved in earlier calculations.

Even more useful was the variation of luminous efficiency with tube bore (Denbigh et al., 1985). In this case, the experiment was done on a single tube with sections having different radii. The theoretical calculation was used to account for the power dissipated in the three sections, while the light output was obtained both theoretically and experimentally. Summing the theoretically obtained power in the three sections and allowing for electrode losses accounted for the observed total power to within 2%, itself an indication of the accuracy of the energy balance calculations. Measurements and calculations also agreed (to better than 10%) with the emitted intensity from each section. The conclusion is that the luminous efficiency increases by 12 lm W^{-1} per mm increase in bore between 4 mm and 7.4 mm. If the wall temperature were

were maintained at the same high value as it was at 4 mm bore, rather than falling as it had to during the experiment, the rate would be 18 lm W^{-1} per mm.

These two examples illustrate the predictive capability of the theory at the pressures ~100 torr used in standard HPS lamps. At higher pressures, the effects of extreme wing line broadening (Wharmby, 1985) must be incorporated to give a satisfactory account of the spectrum, and thus sufficiently accurate values of luminous efficiency and other lamp quantities.

CONCLUSIONS

The properties of the high-pressure sodium arc can be largely understood in terms of a one dimensional radial energy balance. A complete treatment of radiation transport is required, and diffusion and conduction losses must be included. For the lower sodium pressures (<150 torr), standard line broadening theory is adequate for calculation of the spectrum, but at higher pressures, information about extreme wing broadening is required. The theoretical model has proved capable, for example, of predicting the complicated effects which occur when tube bore and wall temperature are changed.

ACKNOWLEDGMENT

I want to thank D. A. J. Mottram and B. F. Jones for unpublished calculations, and P. L. Denbigh for unpublished measurements.

REFERENCES

Akutsu, H., 1984, Lighting Res. & Tech., 16:73.
Cayless, M. A. and Clarke, M. G., 1963, "Improvements to Sodium Vapor Electric Discharge Lamps", British Patent 937938.
Cayless, M. A., 1985, "Radiation Transport in High-Pressure Discharge Lamps", this ASI.
Dakin, J. T., and Rautenberg, T. H., 1984, J. Appl. Phys., 56:118.
Denbigh, P. L., Jones, B. F. and Mottram, D. A. J., 1983, J. Phys. D: Appl. Phys., 16:2167.
Denbigh, P. L., Jones, B. F., and Mottram, D. A. J., 1985, IEE Proc., 132A:99.
Düren, R., 1977, J. Phys. B., 10:3467.
Eardley, G., Jones, B. F., Mottram, D. A. J. and Wharmby, D. O., 1979, J. Phys. D: Appl. Phys., 12:1101.
Griem, H. R., 1964, "Plasma Spectroscopy", McGraw-Hill, New York.
Griem, H. R., 1974, "Spectral Line Broadening in Plasmas", Academic Press, New York.
de Groot, J. J. and Jack, A. G., 1973, J. Quant. Spectrosc. Radiat. Transfer, 13:615.
de Groot, J. J., 1974, "Investigation of the High Pressure Sodium and Mercury/Tin Iodide Arc", Thesis, Technical University, Eindhoven.
de Groot, J. J. and van Vliet, J. A. J. M. 1975, J. Phys. D: Appl. Phys., 8:651.
Hindmarsh, W. R. and Farr, J. M., 1972, Prog. Quantum Electron., 2:143.
Hirayama, C., Andrew, K. F. and Kleinosky, R. L., 1981, Thermochimica Acta, 45:23.
Hirschfelder, J. O., Curtiss, C. F. and Bird, R. B., 1954, "Molecular Theory of Gases and Liquids", Wiley, New York.
Jongerius, M. J., Hollander, Tj. and Alkemade C. Th. J., 1981, J. Quant. Spectrosc. Radiat. Transfer, 26:285.

Froment, N. M., Radmore, P. M. and Stephenson, G., 1981, J. Phys. A: Math. Gen., 14:2201.
Lowke, J. J., 1969, J. Quant. Spectrosc. Radiat. Transfer, 9:839.
Lowke, J. J., Zollweg, R. and Liebermann, R. W., 1975, J. Appl. Phys., 46:650.
Lowke, J. J., 1979, J. Appl. Phys., 50:147.
Lapworth, K. C., 1980, "Development of Bartels' Theory of Radiation from an Inhomogeneous Layer and Extensions to the Theory", NPL Report, Qu55.
McVey, C. I., 1980, IEE Proc., 127A:165.
Mottram, D. A. J., 1980, in: "Industrial Uses of Thermochemical Data", T. I. Barry, ed., Chemical Society, London.
Otani, K., 1983, J. Light & Vis. Env., 7:59.
Reiser, P. A. and Wyner, E. F., 1985, J. Appl. Phys., 57:1623.
Schmidt, K., 1961, "Metal Vapor Lamps", U.S. Patent 2971110.
Shuker, A., Gallagher, A. and Phelps, A. V., 1980, J. Appl. Phys., 51:1306.
Sobel'man, I. I., 1972, "An Introduction to the Theory of Atomic Spectra", Pergamon, Oxford.
Stormbery, H-P., 1980, J. Appl. Phys., 51:1963.
van Vliet, J. A. J. M. and de Groot, J. J., 1981, IEE Proc., 128A:415.
Waszink, J. H., 1973, J. Phys. D: Appl. Phys., 6:1000.
Waszink, J. H., 1975, J. Appl. Phys., 46:3139.
Wiese, W. C., Smith, M. W. and Miles, b. M., 1969, "Atomic Transition Probabilities", Vol 2, NBS, Washington.
Wharmby, D. O., 1980, IEE Proc., 127A:165.
Wharmby, D. O., 1984, J. Phys. D: Appl. Phys., 17:367.
Wharmby, D. O., 1985, "Molecular Spectral Intensities in LTE Plasmas", this ASI.
Woerdman, J. P. and de Groot, J. J., 1982, J. Chem. Phys., 76:5653.
Zollweg, R. J. and Kussmaul, K. L., 1983, Lighting Res. & Tech., 15:179.
Zwicker, H., 1968, in: "Plasma Diagnostics", W. Lochte-Holtgreven, ed., North-Holland, Amsterdam.

METAL HALIDE SOURCES

J. H. Ingold

General Electric Company
Lighting Business Group
Cleveland, OH, USA

INTRODUCTION

The words "metal halide source" are used to describe a family of discharge light sources composed of mercury vapor at a partial pressure of several atmospheres and one or more metal halide compounds at partial pressures generally between 0.1 torr and 10 torr. The main difference between metal halide sources and other discharge sources is that metal halide sources generally have better color-rendering properties. The reason for having better color is that the spectrum of emitted radiation is more balanced, as opposed to the unbalanced spectrum of the high pressure mercury discharge which has no red components. The balanced spectrum is produced by line and band radiation from the metal halides, which may be dissociated in the hot core of the discharge.

The historical development of metal halide light sources can be summarized this way. During the first half of this century, the principle discharge used commercially was the high pressure mercury discharge (Elenbaas, 1951). Although three times more efficient than incandescent tungsten in converting electrical energy into light, the high pressure mercury discharge does not have the desirable color-rendering properties of incandescent tungsten. The reason is that an excited mercury atom emits no radiation between 580 nm and 1000 nm, while emitting strong lines at 435, 546, and 578 nm. Therefore, objects which reflect wavelengths longer than 580 nm are not illuminated very well by the light of a high pressure mercury discharge. Incandescent tungsten, on the other hand, emits a continuous spectrum throughout the visible range, so that all objects are sufficiently illuminated by the light emitted by a tungsten filament.

Applied research directed toward improving the efficiency and color of the high pressure mercury discharge led first to the concept of adding other metals to the discharge. It was soon found that desirable metals, such as sodium, either had insufficient vapor pressure, or reacted strongly with the fused silica container, or both. Early in the 1960's, it was found that the addition of certain metal halides to the mercury discharge produced light with better color more efficiently (Reiling, 1964). The advantage of metal halides over metals is twofold: (1) vapor pressures of metal halides are generally much higher than that of the free metal at a given temperature, allowing a sufficient amount of the vaporized compound to be achieved at a practical temperature; (2) the metal halide compound is dissociated in the core of the discharge, allowing the hot metal atoms to emit characteristic radiation, whereas the compound is associated near the wall of the discharge container, preventing chemical reaction between the metal and silica. This discovery was the birth of the metal halide discharge light source.

Within the discharge light source community, it is customary to refer to "low pressure discharges" and "high pressure discharges." Low pressure discharge light sources are exemplified by the fluorescent lamp, with a gas pressure of a few torr, and high pressure discharge light sources by the high pressure mercury lamp, with a gas pressure of a few atmospheres. Representative values of important parameters in low pressure and high pressure discharge light sources are given in Table 1.

The discharge is contained in a tube of glass or fused silica with electrodes at each end. There are three main parts to the discharge: cathode fall, positive column, and anode fall. These spatial regions have quite different characteristics, as described below.

Cathode fall is characterized by:

   Strong axial variation in charged particle densities
   Strong electric field (high E/p)
   No charge neutrality
   Axial current carried by both electrons and ions
   Electron gas non-Maxwellian
   Little or no light production

Positive column is characterized by:

   Weak or no axial variation in charged particle densities
   Weak electric field (low E/p)
   Charge neutrality (except near wall)
   Axial current carried mainly by electrons
   Radial current = 0 (insulating wall)
   Electron gas pseudo-Maxwellian
   Light production throughout column

Anode fall is characterized by:

   Moderate axial variation in charged particle densities
   Moderate electric field
   No charge neutrality
   Axial current carried by electrons
   Little or no light production

The discussion which follows pertains to the positive column, from which useful radiation emanates, unless otherwise noted.

This paper deals principally with characterization of metal halide sources and description of important physical and chemical processes taking place. New problems caused by the addition of metal halides to the mercury discharge, such as axially non-uniform distributions of radiating species, will be mentioned. Typical metal halide discharges used as light sources will be discussed in terms of energy balance, or "Where the Energy goes." It will be shown, for example, that energy balance considerations suggest that thermal management rather than spectrum-tailoring is the direction that will lead to more efficient metal halide sources. In the next section of this paper, the various types of metal halide sources are described and characterized according to energy balance. After that, physical principles which govern the behavior of metal halide sources are discussed.

Table 1. Discharge Parameters

Parameter	Low Pressure	High Pressure
Electron temperature (K)	10,000–30,000	3000–6000
Gas temperature (K)	300–1,000	3000–6000
Electron (ion) density (m^{-3})	10^{16}–10^{19}	10^{20}–10^{22}
Neutral density (m^{-3})	10^{21}–10^{23}	10^{23}–10^{25}
Excited states	Less than LTE at electron temp	LTE at electron temp
Power density (W/m^3)	10^4–10^5	10^7–10^9

Before proceeding to the next section, however, it is necessary to discuss the qualification of light, because any treaties on light sources must have a basis for comparing one source with another. In this paper, only two of the many parameters commonly used to quantify the "quality" of a light source will be mentioned. The first is called "efficacy". The efficacy of a light source is a measure of the efficiency of light production. The unit of efficacy is the lumen per watt (lm/W), which is defined as the ratio of luminous flux (lumens) emitted by the source to the total power (watts) required to maintain the source. The second quantification parameter used is called "color rendering index," denoted by the symbol Ra. Color rendering index is a dimensionless parameter used to classify sources according to the appearance of an object illuminated by the sources. Black body sources usually have the maximum Ra of 100, where monochromatic sources may have negative values of Ra. For more information on the quantification of light, the reader should consult the IES Lighting Handbook (Kaufman, 1981).

CHARACTERIZATION

In this section, the major types of high pressure metal halide discharge light sources are described, then discussed from the point of view of energy balance. The description of types more or less paraphrases previous review articles. The discussion of energy balance is based partly on published data, and partly on a new interpretation of the data.

Metal halide sources commonly have mercury vapor as the basic ingredient of the discharge. Some of the advantages of mercury have already been mentioned: high vapor pressure and chemical compatibility with silica. When metal halides are added to provide the major emission lines, these two advantages become secondary, because rare gases such as argon and xenon have the same properties. However, mercury has additional advantages which the rare gases do not. One additional advantage is that mercury has a vapor pressure versus temperature relation which is "just right" for practical application--namely, at room temperature, the vapor pressure is several millitorr, providing low ignition voltage. This is usually accomplished by filling the discharge tube with 20 torr of argon so that the argon-mercury Penning effect can be exploited. A second additional advantage is that mercury has an extremely high cross section for electron neutral elastic collisions, providing adjustment of burning voltage.

Common Metal Halide Sources

Metal halide sources fall naturally into two classifications: those which emit mainly line spectra, and those which emit mainly continuous spectra. Line spectra originate from free metal atoms in the hot core of the discharge, whereas continuous spectra originate mostly from molecules. In some cases, there are so many lines that the spectrum has the appearance of being continuous.

Historically, line sources composed of various combinations of sodium iodide, thallium iodide, and indium iodide were investigated first. The visible spectrum of a discharge containing these compounds is dominated by the resonance lines of the metallic elements: yellow (590 nm) for sodium, green (535 nm) for thallium, and blue (451 nm) for indium. Thallium iodide (Larson et al., 1963) and sodium iodide (Martt et al., 1964) were used in the first commercially available metal halide sources designed for general lighting purposes. Subsequently, a combination of all three compounds was found to be the most suitable line source in terms of efficacy and color rendering properties.

Line sources which are presently receiving the most attention from the lighting community are the scandium iodide source (Keeffe, 1980) and the dysprosium iodide source (Drop et al., 1975). The spectra emitted by these sources contain so many lines that considerable overlap occurs due to collision broadening, giving the appearance of continuous spectra.

Continuous sources generally have better color rendering properties than line sources. By definition, "good color rendering properties" implies a continuous spectrum similar to the black body spectrum emitted by the sun, incandescent tungsten, etc. Molecular emission bands, which have been broadened by collisions until they appear continuous, seem to offer the best possibility of simulating the desired continuous spectrum with a discharge light source. Needless to say, prospective candidates for this type of source must be so strongly bound that they are weakly dissociated at high temperature. Perhaps the most widely studied metal halide source with a strong continuum is the tin halide source (Speros et al., 1970; Drop et al., 1975).

Energy Balance

When a gas is heated to a temperature high enough to cause a significant fraction of the atoms and molecules to be excited, the gas emits radiation. In the steady state, the hot gas loses energy mainly by radiation and by thermal conduction. In discussing radiation losses in connection with energy balance, it is useful to define the following wavelength ranges: a) ultraviolet or UV (0 to 400 nm); b) visible or VIS (400 to 700 nm); c) infrared or IR (700 to ∞ nm).

For example, the sun is an optically thick spherical plasma with a surface which radiates like a black body at 6000K. Therefore, 14% of the energy expended is radiated in the UV, 38% in the VIS, and 48% in the IR. The radiation efficiency (watts/watt) of the sun is 100%, meaning there are no energy losses in addition to radiation, and the visible radiation efficiency (watts/watt) is 38%. The efficacy of the sun is 95 lm/W, and the efficacy of the visible spectrum of the sun is 250 lm/W. Efficacy of a light source expressed in units of in lm/W is calculated by integrating over all values of wavelength λ the product of the eye sensitivity function $V(\lambda)$ (Kaufman, 1981), shown in Fig. 1, and the spectral power distribution, then dividing the result by the total power expended. For a black body at temperature T, the spectral power distribution is given by the Planck function $B(\lambda)$:

$$B(\lambda) = 2\pi hc^2/(\exp(hc/kT\lambda)-1)\lambda^5 \quad W/m^3 \tag{1}$$

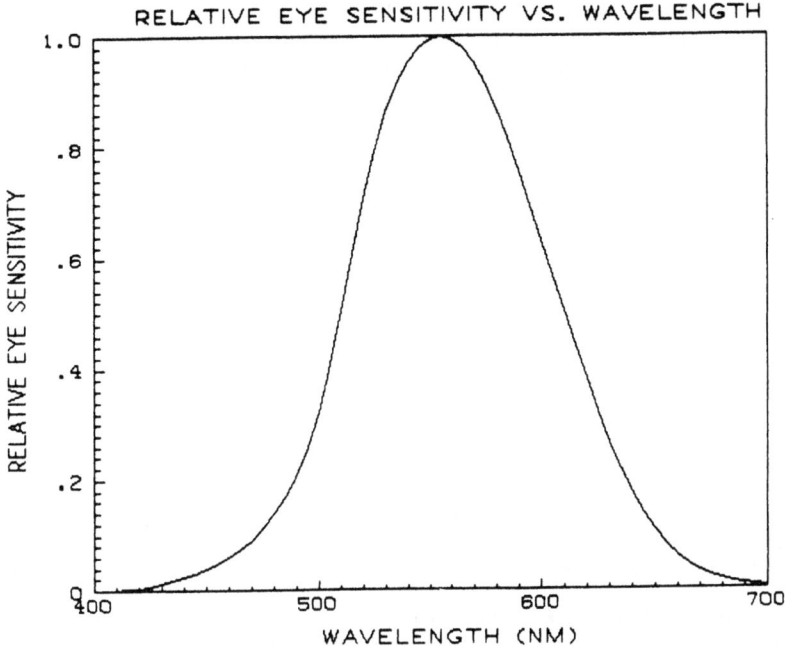

Fig. 1. Relative sensitivity of human eye to radiation. One watt of energy at wavelength of 555 nm is equivalent to 680 lumens. (Kaufman, 1981)

where $\hbar$ is Planck's constant, c the speed of light, k the Boltzmann constant, and T the temperature. The efficacy of the sun's light is

$$680 \int V(\lambda) B(\lambda) d\lambda / \int B(\lambda) d\lambda = 95 \text{ lm/W} \qquad (2)$$

The factor of 680 lm/W in the equation above arises from the fact that one watt of monochromatic radiation at a wavelength of 555 nm, where the human eye is most sensitive, produces 680 lumens. The efficacy of the visible spectrum of the sun light is

$$95/.38 = 250 \text{ lm/W} \qquad (3)$$

In addition to overall efficacy, visible radiation efficiency (dimensionless) and visible radiation efficacy (lm/W) are two important measures of the performance of a light source.

The high pressure gas discharge is a resistive medium with an impedance which varies with gas temperature. The heat generated by Joule heating is transferred from the hot central region to the cool edge of the discharge by flow of radiation, by flow of heat (thermal conduction), and by flow of dissociation energy. Flow of dissociation energy occurs when molecules becomes dissociated in the hot core of the discharge, then flow to the cooler outer regions of the discharge where they recombine, giving up their dissociation energy. Both optically thick radiation flow and dissociation energy flow can be thought of as contributing to the flow of heat by thermal conduction in a high pressure discharge. These various energy transfer processes combine to produce a radial temperature distribution which decreases monotonically from the core of the discharge to the wall. The radial temperature distribution in the high pressure mercury discharge is approximately parabolic, as it is in the high pressure sodium discharge. Narrower radial temperature distributions occur in metal halide discharges. More will be said about this constriction later.

Unlike the sun, a high pressure discharge light source is not optically thick at all wavelengths. As a rule, about 50% of the energy input to the positive column is radiated from the hot core, and the other 50% is thermally conducted down the temperature gradient to the wall. The energy conducted to the wall is dissipated by radiation and convection. The wall temperature is so low, however, that no contribution to the luminous output of the source is made by this energy because it is mostly in the IR region of the spectrum. In discussing energy balance, it is customary to differentiate between IR energy emitted by the discharge and IR energy emitted by the containing wall. In addition to the major energy dissipation mechanisms, about 10% of the input energy goes into the electrodes and does not contribute directly to the radiation output. Consequently, there are five pathways by which energy is lost: UV radiation, visible radiation, IR radiation, thermal conduction, and electrodes. The fraction of input energy going into visible radiation for several light sources, including metal halide, is given in Table 2.

It is evident from the data tabulated above that the most efficient light source converts about 30% of the input energy into light. The relation between performance and visible radiation efficiency of the various light sources is presented graphically in Fig. 2. This graph shows a one-to-one correspondence between lamp performance, measured in lumens per watt, and visible radiation efficiency up to values in the neighborhood of 30%. For higher values of visible radiation efficiency, the curve branches. The branch heading toward 200 lm/W has low Ra (poor color rendering properties) because this is the region of monochromatic light. The branch with good color rendering properties approaches asymptotically to an efficacy of about 100 lm/W. The latter branch includes the sun, which has an efficacy of 95 lm/W and excellent color rendering properties.

Table 2. Fraction of Energy in Visible Spectrum

Type of Source	Visible Fraction	lm/W (Total)	lm/W (vis)
Household Incandescent	0.07	17	250
Fluorescent	.24	80	340
High Pressure Hg	.15	50	356
Sn-I-Cl	.23	60	260
Na-Tl-In-I-Hg	.24	87	360
Dy-I-Hg	.32	81	253
Na-Sc-I-Hg	.33	100	300
High Pressure Na	.30	120	400
Low Pressure Na	.35	180	515
Sun	.38	95	250

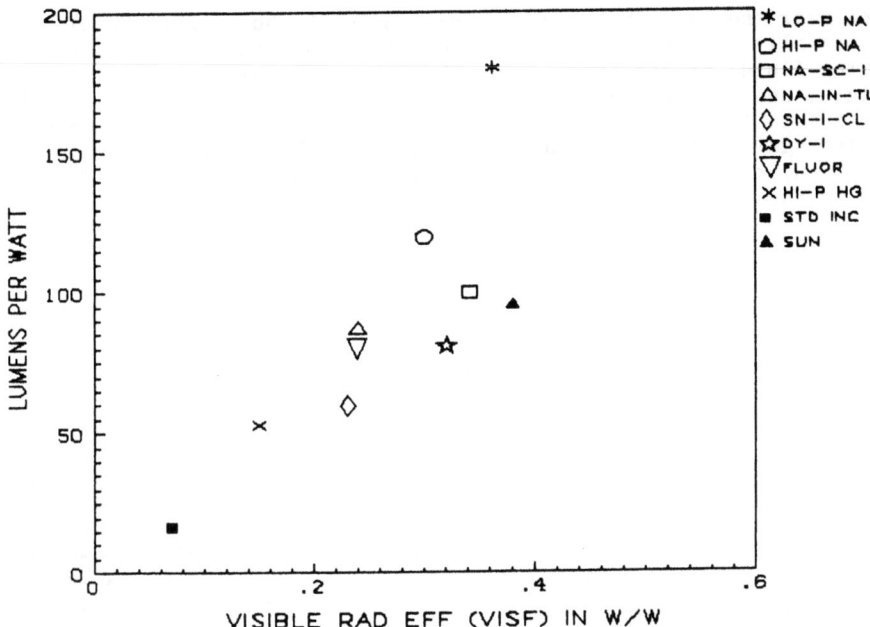

Fig. 2. Efficacy of various light sources plotted versus visible radiation efficiency.

These data can be put into better perspective by adding lines of constant lm/W of the visible spectrum of the various light sources, as shown in Fig. 3. Note that the continuum sources---incandescent, tin halide, and the sun---fall on or near the 250 lm/W curve. Color rendering properties are best for continuum sources, and poorest for monochromatic sources, such as the low pressure sodium source which lies on the curve labelled 515 lm/W. A point to be emphasized here is that an increase in total efficacy caused by an increase in visible efficacy generally is accompanied by a decrease in color rendering capability. However, if the increase in total efficacy is caused by an increase in visible radiation efficiency instead of visible efficacy, then color rendering capability is unaffected. For example, it is evident from this graph that the absolute maximum efficacy that can be achieved with a visible radiation efficiency of 30% is about 200 lm/W, which could be obtained only with a very unsatisfactory monochromatic source emitting green light at 555 nm radiation. By doubling the visible radiation efficiency, however, an efficacy of 200 lm/W with good color rendering properties is possible.

PHYSICAL PRINCIPLES

The high pressure metal halide discharge is a partially ionized plasma. The behavior of this plasma is determined by the combined effects of heat flow, mass flow, and radiation flow, as described below. The cylindrical discharge is maintained by an axial electric current carried mainly by electrons. The electrons gain energy from the applied electric field, and lose energy in elastic and inelastic collisions with other plasma particles. The energy dissipated in elastic collisions eventually reaches the wall by thermal conduction processes, and the energy dissipated in elastic collisions reaches the wall by radiation transfer processes. The metal halides exist as molecules in the cooler region near the wall, but are dissociated in the hot core. Metal atoms are partially ionized in the hot core, providing the electrons for the "working fluid" of the discharge. Throughout the plasma, charged particles and atomic species are diffusing radially outward, where they recombine to form molecules which diffuse radially inward to become dissociated again. In a vertically operating discharge, there is also axial convection, causing an upward mass flow in the hot core, and a downward mass flow near the wall.

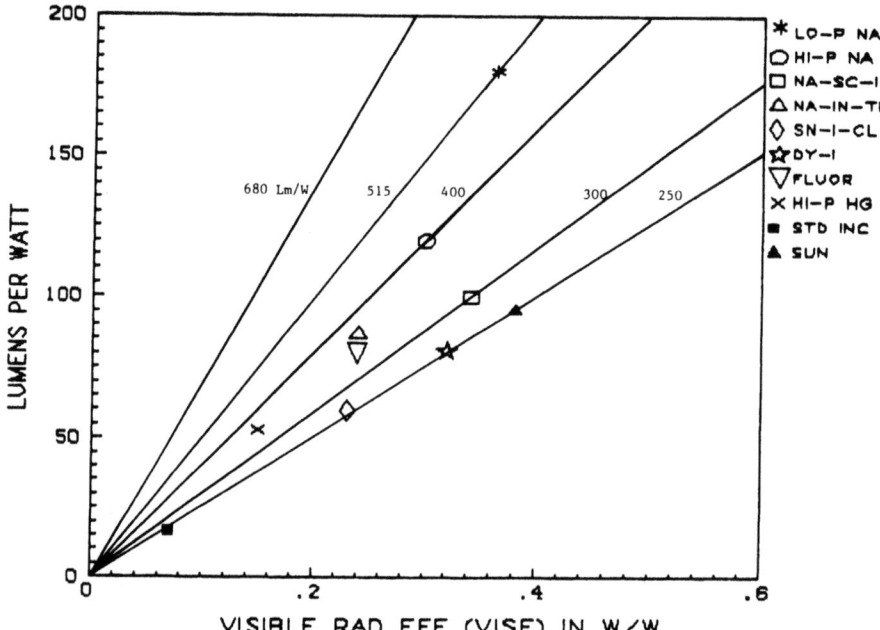

Fig. 3. Efficacy of various light sources showing contours of constant lm/W. Upper limit of 680 lm/W represents monochromatic source at wavelength of 555 nm. Lower limit of 250 lm/W represents continuous source (white light).

A qualitative description of these physical principles is given in this section. The concept of local thermodynamic equilibrium (LTE) is discussed first. When this assumption is invoked, the theoretical analysis of the discharge is simplified. The simplification occurs because fewer equations are required to described the discharge. The basis of the LTE assumption is that all plasma particles have the same temperature, and are related to each other by known functions of temperature, which in the context of this paper will be called equilibrium constants. In other words, the assumption of LTE avoids the necessity of solving rate equations for each species; it is necessary only to solve an energy balance equation for the temperature distribution in the discharge. After the concept of LTE is discussed, the major factors which determine the temperature profile in the metal halide discharge are discussed. Finally, mass flow effects and their influence on the discharge behavior are described.

Local Thermodynamic Equilibrium

A concept which is basic to the discussion of high pressure discharge light source is that of local thermodynamic equilibrium (LTE). Conditions which must obtain for a discharge to be in LTE are the following: 1) electrons and heavy particles (atoms, etc.) have the same temperature T; 2) densities of all species, including excited states, have values approximately equal to the corresponding thermodynamic equilibrium values, which are functions of temperature only, and do not depend on kinetic rate coefficients for population and depopulation.

For Condition 1) to obtain, it is necessary that the rate of energy transfer between electrons and atoms due to elastic collisions be large compared with the Joule heating rate. In mathematical form, this condition is expressed by the relation

$$3(m_e/m_a) \nu_e kT/e\mu_e E^2 \gg 1 \qquad (4)$$

where the symbols have the following meaning: m = mass, ν = elastic energy transfer collision frequency, k = Boltzmann constant, μ = mobility, E = electric field, e = electronic charge, subscript for electron, a = subscript for atom.

It has been shown, for example, that this inequality is satisfied in the core of the high pressure mercury discharge (Elenbaas, 1951) to within about one part in four hundred, i.e., the value of the left hand side of this inequality is about 0.005 for the low pressure Ar-Hg discharge, reflecting the large difference between electron temperature and gas temperature observed for this discharge.

For Condition 2) to obtain, it is necessary that the rate of energy transfer between electrons and atoms due to inelastic collisions (e.g., excitation, deexcitation by electron impact) be large compared with the rate of depopulation of the upper state by other processes, such as ambipolar diffusion when the upper state is a positive ion, or photon emission when the upper state is an excited atom. For example, in the high pressure mercury discharge, the LTE assumption leads to the following relation between electron density, ion density, mercury atom density, and temperature (Fowler, 1955).

$$N_e N_i / N_a = (w_e w_i / w_a)(2\pi mkT/h^2)^{3/2} \exp(-eV/kT) = K(T) \qquad (5)$$

where the symbols have the following meanings: N = density, w = statistical weight, h = Planck constant, V = ionization potential of mercury atom, i = subscript for ion, K = equilibrium constant for reaction $N_e + N_i \longleftrightarrow N_a$.

This relation is valid provided that the loss of ions by ambipolar diffusion is small compared with the loss of recombination, i.e., provided that

$$D/\Lambda^2 \alpha N_e^2 \ll 1 \qquad (6)$$

where the symbols have the following meanings: D = ambipolar diffusion coefficient, Λ = characteristic diffusion length of the discharge, α = recombination coefficient.

This inequality is satisfied in the hot core of the discharge where the electron density is high, but probably is not satisfied in the cooler region near the wall. Similarly, the density of excited state j is related to the gas temperature in equilibrium by the equation

$$N_j = (w_j/w_a)N_a \exp(-eV_j/kT) \tag{7}$$

where V is now the energy of excited level j, provided that the loss of excited species by photon emission and diffusion is small compared with the loss due to superelastic collisions with electrons. As with ionization and recombination, this provision is satisfied in the hot core of the discharge.

Radial Temperature Profile

The radial temperature profile in a high pressure metal halide discharge is somewhat more constricted than that of high pressure discharges in mercury and sodium. In all of these discharges, of course, the temperature profile is established by a balance between power input per unit volume in the form of Joule heating, and power output per unit volume in the forms of thermal conduction, radiation, and, in the case of the metal halide discharge, flow of dissociation energy. When the assumptions of LTE and axial uniformity are made, and the flow of dissociation is neglected, the temperature T(r) in a cylindrical discharge is determined by the energy balance equation

$$\sigma(T)E^2 = -(rK(T)T')'/r + R(T) \tag{8}$$

where the primes mean differentiation with respect to radius r, and where the electrical conductivity σ, thermal conductivity K, and net radiation source R all depend explicitly as well as implicitly on the temperature T through the dependence of these quantities on species densities. The net radiation source R at point r is the difference between the total radiation emitted at point r and the radiation absorbed at point r due to that emitted at all other points in the discharge.

This form of the energy balance equation was used by Elenbaas (1951) to analyze the high pressure mercury discharge. For the net radiation source, Elenbaas assumed that radiation transfer of the resonance lines could be neglected, and that the net radiation source could be approximated by a single line emanating from a lumped energy level near 8 eV above the ground state. In other words, he assumed that R(T) could be expressed by the relation

$$R(T) = (C/T) \exp(-90500/T) \tag{9}$$

where C is a constant with a value depending on the mercury pressure. This form of R(T) is appropriate for optically thin radiation. Solving the energy balance equation with these approximations led Elenbaas to a nearly parabolic temperature profile between an axis temperature of about 6000K and a wall temperature of about 800K for a 0.88 atm mercury discharge 4.1 cm in diameter operating at 35 W/cm, a result which agreed fairly well with measurement.

An interesting conclusion about wall-stabilized discharges can be drawn from solutions of the energy balance equation given above (Waymouth, 1971). Wall-stabilized discharges result when the radiation source R varies more strongly with temperatures than does the electrical conductivity σ, and constructed discharges vice-versa. The high pressure mercury discharge is an example of such a wall-stabilized discharge; the temperature dependence of the electrical conductivity is governed primarily by the factor

$$\sigma \propto \exp(-eV_i/2kT) \text{ where } V_i = 10.4 \text{ eV} \tag{12}$$

and that of the net radiation source primarily by the factor

$$R \propto \exp(-eV_r/kT) \text{ where } V_r = 7.8 \text{ eV} \tag{11}$$

Therefore, the second factor increases more rapidly with T than the first. According to the energy balance given above, the curvature of T(r) on the axis of the discharge, i.e., the second derivative T"(0), is proportional to the difference between power input per unit volume and the net power radiated per unit volume. Now the quantity T"(0) must be negative for a steady state discharge. Therefore, when R increases faster than σ with rising temperature due to increased power input, T"(0) becomes less and less negative, i.e. the temperature profile becomes flat on the axis of the discharge, and the axis temperature increases little or none at all with increasing power input. All that happens is that the discharge becomes "fatter" as power input is increased further. The transition from a parabolic temperature profile to a flattened profile coincides with the minimum in the curve of electric field versus power input noted by Waymouth (1971). In other words, the volt-ampere characteristic of the high pressure mercury discharge becomes positive (Elenbaas, 1951) when the temperature profile begins to flatten, primarily because the axis temperature does not increase. This effect is observed in some metal halide discharges, and not in others, accordingly as the dominant energy level for radiation is greater than or less than one-half the ionization potential (Waymouth, 1971).

While Elenbaas (1951) found acceptable agreement between calculation and measurement by neglecting the optically thick resonance lines of mercury, calculations by Fischer, 1974, led him to the conclusion that radiation transfer of the optically thick resonance lines of mercury should not be neglected. Fischer's calculations result in a profile more closely resembling a parabola than those of Elenbaas. Similarly, calculations of the temperature profile in high pressure sodium discharges result in a parabolic shape (Lowke, 1969; de Groot, 1974).

An important conclusion to be drawn from these calculations is that optically thick radiation, such as that encountered in high pressure mercury and sodium discharges, tends to "fatten" the temperature profile. This "fattening" occurs because part of the radiation emitted from the hot core of the discharge is absorbed in the cooler outer region, creating a heat source there. In other words, the net radiation source R becomes negative, indicating more absorption than emission.

In the metal halide discharge, the effect of absorption in the cooler outer region is much less pronounced, so that the net radiation source does not become negative. This result is consistent with the narrower temperature profile observed for metal halide discharges. The degree of narrowness, often referred to as constriction, is illustrated in Fig. 4, which shows temperature profiles measured for a tin iodide discharge (Fischer, 1974) and for a sodium-scandium iodide discharge (Keeffe et al., 1978), along with a parabola for comparison.

By including in the energy balance equation appropriate terms to describe the radial transport of energy due to dissociation and recombination of molecules, Fischer (1974) was able to show satisfactory agreement of the important parameters in the metal halide discharge, including the temperature profile.

Radial Demixing

Metal halide sources are based on the principle that the metal halide molecule is not dissociated in cooler gas near the wall, and is more or less completely dissociated in the hot core of the discharge, allowing the excited metal species to radiate its characteristic spectrum. In the case of a diatomic molecule AB, each dissociation event results in the production of one atom of A and one atom of B, and the loss of one molecule of AB. In view of the equal production rates of the atomic species, it is tempting to assume that the densities of species A and B are equal everywhere in the discharge. However, because species A

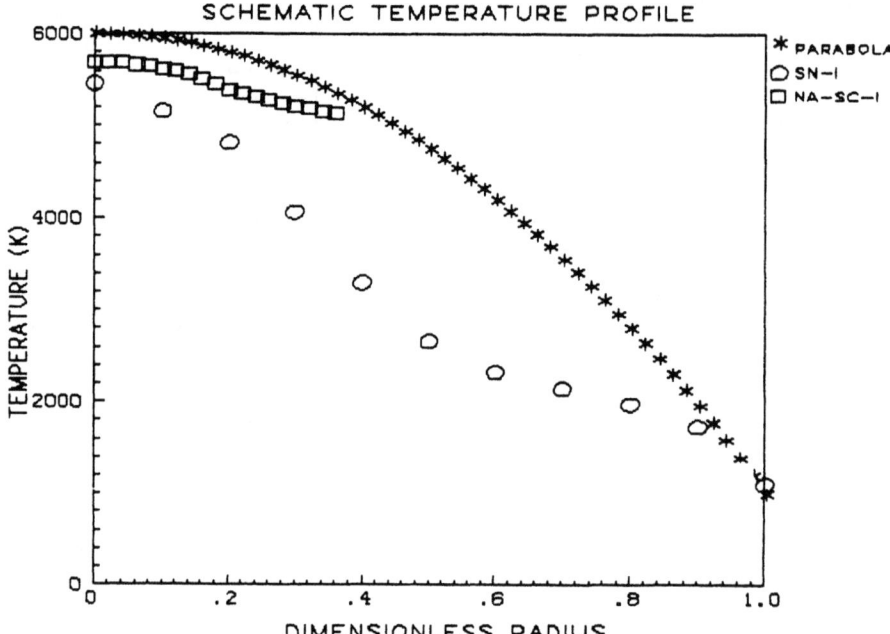

Fig. 4. Temperature profiles typical of metal halide sources: circle---tin iodide (Fischer, 1974); square---sodium-scandium iodide (Keeffe et al., 1978); cross---parabola.

and B generally have unequal diffusion coefficients, the assumption of equal densities is wrong. The term "radial demixing" is commonly used to describe the situation where the radially diffusing species have different diffusion coefficients, causing unequal densities of the dissociated species, even though LTE obtains.

For a steady state to obtain, it is necessary that the two species of dissociated atoms flow toward the wall at the same rate. Consequently, the density gradients of the two species are different if the diffusion coefficients are different. The species with the higher diffusion coefficient will have the smaller gradient in density. A simplified model suffices to illustrate the concept of radial demixing. based on the assumption that the densities of species A, B, and AB are much less than the density of mercury, so that the binary diffusion coefficients depend only on the mercury density, conservation of species mass and momentum dictates the following relation:

$$D_A P'_A + D_{AB} P'_{AB} = 0 \qquad (12)$$

where P is partial pressure. Assuming further that the diffusion coefficients have the same temperature dependence, this equation can be integrated to give the relation

$$P_A(r) + (D_{AB}/D_A) P_{AB}(r) = P_A(a) + (D_{AB}/D_A) P_{AB}(a) \qquad (13)$$

where a is discharge radius. A similar relation can be derived for species B. When the atomic densities are much less than the molecular density at the wall, it can be shown from these equations that the ratio of atomic densities in the hot core of the discharge is given by the relation

$$N_A(0)/N_B(0) = D_B/D_A \tag{14}$$

In addition, it follows that when the species AB is completely dissociated in the core of the discharge, then the partial pressure of the atomic species on the axis is related to the partial pressure of the molecular species at the wall by the equation

$$P_A(0) = (D_{AB}/D_A) \; P_{AB}(a) \tag{15}$$

In halide mixtures, the equations are more complex, but the physical principle is the same: the species diffusion coefficients determine the species densities in the discharge core, even though LTE obtains.

Axial Convection and Segregation of Species

In a high pressure mercury discharge operating in the vertical position, there is an upward flow of gas in the hot core of the discharge, and a downward flow in the cooler region near the wall. This mass flow, an example of natural convection, is due to the fact that the gas density is less in the hot core than near the wall. The convection process in the high pressure mercury discharge was first analyzed theoretically by Elenbaas (1951). Assuming a cylindrical discharge in an infinitely long tube, Elenbaas derived the radial distribution of axial convection velocity by solving the momentum balance equation, or Navier-Stokes equation, for a fluid in the presence of gravity. A dimensional analysis shows that the convection velocity near the axis of the discharge is proportional to the mercury density, the square of the discharge tube radius, and the gravitational constant, and inversely proportional to the viscosity of mercury vapor. The mathematical solution of the momentum balance equation gives a value for the axial convection velocity of 10-100 cm/sec on the axis of the discharge. For example, the radial distribution of axial convection velocity calculated by Elenbaas is shown in Fig. 5.

The convective flow pattern in the metal halide discharge is similar to that in the mercury discharge, because mercury is the most prevalent species in both. There are minor differences in temperature profiles and gas compositions, but the main features which determine the convection velocity are the same for the two discharge types. In the metal halide discharge, however, axial convection can have a profound influence on the axial distribution of the dissociated metal halides, causing axial variation in emission. The convective flow up the center and down the periphery of the discharge sweeps other molecules and atoms along with the mercury atoms. At the lower end of the discharge, metal halide molecules enter the hot core, where they become dissociated into metal and halogen atoms. At the same time, these atoms are diffusing with respect to the upwardly streaming mercury atoms. Different atomic species have different diffusion coefficients, as discussed above in the section on radial demixing, with lighter metallic elements having larger diffusion coefficients than the halogens. Consequently, the upper end of the discharge becomes depleted of the metallic species because radial diffusion carries them out of the upward stream and into the downward stream. This effect can be so pronounced that the radiation coming from the upper end of the discharges composed of mercury lines only, because there are no other metal atoms present (de Groot and Jack, 1973; Zollweg et al., 1975).

A comprehensive model which takes into account these physical principles (Stormberg, 1981) predicts that the density of metallic species falls exponentially in the direction of convective flow along the axis of a discharge containing sodium iodide and thallium iodide, in agreement with measurement. The rate of exponential decrease was found to be small (little segregation) for both small and large convection velocities. For

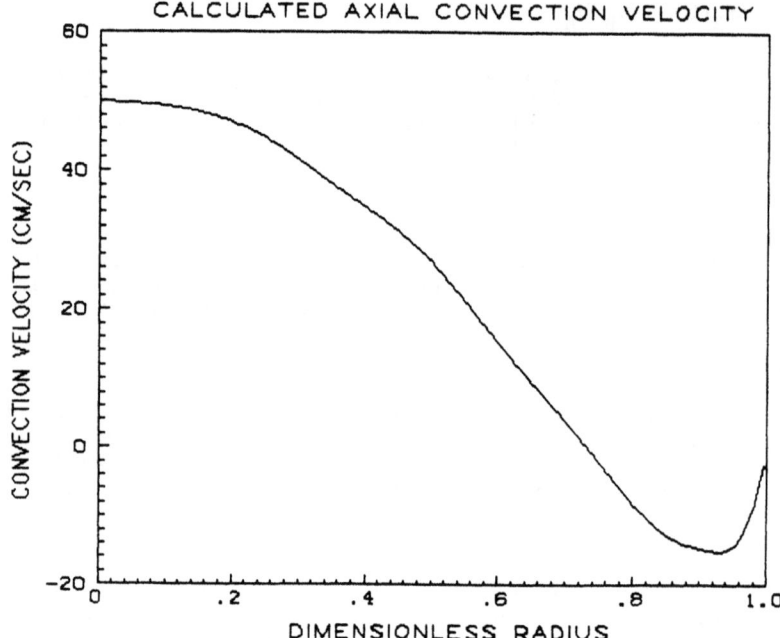

Fig. 5. Axial convection profile calculated for high pressure mercury discharge (Elenbaas, 1951). Mass balance is ensured by this profile, i.e., total mass of mercury flowing up in central region is equal to total mass flowing down in cool gas near wall.

intermediate values, the amount of segregation was found to be larger for lighter atoms. In addition to concentration diffusion, Stormberg found it necessary to include thermal diffusion (Chapman and Cowling, 1970) and ambipolar diffusion to get good agreement between calculation and measurement. Stormberg found it useful to use the following approximation for species densities in his numerical analysis:

$$N_i(r,z) = N_i(r,z^*) \exp(-p_i(r,z)(z-z^*)) \qquad (16)$$

where z^* is the starting value of z on the vertical axis of the discharge, and p is called the segregation parameter. This intuitive form is similar to the rigorous solution of the diffusion equation with axial convection in the special case of plug flow in a pipe with a source of species i at z^*. In the case of plug flow, the segregation parameter has a position-independent value which decreases monotonically with increasing value of the ratio av/D, where a is pipe radius, v is (constant) convection velocity, and D is diffusion coefficient of the impurity species. In other words, there is less segregation when the convection velocity is large due to high pressure or large discharge radius, and when the diffusion coefficient is small due to high pressure or large mass, in qualititative agreement with Stormberg's findings. However, Stormberg also found less segregation at lower pressure and smaller radius, which cannot be explained by simple plug flow analysis, because the analogy breaks down in this limit. The variation of segregation parameter with pressure/radius is shown schematically in Fig. 6.

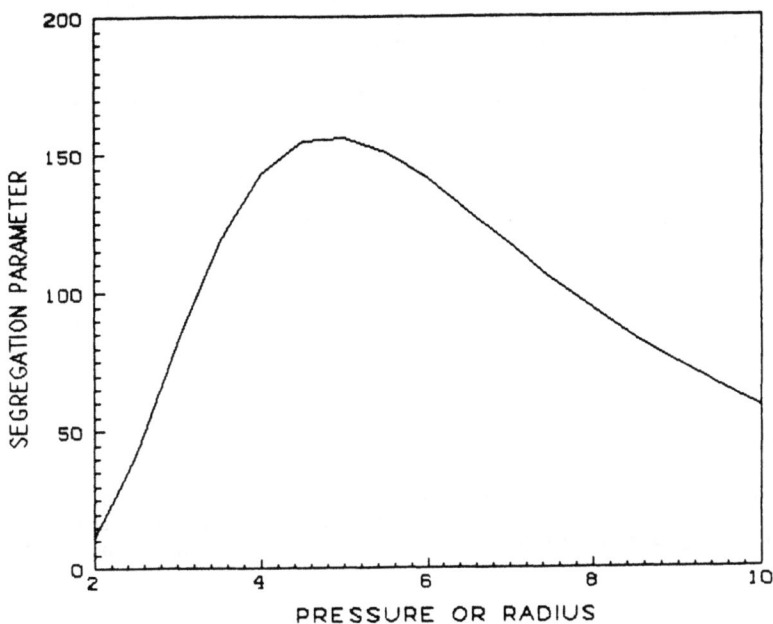

Fig. 6. Schematic variation of axial segregation parameter with pressure or radius of discharge (Stormberg, 1981) in metal halide discharge.

SUMMARY

The metal halide discharge light source is composed of mercury vapor at a partial pressure of several atmospheres, and one or more metal halide compounds at partial pressures in the range of 0.1 to 100 torr. The discovery of this light source with high efficacy and good color rending properties came as a result of a desire to improve the color rendering properties of the high pressure mercury discharge light source. By necessity, the search for means to improve the mercury discharge light source focused on the problem of finding new species to add to the discharge which would radiate in that part of the visible spectrum not filled by mercury, and which would not react chemically with the discharge container. Certain metal halide compounds meet these two important requirements by existing in the molecular state in the vicinity of the discharge container and in the atomic state in the vicinity of the hot core of the discharge where the radiation originates.

The important characteristics which distinguish metal halide discharge light sources from other discharge sources have been reviewed in this paper. The most important of these distinguishing features is the property of color rendition. Of all discharge light sources, the metal halide types have the best color rendition. The price paid for good color is an efficiency of conversion of electrical energy into light which is lower than that of other discharge sources with poorer color rendering properties. Generally speaking, discharges used for light production convert no more than one-third of the input power into visible radiation. It is suggested in this paper that future efforts to improve efficiency of light production by means of gas discharges will be directed to increasing the fraction of input power radiated in the visible spectrum. This suggestion is based on the observations that the most efficient metal halide sources have near-optimum spectral power distribution in the visible spectrum, about 300 m/W, and that further increases can be made only at the expense of color rendition.

In addition to distinguishing characteristics of the light emitted by metal halide discharges, there are many physical and chemical characteristics which differentiate metal halide sources from other sources. Some of the important physical features peculiar to metal halide discharges described in this paper include constricted temperature profile and non-uniform spatial distribution of radiating species caused by axial convection and radial diffusion. A review of published calculations dealing with metal halide discharges leaves the impression that these special discharges are understood qualitatively from a physical point of view, but not quantitatively because of incomplete physical and chemical data, such as diffusion coefficients, thermodynamic properties, especially at high temperature, etc.

ACKNOWLEDGEMENT

The author wishes to acknowledge the support and encouragement of the management of the Lighting Research and Technical Services Operation of the General Electric Lighting Business group during preparation of this work.

REFERENCES

Chapman, S. and Cowling, T. G., 1970, "The Mathematical Theory of Non-Uniform Gases," University Press, Cambridge.
Drop, P. C., Fischer, E., Oostvogels, F. and Wesselink, G. A., 1975, Philip. Tech. Rev.,35:347.
Elenbaas, W., 1951, "The High Pressure Mercury Vapour Discharge," North Holland, Amsterdam.
Fischer, E., 1974, J. Appl. Phys., 45:3365.
Fowler, R. H., 1955, "Statistical Mechanics," 2nd ed., University Press, Cambridge.
de Groot, J. J., 1974, PhD Thesis, Technische Hogeschool, Eindhoven.
de Groot, J. J. and Jack, A. G., 1973, J. Phys. D., 6:1473.
Hirschfelder, R. O., Curtiss, C. F. and Bird, R. B., 1954, "Molecular Theory of Gases and Liquids," Wiley, New York.
Kaufman, J. E., ed, 1981, "IES Lighting Handbook," Illuminating Engineering Society, New York.
Keeffe, W. M., Morris, J. C., and Walter, W., 1978, J. Illum. Engr. Soc., 7:249.
Keeffe, W. M., 1980, Proc. IEE, 127:181.
Larson, D. A., Fraser, H. D., Cushing, W. V. and Unglert, M. C., 1963, Illum. Engr., 58:434.
Lowke, J. J., 1969, J. Quant. Spect. Rad. Trans., 9:839.
Martt, E. C., Smialek, L. J. and Green, A. C., 1964, Illum. Engr., 59:34.
Reiling, G. H., 1964, J. Optical Society of America, 54:532.
Speros, D. M., Caldwell, R. M., Smyser, W. E., Springer, R. H. and Taylor, R. P., 1970, Illum. Engr., 65:641.
Stormberg, H. P., 1981, J. Appl. Phys. 52:3233.
Waymouth, J. F., 1971, "Electric Discharge Lamps," MIT Press, Cambridge, MA.
Zollweg, R. J., 1975, J. Illum. Engr. Soc., 5:12.

LASER-PRODUCED PLASMAS

C. Grey Morgan

Department of Physics
University College of Swansea
Swansea, Wales

INTRODUCTION

The invention of the laser has led to the generation of entirely new plasma phenomena, the first of which to be described was the production of a plasma in a gas by the action of light alone unaided by any other applied electric field.

The discovery of the new plasma form was made almost simultaneously by Marker and his colleagues (1963) and described at the Third International Congress on Quantum Electronics held in Paris. Meyerand and Haught (1963) in the USA, followed quickly by Zeldovich and Raizer (1964) in the USSR, gave accounts of some of the properties of these novel plasmas.

Thus gases initially almost perfectly insulating and transparent to low intensity radiation in the visible part of the electromagnetic spectrum can be transformed into an opaque, highly conducting, self-luminous plasma in times of the order of a few tens of nanoseconds by the action of intense radiation of the same wavelength from a Q-switched ruby laser.

The phenomenon raised some puzzling questions for it was not immediately obvious how the relatively low-energy ruby-laser photons ($h\nu$ = 1.78 eV) could interact with gases having excitation and ionization potentials of about 15 eV, and how they could produce plasmas with electron densities $>10^{13}$ cm^{-3}, with electron energies of several eV or even kV, and provide strong local magnetic fields within the plasmas. Nevertheless, such plasmas were readily produced.

Almost simultaneous reports of electron and ion emission and of plasmas created at the surfaces of solids both in vacuo and in the presence of gases by free-running and Q-switched lasers were made (Honig, 1963; Isenor, 1964). As in the case of plasmas created in gases, strong magnetic fields were detected in these plasmas (Askar'yan et al., 1967; Schwirzke, H., 1974).

These discoveries of new plasma forms and the unravelling of the physical processes responsible for them had significant technological consequences. They have, for example, led to the development of higher power, shorter-flash-duration lasers themselves. They have been used to simplify the rapid switching of very high voltages and large currents by means of laser-triggered spark gaps. They have resulted in novel spectroscopic techniques and have had a seminal and catalytic effect on the development of new kinds of optical reflectors which possess quite astonishing properties, and they offer some promise in achieving the goal of controlled thermonuclear fusion.

It is useful to note some of the orders of magnitude of the electrical vector amplitude E, photon flux F, and radiation pressure R_p which can be generated by a laser beam. The electric field E is related to the beam intensity I by Poynting's theorem as:

$$E \text{ (V/cm)} = 19.4 \, I^{1/2} \quad \text{(when I is expressed in W-cm}^{-2}\text{)}. \tag{1}$$

The flux F is related to I by

$$F = 6 \times 10^{18} \, (I/h\nu) \quad \text{(with } h\nu \text{ in eV)}, \tag{2}$$

and the radiation is given by

$$R_p = 3.33 \times 10^{-11} \, I \, (J/cm^3) = 3.3 \times 10^{-10} \, I \text{ atm}. \tag{3}$$

Table 1 shows values of these parameters assuming a laser with $h\nu \sim 1$ eV.

VISUAL CHARACTERISTICS OF LASER-PRODUCED PLASMAS

The simplest type of laser-produced plasmas is that created in the air by a focused Q-switched laser beam. This plasma is characterized by a brilliant flash of bluish-white light at the lens focus, accompanied by a distinctive cracking noise typical of violent hydrodynamic effects as shock waves propagate in the surrounding air. In certain circumstances more than one discrete plasma may be created by a single laser flash.

The physical appearance of the plasmas in a gas can take many forms depending on the laser beam dimensions, the focusing optics, and the spatial and temporal variation of the laser flash. Fig. 1 shows two distinct plasmas produced in the atmosphere by a single 30-nsec flash from a Q-switched ruby laser focused by a simple biconvex 3-cm lens to create an intensity $I \sim 10^{-11}$ W/cm^2. It was puzzling enough to understand why any plasma should be created at all but it was even more puzzling to explain why there should be more than one.

Intuitively one might guess that a region occupied by a plasma corresponds to a region where a large electric intensity exists capable of causing ionization growth, and a region of no plasma implies low or zero electric intensity. Such regions could correspond to an interference pattern in which constructive interference gives large radiation intensity and destructive interference yields low or zero intensity.

Table 1. Parameters produced by laser beam

I, W-cm^2	E, V-cm^{-1}	F, quanta cm^{-2} sec^{-1}	R_p (Atm.)
10^8	$\sim 2 \times 10^5$	$\sim 5 \times 10^{26}$	3.3×10^{-2}
10^{12}	$\sim 2 \times 10^7$	$\sim 5 \times 10^{30}$	3.3×10^2
10^{14}	$\sim 2 \times 10^8$	$\sim 5 \times 10^{32}$	3.3×10^4

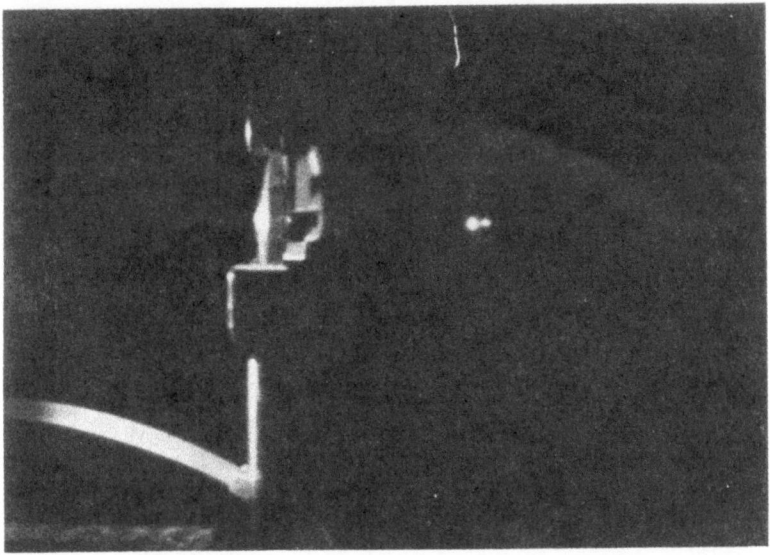

Fig. 1. Two plasmas in air by 30-nsec flash of 694.3-nm radiation.

The dual plasmas shown in Fig. 1 may be obtained using large (one or two cm) diameter laser beams and relatively short focal-length lenses (3 to 5 cm). Thus large amounts of primary spherical aberration were present and this, together with the use of monochromatic coherent laser radiation, leads to interference effects which cause a complex distribution of intensity in the focal region.

Steady-state computations of the interference pattern (Evans and Grey Morgan, 1968, 1969; Ireland et al., 1974), based on the Fresnel-Huygens-Kirchoff equation, for Gaussian beams of various truncation ratios, and with "top hat" profiles focused by a simple biconvex spherical-surfaced lens with varying degrees of aberration function, reveal the presence of several zones of high intensity separated by regions of lower intensity lying along the beam axis. The interference pattern becomes more and more complicated as the aberration function $\phi(\lambda)$ is enlarged by increasing the diameter D of the beam at the lens, or reducing the focal length f, since

$$\phi(\lambda) \propto D^4/f^3. \tag{4}$$

($\phi(\lambda)$ is the distance, measured in wavelengths, between the ideal spherical wavefront and the real front).

Fig. 2 shows the distribution of isophotes (lines of constant intensity) for a Gaussian beam focused by a perfect optical system having zero aberrations. There is one maximum located at the junction of the focal plane and optic axis. The distribution of intensity around the axis at this plane corresponds to the Airy diffraction pattern of the circular aperture formed by the lens. In contrast Fig. 3 shows how the ideal distribution of Fig. 2 becomes distorted by the introduction of six wavelengths of aberration.

The isophotes are drawn in "optical space" with coordinates u and v related to real space coordinates z (along the optic axis) and r (radial, orthogonal to the optic axis) by

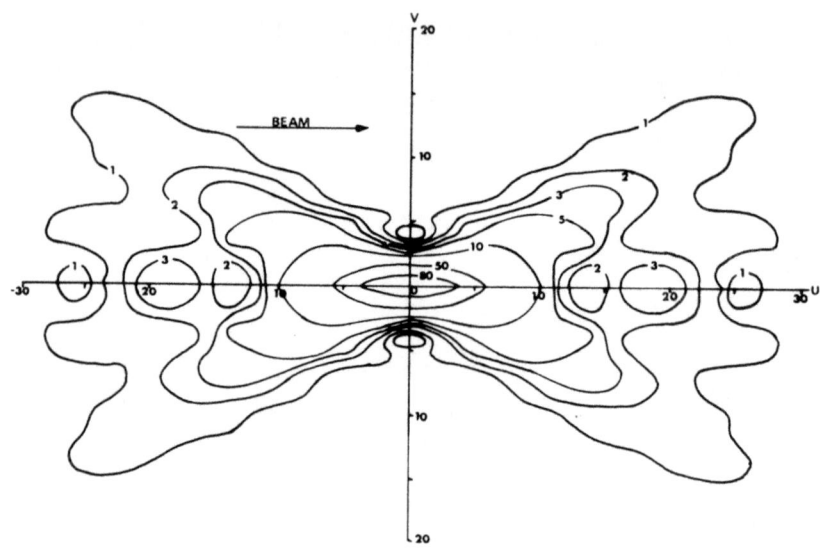

Fig. 2. Isophotes for a gaussian beam focussed by a perfect optical system. $\phi_{(max)}(\lambda) = 0$.

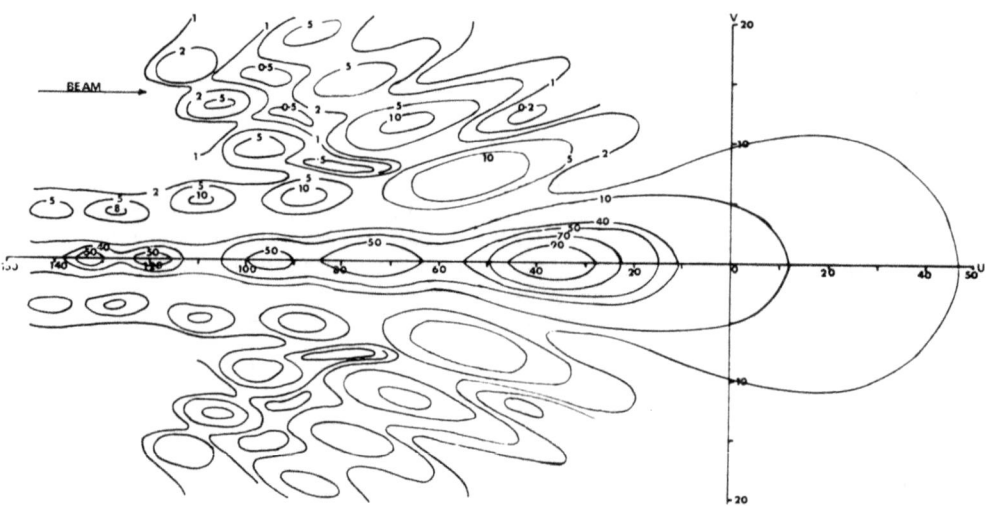

Fig. 3. Isophotes for a gaussian beam focussed by a lens system with $\phi_{max} = 6\lambda$.

$$u = (2\pi/\lambda)(D/f)^2 z \text{ and } v = (2\pi/\lambda)(D/f) r. \tag{5}$$

The v scale is expanded compared to the u scale by the ratio (f/D) for clarity.

As ϕ increases, the degree of concentration of the beam by the lens diminishes, and becomes extended along the optic axis to form multiple foci at each of which plasmas may form if the local intensity is sufficiently large. This is the explanation of the multiple plasmas generated by Q-switched lasers. Similar aberration effects can occur when laser beams are focused using mirrors which are not perfectly spherical, and equally complex energy and intensity distributions are formed along the beam axis. The spatio-temporal development of the interference patterns and isophotes becomes of importance with increasingly shorter duration flashes (~ psec) (Evans, 1976).

Even when the interference effects introduced by spherical aberration and coma are avoided by using well-aligned beams with aspheric lenses it is still possible to produce multiple colinear plasmas if a train of intense pulses from a mode-locked laser is used, as shown in Fig. 4, which illustrates plasmas produced by picosecond flashes from a ruby laser. By interposing a narrow-band filter over the camera lens which transmits only 0.6943μ radiation, individual plasmas created by successive flashes in the train are clearly visible. The millimeter scale shows the size of these plasmas, which is time-resolved to a duration of the length of the pulse train of ~ 250 nsec.

In this succession of plasmas each individual plasma is created by an individual flash in the mode-locked train. The initiatory electron required for each plasma is liberated by the UV radiation from free-free and free-bound transitions in the preceding plasma as Meyer and Stritzke (1976) have shown.

With exceptionally powerful beams, and relatively long wavelength radiation, such as that from TEA carbon dioxide lasers, it is possible to form plasmas over extended lengths - small discrete plasmas distributed irregularly along tens of meters in dusty laboratory air. Each plasma nucleates at microparticulate matter and aerosols in the air.

PLASMA-FORMATION MECHANISMS

Since the first reports on laser-produced plasmas there have been some thousands of papers and several books (e.g., Ready (1971), Hughes (1975), and Raizer (1977)) written on their optical, magnetic, and thermodynamic properties. There is thus a strong body of evidence which explains the initiation and growth of plasmas in terms of multiphoton absorptions leading to a form of stepwise ionization via a series of virtual and allowed optical transitions, and of inverse bremsstrahlung absorption involving absorption of the laser beam photons by a free electron in the presence of a third body - say an atom or an ion. This latter process leads to avalanche growth of the free electrons and ion concentrations in the same manner observed under static and microwave fields applied to gases. It requires a free electron in the lens focus when the flash occurs to initiate the cascade growth process.

MULTIPHOTON ABSORPTION

The laws governing multiphoton ionization processes and their formal dependence on beam intensity, wavelength, polarization, and longitudinal mode structure, and upon the electronic configuration and concentration of the irradiated atoms, have been extensively studied in recent years, mainly at Saclay and the Lebedev Institute, and are the subject of reviews by Bakos (1972), Delone (1975), Grey Morgan (1975), and Eberly and Lambropoulous (1978).

An elementary analysis shows that the probability W that a beam of intensity I or photon F, composed of identical quanta of energy $h\nu$, will ionize an atom of ionization potential E_i in unit time is given by
$W = AF^k = A' I^k$, where A and A' are constants for a given atomic species and radiation wavelength λ. Here k is the next integer larger to $E_i/h\nu$.

Fig. 4. Plasmas created in the laboratory atmosphere by a mode-locked train of pulses at 694.3 nm.

The constants A and A' depend upon the beam polarization and involve the photon absorption cross section σ into a virtual state. In idealized and simplest form A can be expressed as $A = \sigma^k/\nu^{(k-1)} (k-1)!$, so that the number of electron-ion pairs created by irradiating a volume V by a spatially uniform, temporally constant flux F for a time τ is

$$N_i(\tau) = N_o p V \tau F^k \sigma^k / \nu^{k-1} (k-1)!, \qquad (6)$$

where $N_o = 3.56 \times 10^{16}$ and p is the gas pressure in torr. Each atom is ionized individually with no inter-atomic interactions, i.e., the only collisions involved are those between the photons and atoms.

If plasma formation is taken to have occurred when an agreed critical number N_c (arbitrarily taken as $> 10^{13}$ cm^{-3}) of electrons and ions are generated then the threshold flux is

$$F_{th} = (\nu/\sigma) \left[\frac{N_c (k-1)!}{N_o p V \tau \nu} \right]^{1/k}, \qquad (7)$$

and varies with pressure as $p^{-1/k}$. For noble gases illuminated with ruby or glass lasers k lies between 7 for Xe and 21 for He so the pressure dependence is very weak. This is exactly what is observed in practice. The intensities required to produce a single electron or a plasma with $>10^{13}$ electrons in a gas solely by muliphoton absorption differ by very little because of the strong k^{th} power dependence, and are $I_{th} \sim 10^{13}$ w-cm^{-2}.

INVERSE BREMSSTRAHLUNG ABSORPTION - CASCADE IONIZATION

If a free electron in a gas undergoes scattering collisions with atoms during the period of laser irradiation it may gain sufficient energy to excite and ionize the atoms so that plasma formation can proceed by an avalanche or cascade process. However, in this case the rate of ionization growth and the threshold intensity needed to form a plasma are sharply pressure dependent, in contrast to the weak pressure dependence characteristic of multiphoton processes. The ionization frequency ν_i is given by Morgan et al. (1971)

$$\nu_i/N = \left[\frac{377q}{\omega^2}\left(\frac{\nu_m}{N}\right)^2\right] I(t), \tag{8}$$

where ν_m and ω are the electron momentum transfer collision frequency and laser angular frequency, respectively, $N = pN_0$ is the atomic concentration, and q is related to the Townsend primary ionization coefficient α.

As a very rough rule of thumb for the separation of multiphoton ionization from inverse bremsstrahlung absorption regimes we may note that multiphoton effects dominate for values of the product of gas pressure p and flash duration τ less than $\sim 10^{-7}$ torr-sec. The transition from the multiphoton to inverse bremsstrahlung regimes is indicated by the discontinuity in Fig. 5, giving the I_{th}-vs-p dependence for helium, argon, and nitrogen illuminated by $\sim$ 50-psec flashes of ruby laser light (Krasyuk et al., 1970).

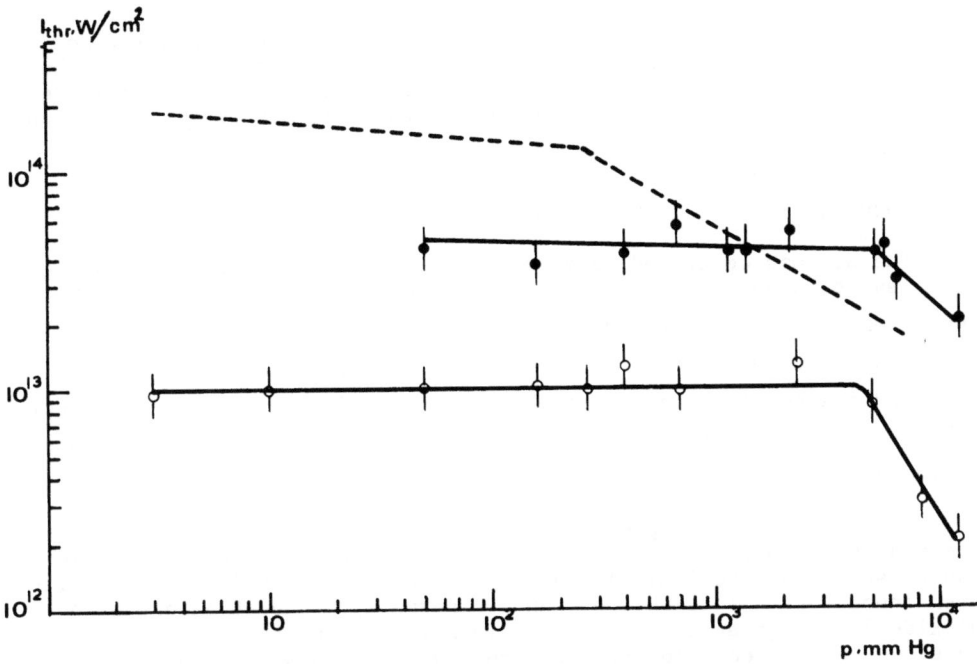

Fig. 5. Subnanosecond breakdown threshold intensities of helium (●), argon (o) and nitrogen (-----).

When plasmas are formed with long, i.e., nanosecond duration, flashes in gases at high (> 0.5 atmospheric) pressure, ionization loss processes compete with the generation processes of free electrons. Thus attachment, recombination, and diffusion effects can become significant and the exact pressure dependence of I_{th} is determined by the relative importance of these electron removal processes.

It is a relatively straightforward matter to derive expressions relating I_{th} to the coefficients describing these production and loss processes (Morgan et al., 1971) by solving an electron concentration continuity equation

$$\frac{dn}{dt} = nN\nu_i - D\nabla^2 n - Rn^2 . \qquad (9)$$

Thus, if diffusion is important the threshold intensity for plasma formation is

$$I_{th} = \frac{pN_o}{377q} \left(\frac{\omega}{\nu_m}\right)^2 \left[\frac{1}{\tau} n(2.3\delta pN_o V) + 2D\left(\frac{4.81}{d}\right)^2\right] , \qquad (10)$$

while at high pressures, when recombination might be the dominant loss process, the threshold intensity is

$$I_{th} = \frac{pN_o}{377q} \left(\frac{\omega}{\nu_m}\right)^2 \left[\frac{1}{\tau} n(\delta pN_o V) + R\delta N_o p \left(\frac{\omega}{\nu_m}\right)\left(\frac{\pi pN_o}{754q I_{th}\tau}\right)^{1/2}\right] . \qquad (11)$$

Here D and R are the electron diffusion and recombination coefficients, d is the focal spot diameter, and 2τ is the flash duration.

These expressions can be replaced by computer modeling solutions to give detailed information about the spatio-temporal development of the plasmas. Nevertheless, these simple equations give order-of-magnitude agreement with experiment and provide insight into the plasma formation mechanisms.

CONTINUOUS OPTICAL DISCHARGES (COD)

The plasmas described above are all of a transient nature lasting some tens of nanoseconds or microseconds, and requiring threshold intensities $> 10^{11}$ w/cm^2 for gases at atmospheric pressure, when generated by flashes from Q-switched ruby or glass lasers, or $> 10^9$ w/cm^2 for infrared flashes from gain-switched carbon dioxide lasers. Laser powers of some tens or hundreds of megawatts are thus needed, largely because of the τ^{-1} dependence of I_{th}.

However, the development of powerful cw carbon dioxide lasers has removed this limitation and another type of laser-induced plasma which can be sustained indefinitely has been observed. This is the so called "continuous optical discharge", and is normally produced using carbon dioxide lasers generating about a kilowatt of 10.6μ radiation - approximately 2 kw is needed to provide a continuous plasma in atmospheric pressure air. While studies of these continuously sustained plasmas are far less numerous than pulsed plasmas they have nevertheless been investigated thoroughly for a variety of gases during the last fifteen years by several workers, for example, Generalov et al. (1970), Smith and

Fowler (1973), and notably by Uhlenbusch and his colleagues at Dusseldorf (Carlhoff et al., 1981; Uhlenbusch, 1983). COD's involve the same energy transfer mechanisms from the laser beam to the gas via inverse bremsstrahlung absorption and electron removal processes due to diffusion, attachment, and recombination, as in pulsed plasmas, but additional energy dissipation processes, such as conduction and convective flow, are involved in these longer-lived plasmas.

Several theoretical treatments have been given to account for the characteristics of COD's, e.g., Muller and Uhlenbusch (1982). Good agreement between the calculated minimum power $P_{L_{min}}$ needed to sustain a CW plasma, based on an analysis of electron creation by inverse bremsstrahlung absorption and removal by ambipolar diffusion, radiative recombination, and three-body recombination, and experiments using argon, helium, and hydrogen have been reported (Uhlenbusch, 1983). A comparison of calculated and measured values of $P_{L_{min}}$ are shown in Fig. 6.

The COD plasmas are generally characterized by a low electron temperature, ~ 1 to 2 eV, but a higher electron concentration, ~10^{18} cm^{-3}, and moderate degree of ionization.

SPECTROSCOPIC APPLICATIONS

Quite apart from being of fundamental interest laser-produced plasmas are of considerable practical importance. Among their many applications their use in spectro-chemical analysis is of growing significance and is particularly noteworthy for the examination of impurities in metals and inspection of fuel elements in-situ in nuclear reactors (Klewe et al., 1972; Tozer, 1975, 1976; Adrain, 1980). Very recently a technique, which has become known as Laser-Induced Breakdown Spectroscopy (LIBS), has successfully been applied to the direct detection of dangerous elements, such as chlorine, fluorine, and beryllium, in the atmosphere. At Los Alamos, for example, beryllium, in as small a concentration as 0.7 $\mu g/m^3$, or 0.6 ng/gm of air, has been detected (Radziemski et al., 1983), in real time, in practical working conditions, using time-resolved spectral analysis of laser-induced air plasmas with a gated diode array and multichannel analyzer. This is a much more attractive and faster method than conventional absorption techniques.

LIBS holds considerable promise for the detection of elements in biomedical fluids such as serum and blood, the latter as an absorbed sample on filter paper. Some nineteen elements present at levels ~ 50 ng/m have been detected in preliminary experiments (Loree, 1983a). These results suggest that LIBS might well be developed into a technique for the rapid detection of trace elements in small samples of body fluids.

In an extension of this work at Los Alamos samples of perspiration are being used to create laser-induced plasmas and preliminary results look very promising for rapid real-time elemental detection for human medical purposes by a totally non-invasive, entirely safe process (Loree, 1983b).

Preliminary studies of COD plasmas as sources for spectrochemical analysis have recently been reported by Los Alamos workers (Cremers et al., 1985), and these plasmas too show considerable promise insofar as they are relatively easy to generate over a wide pressure and temperature range. As the plasma is entirely isolated from electrodes, which are required in some other spectroscopic sources, plasma sample contamination is minimized.

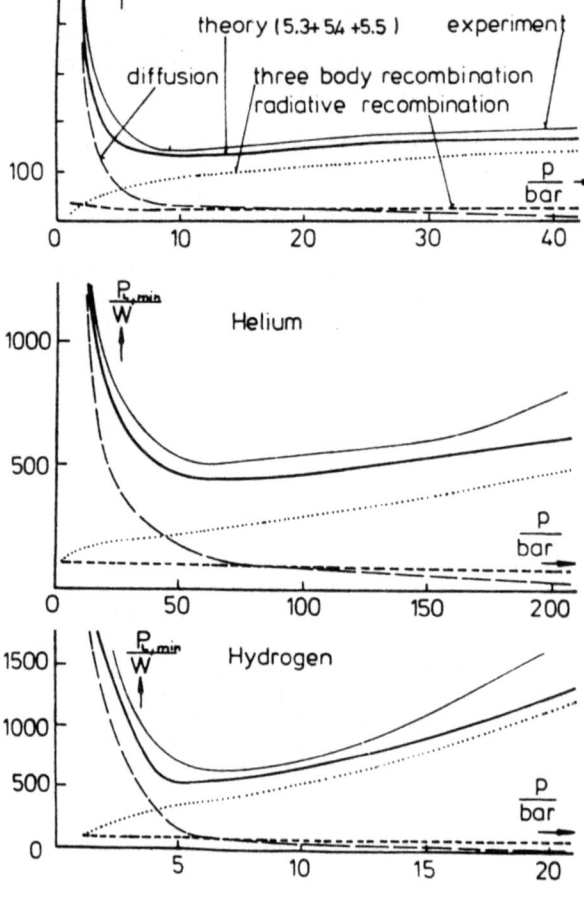

Fig. 6. Comparisons of the calculated and measured minimum plasma sustaining power, $P_{L_{min}}$, for argon, helium, and hydrogen (Uhlenbusch, 1983).

LASER LIGHT SCATTERING - PHASE CONJUGATION - TIME REVERSAL

Knowledge of laser-induced plasmas is based largely on observations of the particles and radiations they emit. However, observations of the laser light and its harmonics scattered by the plasmas is a very useful tool to characterize the plasmas and is widely used.

Among the many important observations made some ten or so years ago in a series of very carefully executed experiments by Eidmann and Sigel at Garching (1972, 1974) was the surprising strength and spatial properties of the radiation back-scattered along the reverse direction of the incident beam which created the plasma at targets of solid deuterium, carbon, copper, and tungsten.

They noted the remarkable fact that quite minute details of the intensity distribution of the incoming beam were reproduced accurately in the slightly red-shifted (~ 2 to 3 Å) back-scattered beam. While one expects scattering when the electron concentration in the plasma is such that the plasma frequency equals the laser angular frequency, the accurate retention of detail implied reflection from a perfectly hemispherical surface whose center of curvature coincided exactly with the lens focus throughout the laser flash duration. Only in these very unlikely circumstances could a light ray be reflected exactly back on itself. Eidmann and Sigel, showing considerable physical insight, favored another explanation of this exact image retention. They suggested that stimulated Brillouin scattering from ion-acoustic waves in the laser-produced plasma was the cause.

Similar observations, including the retention of the original plane of polarization of the incident beam by the reflected light, were later reported by Stamper and colleagues (1974) of the Naval Research Laboratory at Washington, and by workers in the USSR. This remarkable ability of a plasma exactly to reproduce the transverse characteristics of the laser beam is, of course, characteristic of phase-conjugated laser beams, first reported by Nosach and Zeldovich and their co-workers (Nosach et al., 1972; Zeldovich et al., 1972). Undoubtedly these early plasma observations stimulated the current enormous activity on the mechanisms of phase conjugation throughout the world. This intriguing phenomenon in which light is reflected "the wrong way" and in which the beam phase structure undergoes reversal on reflection is likely to have major consequences in the future, especially in attempts to produce matter compression and fusion by lasers, in new types of laser resonators, and in optical communication systems.

To digress for a moment on phase conjugation, we may note that in describing an electromagnetic wave propagating in the z-direction we may write, following Yariv (1978),

$$E_i(\vec{r},t) = \text{Re } [\Psi(\vec{r}) \exp (i (\omega t-kz))] \tag{12}$$

$$= \text{Re } [A_1(\vec{r}) \exp (i\omega t)]$$

as the amplitude E of the wave traveling through a linear lossless but distorting medium. The radial (r) dependence of Ψ allows for spatial modulation by and the effects of distortion by the medium and by diffraction.

The phase conjugate of $E_1(\vec{r},t)$ is defined by

$$E_2(\vec{r},t) = \text{Re } [\Psi^*(r) \exp i(\omega t+kz)] \tag{13}$$

$$= \text{Re } [A_2(\vec{r}) \exp (i\omega t)],$$

and $A_2(\vec{r}) = A_1^*(\vec{r})$ for all $z < z_o$, where z_o is some arbitrary point. E_2 travels in exactly the opposite direction to E_1. To get E_2 from E_1 we take the complex conjugate of the spatial part only, leaving the temporal factor $\exp (i\omega t)$ unchanged. This is equivalent to leaving the spatial part unaltered, but reversing the sign of t. This process is thus known as "wavefront reversal" or "time reversal"*. This is exactly what is observed with laser-produced plasmas which serve as the phase conjugating medium (Basov et al., 1977; Ripin et al., 1977). The practical consequences of phase conjunction are perhaps best illustrated by the following considerations.

---

*For a recent appraisal of optical phase conjugation see "Optical Phase Conjugation", edited by Robert A. Fisher, published by Academic Press, 1983.

While a beam incident at an angle other than the normal to the surface of a plane mirror is reflected at an angle equal to the incident angle, the beam exactly retraces its path if it encounters a phase conjugator for all angles of incidence. A diverging beam incident on a plane mirror is reflected as an even more divergent one. In contrast, reflection at a phase conjugating mirror causes the rays to retrace their path and converge.

Passage of a beam through a medium of variable refractive index introduces phase distortion which is doubled on reflection through the same medium from a conventional mirror. However, automatic collinear reflection from a phase-conjugating mirror through the same medium removes the distortion! We might jocularly say that while the time-reversal reflection may not "shake off the years", it certainly removes the wrinkles!

There are very many processes of a non-linear character, depending on the interaction between coherent beams and matter, which give rise to these unique optical properties, but perhaps the easiest to demonstrate is degenerate four-wave mixing in a saturable absorber. In this a coherent pump beam of sufficient intensity is reflected by a conventional mirror along its own path, after passage through a suitable absorber, to form two counter-propagating pump beams. A third, weaker probe beam, derived from the same laser, is incident on the overlapping region of the pump beams in the medium. These three beams interact non-linearly with the medium effectively to set up interference patterns in their overlap region in such a way as to form thick diffraction gratings which act as an active-volume hologram with its wavefront reversed, and which scatters light along the reverse direction of the incident probe beam.

Four-wave mixing is not the only possible mechanism and, as Eidmann and Sigel (1972, 1974) correctly surmised, stimulated Brillouin scattering is of great importance in this respect.

The ability of plasmas to generate phase-conjugated waves via four-wave mixing has been investigated theoretically by Steel and Lam (1979), who show that as a consequence of the interference of incident beams a spatial modulation of the macroscopic current density develops due to the combined action of convective and Lorentz forces, i.e., the ponderomotive force. The counter-propagating pump wave scatters off this volume perturbation and results in a conjugate wave. They analyzed the non-linear process by treating the plasma as a two-component fluid comprising ions and electrons. In the case of a laser-produced plasma, which is a Brillouin-active medium, the optical wave-plasma inelastic interactions between the incident laser beam and ion-acoustic waves created in the plasma are thought to generate active-volume holograms which back-scatter and conjugate the incident light (Lehmberg, 1978; Lehmberg and Ripin, 1978; Lehmberg and Holder, 1980; Lehmberg et al., 1981).

The highly directional phase-conjugated back-scattered radiation from a laser-produced plasma is a nuisance since it retraces its path into the laser through the amplifiers to the oscillator which can thus suffer irreversible damage unless optical switches (e.g., Faraday rotators) are used to suppress it. However, advantage may be taken of stimulated Brillouin scattering, not from plasmas, but from readily available liquids such as hexane or ethanol, to improve the spatial characteristics of an amplified laser beam. This is likely to have a pronounced influence on progress towards understanding the mechanisms involved in matter compression by powerful laser beams and clarify the route to laser-produced fusion reactors.

Ideally, very well-defined temporally and spatially distributed laser pulses are needed at high intensities (> 10^{14} w-cm^{-2}) and relatively large energies focused on micron-size targets of Li and D_2.

These intensities and energies require complex laser systems with several amplification stages in which the spatial structure of the beam inevitably deteriorates progressively along the amplifying chain even though it may originate as a single-mode gaussian which is reflected into the amplifiers from the target. Phase-conjugated reflection and a second pass through the amplifying system will remove the irregularities introduced on the first pass and restore its well-defined behavior but with greatly enhanced intensity. The exact path reversal property of the phase-conjugated and twice-amplified beam ensures precise focusing on the target to give plasma formation and matter compression. Since there are advantages to be gained in the degree of matter compression which can be achieved by operating at short wavelengths, which give higher ablation pressures and which reduce the undesirable preheating effects of fast electrons, detailed studies of the phase conjugating properties of liquids such as trifluorethanol under 5-nsec flashes at 193-nm ultraviolet light from an ArF laser have been made at the Central Laser Facility at the Rutherford-Appleton Laboratory. These have disclosed that it is possible to achieve high reflectivities which are sufficient to saturate the amplifier on the return path and yield energy fluences and intensities of ~30 mJ/cm^2 and 6 MW/cm^2, respectively, in a defraction-limited beam (Caro and Gower, 1981a,b,c).

Equally exciting is the prospect of using phase-conjugating reflectors as the cavity mirrors of a laser system to remove self-imposed beam distortions in high power excimer or other laser systems to yield idealized outputs for laser-induced plasma, matter compression, and other studies (Auyeung, 1979; Lam and Brown, 1980; Giuliano, 1981; Giuliano et al., 1983). Thus, the original observation of plasma back-scattered laser radiation by Eidmann and Sigel has gone full cycle, with now a much better understanding of its mechanism and the prospect of applying its unique properties to furthering the investigation and use of laser-generated plasmas.

RESONANTLY PRODUCED LASER PLASMAS - CONTINUOUS PLASMAS

The formation of the transient and continuous plasmas described in the preceding sections does not involve absorption of pre-selected wavelength radiation into pre-determined quantum states and, in a sense, the inverse bremsstrahlung and multiphoton absorption processes are "brute force" mechanisms requiring the very large power densities indicated above.

However, large power densities are by no means essential to create laser-induced plasmas. Thus, by taking advantage of the tunable properties of dye lasers it has become possible to produce moderately high-electron-density ($N_e > 10^{14}$ cm^{-3}), low-temperature ($T_e \sim 0.2$ eV) plasmas in alkali metal vapor with only one-tenth of a watt (Tam and Happer, 1977) or even a milliwatt (Koch et al., 1980).

As long ago as 1970 it had been suggested (Measures, 1970) that efficient laser-produced ionization could be generated if the laser quantum energy was in resonance with the energy difference between allowed atomic states. Calculations for the case of a potassium-seeded argon plasma showed that it should be possible to sustain a steady-state plasma with an electron concentration of ~ 2.5×10^{14} cm^{-3} at ~ 3600 K by the expenditure of only 15 W-cm^{-3} of resonant radiation. Suitable tunable lasers were, however, not available at the time to conduct experiments. Nevertheless, the principle of plasma formation by resonance radiation had been established in work by Morgulis et al (1968, 1970) and Yamada and Okuda (1977a,b) for the case of cesium irradiated by resonance radiation to excite the doublet $6P_{1/2}$ and $6P_{3/2}$ levels, which are then ionized by collisional interactions involving the formation of excited cesium atoms and molecules. A quiescent plasma with an electron concentration of $\sim 10^{12}$ cm^{-3} was obtained. With the availability of sufficiently powerful pulsed tunable dye lasers a resonantly excited laser-produced plasma was demonstrated by Lucatorto and McIlrath (1976, 1977), who irradiated sodium vapor at a relatively high density ($\sim 10^{16}$ cm^{-3}) with 589.6-nm radiation and fully ionized it. Again by using a 1-MW pulsed tunable dye laser generating 670.8-nm flashes, i.e., tuned to the $1s^2 2s \rightarrow 1s^2 2p$ transition of ground-state lithium, they were able to create plasmas with $\sim 10^{16}$ cm^{13}.

In these dense vapors collisions between electrons and the resonantly excited atoms, which are produced from the ground state by laser photon absorption, rapidly transfers the laser energy into plasma energy. An initially small number of free electrons are heated by superelastic collisions with the excited atoms. The resulting fast electrons further excite and ionize the atoms to form a plasma. The role of atom-atom collisions in resonantly produced plasmas has been considered by Bobin and Zaibi (1985).

Fig. 7 illustrates possible sequences. Here A, A* and A** indicate the ground, the lowest, and higher excited states, respectively. The very small initial population N_2 of A* may be formed by dissociative excitation of molecules of the alkali metal vapor (Tam and Zapka, 1982). When both I and Ng are low, atoms in the upper excited level A** decay largely by radiative transitions into lower levels indicated by A' in Fig. 7(a).

If I is small but Ng is large, interactions between A** and A atoms, i.e., collisional ionization, become significant, either by the process of association and molecular ion formation

$$A^{**} + A \rightarrow A_2^+ + e ,$$

or by the process of ion-pair formation yielding a positive and negative ion

$$A^{**} + A \rightarrow A^- + A^+ .$$

If a plasma is to be formed one or both of these reactions must occur at a rate which is large compared to the spontaneous decay rate of A**, i.e.,

$$Ng\, \alpha_i \tau_2 > 1 ,$$

where α_i is the ionization rate coefficient for A** + A collisions, and τ_2 is the radiative lifetime of A**.

The electrons liberated in this process have low energies (< 1 eV) which are insufficient to create A* atoms from the ground state, and

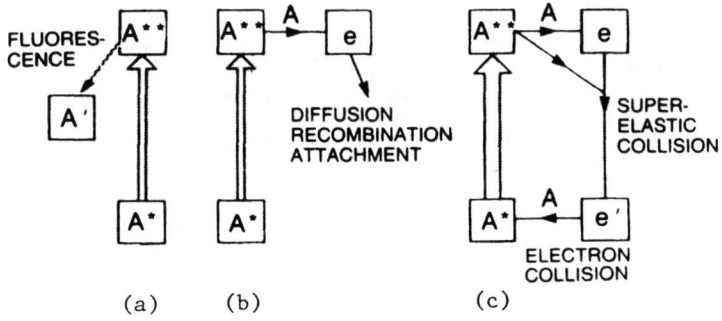

Fig. 7. Energy level diagram showing quasi-resonant laser-produced plasma formation.

(a) At low beam intensity and low concentration of ground state atoms radiative decay of A** dominates.

(b) At high ground state concentration and low intensity A** undergoes collisional ionization, but electrons liberated do not cause ionization and are lost.

(c) At high density and high intensity electrons are heated by superelastic collisions and form a self-propagating loop and generates a plasma. (Tam and Zapka, 1982).

they are lost by recombination, attachment, and diffusion, as shown in Fig. 7(b). However, if I is large enough, together with a big enough population of A** as shown in Figure 7(c), then the slow electrons can undergo superelastic collisions with A** and gain sufficient energy to excite ground state A atoms to the A* state.

The net result is that for each A* atom transferred to A** by the resonant laser light, two are produced. The interactions are

$$A^{**} + e_{slow} \rightarrow A^* + e_{fast} \quad \text{- electron heating}$$

$$A + e_{fast} \rightarrow A^* + e_{slow} \quad \text{- replenishing } A^*.$$

Adding these we see that for each A* atom destroyed by laser radiation, two are regenerated, thus setting up a regenerative loop which enhances the A** population, ultimately resulting in breakdown and plasma formation. The intensity required is only $\sim 10^4$ W-cm^{-2}, i.e., about ten orders-of-magnitude less than the threshold for non-resonant ionization of rare gases.

Detailed computer modeling of the processes of resonance laser beam absorption for ionization and plasma formation has been successfully undertaken by Measures and his colleagues (Measures et al., 1979), which show the effectiveness of resonance ionization plasma formation for the creation of long plasma channels needed in electron or ion-beam transportation in inertial fusion schemes.

REFERENCES

Adrain, R. S., 1980, Optics and Laser Technology June, 137.
Auyeung, J. et al., 1979, IEEE J.Qu.E., 15:1180.
Askar'yan, G. A., et al, 1967, Sov. Phys. JETP Lett., 5:93.
Bakos, J. S., 1974, Adv. Electronics & Electron Phys., 36:58.
Basov, N. G., et al., 1977, Sov. J. Quant. Electr. 7:1300.
Bobin, J. L., and Zaibi, M. A., 1985, Proc. XVII ICPIG, 2:860.
Caro, R. G., and Gower, M. C., 1981a, IEEE, J.Qu.E., 17:225.
           1981b, Optics Lett. 6:557.
           1981c, Appl. Phys. Lett., 39:855.
Carlhoff, C., Krametz, E., Schafer, J. H., Schildbach, K., Uhlenbusch, J., and Wroblewski, D., 1981, Physica, 103C:439.
Cremers, D. A., Archuleta, F. L., and Martinez, R. J., 1985, Spectrochimica Acta B (in the press).
Delone, N. B., 1975, Sov. Phys. Usp. 18:169.
Eberly, J. H., and Lambropoulous, P., 1978, "Multiphoton Processes," Wiley, New York.
Eidmann, K., and Sigel, R., 1972, Int. Conf. on Interaction of Radiation with Matter.
Eidmann, K., and Sigel, R., 1974, 3B:667 in "Laser Interactions and Related Plasma Phenomena," Schwarz, H. J., and Hora, H., ed., Plenum Press, New York.
Evans, C. J., 1976, Optics Comm., 16:218.
Evans, L. R., and Grey Morgan, C., 1968, Nature, London, 219:712.
           1969, J. Phys. Med. Biol., 14:205.
Generalov, N. A., et al., 1970, JETP Lett. 11:302.
Giuliano, C. R., 1981, Physics Today, 34:27.
Giuliano, C. R., et al., 1983, "Laser Focus," (Feb.), 55.
Grey Morgan, C., 1975, Rep. Prog. Phys. 38:621.
Hellwarth, R. W., 1982, Optical Engineering, 21:257.
Honig, R., 1963, Appl. Phys. Lett. 3:8.
Hughes, T. P., 1975, "Plasmas and Laser Light," Adam Hilger.
Ireland, C. L. M., Yi, A., Aaron, J. M., and Grey Morgan, C., 1974, Appl. Phys. Lett. 24:4.
Isenor, N. 1964, Can. J. Phys., 42:1413.
Klewe, R. C., et al., 1972, J. Phys. E. Sci. Inst., 5:203.
Koch, M. E., Verma, K. K., Stwalley, W. C., 1980, J. Opt. Soc. Amer., 70:627.
Koch, M. E., Verma, K. K., Bohns, J. T., and Stwalley, W. C., 1982, Proc. Int. Conf. Lasers, 1982.
Krasyuk et al, 1970, Sov. Phys. JETP, 31:860.
Lam, J. F., and Brown, W. P., 1980, Optics Lett. 5:61.
Lehmberg, R. H., 1978, Phys. Rev. Lett., 41:863.
Lehmberg, R. H., and Holder, K. A., 1980, Phys. Rev. A, 22:2156.
Lehmberg, R. H., and Ripin, B. H., 1978, Proc. Int. Conf. on Lasers,
Lehmberg, R. H., Tripathi, V. K., and Liu, C. S., 1981, Phys. Fluids, 24:703.
Loree, T. R., 1983a, ICALEO Conference.
       1983b, Private Communication.
Lucatorto, T. B., and McIlrath, J. J., 1976, Phys. Rev. Lett., 37:428.
Lucatorto, T. B., and McIlrath, J. J., 1977, Phys. Rev. Lett., 38:1383.
Maker, P. D., Terhune, R. W., and Savage, C. M., 1963, Proc. 3rd Int. Conf. on Quantum Electronics, P. Grivet and N. Bloembergen, ed., Columbia U.P., New York.
Measures, R. M., 1970, J. Quant. Spect. Radiot. Trans. 10:107.
Measures, R. M., Drewell, N., and Cardinal, P. 1979, J. Appl. Phys., 50:2662.
Meyer, I., and Stritzke, P., 1976, Applied Physics, 10:125.
Meyerand, R. G., and Haught, A. F., 1963, Phys. Rev. Lett. 11:401.
Morgulis, N. D., et al, 1968, Sov. Phys. JETP 26:279.
        1970, ibid 31:1005.
Morgan, F., Evans, L. R., and Grey Morgan, C., 1971, J. Phys. D., 4:225.
Muller, S., and Uhlenbusch, J., 1982, Physica, 112C:259.

Nosach, O. Y., et al., 1972, Sov. Phys. JETP, 16:435.
Payne, M. G., et al., 1975, Phys. Rev. Lett., 35:1154.
Radziemski, L. J., Cremers, D. A., and Loree, T. R., 1983, Spectrochimica Acta, 38B:349.
Raizer, Yu. P., 1977, "Laser-Induced Discharge Phenomena," Plenum Press, New York.
Ready, J. E., 1971, "Effect of High power Laser Radiation," Academic Press, New York.
Ripin, B. H., et al., 1977, Phys. Rev. Lett., 39:611.
Schwirzke, F., 1974, "Laser Interaction and Related Plasma Phenomena," 3A, 213, H. J. Schwarz and H. Hora, ed., Plenum Press, New York.
Smith, D. C., and Fowler, M. C., 1973, Appl. Phys. Lett., 22;500.
Stamper, J. A., et al., 1974, "Laser Interaction and Related Plasma Phenomena," 3B, 713, H. J. Schwarz and H. Hora, ed., Plenum Press, New York.
Steel, D. G., and Lam, J. F., 1979, Optics Lett., 11:363.
Tam, A. C., 1979, App. Phys. Phys. Lett., 35:683.
    1980, J. Appl. Phys., 51:4682.
Tam, A. C., and Happer, W. 1977, Optics. Comm. 21:403.
Tam, A. C., and Zapka, W., 1982, Laser Focus, March, 1969.
Tozer, B. A., 1975, Physics in Technology, Nov., 251.
    1976, Optics and Laser Tech., April, 1957.
Uhlenbusch, J., 1983, Proc. XVI ICPIG, 119, (Vol. Invited Lectures).
Yariv, A., 1978, IEEE, J. Qu, E., 14:650.
Yamada and Okuda, 1972a, J. Phys. Soc., Japan, 32:1162.
    1972b, Phys. Rev Lett., 39A:223.
Zeldovich, Y. B., and Raizer, Yu P., 1964, J. Exp. Theor. Phys., 47:1130.
Zeldovich, Y. B., et al, 1972, Sov. Phys. JETP, 13:109.

PLASMAS SUSTAINED BY SURFACE WAVES AT MICROWAVE AND RF FREQUENCIES:

EXPERIMENTAL INVESTIGATION AND APPLICATIONS

M. Moisan and Z. Zakrzewski

Department de Physique,
Universite de Montreal,
Montreal, Quebec, Canada

INTRODUCTION

The search for plasma sources that have the necessary requirements for given applications has a history as long as that of plasma physics. From the beginning, interesting plasma applications have been proposed but often their practical realization has been limited by the lack of adequate plasma sources, sources that would, for example, yield quiescent, reproducible, stable, low contamination plasmas and that would be simple to build, easy to operate, and inexpensive (Crawford, 1971). Some additional features that may be desired are a high electron density or a high ionization degree, a large plasma volume, the possibility of operating at a relatively low gas pressure ($\simeq 10^{-5}$ torr) or at many times the atmospheric pressures, the possibility of using corrosive gases, etc. Though it is far from meeting all these specifications, the most widely used gas discharge is still that obtained in a dc or ac ($\simeq 60$ Hz) electric field, using electrodes within the plasma tube. This observed preference is explained by the fact that the high-frequency-produced plasmas available up to now are not that flexible, are often more costly, and their modeling is far from complete. One reason for this poor modeling situation is that, very often, the various HF sources have been developed while pursuing specific applications, and the desire to achieve such immediate useful goals (e.g., making a laser) did not allow for longer research time. Moreover, in many cases, such works were conducted by specialists from fields other than plasma physics. The impression that arises is that there does not exist any coherent description, common to all these HF discharges.

This paper presents a recently developed plasma source that uses the propagation of an electromagnetic surface wave. The surface-wave discharges emerged in the early seventies as a new kind of microwave-produced plasma with many advantages. Recently made progress in terms of surface-wave launchers allows one to sustain surface-wave discharges in the rf range as well, now covering a frequency band extending from a few MHz up to 10 GHz. The fact that the same field configuration can be used to produce plasmas in the rf as well as in the microwave range is a unique feature of surface-wave plasmas. A large amount of experimental data has already been accumulated on these plasmas and their modeling is already well developed (Ferreira, this Proceedings). This enables one to have a rather unified and complete picture of their properties. In addition, it has led to the elaboration of a simplified general model for rf- and microwave-produced electrodeless (no internal electrodes) discharges in general. These are designated as high frequency (HF) plasmas in this work.

The fact that an electromagnetic surface wave can propagate along the interface between a positive-column plasma and the surrounding dielectric had been known for years (Trivelpiece and Gould, 1959; Trivelpiece, 1967). The term surface wave is justified by the fact that the wave energy is concentrated along the propagating interface. For surface-wave sustained plasmas, one can find many reports that present microwave-produced plasmas that are obviously the result of surface-wave propagation but that were not recognized as such (e.g., Fehsenfeld et al., 1965; Vidal and Dupret, 1976). Tuma (1970) was probably the first one to clearly identify a surface-wave produced discharge. Moisan et al. (1974, 1975, 1979) developed simple, compact, and efficient surface-wave launchers, based on coaxial or waveguide components, suitable for the generation of long plasma columns at microwave frequencies. The nature of the wave sustaining such plasma columns has been explicitly confirmed by Zakrzewski et al. (1977), who measured its dispersion and attenuation characteristics.

The surface-wave plasmas have interesting applications. Some applications are well documented and interesting developments are foreseen in the fields of surface treatment, lasers, and elemental analysis. One advantage of these plasmas is connected with the propagation properties of surface waves: a plasma source using this principle can yield long plasma columns with wave launchers of dimensions that are not larger than those of a conventional resonant cavity. Moreover, the range of their discharge conditions and plasma parameters is broader than that of dc discharges. Furthermore, the plasma obtained is stable, reproducible, and quiescent.

There are two main objectives to this paper. One is to provide the theoretical and experimental background necessary to understand the physical phenomena governing the maintenance and determining the plasma parameters of hf discharges in general. The other is to present the properties of surface-wave sustained plasmas and their applications, with particular attention to radiative processes. In this presentation, we try to strike a balance between a sufficient introduction to the subject for scientists who are not directly involved in the field of hf gaseous discharges, and the presentation of our view on the actual state of the art.

The content of the paper is organized as follows. The next section contains a short review of the physical processes occurring in hf discharges in general, followed by the presentation of our simplified general model for these discharges. Then, the foundations of similarity relations for hf discharges are discussed and a classification of these discharges introduced. A simplified general model is then applied to the particular case of a surface-wave discharge. In the last section, the properties of surface-wave discharges are presented followed by a review of their applications.

PHYSICAL CHARACTERISTICS AND SIMPLIFIED MODEL OF RF AND MICROWAVE DISCHARGES IN GENERAL

Outline of a Simplified Model for HF Discharges

An hf-produced gaseous discharge is a complex physical phenomenon. Besides the rich variety of physical processes that is characteristic of plasmas, there is additionally the interaction between the plasma and the electromagnetic field that sustains the discharge. In that respect, an rf- or microwave-driven plasma is more complicated than a plasma sustained by a dc field, in particular when the time constants that characterize the various processes occurring in the plasma become comparable with the oscillation period of the field.

Clearly, to take into account the interaction between the hf field and the plasma, a fully self-consistent theoretical model of the discharge seems needed. It should be based on a set of equations, including Maxwell's equations and the equations describing the plasma maintenance processes within the discharge. The proper boundary conditions for the electromagnetic field at the media interfaces as well as those for the charged particle density at the surface surrounding the plasma should also be included. Unfortunately, the problem formulated in that way seems to elude any successful analytical treatment. To make it tractable, some physically justified simplifying assumptions have to be sought.

In the present work, we introduce a simpler model for rf and microwave discharges. It is based on the assumption that the formal treatment of the physical processes occurring in the discharge setup can be divided into two separate parts: the first concerns the maintenance processes within the discharge; the second, the electromagnetic behavior of the hf circuit containing the plasma. Thus, we shall first investigate what are the physical processes governing the discharge. It turns out that they are determined by the discharge conditions (nature and pressure of the gas; shape, size, and wall material of the vessel; frequency of the electromagnetic field) and the amount of power transferred to the plasma. The way that the electromagnetic field is imposed on the plasma is assumed, to a first approximation, not to affect the discharge processes*. From this investigation follows the knowledge of the amount of power that is lost on the average by an electron under given discharge conditions. Then, in a second step, we determine the electrodynamic properties of the hf circuit-plasma ensemble using standard methods from electromagnetic field theory and circuit analysis. The influence of the discharge plasma on the properties of the hf circuit is treated phenomenologically by considering the plasma as a dielectric with a complex conductivity determined by the electron density and the collision frequency for momentum transfer. This treatment yields the fraction of the electromagnetic energy, incident on the discharge, that is absorbed in the plasma.

Finally, with the above results at hand, we can address and solve the problem of the charged particle balance as well as that of the energy balance in a steady-state discharge. The quantitites derived are: from the charged particle balance, the spatial distribution of the electron density and the average value of the electron energy; and from the energy balance, the amplitude of the electric field within the plasma that is necessary to maintain the discharge, and the average value of the electron density.

We shall now proceed with the description of the discharge, as outlined above. The justification for this procedure and the discussion of its range of application will be dealt with throughout the next paragraphs and summarized at the end of this section.

Principal Physical Processes in the Discharge

There are always some free electrons and ions present, for various reasons, in any given finite volume of gas. When an electric field is applied across such a volume, charged particles are accelerated. In an

---

* This assumption is supported by the experimental results obtained with surface-wave produced plasmas, as we shall see further.

rf or microwave field, the movement of particles differs distinctly from that in a constant or low frequency field: a) The ions, because of their large mass, can be considered immobile; and b) the electrons oscillate in the gas volume, usually with an amplitude that is small in comparison with the plasma dimensions. If, during this oscillation process, the electron collides with another particle, it gains some energy at the expense of the field. Otherwise, its net average energy gain over one period is zero. The electron gas is the only medium able to effectively absorb energy from the high frequency field and to transfer it to the heavy particles and to the environment. The randomization of the electron movement due to collisions leads to the establishment of an electron energy distribution with a mean energy much larger than the energy available from the organized motion in the HF field. The fast electrons from the tail of the distribution function are able to ionize the gas molecules. In that way, new electrons can be produced. At the same time, however, electrons are being lost due to processes such as diffusion, volume recombination, and attachment.

In this paper, we assume that direct ionization from the ground state and diffusion are the predominant particle creation and loss processes, respectively, determining the charged particle balance in the discharge. More complicated cases such as step-wise ionization and volume recombination are not considered in this presentation. Let us start by briefly reviewing the physical processes directly related to the maintenance of the diffusion-controlled discharge. These processes are: elastic collisions, inelastic collisions, and diffusion.

The various collision processes (elastic, excitation, ionization) are conveniently described in an hf discharge by using specific collision frequencies for each given process. When considering, for example, the energy loss by electrons in elastic collisions, one should use the collision frequency for momentum transfer. The dependence of this frequency on the electron energy, in the case of helium and argon, is illustrated in Fig. 1. The fraction of the total number of collisions that results in a given non-elastic process is called the efficiency of this process. Fig. 2 shows the efficiency of ionization, both for helium and argon. Similar curves can be plotted that represent the excitation efficiency to a given level.

The total power loss, suffered by an average electron in collisions of all kinds, can be expressed as

$$\Theta \equiv \frac{2m}{M} <\nu_m \varepsilon> + \sum_k <h_k \nu_m> eV_k \; ; \; \varepsilon = \frac{mv^2}{2}, \qquad (1)$$

where m and M are the electron and ion mass, ε is the electron energy, ν_m the collision frequency for momentum transfer, and h_k and V_k are, respectively, the efficiency and the threshold potential for excitation by collisions leading to the k-th level. The brackets < > denote the averaging over the electron energy distribution function (EEDF).

In the discharges considered in the present work, the EEDF, and thus Θ, are assumed electron density independent. Also, the dependence of these quantities on the shape of the electromagnetic field distribution within the plasma is, as a first approximation, neglected. As we shall show, under these conditions, the value of Θ is a function of the discharge conditions only. Furthermore, provided that the collision frequency ν_m does not depend on the electron velocity (or, if it does, provided that an effective value of the collision frequency can be introduced), the total energy loss due to collisions, per electron, can be written in a form that will prove useful later:

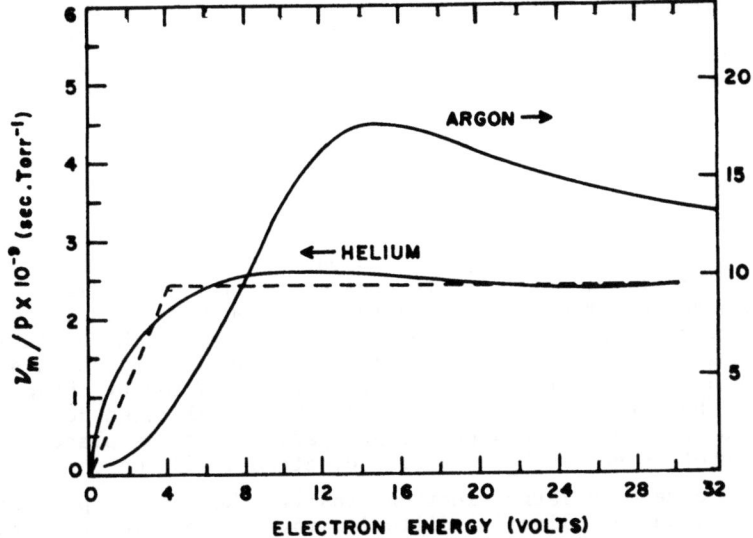

Fig. 1. Reduced value of electron collision frequency for momentum transfer in argon and helium (after MacDonald, 1966; MacDonald and Tetenbaum, 1978).

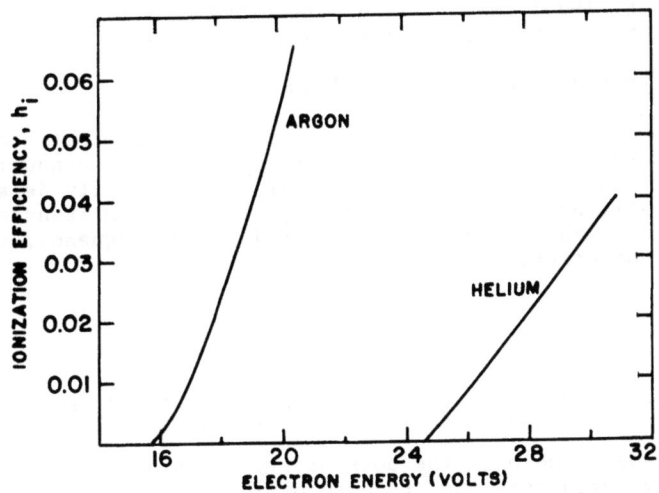

Fig. 2. Ionization efficiency in argon and helium (after MacDonald and Tetenbaum, 1978).

$$\Theta/\nu_m = \frac{2m}{M} <\varepsilon> + \sum_k <h_k> eV_k \qquad (2)$$

We shall see that the assumption of Θ being independent of the electron density is physically justified in the case of the diffusion-controlled discharge, and it is used in this work mainly for reasons of simplicity. However, the present model applies directly to the more general case of a density-dependent Θ, provided that the dependence of Θ on density is known.

We now turn to the diffusion of electrons. The diffusion results from the random motion of electrons and causes a net directed flow of electrons to occur because of the existence of a density gradient. This electron flux is directed opposite to the gradient. The proportionality factor between the flux value and the gradient value is called the diffusion coefficient. In the most general case, this coefficient depends on both the electron energy and the electron density. There exists, however, two important asymptotic cases for which the diffusion coefficient does not depend directly on the electron density (such a dependence could occur indirectly if the EEDF was affected by changes in the density). The first case corresponds to the low electron density limit, where no interaction between charged particles takes place, and the diffusion coefficient is D_e, the free electron diffusion coefficient.

In the opposite case of a large electron density, the charged particles interact strongly and a collective diffusion movement of electrons and ions occurs. This so-called ambipolar diffusion is slower than the free diffusion of electrons, and the corresponding diffusion coefficient D_a is much smaller that D_e. In the intermediate range of electron density values, the diffusion coefficient is density dependent. Thus, it always varies spatially within any bounded plasma. To cope with this problem, an effective diffusion coefficient D_s can be introduced. The theoretical dependence of D_s on the electron density is shown in Fig. 3 (Muller and Phelps, 1980; remember that the Debye length λ_D is inversely proportional to the square root of the electron density).

Transfer of Electromagnetic Energy to Plasma

The description of the plasma interaction with the electromagnetic field requires a knowledge of the plasma complex electron conductivity (or an equivalent quantity). The value of this conductivity depends on the electron density and collision frequency for momentum transfer. In the simplest case of an energy independent collision frequency, the so-called Lorentz conductivity is

$$\sigma = \frac{ne^2}{m} /(\nu_m + j\omega), \qquad (3)$$

where n, the electron density, is in general a function of position. This relation, however, does not hold for most real gases, where ν_m is not a constant. Nevertheless, the possibility of using the Lorentz formula has always been tempting, because of its simplicity. For that purpose, effective electron collision frequencies have been defined (Heald and Wharton, 1965) that can be directly used in Eq. (3). Fortunately, this is possible in most practical cases. Thus, we assume that ν_m is either an energy independent collision frequency or it represents an effective value of the collision frequency.

The transfer of energy from the electromagnetic field to the plasma can be expressed, using the real part of the plasma conductivity (Reσ) as follows. In a harmonically varying electric field of angular frequency ω

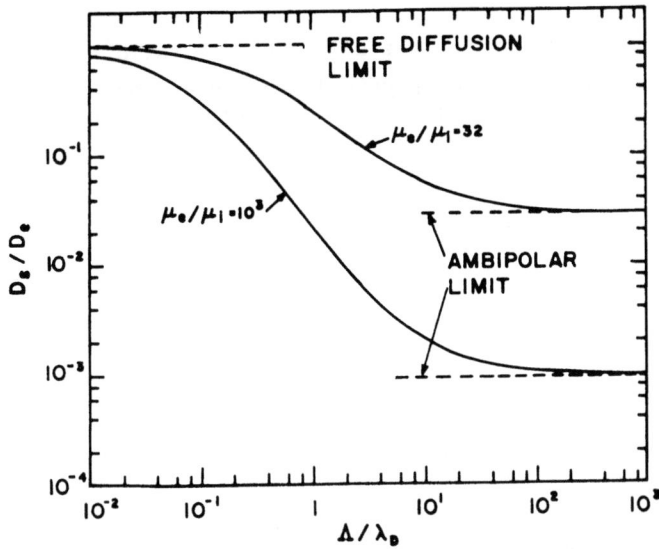

Fig. 3. Theoretical values of the effective diffusion coefficient D_s for electrons (normalized with respect to the free diffusion coefficient D_e), for various ratios of electron-to-ion mobility, μ_e/μ_i (Data from Phelps, 1982; Muller and Phelps, 1980).

and intensity E, the power acquired by electrons from the field and lost in collisions is

$$P_L = \frac{1}{2} \int_{V_p} E^2 \, \text{Re}\sigma \, dV = \frac{1}{2} \frac{e^2 \bar{n}}{m} \frac{\nu_m}{\omega^2 + \nu_m^2} \bar{E}^2 V_p \tag{4}$$

with $\bar{n} \equiv \frac{1}{V_p} \int_{V_p} n \, dV$,

where V_p denotes the integration over the active zone* of the discharge and $\bar{E}^2$ is a spatially averaged value of the field intensity. An effective value E_e for the electric field intensity can be defined, such that

$$E_e^2 = \frac{1}{2} \frac{\nu_m^2}{\omega^2 + \nu_m^2} \bar{E}^2, \tag{5}$$

and then

$$P_L = \frac{e^2 \bar{n}}{m \nu_m} E_e^2 V_p. \tag{6}$$

---

* The active zone of the discharge is the volume area in which the energy transfer from the electromagnetic field to the plasma occurs.

The value of E_e incorporates the effect of the ratio ν_m/ω on the power transfer to the plasma. The physical meaning of the effective field becomes clear when one notices that the power dissipated in a plasma placed in a dc and an hf field is the same, providing that $\bar{n}$ has the same value and $E_e = E_{dc}$.

In a practical discharge setup, the plasma is sustained by a given type of field applicator (e.g., hf rings, coil, cavity, waveguiding structure) having specific properties. Thus, the power transfer from the generator to the plasma will be determined by the electrodynamic characteristics of the specific applicator used, as loaded with the plasma produced. Obviously, in contrary to the problem of the energy loss by electrons, the power transfer depends strongly on the kind of applicator used and on the way that the plasma is located within it. Nevertheless, it is possible to define, in a general form, a power absorption coefficient that represents the results of the interaction between the plasma and the electromagnetic field.

Consider a discharge setup consisting, in general, of a plasma and of an rf or microwave field applicator and matching circuit. Let us choose any arbitrary surface enclosing the plasma together with, if any, only the lossless elements of the setup. The power absorption coefficient $A(\bar{n})$ is defined as

$$A(\bar{n}) = P_A/P_I, \tag{7}$$

where P_I is the total flux of power directed inward to the surface considered, and $(P_I - P_A)$ is the flux of power directed outward as a result of reflections, transmission to other areas in the discharge setup, and electromagnetic radiation into the surrounding space.

As seen from (7), it is assumed that, under given discharge conditions, $A(\bar{n})$ depends only on the average electron density. This is strictly true only in the case of a uniform distribution of the field within the plasma. However, even with a nonuniform field distribution, this approximation remains valid, providing an appropriate spatial average field can be used.

Charged Particle Balance in a Steady-State Discharge

Within the present model, it is assumed that the charged particles are created in the gas volume by direct electron impact on the atom in the ground state, and that they disappear through diffusion toward the walls where they recombine. Thus, the continuity equation can be written

$$\frac{\partial n}{\partial t} = \nabla^2 (Dn) + \nu_i n, \tag{8}$$

where ν_i is the ionization frequency and t denotes time. For the reasons already discussed, the EEDF can be assumed spatially uniform in a diffusion-controlled discharge. As a consequence, the ionization frequency ν_i, as well as the diffusion coefficient D_e or D_a, can be considered constant within the active zone of the discharge.

For plasma vessels with a diffusion length Λ of the order of centimeters, the diffusion becomes ambipolar when $n \geq 10^8$ cm^{-3}. This condition is always met in rf and microwave discharges of any practical use. The steady state in such a discharge is characterized by the following set of conditions:

$$\partial n/\partial t = 0; \quad n\Lambda^2 \to \infty; \quad D = D_a. \tag{9}$$

Therefore, Eq. (8) becomes

$$D_a \nabla^2 n + \nu_i n = 0. \tag{10}$$

The solution to this linear equation yields the distribution of the electron density in a steady-state discharge. Furthermore, as an eigenvalue equation, it imposes the condition

$$\nu_i/D_a = \Lambda^{-2}. \tag{11}$$

For a cylindrical tube of radius a and length ℓ, the radial density profile is

$$n(r) = n_0 J_0(r/\Lambda), \tag{12}$$

with $\Lambda^{-2} = (\pi/\ell)^2 + (2.405/a)^2$, which becomes $\Lambda \equiv a/2.405$ for $\ell \gg a$.

The ratio (11) represents the condition for the charged particle balance. It denotes the equality that has to exist between the rate of production of electrons and ions in the plasma volume and their removal from this volume due to ambipolar diffusion. From (11), it is possible to determine the average electron energy if the dependence of ν_i and D_a on $\langle\varepsilon\rangle$ is known. The Schottky theory of a diffusion-controlled plasma column, for example, yields such a dependence of ν_i on the electron temperature for a plasmas with a Maxwellian EEDF.

So far, we have shown that the average electron energy and the spatial distribution of the electron density can be found from the charged particle balance. To determine the absolute value of the electron density and the absolute value of the maintenance field intensity in the discharge, we shall now address the balance of electron energy.

Energy Balance in a Steady-State Discharge

In a steady-state discharge, the energy balance requires that the energy acquired by the electrons from the electromagnetic (em) field be equal to the energy lost by them due to interactions with heavy particles and with the tube wall. Since the amount of power transferred to the electrons from the em field depends on the electrodynamic characteristics of the hf applicator-plasma ensemble, one has to deal with a specific energy balance relation for any practical realization of the discharge setup. However, we can formulate the problem of the power balance in a general way, using the absorption coefficient $A(\bar{n})$.

By equating the power absorbed from the applied field by the plasma with the power lost by these electrons into various interaction processes (left-hand side and right-hand side of the following equation, respectively), we have:

$$A(\bar{n}) P_I = \bar{n} \Theta V_p. \tag{13}$$

Fig. 4 presents a flow chart of the different states of the power transfer from the applicator to the plasma. One notices that the interactions of the electrons with the ions and with the tube walls are neglected in our definition of Θ. An interested reader may find a detailed discussion of the energy balance in a steady-state discharge in Golant et

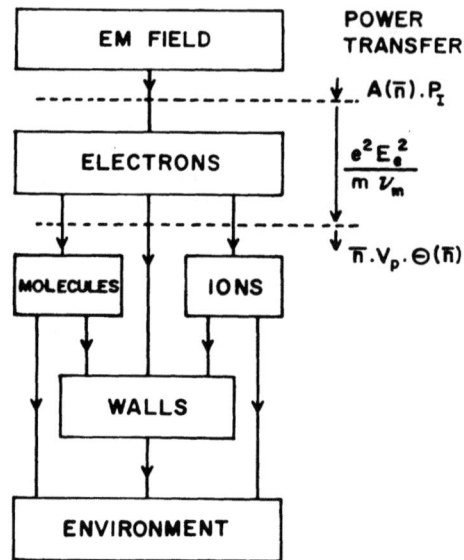

Fig. 4. Simplified diagram of the power flow in hf discharges.

al. (1980); although it concerns a discharge in a dc field, it can be applied directly to the case considered here.

From the energy balance condition, the average electron density in the discharge is

$$\bar{n} = \frac{A(\bar{n})}{\Theta V_p} P_I. \tag{14}$$

Thus, when the EEDF (and Θ) does not depend on the electron density, the average density is directly proportional to $A(\bar{n}) P_I$, the power absorbed in the plasma.

The intensity of the electric field within the plasma sustaining a steady-state discharge is called the maintenance field intensity. It has to acquire a value complying with the balance condition (13). To calculate the effective value of this maintenance field, let us notice that the power lost by the electrons can be written either in the form of Eq. (6) or in the form of the right-hand side of Eq. (13); thus

$$\frac{e^2 E_e^2}{m \nu_m} \bar{n} V_p = \bar{n} \Theta V_p. \tag{15}$$

From this, it follows directly that

$$\left(\frac{E_e}{\nu_m}\right)^2 = \frac{m}{e^2} \frac{\Theta}{\nu_m}. \tag{16}$$

Let us dwell on this relation. First, we note that the effective value of the maintenance field intensity is determined only by the physical processes occurring within the plasma and not by the absorbed power or by the applicator characteristics. This is true in so far as we can assume that the EEDF, and consequently Θ, does not depend on the electron density. This fact seems to have escaped the attention of many researchers, so we are emphasizing it: Once the discharge conditions are set, in an ambipolar diffusion-controlled plasma, the value of the field intensity in the plasma is not (to a first approximation) affected, either by increasing the power delivered to the plasma or by changing applicators. The result of any increase in the absorbed power density (proportional to $\bar{n} E_e^2$) is an increase in the average electron density value and not in the maintenance field intensity.

A further conclusion following from (16) is that Θ/ν_m and E_e/ν_m are equivalent parameters* in a steady-state discharge. The latter, or more often E_e/p, has been commonly used to characterize dc discharges. It is worth pointing out that there exists ample experimental data, particularly for positive-column plasmas, linking E_e/p with the other discharge parameters. These results could thus be readily used in the modeling of rf and microwave discharges, as they provide Θ/ν_m.

In Fig. 5, one can find the calculated values of E_e/p as a function of $p\Lambda$, for various values of $n_o \Lambda^2$. The $n_o \Lambda^2 \to 0$ curve shows the breakdown field value, whereas the $n_o \Lambda^2 \to \infty$ curve corresponds to the case of a steady-state ambipolar diffusion-controlled discharge. Note that the maintenance field intensity is lower than the breakdown value. These data were first obtained theoretically and then experimentally verified by Rose and Brown (1955). Since the combined effects of ω and ν_m on the value of the maintenance field have been included within the value of E_e, the E_e/p-vs-$p\Lambda$ curves behave exactly like the E_{dc}/p-versus-$p\Lambda$ curves for a dc discharge - they decrease monotonically with $p\Lambda$. To show explicitly the influence of the radio ν/ω on the value of the maintenance field, we present in Fig. 6 the breakdown and steady-state E-versus-p curves for hydrogen (according to Brown, 1961). The minimum of both these curves corresponds approximately to $\nu_m = \omega$.

The energy balance condition (13) is necessary, but not sufficient, for a steady-state discharge to exist. Indeed, not only the power lost by electrons has to be compensated by the power transferred to the plasma from the electromagnetic field, but this power equilibrium has to be stable. Let us have a closer look at the problem. In Fig. 7, a sketch illustrating the balance of power in a microwave discharge is shown. The absorbed power curves, $P_A = A(\bar{n})P_I$, all have the same shape, with P_I as a multiplying constant. The P_L curve for given discharge conditions, and with Θ independent of density, is a straight line. For each value of the power delivered to the discharge setup, the average electron density in the plasma assumes the value corresponding to the crossing point of the P_L and P_A curves. No discharge exists if the power delivered is too small (P_{I1}). To curves P_{I2} and P_{I3} there corresponds a steady-state

---

* The same conclusion is true in the more general case, where the EEDF depends on the electron density. Then, however, the values of Θ and of E_e also depend on the density.

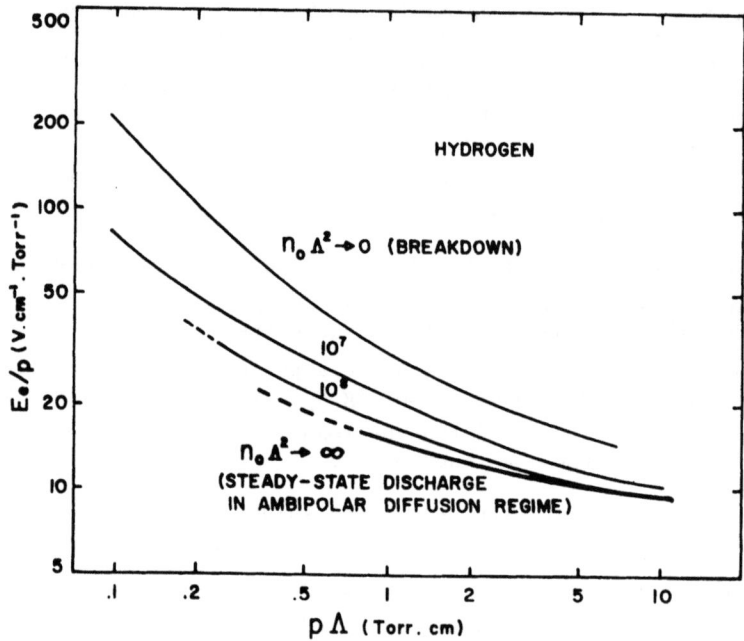

Fig. 5. Reduced value of the effective maintenance field intensity (after Rose and Brown, 1955).

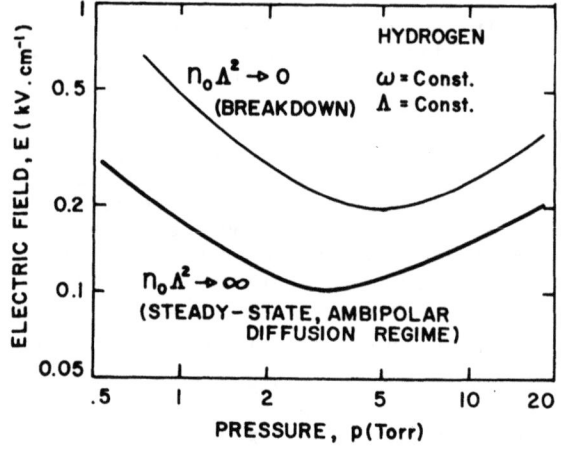

Fig. 6. Dependence of the maintenance field intensity on the gas pressure in a hydrogen discharge at 3.2 GHz (after Brown, 1961).

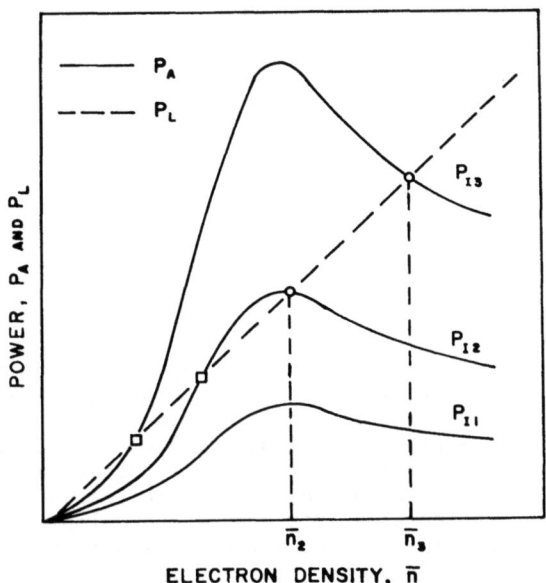

Fig. 7. Sketch illustrating the stable (o) balance of power in a steady-state HF discharge. P_I, P_A, P_L are the incident wave power, the absorbed power, and the power lost by electrons, respectively.

discharge with densities $\bar{n}_2$ and $\bar{n}_3$ (circles), respectively. A stable equilibrium of power does not, however, exist for the other crossing points (squares). The reasons are now examined.

The discharge is stable if any random fluctuation of the electron density triggers a sequence of events which leads to the restoration of the original density value. Consider, for example, a situation in which a random decrease of the electron density occurred. This, in turn, causes the electrodynamic characteristics of the whole plasma-hf circuit to change, i.e., $A(\bar{n})$ changes (also $\Theta(\bar{n})$ may change, if it is not density independent). If all this led to a reduction of the power gained by the electrons relative to the power lost in collisions, a further decrease of the density would occur that would ultimately lead to the extinction of the discharge. On the other hand, the original value of the electron density is restored when any decrease of the electron density causes the power absorbed from the electromagnetic field to prevail over the power loss due to electron collisions and, conversely, if the electron density was to increase from its stable point value. Thus, the following condition has to be fulfilled:

$$dP_L/d\bar{n} > dP_A/d\bar{n} . \qquad (17)$$

This condition was first formulated by Taillet (1969) and Leprince et al. (1971) on the basis of the stability of resonantly sustained plasmas. Within the present model, it becomes

$$\frac{dA(\bar{n})}{A(\bar{n})} - \frac{d\Theta(\bar{n})}{\Theta(\bar{n})} < \frac{d\bar{n}}{\bar{n}} . \qquad (18)$$

The criterion (18) determines the range of electron density (and thus that of absorbed power) over which a steady-state discharge can be operated in a given setup and under given discharge conditions. In the particular case of a density independent Θ, the condition (18) reduces to

$$\frac{dA(\bar{n})}{A(\bar{n})} < \frac{d\bar{n}}{\bar{n}}, \qquad (19)$$

and its physical meaning becomes more straightforward. A stable steady-state discharge can be operated in a range of electron densities for which the relative changes of $A(\bar{n})$ due to the density fluctuations are either smaller than the relative changes in density or of opposite signs.

Similarity of Discharges

The behavior of a gaseous discharge depends on many variables. In order to organize and to simplify the description of the discharge, it has proven useful to reduce the number of variables by introducing properly chosen combinations of these variables (e.g., Golant, 1958; MacDonald, 1966). This procedure cannot replace the solving of a full set of the appropriate equations describing a particular physical problem. However, it simplifies the analysis of the discharge processes. Moreover, it has a considerable impact on the development of various concepts connected with the discharge theory, and it facilitates the way that the results of the discharge investigations are presented; but there is even more to it. The introduction of reduced parameters leads to the concept of similarity of discharges. This problem has been under consideration for decades (see, e.g., Francis, 1956; MacDonald, 1966) and is still far from being fully solved. In the recent literature, it is pursued in relation to microwave (e.g., Lebedev and Polark, 1979) as well as to dc (e.g., Pfau et al., 1969; Ferreira, 1983b) discharges. There are reasons for this continuing interest. First, once the legitimacy of some similarity law has been established, the quantities connected with this law can be readily determined for any discharge provided they are known from experiment for at least another similar discharge. This then increases by many times the value of the results gathered over the years from the investigations of various kinds of discharges. These results can be put directly into practical use for similar discharges, without the necessity of performing experiments for any particular case. The similarity principle also has merits of a more general nature. Its application helps to bring into the open and to understand general rules governing the behavior of the discharges.

We have already pointed out the key role played by the electrons of the plasma in the energy transfer to the microwave discharge. The shape of the electron energy distribution function (EEDF) determines the physical processes in the discharge. In the present case, the EEDF is assumed spatially uniform and the similarity of discharges can be defined as follows. Two discharges are similar if they occur in the same gas, in geometrically similar vessels, and if their EEDF is identical, up to a multiplication factor (electron density), in both discharges.

Consider two similar rf or microwave discharges. The EEDF is a solution of the Boltzmann equation. From the definition of similarity, it follows that the quantities characterizing these two discharges may only vary in such a way that the corresponding terms in the Boltzmann equation describing these discharges remain identical. In the case of discharges in long cylindrical tubes, the radius a is the only dimension affecting the maintenance of the discharge. It may be shown (e.g., Francis, 1956; Golant, 1958) that two discharges, sustained in tubes of radii a_1 and $a_2 = k\, a_1$, are similar if

The similarity relation of interest with respect to the present model for hf discharges is the dependence of Θ/p on pa. Fig. 8 shows the theoretical (Ferreira, 1983) and experimental (Chaker et al., 1982; Chaker and Moisan, 1985) dependence of Θ/p on pa in a surface-wave discharge in argon. The following conclusions can be drawn from it. First, the experimental points obtained under very different discharge conditions (ω, p, a) all fall on the same curve, i.e., Θ/p-versus-pa is a similarity relation in the range investigated. Second, these points are in good agreement with the available theoretical results. Third, since the density varies considerably under these conditions, it further proves that the value of Θ is independent of density. Finally, the fact that the radial distribution of the electric field intensity of surface waves varies considerably in shape with frequency and electron density (see further) does not seem to play any significant role on the value of Θ. This justifies, a posteriori, our assumption about being independent of the field applicator.

The fact that Θ/p-versus-pa defines a similarity relation for argon in a very wide range of operating conditions is surprising because the electron collision frequency in argon, due to a pronounced Ramsauer effect, is far from being velocity independent as required in our development of the similarity law. To date, there are no similar experimental data available for other gases. However, the fact that such a similarity law applies even in the case of argon sets up an optimistic precedent.

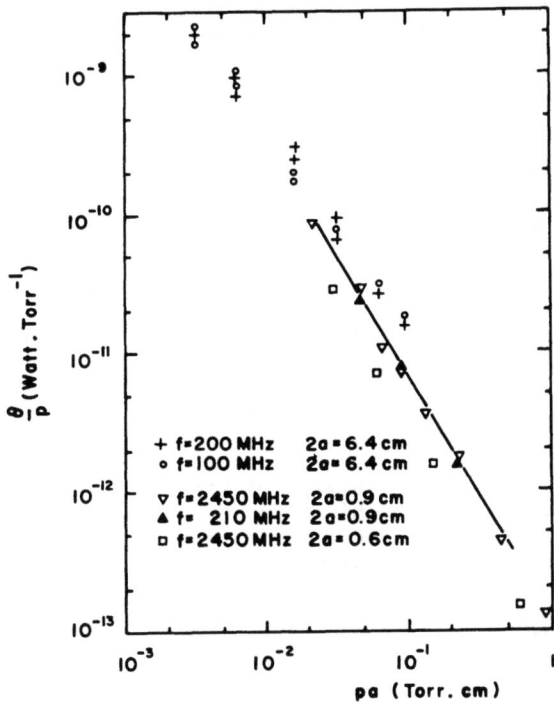

Fig. 8. Theoretical and experimental values of the ratio of power loss per electron to the gas pressure in an argon discharge (from Chaker and Moisan, 1985).

$$E_{e_1}/E_{e_2} = \tau_2/\tau_1 = k, \tag{20}$$

where τ is the time interval characterizing any plasma process. Notice that the radii of the discharge tubes do not appear directly in the Boltzmann equation. However, from the fact that $v = dx/dt$ (x is an arbitrary longitudinal coordinate), it may be inferred that all time-related and distance-related quantities have to vary proportionally in similar discharges. All the parameters related to those appearing in (20) should obey the same proportionality requirement. For example,

$$N_1/N_2 = p_1/p_2 = \ell_{e_2}/\ell_{e_1} = \nu_{m_1}/\nu_{m_2} = k, \tag{21}$$

where N and ℓ_e are the neutral particle density and the electron mean free path, respectively. As can be seen, proper combination of variables for HF discharges are, for example, E_e/N, E_e/p, $E_e a$, pa, ωa, ν/ω, and so on. If the similarity principle applies, this means that there exists definite functional relations (for a given gas) between such combined variables (an example is presented further).

At this stage, we would like to dwell in more detail on the question of applicability of the similarity laws. It follows from the above discussion that some general conditions should be met. The main conditions are the following. Since the EEDF has to be independent of the electron density, only processes involving interactions between electrons and neutral particles may take place in similar discharges.* This leads to a whole spectrum of processes which are, according to this, either "permitted" or "forbidden" in such discharges (e.g., Francis, 1956; Lebedev and Polark, 1979). The processes dominating in the discharge considered here, namely the direct ionization and the ambipolar diffusion, are both "permitted" (both ν_i and D_a are density independent). Notice, also, that the similarity in the sense discussed above requires the introduction of an effective maintenance field E_e or an equivalent parameter, like Θ. In principle, this can be done only for either $\omega^2 \ll \nu^2$ or for the case of a velocity-independent electron collision frequency. For this reason, strictly speaking, the similarity principle is applicable only to a small fraction of hf discharges. On the other hand, there does not seem to exist any reasonably tractable alternate way to describe these discharges. Thus, various attempts to extend the range of applicability of the similarity laws have been undertaken recently. The most promising solution to this problem is to introduce an effective electron collision frequency applicable over a specified range of discharge conditions. Examples may be found in the work by Lebedev and Polak (1979) and by Ferreira and Loureiro (1983).

It follows from the definition of similarity that it can include dc as well as hf discharges. A high frequency and a dc discharge are similar when all the above conditions are met, the effective field value in the dc case being the actual maintenance field of the dc discharge. Such a similarity situation is sometimes called the direct-current analogy (e.g., Lebedev and Polak, 1979). It makes available, for use in modeling of hf discharges, the wealth of the various experimental data accumulated during decades of investigation of dc discharges.

---

* Alternate definitions of similarity can be introduced. However, this always limits the range of application of similarity laws because one has to introduce an additional requirement on the values of electron density in such similar discharges (e.g., Golant, 1958).

Classification of RF and Microwave Discharges

Though a large variety of setups in which rf and microwave plasmas can be sustained have been described in the literature, there does not seem to exist any established and widely used classification of these discharges. When undertaking such a task, various criteria can be chosen (e.g., Kundel, 1966). In this presentation, we systematize the discharges according to the way by which the electromagnetic field is imposed on the plasma. In contrary to some earlier classifications (see, e.g., Marec et al., 1982), we feel that all the rf and microwave discharges fall into one of two broad categories.

The first category consists of discharges which are sustained within an hf circuit and for which the active zone of the discharge can be considered as a localized one. This is possible when the time difference between field variations at any two points within the discharge is small in comparison with the oscillation period. In this case, as a rule, the whole active zone of the discharge is confined within the field applicator and the influence of the plasma on the electrodynamic properties of the hf circuit can be represented by an equivalent lumped impedance. We propose to use the term "discharge with a localized active zone" to designate such discharges.

The second category includes all the remaining discharges. However, we shall reduce it to long rf and microwave discharges. These are discharges, long with respect to the tube diameter, sustained by the electric field of a wave propagating along the discharge. They have been labeled traveling-wave discharges (Zakrzewski, 1983). In this case, the active zone of the discharge extends in the direction of the wave propagation, and the time difference between field variations at different points within the active zone cannot be neglected in comparison with the wave period. In fact, it may exceed many times the wave period when the discharge length is large in comparison with the traveling wave length. The field applicator may, or may not, extend along the discharge tube. In the latter case, corresponding to the surface-wave discharge situation, the energy is carried out from the localized applicator (wave launcher) by a wave propagating along the plasma-dielectric interface.

Some typical examples of discharges of the first category are: the capacitively or inductively coupled rf discharges, and the microwave discharges that are sustained in resonant cavities or in waveguides (the latter case includes only such situations where the axes of the discharge tube and of the waveguide are perpendicular to each other). The surface-wave discharge, to be treated in detail in the present chapter, is a typical representative of the second category.

Let us now examine how the previously described physical model can be applied to each category of discharges. To this end, we have to know how the coefficients $\Theta(\bar{n})$ and $A(\bar{n})$ are affected by the way the field is imposed on the plasma. As far as the value of $\Theta(\bar{n})$ is concerned, we know that it is determined predominantly by the discharge conditions. It may vary from one setup to the other only because of a possible difference in the field distribution within the plasma. This effect, however, has usually little influence on the charged particle balance and power balance in the discharge and it can be neglected in practice. Thus, $\Theta(\bar{n})$ is the same for the discharges of the two categories, providing the discharge conditions are the same. On the other hand, the method of determining the absorption coefficient $A(\bar{n})$ depends not only on the category to which the discharge in question belongs, but it also depends on each particular setup considered. For discharges with a localized active zone, $A(\bar{n})$ can be calculated using the methods of circuit analysis. Note that the part of the equivalent circuit representing the plasma should not depend on the kind of setup considered. $A(\bar{n})$, however, once again is different for various discharge setups, even if the discharge conditions, and the plasma parameters, are the same. Sketches illustrating the power balance for a stable operation in a waveguide discharge and in a resonant cavity are given as examples in Fig. 9. Identical discharge conditions have been assumed in both cases as well as Θ being independent of the electron density; thus, the electron density, the absorbed power, and Θ are identical in both discharges. Nevertheless, $A(\bar{n})$ and the dynamical behavior of the discharge are radically different in the two cases shown.

We now turn to the discharges sustained by traveling waves. In this case, the plasma parameters may vary along the axis of the discharge. Thus, the balance of power has to be considered locally in each cross-section of the plasma column. We shall show further that, in this case, the power absorption coefficient $A(\bar{n})$, appearing in the present discharge model, can be expressed in terms of the wave attenuation.

Conclusion

The full set of equations describing the physical processes in a gaseous discharge, and the interaction between the plasma and the electromagnetic field sustaining the discharge, appears to be, in most instances, analytically intractable. The task of discharge modeling may be simplified by dividing it into two parts. First, one has to determine the dependence of the power loss per electron $\Theta(\bar{n})$ on the discharge conditions. This implies that the concept of the effective maintenance field E_e can be introduced. This can be done either theoretically or experimentally providing that some requirements, discussed in this section, are met. Second, one seeks the power absorption coefficient $A(\bar{n})$ which characterizes the electrodynamic behavior of the system consisting of the hf circuit and the plasma. This can be arrived at by using conventional methods from electromagnetic field theory and circuit analysis. The above formulation presents a complete, although not self-consistent, physical model of the discharge. With $\Theta(\bar{n})$ and $A(\bar{n})$ known, the description of the discharge behavior becomes a relatively simple and standarized procedure.

The analysis of the charged particle balance and energy balance makes it possible to obtain the main properties of the discharge. In the case of diffusion-controlled plasmas, this leads to a set of equations describing a steady-state condition, consisting of Eqs. (10) and (13). The first of these equations yields the spatial distribution, but not the

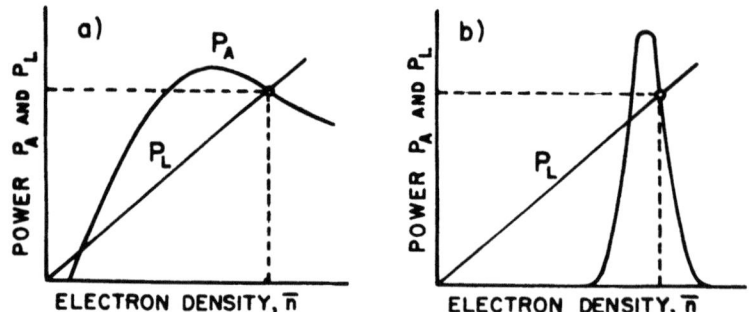

Fig. 9. Sketch presenting a comparison of the stable power equilibrium condition in two discharges under the same operational conditions, but with different field applicators: a) discharge in a waveguide, b) discharge in a cavity.

absolute value, of the electron density. With Θ and $A(\bar{n})$ known beforehand, the solution of the second equation provides the average electron density. If the EEDF does not depend on the electron density, the effective maintenance field intensity neither depends on the amount of power delivered to the plasma nor on the way the coupling between the plasma and the electromagnetic field is realized. It is determined exclusively by the discharge conditions: nature and pressure of the gas, shape and dimensions of the vessel, and wave frequency.

As far as the average electron density is concerned, it also depends on the amount of microwave power delivered to the discharge setup, as well as the discharge conditons, and on the electrodynamic characteristics of this setup, namely the absorption coefficient $A(\bar{n})$. For constant Θ, the density is proportional to the absorbed power density. In the simplest case where both Θ and A are density independent, the electron density is proportional to the input power.

The processes occurring in the plasma are predominantly governed by the discharge conditions and not by the way the HF field is imposed on the plasma. Therefore, the results obtained from the investigation of some particular kind of discharges may be applied to the modeling of other discharges. This proves the usefulness and importance of the similarity principle.

PRINCIPLE AND PHYSICAL MODEL OF SURFACE-WAVE DISCHARGES

Concept and Model of Traveling-Wave Discharges in General

Consider a long cylindrical tube filled with gas. An appropriate launcher excites an electromagnetic wave that propagates in the z-direction (z is a linear coordinate coinciding with the tube axis). Free electrons and ions are created under the action of the electric field, in the gas volume, and they recombine at the tube wall after reaching it due to diffusion. The wave power decreases gradually along z as it is used up in sustaining the plasma. The length of the active zone of the discharge is much larger than its diameter. The effective collision frequency ν_m for momentum transfer is constant in the plasma volume. The electron density may vary axially, but the axial density gradient is small with respect to the radially averaged transverse gradient.

Let us now concentrate on the application of our general model of an hf discharge to the case of traveling-wave discharges, of which the surface-wave discharge is a particular case. The axial density gradient can be neglected when writing down the balance equations. Therefore, the value of Θ (and that of the effective maintenance field) is the same in the traveling-wave discharge as in any uniform discharge under the same operational conditions (see the comparative analysis by Ferreira and Loureiro, 1984; also Ferreira, this Proceedings). The actual value of Θ may either be calculated when all the cross-section data are available, or determined from the existing experimental data if the similarity principle can be applied. Consequently, we shall assume that Θ is known for given discharge conditions.

Recall that to write the power balance equation, an absorption coefficient characterizing the interaction between the plasma and the electromagnetic field also needs to be known. Because of the axial non-uniformity of the discharge, the balance of power has to be considered separately for any elementary length of the plasma column. Let us denote by $dA(\bar{n})$ the fraction of the wave power absorbed in the plasma as the wave travels over a distance dz between the z and z = dz planes (Fig. 10). The amount of power absorbed is then:

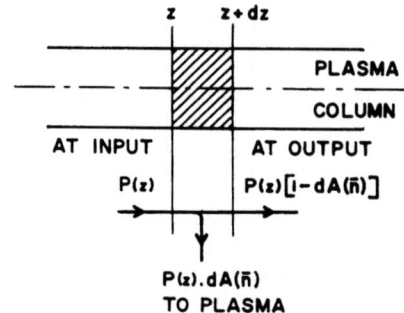

Fig. 10. Flow of power within an elementary slab for a traveling-wave discharge.

$$dP_A = dA(\bar{n}) P(z), \qquad (22)$$

where $P(z)$ is the total flux of the wave power in the z-plane.

To determine $dA(\bar{n})$, we need to consider the propagation of the wave sustaining the discharge. We will assume that the radial inhomogeneity of the plasma has a negligible influence on the propagation of the wave. This assumption has been verified experimentally in the case of a surface wave (e.g., Zakrzewski et al., 1977; Glaude et al., 1980; Moisan et al., 1982a; Shivarova and Zhelyazkov, 1982). The phase coefficient $\beta(\bar{n})$ and the attenuation coefficient $\alpha(\bar{n})$ of the wave, at each z = constant plane, depend only on the local value of the cross-section average electron density $\bar{n}(z)$. This assumption is, in fact, justified if the variation of the electron density over one wavelength distance is small, i.e.,

$$\frac{d\bar{n}(z)}{\bar{n}(z)} \ll \beta(n)dz. \qquad (23)$$

By definition, the attenuation coefficient links the amount of power $dP(z)$, lost by the wave when propagating along an elementary distance dz, with the total power flux:

$$dP(z) \equiv -2\alpha(\bar{n}) P(z) dz. \qquad (24)$$

Notice that $dP_A = -dP(z)$; thus from (22) and (24),

$$dA(\bar{n}) = 2\alpha(\bar{n})dz. \qquad (25)$$

One notices, therefore, that the macroscopic quantities $A(\bar{n})$ and P_I, which appear in the general model, in the traveling wave case, take the form of the local absorption coefficient $dA(\bar{n})$ and of the power flux $P(z)$. The graph of the power flow within an elementary section of the plasma column is shown in Fig. 10.

It follows from the above that, in the case of traveling-wave discharges, the attenuation coefficient $\alpha(\bar{n})$ of the wave sustaining the plasma becomes, together with Θ, the key element of the discharge model. Thus, to proceed further with the description of surface-wave discharges, we need to determine the attenuation characteristics of the surface propagating along a cylindrical plasma column.

Application to the Case of a Surface Wave Discharge

Now we shall concentrate on the application of our model to surface-wave-produced discharges. As they are a particular case of the traveling-wave discharges, the only quantity to be determined is the attenuation characteristics $\alpha(\bar{n})$. This relation is easily accessible experimentally (Zakrzewski et al., 1977; Glaude et al., 1980; Moisan et al., 1982b). As concerns its calculation, different methods may be used (Glaude et al., 1980). We will present one that is straightforward and easy to perform with present computer techniques, and consists of solving the dispersion relation using complex algebra. In this method, it is implicitly assumed that, for each value of $\bar{n}$ for which we want to determine α, we are dealing with an infinitely long, homogeneous plasma column, i.e., the axial gradient of the electron density is neglected and the radial density variation is taken care of by using the cross-section averaged density $\bar{n}$. This latter assumption is fully justified in the case of $\beta a < 1$ (Trivelpiece, 1967; Ferreira, this Proceedings) and has been verified experimentally (e.g., Moisan et al., 1982b).

In this method, the field components of the surface wave are complex functions of complex arguments. Using a complex wavenumber $h = \beta + j\alpha$, the axial component of the electric field for the $m = 0$ surface-wave mode, in the k medium of permittivity ε_k, is expressed as

$$E_{zk} = A_k J_0[(\beta_o^2 \varepsilon_k - h^2)^{1/2} r] + B_k H_0^{(1)}[(\beta_o^2 \varepsilon_k - h^2)^{1/2} r], \qquad (26)$$

where $H_0^{(1)}$ is the zero-order Hankel function of the first kind. The sign of the square roots is chosen so the imaginary part of the argument is positive. This restriction is necessary to obtain a zero field value with $H_0^{(1)}$ when r tends toward infinity. For example, in the case of a cylindrical plasma tube surrounded by air (vacuum), $B_k = 0$ in the plasma and $A_k = 0$ in air. The remaining field components E_{rk} and $H_{\phi k}$ are directly related to E_{zk} by the usual em field relations. From the boundary conditions for the field components E_{zk} and $H_{\phi k}$ at the various medium interfaces, one obtains a set of linear equations which has a nontrivial solution only if their corresponding determinant D is zero. The condition $D(\omega/\omega_{pe}, \nu, a; \alpha + j\beta) = 0$, for ω, ν, and a fixed, and ω_{pe} varying, yields two relations: ω/ω_{pe}-versus-β, called the phase characteristics (we use this term for the case of fixed ω, in contrary to the customary dispersion characteristics where ω varies and ω_{pe} is fixed), and ω/ω_{pe}-versus-α, called the attenuation characteristics. This method is valid whatever the plasma parameters are and, thus, whatever the value of α as compared to β. Figs. 11a and 11b show an example of such a calculated set of curves.

SURFACE-WAVE SUSTAINED PLASMAS

Experimental Apparatus and Procedures

A simplified diagram of the setup with which the experimental investigations are usually performed is shown in Fig. 12. The azimuthally symmetric surface wave is launched by a localized field-shaping and

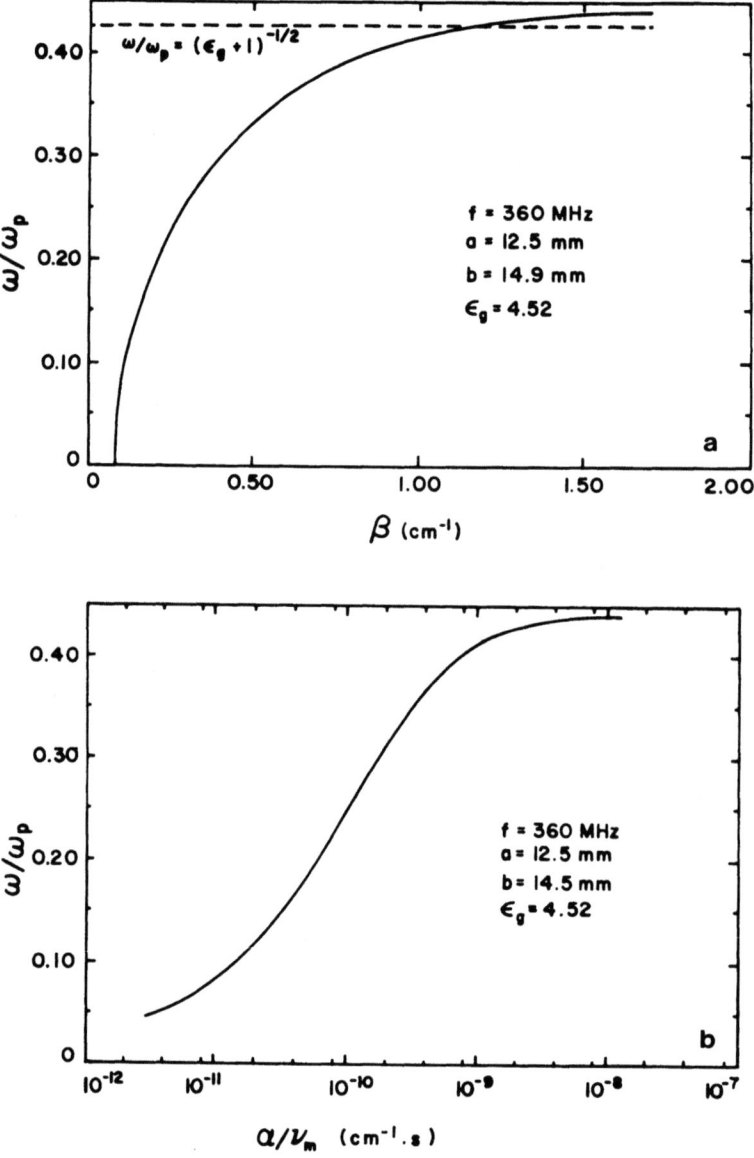

Fig. 11. Phase coefficient β (a) and attenuation coefficient α (b) for an axially symmetric surface wave propagating over a cylindrical plasma column enclosed in a glass tube (ω_p is the plasma angular frequency).

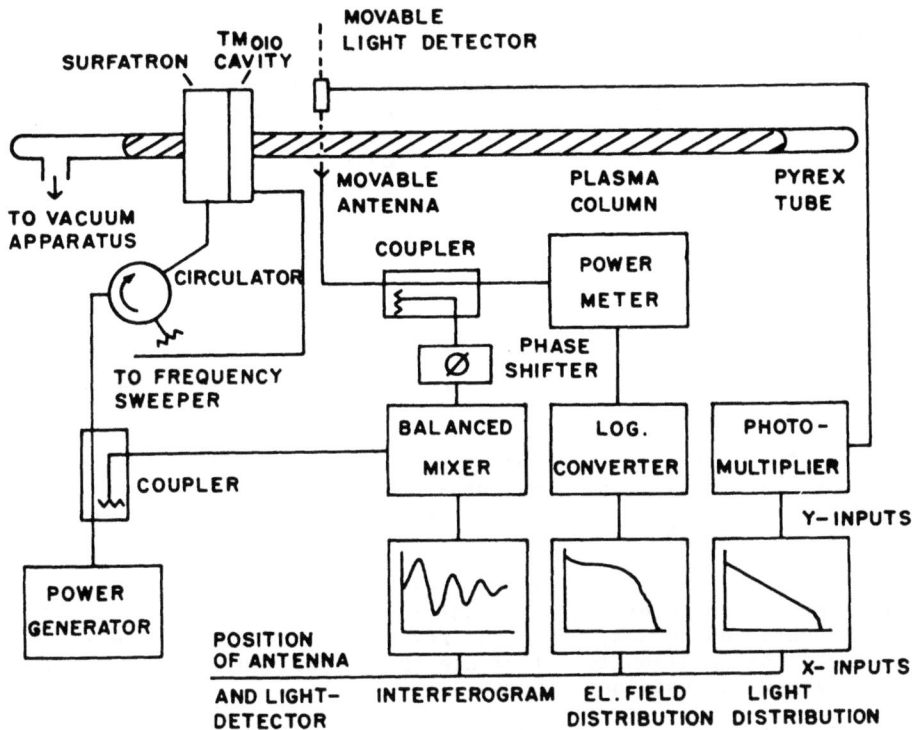

Fig. 12. Diagram of the experimental setup.

impedance matching device.* This wave creates a plasma column that extends from the launching structure. The interface between the plasma and the surrounding dielectric serves as a wave-guiding structure. Both the total flux of the wave power and the electron density in the plasma decrease gradually with increasing distance from the launcher. In the low pressure limit ($\nu_m^2 \ll \omega^2$), the column ends where the electron density reaches the value corresponding approximately to the dipole resonance for a cylindrical plasma column: $n_D = n_c (\varepsilon_g + 1)$, where n_c is the critical density, $\omega_{pe}(n_c) \equiv \omega$, and ε_g is the plasma tube relative permittivity.

The experimental setup shown in Fig. 12 allows the use of a TM_{010} resonant cavity for the measurement of the electron density. This method yields an average value for the plasma contained within the cavity. The cavity is moved axially along the plasma column to probe it. To minimize the perturbation that it causes to the propagation of the wave, the passage holes through which the plasma column crosses the cavity are made 1-1/2 to 2 times larger than the plasma tube diameter. The resulting edge effects on the TM_{010} mode are taken into account by a proper calibration using conventional dielectrics with known values. An alternate

---

* A variety of surface wave launchers has been described in the literature in connection with experimental investigations of wave propagation (Moisan et al., 1982b) and plasma production. Some of these wave launchers, specifically intended for the efficient generation of long plasma columns, have been presented by the authors and colleagues (Moisan et al., 1979, 1982a).

way to obtain the axial distribution of the electron density is to use the dispersive properties of the wave, i.e., the fact that its wavelength depends on the electron density. This is done by taking interferograms of the wave along the plasma column, and using a double-balanced mixer and a reference signal from the generator output. This yields $\beta(z) = 2\pi/\lambda(z)$ and, from the dispersion relation, the value $\bar{n}(z)$. This latter method perturbs the plasma less and can be used at higher gas pressures and higher electron densities than the cavity method. It requires, however, that the validity of the calculated phase relation has been demonstrated experimentally.

The total power flux carried by a surface wave is approximately proportional to E_r^2, the radial component of the electric field outside the tube, everywhere, but near the end of the column. When this assumption can be made, this provides a convenient way of measuring the axial distribution of wave power, using a movable radially oriented antenna connected to a bolometer.

The above experimental apparatus and procedure were first described by Zakrzewski et al., (1977) and became a standard way of investigating surface-wave discharges (e.g. Glaude et al., 1980; Chaker et al., 1982; Nghiem et al., 1982).

There exists ample experimental data concerning surface-wave discharges sustained under various conditions (see, e.g., Moisan et al., 1982a; Chaker and Moisan, 1985, and references therein). Some are used further in this section to illustrate the properties and applications of the surface-wave sustained plasmas.

Properties of Surface-Wave Produced Plasmas

Theoretical summary on the radial variation of the surface-wave field component intensity. Some properties of the surface-wave-produced plasmas are dependent on the spatial distribution of the electric field within the plasma, as we shall see. This field distribution is different for the various surface-wave modes. These are defined by the value of an integer m appearing in the argument of the phase term, exp (jmΦ), affecting the field components and representing their azimuthal dependence. Although various surface-wave modes could eventually be used for plasma generation, in what follows we only consider the azimuthally symmetric mode (m = 0). In this case, the components of the electric and magnetic field of the wave are, using cylindrical coordinates, E_r, E_z, and H_Φ, respectively. This defines a transverse magnetic (TM) wave since the wave-vector is directed axially, along the z coordinate.

Consider a homogeneous plasma column of relative permittivity ε_p located in a dielectric tube of relative permittivity ε_g, the latter being surrounded by air (vacuum). Using the full set of Maxwell's equations and the usual boundary conditions at the plasma-tube and tube-air interface, one obtains the radial distribution of the electric field components illustrated in Fig. 13. The electric field intensity increases toward the interface between the plasma and the surrounding dielectrics (glass and air), hence the name of the wave. The electric field intensity, within the plasma, assuming that the plasma is homogeneous, is given by:

$$E_z(r) = A_1 I_0[(\beta^2 - \beta_0^2 \varepsilon_p)^{1/2} r], \quad \text{and} \tag{27}$$

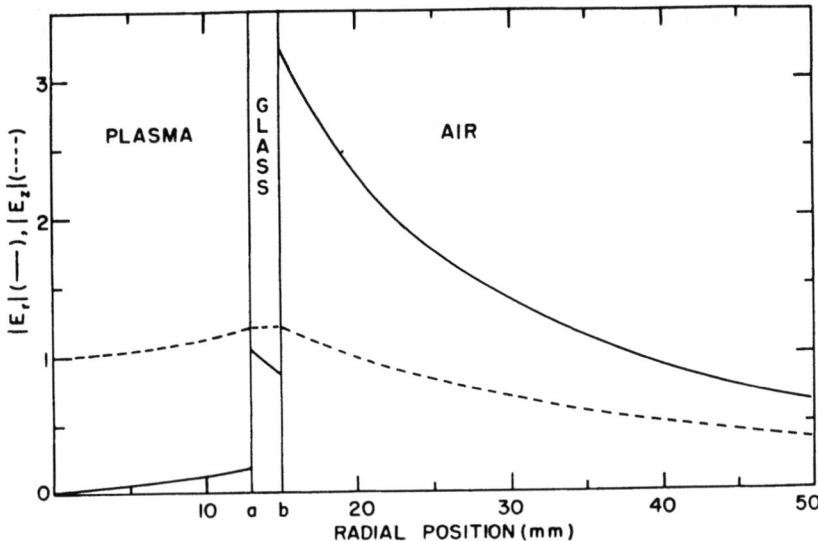

Fig. 13. Theoretical radial distribution of the intensity of the electric field components for an axially symmetric surface wave in a homogeneous plasma (from Moisan et al., 1982b).

$$E_r(r) = A_2 \, I_1[(\beta^2 - \beta_0^2 \, \varepsilon_p)^{1/2} r], \tag{28}$$

where A_1 and A_2 are constants, I_0 and I_1 are modified Bessel functions of the first kind, of order zero and one, respectively, and $\beta_0 = 2\pi/\lambda_0$, where λ_0 is the vacuum wavelength. Fig. 14 shows the radial distribution, within the plasma, of the two electric field components, for two different values of the electron density. From these two figures, one notes two important points: The main field component within the plasma is E_z (whereas, outside the plasma, in air, it is E_r), and the radial variation of $E_z(r)$ presents a deeper and deeper minimum at the axis as the electron density increases (the same is true for increasing the wave frequency, as could be shown).

In reality, the electron density varies both radially and axially in a surface-wave-produced plasma column. One can neglect the axial variation when looking for the radial distribution of the field components. This amounts to numerically solving a differential equation coupled to the equations yielding the electron density radial distribution (this is a self-consistent problem, as mentioned earlier). The overall behavior of the field distributions obtained remains the same and, to a first approximation, the radial distribution of the total electric field intensity can be fitted with:

$$E_T(r) = A_0 \, I_0 \, (\sigma r), \tag{29}$$

where A_0 is a constant and σ is a fitting parameter that increases with the cross-section average value of the electron density and with the wave frequency (see Ferreira, this Proceedings, for an exact solution).

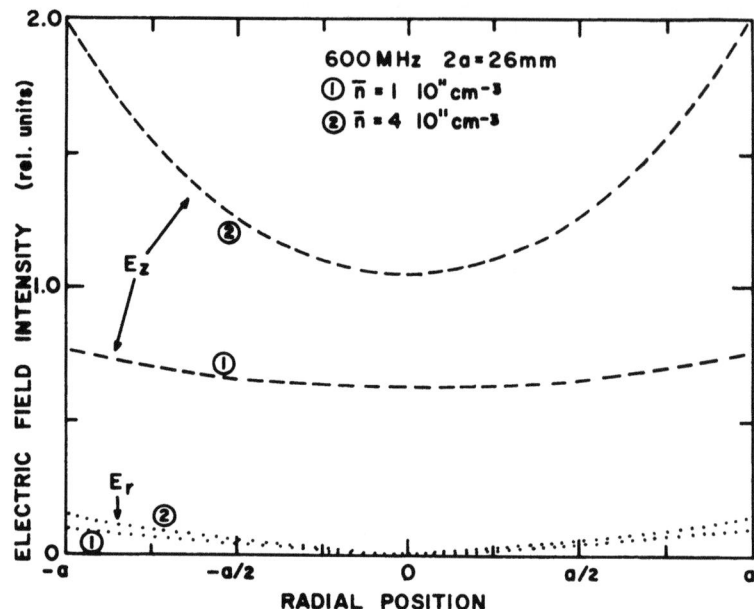

Fig. 14. Theoretical radial distribution of the components of the electric field intensity for an axially symmetric surface wave within a homogeneous plasma column (from Moisan et al., 1980).

Range of the discharge conditions in surface-wave-produced plasmas. The characteristics of the azimuthally symmetric surface-wave-produced plasma do not depend on the particular type of launcher used. In fact, they are influenced solely by the discharge conditions (nature and pressure of the gas, wave frequency, diameter and dielectric permittivity of the tube) and by the wave power that is absorbed in the plasma.

As concerns the nature of the gas, the most systematic set of results obtained until now are for argon, though discharges were realized in a large variety of gases and mixtures of gases (see under Applications). The pressure range over which a stable discharge can be operated depends on the plasma tube diameter, as shown in Table 1 for argon. Note that a stable discharge can be obtained at pressures as low as those attainable with dc discharges, but as a rule this is more easily achieved with surface-wave plasmas. As for the high pressure limit, stable argon plasmas were recently obtained, in capillary tubes of i.d. from 1 to 4 mm, at pressures up to about ten times the atmospheric pressure, this limit being determined by technical considerations. As a rule, the plasma obtained at atmospheric pressures is stable, provided that the plasma tube i.d. is less than 5 to 7 mm, but it is constricted, i.e., it does not fill the entire cross-section of the tube. It appears as a bright, straight filament, centered on the axis, with a diameter smaller than the tube bore. This filament is not in contact with the walls (such a constriction is commonly observed with other hf plasmas as well as with dc arcs, operated at atmospheric pressures).

Concerning the frequency range, the surface-wave plasmas investigated have frequencies from 27 MHz up to 10 GHz, although such plasmas have been produced at still lower frequencies ($\simeq$ 7 MHz). It is believed that the new launchers now under development will allow operation below the 1-MHz mark. The possibility of producing an hf plasma over such a large frequency band, using the same field configuration throughout, is a unique feature of surface-wave-produced plasmas.

Table 1. Argon pressure range for stable operation with different plasma tube diameters.

Tube i.d. (mm)	Minimum Pressure (torr)	Maximum Pressure (torr)
0.5 to 7	~10^{-2}	~ 7000 (maximum tested value)
25	~10^{-3}	~ 20
150	~10^{-4}	~ .35[a]

[a]This value increases by more than two orders-of-magnitude when gas flow is permitted.

A large variety of plasma column diameters have been obtained (0.5 mm to 150 mm demonstrated) as well as column lengths up to 6 meters. These limits are a result of technical constraints. The flexibility of these discharges as far as operating conditions are concerned thus appears to be extremely large as compared to other hf-produced and dc discharges.

Plasma parameters and characteristics of surface-wave plasma columns. The parameters of interest are the electron density, the density of excited atoms, and the average electron energy as well as the shape of the electron energy distribution.

When the diffusion is the main loss mechanism, the shape of the electron density radial distribution is determined mainly by the shape of the vessel and is not much affected by the radial variation of the surface-wave electric field intensity (Ferreira, 1981). The radial profile of the electron density, as obtained from calculations (Ferreira, 1981), is only slightly flatter than that given by the Bessel function, $J_o(2.4\ r/a)$, that describes the electron density radial distribution in the positive-column plasma. On the contrary, the electron density axial distribution in surface-wave plasma columns shows a distinctive behavior. We thus turn to examine this problem in detail, determining the role played by the amount of power absorbed in the plasma column and by the discharge conditions. In such a study, the radial distribution of the electron density can be ignored and the electron density in a thin slab transverse to the axis can be expressed by the cross-section average value:

$$\bar{n}(z) = 2\pi \int_0^a n(r,z)\ r\ dr/(\pi a^2). \tag{30}$$

Fig. 15 is a typical example of the axial distribution of the electron density, $\bar{n}(Z)$, recorded along a surface-wave-produced plasma column. The electron density decreases linearly away from the launcher, and the column ends abruptly where the wave ceases to propagate. There are two essential characteristics of surface-wave discharges. First, the plasma column length increases with the power transferred to the plasma and, in most instances, the column is much longer than the launching structure. Second, the electron density axial gradient is completely determined when

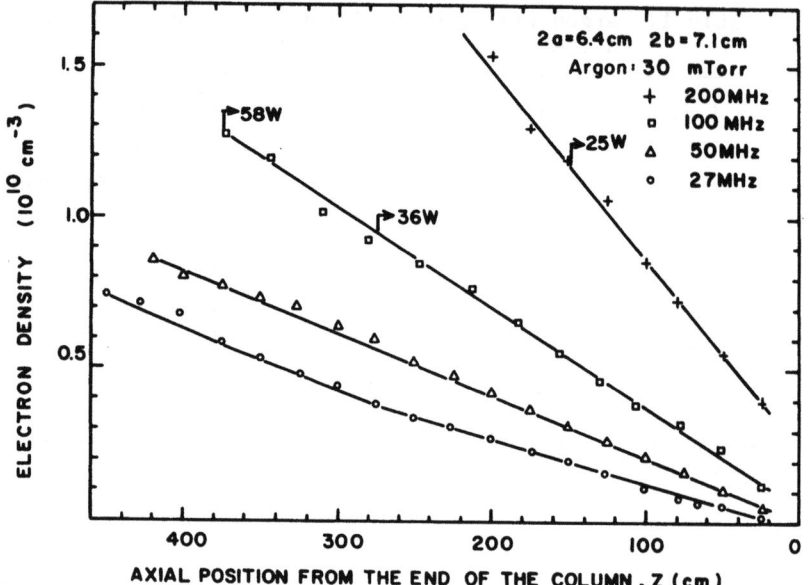

Fig. 15. Axial distribution of the electron density at various wave frequencies (from Chaker et al., 1984).

the wave frequency, the gas pressure, and the tube diameter are given, and it does not depend on the HF power level. Thus, when the power is raised from P_1 to P_2 (say from 36 W to 58 W in Fig. 15), the plasma column that existed at power P_1 is translated, without being modified, away from the launcher. An additional plasma length is created that extends from the launcher gap and connects with the beginning of the P_1 column. The electron density plot of this additional plasma length shows that it is the continuation of the density plot of the plasma column obtained for power P_1 (see the arrows on the density plot in Fig. 15; they indicate the position of the launcher aperture (gap) for the corresponding total absorbed power in the column). As a result of this behavior, the axial position along the plasma column is better referenced from the end of the column rather than from the surfatron gap region.

The power absorbed in the entire plasma column is the difference between the power of the incident wave and that of the reflected wave in the feeder line, as measured with a directional coupler line. Most of the experimental results that we are reporting have been obtained in diffusion-controlled discharges. In such a case, Θ is approximately independent of the electron density so that, for given discharge conditions, the cross-section average density $\bar{n}(z)$ at any position can be assumed to be proportional to the local value of the power absorbed per unit length of the column. Therefore, the axial density distributions shown further also represent the axial distribution of the absorbed power.

Fig. 15 shows that the value of the axial gradient of electron density along a surface-wave plasma column, for a given tube diameter and a given pressure, increases with the wave frequency. Also, it can be seen that the density value at the end of the column ($z = 0$) increases with the wave frequency. This density value can be calculated from the approximate dispersion relation that characterizes the end of the plasma column and is given further in Eq. (32). Finally, it can be seen in Fig. 15 that a plasma column length in excess of 4 meters has been achieved.

Fig. 16 shows that, for a given wave frequency ($f = \omega/2\pi$) and plasma diameter, the axial gradient of the electron density increases with the gas pressure. Fig. 17 shows that, for a given wave frequency and gas pressure, the electron density gradient decreases as the plasma diameter increases. As can be seen from Figure 18, it is possible to sustain a surface-wave plasma column of similar properties by waves of greatly differing frequencies.

The observed influence of the discharge conditions on the plasma parameters (Figs. 15 to 18) can be retrieved from a single theoretical relation (Mateev et al., 1983). For a cold, low collision ($\nu_m^2 \ll \omega^2$) plasma in the ambipolar diffusion regime and in the electrostatic approximation, the axial gradient of the electron density can be expressed as

$$d\bar{n}/dz = 0.73 \times 10^{-2} \, f \, \nu/a \quad (cm^{-4}), \qquad (31)$$

where the wave frequency is in MHz. This equation yields the same $\bar{n}(z)$ variations as observed in Figs. 15, 16, 17, and in Fig. 18, except for the plasma column obtained at 210 MHz and 0.2 torr. This latter situation corresponds to $\nu = \omega$, i.e., it is no longer possible to assume a low collision plasma as in Eq. (31).

Fig. 19 shows that it is possible to achieve a surface-wave plasma in capillary tubes at pressures exceeding the atmospheric pressure. The gas used is argon, but supra atmospheric plasmas have also been achieved in helium, hydrogen, and nitrogen (Abdallah et al., 1982; Chevrier et al., 1982). Note that these atmospheric pressure plasmas are not covered by the existing theoretical modeling of surface-wave plasmas.

In summary, we observe in surface-wave plasmas a large range of electron densities (0.5×10^8 cm^{-3} to 1.5×10^{15} cm^{-3}), depending on the discharge conditions. The electron density, in the absence of a reflected wave, is always maximum at (or very close to) the launcher gap, and this value increases with the power transmitted to the launcher. Along the plasma column, the electron density decreases away from the launcher (according to Eq. (31), if $\nu^2 \ll \omega^2$) and, finally, reaches a minimum value, determined by wave propagation conditions. This minimum value is given approximately by

$$\bar{n}(0) \simeq 1.2 \times 10^4 \, (1 + \varepsilon_g) \, f^2 \quad (cm^{-3}), \qquad (32)$$

with the wave frequency expressed in MHz (the relative permittivity ε_g of the tube wall is 4.52 and 3.78 for pyrex and fused silica, respectively). This relation is valid for $\nu/\omega \ll 1$; as ν/ω increases above this condition, the density value at the end of the column grows larger than that given by Eq. (32), but it still increases with the wave frequency. However, when ν/ω becomes larger than unity, the plasma parameters can no longer be controlled by changing the wave frequency; such is the case at atmospheric pressures (Hubert et al., 1979).

It is clear from the above that surface-wave plasmas are axially inhomogeneous media. This inhomogeneity may be a shortcoming in some applications. There exists possible remedies to this situation (see under <u>Applications</u>).

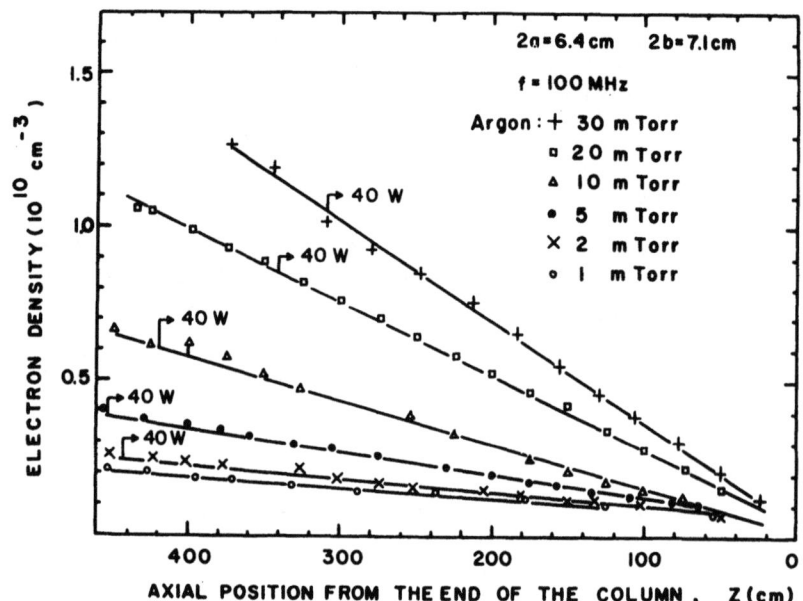

Fig. 16. Axial distribution of the electron density at various gas pressures (from Chaker et al., 1985).

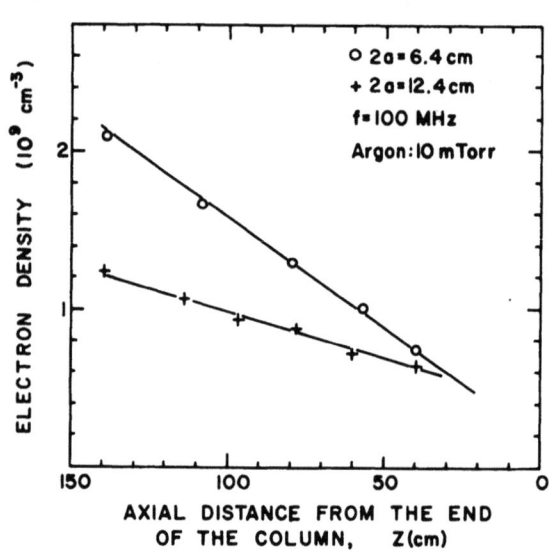

Fig. 17. Axial distribution of the electron density in tubes of different diameters (from Chaker et al., 1985).

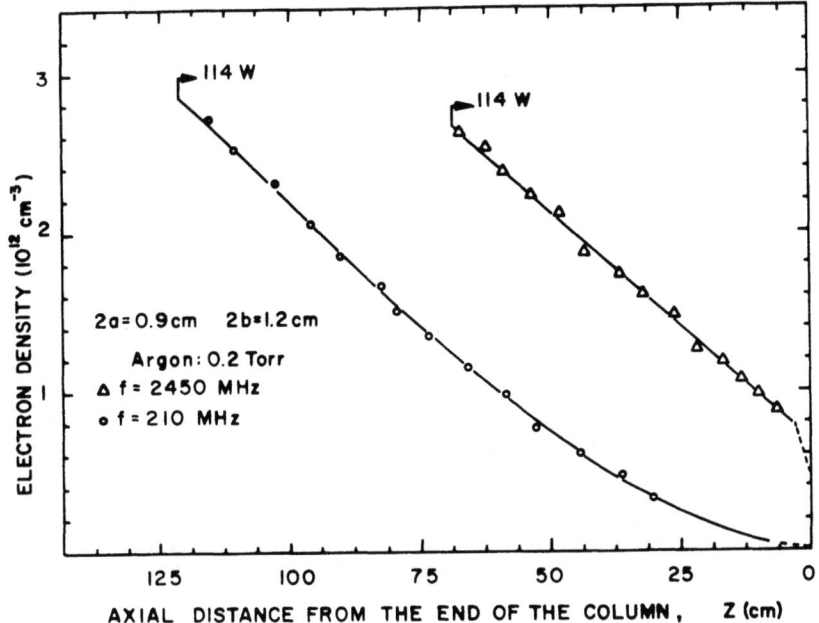

Fig. 18. Axial distribution of the electron density in plasma columns sustained by surface waves of greatly differing frequencies (from Chaker et al., 1984).

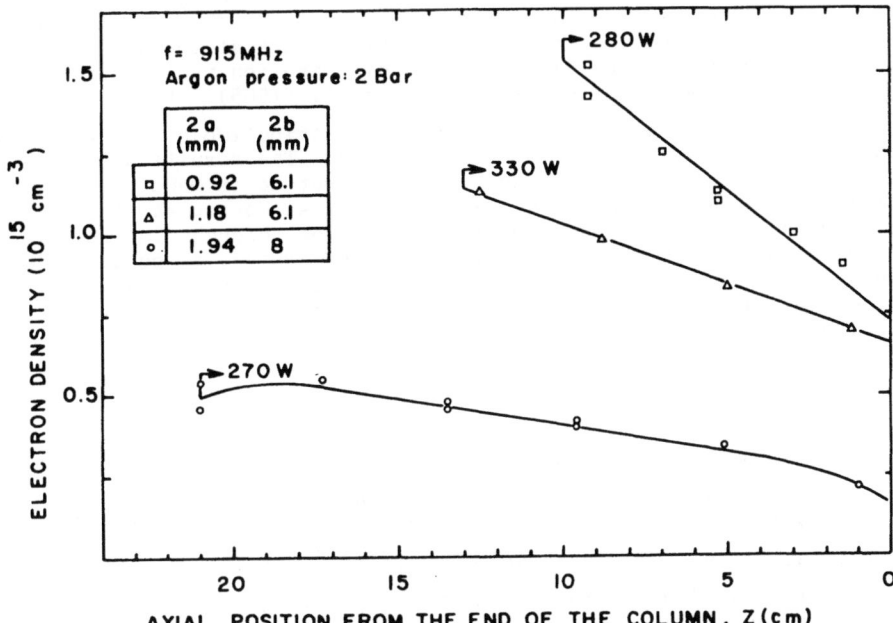

Fig. 19. Axial distribution of the electron density in an atmospheric pressure plasma sustained in capillary tubes of various diameters (Moisan, Pantel and Hubert, unpublished).

In surface-wave-produced plasmas, the shape of the radial density distributions of excited atoms can differ considerably from those observed in a positive-column plasma, where the radial distributions of all species are close to that given by J_o (2.4 r/a). The shape of these density distributions depend considerably on the gas pressure, the wave frequency, and the plasma diameter. Moreover, they show a strong dependence on whether the excited atoms concerned are in a radiative state or in a long-lived state (metastable and resonant states). These differences are so marked that they can be used to determine the predominant channel among a series of possible kinetic channels populating a given level (Ricard et al., 1983).

All the experimental results presented in this section were obtained by an end-on measurement method, both tube ends being terminated by windows (Moisan et al., 1980, 1982a; Pantel et al., 1983). With this technique, the light emitted (or absorbed), integrated along the axis, is measured at a given radial position. A radially movable and axially oriented collimator, positioned at the end of the tube, is used for this purpose. As a rule, the plasma tubes used are short (10-20 cm long). Thus, by means of optical emission measurements, the relative radial density distribution for the excited atoms in a radiative state may be determined. On the other hand, optical absorption measurements yield the radial density distribution of the atoms in a metastable or resonant state.

Radiative States

Figures 20a, 20b, 20c, and 20d show the relative radial density distribution of excited atoms in a given radiative state (ArI, 549.6 nm, is optically thin over the tube length). As can be seen from Fig. 20a, in tubes of large enough diameters, the radial distribution shows a maximum near the wall and a minimum at the axis. This radio of the maximum to the minimum density increases with increasing tube diameter. On the other hand, as the tube diameter is reduced, the distribution tends to flatten, and the relative maximum appears at the axis. By further reducing the tube diameter, the distribution would approach the J_o (2.4 r/a) Bessel function (Moutoulas et al., 1985). Figs. 20b and 20c show the dependence of the radial density distribution on the wave frequency and on the gas pressure, respectively. The results presented here have been obtained for argon, but recent measurements made with neon, helium, and He-Ne mixtures gave similar results (Moutoulas et al., 1985; Ricard et al., 1985). When one-step excitation through electron collision with an atom in the ground level can be assumed, these distributions depend on the radial distribution of the electric field intensity and on that of the electron density, through the relation (Ferreira and Loureiro, 1983):

$$n_j(r) = A\, n(r)\, E_T^k(r), \qquad (33)$$

where $n_j(r)$ is the population density of the radiative atoms in level j, A is a constant independent of position, and k is a value that can be determined from theory (Ferreira and Loureiro, 1983). For argon, it is 1.7 for the $^3P_2$ and $^3P_0$ metastable levels, 2.0 for the $^3P_1$ and $^1P_1$ resonant levels, 2.9 for the group of higher-lying levels, and, finally, 4.1 for ionization.

The above results can be unified by noting that the same sequence of radial profiles (i.e., Bessel shape → flat shape → minimum at the axis)

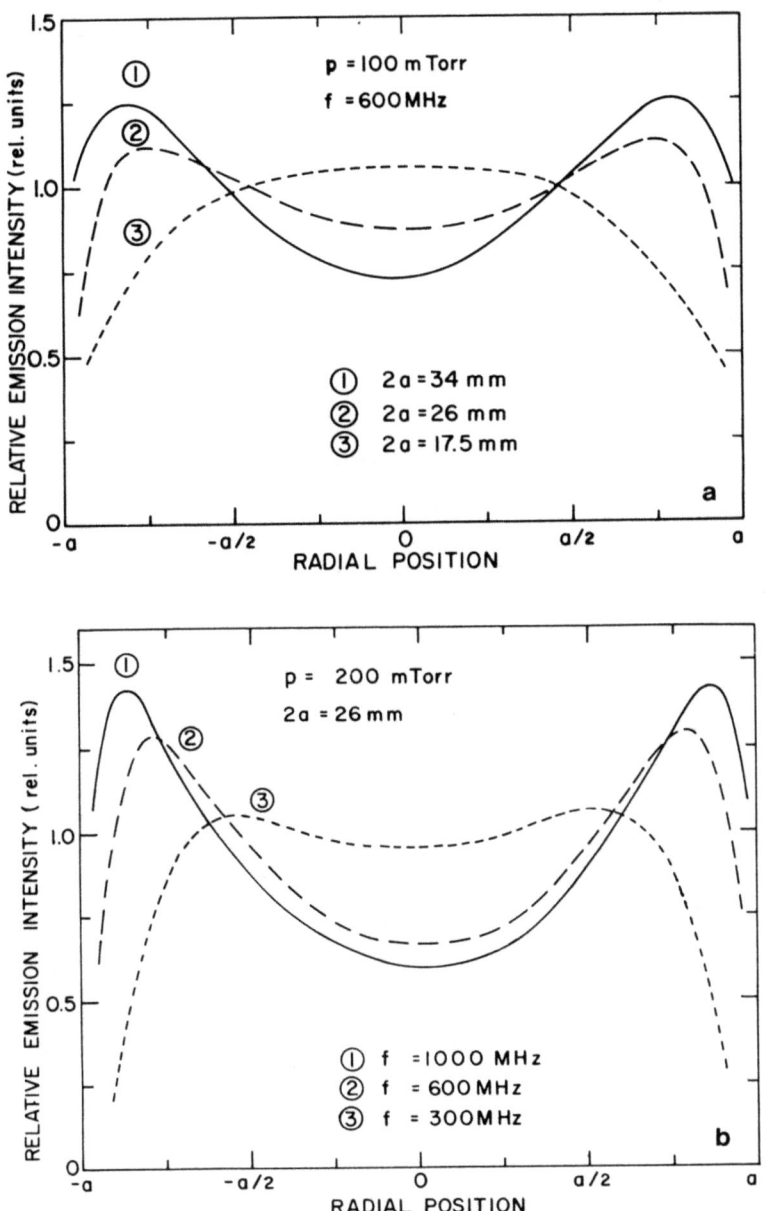

Fig. 20. Observed dependence of the emission intensity of the ArI, 549.6-nm thin line on tube diameter (a), wave frequency (b), gas pressure (c), and electron cyclotron frequency f_{ce} (d). The data for (c) are from Moisan et al., 1982a and the remaining from Moisan et al., 1982c.

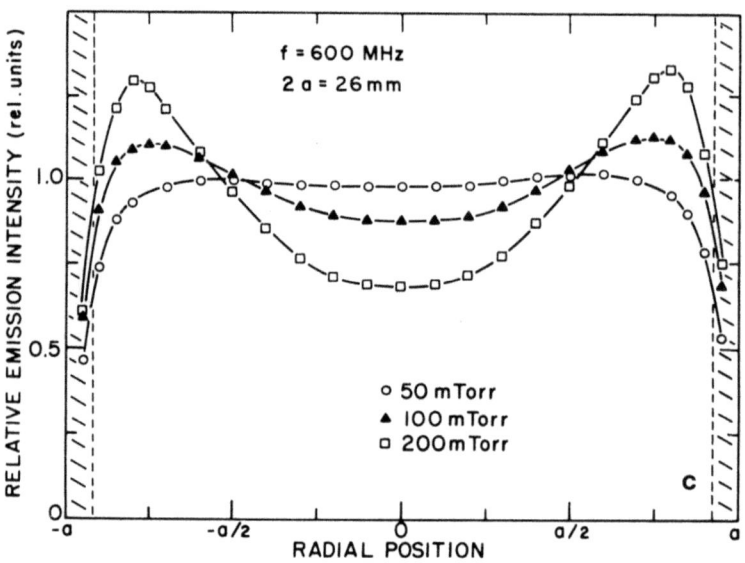

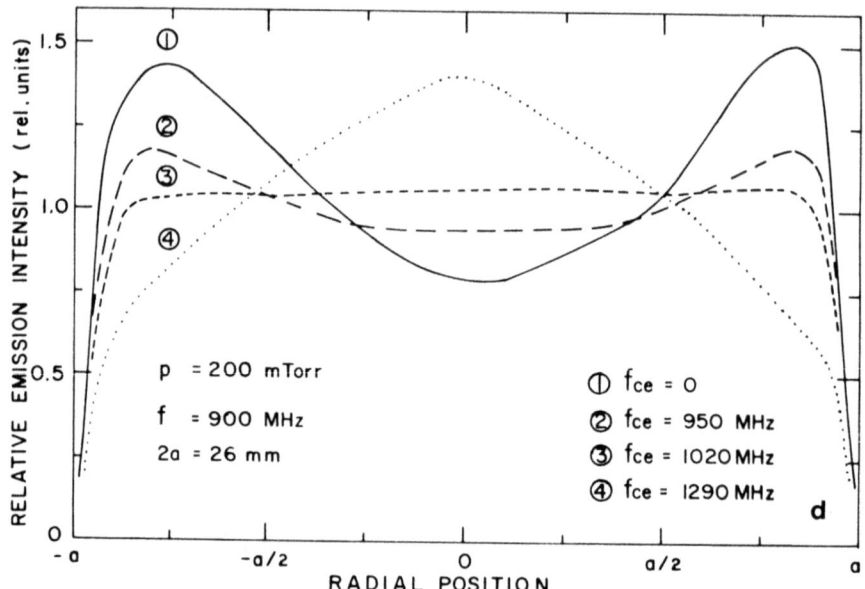

Fig. 20. (cont'd) Observed dependence of the emission intensity of the ArI, 549.6-nm thin line on tube diameter (a), wave frequency (b), gas pressure (c), and electron cyclotron frequency f_{ce} (d). The data for (c) are from Moisan et al., 1982a and the remaining from Moisan et al., 1982c.

can be observed when either increasing the wave frequency, the gas pressure, or the tube diameter. This sequence is determined by the behavior of the electric field and it can be followed by combining Eq. (33) with the approximate relation (29) for E_T: as the argument (σr) grows, σ increases with frequency and pressure (i.e., with electron density) and r becomes larger with increasing tube diameter. Such a dependence explains that, even in small bore tubes (< 5 mm) but at much higher electron density ($\simeq 10^{15}$ cm^{-3}, at atmospheric pressures), one can observe a maximum excited atom density close to the tube wall (Waynant et al., 1985; see under <u>Applications</u>).

Finally, Fig. 20d shows the influence of an axially uniform dc magnetic field on the radial density distribution of atoms in a radiative state (the electron-cyclotron frequency is related to the magnetic field intensity B expressed in gauss by $f_{ce} = 2.7 \times 10^6$ B (Hz)). The gradual increase of the relative density at the axis reflects the transformation of the surface wave into a volume wave (Moisan et al., 1982b), as the intensity of the magnetic field is increased (in the present case, $\nu \ll \omega_{pe}$).

In summary, when decreasing either the tube diameter, the gas pressure, or the wave frequency, the radial density distributions of atoms in a radiative state will tend toward that observed in the positive column plasma of a dc discharge. By using the opposite conditions, and eventually applying an axially constant magnetic field, a large variety of shapes (e.g. square, hollow, bell shaped) can be obtained for these distributions.

<u>Long-Lived States</u>

Figure 21 shows the radial density distribution of argon atoms in the $^3P_2$ metastable state, for three different gas pressures. The tube diameter and the wave frequency are the same as in Fig. 20c. One notes that the density distribution of the $^3P_2$ metastable level, in the range 50-350 mtorr, is flatter than the density distribution observed in Fig. 20c for atoms that are in a radiative state. For example, the relative minimum in the $^3P_2$ density distribution at 350 mtorr is less pronounced than the one at 200 mtorr for the radiative state density. The reason for this is that the atoms in a metastable state, being long-lived, have time to diffuse radially before going, as a result of a collision, into another state. The diffusion tends to level the density profiles. The radial density of atoms in resonant states (e.g., $^1P_1$) are similar to that of the $^3P_2$ level. The trapping of photons connected with these resonant levels makes them appear as long-lived species.

To prove that the flatter density profile of the $^3P_2$ level is due to diffusion, we look in Fig. 22 to what happens when 10% of nitrogen is introduced in the argon plasma. The presence of nitrogen quenches the argon metastable state so rapidly that its density distribution resembles that of a radiative state; the density profile is that given by Eq. (33). One further notes that the density distributions of the $^3P_2$ level obtained through emission and absorption measurements are very close to each other, which proves that the ArI, 696.5-nm line (its lower level is

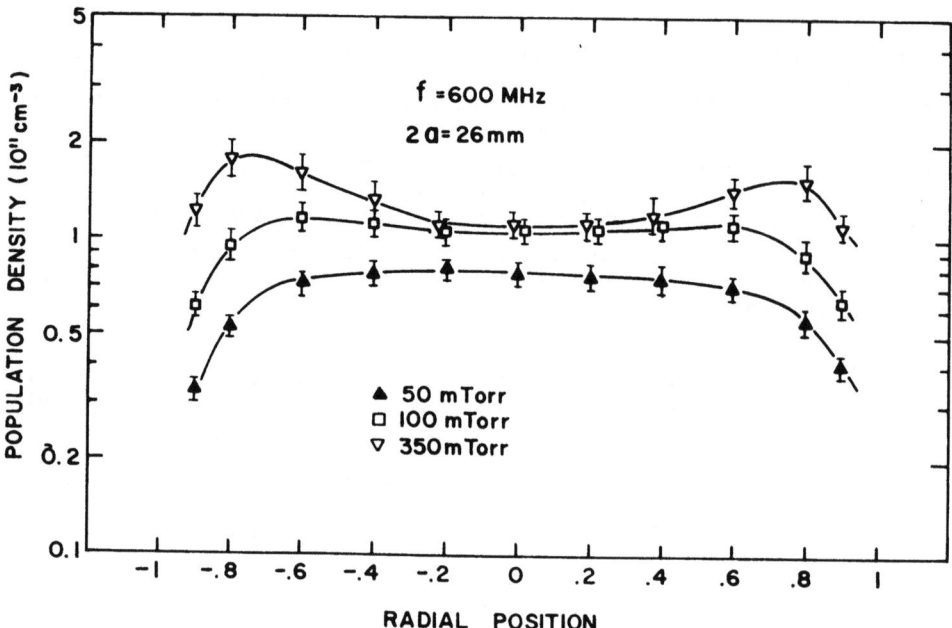

Fig. 21. Observed radial distribution of the population density of $^3P_2$ argon atoms in a metastable state (from Moisan et al., 1980).

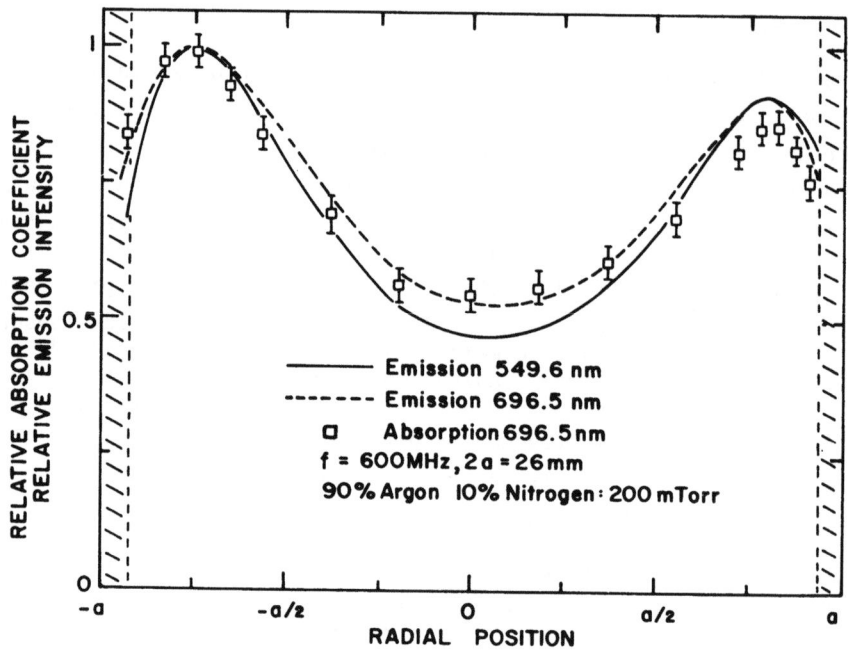

Fig. 22. Influence of the addition of nitrogen on the radial distribution of the $^3P_2$ argon density (compare with Fig. 21). The $^3P_2$ level population density is obtained through the ArI 696.5-nm line (from Moisan et al., 1980).

the $^3P_2$ level) is optically thin under such heavy quenching. Finally, the fact that the density profile of the radiative state yielding the ArI, 549.6-nm emission line has a comparatively slightly deeper relative minimum than that of the level yielding the ArI, 696.5-nm emission line can be attributed to the fact that this state lies higher in the energy diagram. The value of the exponent k in Eq. (33) in this case should be slightly larger.

It is interesting to compare the radial density distributions for the Ne($^3P_2$) metastable and for the Ne($^1P_1$) resonant levels in a surface-wave plasma with those in a positive-column plasma, in tubes of the same diameter and length, at the same gas pressure. This is shown in Fig. 23. It can be seen that:

(1) The radial distributions for the surface-wave plasma are comparatively flatter.

(2) The maximum density value attained radially, in both types of discharges, is not very different. However, it can be noted that the maximum density value is slightly larger for the $^3P_2$ level in the positive-column plasma, while it is the opposite for the $^1P_1$ level. The same behavior, with respect to the shape of the density distributions and to the relative values of the $^3P_2$ and $^1P_1$ density maximum, in these two types of discharges, had been observed in argon (Pantel et al., 1983).

(3) For the gas pressures reported, the surface-wave plasma has a larger cross-section density $\bar{n}_j$, where $\bar{n}_j$ is defined in a similar way as $\bar{n}$ (Eq. (30)). The numerical integration of the distributions in Fig. 23 shows that $\bar{n}(^3P_2)$ and $\bar{n}(^1P_1)$ are 2.5 and 5.2 larger, respectively, in the surface-wave plasma as compared to the positive-column plasma (Ricard et al., 1985). Similar results are also reported with the $2\,^3S$ and $2\,^1S$ metastable states in helium (Ricard et al., 1985).

A larger yield of metastable atoms can, under certain circumstances, increase the reaction rate in plasma chemistry, where a large population density for resonant states may be of interest in realizing intense uv lamps. In the latter case, the fact that the $^1P_1$ radial distribution extends to the wall with an appreciable density value means that a still larger number of photons, as compared with a positive-column plasma, will leave the tube since the trapping of photons in the plasma volume is less probable for those created close to the wall.

As discussed earlier, under ambipolar diffusion conditions, the average electron energy is determined almost exclusively by the discharge conditions. As a consequence, the average electron energy is not much different from that in a positive-column plasma under the same operating conditions, though it shows some dependence on the ratio ν/ω (Ferreira, 1981; Ferreira and Loureiro, 1984). Ferreira and Loureiro have shown theoretically that, in argon, the EEDF is not Maxwellian for $\nu/\omega < 1$ (rf or dc discharge), whereas it tends toward a Maxwellian distribution as ω becomes large ($\nu/\omega \ll 1$), and more rapidly as the electron density exceeds some 10^{12} cm^{-3}.

One consequence of this influence of ν/ω on the EEDF is that, at values $\nu/\omega > 1$, the distribution contains comparatively more low energy

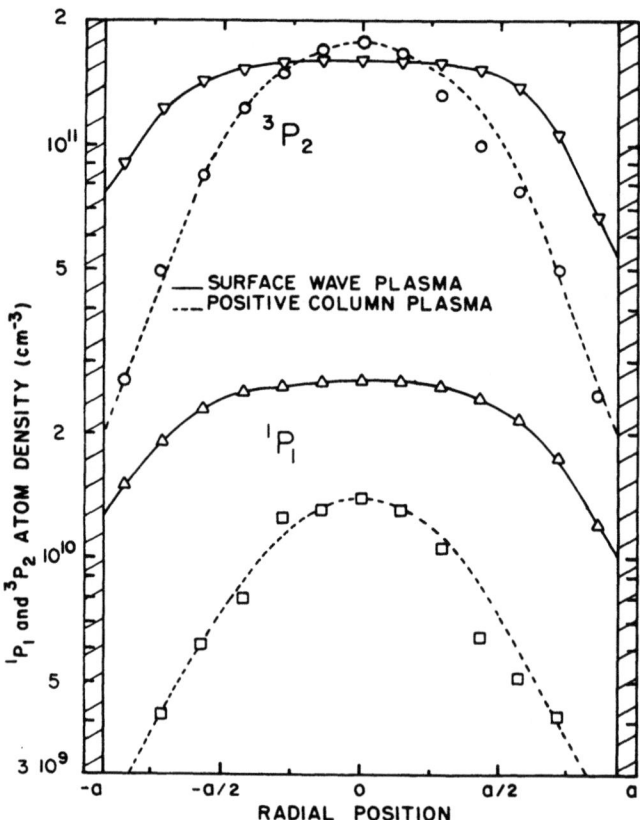

Fig. 23. Measured atom density of the $^3P_2$ metastable and $^1P_1$ resonant states in neon as a function of radial position in a surface-wave-produced plasma (f = 900 MHz) and in a positive-column plasma, in the same plasma tube (2a = 26 mm), at the same gas pressure (p = 400 mtorr). The column length is 150 mm (after Ricard et al., 1985).

electrons than in a Maxwellian distribution with the same average electron energy. On the other hand, these authors have also obtained that the average electron energy, for a given gas pressure, decreases with increasing frequency (see Fig. 7, in Ferreira and Loureiro, 1984), which means that, for the same hf power density deposited into the plasma, the electron density increases with frequency. In the microwave-sustained plasmas, the combined effects of the increased electron density and of the EEDF tending toward a Maxwellian distribution prevail over the decrease of the average electron energy. Thus, the higher levels (including ionization levels) will be more densely populated as compared to dc and rf discharges (Ferreira and Loureiro, 1984) for the same density of power deposited into the plasma. These predictions seem to agree very well with experimental observations (Lebedev and Polark, 1979). They also resolve that apparent paradox that the lines emitted from high lying levels are more intense in microwave discharges, even if the average electron energy is lower, as compared with rf and dc discharges. Due to the wide range of wave frequencies and of gas pressures in surface-wave discharges, this dependence on ν/ω can be used as a convenient means of achieving an external control of the EEDF to increase the efficiency and the selectivity of chemical reactions in plasma (Bell, 1974; Lebedev and Polark, 1979). Further analysis of the dependence of the EEDF on ν/ω can be found in the work by Ferreira and colleagues.

Applications

General considerations.
Surface-wave-produced plasmas can be used in many applications to replace, with definite advantages, either dc discharges or other kinds of rf- and microwave-produced plasmas. For some other applications, certain of the properties of surface-wave-produced plasmas may constitute shortcomings.

The advantages of surface-wave-produced plasmas are first of all those in general of rf and microwave plasmas that have no electrodes of any kind within the plasma vessel: There are no electrode erosion problems and the gas contamination induced by the operation of the discharge is comparatively low, resulting in longer lifetimes for sealed off tubes. Such electrodeless hf plasmas, as compared to the positive-column of dc discharges, are quiescent, i.e., they have a much lower rate of electron density fluctuations, and, as a rule, they do not show moving striations. As compared to some other hf plasmas (e.g., these produced in resonant cavities), a distinctive feature of surface-wave-produced plasmas is that they are perfectly reproducible for given discharge conditions, since only one mode of propagation can be assured.

The stability of surface-wave-produced plasmas in response to changes in the gas composition, the wave frequency, or the power value, is remarkably high as compared, for example, to plasmas produced in resonant cavities. In the latter case, a minute change in the discharge conditions affects the value of the electron density and the system goes off resonance, which often results in the extinction of the plasma. The same changes in the discharge conditions do not lead to such an unstable behavior with surface-wave plasmas, provided that the plasma column length exceeds some minimal length. This is because such changes primarily affect the length of the plasma column. It can be shown that as long as the electron density remains a few times larger than the cut-off density for the wave propagation, the impedance of the surface wave seen from the launcher is almost constant. This is the reason why surface-wave launchers tend to be broad-band devices. This stability feature is particularly important in applications such as elemental analysis by optical spectroscopy, where samples are introduced in the carrier gas, perturbing the plasma. Our experience shows that under constant discharge conditions, after a warm-up period, any variation observed in the surface-wave plasma parameters (e.g., electron density, emitted line intensities) is in fact governed by external factors such as the generator frequency and power stability, and the gas flow and pressure stability.

The main shortcoming of surface-wave plasmas is the axial inhomogeneity of the plasma column connected with the gradient of electron density that results from the decrease of the wave power flow along the plasma column. One solution to this problem is to make use of two launchers, each located at one end of the plasma tube, supplying them from two separate hf generators that are not phase related (to avoid standing waves along the column). A good axial homogeneity can be obtained by that method but the proper power setting of the two generators is not straightforward. Another solution, currently being examined, consists of progressively reducing the plasma tube diameter in the direction of the wave propagation, to compensate for the axial decrease in the wave power flow.

Surface-wave-produced plasmas have been used in a series of applications. Some of them, previously reported in the review paper by Moisan et al. (1982a), are only briefly recalled (they are indicated by the letter "R"). The emphasis is on new results.

Ion source (R). A space-qualified ion source has been realized by Hajlaoui, Henry, and Arnal. It is based on a surfatron launcher that weights less than 150 g. It yields a proton current of about 20 mA with a microwave power value in the range 50-65W (870 MHz). It is a compact, low weight, efficient ion source.

Hydrogen fluoride laser (R). In this application, the plasma is used to provide fluorine atoms from a mixture of SF_6 with a rare gas and oxygen. The F atoms are mixed with hydrogen in the laser chamber to form a vibrationally excited FH molecule. This creates the necessary population inversion for lasing action in the range 2.6 to 3 μm. The multiline output power, for example at 800 W of microwave power, can attain 9W, provided sufficiently fast flow rates are realized. This is presumably one of the most compact FH laser that has been realized in that power range (Bertrand et al., 1978, 1979).

Elemental analyses by optical emission spectroscopy. Surface-wave plasmas can be efficiently produced in capillary quartz tubes at atmospheric pressures in various gases, including argon and helium. Helium is more efficient than argon in terms of excitation of the elements to be analyzed. As compared to other microwave-induced plasmas (MIP) used in that field (e.g., TM_{010} cavity, Evenson cavity), the surface-wave plasma generator is easier to tune and operate and it is more stable in the sense discussed earlier. As for the inductively coupled plasma (ICP), an RF device (usually 27 MHz) used in most commercially available elemental analysis equipment, it is not routinely operated in gases other than argon or argon-based mixtures. Further, it requires flow rates of more than 15 l/min, whereas the surface-wave plasma (typically in a 2-3 mm bore tube) is operated with a flow rate of less than 0.5 l/min, providing an appreciable gas economy. The surface-wave plasma is now being evaluated for such an application by various groups and it is already considered a competitive MIP device for some types of analyses. For example, detection limits of a few pg/s were recently observed with C,Cl,Br, and I in halogen-containing compounds, with 90 W (2.45 GHz), in a 2.8-mm bore tube, with total helium gas flow between 20 and 30 ml/min (Hanai et al, 1981; Chevrier et al., 1982; Abdallah et al., 1982).

Small plasma-jet (R). By using a short open-ended capillary tube and enough microwave power, the atmospheric-pressure surface-wave produced plasma can be made to come out from the plasma tube into the room atmosphere. Under laminar flow conditions (< 1 l/m in argon), it yields a stable, well-defined plasma tip of 0.5 to 2 mm diameter (depending on the capillary tube bore and the microwave power level). The extent of this torch in air is 10 to 30 mm. This small diameter plasma-jet could be used, for example, in microelectronics, for precision and low-heat capacity brazing (> 425 C) on thin metals or tabs, without affecting the other elements of a circuit.

Spectral lamps. The fact that very stable and reproducible plasmas can be achieved with the surface-wave technique, that there are no electrodes in contact with the plasma, and that large densities of excited atoms can be obtained, are interesting features for the realization of stable, low-noise spectral lamps.

Surface-wave-excited lamps were tested by our group and that of Prof. Hubert in view of replacing, for optical absorption measurements, some commercially available spectral lamps that drift with time and/or are too noisy. Two types of lamps were made: one, destined to reduced pressure operation, to provide pure Doppler line profiles; and the other, operated above atmospheric pressures, to tentatively provide a continuous spectrum emission.

Figure 24 shows the first type lamp (R). It is operated in the tenths of a torr to a few torr pressure range with argon or argon-oxygen. It is made of a 1.5-mm i.d. pyrex tube about 150 mm long, connected to a comparatively large gas reservoir. The tube is out-gassed and pumped out at 10^{-7} torr, whereas the corresponding electroded commercial lamps are filled at higher pressures, usually a few torr, to ensure long-term gas purity. The lamp is excited with a surfatron at 900 MHz, modulated at one kHz to allow for lock-in detection, using a maximum of 65 W average power. The line intensities were found to be much more stable as a function of time than those recorded from the commercial lamps. Viewing direction 1 presents lines that can be self-absorbed, while viewing direction 2 yields a pure Doppler profile (argon at 0.4 Torr). Our absorption measurements are facilitated when Doppler-broadened line profiles are available. This type of lamp is rather easy to make but it has a shorter life time than the commercial lamps we used. After a few hundreds of hours of operation, spurious lines due to gas contamination appear, due to a large extent from our lack of technological knowledge in the preparation of sealed-off tubes (Moisan et al., 1982a).

Beauchemin et al. (1986) examined the characteristics of sealed-off xenon lamps filled at pressures ranging from 0.5 to 5 times the atmospheric pressures. The lamps have an i.d. of 4-5 mm, are 10 cm long, and have no gas reservoir. As compared to a short-arc xenon lamp, filled at a few times the atmospheric pressures, the surfatron-excited lamp is less noisy. However, the optical emission, which had to be observed transversally to the tube axis, shows a comparatively stronger line emission intensity than in the short-arc plasma, for comparable dissipated powers. This result is far from intended, which is to obtain a quiescent continuous spectrum emission with little or no line emission, to replace the short-arc lamp for absorption measurements.

He-Ne laser. A He-Ne laser pumped by a surface-wave plasma has been recently realized (Moutoulas et al., 1985). Comparing the characteristics of this laser with those of the classical dc-operated He-Ne laser helps to evaluate the potential of the surface-wave plasma technique for gas laser operation in general. One unique advantage of this pumping method, as compared to other hf pumping schemes, is that it allows

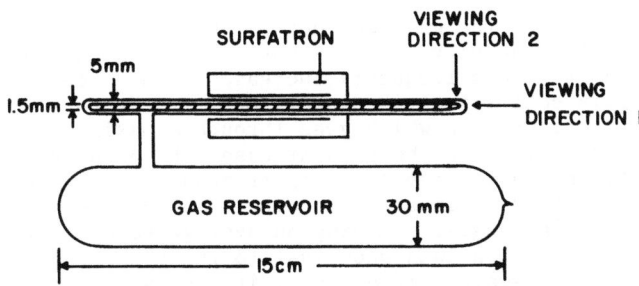

Fig. 24. Simple spectral lamp designed for optical absorption measurements. Viewing direction 1 presents lines that can be self-absorbed, while viewing direction 2 yields pure Doppler line profiles (from Moisan et al., 1982a).

optimization of the laser operation in terms of the pumping field frequency over both the rf and the microwave range.

The experimental arrangement is shown in Fig. 25. The plasma is produced by a surfatron in a 5-mm i.d., T-shaped quartz tube. The wave emerges from the launcher at the base of the T, where it is divided into two identical waves, propagating in opposite directions in the two arms of the T. The amount of hf power supplied to the launcher determines the plasma column length in the active region. The T-configuration was chosen instead of a straight tube because it permits more accurate measurements of the electron density at every point along the active region. A small gas flow is maintained in order to preserve gas purity.

When mounted as an oscillator, the laser tube has been inserted in an optical cavity formed by a curved (R_1 = 5 m) and a flat mirror having transmissions of 0.015% and 0.01%, respectively. The dependence of the laser output intensity on the gas pressure was investigated for various He:Ne mixtures. The best results were obtained for a 7:1 ratio, at a total gas pressure of 0.7 torr (pd = 3.5 torr-mm). The (average) optical gain corresponding to a fixed active length of 20 cm was determined for various surface wave frequencies in the 100-915 MHz range. Figure 26 shows this value normalized in terms of a percentage average gain per meter. Operation above 1 GHz was not possible because the HF generator, due to the increase in electron density with frequency, could not provide enough power to produce the 20-cm reference length. As for the lower frequency limit, it is set by technical problems encountered with the launcher used.

The average gain value $\bar{g}_0$, reported in Fig. 26, is related to the incident intensity I_0 of an incident beam and to the intensity I obtained after one passage by

$$I = I_0 \exp(\bar{g}_0 \ell),$$

where ℓ is the length of the active medium transversed by the beam. This average gain value in an axially inhomogeneous medium is in fact given by:

$$\bar{g}_0 = \int_{Z_1}^{Z_2} g_0(z) \, dz / \int_0^\ell dz,$$

where $g_0(z)$ is the gain coefficient along the axis. Since the electron density varies axially, the gain depends on the cross-section average value of the electron density at z. Figure 26 indicates that the average gain decreases as the pump frequency increases. This results from the fact that the gain coefficient is proportional to the population inversion of the laser transition which goes through a maximum as a function of the electron density. In the present case, the axial gradient of electron density increases with the pump frequency. The average gain value thus results from an integration over a larger electron density range as frequency increases, yielding an average value that tends to decrease. These considerations show that to fully understand the behavior of this surface-wave-pumped laser, it proves necessary to determine the gain coefficient as a function of the electron density.

Taking advantage of the fact that the active length of the laser could be varied by changing the hf power level sent to the launcher, a differential measurement of the average gain was performed. It leads to

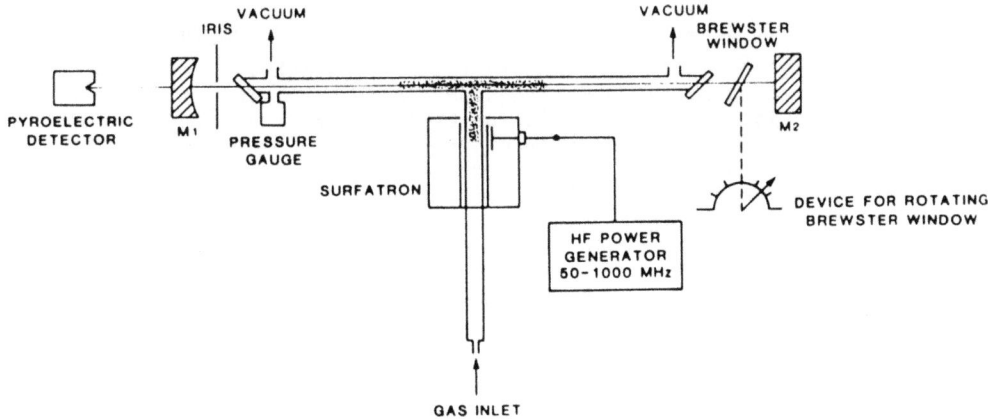

Fig. 25. Schematic diagram of the laser showing how the plasma is generated (from Moutoulas et al., 1985).

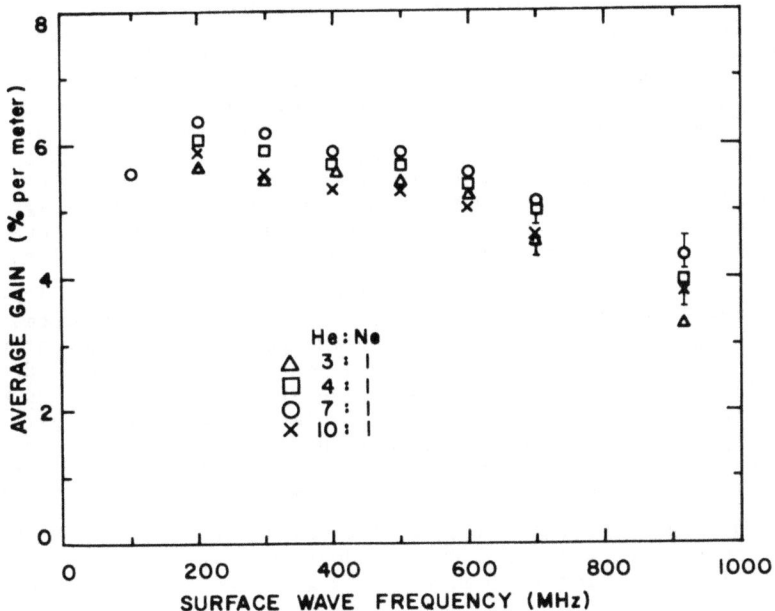

Fig. 26. Average percentage gain per meter, as a function of the surface wave frequency, for four different He:Ne mixtures (from Moutoulas et al., 1985).

the value of g_o as a function of $\bar{n}$ (as with an axially homogeneous plasma), provided the electron density axial distribution is known. It is observed that the maximum attainable gain is 7.5%/m for a wave frequency of 915 MHz and $\bar{n} \simeq 0.6 \times 10^{11}$ cm^{-3}, as compared to less than 6%/m for dc discharge lasers in a tube having the same diameter (Willett, 1974). An important result from the plot of $g_o(\bar{n})$, for various pumping frequency values, is that the electron density value for which the maximum gain is obtained decreases as the pump frequency increases. This frequency effect on the gain coefficient can be explained as resulting from a modification of the electron energy distribution, connected with the fact that the ratio ν/ω varies from 2 (100 MHz) to 0.22 (915 MHz). This explanation is based on a qualitative extension of Ferreira's theory for argon to neon, noting that neon, like argon, has an electron-neutral collision frequency for momentum transfer that varies with the electron energy. The fact that it is the neon gas that controls the He-Ne discharge, though there are much more helium atoms, is due to the lower ionization potential of neon. (The details of this experiment on $g_o(n)$, the corresponding results, and the population inversion relation on which the previous discussion is based, shall be presented in an article to be submitted by Moutoulas et al.)

Knowing that in surface-wave-produced plasmas the radial distribution of excited atoms can take various shapes depending on discharge conditions, it had to be checked to what extent such variations in the radial distribution could not be responsible for the observed gain coefficient dependence on pumping frequency. In order to obtain the radial distribution of the upper level population density, the spontaneous emission of NeI (632.8 nm) was recorded, and integrated along the tube axis (20 cm long), as a function of radius as seen from one of the tube ends. These measurements were performed without the mirrors. The signal was spectrally filtered by a monochromator and the spatial resolution was ensured by a collimator system coupled to a scanning aperture. The spatial resolution was at best 1.2 mm, so the smallest inner diameter plasma tube that could reasonably be used for this series of measurements was 10 mm, whereas the laser operated with a 5-mm i.d. tube. Fig. 27a shows that increasing the wave frequency from 90 to 900 MHz slightly flattens the radial density distribution of the upper-level atoms. This can be seen by comparison with a J_o Bessel function (dashed line); curve # 1 at 0.3 torr corresponds to the optimum pd-product for the laser gain. Figure 27b presents results for the same type of measurement but performed on a 25-mm i.d. tube (the optimum pd-product is now at 0.14 torr). By referring to our observations concerning the radial density distribution of argon atoms in a radiative state (see Properties of Surface Wave-Produced Plasmas), in particular those presenting the tube diameter dependence, we can expect that the radial density distribution of the upper level atoms in the 5-mm i.d. tube will be still closer to a J_o Bessel function function than in the 10-mm i.d. tube (Fig. 27a). We also conclude that the flattening of the radial distribution with the pumping frequency is minor and that it does not explain the observed influence of the pumping frequency on the gain coefficient curve $g_o(\bar{n})$.

The gain values obtained with this He-Ne laser compare favorably with results for a laser excited by dc. As concerns the potential of the surface-wave plasma technique for laser operation in general, we note that the high efficiency of the hf power coupling appears attractive for high power operation as needed with ion and excimer lasers.

Recent results obtained at the Naval Research Laboratory in Washington by Waynant et al. (1985), with a high power surface-wave-pumped excimer laser, are useful in evaluating further this pumping technique. The plasma is produced at atmospheric pressures in a 5-mm i.d. tube with pulses of 1.5 MW at 10 GHz. The electron density value achieved in this case is, however, so large that the radial density distribution of excited atoms presents a relative maximum very close to the tube wall (at 0.3 mm from the wall with pure helium) and a deep

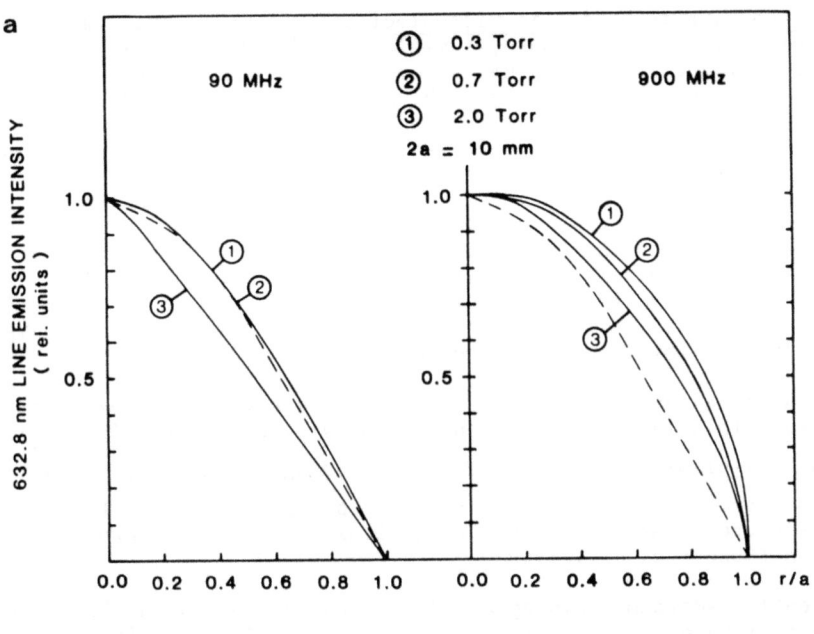
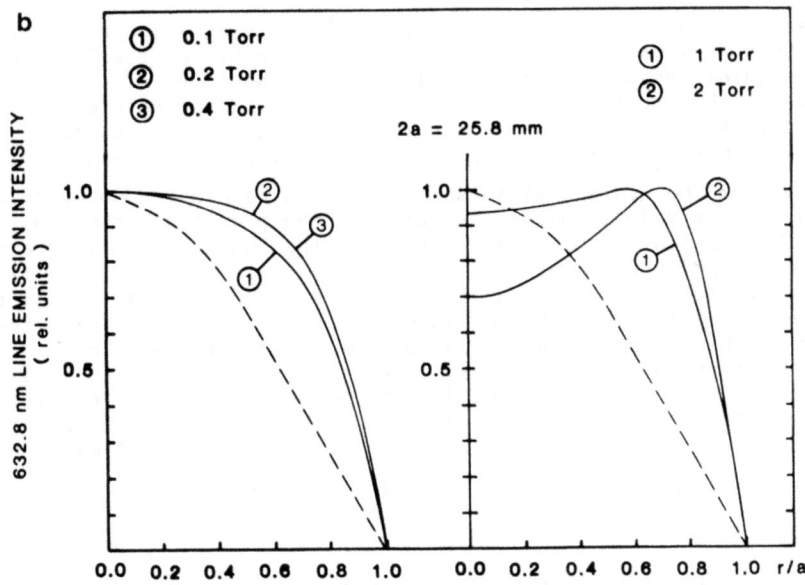

Fig. 27. Radial distribution of the NeI (632.8 nm) emission line intensity in a 5:1 He-Ne mixture: (a) versus wave frequency, (b) 25-mm i.d. tube. The dashed line is the Bessel function, J_0 (2.4 r/a) (from Moutoulas et al., 1985).

minimum on the axis. This kind of a radial distribution is disadvantageous for the operation of the laser in the fundamental mode which requires a gaussian radial profile for the laser beam.

<u>Surface wave technique in microelectronics</u>. In recent years, the use of cold (i.e., low-pressure, low-electron density) plasmas, sustained by hf electromagnetic fields, has become commonplace in the semiconductor industry, both for dry etching and for thin-film deposition.

As compared to the rf capacitive discharge, the behavior of ions in the surface-wave plasma is different. In the rf capacitive discharge, ions can gain energies up to a few hundreds of eV in the self-biasing electrode sheath and thus show anisotropy in their action (e.g., etching) on samples located on the electrodes. Within surface-wave-produced plasmas, the movement of ions hitting a sample is comparatively isotropic. Recently, Paraszczak et al. (1984) have shown that anistropy could be introduced in a microwave plasma with rf biasing of the sample (relatively very low rf power is needed).

The surface-wave plasma is used to provide excited nitrogen that is then mixed with silane to yield deposits of silicon nitride (Si_3N_4). This reactive-plasma deposition technique, using surfatrons, can be carried out on an extremely broad range of pressures, namely from 10^{-5} to atmospheric pressures, allowing a great flexibility in the experimental conditions (Loncar et al., 1980).

Recently, amorphous hydrogenated silicon (a - Si:H) has been prepared by microwave-produced plasma at 2.45 GHz in $Ar-SiH_4$ mixtures using two different deposition systems: a surface-wave plasma and a plasma produced by a slow-wave structure (a ladder-type applicator called LMP) that accompanies the plasma column all the way (Paquin et al., 1985). Fig. 28 shows the experimental setup (a) with the LMP structure, and (b) for the surface wave scheme using a surfatron. In both cases, the plasma is generated in the $Ar-SiH_4$ feeder gas in a portion of the reactor vessel downstream from the heated sample substrate. In general, a non-glowing region about 2 cm wide is maintained between the plasma and the substrate. The gas flow comes from the back of the substrate. Films are thus formed by a "counter-current" diffusion of plasma-generated active species toward the substrate. Although this procedure tends to give rise to lower growth rates than are possible with other configurations, resulting films have higher quality and better reproducibility under a given set of fabrication conditions. The reactor diameters are similar in the two systems ($\simeq$ 8 cm), resulting in comparable gas flow conditions with respect to the two identical sample holders. The total flow rate ranges from 10 to 100 sccm, with about 6% of SiH_4. The total gas pressure is in the range 0.1 - 0.3 torr. The substrate temperature is held at 300 $\pm$ 10 C. Comparable microwave power densities are maintained in the plasma.

Deposition in the surfatron system gives rise to device-grade a-Si:H, as demonstrated by Schottky cell efficiencies exceeding 3%. It was not possible to duplicate this in the LMP system in spite of nominally identical fabrication conditions: the LMP films have gross columnar morphology, and they react with atmospheric constituents to give a-Si:(H,C,O) alloys. More pronounced ion bombardment of the sustrate during deposition is thought to account for the better quality of the surface-wave-produced films. This explanation is based on several reports in the literature which show that ion bombardment during a-Si:H film growth eliminates rough surface morphology, presumably through the mobilization and rearrangement of reactive species on the surface during film growth. Further, there is also much evidence that "smooth" films

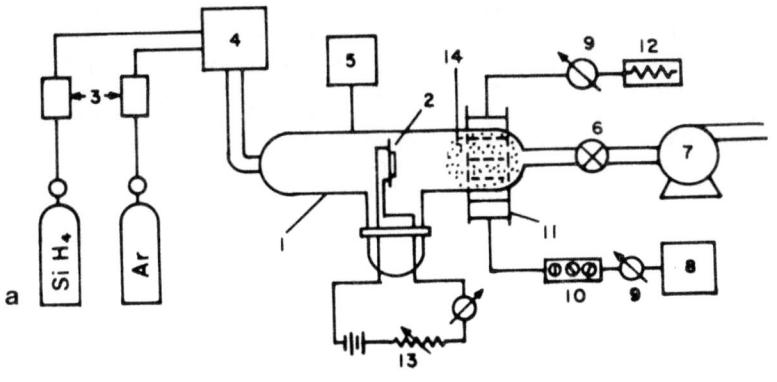

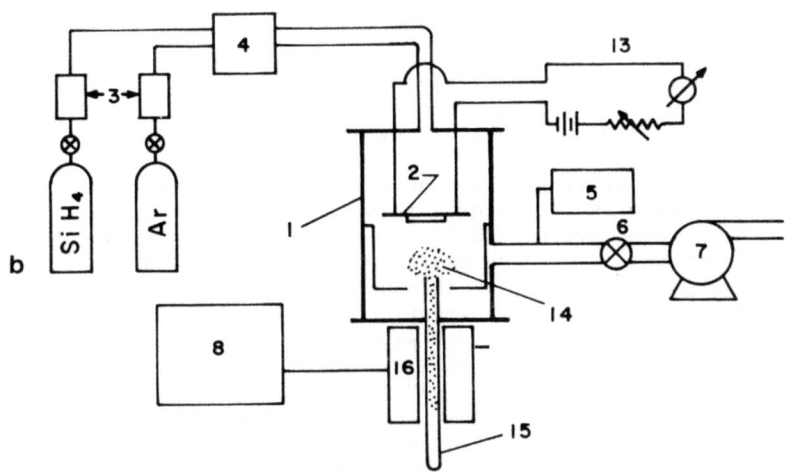

Fig. 28. Experimental setup for the deposition of amorphous hydrogenated silicon: (a) using a slow-wave applicator (LMP), (b) using a surface-wave plasma (surfatron). 1: reactor vessesl; 2: heated sample holder; 3: electronic flowmeter/controller; 4: mixing manifold; 5: capacitive manometer; 6: automatic throttling valve; 7: turbomolecular and mechanical backing pumps; 8: 2.45-GHz power supply; 9: power monitors; 10: triple-stub tuner; 11: slow-wave microwave applicator; 12: dummy load; 13: substrate heater supply; 14: glow region of plasma; 15: quartz tube; 16: surfatron (from Paquin et al., 1985).

tend to have good electro-optical properties, whereas rough (columnar) films do not.

The origin of the ion bombardment has not been demonstrated directly in the present experiment. It is suggested that the large electric field discontinuity that occurs at the end of the surface-wave plasma (where the surface wave ceases to propagate) creates a ponderomotive force that accelerates the electrons and, subsequently, through the space-charge field, the ions. There is no such sharp E-field break in the plasma produced with the LMP system, since the electromagnetic energy is carried by the slow-wave structure (even in the absence of plasma). Comparatively, in the LMP scheme, the E-field is homogeneous. In such a case, no ion acceleration results, as the ions are too heavy to move in a microwave field period. The existence and action of a ponderomotive force in surface-wave-produced plasmas has been clearly demonstrated by Bloyet et al., (1981). It has been shown to be responsible for the movement of the ionization front that opens the way to surface wave propagation as the hf power is initially applied (transient regime).

CONCLUSION

The electrodeless microwave and rf discharges have been, for decades, attractive sources of plasmas. However, two shortcomings always prevented their wider recognition. First, the understanding of the complex physical processes occurring in such discharges has been insufficient. This made their modeling difficult and restricted the possibility of predicting their behavior under various operational conditions. Second, the plasma volume attainable was limited by the size of the field applicator within which, as a rule, the plasma had to be contained. This was particularly the case with microwave discharges. As a result, the application of such plasmas, having potentially attractive properties, has been limited mainly to laboratory studies.

In the present paper, we have shown that the surface-wave discharge is free of both these shortcomings. This results, first of all, from the fact that the surface-wave plasma is a particularly convenient object for experimental studies. It can be sustained under a wide range of wave frequencies and gas pressures, it is easy to operate, stable, and reproducible. Due to the traveling wave principle of operation and to the absence of any external hf structure along the plasma column, the balance processes may be investigated locally along the discharge axis. This provides a better insight into the physical processes occurring in the plasma than can be arrived at with other kinds of discharges. For the same reason, the verification of the discharge model becomes straightforward. This, in connection with the intensive theoretical and numerical work performed, explains why the modeling of surface-wave discharges has reached such a highly advanced stage. Many results obtained in that way are of a general character and allows for a better understanding of the hf discharges in general.

Among the features of surface-wave discharges, probably the most outstanding one is the possibility of sustaining long plasma columns (and, therefore, to obtain large volume plasmas) using a localized wave launcher of relatively small dimensions. This, together with the advantages already mentioned, make these discharges interesting for various applications.

REFERENCE

Abdallah, M. H., Coulomb, S., Mermet, J. M., and Hubert, J., 1982, Spectrochim. Acta, 37B:583.
Beauchemin, D., Hubert, J., and Moisan, M., 1986, To appear in Appl. Spectroscopy, April.
Bell, A. T., 1974, in: "Techniques and Applications of Plasma Chemistry", J. R. Hollahan and A. T. Bell, eds., Wiley, New York, Ch. 1 and 10.
Bertrand, L., Gagne, J. M., Bosisio, R., and Moisan, M., 1978, IEEE J. Quantum Electron., QE14:8.
Bertrand, L., Monchalin, J. P., Pitre, R., Meyer, M. L., Gagne, J. M., and Moisan, M., 1979, Rev. Sci. Instrum., 50:708.
Bloyet, E., Leprince, P., Llamas Blasco, M., and Marec, J., 1981, J. Physique-Lettres, 83A:391.

Bollen, W. M., Waynant, R. W., and Christensen, C. P., 1985, Naval Research Lab., Memo. Report 5432.
Brown, S. C., 1961, "Basic Data of Plasma Physics", M.I.T. Press, Cambridge, Mass.
Chaker, M., Nghiem, P., Bloyet, E., Leprince P., and Marec, J., 1982, J. Physique-Lettres, 43:L-71.
Chaker, M., Moisan, M., and Sauve, G., 1984, in "Proceedings of the 34th Canadian Chemical Engineering Conference, Quebec", pp. 47-50.
Chaker, M. and Moisan, M., 1985, J. Appl. Phys., 57:91.
Chaker, M., Moisan, and Zakrzewski, Z., 1986, Plasma Chemistry and Plasma Processing, 6(1).
Chevrier, G., Hanai, T., Tran, K. C., and Hubert, J., 1982, Can. J. Chem., 60:898.
Crawford, F. W., 1971, Proc. IEEE, 59:4.
Ferreira, C. M., 1981, J. Phys. D: Appl. Phys., 14:1811.
Ferreira, C. M., 1983, J. Phys. D: Appl. Phys., 16:1673.
Ferreira, C. M., 1986, in this volume.
Ferreira, C. M., and Loureiro, J., 1983, J. Phys. D: Appl. Phys., 16:2471.
Ferreira, C. M., and Loureiro, J., 1984, J. Phys. D: Appl. Phys., 17:1175.
Francis, G., 1956, in: "Handbuch der Physik", vol. 22, S. Flugge, ed., Springer, Berlin.
Glaude, V. M. M., Moisah, M., Pantle, R., Leprince, P. and Marec, J., 1980, J. Appl. Phys., 51:5693.
Golant, V. E., 1958, Uspiekhi Fiz. Nauk, 45:40.
Golant, V. E., Zhilinsky, E. P., Sakharov, I. E., and Brown, S. C., 1980, "Fundamentals of Plasma Physics", Wiley, New York.
Hanai, T., Coulombe, S., Moisan, M., and Hubert, J., 1981, in: "Developments in Atomic Plasma Spectrochemical Analysis", R. Barnes, ed., Heyden, London.
Heald, M. A., and Wharton, C. B., 1965, "Plasma Diagnostics with Micowaves", Wiley, New York.
Hubert, J., Moisan, M., and Ricard, A., 1979, Spectrochim. Acta, 33B:1.
Kunkel, W. B., 1966, in: "Plasma Physics in Theory and Appliation", W. B. Kunkel, ed., Chapter 10, McGraw Hill, New York.
Lebedev, Yu, A., and Polak, L. S., 1979, Khimiya Vysokikh Energii, 13:387 [High Energy Chemistry, 13:331].
Leprince, P., Matthieussent, G., and Allis, W. P., 1971, J. Appl. Phys., 42:412.
Loncar, G., Musil, J., and Bardos, L., 1980, Czech. J. Phys., B30:688.
MacDonald, A. D., 1966, "Microwave Breakdown in Gases", Wiley, New York.
MacDonald, A. D., and Tetenbaum, S. J., 1978, in: "Gaseous Electronics", M. N. Hirsh, and H. J. Oskam, eds., Chapter 3, Academic Press, New York.
Marec, J., Bloyet, E., Chaker, M., Leprince, P., and Nghiem, P., 1982, in "Electrical Breakdown and Discharges in Gases", Pt. B, E. E., Kunhardt and L. H. Luessen, eds., Plenum Press, New York.
Mateev, E., Zhelyazkov, I., and Atanassov, V., 1983, J. Appl. Phys., 54:3049.
Moisan, M., Beaudry, C., and Leprince, P., 1974, Phys. Lett., 50A:125.
Moisan, M., Beaudry, C., and Leprince, P., 1975, IEEE Trans. Plasma Sci., PS-3:55.
Moisan, M., Zakrzewski, Z., and Pantel, R., 1979, J. Phys. D: Appl. Phys., 12:219.
Moisan, M., Pantel, R., Ricard, A., Glaude, V. M. M., Leprince, P., and Allis, W. P., 1980, Rev. Physique App., 15:1383.
Moisan, M., Ferreira, C. M., Hajlaoui, Y., Henry, D., Hubert, J., Pantel, R., Ricard, A., and Zakrzewski, Z., 1982a, Rev. Physique Appl., 17:707.
Moisan, M., Shivarova, A., and Trivelpiece, A. W., 1982b, Plasma Phys., 24:1331.
Moisan, M., Pantel, R., and Ricard, A., 1982c, Can. J. Phys., 60:379.

Moutoulas, C., Moisan, M., Bertrand, L., Hubert, J., Lachambre, J. L., and Ricard, A., 1985, Appl. Phys. Lett., 46:323.
Muller, Ch. H. III, and Phelps, A. V., 1980, J. Appl. Phys., 51:6141.
Nghiem, P., Chaker, M., Bloyet, E., Leprince, P., and Marec, J., 1982, J. Appl. Phys., 53:2920.
Paraszczak, J., Hatzakis, M., Babich, E., Shaw, J., Arthur, E., Grenon, B. and DePaul, M., 1984, in: "Proceedings, Microcircuit Engineering Conference, Berlin".
Pantel, R., Ricard, A., and Moisan, M., 1983, Beitr. Plasmaphys., 23:561.
Paquin, L., Masson, D., Wertheimer, M., and Moisan, M., 1985, Can. J. Phys., 63:831.
Pfau, S., Rutscher, A., and Wojaczek, K., 1969, Beitrage Plasmaphys., 9:333.
Phelps, 1982
Ricard, A., Collobert, D., and Moisan, M., 1983, J. Phys. B: At. Mol. Phys., 16:1657.
Ricard, A., Hubert, J., and Moisan, M., 1985, Proc. 17th Int. Conf. on Phenomena in Ionized Gases, Budapest, Contributed papers, p. 741.
Rose, D. J., and Brown, S. C., 1955, Phys. Rev., 98:310.
Taillet, J., 1969, Am. J. Phys., 37:423.
Trivelpiece, A. W. and Gould, R. W., 1959, J. Appl. Phys., 30:1784.
Trivelpiece, A. W., 1967, "Slow-Wave Propagation in Plasma Waveguides," San Francisco University Press, San Francisco.
Waynant, R. W., BOllen, W. M., and Christensen, C. P., 1985, Bull. Am. Phys. Soc., 30:136.
Zakrzewski, Z., Moisan, M., Glaude, V. M. M., Beaudry, C., and Leprince, P., 1977, Plasma Phys., 19:77.
Zakrzewski, Z., 1983, J. Phys. D: Appl. Phys., 16:171.

PLASMAS SUSTAINED BY SURFACE WAVES AT RADIO AND

MICROWAVE FREQUENCIES: BASIC PROCESSES AND MODELING

C. M. Ferreira

Departamento de Fisica and Centro de Electrodinamica
da Universidade Tecnica de Lisboa (I.N.I.C),
Instituto Superior Tecnico, Lisbon, Portugal

INTRODUCTION

The propagation of surface waves (SW) along plasma columns has long been a subject of wide interest motivating numerous investigations (for a recent review of the subject see, e.g., Moisan et al., 1982a). In fact, surface waves can be involved in such fundamental problems as plasma heating by electromagnetic fields, absorption of laser power by dense plasmas, and nonlinear interactions such as parametric instabilities. Furthermore, surface waves can be used as a diagnostic tool for low-temperature plasmas, e.g. to determine the electron density, collision frequency, and drift velocity in a positive column (Trivelpiece and Gould, 1959; Carlile, 1964; Akao and Ida, 1963, 1964). There exist as well some promising applications of surface waves in the field of microwave devices employing plasmas (Clarricoats et al., 1966; Aronov et al., 1976; Burykin et al., 1975).

The production of plasma columns sustained by the propagation of surface waves has stimulated intensive experimental and theoretical research in recent years (see, e.g., Moisan et al., 1982b; Marec et al., 1983). Gas discharges of this type may be operated over an extremely large range of gas pressures, from a few mtorr to atmospheric pressures, and frequencies, from a few MHz to some 10 GHz. The electron density in these discharges may vary from 10^{10} to 10^{14} cm^{-3}, depending on HF power and tube diameter. Typically, a 2.5-cm diameter, 1.8-m long plasma column with an electron density of a few x 10^{10} cm^{-3} can be produced at a pressure of a few tens of mtorr with 80W of HF power at 500 MHz. The properties of these plasmas make them interesting for numerous applications such as plasma chemistry, laser excitation, spectroscopic sources, etc.

In this work we analyse the basic processes of SW-produced plasmas and the present degree of achievement of their modeling. The experimental aspects will not be discussed here as they are reviewed in a complementary work presented in this Proceedings (Moisan and Zakrzewski, 1985). First, we will review basic phenomena associated with the propagation of surface waves, such as the distribution of the wave fields, the dispersion characteristics in the case of a homogeneous plasma, the effects of plasma inhomogeneity, and the absorption of the wave power in the plasma. The second part is concerned with the modeling of SW-produced plasmas, but we have chosen to insert the analysis in a wider context: that is, the general problem of the maintenance of a stationary, diffusion-controlled discharged sustained by a HF field. The basic theory is first discussed and applied in the simple case of a homogeneous field. Then, we extend the analysis to situations where the field is inhomogeneous, which implies that its spatial distribution needs to be determined self-consistently in the framework of the theory itself. We discuss two examples of such situations: the HF discharge between parallel plates, and the cylindrical plasma column sustained by an axially symmetric surface wave. In this way one readily recognizes that these two types of discharges possess many features in common.

PROPERTIES OF SURFACE WAVES PROPAGATING ALONG A PLASMA COLUMN

The Concept of a Surface Wave (SW)

With regard to wave propagation, a cold plasma with negligible losses may be described as a dielectric medium of relative permittivity ε_p given by

$$\varepsilon_p = 1 - (\frac{\omega_{pe}}{\omega})^2 \qquad (1)$$

where ω is the wave angular frequency and ω_{pe} is the angular electron plasma frequency. For $\omega < \omega_{pe}$, the plasma permittivity is negative, the only dielectric medium known to have this property. This medium, when bounded by a dielectric of positive permittivity ε_g, such that $\varepsilon_g < |\varepsilon_p|$, can sustain the propagation of pure surface waves characterised by evanescent fields on both sides of the boundary if both media are semi-infinite. For closed plasma configurations (e.g., slabs, cylindrical columns) the field intensity in the plasma decreases from the boundary towards the axis, so that, in some sense, surface waves carry their energy predominantly near the boundary along which they propagate.

In general, surface waves can be expressed as a superposition of TM and TE waves. However, only TM modes can satisfy the field continuity relations across the boundary for axially symmetric surface waves, whatever the configuration. For instance, this means that in cylindrical geometry the non-zero electric and magnetic field components associated with a symmetric surface wave propagating along the axis (z-axis) are E_r, E_z, and H_ϕ, respectively. This corresponds to m=0, where m is the wave number corresponding to the azimuthal angle ϕ. For higher modes (m$\geq$1) the surface waves are hybrid, i.e., they possess non-zero axial components of both $\vec{E}$ and $\vec{H}$.

The SW Field Equations

Consider a SW propagating along the z-axis and assume that the variation of $\vec{E}$ and $\vec{H}$ in time and along z is of the form exp j(ωt-kz), where k= β-jα is the wave vector, β=2π/λ is the wave number, and α is the attenuation coefficient. Describing the plasma as a dielectric, the complex field vectors in the plasma satisfy the equations

$$\vec{\nabla} \times \vec{E} = -j\omega\mu_o \vec{H} \qquad (2)$$

$$\vec{\nabla} \times \vec{H} = j\omega \varepsilon_0 \varepsilon_p \vec{E} \tag{3}$$

$$\vec{\nabla} \cdot \vec{H} = 0 \tag{4}$$

$$\vec{\nabla} \cdot (\varepsilon_p \vec{E}) = 0 . \tag{5}$$

The relative plasma permittivity ε_p is, in general, a complex number. It is related to the complex plasma conductivity, σ, by the expression

$$\varepsilon_p = 1 - j \frac{\sigma}{\omega \varepsilon_0} \tag{6}$$

σ being given by (see, e.g., Allis, 1956)

$$\sigma = -\frac{2}{3} \frac{e^2 n_e}{m} \int_0^\infty \frac{u^{3/2}}{\nu_c + j\omega} \frac{df}{du} du . \tag{7}$$

In the above expression, e and m denote the charge and mass of an electron, ε_0 is the electric permittivity of free space, n_e is the electron number density, ν_c is the collision frequency for momentum transfer between the electrons and the molecules of the gas, $u = mv^2/2e$ is the electron energy expressed in volts, and $f(u)$ is the electron energy distribution function with the normalization $\int_0^\infty u^{1/2} f(u) du = 1$. For ν_c independent of u, Eq. (7) reduces to the simpler form $\sigma = e^2 n_e/m(\nu_c + j\omega)$. In the absence of collisions, i.e., for $\nu_c = 0$, σ is purely imaginary and Eq. (6) for ε_p reduces to the form of Eq. (1), with $\omega_{pe}^2 = e^2 n_e/m\varepsilon_0$.

Noting that $\vec{\nabla} \equiv \vec{\nabla}_T - jk\vec{\ell}_z$, where $\vec{\ell}_z$ is the unit vector along the z-axis and $\vec{\nabla}_T$ is an operator involving only derivatives with respect to the transverse coordinates, we easily obtain from Eqs. (2) and (3)

$$\vec{\nabla}_T \times \vec{E}_T = -j\omega \mu_0 \vec{H}_z \tag{8}$$

$$\vec{\nabla}_T \times \vec{E}_z - jk\vec{\ell}_z \times \vec{E}_T = -j\omega \mu_0 \vec{H}_T \tag{9}$$

$$\vec{\nabla}_T \times \vec{H}_T = j\omega \varepsilon_0 \varepsilon_p \vec{E}_z \tag{10}$$

$$\vec{\nabla}_T \times \vec{H}_z - jk\vec{\ell}_z \times \vec{H}_T = j\omega \varepsilon_0 \varepsilon_p \vec{E}_T . \tag{11}$$

Here, $\vec{E}_T$ and $\vec{H}_T$ denote the transverse components of the complex field vectors. Now, cross-multiplying Eq. (9) by $\vec{\ell}_z$ and combining the resulting equation with Eq. (11), we readily obtain

$$\vec{E}_T = -\frac{jk}{\frac{\omega^2}{c^2} \varepsilon_p - k^2} \vec{\nabla}_T E_z + \frac{j\omega \mu_0}{\frac{\omega^2}{c^2} \varepsilon_p - k^2} (\vec{\ell}_z \times \vec{\nabla}_T H_z) \tag{12}$$

where use was made of the identities of $(\vec{\nabla}_T \times \vec{E}_z) \times \vec{\ell}_z \equiv -\vec{\nabla}_T E_z$ and

$\vec{\nabla}_T \times \vec{H}_z \equiv -\vec{\ell}_z \times \vec{\nabla}_T H_z$; $c = (\varepsilon_o \mu_o)^{-1/2}$ denotes the speed of light in vacuum.

Using a similar procedure, we obtain from Eqs. (8) and (10)

$$\vec{H}_T = -\frac{jk}{\frac{\omega^2}{c^2}\varepsilon_p - k^2} \vec{\nabla}_T H_z - \frac{j\omega \varepsilon_o \varepsilon_p}{\frac{\omega^2}{c^2}\varepsilon_p - k^2} (\vec{\ell}_z \times \vec{\nabla}_T E_z). \tag{13}$$

Eqs. (12) and (13) constitute general expressions for the transverse field components as a function of the transverse gradients of the axial components. They reduce to simpler forms in the case of TE waves ($E_z = 0$) or TM waves ($H_z = 0$). In the latter case we simply have

$$\vec{E}_T = -\frac{jk}{\frac{\omega^2}{c^2}\varepsilon_p - k^2} \vec{\nabla}_T E_z \tag{14}$$

$$\vec{H}_T = \frac{\omega \varepsilon_o \varepsilon_p}{k} (\vec{\ell}_z \times \vec{E}_T). \tag{15}$$

Now we can obtain two equations for the axial components E_z and H_z from Eqs. (2) - (4) by making use of the identity $\vec{\nabla} \times (\vec{\nabla} \times \vec{A}) \equiv \vec{\nabla}(\vec{\nabla} \cdot \vec{A}) - \nabla^2 \vec{A}$. The case of a homogeneous plasma is particularly simple because Eq. (5) yields $\vec{\nabla} \cdot \vec{E} = 0$ when $\varepsilon_p = $ const. A straightforward calculation then yields

$$\nabla_T^2 \vec{E} + (\frac{\omega^2}{c^2}\varepsilon_p - k^2)\vec{E} = 0 \tag{16}$$

$$\nabla_T^2 \vec{H} + (\frac{\omega^2}{c^2}\varepsilon_p - k^2)\vec{H} = 0 \tag{17}$$

where we have used $\nabla^2 \equiv \nabla_T^2 - k^2$. The axial components E_z and H_z in a homogeneous plasma satisfy, therefore, the equations

$$\nabla_T^2 E_z + (\frac{\omega^2}{c^2}\varepsilon_p - k^2) E_z = 0 \tag{18}$$

$$\nabla_T^2 H_z + (\frac{\omega^2}{c^2}\varepsilon_p - k^2) H_z = 0. \tag{19}$$

However, plasma inhomogeneity in transverse directions must often be taken into account (e.g., cylindrical plasma column) so that these equations are not applicable, in general. From Eq. (5) we obtain $\vec{\nabla} \cdot \vec{E} = -\vec{E} \cdot \vec{\nabla} \varepsilon_p/\varepsilon_p$ and, instead of Eq. (16), we now have

$$\nabla_T^2 \vec{E} + (\frac{\omega^2}{c^2}\varepsilon_p - k^2) \vec{E} + \vec{\nabla}(\vec{E}_T \cdot \frac{\vec{\nabla}_T \varepsilon_p}{\varepsilon_p}) = 0. \tag{20}$$

For a TM wave E_T is given by Eq. (14), and substituting into Eq. (20) we obtain the following equation for E_z

$$\nabla_T^2 E_z + (\frac{\omega^2}{c^2}\varepsilon_p - k^2) E_z - \frac{k^2}{\frac{\omega^2}{c^2}\varepsilon_p - k^2} \vec{\nabla}_T E_z \cdot \frac{\vec{\nabla}_T \varepsilon_p}{\varepsilon_p} = 0 . \qquad (21)$$

Similarly, Eq. (19) is replaced by the equation

$$\nabla_T^2 \vec{H} + (\frac{\omega^2}{c^2}\varepsilon_p - k^2) \vec{H} + j\omega \varepsilon_o \vec{\nabla}_T \varepsilon_p \times \vec{E} = 0 \qquad (22)$$

which may be combined with Eq. (12) for $\vec{E}_T$ to provide an equation for H_z.

Properties of Axially Symmetric Surface Waves Propagating Along a Cylindrical Plasma

Let us consider the case of a TM wave propagating down a cylindrical plasma column enclosed in a glass tube. Figure 1 represents schematically an apparatus usually employed to produce a plasma column by the propagation of a surface wave (Marec et al., 1983; Moisan et al., 1982b). The wave is launched by an HF structure called a "Surfatron".

From Eqs. (14), (15), and (21) we obtain the following equations for the field components E_z, E_r, and H_ϕ

$$\frac{d^2 E_z}{dr^2} + (\frac{1}{r} - \frac{k^2}{\frac{\omega^2}{c^2}\varepsilon_p - k^2} \frac{1}{\varepsilon_p} \frac{d\varepsilon_p}{dr}) \frac{dE_z}{dr} + (\frac{\omega^2}{c^2}\varepsilon_p - k^2) E_z = 0 \qquad (23)$$

$$E_r = -j \frac{k}{\frac{\omega^2}{c^2}\varepsilon_p - k^2} \frac{dE_z}{dr} \qquad (24)$$

$$H_\phi = \frac{\omega \varepsilon_o \varepsilon_p}{k} E_r . \qquad (25)$$

In the glass tube and in the outer free space the field components are given by similar equations, but with ε_p replaced by the relative permittivity of the corresponding medium, i.e., ε_g in the glass, and $\varepsilon_{air}=1$ in the outer space. The dispersion characteristics follow from the continuity of the components E_z and H_ϕ at the plasma-glass (r=a) and the glass-air (r=b) interfaces. In the nearly collisionless case, i.e., for $\nu_c \ll \omega$, the attenuation coefficient may be neglected ($\alpha \ll \beta$) and the dispersion characteristics may be obtained using real algebra only. The problem becomes particularly simple in this case if one further assumes that the plasma is homogeneous (Zakrzewski et al., 1977; Glaude et al., 1980; Chaker et al., 1981, 1982; Nghiem et al., 1982; Moisan et al., 1982b). The solutions of Eqs. (23) and (24) are then

$$E_z(r) = E(0) I_0(\Gamma r) \qquad (26)$$

$$E_r(r) = E(0) \frac{\beta}{\Gamma} I_1(\Gamma r) \qquad (27)$$

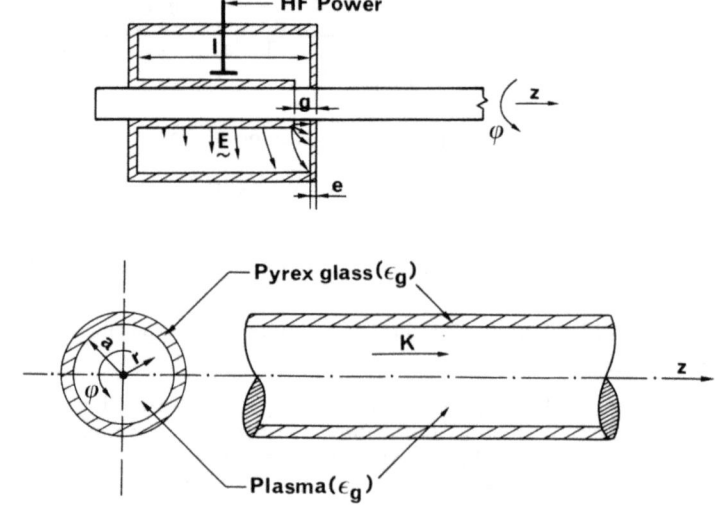

Fig. 1. Schematic diagrams of the HF structure and the coupler in a Surfatron, and geometry of the discharge tube.

where I_0 and I_1 denote the zeroth and first-order modified Bessel functions, respectively, E(0) is the field intensity at the axis, and

$$\Gamma^2 = \beta^2 + \frac{\omega_{pe}^2 - \omega^2}{c^2}. \tag{28}$$

Figure 2 shows a phase diagram of ω/ω_{pe}-vs-βa calculated assuming a homogeneous, cold, collisionless plasma, along with data from experiments (here, the term phase diagram is used to emphasize that, in the experiments, ω is kept fixed while ω_{pe} is varied). The agreement is rather satisfactory except at high βa values.

We note the following important limits in the phase diagram of surface waves. As $\omega/\omega_{pe} \to 0$, we have $\beta \to (\omega/c)\sqrt{\varepsilon}$, where $1 < \varepsilon < \varepsilon_g$. Then, for $\omega_{pe} \gg \omega$ the SW wavelength remains close to that of an ordinary electromagnetic wave propagating in a dielectric of relative permittivity ε. In the opposite limit, i.e., for $\beta \to \infty$, we have $\omega/\omega_{pe} \to (1 + \varepsilon_g)^{-1/2}$ which is the same as $|\varepsilon_p| \to \varepsilon_g$. This condition defines the so-called cut-off limit; the SW cannot propagate if the plasma density is lower than a critical value which is given by

$$n_c(\text{cm}^{-3}) = 1.245 \times 10^{-8} f^2 (1 + \varepsilon_g) \tag{29}$$

where f is the wave frequency in Hz.

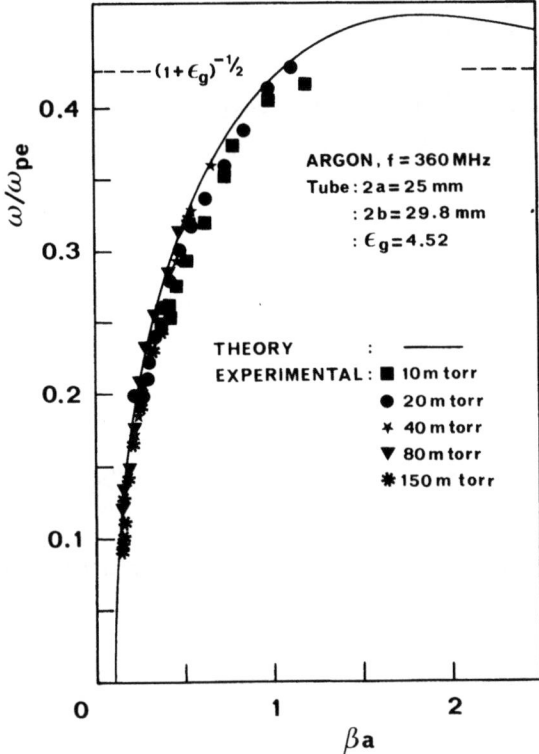

Fig. 2. Dispersion characteristics of a SW with f=360 MHz, propagating along a homogeneous argon plasma.

Close to the cut-off limit we have $\Gamma \sim \beta$ and Γa becomes large, so that the electric field is mostly concentrated at the plasma boundary. A similar situation arises at high plasma densities, as $\Gamma a \to \omega_{pe} a/c$ for $\omega_{pe} \gg \omega$. These limits are well illustrated in the curves of Γa and βa vs $-\omega_{pe} a/c$ given in Fig. 3, calculated for a homogeneous plasma.

Figure 4 shows typical radial profiles of the components E_z and E_r in the plasma, the glass, and the outer free space, for a surface wave with f = 600 MHz and assuming $\omega_{pe}/\omega = 5$.

Calculations of the attenuation coefficient due to collisions have also been performed by various authors for a cold, homogeneous plasma in the limit $\nu_c \ll \omega$. The attenuation coefficient is given by (see, e.g., Bekefi, 1966)

$$\alpha = \frac{1}{2} \frac{d\,P_{abs}/dz}{v_g W_t} \qquad (30)$$

where v_g is the wave group velocity, W_t is the time-averaged total energy (i.e., magnetic + electric + kinetic energy of charged particles) per unit axial length, and dP_{abs}/dz is the time-averaged absorbed power in the plasma per unit axial length. The latter is given by

$$\frac{dP_{abs}}{dz} = \frac{1}{2} \int_0^a \text{Re}(\sigma) \, E^2 \, 2\pi r \, dr \qquad (31)$$

where Re(α) is the real part of the complex plasma conductivity given by Eq. (7), and $E^2 = E_z^2 + E_r^2$. For $\nu_c \ll \omega$, Eq. (7) yields

$$\text{Re}(\sigma) = \frac{e^2 n_e}{m\omega^2} \bar{\nu}_c \qquad (32)$$

where we have defined an effective collision frequency

$$\bar{\nu}_c = -\frac{2}{3} \int_0^\infty u^{3/2} \, \nu_c \, \frac{df}{du} \, du \, . \qquad (33)$$

We obtain, therefore

$$\frac{dP_{abs}}{dz} = \int_0^a \frac{e^2}{2m} \left(\frac{E}{\omega}\right)^2 \bar{\nu}_c \, n_e \, 2\pi r \, dr = \frac{\bar{\nu}_c \varepsilon_0}{2} \left(\frac{\omega_{pe}}{\omega}\right)^2 \int_0^a E^2 \, 2\rho r \, dr \qquad (34)$$

the second equality holding only in the case of a homogeneous plasma.

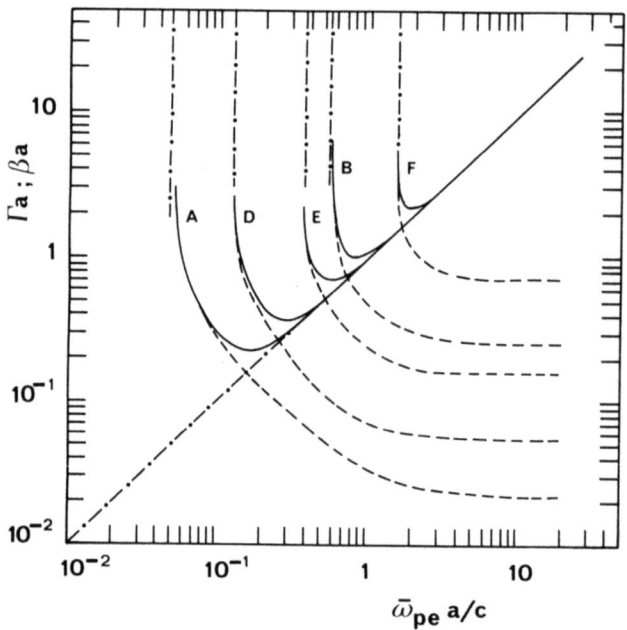

Fig. 3. Values of βa (dashed curves) and Γa (solid curves) vs. $\omega_{pe} a/c$ in the case of a radially homogeneous plasma for: a=0.45 cm, b=0.60 cm, f=210 MHz (A) and 2450 MHz (B); a=1.25 cm, b=1.49 cm, f=210 MHz (D), 600 MHz (E), and 2450 MHz (F).

In a cold plasma, under low absorption conditions, the term $v_g W_t$ in Eq. (30) is equal to the total flow of electromagnetic energy across a plane perpendicular to the z-axis, i.e., the total flux across this plane of the real component of the complex Poynting vector $\vec{E} \times \vec{H}*/2$ (the asterisk denotes complex conjugate).

Figure 5 shows calculated values of $\alpha \omega/\bar{\nu}_c$-vs-$\omega_{pe}/\omega$ for various values of the wave frequency and geometrical parameters. The attenuation is small at high electron densities, such that $\omega_{pe} \gg \omega$, and becomes large close to the cut-off limit, i.e., when $\omega_{pe}/\omega \to (1+\varepsilon_g)^{1/2}$.

The calculations of the attenuation coefficient from Poynting's theorem, in the approximation $\nu_c \ll \omega$ and using real algebra only, have been shown to agree remarkably well with calculations based on complex algebra solutions of the dispersion equation (Glaude et al., 1980).

The effects of radially inhomogeneous electron density on the dispersion relation have also been investigated by various authors using the quasistatic approximation ($\vec{\nabla} \times \vec{E} = 0 \to \vec{E} = -\vec{\nabla}F$). Trivelpiece (1967) and Ilic (1968) assumed a parabolic radial electron density distribution, that is

$$n_e(r) = n_e(0) \left[1 - K\left(\frac{r}{a}\right)^2\right] , \qquad (35)$$

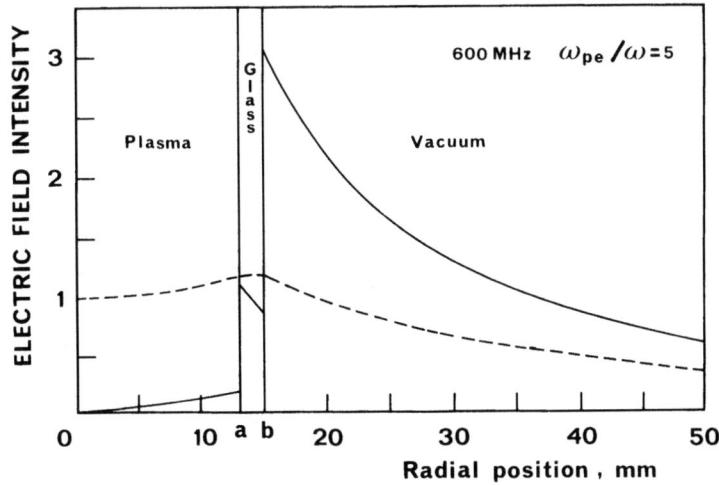

Fig. 4. Radial distribution of the radial (solid curves) and axial (dashed curves) electric field components of an azymuthally symmetric SW propagating along a homogeneous cylindrical plasma column (from Moisan et al., 1982a).

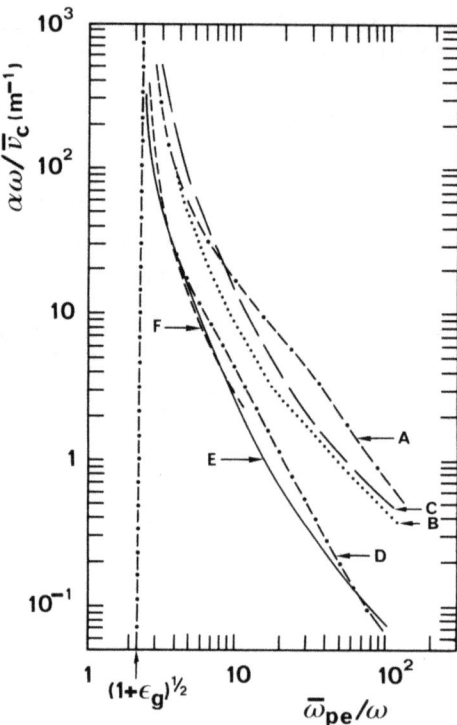

Fig. 5. Attenuation characteristics of azimuthally symmetric SW's propagating along a radially homogeneous plasma for: a=0.45 cm, b=0.60 cm, f=210 MHz (A) and 2450 MHz (B); a=0.30 cm, b=0.45 cm, f=2450 MHz (C); a=1.25 cm, b=1.49 cm, f=210 MHz (D), 600 MHz (E), and 2450 MHz (F).

where K measures the rate of radial decrease. Figure 6 shows curves of $\omega/\bar{\omega}_{pe}$-vs.-βa calculated by Ilic for various values of K ranging form 0 (homogeneous plasma) to 0.9. Here $\bar{\omega}_{pe}$ denotes the plasma frequency that corresponds to the radially averaged electron number density, that is, $\bar{n}_e = n_e(0)(1 - K/2)$. It is seen that for $\beta a \geq 1$ the inhomogeneous distribution gives lower values of $\omega/\bar{\omega}_{pe}$ than in the homogeneous case. This may be easily understood since for high βa the field concentrates near the plasma boundary, so that the asymptote is determined by $n_e(a)$, i.e., $\omega/\omega_{pe}(a) \to (1 + \varepsilon_g)^{-1/2}$ as $\beta a \to \infty$.

Hence, $\omega/\bar{\omega}_{pe} \to (1 + \varepsilon_g)^{-1/2} [(1 - K)/(1 - K/2)]^{1/2}$, which is lower than the asymptote corresponding to the homogeneous case since $0 < K < 1$. For $\beta a < 1$, however, the SW propagation is determined only by the average electron density.

An interesting feature resulting from these calculations is the appearance of a region of backward wave propagation. However, by considering simultaneously the combined effects of radial inhomogeneity and of a sheath region near the wall (Ilic, 1986) this backward behavior

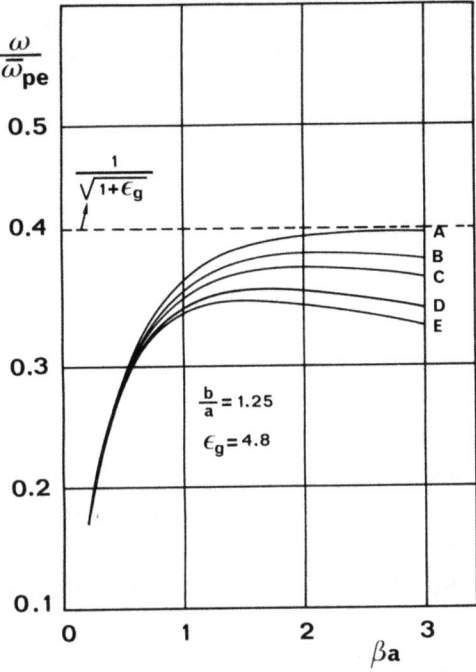

Fig. 6. The effect of the parabolic density $n_e = n_{eo}[1-K(r/a)^2]$ on the dispersion characteristics of the symmetric mode for K=0 (A); 0.4 (B); 0.6 (C); 0.8 (D); and 0.9 (E) (from Ilic, 1968).

may eventually be eliminated. It is worth pointing out that Ilic's calculations assume that the sheath is an electron-free region and that the plasma density remains over-critical at the plasma-sheath boundary, thus avoiding resonance problems at $\omega_{pe} = \omega$ in the calculations. In fact, one should account for a transition region near the boundary along which the electron density decreases to zero, thus reaching somewhere in this layer the resonance value. The excitation of local Langmuir waves by the SW in the resonance region causes a damping of the surface modes (Gradov and Stenflo, 1983) in addition to that resulting from collisions. In this paper, however, only collisional damping will be taken into account for the purposes of modeling SW-produced plasmas. The good agreement of the calculations with experiment gives an a posteriori justification for this assumption.

THE MAINTENANCE FIELD FOR DIFFUSION-CONTROLLED HF DISCHARGES

The Two-Moment Equations

It is well known that the maintenance electric fields of steady-state, diffusion-controlled discharge plasmas, i.e., the discharge characteristics, are the result of a balance between collisional ionization of the gas and the loss of electrons to the wall at a rate determined by the collisions of ions and atoms, and by the space-charge fields. For a quasi-neutral discharge (electron Debye length, λ_D, much smaller than any dimension of the discharge vessel), the diffusion of the charged particles to the wall may be approximately described by a two-moment theory (Self and Ewald, 1966), i.e., by the continuity and the momentum transport equations for the electrons and the ions. These equations may be written

$$\vec{\nabla} \cdot (n\vec{v}) = \nu_i n \tag{36}$$

$$\vec{v} = -\mu_e \vec{E}_A - (1/n) \vec{\nabla}(D_e n) \tag{37}$$

$$(\vec{v} \cdot \vec{\nabla})\vec{v} + \nu_i \vec{v} = (e/M)\vec{E}_A - (K T_i/M)(1/n)\vec{\nabla}n - \nu_{in} \vec{v}. \tag{38}$$

Here, $n = n_e = n_i$ is the plasma density, $\vec{v}$ is the mean velocity of the electrons and the ions under the action of the space-charge field $\vec{E}_A$ and of the spatial density gradients, ν_i is the ionization rate per electron, μ_e and D_e are the electron dc mobility and free-diffusion coefficient, respectively, T_i is the ion temperature, M is the mass of an ion, and ν_{in} is the effective ion-neutral momentum transfer collision frequency including the effects of both elastic and charge transfer collisions.

The Isothermal Discharge

Let us consider first, as an illustrative example, the simple case of a discharge where the electron energy distribution function (EEDF) is spatially homogeneous and, say, Maxwellian with temperature T_e. It has been shown that, in this case, the presence of the nonlinear inertia term in Eq. (38) makes the plasma-sheath boundary appear naturally when the Eqs. (36) to (38) are solved (Persson, 1962; Self and Ewald, 1966; Forrest and Franklin, 1966). In fact, the solutions of this system show that the ion drift velocity converges to the isothermal ion sound speed, i.e., $v \rightarrow [K(T_e + T_i)/M]^{1/2}$, the plasma-sheath boundary being just the point where this velocity is reached (Bohm criterion). Since the sheath thickness is negligibly small compared to the dimensions of the vessel, it can be shown that this condition determines the maintenance temperature T_e as a function of $N\Lambda$, N being the gas density and Λ the characteristic diffusion length for the discharge vessel (for example, $\Lambda = R/2.405$ for infinite cylindrical geometry and $\Lambda = L/\pi$ for infinite parallel plane geometry, L being the distance between the walls).

The results of this theory may be conveniently expressed in the simple form (Ingold, 1978; Muller and Phelps, 1980; Ferreira and Loureiro, 1984)

$$\nu_w = \frac{D_{se}}{\Lambda^2} \tag{39}$$

where ν_w is the rate of electron loss by diffusion to the wall, and D_{se} is an effective electron-diffusion coefficient. The maintenance condition for a steady-state discharge is $\nu_i = \nu_w$, hence

$$\nu_i = \frac{D_{se}}{\Lambda^2}. \tag{40}$$

The ratio of D_{se} to the ambipolar diffusion coefficient D_a ($D_a = D_e/(\sigma+1)$, where $\sigma = \mu_e/\mu_i$ is the mobility ratio) is a function only of Λ/λ_i, where

$$\lambda_i(T_e) = (M \mu_i/e)(3 e T_e/M)^{1/2} \tag{41}$$

denotes an effective ion mean-free-path (Ingold, 1972). At sufficiently high pressures, i.e., for $\lambda_i \ll \Lambda$, we have $D_{se} \to D_a$, so that $\nu_i = D_a/\Lambda^2$, which is the well-known result of Schottky's ambipolar diffusion theory with $n_e = 0$ at the wall (Schottky, 1924). In the low-pressure limit, i.e., when $\lambda_i \gg \Lambda$, the results are similar to those of the free-fall theory (Self, 1965). Figure 7 shows curves of D_{se}/D_a-vs-Λ/λ_i for argon discharges in plane-parallel and cylindrical geometries.

The fact that Eq. (40) determines T_e as a function of $N\Lambda$ may then be readily understood by rewriting it in the form

$$C_i(T_e) \equiv \frac{\nu_i}{N} = \frac{D_{se}}{D} \frac{D_a N}{(N\Lambda)^2}, \tag{42}$$

$C_i(T_e)$ being the electron rate coefficient for ionization.

For a non-Maxwellian electron energy distribution the theory applies as well, but with T_e replaced by the characteristic energy $u_k = D_e/\mu_e$ in Eq. (41). In this case C_i, D_e, and μ_e must be determined from the solutions to the electron Boltzmann equation. For $\omega \ll \nu_c$, the electron energy distribution function (EEDF) associated with an HF field of RMS amplitude E is identical to that associated with a dc field of the same amplitude* (Holstein, 1946), so that Eq. (40) provides discharge characteristics of E/N as a function of $N\Lambda$. In the opposite limit of $\omega \gg \nu_c$, the EEDF depends only on the parameter E/ω (see, e.g., Brown, 1956; Ferreira and Loureiro, 1984), so that the discharge characteristics may be expressed as curves of E/ω-vs-$N\Lambda$. The intermediate situations are more complicated except when ν_c is independent of the electron velocity. Then the EEDF depends only on E_e/N, where

$$E_e = E \frac{\nu_c}{(\nu_c^2 + \omega^2)^{1/2}} \tag{43}$$

represents an effective field (Rose and Brown, 1955; Brown 1956). In this case the discharge characteristics can be expressed as curves of E_e/N-vs-$N\Lambda$ for all frequencies.

---

* Provided ω is sufficiently high so that the isotropic part of the distribution function does not change appreciably during a cycle of field oscillation. This requires that $\omega \gg$ v/smallest dimension of the discharge vessel, and $\omega \gg \tau_e^{-1}$, where τ_e is the characteristic time for energy relaxation by collisions with the atoms.

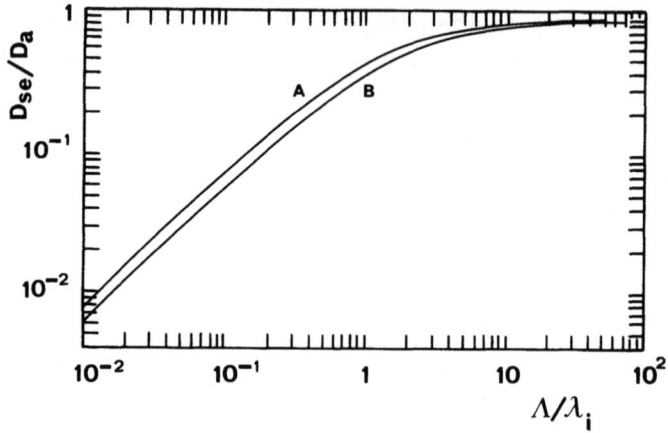

Fig. 7. Ratio of the effective diffusion coefficient to the ambipolar diffusion coefficient vs. the ratio of the diffusion length to the effective ion mean-free-path for cylindrical (A) and plane-parallel (B) argon discharges (from Ferreira and Loureiro, 1984).

When ν_c is velocity dependent (e.g., in argon), the concept of effective field does not generally apply and the discharge characteristics depend, in principle, on an additional parameter proportional to N/ω (e.g., ν_{ce}/ω, where ν_{ce} is the collision frequency for the electrons of some representative velocity). This follows from the fact that, for $\omega \sim \nu_c$, the EEDF depends on E/N and E/ω, or some other combination of these parameter such as, E/N and N/ω.

It is interesting to classify the various frequency operating regimes with the help of a diagram of E/N-vs-E/ω, such as that shown in Fig. 8 for argon (Ferreira and Loureiro, 1983). The straight line A corresponds to the condition $R \equiv \dfrac{\omega}{\nu_{cM}} = 1$, where ν_{cM} is the collision frequency for the electrons with the energy corresponding to the peak of the momentum-transfer cross section ($\nu_{cM}/N = 3.2 \times 10^{-7} \mathrm{cm}^3 \mathrm{s}^{-1}$). In the upper ($R > 1$) half-planes the distribution becomes a function only of E/ω (HF regime) or E/N (dc regime), respectively, except in a transition region where it depends on both parameters. The condition $\omega \gg \nu_{cM}$ may be, however, too severe for defining the onset of the HF regime and could be replaced, by $\omega \gg \bar{\nu}_c$, with $\bar{\nu}_c$ given by Eq. (33). Curve B corresponds to $\omega = 10\bar{\nu}_c$; the HF regime is certainly established along this line and in the left half-plane. The vertical line marked on the diagram (curve C) corresponds to the value of E/ω for which the elastic and the inelastic electron energy losses are of the same magnitude, in the case of the HF regime. To the left of this line the body of the EEDF becomes Maxwellian with temperature $T_e = (e M/3m^2)(E/\omega)^2$ (see, e.g., Allis, 1956). A scale for

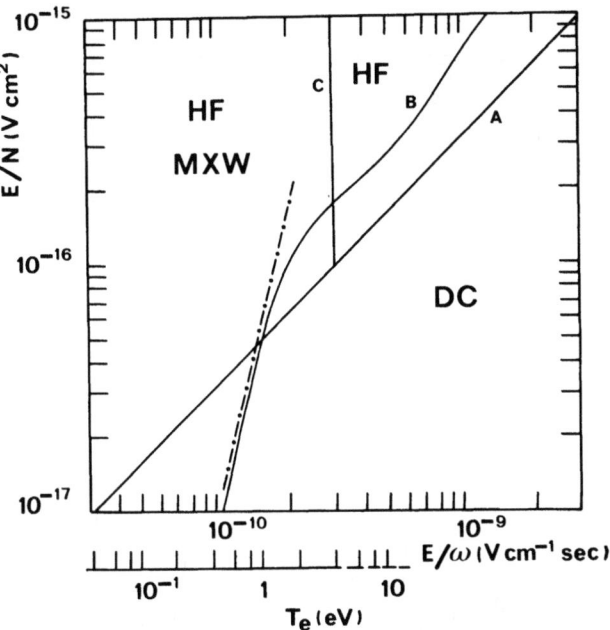

Fig. 8. E/N-vs-E/ω diagram: (A) $r \equiv \omega/\nu_{cM}=1$; (B) $R' \equiv \omega/\bar{\nu}_c=10$; (C) $E/\omega = 3 \times 10^{-10}$ Vcm^{-1}s (equal elastic and inelastic loss rates in the case R>>1). The chain curve is the asymptotic limit of the line R'=10 at low E/ω (from Ferreira and Loureiro, 1983).

these T_e values is also provided in the diagram of Fig. 8.

The shape of the EEDF in argon is quite different in the two limiting cases of $\omega \ll \nu_c$ and $\omega \gg \nu_c$. This is illustrated in Fig. 9 where we have represented the EEDF for a constant value of $E/N = 6.5 \times 10^{-16}$ Vcm^2 and various values of $R \equiv \omega/\nu_{cM}$ ranging from zero (dc case) to 31.2. As R increases at constant E/N the representative point on the diagram of Fig. 8 moves along a horizontal line from the right to the left, i.e., from the dc range to the HF one. We see that the body of the EEDF becomes a steeply decreasing function of the electron energy in the HF range, this behavior being a consequence only of the fact that $\nu_c(u)$ increases with u in the energy range of the bulk electrons (this can be understood from the approximate analytical solution given by Eq. (7) in Ferreira and Loureiro, 1983). Therefore, we may expect a similar behavior for all gases where $\nu_c(u)$ is an increasing function.

A direct consequence of this deep change in the shape of the EDDF is that, for a similar average energy, the number of electrons in the high energy tail is much higher, therefore the excitation and the ionization

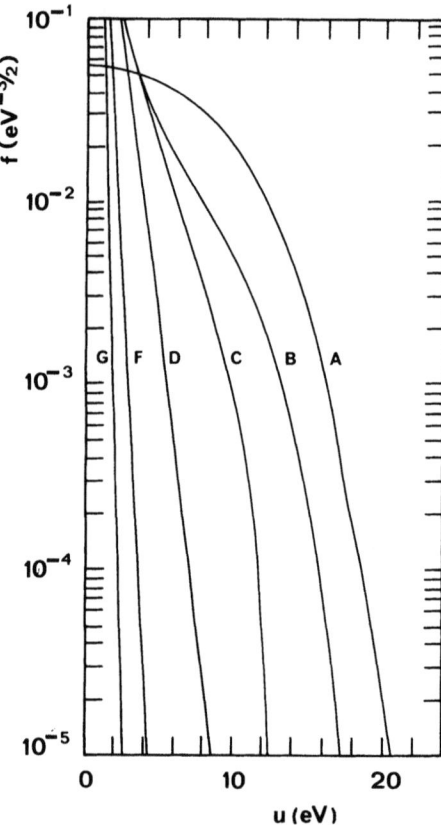

Fig. 9. Electron energy distribution functions in argon for $E/N = 6.5 \times 10^{-16}$ Vcm^2 and $R \equiv \omega/\nu_{cM}=0$ (A); 1.25 (B); 9.35 (C); 15.6 (D); 23.4 (F); 31.2 (G) (from Ferreira and Loureiro, 1983).

are much more efficient for $\omega \gg \nu_c$ than for $\omega \ll \nu_c$, as illustrated in Fig. 10. This is well reflected in a plot of the fractional power transferred by the electrons via different collisional channels versus the average absorbed power per electron at unit gas density, $\theta/N=(e\, n_e)^{-1} Re(\sigma)E^2$ *, shown in Fig. 11. The curves for $\omega \gg \nu_c$ are shifted towards lower θ/N values with respect to those for $\omega \ll \nu_c$; this means, in particular, that the average absorbed power per electron required to sustain on HF discharge is lower when $\omega \gg \nu_c$ than when $\omega \ll \nu_c$.

---

* Note that θ/N is a unique function of E/N for $\omega \ll \nu_c$ and of E/ω for $\omega \gg \nu_c$.

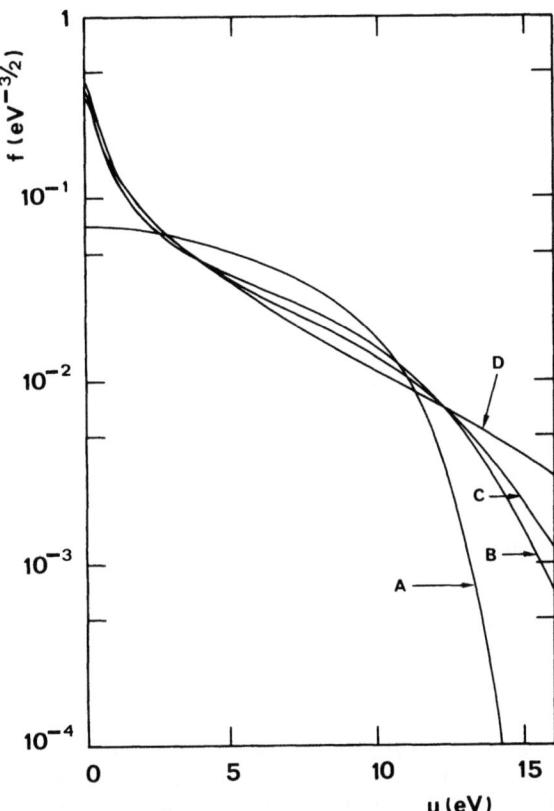

Fig. 10. Electron energy distribution functions in argon with the same mean energy of 3.5eV for the following values of ω/ν_{ce}: 0 (A); 0.5 (B); 0.8 (C); >>1 (D).

Though the concept of effective field is not applicable in argon, it was shown that the ionization rate coefficient can be approximately expressed as a function of $E_e/N = (E/N)\nu_{ce}/(\nu_c^2 + \omega^2)^{1/2}$, with $\nu_{ce}/N = 2.0 \times 10^{-7}$ cm^3 s^{-1}, for all frequencies (Ferreira and Loureiro, 1984). Figure 12 shows that the calculated values of C_i as a function of E_e/N for various ratios of ω/ν_{ce} may be well fitted by a sole curve. Hence, E_e constitutes, in some sense, an effective field for the ionization.

The practical importance of this is that the discharge characteristics derived from Eq. (40) can then be expressed as curves of E_e/N-vs.-$N\lambda_i$ for all frequencies, as illustrated in Fig. 13 for plane-parallel and cylindrical geometry.

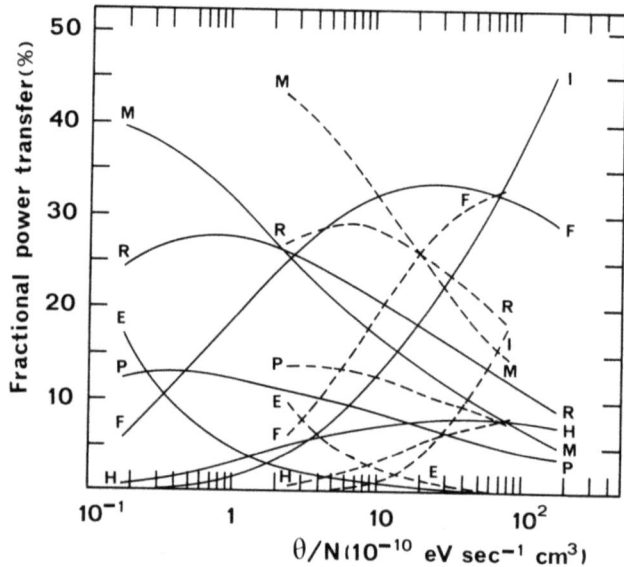

Fig. 11. Percentage electron energy losses in argon vs. the average input power per electron at unit gas density for $\omega \gg \nu_{ce}$ (solid curves) and $\omega \ll \nu_{ce}$ (dashed curves): E = elastic; M = $^3P_0 + ^3P_2$; R = $^1P_1$; P = $^3P_1$; F = forbidden transitions; H = higher-lying allowed levels; I = ionization (from Ferreira and Loureiro, 1984).

Discharges Sustained by Inhomogeneous HF Fields

The steady-state condition for a discharge described by Eqs. (36) to (38) constitutes always a boundary value problem, whether one assumes that the discharge is isothermal or not. The isothermal case is particularly simple because we only need Eqs. (36) to (38) plus the electron Boltzmann equation. When solved for a given geometry, this eigenvalue problem yields, as we have seen, the maintenance reduced field, E_e/N, as a function of the product $N\Lambda$.

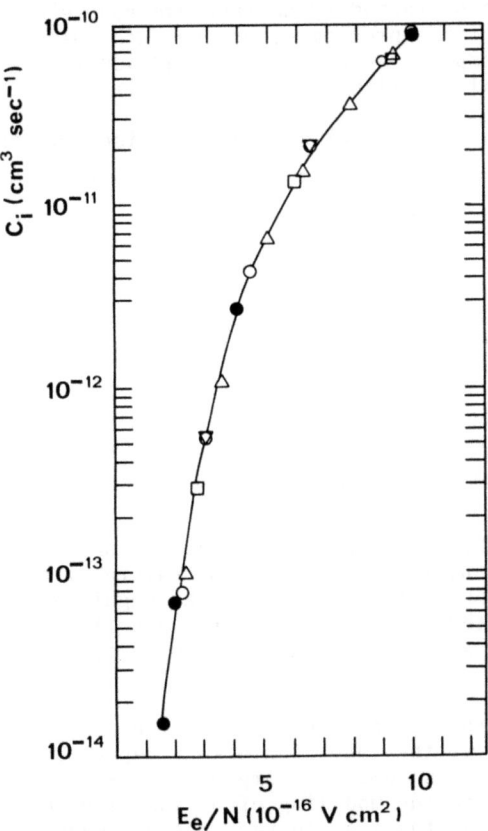

Fig. 12. Ionization coefficient in argon vs. the ratio of the effective field to the gas density for various values of ω/ν_{ce} ranging from zero to ∞ (data points). The curve is the best fit to these data (from Ferreira and Loureiro, 1984).

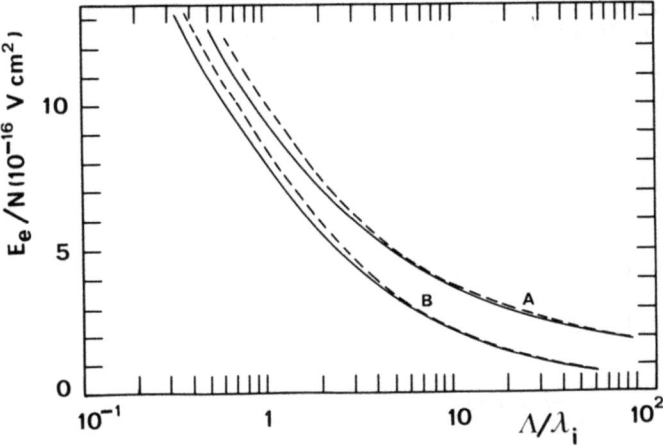

Fig. 13. Characteristics of E_e/N-vs-Λ/λ_i for plane parallel (solid curves) and cylindrical (dashed curves) argon discharges: A = EEDF calculated from Boltzmann equation; B = Maxwellian EEDF (from Ferreira and Loureiro, 1984).

For a discharge sustained by an inhomogeneous electric field the fundamental nature of the problem is not so much changed, though the situation is more complicated because we need additional equations to determine the spatial distribution of the HF field.

For example, in the case of a planar HF discharge the additional equation needed for self-consistency is simply $\vec{\nabla} \cdot \vec{J} = 0$, i.e., J=const., where $\vec{J}$ is the total current density, including the electron current and displacement current. This problem has been treated by Allis, Brown and Everhart in a famous paper (Allis et al., 1951).

In the case of a cylindrical plasma sustained by, say, a TM traveling wave propagating along the axis the additional equations are $\vec{\nabla} \cdot \vec{J} = 0$, Eq. (14), which relates the transverse and the axial components of $\vec{E}$ for any TM traveling wave, and the dispersion relation *. The set of Eqs. (14) and $\vec{\nabla} \cdot \vec{J} = 0$ yields, of course, Eqs. (23) and (24) for the components E_z and E_r in the case of cylindrical geometry.

In both cases the additional equations are necessary to determine the field profiles, but the boundary value nature of the diffusion problem remains unchanged and determines the magnitude of the field required to sustain the discharge. The reduced field intensity at the axis constitutes, in some sense, the eigenvalue for this boundary value problem.

We now proceed to a deeper analysis of the effects of inhomogeneous fields. We first treat the case of a planar capacitive discharge as it serves to illustrate in a somewhat simpler context the fundamental problems encountered in the modeling of SW sustained discharges. In fact, we shall see that these discharges possess many features in common.

HF DISCHARGE BETWEEN PARALLEL PLATES

Let us consider a gas discharge taking place between two parallel-plate electrodes separated from the gas by a dielectric barrier. The total current density is a constant across the discharge (as $\nabla \cdot \vec{J} = 0$) and, assuming ν_c independent of the electron velocity, is given by

$$J = \left(\frac{n_e e^2}{m} \frac{1}{\nu_c + j\omega} + j\omega \varepsilon_0 \right) E . \tag{44}$$

The magnitude of the resistivity is therefore

$$\left| \frac{E}{J} \right| = \frac{1}{\omega \varepsilon_0} \frac{(\delta^2 + 1)^{1/2}}{[(y - 1)^2 + \delta^2]^{1/2}} , \tag{45}$$

where we have defined

$$y \equiv \left(\frac{\omega_{pe}}{\omega} \right)^2 = \frac{n_e e^2}{\varepsilon_0 m \omega^2} , \text{ and} \tag{46}$$

---

* In the case of a surface wave the problem may be simplified by assuming that the dispersion properties of the inhomogeneous plasma are approximately the same as those of a homogeneous plasma with the same average density. This constitutes a good approximation for $\beta a \lesssim 1$ as we have seen.

$$\delta \equiv \frac{\omega_c}{\omega} . \tag{47}$$

It follows that the magnitude of the field varies according to the law

$$\frac{E}{E_o} = \left[\frac{(y_o - 1)^2 + \delta^2}{(y - 1)^2 + \delta^2} \right]^{1/2} \tag{48}$$

where the subscript zero denotes the values at the axis. As the plasma density decreases, the magnitude of the field increases from the axis towards the walls and passes through a maximum at the plasma resonance where y=1 (if the density at the plasma-sheath boundary is below the resonance value).

The set of Eqs. (48) and (36) to (38) together with the electron data obtained from solutions to the Boltzmann equation determine self-consistently the spatial distributions of the density and of the field intensity, and the values of E_o and n_{eo} for specified discharge conditions (i.e., ω, N, electrode separation L, and total current density). In practice, for ω, N, and L given one may in principle solve numerically the boundary value problem and obtain the spatial distributions of y and E, as well as E_o (determined as a characteristic value for the problem), for each value of y_o. One may then calculate the average normalized density $\bar{y}$ or the volume-averaged power density and express the solutions as a function of any of these two quantities in place of y_o. As we have assumed that ν_c is velocity-independent, the number of independent variables may be further reduced using the effective field $E_e = E \delta (1 + \delta^2)^{-1/2}$; the spatial distributions and E_{eo}/N are unique functions of NL, δ, and $\bar{y}$.

Allis et al. (1951) and later Bell (1970) treated this problem using the ambipolar diffusion equation

$$\frac{d^2 n_e}{dx^2} + \frac{\nu_i}{D_a} n_e = 0 \tag{49}$$

with the boundary condition $n_e(L/2)=0$, instead of the two-moment equations (36) to (38), which is justified at sufficiently high pressures. Allis et al assumed that the ionization frequency varies as a power h of the field, i.e.,

$$\frac{\nu_i}{\nu_{io}} = \left(\frac{E}{E_o}\right)^h , \tag{50}$$

and carried out approximate calculations of the spatial distributions in the discharge for various ranges of parameters. For sufficiently high pressures ($\delta^2 \gg 1$), or for electron concentrations well below resonance, the ac electron current is negligible relative to the displacement current and the field is homogeneous. The solution of the boundary value problem is then

$$n_e = n_{eo} \cos\left(\frac{\pi x}{L}\right) \text{ and} \tag{51}$$

$$\frac{\nu_i}{D_a} = \left(\frac{\pi}{L}\right)^2 = \frac{1}{\Lambda^2}. \tag{52}$$

Since $\nu_i/(D_a N^2)$ is an unique function of E_e/N, Eq. (52) determines the sustaining reduced field E_e/N as a function of $N\Lambda$.

For $y^2 + 1 + \delta^2 \gg 2y$, the resonance effects on the electron spatial distribution are eliminated. One may then neglect the negative terms in Eq. (48) and determine the distribution of electrons for various values of the parameter $S \equiv y_o (\delta^2 + 1)^{-1/2}$. As $S \to 0$ we obtain the cosine solution (51); as $S \to \infty$, i.e., for high electron concentrations, the diffusion equation reduces to the form

$$\frac{d^2}{dx^2}\left(\frac{y}{y_o}\right) + \frac{\nu_{io}}{D_a}\left(\frac{y_o}{y}\right)^{h-1} = 0 \tag{53}$$

whose solution for the specified boundary conditions is simply a circle when $h=4$, that is

$$y = y_o \left[1 - \left(\frac{2x}{L}\right)^2\right]^{1/2}. \tag{54}$$

The characteristic value problem for this case corresponds to the condition

$$\frac{\nu_{io}}{D_a} = \frac{4}{L^2} \tag{55}$$

which determines the reduced field at the axis E_{eo}/N as a function of $N\Lambda$. It is smaller than the reduced field by a factor of $(4/\pi^2)^{1/4} \sim 0.80$ for the homogeneous case.

Bell (1970) calculated numerically the exact solutions in this limit of high electron concentrations for a helium discharge operating at 20 MHz, using available experimental data for ν_i/N and D_a/N. His solutions, illustrated in Fig. 14, show that the electron density profiles are approximately circular-shaped and in agreement with the predictions of Allis et al. (1951). They depend, however, on the parameter $N\Lambda$ and become more and more circular for the highest $N\Lambda$ values of his work because the power law (50) with $h=4$ is then better verified (see Fig. 6 in Allis et al., 1951).

The electric field intensity rises from low values at the axis up to a sharp maximum value at the position where $y=1$[*]. Bell's calculation of the spatially averaged and axial-reduced field strength are represented in Fig. 15 as a function of $N\Lambda$. There is, however, a slight dependence of the spatially averaged field on the axial density n_{eo}, which is a

---

[*] Note that we obtain from Eq. (48) $E_{MAX}/E_o = \left[1 + \left(\frac{y_o - 1}{\delta}\right)^2\right]^{1/2}$

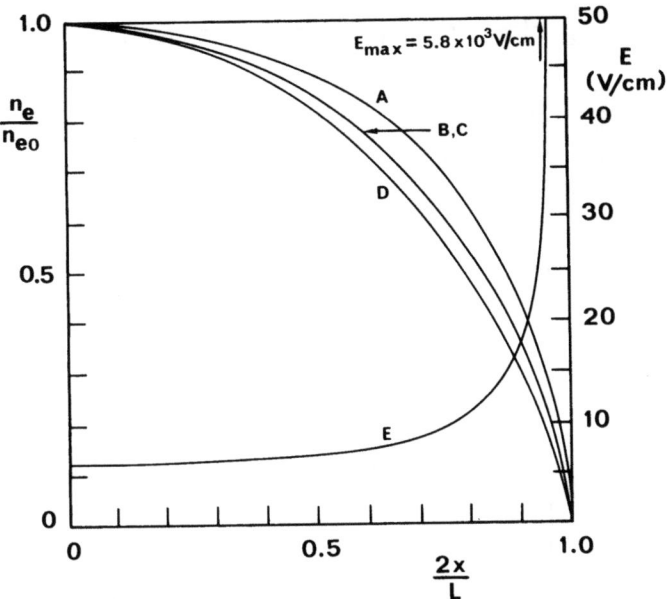

Fig. 14. Spatial variations of electron density (curves A,B,C, and D) and electric field intensity (curve E) for a helium discharge operating at 20 MHz between parallel plates and for the following values of the pressure and plate separation: 1 torr, 10 cm (A); 1 torr, 2 cm (B); 10 torr, 0.2 cm (C); 1 torr, 0.5 cm (D); 1 torr, 2 cm (E) (from Bell, 1970).

result of the contribution from regions very close to the wall. The points on Fig. 15 correspond to the approximate solution (55) and agree very closely with the exact calculation.

Fig. 16 illustrates Bell's calculations of the electron density at the axis as a function of the volume averaged power density at unit gas pressure for various values of pL. These plots are essentially linear, which is indicative of the fact that the average absorbed power per electron at unit gas density, Θ/N, is practically a unique function of $N\Lambda$

$$\frac{\Theta}{N} = f(N\Lambda) . \tag{56}$$

We will obtain a similar conclusion in the case of SW-sustained discharges.

The ionization and the excitation rates are the product of the corresponding rates per electron and the electron concentration so that their distribution across the discharge will be given by expressions of the type

$$\frac{y}{y_o} [\frac{(y_o - 1)^2 + \delta^2}{(y - 1)^2 + \delta^2}]^{h/2}$$

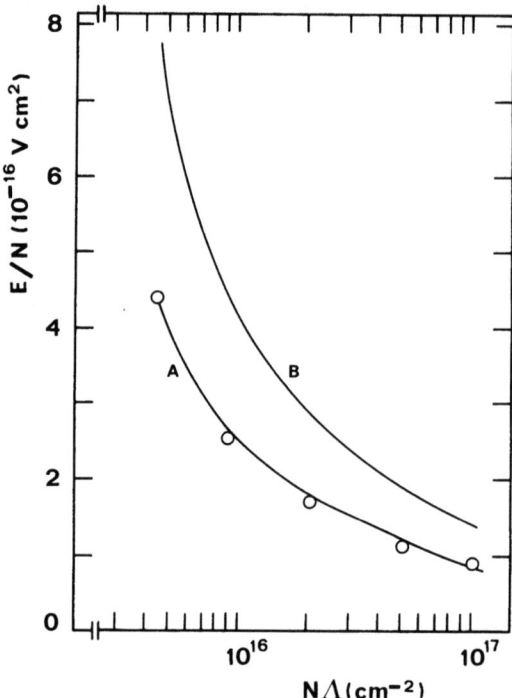

Fig. 15. Axial (curve A and points) and spatially averaged (curve B) values of the reduced sustaining field for a plane-parallel helium discharge operating at 20 MHz as a function of NΛ. The points correspond to the approximation of a circular-shaped electron density distribution (from Bell, 1970).

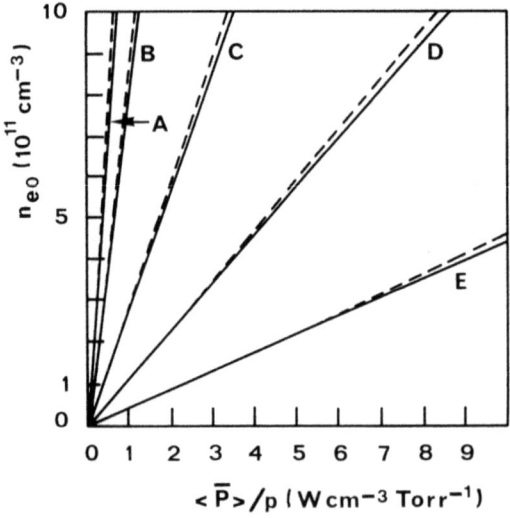

Fig. 16. Variation of the electron density at the axis vs. the volume-averaged power density at unit pressure in a plane-parallel helium discharge operating at 20 MHz for p=1 torr (solid curves) and p=10 torr (dashed curves) and the following values of pL in torr-cm: 10 (A); 5 (B); 2 (C); 1 (D); 0.5 (E) (from Bell, 1970).

if one assumes a power law of the type (50) with a specific value of h (>1) for each process. These distributions will generally present a maximum at some position except when the field intensity is nearly homogeneous. The maximum occurs at the point where the normalized plasma concentration verifies the condition

$$y = y_M \equiv 1 + \frac{hA}{2(h-1)}, \quad (57)$$

where $1 + A \equiv [1 + 4(\delta/h)^2(h-1)]^{1/2}$. There exists a maximum provided $y_M < y_o$. For $\delta \ll 1$, we have $y_M \sim 1 + \delta^2/h$, so that the maximum is nearly coincident with the plasma resonance. For $\delta \gg 1$, we have $y_M \sim 1 + \delta(h-1)^{-1/2}$, so that the maximum may occur much closer to the axis than the resonance point.

Figure 17 illustrates the results of Allis et al. (1951) for a helium discharge in the case pL = 23 torr-cm (which corresponds to h ~ 7), $\delta=1$, and $y_o=2$. The distribution of the light intensity emitted by the discharge should follow approximately the distribution of the ionization rate.

DISCHARGE SUSTAINED BY A SURFACE WAVE

A model of a cylindrical plasma column sustained by a weakly damped ($\omega \gg \nu_c$), axially symmetric SW at low pressures has recently been proposed by this author (Ferreira, 1981, 1983). The basic equations used are the continuity and the momentum transfer Eqs. (36) to (38), Eqs. (23) and (24) for the field components E_z and E_r, and a local electron power balance equation which determines T_e as a function of E/ω (a Maxwellian EEDF was assumed which is justified only for degrees of ionization $>10^{-4}$; see the analysis given in Ferreira and Loureiro 1983). The dispersion characteristics used were those corresponding to a homogeneous plasma column, which is a good approximation for $\beta a \leq 1$, as we have seen.

This formulation determines the electron density, the electron temperature, and the magnitude of the field as a function of the radial position for specified discharge conditions. In the first paper (Ferreira, 1981) a detailed numerical application was made for an argon discharge sustained by a SW with f=600 MHz in a 2.98-cm o.d., 2.50-cm i.d. Pyrex glass tube ($\varepsilon_g=4.52$). The radial distributions of n_e, E, and T_e, the average absorbed power per unit axial length, and the values of the sustaining field at the axis were computed for various values of the gas pressure in the range $0.05 \leq p \leq 2$ torr, and of the average plasma density in the range 5×10^{10} cm$^{-3} \leq \bar{n}_e \leq 5 \times 10^{11}$ cm^{-3}.

Figures 18 and 19 show calculated radial profiles of the electron concentration and of the intensity of the two field components for p=0.1 torr, and various values of the average electron concentration, $\bar{n}_e$. It is seen that the increase in $\bar{n}_e$ results in a steeper increase of the field with radius. A high ionization rate near the boundary makes the density profile flatter in the central region and more steeply decreasing near the boundary. Figure 19 also shows that E_r is negligible as compared with E_z for $\bar{n}_e$ well above the cut-off limit ($n_{ec} = 2.47 \times 10^{10}$ cm^{-3}

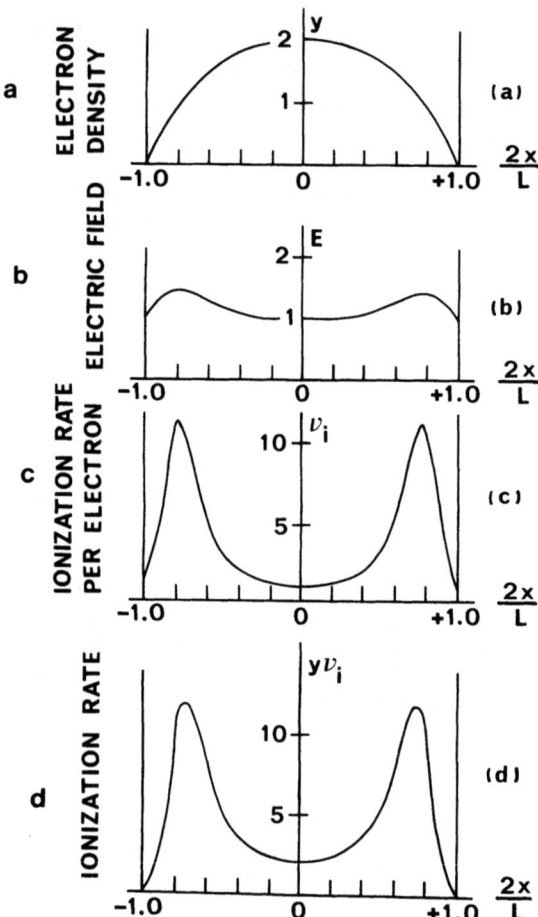

Fig. 17. Variation of electron density (a), electric field (b), ionization rate per electron (c), and ionization rate (d) across a plane-parallel discharge in helium for pL=23 torr-cm, $\omega_{pe}^2(0)/\omega^2=2$, and $\omega/\nu_c=1$ (Allis et al., 1951).

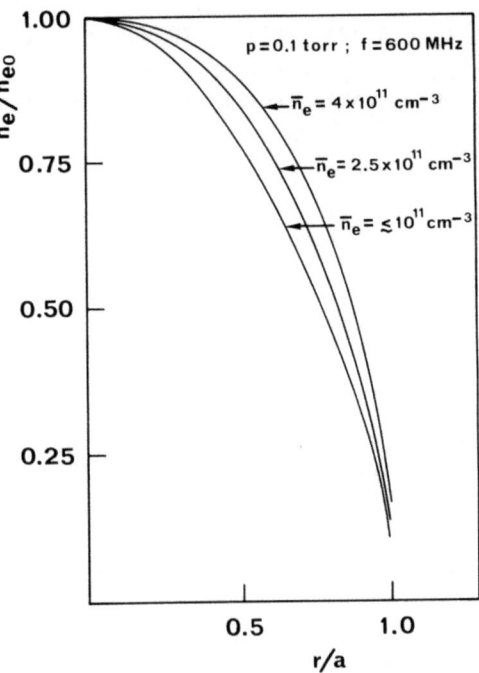

Fig. 18. Radial distribution of electron density in a SW-produced argon plasma column at 0.1 torr and for various values of the average electron density (from Ferreira, 1981).

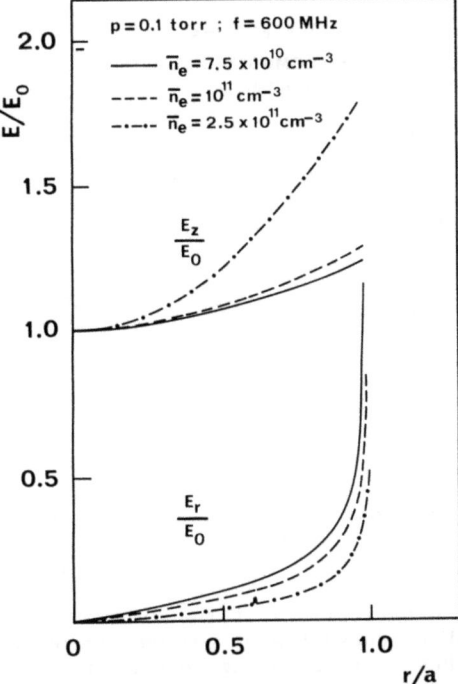

Fig. 19. Radial variation of the axial and radial components of the SW electric field in a SW-produced argon plasma column for various values of the average electron density (from Ferreira, 1981).

for $f = 600$ MHz and $\varepsilon_g = 4.52$). Figure 20 shows that the radial distribution of the field intensity is strongly affected by the plasma inhomogeneity, differing considerably from that corresponding to a homogeneous plasma with a density equal to $\bar{n}_e$.

Calculated radial profiles of excited argon atoms in the $3p^5 6d$ configuration are shown in Fig. 21 for various discharge conditions. For the higher electron densities the distributions have a sharp peak near the boundary which is a consequence of the increase in the field intensity (hence, of the excitation rate) with radius. The situation is quite similar to that found in HF discharges between parallel plates as discussed above. The basic trends exhibited by the calculations are well confirmed by measurements of the radial distribution of the light intensity emitted by these states (Moisan et al., 1980, 1982b).

In the second paper (Ferreira, 1983) it was proved that this discharge obeys simple similarity laws that enable the extension of those specific calculations to other experimental conditions. In fact, the basic equations may be written in this case:

(a) Continuity and momentum transfer equations for the electrons and the ions:

$$\frac{1}{x}\frac{d}{dx}(nvx) = (Na) C_i n \tag{58}$$

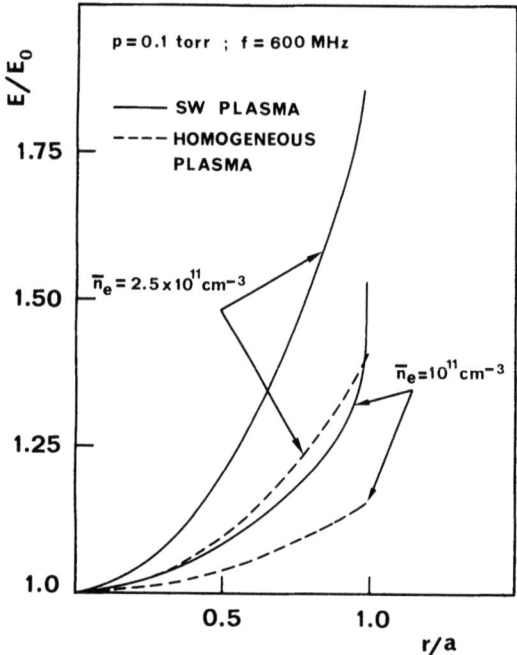

Fig. 20. Radial distributions of the electric-field strength in a SW-produced plasma column (solid curves) compared with those corresponding to a homogeneous plasma with the same average electron density (from Ferreira, 1981).

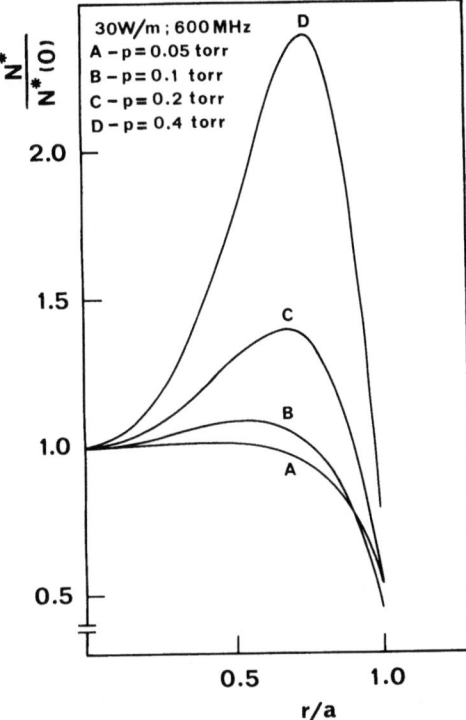

Fig. 21. Radial distributions of excited argon atoms in a SW-produced plasma column (from Ferreira, 1981).

$$v = \frac{D_e N}{Na} \left(\frac{e}{KT_e} \frac{d\Phi}{dx} - \frac{1}{n} \frac{dn}{dx} \right) \tag{59}$$

$$v \frac{dv}{dx} = - \frac{e}{M} \frac{d\Phi}{dx} - \frac{KT_i}{M} \frac{1}{n} \frac{dn}{dx} - Na \left(C_i + \frac{\nu_{in}}{N} \right) v \tag{60}$$

(b) Equations for the components of the electric field:

$$\frac{d^2 E_z}{dx^2} + \left(\frac{1}{x} + \frac{\beta^2}{\Gamma^2} \frac{1}{n} \frac{dn}{dx} \right) \frac{dE_z}{dx} - (\Gamma a)^2 E_z = 0 \tag{61}$$

$$E_r = \frac{\beta a}{(\Gamma a)^2} \frac{dE_z}{dx} \tag{62}$$

(c) Local power balance equation for the electrons:

$$\frac{e^2}{2m} \bar{\nu}_c \left(\frac{E}{\omega} \right)^2 = \frac{3m}{M} KT_e \bar{\nu}_c + \sum_j eV_j NC_j + eV_i NC_i \; . \tag{63}$$

In these equations x=r/a is the normalized radial coordinate, $E_z(x)=E_z(x)/E_0$ and $E_r(x)= E_r(x)/E_0$ are the magnitudes of the field components normalized to the magnitude of the field at the axis, $\Phi(x)$ is the plasma potential, $\bar{\nu}_c$ and Γ are given by Eqs. (33) and (28), respectively, and C_j is the electron rate coefficient for the excitation of the atomic level j of potential V_j (the subscript i holds for the ionization process).

As we have seen, $\Gamma \sim \omega_{pe}^2 a/c \gg \beta$ for $\omega_{pe}^2 \gg \bar{\omega}^2$ (Fig. 3). In practice this limit is reached provided that $\bar{\omega}_{pe}$ slightly exceeds the cut-off value $\bar{\omega}_{pe} = \omega(1 + \varepsilon_g)^{1/2}$, so that this approximation can be used along a major portion of the plasma column. It only fails at the end of the column where n_e approaches the critical value and near the plasma-sheath boundary where n_e becomes small. Eq. (61) and (62) may then be written in the simpler form

$$\frac{d^2 E_z}{dx^2} + \frac{1}{x}\frac{dE_z}{dx} - \left(\frac{\bar{\omega}_{pe} a}{c}\right)^2 \frac{n_e}{\bar{n}_e} E_z = 0 \tag{64}$$

$$E_r \ll E_z . \tag{65}$$

When this approximation is valid the boundary value problem depends only on two independent parameters, Na and $\bar{\omega}_{pe} a/c$. This means that $n_e(x)/\bar{n}_e$, $E(x)$, $T_e(x)$, and E_o/ω are unique functions of these two parameters. Moreover, the variation of ν_c/N with radius is also uniquely determined by these parameters because this quantity is a function only of T_e.

The average absorbed power per unit length of the column may be written (see Eq. 34) in the present case

$$\frac{d P_{abs}}{dz} = \pi \varepsilon_o c^2 N \left(\frac{\bar{\omega}_{pe} a}{c}\right)^2 \left(\frac{E_o}{\omega}\right)^2 \int_0^1 \frac{\bar{\nu}_c}{N} E^2(x) \frac{n_e(x)}{\bar{n}_e} x\, dx \tag{66}$$

This may be expressed in the form

$$\frac{d P_{abs}}{dz} = \Theta \pi a^2 \bar{n}_e , \tag{67}$$

where Θ represents the average absorbed power per electron required to sustain the discharge. It follows from Eq. (66) and the discussion above that Θ/N is a unique function of the parameters $\bar{\omega}_{pe} a/c$ and Na, i.e.,

$$\frac{\Theta}{N} = f_\Theta \left(\frac{\bar{\omega}_{pe} a}{c}, Na\right) . \tag{68}$$

The dependence on $\bar{\omega}_{pe} a/c$ is, however, quite small as shown by the numerical calculations, so that Θ/N is, in practice, a unique function of Na.

Figure 22 shows the reduced effective field at the axis $E_{eo}/N = (E_o/N)(\nu_{ce}/\omega)$, with $\nu_{ce}/N = 2.0 \times 10^{-7}$ cm^3 s^{-1}, as a function of Na for various values of $\bar{\omega}_{pe} a/c$. Also shown for comparison are the results obtained assuming a homogeneous field in the cases of a Maxwellian distribution and exact calculations based on the Boltzmann equation. The

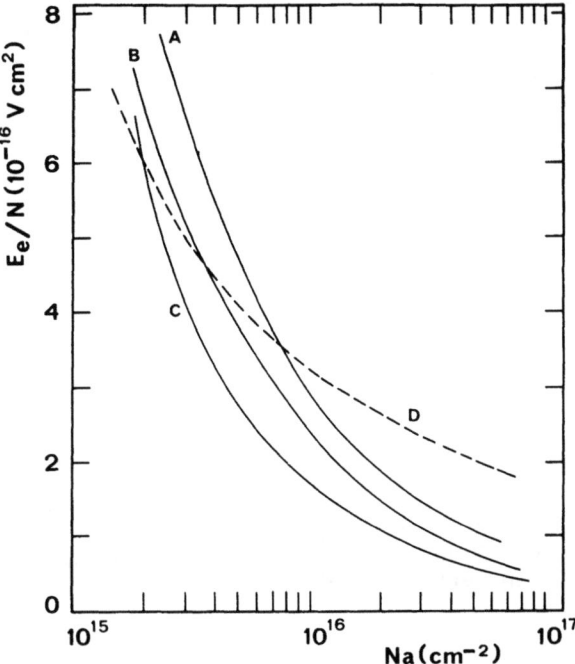

Fig. 22. Reduced sustaining field at the axis for a SW-produced argon plasma vs. Na for $\bar{\omega}_{pe} a/c = 0.75$ (B) and $\bar{\omega}_{pe} a/c = 1.5$ (C) assuming a Maxwellian EEDF. Curve A = homogeneous field, Maxwellian EEDF; Curve D = homogeneous field, EEDF calculated from the Boltzmann equation.

magnitudes of the sustaining field are quite comparable in all these situations.

Figure 23 shows a comparison of computed radially averaged values of $\bar{\nu}_c/p^*$ with data obtained from various experiments (Glaude et al., 1980; Chaker et al., 1982). The agreement between the theoretical predictions and the measurements is quite satisfactory indicating, in particular, that the attenuation of the SW as it propagates along the column is caused by the electron collisions.

The calculated values of Θ/p-vs-pa^* are compared in Fig. 24 with experimental data obtained by Dervisevic et al. (1983) in plasma columns sustained by surface microwaves of various frequencies in capillary tubes with different diameters. The good agreement between theory and experiment confirms the validity of the similarity law $\Theta/N \sim f_\Theta(Na)$. Also shown in this figure are calculations of Θ/p-vs.-pa, assuming a homogeneous field for the following cases: Maxwellian EEDF and $\omega \gg \nu_c$; EEDF calculated from the Boltzmann equation for $\omega \gg \nu_c$ and $\omega \ll \nu_c$. Comparison

---

* Here, the pressure refers to a gas temperature of 300 K, i.e., a pressure of 1 torr actually means a number density of 3.22×10^{16} cm^{-3}.

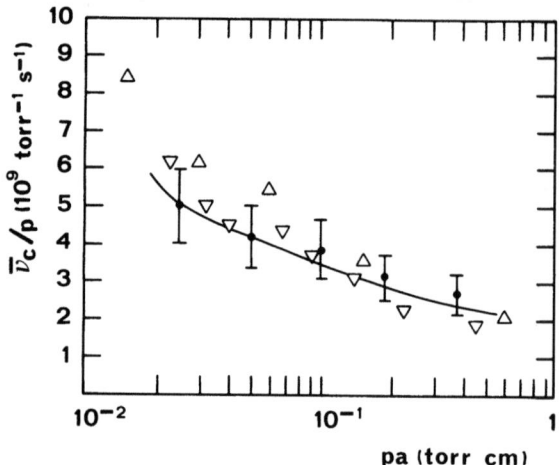

Fig. 23. Effective electron collision frequency at unit pressure vs. pa in a SW-produced argon plasma. Solid curve = Theory. Points = experimental data by Glaude et al., 1980 and by Chaker et al., 1982 (from Ferreira, 1983).

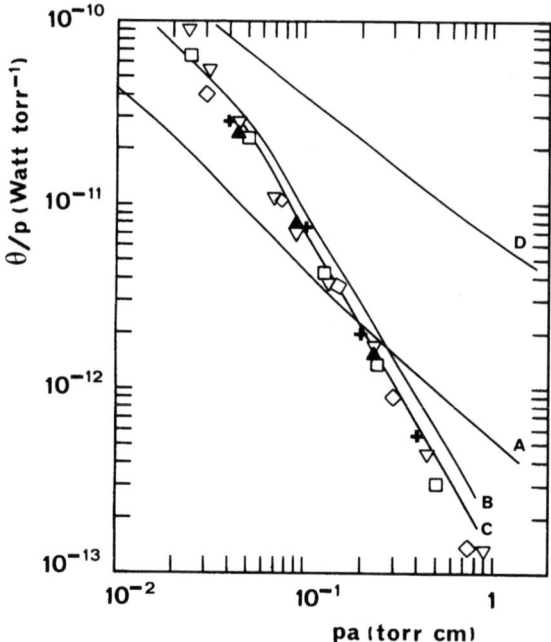

Fig. 24. Average input power per electron at unit pressure required to sustain a cylindrical argon plasma as a function of pa. Homogeneous field: A = $\omega \gg \nu_{ce}$, EEDF from Boltzmann code; B = $\omega \gg \nu_{ce}$, Maxwellian EEDF; D = $\omega \ll \nu_{ce}$, EEDF from Boltzmann code. SW-produced plasma: C = Theory with Maxwellian EEDF; Data points = experiments by Dervisevic et al., 1983 (from Ferreira and Loureiro, 1984).

of curves B and C reveals that the effects of the field inhomogeneity are quite small in determining the values of Θ/p required for the steady-state operation of the discharge. This conclusion could have been anticipated on the basis of the results shown in Fig. 22.

Finally, let us determine the axial profile of the radially averaged electron density. This can be easily determined using the fact that Θ is approximately independent of $\bar{n}_e$ and combining Eqs. (67) and (30). We then obtain

$$\bar{n}_e = \frac{2\alpha P_i}{\Theta \pi a^2} \tag{69}$$

where $P_i = v_g W_t$ is the incident power at the abscissa z. Differentiating this equation and again using Eq. (30) one readily obtains

$$\frac{d\bar{n}_e}{dz} = -2\alpha \bar{n}_e \left(1 - \frac{\bar{n}_e}{\alpha} \frac{d\alpha}{d\bar{n}_e}\right)^{-1} . \tag{70}$$

Using the attenuation characteristics (Fig. 5), this equation can be easily integrated by numerical methods to yield $\bar{n}_e(z)$. It is, however, instructive to make some approximations which provide simple, closed formulas for the distribution $\bar{n}_e(z)$ and for the length of the plasma column as a function of the operating parameters. As shown in Fig. 5, $\log(\alpha\omega/\bar{v}_c)$ varies more or less linearly with $\log(\bar{\omega}_{pe}/\omega)$, except close to the cut-off (which corresponds to the end of the column). Therefore, one may use the approximation

$$\alpha \frac{\omega}{\bar{v}_c} = g(\omega,a)\left(\frac{\bar{\omega}_{pe}}{\omega}\right)^{-K} \tag{71}$$

where $g(\omega,a)$ and K are determined by fitting this equation to the curves given in Fig. 5. This approximation is satisfactory for $\bar{\omega}_{pe} \gg \omega$, i.e., when the above similarity laws also apply. The integration of Eq. (70) is then straightforward and yields (Zakrzewski, 1983; Ferreira, 1983)

$$\frac{\bar{n}_e(z)}{\bar{n}_e(0)} = \left(1 - \frac{K}{1+(K/2)} \alpha_0 z\right)^{2/K} \tag{72}$$

where α_0 is the value of the attenuation at z=0. Here, the axial distance z is measured from the launcher. The initial values $\bar{n}_e(0)$ and α_0 are fully determined by Eq. (69), the attenuation curves, and the theoretical values of Θ and ν_c for given operating conditions (ω, p, geometrical parameters, and incident power at z=0). The column ends when $\bar{n}_e(L)$ is approximately equal to the cut-off value so that its length is approximately given by

$$L = \frac{1+(K/2)}{K} \frac{1}{\alpha_0} \left[1 - \left(\frac{1+\varepsilon_g}{y_0}\right)^{K/2}\right] \tag{73}$$

where $y_0 \equiv (\bar{\omega}_{pe}(0)/\omega)^2$.

Fig. 25 shows calculated and measured axial distributions of the electron density for different values of the SW frequency, gas pressure, and incident power. The agreement is very satisfactory except near the end of the column where the approximations used no longer apply, especially for the lower frequency (210 MHz). Note, however, that the low-attenuation conditions are but marginally satisfied in this case since $\omega/\nu_c \sim 2$ for the higher pressure values of this calculation.

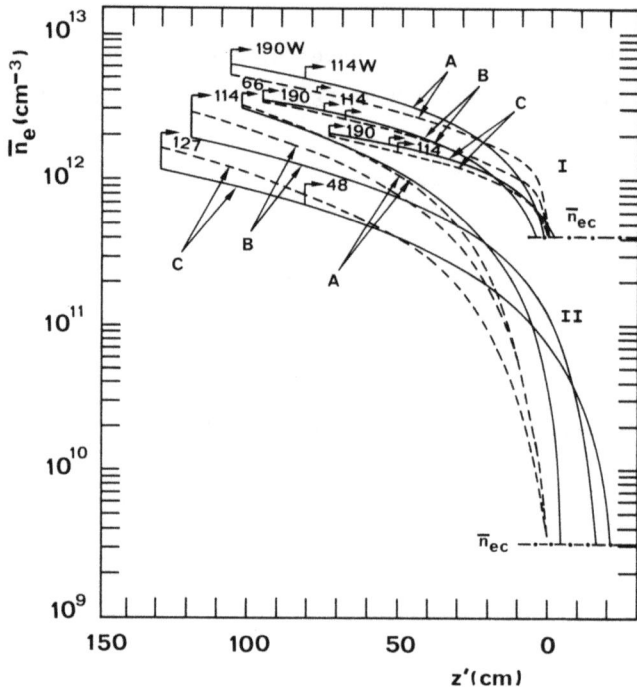

Fig. 25. Calculated (solid curves) and measured (dashed curves) axial distributions of electron density along a SW-produced argon plasma for various values of frequency (I = 2450MH, II = 210MHz), wave incident power in W, and pressure (A = 0.5 torr; B = 0.2 torr; C = 0.1 torr). The abscissa scale is the axial distance measured from the point where the column was experimentally observed to end (from Ferreira, 1983).

CONCLUDING REMARKS

The present degree of achievement of the modeling of surface-wave-produced plasmas is rather high as compared to that of other rf and microwave plasmas. The model developed thus far treats in a nearly self-consistent way the radial distributions of the plasma density and SW field intensity and provides detailed predictions concerning the radial and the axial variations in the plasma properties as a function of the operating parameters. It also predicts the existence of simple similarity laws for this type of discharge which enable the extension of the results to different experimental circumstances. The agreement between theory and experiment is, in general, quite satisfactory.

One may expect that other rf and microwave plasmas possess many features in common with SW-produced plasmas. The latter, however, are particularly suitable for the investigation of local plasma properties as related to the local absorbed power, a unique feature that enables detailed comparisons between theory and experiments. Hence, the study of these discharges opens the way to the understanding of plasmas produced by other HF sources.

ACKNOWLEDGMENTS

The author is greatly indebted to Professors M. Moisan and Z. Zakrzewski for stimulating discussions and valuable suggestions.

REFERENCES

Akao, Y., and Ida, Y., 1963, J. Appl. Phys., 34:2119.
Akao, Y., and Ida, Y., 1964, J. Appl. Phys., 35:2565.
Allis, W. P., Brown, S. C., and Everhart, E., 1951, Phys. Rev., 84:519.
Allis, W. P., 1956, Handb. of Physics, 21:383.
Aronov, B. I., Bogdankevich, L. S., and Kukhadze, A. A., 1976, Plasma Phys., 18:101.
Bekefi, G., 1966 "Radiation Processes in Plasmas", Wiley, New York.
Bell, A. T., 1970, Ind. Engng. Chem. Fundam., 9:160.
Brown, S. C., 1956, Handb. of Physics, 22:531.
Burykin, Yu., I., Levitskiy, S. M., and Martynenko, V. G., 1975, Radio Engng. Electron. Phys., 22:86.
Carlile, R. N., 1964, J. Appl. Phys., 22:531.
Chaker, M., Nghiem, P., Bloyet, E., Leprince, P., and Marec, J., 1981, Rapport Interne L. P. 190, Universite de Paris-Sud (Orsay); 1982, J. Physique-Lett., 43:L71.
Clarricoats, P. J. B., Oliver, A. D., and Wong, J. S. L., 1966, Proc. IEEE, 113:755.
Dervisevic, E., Bloyet, E., Laporte, C., Leprince, P., Marec, J., Pouey, M., and Saada, S., 1983, Proc. XVI th Int. Conf. Phenomena in Ionized Gases, Dusseldorf, p. 468.
Ferreira, C. M., 1981, J. Phys. D: Appl. Phys., 14:1811.
Ferreira, C. M., 1983, J. Phys. D: Appl. Phys., 16:1673.
Ferreira, C. M., and Loureiro, J., 1983, J. Phys. D: Appl. Phys., 16:2471.
Ferreira, C. M., and Loureiro, J., 1984, J. Phys. D: Appl. Phys., 17:1175.
Forrest, J. R., and Franklin, R. N., 1966, Br. J. Appl. Phys., 17:1061.
Glaude, V. M. M., Moisan, M., Pantel, R., Leprince, P., and Marec, J., 1980, J. Appl. Phys., 51:5693.
Gradov, O. M., and Stenflo, L., 1983, Physics Reports, 94:111.
Holstein, T., 1946, Phys. Rev., 70:367.
Ilic, D., 1968, Int. J. Electronics, 24:439.
Ingold, J. H., 1972, Phys. Fluids, 15:75.
Ingold, J. H., in "Gaseous Electronics", M. N. Hirsh and H. J. Oskam, eds., Academic, New York, Chap. 2.
Marec, J., Bloyet, E., Chaker, M., Leprince, P., and Nghiem, P., 1983, in "Electrical Breakdown and Discharges in Gases", E. E. Kunhardt and L. H. Luessen, eds., NATO ASI Series Vol. 89b, Plenum, New York.
Moisan, M., Pantel, R., Ricard, A., Glaude, V. M. M., Leprince, P., and Allis, W. P., 1980 Revue Phys. Appl., 15:1383.
Moisan, M., Shivarova, A., and Trivelpiece, A. W., 1982a, Plasma Physics, 24:1331.
Moisan, M., Ferreira, C. M., Hajlaoui, Y., Henry, D., Hubert, J., Pantel, R., Ricard, A., and Zakrzewski, Z., 1982b, Revue Phys. Appl., 17:707.

Moisan, M., and Zakrzewski, Z., 1985, in this Proceedings.
Muller, C. H. III, and Phelps, A. V., 1980, J. Appl. Phys. 51:6141.
Nghiem, P., Chaker, M., Bloyet, E., Leprince, P., and Marec. J., 1982, J. Appl. Phys., 53:2920.
Persson, K. B., 1962 Phys. Fluids, 5:1625.
Rose, D. J., and Brown, S. C., 1955, Phys. Rev., 98:310.
Schottky, W., 1924, Z, Phys., 25:635.
Self, S. A., 1965, J. Appl. Phys., 36:456.
Self, S. A., and Ewald, H. N., 1966, Phys. Fluids, 9:2486.
Trivelpiece, A. W., and Gould, R. W., 1959, J. Appl. Phys., 30:1784.
Trivelpiece, A. W., 1967 "Slow-Wave Propagation in Plasma Waveguides", San Francisco University Press, San Francisco.
Zakrzewski, Z., Moisan, M., Glaude, V. M. M., Beaudry, C., and Leprince, P., 1977, Plasma Physics, 19:77.
Zakrzewski, Z., 1983, J. Phys. D: Appl. Phys., 16:171.

GAS DISCHARGE LASER DIAGNOSTICS UPDATE

A. Garscadden

Air Force Wright Aeronautical Laboratories
Wright-Patterson AFB, OH, USA

INTRODUCTION

The investigation of gas discharges dates back to the time of Faraday and other notables of the nineteenth century. Many of the studies were very ambitious in that quite complex gases and complicated geometries were used. Naturally, these early results were qualitative. Later, as vacuum techniques improved, the beautiful results of the Cavendish Laboratory led to many of the most fundamental advances in atomic physics. Electrical discharge studies gradually became refined and the importance of gas purity to achieve reproducible results was recognized. Hence, many discharge experiments were performed using the rare gases, the simpler diatomic gases and some metal vapors. Thus it seems that Irving Langmuir's favorite discharge was that in mercury vapor because of its compatibility with mercury diffusion pumps. After 1940 the interest in discharges in gas mixtures and in complex molecular gases declined because it was difficult to make quantitative analyses or even to obtain reproducible results. At that time film depositions and sputtering were generally considered nuisances.

However, during the past twenty-five years there were two parallel synergistic activities. The first was the discovery and rediscovery of applications for discharges in gases that produce chemical reactions, thin-film depositions and etching of substrate materials. The second activity was the development of reliable laser sources, especially tunable laser sources that permitted new measurement techniques for complex discharges. This lecture will provide a concise update on some of these techniques, especially those applicable to low-energy, low-density plasma discharges.

It is now possible, using laser diagnostics, to measure the electron, positive ion and negative ion densities, their characteristic temperatures, gas temperature and the vibrational and rotational temperatures or level population distributions. It is also possible to measure collective properties of the plasma. More recently, measurements of the local electric field have been demonstrated for relatively "cooperative" experimental configurations. These laser techniques are complemented by microwave and probe methods at lower electron densities ($< 10^{12}$ cm^{-3}) and by emission spectroscopy at high electron densities ($< 10^{15}$ cm^{-3}). In general, however, it is still difficult to achieve short time resolution and to measure the actual energy distributions functions (e.g., for electrons) for plasmas of interest.

In this update we will provide laser diagnostics examples especially for low pressure discharges used in thin film deposition and etching.

ELECTRON DENSITY MEASUREMENT WITH INTERFEROMETERS

The electron density in a plasma can be measured using a laser source and a Michelson, Jamin or Mach-Zender interferometer in a configuration similar to that used in microwave diagnostics (Johnson, 1967; Gerardo and Verdeyen, 1964; Musal, 1969). The plasma refractivity is described by the solution of Poisson's and Maxwell's equations. When electron-neutral collisions can be neglected, the resultant index of refraction is

$$n = \left(1 - \frac{\omega_p^2}{\omega^2}\right)^{1/2} \quad (1)$$

where ω_p is the electron plasma frequency,

$$\frac{\omega_p}{2\pi} = f_p = 8980 \, n_e^{1/2} \quad (2)$$

and ω is the probing frequency. For visible or infrared lasers and electron densities of interest $\omega \gg \omega_p$ and thus

$$n = 1 - \frac{\omega_p^2}{2\omega^2} \quad (3)$$

The fringe shift for a typical Michelson interferometer (Fig. 1) is then, for a homogeneous plasma

$$\Delta\phi = -\frac{\pi f_p^2}{c^2} \lambda L \quad (4)$$

or for an inhomogeneous plasma

$$\Delta\phi = -2.82 \times 10^{-13} \int_{z_1}^{z_2} n_e(z) dz \quad (5)$$

where L is the length of plasma = $(z_2 - z_1)$, and λ is the wavelength of the laser. Regular interferometer techniques permit fringe shifts of about $10^{-2} \lambda$ to be detected so that for $\lambda = 5000$ Å a minimum detectable electron density is approximately 10^{16} cm^{-3}.

The phase shift is directly proportional to the wavelength of the laser, so that higher sensitivity is obtained by using far infrared lasers. The penalty paid is that spatial resolution, given approximately by the laser waist size, is reduced. It is also necessary to stay below the plasma (or cut-off) frequency to avoid attenuation of the signal. It is possible to enhance the phase shift by using a resonant cavity. However, for practical use, this effectively reduces the cut-off frequency. Figure 1 (after Musal, 1969) shows $\frac{\Delta\phi}{2\pi L}$, the cycles of phase shift per cm of path length for typical laser sources as a function of electron

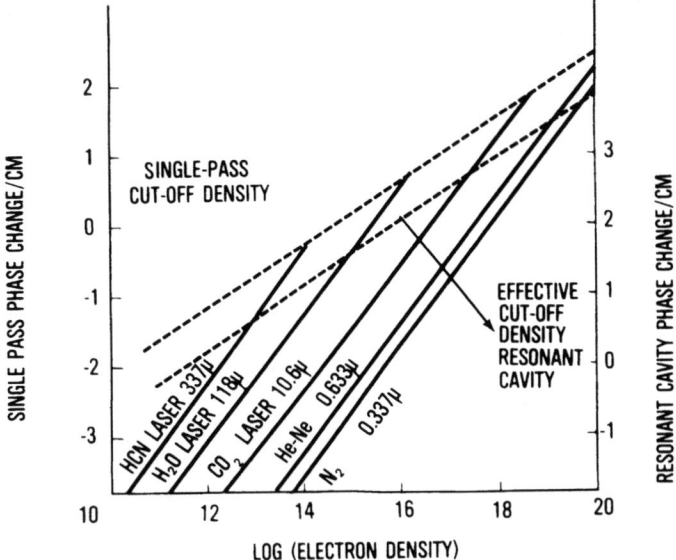

Fig. 1. Nomograph for interferometer phase shifts: LH ordinate single-pass; RH ordinate resonant cavity multi-pass (after Musal, 1969).

density for a single pass interferometer and for a resonant cavity interferometer (RCI) with an enhancement factor K = 100 (where $K = \frac{Q\lambda}{4\pi L}$; Q = cavity quality factor). Also shown are the densities for two interferometers. The RCI gives a larger phase shift than the single pass interferometer but is unsuitable at higher densities. A further constraint for extended plasmas is to avoid too much probe laser deflection if there is an electron density profile.

This straightforward method is used for higher density plasmas such as a pulsed electron beam diode. Recently Hinshelwood (1985) in a study of cold cathode electron beams used several pulsed nitrogen lasers triggered sequentially to permit framing operation of the interferometer method. This gave interferograms separated by only a few nanoseconds. However, these approaches are not so popular for lower electron densities or for longer duration plasmas. In these interferometers the output signal is proportional to $I_1 I_2 \cos\theta$ where $I_1 I_2$ are the amplitudes of the probe and reference and θ is the plasma shift. It is difficult to keep I_1 and I_2 exactly stable and one does not know the sign of θ. A normalized signal (I_1/I_2) can be generated, however. These difficulties led to the development of the phase modulation interferometer (PMI).

The schematic of a PMI is shown in Fig. 2. The laser is frequency-shifted from f_o to $f_o + f_m$ by either mechanical means for the far infrared or by an optoacoustic cell for 12 microns and shorter wavelengths. This probing signal is then phase modulated by the temporal changes in the electron density of the discharge of interest. The signal is downconverted to $f_m + \theta$ and enters a digital phase comparator with f_m as its reference, providing an output proportional to the phase shift $\theta(t)$.

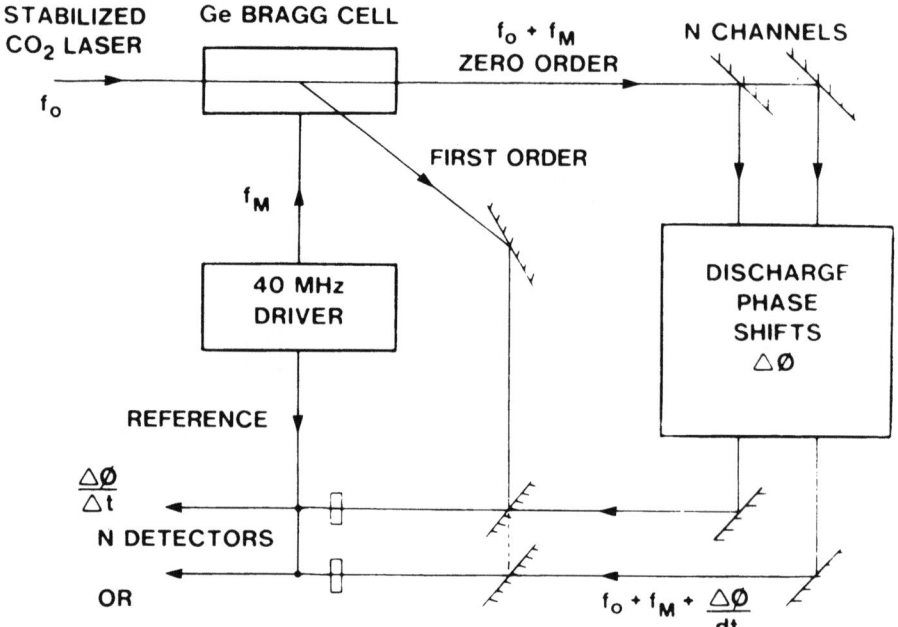

Fig. 2. Schematic of phase-modulation interferometer.

At the longer laser wavelengths desirable for higher sensitivity, two optically pumped molecular lasers (CH_3OH) tuned to slightly different frequencies under the gain curve have been used. The frequency shift is calibrated by a visible interferometer measurement of the cavity length (Luhmann and Peebles, 1984).

The phase shift is that due to all effects in the probe traversal so that care must be taken to avoid near-resonances with any absorbing lines. The previously developed techniques of making measurements at several frequencies are of course applicable. Strictly speaking, the refractive index is

$$n = 1 - \frac{n_e e^2}{2m\omega^2} + 2\pi\alpha_o N_o + 2\pi \sum_j \alpha_j N_j \qquad (6)$$

(electrons)   (ground state)   (excited stated and ions)

where α = polarizability per cm^3 at the laser frequency and N = number density.

Provided the laser frequency is not near an absorption resonance, only the electronic term varies rapidly with frequency. Hence, the electronic contribution to the refractive index is resolved by making the measurements at several different laser frequencies. In this case

$$\lambda_1 \Delta\phi_1 - \lambda_2 \Delta\phi_2 = -\frac{1}{2} \frac{f_p^2}{c^2} L (\lambda_1^2 - \lambda_2^2) \qquad (7)$$

where λ_1, λ_2 are the probe wavelengths and $\Delta\phi_1$, $\Delta\phi_2$ are the fringe shifts at the respective wavelengths.

The available far infrared lasers are not very convenient to use. Therefore, research requirements for easier diagnostics are for more intense, stable, long lived far infrared lasers. As a result of budget considerations the most elaborate multiplex laser interferometers are those used at the fusion centers. A recent comprehensive review of techniques used in the fusion plasma community has been given by Luhmann and Pebbles (1984).

MEASUREMENT OF NEUTRAL SPECIES: LASER INDUCED FLUORESCENCE (LIF)

The development of techniques to measure neutral species by laser methods has proceeded very rapidly because the topic is of great interest to chemists and to combustion engineers. Some of the approaches that have been applied are summarized in Table 1. These methods have benefited from the improvements in intense tunable UV sources that now permit linear and non-linear methods.

Laser induced fluorescence (LIF) is one of the more versatile methods. This discussion will attempt to transfer some of the combustion techniques, notably by Crosley (1983) and by Dailey (1977), to the low pressure discharge community. As indicated in Fig. 3, LIF can be used in two modes. The first is a scanning pump mode where the pump laser is tuned to excite different excitation lines and the fluorescence in one line or band is observed. The second mode is where a fixed laser pump wavelength is applied and the resulting fluorescence is scanned in wavelength. The fluorescence signal is dependent on the number of atoms excited, the transition probability of the fluorescence line, the collisional mixing of the upper state and the collisional quenching of the upper state. The advantage of pulsed LIF is that ideally one detects a signal on a null background. In practice, good laser beam sinks (Wood's horns) and line selectivity are needed. If a laser pulsewidth much less than the fluorescence lifetime and gated detection are used, then total absorptions of less than 10^{-6} produce signals that are easily analyzed. Some of the interesting discharge species that have been addressed by LIF are listed in Table 2.

Table 1. Plasma parameters accessible with lasers

Electron density	$n_e(\bar{r},t)$	$f_e(\bar{v},\bar{r},t)$	ALSO
Positive ion density	$n_p(\bar{r},t)$	$f_p(\bar{v},\bar{r},t)$	COLLECTIVE
Negative ion density	$n_n(\bar{r},t)$	$f_n(\bar{v},\bar{r},t)$	PROPERTIES
Electric field	$E(\bar{r},t)$		

Excited state densities

Total gas densities

Neutral gas energy distributions, e.g., N_v, N_{ROT}

Dissociation, e.g., N(atom) / N(molecule)

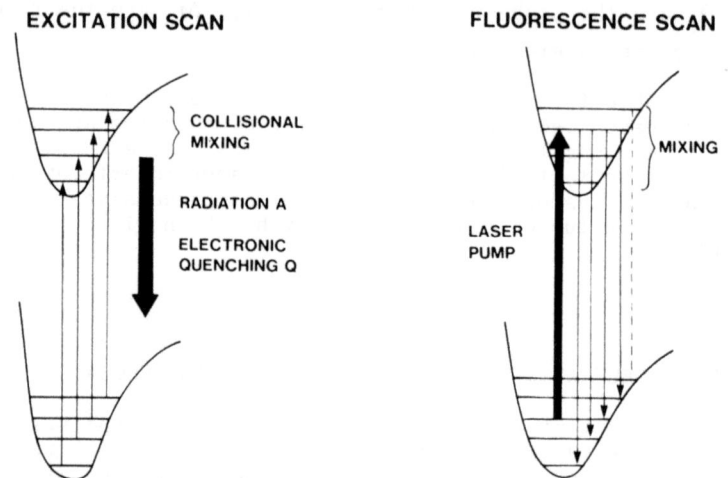

Fig. 3. Laser induced fluouresence experimental scans.

The rate equation approach to LIF has been applied by Crosley and Smith (1983) and by Bontilier et al., (1978). The simplified equation for the total integrated fluorescence of a two level system is

$$I_F = c\ BIN_o \left(\frac{A}{A + Q} \right) \qquad (8)$$

Here c is a collection-detection efficiency factor, A is the spontaneous emission coefficient for the observed transition, N_o is the ground state

Table 2. Examples of discharge species accessible by laser induced fluorescence

O, N, H	S_2, SH, SO, SO_2
Si, Fe, Ti, Li, Li_2	CS, CS_2
CH, CH, C_2, CO	HCN, NCO
NH, NH_2	C_2O, C_3
NO, NO_2, HNO	CH_2O, CH_3O
PO, BO, OH	N_2^+, CL_2^+

population, I is the laser intensity, Q is the quenching rate of the upper level, and BI is the excitation rate due to absorption of the laser (actually a convolution over line shapes).

A particularly nice example of the effect of collisions in S_2 has been provided by Crosley and Smith (1983). The fluorescence spectra of S_2 B^3 initially excited in the $v^1 = 4$, $J^1 = 4^1$ level is shown in Fig. 4. The S_2 is contained in a small heated cell. The liquid reservoir is about 100 C which provides a total sulfur pressure of 50 mtorr (sum of the partial pressures of the molecules S_2, S_4, S_6, S_8). The center of the cell is about 600 C giving predominantly S_2. The initial LIF scan for no buffer gas is shown on the lowest trace of Fig. 4. Successive scans are at the indicated pressures of argon buffer gas keeping other conditions constant. First of all, collisional rotational-level mixing occurs within the pumped vibrational level. Then vibration level relaxation is observed with the $v^1 = 4$ level eventually (at argon pressures above 50 torr) relaxing almost totally to the $v^1 = 0$, 1 levels before fluorescing. Gases other than inert gases often also quench the electronic excitation rapidly. These results illustrate that it is relatively easy to study collisional energy transfer and effective lifetimes using LIF if a convenient absorption to the upper state is available. It would be of interest to repeat these LIF measurements with discharge excitation superimposed, as it appears that electron impact dissociation could be examined for clustering type molecules such as S_2.

The OH radical is exceptionally well characterized due to its importance in combustion and its strong absorbing transitions in the near ultraviolet. For OH, the A coefficient is about 1.4×10^6 s^{-1} so that with a suitable 10 ns laser, OH can be detected at less than 10^{-9} concentration levels in an atmospheric pressure flame with 1 mm^3 spatial resolution. In fact the question of data retrieval and assimilation is made easier by the sensitivity of LIF. Dyer and Crosley (1982) and, independently, Kychakoff et al. (1982) have obtained two dimensional images of OH radicals in a flame. It appears that optical multichannel analyzers and vidicons coupled with pulsed LIF can be used to similar advantage in discharges. The radical concentrations will be of the same order, however the atom collisional quenching in a low pressure discharge is usually less of a problem. On the other hand, collisional mixing and quenching by electrons can also occur.

A very important experiment on imaging of atomic hydrogen in flames has been performed recently by Goldsmith and Anderson (1985). Two-step saturated fluorescence spectroscopy was used to excite the hydrogen atoms to the n = 3 state along the line of a laser intersecting the flame. The spontaneous Balmer- β emission at 656 nm was recorded on a linear array detector to produce a line image of the ground state atom concentration. Successive lines are assembled to give a two dimensional mapping of concentration at a sensitivity better than 10^{15} atoms cm^{-3}. Carbon monoxide and atomic oxygen have been observed in a similar manner. These two-dimensional methods can aid the interpretation of phenomena in flowing plasmas and around electrodes, grids, and probes in discharges. The removal of the symmetries demanded by older spectroscopic techniques should lead to new discharge configurations and applications.

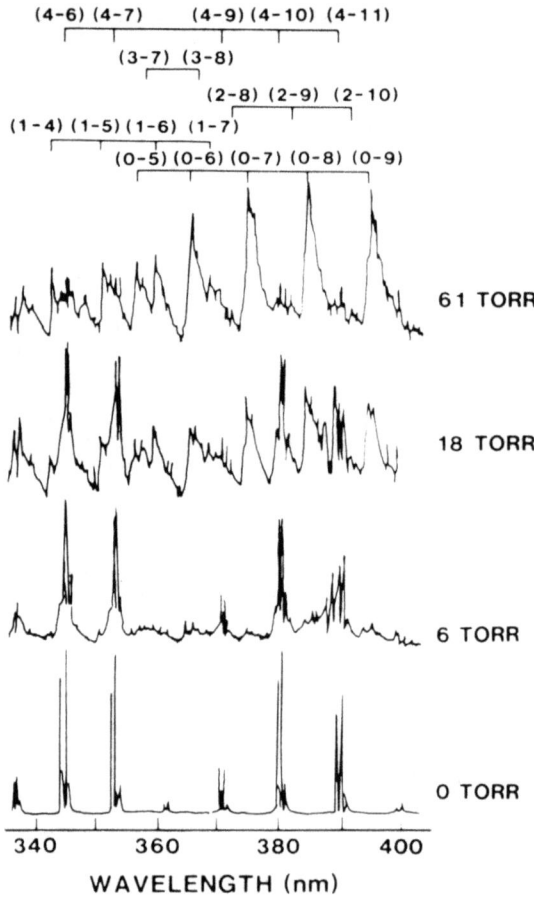

Fig. 4. Laser induced fluorescence scans in sulfur as function of argon buffer gas pressure.

Time resolved LIF measurements in etching plasmas have been applied by Donnelly et al., (1982). As examples, nitrogen (Gottscho et al., 1984) and chlorine molecular ions are accessible. These methods will be discussed by R. Gottscho.

NONLINEAR SPECTROSCOPY: COHERENT ANTI-STOKES RAMAN SCATTERING (CARS)

Four-wave mixing with resonant excitations is an ubiquitous and powerful diagnostic technique for many phases of materials including plasmas. A representation of one four-wave process, CARS, is shown in Fig. 5. Two laser beams with frequencies ω_p (pump) and ω_S (idler) and powers P_p and P_S interact through the third-order nonlinear susceptibility $\chi^{(3)}$ to produce a coherent emission at the anti-Stokes frequency ω_{AS} when

$$\omega_{AS} = 2\omega_p - \omega_S \tag{9}$$

The advantages of CARS spectroscopy are its ability to measure molecules that do not have easily accessible strong absorption lines, its spatial resolution (mm^3) and its inherent time-resolution (10^{-12} to 10^{-8} s).

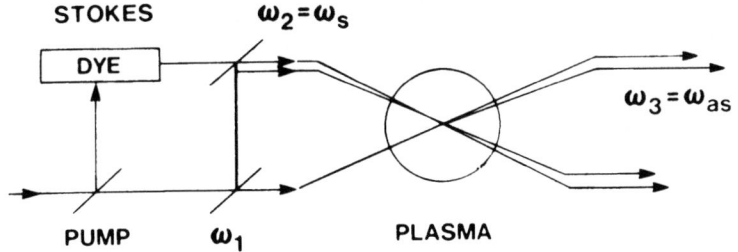

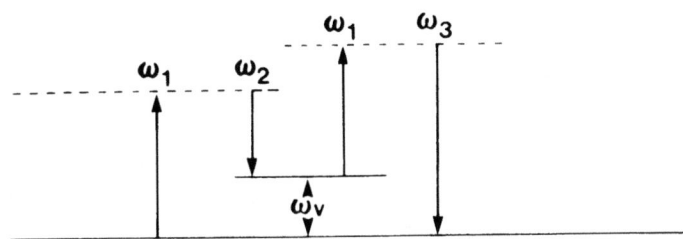

Fig. 5. Diagram of CARS interaction.

Most of the CARS development has been for combustion studies where the time resolution is less important than the data rate. Although CARS has a conversion efficiency up to 1% it has been generally applied to gases at pressures above 1 torr. However polarization sensitive CARS that reduces the nonresonant background now permits measurements down to 10^{-3} torr. It is also possible to measure higher vibrational level densities. The full expression for $\chi^{(3)}$ shows that it can have single, double or even triple resonances depending on the medium resonances. Normally, one works with the single vibrational resonance, $W_P - W_S = W_V$ which gives an enhancement of the power into the anti-Stokes beam P_{AS}. There is also the boundary condition on the wave vectors

$$2\,\underline{k}_P = \underline{k}_{AS} + \underline{k}_S \tag{10}$$

so that the anti-Stokes signal is in a predetermined direction and not isotropic as in Raman scattering. This feature alone can assist the experimental detection of the CARS signal several orders of magnitude over Raman scattering.

It has been shown that for a single isolated line with Raman frequency and intersecting Gaussian beams (Tolles et al., 1977)

$$P_{AS} \propto = \left(\frac{16\,\pi^3}{\lambda_P \lambda_{AS}}\right)^2 \left|\chi^{(3)}\right|^2 P_P^2 P_S \tag{11}$$

where λ_P, λ_{AS} are the wavelengths of the pump and anti-Stokes beams, and

$$\chi^{(3)} = \frac{2N_j e^4}{\hbar\,\omega_S^4}\left(\frac{\partial\sigma}{\partial\Omega}\right)_j \left(\frac{g_j\,\omega_j\,\Delta_j}{\omega_j^2 - (\omega_P - \omega_S)^2 - i\,\Gamma_j\,(\omega_P - \omega_S)}\right) \tag{12}$$

where N = molecular number density, j = average population difference per molecule between the lower and upper energy levels of the Raman transition, g_j = statistical weighting factor, and Δ_j = linewidth of the Raman transition.

Hence, the power in the anti-Stokes beam is proportional to the square of the differential Raman cross section. Generally $\chi^{(3)}$ can be expressed as the sum of three terms

$$\left|\chi^{(3)}\right| = \chi' + i\chi'' + \chi^{NR} \tag{13}$$

where χ^{NR} (real) is the slowly varying (with frequency) non-resonant background due to contributions from other remote lines and from electronic transitions. Careful high resolution tests (resolution 0.03 cm^{-1}) of the theory have been made by Nitsch and Kiefer (1971). As the CARS signal depends on the product of the density squared and the pump power squared, it is possible to achieve 1% conversion and thus to obtain very strong signals (easily visible as contrasted to Raman scattering). Also the frequency ω_S may be made broadband (~50Å), in which case it is possible, using multichannel detection, to obtain simultaneous vibrational-rotational resonances or, in effect, a 10 ns snapshot of the rotational temperature.

It is noted that because the process is non-linear, the interferences of neighboring vibrational lines give rise to asymmetries and apparent line shifts. The shape of the spectral lines of the CARS signals are different from those of the spontaneous Raman lines. Thus in CARS spectra it is absolutely necessary to couple theory and experiment in order to achieve accurate data. A valuable library of CARS spectra has been calculated for many species and conditions by Verdieck et al., (1982).

The effect of the non-resonant background shows up in limiting the detection of impurity concentrations. Thus the limiting detectable concentration of hydrogen in nitrogen at 1 amagat is ~ 60 ppm determined by the ratio $\chi^{(3)}_{H_2}$ (resonant) /$\chi^{(3)}_{N_2}$ (non-resonant). Some of this problem is overcome by applying the difference in the symmetry properties of the resonant and non-resonant components of $\chi^{(3)}$. Whenever the pump and Stokes beams are linearly polarized and their relative polarization angle is chosen appropriately in the range 40 - 70°, the polarizations of the anti-Stokes and the non-resonant signals are orthogonal. This technique often permits a significant improvement in the signal/noise, typically by several orders of magnitude. [Note that the signal is actually attenuated; the difference is that the signal/noise is much improved]. The non-resonant background, however, can be used to advantage to provide an absolute calibration for the signal levels. In fact, Eckbreath and Hall (1981) found that for some circumstances in flame studies polarization CARS is not an advantage. Relating to deposition studies, the CARS spectrum of silane has been studied by Hata et al. (1983) and by Bulatov (1980).

Several groups have examined the discharge vibrational excitation of nitrogen using CARS. As the vibrational upper state v_i is populated significantly the anti-Stokes power is proportional to the square of the density difference ΔN_{ij}

$$P_{AS} \propto (\Delta N_{ij})^2 P_P^2 P_S \tag{14}$$

The finite anharmonicity of the diatomic nitrogen provides a spectral separation of the CARS signals. The v = 0, 1, 2 are separated by 30 cm^{-1} and 28 cm^{-1} intervals. Figure 6 shows data obtained by Smirnov and Gabelinskii (1978). The "vibrational temperature" is measured to be around 300 K and appears to be slowly increasing with vibrational level. However, the rotational temperature is 395 K. This method can demonstrate v-v pumping in homopolar diatomics excited by discharge excitation or by chemical pumping. Dreier (1982) checked the validity of the CARS method using CO titration and infrared fluorescence, and found excellent agreement between the different methods. Shaub et al. (1977) obtained unusual deviations from the Boltzmann distribution in discharged nitrogen. However, as $(\Delta N_{ij})^2$ occurs in Eq. (14), it was not possible to unambiguously measure population inversions.

The most sensitive application of CARS without crossing the beam polarizations, has been by Pealat et al. (1981) in their investigation of hydrogen vibrational populations in a magnetic multipole low pressure plasma (0.1 torr). A sensitivity of approximately 10^{12} cm^{-3} was achieved by careful attention to detail. It must be emphasized that this was an exceptionally well-performed experiment and that the CARS method is more appropriate to densities of 10^{15} or higher.

Higher sensitivity can be obtained if more than one resonance is obtained simultaneously for a single isolated Raman resonance where the lower vibrational state population is large:

$$\chi_{ij}^{13} = \frac{N\, p_i^o}{h^3(\omega_{ji} - \omega_1 + \omega_2 - i\Gamma_{ji})}$$

$$\times \frac{\mu_{in'}\, j_{n'}}{(\omega_{ni} - \omega_3 - i\,\Gamma_{n'})}$$

$$\times \frac{\mu_{in}\, \mu_{jn}}{(\omega_{ni} - \omega_1 - i\Gamma_{ni})} \quad (15)$$

where N = the number density, Γ is the pressure-broadened linewidth, μ is the electric dipole matrix element, p is the Boltzmann population, and i, j are the initial and final vibrational quantum numbers of a Raman transition and n, n' are excited electronic states.

In CARS, normally one is addressing the resonance of $(\omega_{ji} - \omega_1 + \omega_2 - i\Gamma_{ji})$; however, electronic resonant enhancement will also occur if ω_1, ω_2 or ω_3 coincide with an allowed electronic transition. This usually means using two dye lasers so that ω_1, ω_2 are separately variable. The ω_2 resonance has been omitted in Eq. (15) as it is proportional to the upper vibrational level population, which is assumed small. If the pump frequency is fixed at $\omega_1 = \omega_{in}$, then the Stokes frequency sweeps out the Raman frequencies and double resonance occurs when $(\omega_1 - \omega_2) = \omega_{ji} = \omega_v$ and when $\omega_3 = \omega_n 1_i$. (Triple resonances could, in principle, occur if the latter are simultaneous.) Studies of double resonances have been made on I_2 (Druet et al., 1978), N_2O (Guthals et al., 1979), and C_2 (Attal et al., 1984). While most of this work originated in combustion diagnostics, the results and conclusions are applicable to

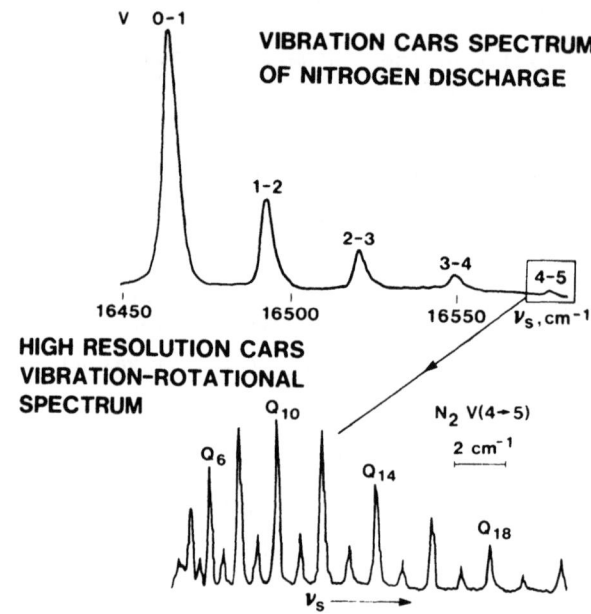

Fig. 6. CARS results on vibrationally excited nitrogen (Smirnov and Gabelinskii, 1978).

discharges, as the double resonance increases the anti-Stokes power yet further.

LASER DIAGNOSTICS OF ELECTRIC FIELDS IN PLASMAS

The local measurement of the electric field in non-equilibrium, low pressure discharges is one of the most interesting generic challenges to the experimenter. While there have been advances in various active probe and spectroscopic methods, we will concentrate on techniques applying lasers.

The laser or spectroscopic measurement of the electric field are based on some aspect of the Stark effect. The electric field induces a dipole moment in the atomic or molecular species. The interaction between the dipole and the field causes a precession of J, the angular momentum vector, about the applied field such that the m_j components of angular momentum are a constant of the motion. Results of such calculations show that hydrogen is a special case. The shifts of the Stark components are linear with the applied field and quite large. The electric field is added to the Coulomb potential and causes a lowering of the ionization potential (in parabolic coordinates) to $E_i = Ze(eZF)^{1/2}$. In addition to hydrogen, the molecular species in electronic states other than a Σ state and possessing a permanent dipole moment also show a linear Stark shift. Other species at low-applied fields show a quadratic Stark effect. The Stark splitting of the energy level depends on the square of the applied electric field. The energy level diagrams for hydrogen have been calculated by many authors including Littman et al. (1978) and Harmin (1985). The shifts for helium have been calculated by Foster (1927) who also performed quality experiments on the shift in high applied fields.

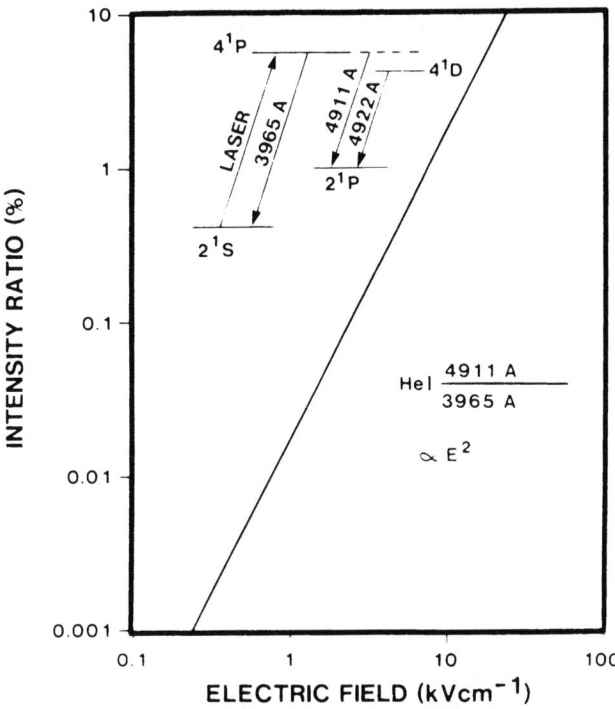

Fig. 7. Transitions in helium used by Oda et al., for forbidden line intensity method of measuring electric field.

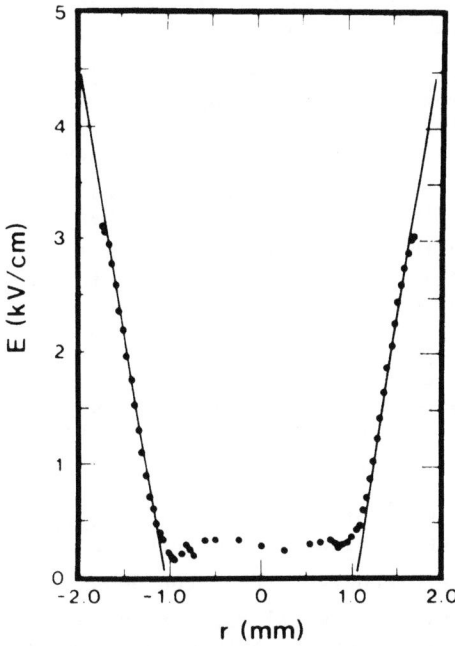

Fig. 8. Experimental data on electric fields in hollow cathode (Oda et al., 1984) (O = 4mm, length = 18mm; helium, 6 torr; discharge current, 15mA; voltage, 220V).

The measurement of the line shift or Stark splitting in a discharge will be where there is also Doppler broadening, Holtsmark field induced effects and pressure broadening. The first can be overcome by using counterpropagating beams and hence Doppler-free spectroscopy. The other effects pose fundamental limits to the sensitivity of the diagnostic. New results on the measurement of electric fields by Stark effects are included elsewhere in the discussion of Rydberg states.

The basic technique of laser induced fluorescence has been combined very successfully with field induced forbidden transitions to measure the electric field in rf discharges by Moore et al. (1984). The method has good time resolution and fair sensitivity. The technique has been used with BCl transitions and detailed spatial and temporal mapping of the plasma sheath fields have been obtained. These are discussed elsewhere by Gottscho.

It is noted that the LIF field measurement technique can be applied also to helium which is less disturbing to many discharges than BCl_3. T. Oda et al. (1984) have used the ratio of the He I 491.1-nm and 396.5-nm transitions to measure the electric field in a hollow cathode discharge. The calculated ratio versus electric field is given in Fig. 7. The estimated sensitivity is 200 volts cm^{-1}. Results for a hollow cathode discharge in helium are shown in Fig. 8. Generally, this method will tend to overestimate the field due to the non-linearity in the response (Fig. 7) and the possible errors from scattered radiation or from collisional mixing by electrons. Higher principal quantum number transitions can be used to improve sensitivity but these levels will be more sensitive to collisional quenching and hence are more appropriate to low pressures and lower electron densities.

The experiment requires care when probing near the electrodes to avoid direct scattering of the probe beam. Use of the technique at higher pressures requires knowledge of the quenching rates of both the allowed and forbidden transitions.

A higher accuracy is obtained if the individual fine-structure components can be measured using Doppler-free spectroscopy. In one such technique two counterpropagating beams modulated at different high frequencies ω_1 and ω_2, are passed through the discharge. The fluorescence at the sum or difference frequency is detected as the beam carrier frequency is tuned to line center. The method has been used by Freeman et al. (1980) to study the $2^3P - 3^3D$ isotopic shift of helium in a glow discharge. Extrapolation of the sensitivity indicates that fields as low as 10V/cm can be measured. Optogalvanic spectroscopy basically uses the same techniques to address the same transitions. However, it gains several orders of magnitude in sensitivity over LIF methods. This technique has been applied very successfully by Doughty, Salih and Lawler (1984) to measure sheath fields and to achieve point resolution in a discharge. Recently, by addressing high Rydberg states, we have been able to measure electric fields to 2V cm^{-1}.

Doppler-free, two-photon spectroscopy (Fig. 9) has been used by Cornelissen and Burgmans (1982) to measure the electron density in a low pressure sodium-neon discharge. Electron densities from 10^{12} to 10^{13} cm^{-3} were measured with a spatial resolution better than 0.5 mm. The experiment measured the Stark broadening and the Stark shift of the sodium 3S - 4D transition. The power of the technique is illustrated by the fact that the Stark broadening and shift were of the order of 100 MHz whereas the Doppler width was a few GHz. The laser was stabilized to

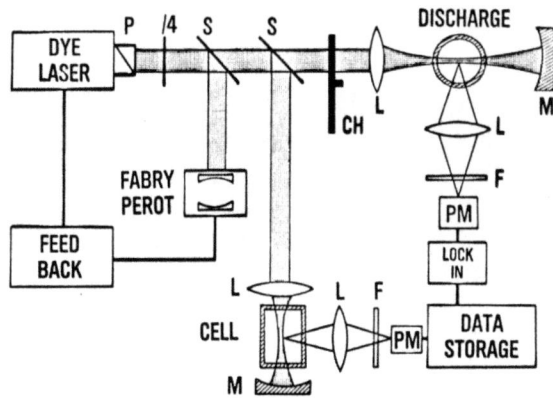

Fig. 9. Experimental arrangement of Cornelissen and Burgmanns (1982) for Doppler-free measurement of electron density.

less than 8 MHz by an external Fabry-Perot and feedback loop, and the experiment also used a heated reference cell. The advantage of the method over microwave resonant cavity interferometers is its high spatial resolution (0.5 mm).

Another Doppler-free technique that permits high resolution is polarization spectroscopy (Wieman and Hansch, 1976). The method works as follows: a circularly polarized pump beam is passed through the discharge. A probe beam, linearly polarized is counterpropagated through the same region. In the field-free situation, the linearly polarized beam, considered as two circularly polarized beams, +ve and -ve, retain the same phase and amplitude and no signal is detected through the orthogonally polarized detector. In the presence of the pump beam the medium is depleted of the same absorbers that interact with the +ve component, and a signal is generated. In the presence of a pump beam and an electric field the signal generated is dependent also on the electric field "broadening" of the transition and hence fields can be measured. The sensitivity using the H_β transition in a hydrogen glow discharge was estimated at about 5 volts per cm.

As an example of one of these new methods, Fig. 10 shows the electric field in the cathode sheath of a glow discharge. Goss (1985) obtained this profile using the optogalvanic effect on appropriate high principal quantum numbers in helium. The sensitivity increases at higher n but there is also more overlap so there are optimum levels to be examined depending on the field to be measured. Figure 11 shows the calculated sheath field profile treating the electrode as a biased probe in a homogeneous plasma. The comparison is considered reasonable. Note that we expect the linear approximation to be valid only close to the electrode. The plasma-sheath transition region will be quadratic at low discharge currents when gas heating is assumed to be a small perturbation in the sheath.

ABSORPTION MEASUREMENTS

Conventional absorption measurements are well known so we have selected some improved or new techniques that are receiving wide acceptance.

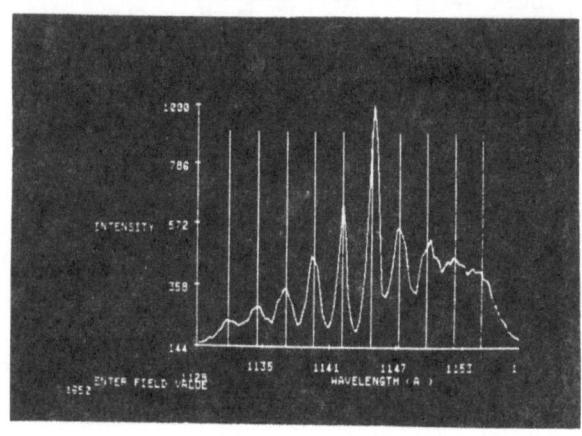

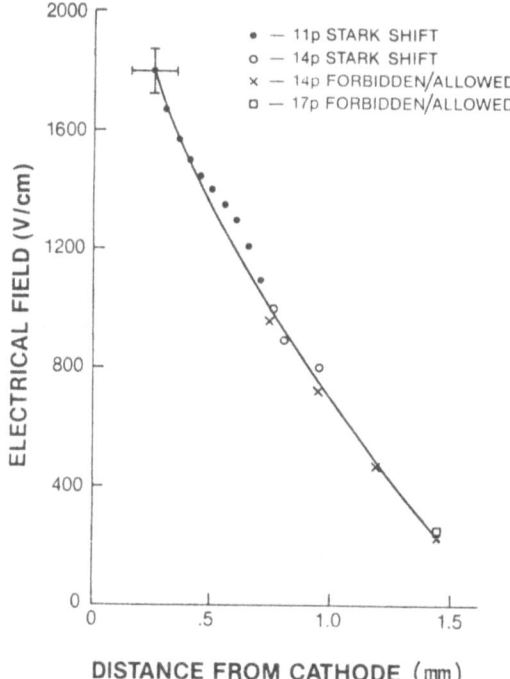

Fig. 10. Experimental results of Goss (1985) on electric field in hollow cathode discharge.

Tunable Laser Diode Spectrometry

The discharges that are used for plasma enhanced deposition and etching are run in gas mixtures where one or more components dissociate. The emission and absorption spectra are often very complicated. Fortunately the discharges are at low pressures, so high resolution spectroscopy can be used to identify unambiguously lines of a particular species. Absorption spectroscopy usually probes the ground electronic state of the molecules and in principle, for the same oscillator strengths, is more sensitive than emission spectroscopy.

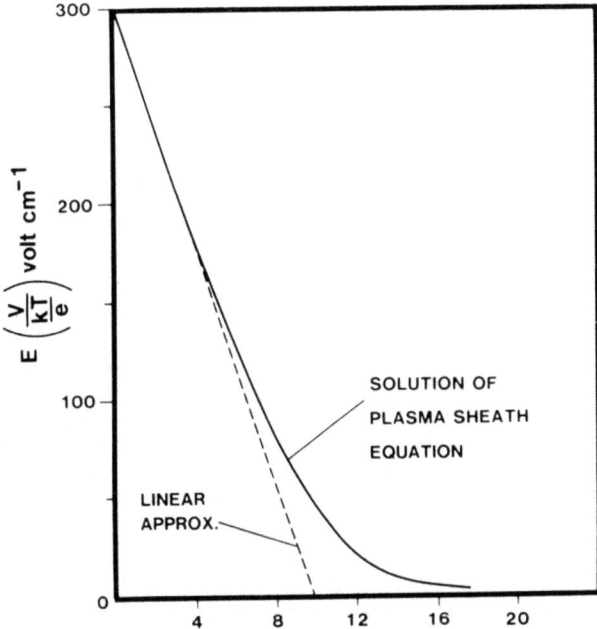

Fig. 11. Calculated cathode sheath field (y-axis normalized with respect to electron temperature; x-axis normalized with respect to Debye length.)

Broadband sources and Fourier Transform Spectroscopy (FTS) are very useful for surveys and calibration of even higher resolution methods. The development of lead salt ternary compound tunable laser diodes (TLD) permits high sensitivity and the detection of a wide variety of etching and deposition species. Reviews of the various applications have been given by Schlossberg and Kelley (1981) and by Eng and Ku (1982). A comprehensive analysis directed to the detection of gas phase species in plasma chemistry reactors has been published by Wormhoudt et al. (1983).

Figure 12 shows the experimental layout used by DeJoseph et al. (1983) in our laboratory. The laser diode system is from a commercial

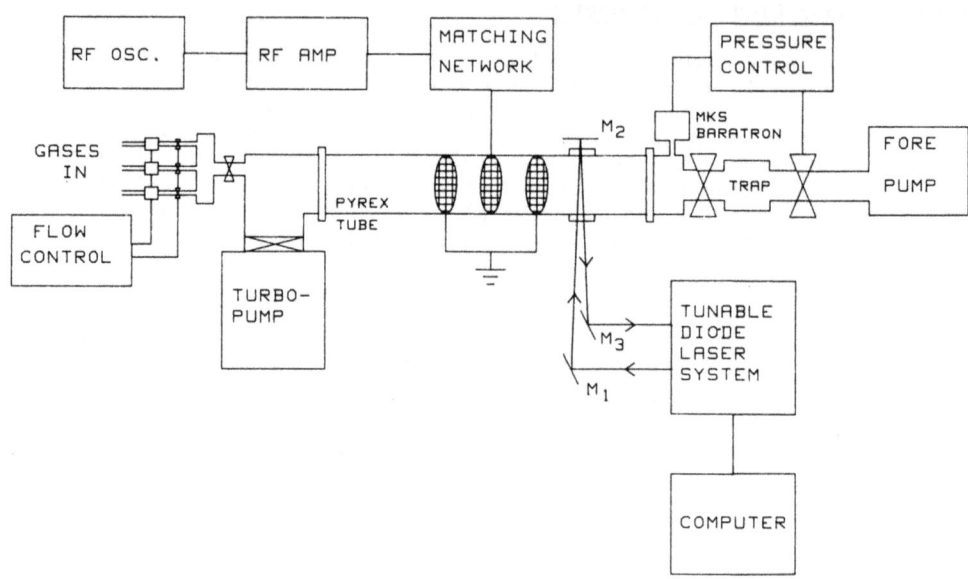

Fig. 12. Schematic of tunable laser diode experiments (DeJoseph et al., 1983).

source. It consists of diode laser mounted on a carousel in a temperature-stabilized and to some extent vibration-stabilized, cryogenic closed cycle compressor-refrigerator which provides cooling in the range 10 - 60K. The laser wavelength is controlled by the temperature (coarse control) and the injected current (fine control). A 1/2-meter monochromator is used to separate the longitudinal modes of the laser (spaced by about 1 cm^{-1}) and to provide wavelength calibration. An etalon gives a fine calibration. A single longitudinal-mode laser beam is sent through the discharge or through an absorption cell downstream of the discharge. Changes in absorption are detected by cooled detectors (copper doped germanium if the wavelength is greater than 14 microns, mercury-cadmium telluride if the wavelength is between 5 and 14 microns and indium antimonide if the wavelength is less than 5 microns). The laser tunability is usually quoted as being over a range of several tens of wave numbers, however the prospective user should be cautioned that because of mode hopping, the continuous tuning range is 1 - 3 cm^{-1}, while the Doppler width of the silane lines, for example, is about 2×10^{-3} cm^{-1}. The high resolution therefore permits absorption measurements without having to make line profile corrections. An example of a high resolution scan and the etalon reference is given in Fig. 13.

Table 3 lists some of the methods used to measure some of the silane discharge species and the estimated or demonstrated sensitivities. A discussion of many radical species that are found or anticipated in plasma reactors is given by Wormhoudt et al. (1983) in terms of TLD absorption spectroscopy and LIF detection. Generally, the minimum detectable densities are of the order of 10^{12} cm^{-3} by absorption spectroscopy and where LIF is possible, a higher sensitivity of $10^8 - 10^9$ cm^{-3} is achievable.

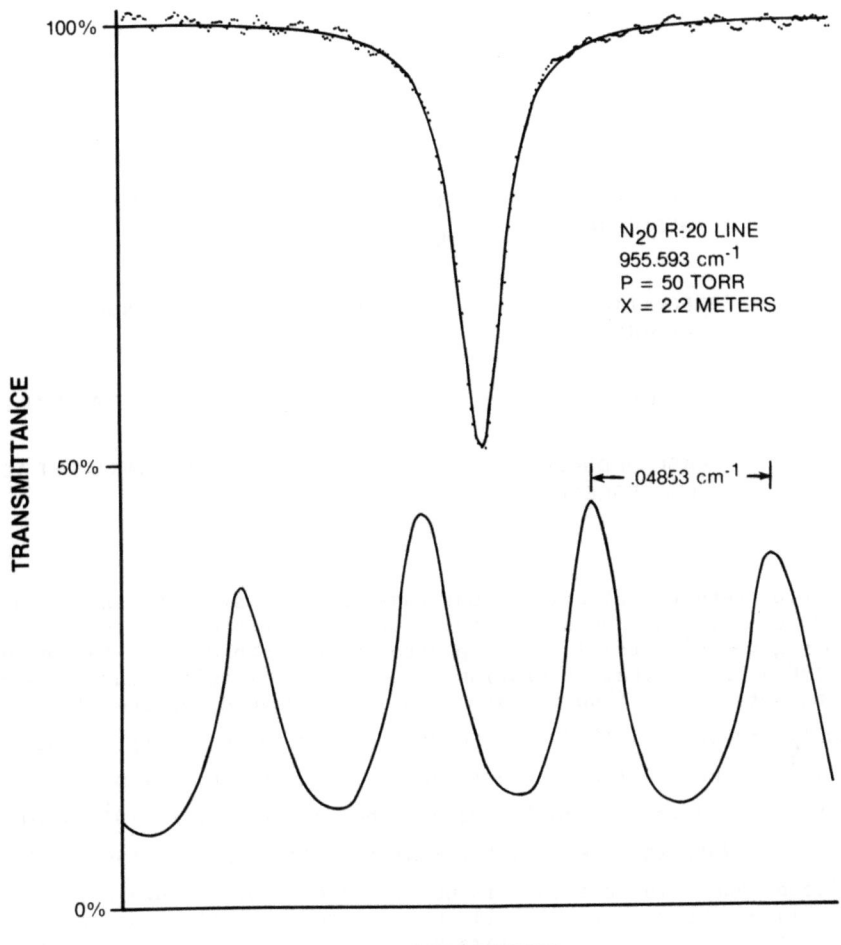

Fig. 13. Tunable laser diode scan of N_2O.

Table 3. Measurement techniques and sensitivities for species in silane discharges.

SPECIES	METHOD	COMMENT/SENSITIVITY	REFERENCE
Si_2H_6	TUNABLE LASER DIODE	$NL = 4 \times 10^{14} cm^{-2}$	DEJOSEPH (1985)
SiH_4	PULSED UV RAMAN	$NL = 3 \times 10^{14} cm^{-2}$	BREILAND & KUSHNER (1983)
SiH_4	TUNABLE LASER DIODE	$NL = 6 \times 10^{12} cm^{-2}$	DEJOSEPH ET AL (1982)
SiH_4	CARS	$\nu_1 = 2190\ cm^{-1}$ ESTD $N = 3 \times 10^{14} cm^{-3}$	N. HATA ET AL (1983)
SiH_3	LASER MAGNETIC RESONANCE	$P(10)\ CO_2$ LINE $NL = 4 \times 10^{12} cm^{-2}$	BRAUN ET AL (1981)
SiH_2	FREQUENCY-MOD ABSORPTION	$\frac{\Delta I}{I}\ 10^{-6}$	JASINSKI ET AL (1984)
SiH_2	CARS		N. HATA ET AL (1983)
SiH	LASER INDUCED FLUORESCENCE		DREVILLON ET AL (1981)

Schlossberg (1976) pointed out that the fine structure splitting of the ground state halogen atoms would permit measurement of their concentrations given high resolution, spectroscopic techniques. Stanton and Kolb (1980) showed that absorption at 404 cm^{-1} was sufficient to permit their measurement of fluorine atoms in a microwave discharge afterglow at a sensitivity of $NL = 10^{16}\ cm^{-2}$ (corresponding to an absorption of 1.5 x $10^{-3}\ cm^{-1}$). They further suggest that refinements such as multipass cell and derivative detection could improve the sensitivity to $10^{12} - 10^{13}$ atoms cm^{-3}. Note that perhaps the most important radical species SiH_3, in silicon deposition reactors is not yet accessible. However, Braun et al. (1981) and Krashoperov (1981), using laser magnetic resonance (LMR) applied to a rarefied silane flame and to silane reactions with chlorine and fluorine atoms, have measured the SiH_3 concentrations. Using SiH_3 996 cm^{-1} and 925 cm^{-1} transitions, LMR spectra were obtained on many 10.4 micron P and R transitions. The magnetic field had a maximum value of 0.55 Tesla and in order to improve detection, resonance modulation at 150 kHz with amplitude 0.25 mTesla. The presence of absorption by the SiH_3 radical on some lines of the CO_2 laser in zero magnetic field indicates the possibility of active discharge measurements.

A major disadvantage of line of sight absorption techniques is that they measure the line integral value and do not have spatial resolution for discharge configurations of interest. Knapp and Hanson (1983) have reported the first successful method to circumvent this limitation by Stark modulation of tunable laser diode absorption. A high power off-resonant laser can produce a Stark shift of a line into absorption in the intersection volume of the TLD and the high power laser. The experiment measured spatially resolved concentrations of carbon monoxide in a pre-mixed methane/air flame. It appears that the technique can be adapted for plasma measurements when the high power laser will not perturb the study. The method has an inherent fast time resolution of the order of the pulsewidth of the pulsed laser (8 ns or less).

Frequency-Modulated Absorption Spectroscopy

A new and sensitive method of absorption spectroscopy was proposed by Bjorklund (1980) and rapidly implemented (Jasinski et al., 1984). A single axial mode laser is driven at low modulation index at a frequency that is large compared to the width of the absorbing transition. The modulated beam has a pure frequency-modulated spectrum described by

$$E(t) = \frac{E_o}{2} \sum_{n=-\infty}^{\infty} J_n(M) \exp\left\{i(\omega_o + n\omega_m)t\right\} \quad (16)$$

where M is the modulation index and J_n refers to the Bessel functions of order n. When $M \ll 1, J_o(m) \simeq 1$, $J_{\pm 1}(M) = \pm M/2$, and all other terms vanish. This gives a strong carrier frequency at ω_o and two very weak sidebands at $\omega_o \pm \omega_m$. The beam is passed through the discharge or afterglow of interest. The dye laser frequency is adjusted so that one of the sidebands suffers absorption but not the other one. The sideband at the absorption feature is tuned across the band by varying either the carrier frequency or the radio-frequency. The losses and the phase shift of the other sideband are assumed to be constant. The signals are recombined at a photodetector. A beat signal at the modulation frequency will appear if the phase shifts and the attenuation of the sidebands are not equal. The in-phase component of the beat signal is proportional to the loss suffered by the probing sideband. The quadrature component is proportional to dispersion caused by the discharge species of interest. Fig. 14 shows a circuit arrangement for this method that was used to measure the absorption due to the SiH_2 radical in silane and disilane. The beam from a stabilized ring dye laser is rf modulated by an electro-optic modulator at 800 MHz. The differential absorption of the sidebands produces the rf beat frequency on the detector. Further sensitivity is achieved by chopping the beam at 2 kHz and modulating the discharge at 1.1 kHz. After demodulation, the final signal is detected by a lock-in amplifier at 3.1 kHz. Using normalizing techniques and the same photodiode to measure both the frequency modulated signal and the laser power, a detection limit of 10^{-6} was obtained (Jasinski et al., 1984) for in-situ absorption by SiH_2. Calculation of the absolute cross-sections was not possible as the absorption cross sections for the transitions are not known.

Survey Techniques: TRISP

The mid-infrared range is the region of choice for the detection of many molecules and radicals. However, if the species are short lived, as the interesting reactive ones often are, the spectroscopy is difficult or, at best, time-consuming. A new technique: time-resolved infrared

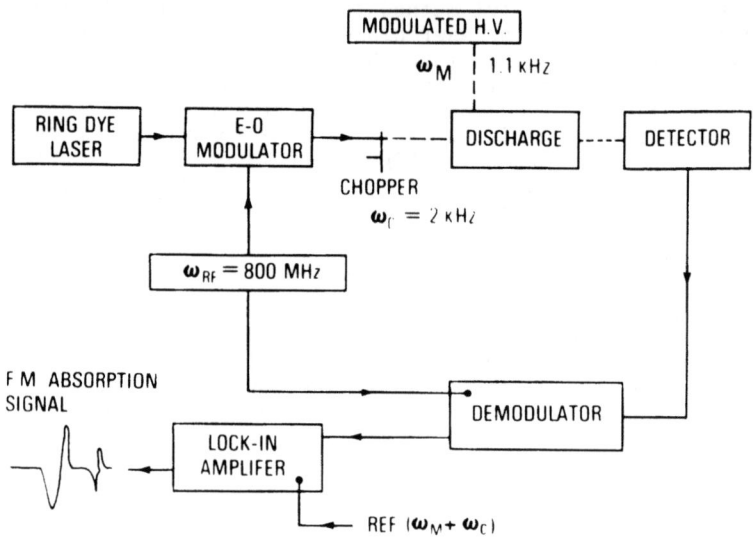

Fig. 14. Experimental arrangement for frequency modulation spectroscopy.

absorption spectroscopy (TRISP) has been introduced by Bethune et al. (1983). The method is illustrated in Fig. 15. A broadband, pulsed infrared continuum is generated by stimulated electronic Raman scattering of a broadband dye laser in a cesium heat pipe cell. The infrared beam is passed through the reaction zone (or plasma) and suffers absorption. The resulting output is upconverted back to the visible regime by four-wave mixing in another cesium cell. The visible beam carries the record of the infrared absorption and it can be easily recorded photographically or with an optical multichannel analyzer. The time resolution of the technique is determined by the pulse length (10 ns) of the dye laser-cesium converter output. It appears that picosecond response is in fact possible. The IBM group has applied the method to photoinitiated chemical reactions using time-resolved absorption in the ranges 7 - 11 μm and 2.5 - 4 μm. The method is very promising for transient discharge investigations, at least until fast infrared detector arrays are more available.

Detection of Excited Species: Laser Source Hook Interferometry

The hook interferometer method to measure excited state concentrations was introduced a long time ago by Rozhdestvenskii (1912). The method is based on the rapidly varying refractive index (anomalous dispersion) in the neighborhood of an absorption line. The real and imaginary parts of the refractive index are related by the Kramers-Kronig equations. Thus the refractive index changes sign across a spectral line. This leads to the appearance of hooks in the fringe pattern of an interferometer if the discharge arm contains line absorption corresponding to the source illumination wavelength. The hook interferometer has several advantages over straightforward absorption techniques: the linewidth and lineshape need not be known for many experiments, the method is applicable to high concentrations of excited states (i.e., optically thick plasmas), and the dynamic range is several orders of magnitude. With the availability of intense pulsed tunable lasers, another advantage is the fast time response determined by the laser pulsewidth. The experimental arrangement used by Kushner (1984) and

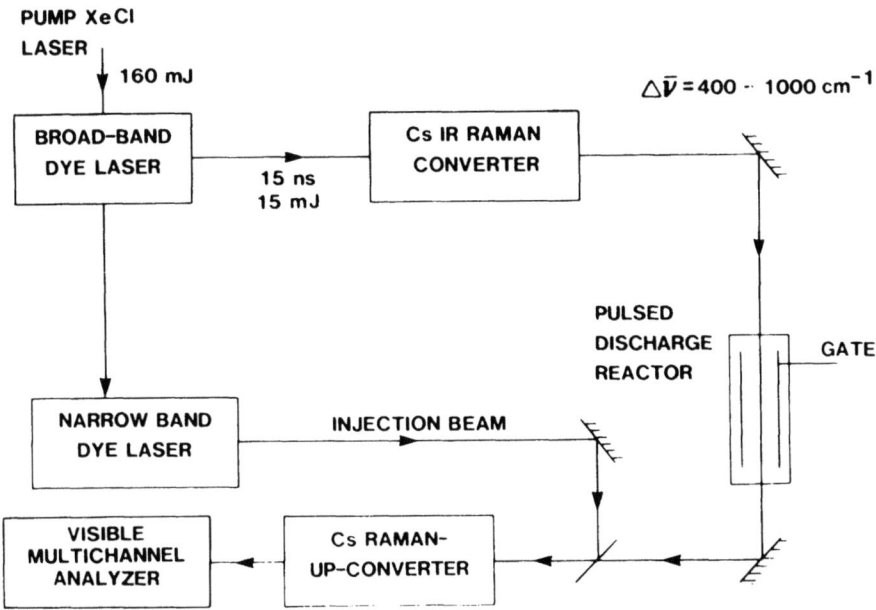

Fig. 15. Experimental arrangement for time-resolved infrared spectroscopy (TRISP). The first Cs cell generates wide band infrared radiation which after absorption in the discharge cell is up-converted to the visible in the second Cs cell.

colleagues to investigate thyratron switching is shown in Fig. 16. The thyratron is placed in one arm of a Mach-Zender interferometer and compensating optics are placed in the other arm. A wide bandwidth dye laser beam is split, passed through both arms of the interferometer, recombined and imaged on the slit of a spectrometer. The wavelength dispersed image at the output of the spectrometer resembles Fig. 17 (adapted from van de Weijer and Cremers, 1982). The feature of interest is the separation in wavelength, of the upper and lower hooks. It can be shown that the density of excited states causing the dispersion is

$$N^* = \frac{\pi k}{\lambda_o^3 \, a_o \, fL} \frac{(\Delta^2 + (\Delta\lambda)^2)^2}{(\Delta^2 - (\Delta\lambda)^2)} \qquad (17)$$

where a_o is the classical electron radius, k equals the order of the interference pattern, f is the oscillator strength of the transition, L is the beam path length through the plasma, and $\Delta\lambda$ is the spectral width of the transition at λ_o. At low pressures $\Delta\lambda$ is determined by Doppler and Stark broadening. If the excited state density is greater than about 5×10^{11} cm^{-3}, $\Delta \gg \Delta\lambda$, hence

$$N^* = \frac{\pi \, k \, \Delta^2}{\lambda_o^3 \, a_o \, fL} \qquad (18)$$

When experimental conditions are such that Stark broadening is the dominant line broadening mechanism (>10^{13} ions cm^{-3}) the slope of the fringes adjacent to line center can be used to determine the electron

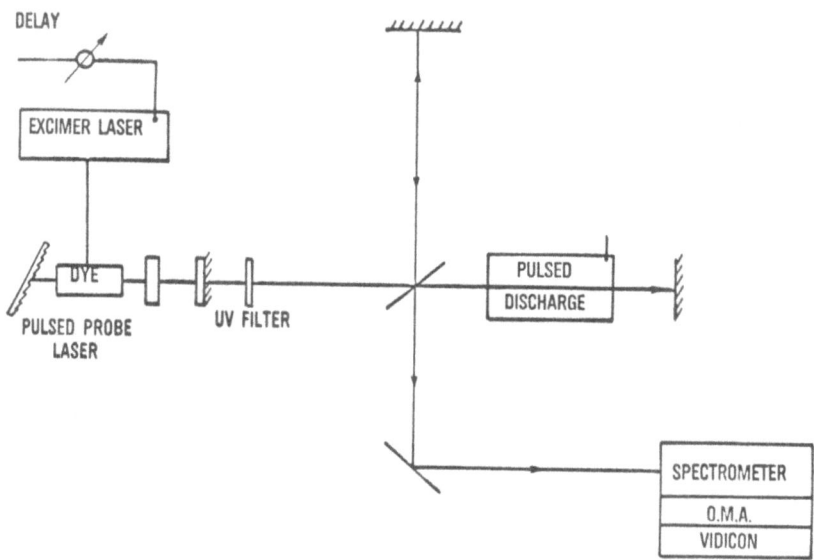

Fig. 16. Schematic of time-resolved 'hook spectroscopy' experiments.

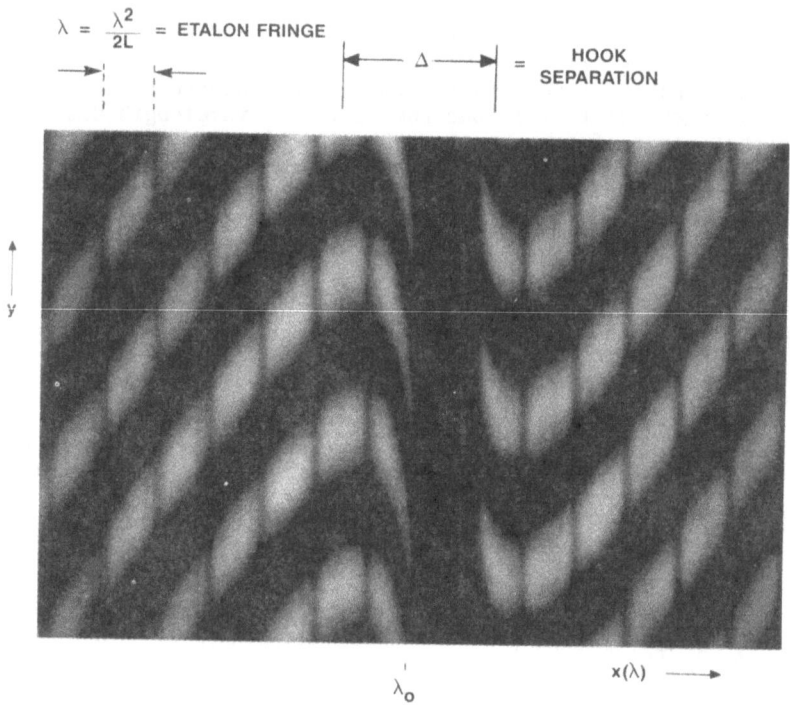

$\lambda = \frac{\lambda^2}{2L}$ = ETALON FRINGE Δ = HOOK SEPARATION

Fig. 17. Experimental results of van de Weijer and Cremers on the 546.1 nm Hg line. Distance between two vertical fringes is 3×10^{-2} nm.

density. Away from any resonances the interferometer fringe shift will record gas density changes. Van de Weijer and Cremers (1982) have replaced the spectrometer with a grating and a camera. The grating was used at diffraction angles close to $\frac{\pi}{2}$ and the hook spectrum was projected a distance of 4 meters, thanks to the use of an intense laser. High quality images were obtained in this way for mercury discharges. The distance between the hooks in Fig. 17 is about 40 mm permitting accurate measurements to be easily obtained from instant photographs. The method permits measurement of transient discharges by using a delayed trigger of the laser pulse with respect to the discharge pulse. The data can be video recorded and the images processed using a frame grabber and a minicomputer. Simultaneous measurement of the excited state densities of two or more high-lying levels can be used to estimate bulk "electron temperatures" in the discharge.

ION DETECTION

Velocity Modulated IR Laser Spectroscopy

The detection of ion densities by LIF and other methods are discussed elsewhere. Here we introduce a technique that is suitable for complex ions that may be found in discharge reactors. The method is based on pioneering spectroscopic studies by Wing et al. (1976) and by Oka (1980). The vibrational HD^+, HeH^+ and D_3^+ in fast ion beams were observed by measurement of the changes in charge-transfer cross-sections when vibrational-rotational transitions of the ions were Doppler-tuned into coincidence with a carbon monoxide laser line. If the measurements are now made on minority ions in a glow discharge, the ions will have a reasonably defined drift velocity in the direction anode to cathode. For HCO^+ in hydrogen at 1 torr in a wide tube, the drift velocity is about 300 m/s. This is equivalent to a Doppler shift of several parts per million. If the discharge is run ac at low frequencies, an effective frequency modulation of the laser is obtained in the ion frame of reference. Then the use of lock-in detection will only give signals due to the ions and not due to any neutral absorption, even though the neutrals are present in much higher concentrations than the minority ions. A schematic of the experiment is shown in Fig. 18. The sensitivity of the method has been estimated (Gudeman et al., 1983), at 10^{-5} cm^{-1} absorption, equivalent to a density of 10^{11} cm^{-3} HCO^+ ions. Gudeman et al. (1983) used a color-center laser pumped by a krypton ion laser to detect HCO^+, H_3O^+, NH_4^+, H_2F^+, HNN^+, H_3^+ and in visible absorption CO^+ and N_2^+. Haese et al. (1983) employed tunable diode lasers to measure ion mobilities of ArH^+, NH_4^+, $HCNH^+$ and $HCOP^+$. The combination of these techniques with microwave spectroscopy to determine the molecular structure represents a new powerful method to measure mobilities and their rotation or vibrational energy dependence.

Negative Ion Measurements

Photodetachment techniques for precision measurements of electron affinities are now well developed. Bacal (1982) have used the photo detachment step as a sort of opto-galvanic process to measure the densities of H^- and D^-. Previously, some measurements were made by Taillet (1969) on O^-. The principle of the method is to photodetach the electrons from the negative ions in a defined volume and to measure the resultant current to a nearby Langmuir probe or to measure the change in

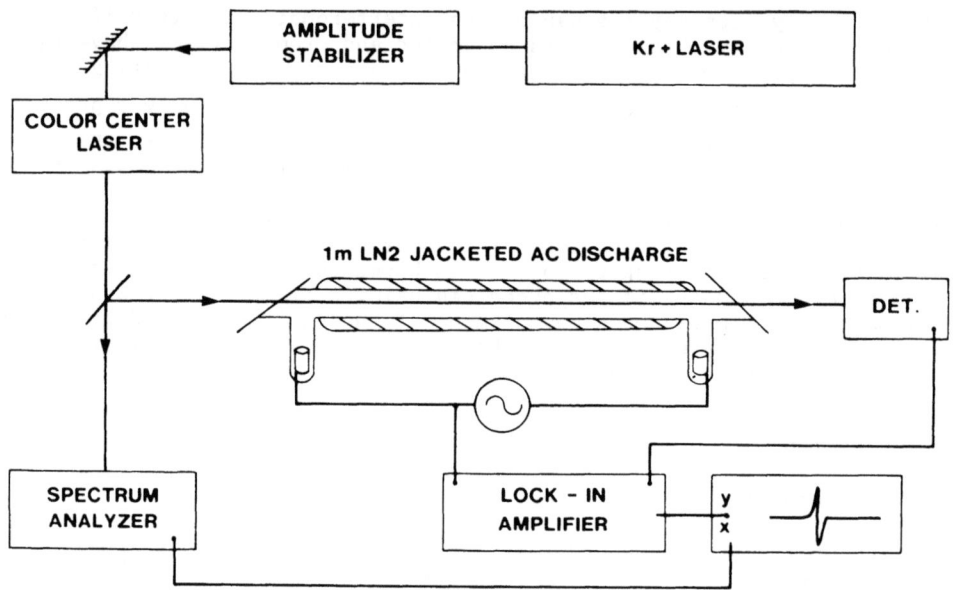

Fig. 18. Schematic of velocity modulation spectrometer by ion drift modulation.

conductivity of the discharge. The photon energy required is not selective, provided it is above threshold. It is necessary to avoid any other photoelectric effects that might cause anomalous signals. Therefore, to measure H^-, a ruby laser was used because its wavelength at 6943Å (1.8 eV) is reasonably close to the maximum in the photodetachment cross section (4×10^{-17} cm^2). The fraction of negative ions destroyed by photodetachment is

$$\frac{\Delta n_-}{n_-} = 1 - \exp\left(-\frac{I\sigma}{h\nu}\right) \qquad (19)$$

where σ is the photodetachment cross section and I is the laser pulse energy per cm^2. For H^-, $\sigma/h\nu = 139$ cm^2 J^{-1} at 6943Å, 182.3 cm^2 J^{-1} for Nd:YAG at 1.06 µm (1.2 eV). This means we need > 100 millijoules cm^{-2} to saturate the detachment. The pulse length requirement that reverse processes should be slow is easily satisfied. Bacal (1982) has used the outlined technique to measure hydrogen negative ion densities in the range $10^7 - 10^{10}$ cm^{-3}. It therefore appears that the method can be applied to plasma reactors if there is a dominant negative ion. If several negative ions are present, the deconvolution of the signal will require a tunable near IR laser.

This talk has briefly reviewed some of the laser techniques which are available for discharge measurements. The laser methods permit very high spatial and fast temporal resolutions. Many atomic, molecular and ionic species can be measured without the ambiguities of mass-spectrometric sampling. While very impressive progress has been made, it is also recognized that many opportunities exist for further laser based measurements in this area, especially on the interactions between the substrate and the reactive plasma.

ACKNOWLEDGEMENT

It is a pleasure to recognize the consultations of C. A. DeJoseph, B. N. Ganguly, L. Goss, P. Haaland and M. J. Kushner.

REFERENCES

Attal, B., Debarre, D., Muller-Dethlefs, K., and Taran, J. P. E., 1984, Revues des Physique Appliquee, 18:39.
Bacal, M., 1982, Physics Scripta, T2:467.
Bethune, D. S., Schell-Sorokin, A. J., Lankard, J. R., Loy, M. M. T., and Sorokin, P. P., 1983, in: "Adv. in Laser Spectroscopy," Vol 2, eds B. A. Garity and J. R. Lombardi, John Wiley & Sons, New York.
Bjorklund, G. C., 1980, Opt. Lett., 5:15.
Boutilier, G. D., Blackburn, M. B., Mermet, J. M., Weeks, S. J., Haraguchi, H., and Winefordner, J. D., 1978, App. Optics, 17:2291.
Braun, V. R., Krasnoperov, L. N., and Panfilov, V. N., 1981, Doklady Akademii Nauk SSSR, 260:901. Catalysts, 22:1332.
Bulatov, E. D., Kozlov, D. N., Otlivanchik, E. A., Pashinin, P. P. Prokhorov, A. M., Sisakyan, I. N., and Smirnov, V. V., 1980, Sov. J. Quantum Electron, 10:740.
Cornelissen, H. J. and Burgmans, A. L. J., 1982, Opt. Commun., 41:187.
Crosley, D. R. and Smith, G. P., 1983, Optical Engineering, 22:545.
Dailey, J. W., 1977, Appl. Opt., 15:568.
DeJoseph, C. A., Garscadden, A., and Pond, D. R., 1983, in: "Proceedings, International Conference on Lasers," V. Corcoran, ed., 1982:738.
Donnelly, V. M., Flamm, D. L., and Collins, G. J., 1982, J. Vac. Sci. Technology, 21:817.
Doughty, D. K., Salih, S., and Lawler, J. E., 1984, Phys. Lett., 103A:41.
Dreier, T., Weilhausen, U., Wolfrum, J., and Marowsky, G., 1982, Appl. Phys., B29:31.
Druet, S., Attal, B., Gustafson, T. K., and Taran, J. P. E., 1978, Phys. Rev., A18:1529.
Dyer, M. J. and Crosley, D. R., 1982, Optics Lett., 7:382.
Eckbreth, A. C., and Hall, R. J., 1981, Combustion Science and Technology, 25:175.
Eng, R. S. and Ku, R. T., 1982, Spectroscopy Letters, 15:803.
Foster, J. S., 1927, Proc. Roy. Soc., A114:47.
Freeman, R. R., Liao, P. F., Panock, R. and Hymphrey, L. M., 1980, Phys. Rev. A22:1510.
Gerardo, J. B. and Verdeyen, J. T., 1964, Proc. IEEE, 52:690.
Goldsmith, J. E. M. and Anderson, R. J. M., 1985, Appl. Optics, 24:607.
Goss, L., 1985, private communication.
Gottscho, R. A., Burton, R. H., Flamm, D. L., Donnelly, V. M., and Davis, G. P., 1984, J. Appl. Phys., 55:2707.
Gudeman, C. S., Begemann, M. H., Pfaff, J., and Saykally, R. J., 1983, Phys. Rev. Lett., 50:727.
Guthals, D. M., Gross, K. P., and Nibler, J. W., 1979, J. Chem. Phys., 70:2393.
Haese, N. H., Pan, F-S, and Oka, T., 1983, Phys. Rev. Lett., 50:1575.
Harmin, D. A., 1985 in: "Atomic Excitation and Recombination in External Fields," M. H. Nayfeh and C. W. Clark, eds., Harwood Press, New York.
Hata, N., Matsuda, A., Tanaka, K., Kajiyama, K., Moro, ., and Sajiki, K., 1983, Japan J. Appl. Phys., 22:L1.
Hinshelwood, D., 1985, NRL Memorandum Report, No. 4592.

Jasinski, J. M., Whittaker, E. A., Bjorklund, G. C., Dreyfus, R. W., and Estes, R. D., 1984, Appl. Phys. Lett., 44:1155.
Johnson, W. B., 1967, IEEE Trans. Antennas and Propagation, AP-15:152.
Knapp, K. and Hanson, R. K., 1983, Appl. Optics, 22:1980.
Krasnoperov, L. N., Braun, V. R., Nosov, V. V., and Panfilov, V. N., 1981, Kinetics and Catalysts, 22:1332.
Kushner, M. J., private communication.
Kychakoff, G., Howe, R. D., Hanson, R. K., and McDaniel, J. C., 1982, Appl. Optics, 21:3225.
Littman, M. G., Kash, M. M. and Kleppner, D., 1978, Phys. Rev. Lett., 41:103.
Luhmann, N. C. and Peebles, W. A., 1984, Rev. Sci. Instrum., 55:279.
Moore, C. A., Davis, G. P., and Gottscho, R. A., 1984, Phys. Rev. Lett., 52:538.
Musal, H. M., 1969, Proc. IEEE, 57:98.
Nitsch, W. and Kiefer, W., 1977, Optics Comm., 23:240.
Oda, T., Usia, T., Takiyama, K., Fujita, T., Kamiura, Y., and Kawasaki, K., 1984, in: "Proceedings, Japan Workshop on Tokomak Diagnostics," Inst. Plasma Physics Report IPPJ-703, Nagoya, Japan.
Oka, T., 1980, Phys. Rev. Lett., 45:431.
Pealat, M., Taran, J. P. E., Taillet, J. Bacal, M., and Bruneteau, A. M., 1981, J. Appl. Phys., 52:2687.
Rozhdestvenskii, D. S., 1912, Ann. Phys., 39:307.
Schlossberg, H. R., 1976, J. Appl. Phys., 47:2044.
Schlossberg, H. R. and Kelley, P. L., 1981 in: "Spectrometric Techniques, II," Ch 4, G. Vanesse, ed., Academic Press, New York.
Shaub, W. M., Nibler, J. W., and Harvey, A. B., 1977, J. Chem. Phys., 67:1883.
Smirnov, V. V. and Gabelinskii, V. I., 1978, JETP Lett., 28:427.
Stanton, A. C. and Kolb, C. E., 1980, J. Chem. Phys., 72:6637.
Taillet, J., 1969, Compt. Rend., 269:352.
Tolles, W. M., Niblr, J. W., McDonald, J. R., and Harvey, A. B., 1977, Appl. Spectroscopy, 31:253.
van der Weijer, P. and Cremers, R. M. M., 1982, J. Appl. Phys., 53:1401.
Verdieck, J. F., Shirley, J. A., Hall, R. J., and Eckbreth, A. C., 1982 in: "Temperature: Measurements and Control in Science and Industry," 5:595, J. F., Schooley, ed., American Inst. of Physics, New York.
Wieman, C. and Hansch, T. W., 1976, Phys. Rev. Lett., 361170.
Wing, W. H., Ruff, G. A., Lamb Jr., W. E., and Spezeki, J. J., 1976, Phys. Rev. Lett., 36:1488.
Wormhoudt, J., Stanton, A. C., and Silver, J., 1983, in: "Proceedings, SPI," Vol. 452.

LASER (AND OTHER) DIAGNOSTICS OF RF DISCHARGES

Carl E. Gaebe and Richard A. Gottscho

AT&T Bell Laboratories
Murray Hill, NJ, USA

INTRODUCTION

Interest in radio frequency glow discharges has burgeoned in recent years because of the wide-spread use of plasmas in the electronics and photonics industries for fabrication of microscopic circuit components. Plasmas are used because: (1) they provide a means for achieving directional surface chemistry, through the directional transport of ions and electrons across the sheaths to device surfaces; and, (2) because they provide a means for achieving high temperature chemistry without the use of high temperatures (Tang and Hess, 1984), through the formation of radicals by electron impact dissociation. RF plasmas are used primarily to avoid surface charging when etching or depositing insulating thin films. Despite the complexity of plasma reactions and past emphasis on understanding dc glow discharges, radio frequency plasmas have been used successfully in producing desired device structures. However, it is probably an understatement to say that few processes have been optimized for the range of desired results, for this can only be realized by accident or extensive empirical investigation of a multi-dimensional, non-linear parameter space.

One of the reasons for developing plasma diagnostics is to develop reliable plasma models with which current processes can be optimized computationally and alternative processes or entirely new processes can be created. Diagnostic data are essential to test the output of such models as well as to provide critical input data. Of course, to the extent that a diagnostic technique is simple to implement, it may also be useful in real-time process monitoring. The focus here, however, is on diagnostic techniques which elucidate some of the fundamental physics and chemistry of rf glow discharges. Three techniques will be discussed in terms of their experimental implementation and the information they can provide: laser-induced fluorescence spectroscopy (LIF), plasma-induced emission (PIE), and optogalvanic spectroscopy (OGS). The emphasis is on application to rf discharges where it is desirable to monitor events on the time scale of a single rf cycle. Optogalvanic spectroscopy is considered in greater detail by Garscadden and Lawler (this Proceedings).

The organization of this paper is centered about the quantities of interest which are to be measured. That is, given a particular quantity of interest, which of the above techniques is best suited for its determination, how reliable are these techniques, and from what limitations do the techniques suffer? In the next section we address the question of what should be measured within the context of fundamental plasma equations. The experimental techniques are then outlined in brief; the reader is referred to original references for further detail. We then discuss the application of the above techniques to measurement of particle concentrations, electric fields, energy distributions, and kinetic parameters.

WHAT SHOULD BE MEASURED?

To answer this question it is best to consider the fundamental equations which govern plasma physics and plasma chemistry (Chen, 1974). The Boltzmann equation relates changes in six-dimensional (three space and three velocity coordinates) phase space to the rate of production and loss by collisions (in cgs electrostatic units):

$$\frac{\partial(nf)}{\partial t} + v \cdot \nabla(nf) + \frac{F}{m} \cdot \nabla_v(nf) = \left[\frac{\partial(nf)}{\partial t}\right]_c , \qquad (1)$$

where n is the density, f is the six-dimensional distribution function, v is the particle velocity, F is the force field acting upon the particle (zero for neutral particles), ∇_v is the gradient with respect to velocity, and m is the particle mass. The left hand side describes the flow of flux, nf, into a volume element by virtue of the flux gradient (second term) and force fields (third term). In the absence of collisions, the right-hand side of Eq. (1) and the net flux in phase space both go to zero. There is an equation for each particle in the plasma and, in general, all of the functions are coupled through the collisional term on the right-hand side and through the self-consistent field, which results from the application of an external field as well as the presence of bound and free charges in the plasma. Note that the collision term on the right not only includes gas-phase interactions but also plasma-surface interactions, of which etching and deposition are only two examples. In an rf discharge, Eq. (1) is further complicated by the introduction of a time-dependent force field.

For capacitively coupled glow discharges in the absence of an external magnetic field and for frequencies below ~ 100 MHz, the only field equation which need be considered is Poisson's equation, which relates the gradient in the local field to the charge density (in cgs electrostatic units):

$$\Delta \cdot E = 4\pi e (n_+ - n_e - n_-), \qquad (2)$$

where e is the electronic charge, n_e is the electron density, n_- is the negative ion density, and n_+ is the positive ion density. The force field in Eq. (1) for singly charged species, given by

$$F = eE, \qquad (3)$$

couples Poisson's equation to the set of Boltzmann equations.

It should now be obvious what to measure in any diagnostic endeavor: the densities and distribution functions for all particles as well as the self-consistent electric field which drives charged particle transport. These data could then be used to test the validity of Boltzmann computer codes or to provide input data for semi-empirical solutions to the equations. Such measurements also provide immediate insight into the effects of varying plasma parameters such as power density, frequency, pressure, electrode gap, gas composition, flow rate, etc., on the discharge chemistry. Needless to say, we are a long way from achieving this diagnostic goal. However, many of these quantities are measurable and concentrations of the most important species, such as ions and free radicals, are often amenable to spectroscopic interrogation.

TECHNIQUES

The Ideal Plasma Diagnostic

Before considering the measurement of a particular component of the Boltzmann and field equations or a specific technique, it is worthwhile defining the ideal characteristics of any plasma diagnostic. We want "in situ" diagnostics so that we do not have to infer what is happening in the plasma from downstream measurements or extracted samples, and we want non-intrusive diagnostics so that our measurements do not perturb the system.

We also desire techniques which have high spatial and temporal resolution compared to the characteristic dimensions and times of interest, which are in turn a function of the quantity to be measured. For example, if we want to measure ion densities as a function of rf voltage at an rf frequency of 10 kHz, the temporal resolution should be at least 1 µs in order to avoid deconvolution of the time-dependent ion density from the apparatus time response. If the same measurement were made an an rf frequency of 13 MHz, one might suppose, by a similar line of reasoning, that the temporal resolution required would be 700 ps. However, because only the lightest ions at low pressure can respond to the instantaneous field above a few MHz, time-resolved ion density diagnostics at 13 MHz are not necessary and a time-averaged technique should suffice. Spatial resolution should be high enough to determine the magnitudes of ion and radical concentration gradients. For example, if the concentration of a reactant is monitored near a reactive surface, a resolution of 1 mm or 0.1 mm might be needed, depending on the pressure, temperature, and reaction rate.

The ultimate temporal resolution for laser-based spectroscopies is determined by the laser pulse width, although other considerations may preclude attaining this limit. For example, if the excited state radiative decay time is much longer than the laser pulse width, the ultimate temporal resolution in a laser-induced fluorescence experiment can be achieved only at the expense of signal intensity. The temporal resolution limit for plasma-induced emission is ultimately limited by the radiative decay rate. For laser-based techniques, the spatial resolution limit is determined by the diffraction of the beam when it is focussed. Similarly, the ultimate spatial resolution achievable in an emission experiment is determined by the extent to which the incoherent light from the plasma source can be focussed.

Laser-Induced Fluorescence

The idea behind laser-induced fluorescence is to excite a specific atom or molecule from a specific quantum state to an excited state and then detect fluorescence from the excited state. If the transition probabilities for excitation and fluorescence, which depend upon laser polarization and intensity as well as the quantum numbers for both lower and upper states, and the interaction volume are known, the density in a particular quantum state can be determined from an absolute measurement of the fluorescence intensity (Zare and Dagdigian, 1974; Kinsey, 1977). In practice, this is rarely done because the measurement of absolute fluorescence intensities is tedious when an accuracy of 10% or better is desired. On the other hand, relative fluorescence intensity measurements are straightforward and often yield a wealth of information. For example, measurement of the relative change in concentration as a function of frequency in an rf discharge provides insight into the competitive modes of power dissipation (see CONCENTRATIONS).

To obtain total particle concentrations it is important to characterize internal energy distributions of the atom or molecule. For example, the rotational temperatures of molecules may depend upon position in the discharge (Davis and Gottscho, 1983), in which case the entire rotational energy distribution should be measured in order to obtain the total concentration in a particular vibrational level at any given position. Similarly, the vibrational and electronic energy distributions should be measured to obtain an accurate total number density. The translational energy distribution might also have to be characterized, particularly for ions accelerated by sheath fields, if the Doppler width is larger than the exciting laser linewidth. Fortunately, the situation is not quite as bad as it might first seem. Rotational transitions can be chosen which correspond to levels whose populations are relatively insensitive to temperature over temperature ranges of a few hundred degrees (Gottscho et al., 1983). On the other hand, measurement of these energy distributions provides useful insight into the mechanisms for power dissipation and energy balance in glow discharges.

A typical LIF diagnostic setup is shown in Fig. 1 (Donnelly et al., 1982). The laser beam propagates parallel to the electrode surfaces and creates a linear fluorescence region which is collected perpendicularly with a collimating lens and then focussed onto the slit of a scanning monochromator. Spatial resolution is determined by the laser beam diameter, the monochromator slit dimensions, and the fluorescence collimating and focusing lenses. Typically, the spatial resolution is 0.01 cm in the axial dimension (i.e., perpendicular to the electrode surface) and 0.1 to 1 cm in the radial dimension. The monochromator not only provides spatial resolution but also spectral resolution. Both are useful in

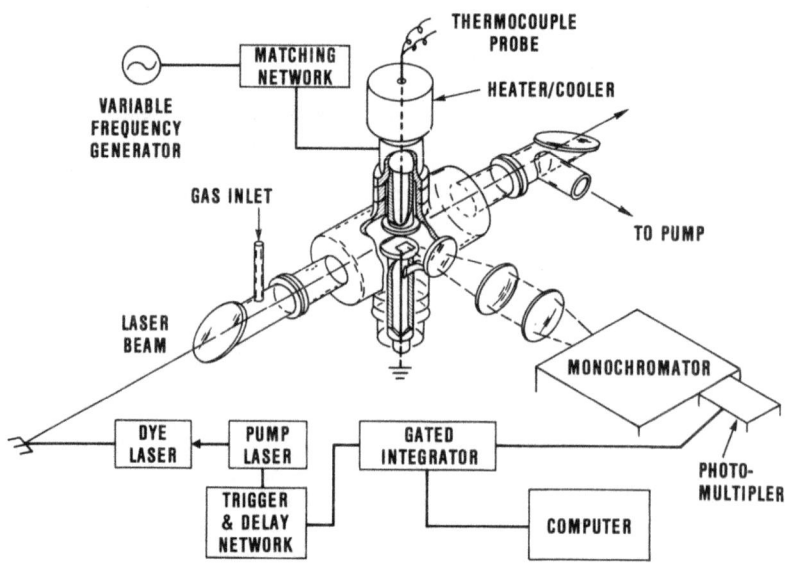

Fig. 1. Illustration of apparatus used for laser diagnostics of a parallel-plate rf discharge. Reproduced with the kind permission of V. M. Donnelly et al. (1982).

discriminating against PIE, which is the primary source of noise. Fluorescence is usually detected by either a photomultiplier tube or a diode array; the latter is useful for simultaneously recording a larger part of the LIF spectrum but generally suffers from lower sensitivity.

Almost all LIF diagnostic experiments performed to date have been with pulsed lasers: the most common systems use a nitrogen, YAG, or excimer laser to pump a tunable dye laser. Pulsed lasers are used widely because they provide: (1) a broader tuning range (from the vacuum ultraviolet to the infrared); and (2) sufficient power to saturate transitions and thereby maximize the LIF intensity with respect to PIE. Also, pulsed excitation of a short-lived transition permits the use of a gated integrator to discriminate against PIE. In studying rf discharges, the laser pulse and detector gate can be synchronized with a particular point on the rf cycle in order to record changes which occur on a time scale shorter than a single cycle (Gottscho et al, 1984).

An example of the timing and discrimination achievable in a two-photon LIF experiment is shown in Fig. 2, where O atoms have been excited by pumping at 226 nm (DiMauro et al., 1984). An O_2 discharge operating at 80 kHz produces both ground and excited state atoms. The excited state density is directly proportional to the PIE intensity, which is seen to be 100% modulated and in phase with the driving voltage. The ground state O atom density is measured by promoting population to the

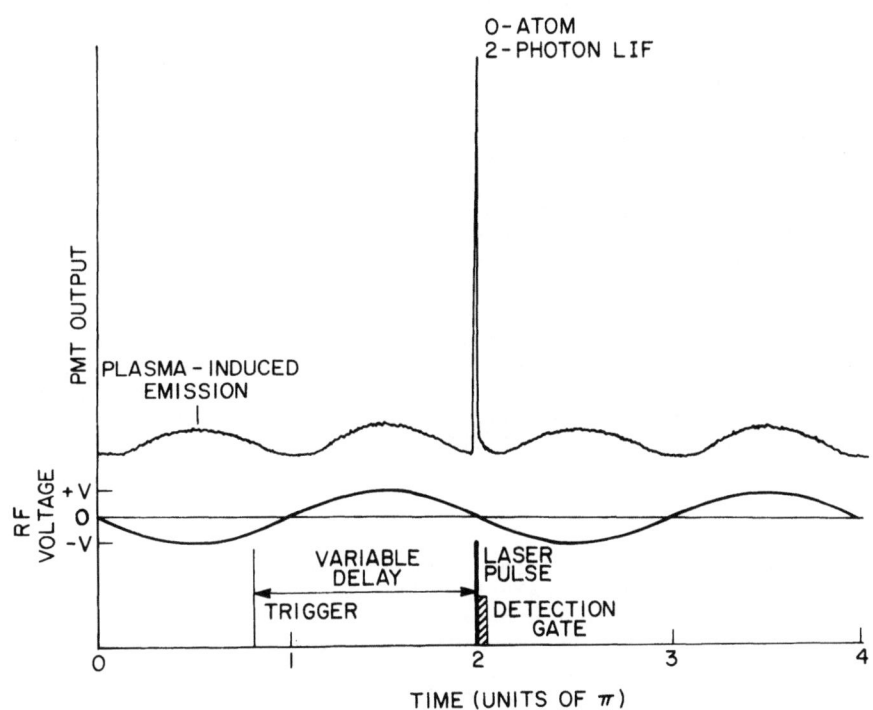

Fig. 2. Timing scheme for detection of two-photon LIF (D. Mauro et al., 1984). Upper trace is the photomultiplier output averaged over 400 laser shots. The middle trace shows the reference rf voltage waveform, and the lower trace illustrates the triggering scheme which fires the laser and detection gate at a PIE minimum. The detection gate duration is 100 nsec.

$3p\,^3P$ state and detecting the induced $3p\,^3P \to 3s\,^3S$ fluorescence at 845 nm. At 80 kHz, the ground state O atom concentration is found to be time-independent. Thus, the LIF signal-to-PIE-noise ratio is optimized by firing the laser at a zero crossing of the rf voltage where PIE is a minimum.

Optogalvanic Spectroscopy

Optogalvanic spectroscopy (OGS) entails the use of a laser to create a change in the discharge impedance and the measurement of this change by monitoring the change in voltage between anode and cathode or the change in current through the discharge circuit. When monitoring the transient voltage induced optogalvanically in a dc discharge, the signal is capacitively coupled to the detection electronics in order to discriminate against the background dc current. In rf discharges, a similar approach has been taken by modulating a cw laser and detecting the OGS signal with a phase-sensitive amplifier (Suzuki, 1981).

One of the difficulties in using OGS as a quantitative plasma diagnostic is that one usually needs to understand the mechanism for the optogalvanic effect. A notable exception is the measurement of electric fields by exciting Rydberg states of a rare gas atom (see Lawler, Garscadden, this Proceedings; Doughty and Lawler, 1984; Doughty et al., 1984; Ganguly and Garscadden, 1985); (see also ELECTRIC FIELD MEASUREMENTS). It is often advantageous to measure the full time-dependence of the optogalvanic signal in order to determine the mechanism (Walkup et al., 1983). Therefore, one must be careful with the choice of detection circuitry so as not to discriminate against fast or slow processes. When capacitive coupling is employed in rf discharges, a notch filter, needed to discriminate against the applied rf voltage, could cause distortion of the optogalvanic signal and misinterpretation of the mechanism. An alternative is direct coupling, as illustrated in Fig. 3. A 50 ohm resistor is placed in series with the discharge and the voltage across this resistor is sampled by a fast transient digitizer (Gaebe and Gottscho, in preparation). To isolate the optogalvanic signal, the laser is synchronously fired with the applied voltage waveform. The "laser-off" signal is subtracted from the "laser-on" signal. The major sources of noise in OGS experiments are the same in rf and dc discharges, namely, discharge noise (e.g., current instabilities) and electromagnetic interference generated by the pulsed laser.

In a one-photon process, OGS is necessarily a two dimensional, line-of-sight probe. However, three-dimensional spatial resolution is obtainable in a two-step process with crossed laser beams. This has been elegantly demonstrated by Doughty and Lawler (1984) in their measurements of spatially resolved electric fields in the cathode fall regions of a dc discharge.

Plasma-Induced Emission

As a diagnostic tool, plasma-induced emission (PIE) has been used most often for qualitative analysis. Its use in quantitative concentration measurement is limited because of the myriad of processes giving rise to the emission and the convolution of excitation functions with electron and ion energy distributions. On the other hand, emission from a short-lived state created in an optically thin plasma can be useful as a measure of the rate of formation of excited states and, therefore, of the impact processes which are responsible for this emission.

There are many mechanisms for the production of PIE: electron-impact excitation, electron-impact dissociation, radiative recombination, chemiluminescence, radiative and collisional transfer from metastable states, ion-impact excitation, sputtering, and surface neutralization.

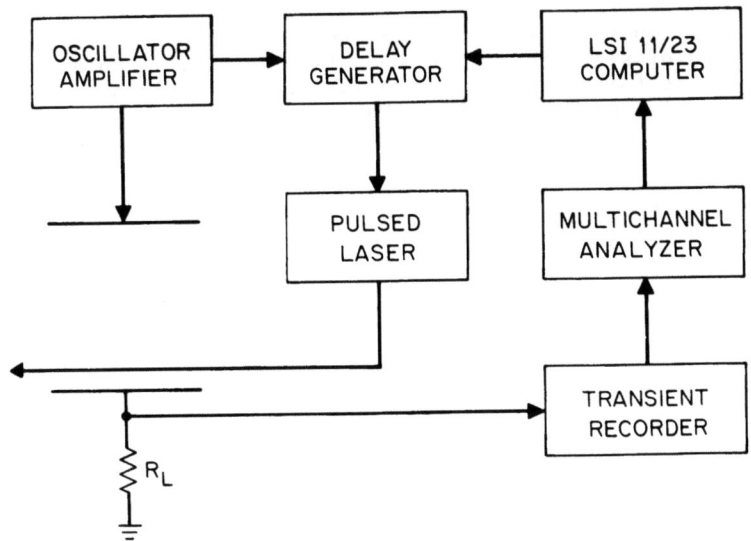

Fig. 3. Schematic illustration of apparatus used to detect time-resolved photodetachment optogalvanic signals from rf discharges through electronegative gases. The value of R_L is 50 Ω. The laser is fired synchronously with the applied voltage (Gaebe and Gottscho, 1985).

Temporal, spatial, and spectral resolution can all be used to distinguish between various excitation mechanisms. For example, because sputtering and surface neutralization are confined to the sheath regions of the discharge and result in greatly broadened or shifted Doppler profiles, the importance of these excitation mechanisms can be discerned by careful linewidth measurements of emission collected from the sheath regions of the discharge (Cappelli et al, 1985; Benesch and Li, 1984). Similarly, chemiluminescence, ion-impact excitation, and electron-impact dissociation produce excess translational energy which can also be detected in the form of Doppler broadening or shifts (Gottscho and Donnelly, 1984; Cappelli et al., 1985; Benesch and Li, 1984). Collisional transfer from metastable states often occurs at slower rates than electron-impact excitation, so measurement of the emission time dependence in an rf discharge can be a useful means to distinguish between excitation processes (Gottscho et al., 1984).

Once it has been established that a particular emission line or band is produced by a specific mechanism, the intensity of that line can be used for diagnostic purposes. For example, if it is known that a particular line is induced by electron-impact excitation, then the intensity of that line can be used to monitor either one portion of the electron energy distribution (if the lower state concentration is known), or the lower state concentration (if the electron energy distribution is known). As discussed below (see <u>Neutral Intermediates</u>), this latter application is the basis for emission actinometry.

PIE can also be detected synchronously with the applied rf field (Flamm and Donnelly, 1985; Gottscho et al., 1984; Rosny et al., 1983) by either scanning the gate position of a gated integrator or by using a transient digitizer. Spatial resolution in a PIE experiment is limited to two dimensions unless a series of line-of-sight measurements are made and an inversion procedure is used to convert the cross sectional measurements to radial profiles (Choi and Kim, 1982; Blades and Hozlick,

1980). It is important to note that many PIE measurements are not strictly line-of-sight in that a cone of emission is sampled. Only when small apertures (high f numbers) are employed is the measurement truly two-dimensional (Gottscho et al., 1983).

CONCENTRATIONS

Unfortunately, a single spectroscopic diagnostic technique cannot be used to measure all species concentrations. In general, different techniques must be used for different species. Moreover, there are still many species for which no suitable spectroscopic diagnostic technique is currently available. A summary of species which can be measured by diode laser and dye laser spectroscopy has been compiled recently (Wormhoudt et al., 1983). The three techniques described above are best suited for detection of reactive intermediates, upon which we focus our attention. By reactive intermediates, we mean short-lived species which play an important role in plasma chemistry, i.e., positive and negative ions and many reactive neutrals.

Positive Ions

Although the spectroscopy and chemistry of positive ions has been studied extensively and there has been much recent progress (Miller and Bondybey, 1983), relatively few diagnostic studies have been reported. Two techniques which have been applied successfully to the study of diatomic molecular ion concentrations in glow discharges are laser-induced fluorescence (LIF) and opto-galvanic spectroscopy (OGS). Recent applications of microwave and infrared spectroscopy to the study of molecular ions have been impressive but the emphasis has been more on molecular structure and less on plasma chemistry. These methods should prove useful for diagnostic purposes in the future (Gudeman and Saykally, 1984; Woods, 1981, 1983; Yamada et al., 1981).

An example of the sort of information which can be obtained from LIF is shown in Fig. 4, where the concentration of Cl_2^+ is plotted as a function of position between the electrodes for two times during a low frequency (55kHz) rf cycle (Gottscho et al., 1984). In the upper trace, the laser is fired when the powered electrode (on the left) is the momentary anode; that is, the applied voltage is a maximum. The lower trace is recorded when the powered electrode is the momentary cathode. Note that the two traces are mirror images of one another, which is indicative of the two electrodes having equal areas and there being no dc bias. During the anodic part of the cycle, ion density builds in the sheath primarily because of electron-impact ionization (Gottscho, 1985). When the applied voltage crosses zero, however, anode and cathode switch, cations feel a strong extraction force, and the ion density decreases to an unmeasurable level ($< 10^7$ ions cm^{-3}). Note that ion density in the plasma center is the same for both cathodic and anodic parts of the cycle. This again results from the symmetric nature of the reactor. However, at other times during the rf cycle, the ion density in the center of the plasma is different because of periodic formation and loss. This will be discussed further (see KINETIC MEASUREMENTS) when we consider kinetic diagnostics.

Periodic high-energy extraction and build-up of cations in the sheath is unique to low frequency ac discharges. In the dc cathode sheath the cation density is always low because of the ever-present, strong electric field which results in steady, high energy ion extraction. In the dc anode sheath, the cation density is always higher than in the cathode sheath because the anode sheath field is smaller and the ions are extracted with lower energy. In a high frequency rf discharge (i.e., above a few MHz), ions do not respond to the instantaneous field

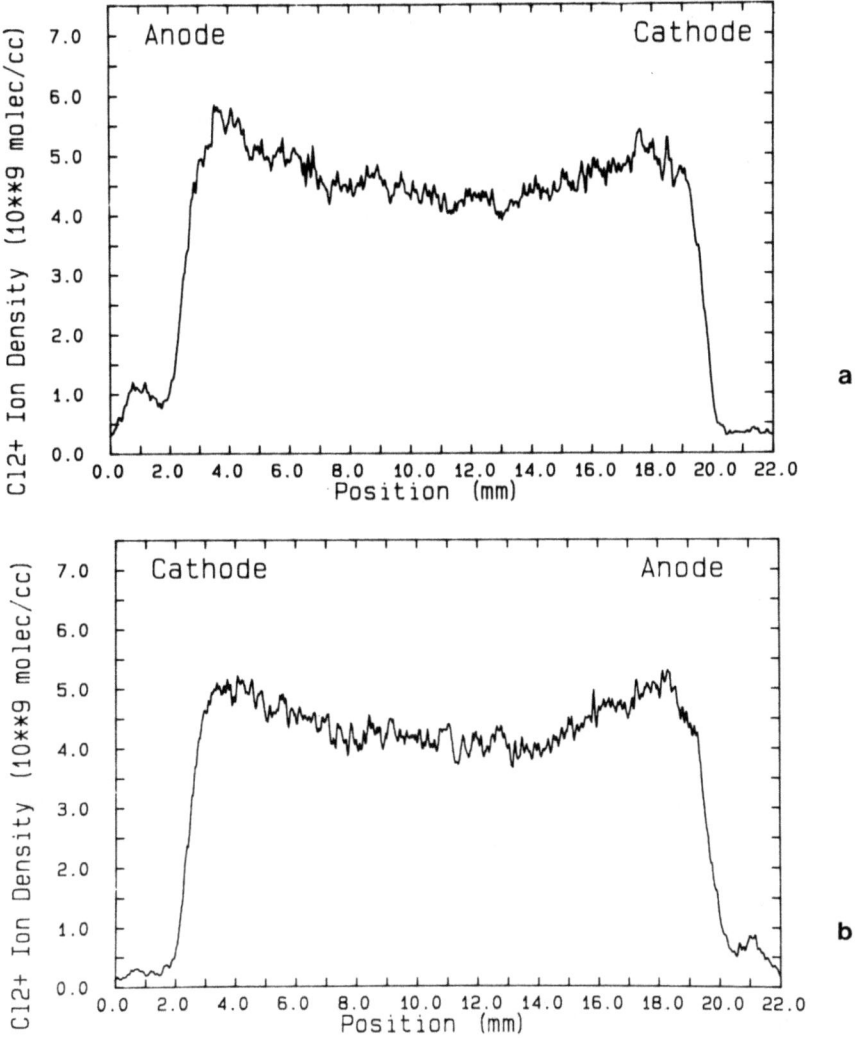

Fig. 4. Spatial profiles of Cl_2^+ concentration at two different times during an rf cycle in a 55 kHz discharge through Cl_2 at 0.3 Torr and 0.6 W cm^{-3}. (a) Powered electrode, at 0 mm, is momentary anode and grounded electrode, at 22 mm, is momentary cathode. (b) Powered electrode is momentary cathode and grounded electrode is momentary anode. The absolute concentrations should be accurate to within an order of magnitude and are lower bounds.

but rather to the time-averaged field. Therefore, the cation density builds to a steady-state level higher than in the low-frequency cathode sheath but lower than in the low-frequency anode sheath. Because the time-averaged field is smaller in the high frequency plasma for the same power, ions do not impact the electrode surfaces with as large an energy as in the low frequency plasma (Bruce, 1981a, 1981b).

One of the consequences of high energy ion bombardment of electrode surfaces in low frequency as well as dc discharges is the production of secondary electrons which help to sustain the discharge. At sufficiently high pressures these secondaries can initiate an ionization and excitation cascade as they are accelerated across the cathode sheath. This is evident in Fig. 5 where the emission intensity from excited BCl radicals is plotted as a function of both time and position from the center of the plasma into the sheath of a parallel plate discharge through BCl_3 operated at 750 kHz. At this frequency, the ion transit time across the sheath is a substantial fraction of the rf period so that a delay (~ 250 ns) is apparent in the emission intensity in the cathode sheath. The onset of emission is retarded by the time it takes ions to traverse the sheath, impact the electrode, and eject secondary electrons, which presumably initiate the excitation cascade (Flamm and Donnelly, 1985). Also note that emission throughout the discharge is 100% modulated: as the applied voltage crosses zero, the local field decreases, and electrons lose energy rapidly enough for excitation to cease briefly before the voltage increases once more (Figs. 2 and 5) (Flamm et al., 1973).

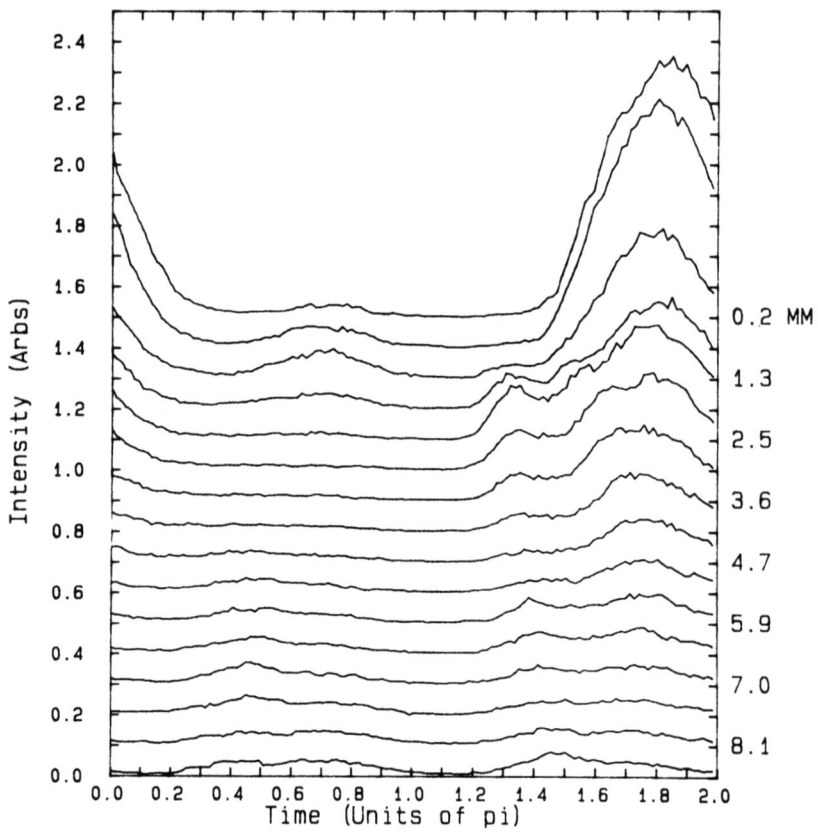

Fig. 5. Plasma-induced emission from BCl radicals ($A^1\Pi \rightarrow X^1\Sigma^+$, $\lambda = 272$ nm) formed in a 750 kHz discharge through 0.3 Torr of BCl_3. Power density 0.1 W cm^{-3}, electrode gap 16 mm, flow rate 10 sccm. Time in units of pi refers to phase of voltage waveform on electrode located at 0 mm (2π corresponds to 1.33 μsec). Each trace is taken at a different position (given on the right) from the electrode and is displaced vertically by one intensity unit.

Optogalvanic spectroscopy has also been employed to measure positive ion densities in situ (Walkup et al., 1983). The experiments to date have been performed only in dc glows but there is no obvious reason why the technique could not be applied equally well in an rf discharge (see next section). Often, the optogalvanic signal is difficult to interpret because it can involve secondary processes. For example, if the atomic concentration in a metastable state is transferred to a higher lying state by absorption of radiation, collisional ionization of this higher lying state may be required in order for a change in the discharge impedance to occur. The mechanism responsible for the optogalvanic signal and whether or not secondary, collisional processes are involved may often be discerned from the time-dependence of the optogalvanic signal. In this fashion, Walkup et al., 1983 were able to show that the optogalvanic signals resulting from excitation of N_2^+ and CO^+ ions in N_2 and CO discharges, respectively, are due to larger ion mobilities in the excited electronic state. Given this mechanism, they were then able to determine the absolute ion concentration from the optogalvanic signal amplitude. Because the optogalvanic signal results from a change in ion mobility, the strongest signals occur in the sheath regions of the discharge where the ions are accelerated by the local field (Fig. 6). This is in marked contrast with the results of LIF experiments, where signals are strongest in the weak-field plasma body (Figs. 4 and 6). Thus, the two techniques are complementary.

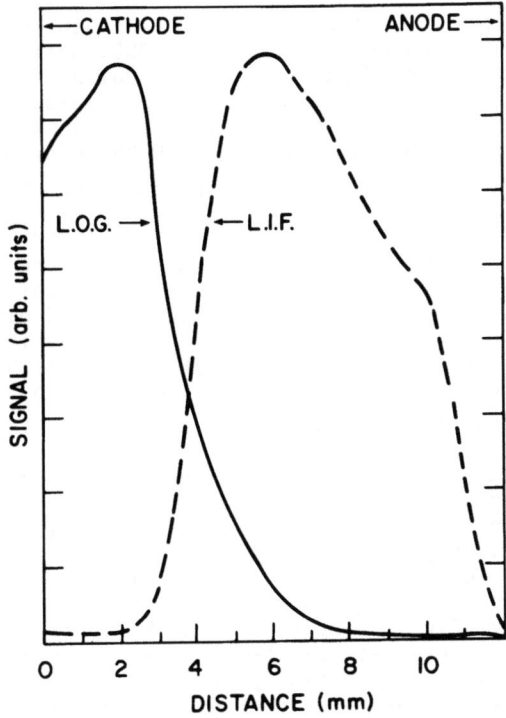

Fig. 6. N_2^+ optogalvanic (LOG) and laser-induced fluorescence (LIF) signal as a function of position between cathode and anode of a dc discharge through N_2 at a pressure of 1.0 Torr. Reproduced with the kind permission of R. Walkup (from Walkup et al., 1983).

Negative Ions

The effects of negative ions on dc glow discharges have been discussed in the past (Massey, 1976; Thompson, 1959; Garscadden, 1978; Emeleus and Woolsey, 1970), but their effects on rf discharges have not been explored until recently. Because of ion and electron energy modulation one expects negative ion kinetics to be different in an rf discharge. One of the reasons for the lack of data on negative ions has been the lack of good diagnostic techniques.

Recently, several groups have reported photodetaching negative ions in dc glow discharges (Taillet and Legendre, 1969; Bacal et al., 1979; Greenberg et al., 1984; Webster et al., 1983, Klein et al., 1983). The subsequent change in discharge impedance resulting from the increase in negative charge mobility is readily detected. Because of the method's high sensitivity and signal-to-noise ratio, Webster et al., (1983) were able to use this technique to extend by an order of magnitude the precision with which the I^- photodetachment threshold had been measured. Greenberg et al., (1984) used microwave interferometry, albeit without very high spatial resolution, to measure the transient increase in electron density resulting from excimer laser detachment of F^- in a dc discharge through NF_3. From the dependence of the detached electron density on laser power, the absolute negative ion density was found to be two orders of magnitude greater than the electron density. The findings are consistent with earlier determinations (Emeleus and Woolsey, 1970; Thompson, 1959). Taillet and Legendre (1969) and Bacal et al. (1979) used a ruby laser to measure densities of O^- in O_2 and H^- and H_2 dc discharges and found similar anion densities.

We have extended these techniques to the measurement of negative ions in rf discharges (Gaebe and Gottscho, 1985; Gottscho and Gaebe, 1985). The full time-dependence of the opto-galvanic signal in an rf discharge is measured by synchronously firing a pulsed laser with respect to the applied rf voltage. A typical transient signal resulting from photodetachment in the sheath region of a 50-kHz discharge through BCl_3 is shown in Fig. 7. Similar current transients are obtained when pure Cl_2 discharges are examined. The oscillations evident in Fig. 7 most likely result from a combination of ion waves and secondary ionization, although they may depend also upon the external circuitry. We associate the first peak with initially detached electrons and plot its amplitude as a function of the applied voltage in Fig. 8. The largest signals are observed in the vicinity of the rf voltage zero crossings for two reasons: (1) dissociative attachment of electrons to BCl_3 or Cl_2 is most rapid at electron energies below 1 eV (Kurepa and Belic, 1978; Stockdale et al., 1972; Christophorou, 1980) and (2) detachment of anions by collisions with neutrals is most rapid at energies above 10 eV (Huq et al., 1984). When the applied voltage crosses zero in a low frequency rf cycle, the electron energy distribution relaxes (Figs. 2 and 5) and electrons dissociatively attach to form negative ions. At 50 kHz, there is sufficient time for virtually all of the electrons to attach and be converted into anions, and large optogalvanic signals result from the large anion density (Figs. 7 and 8). When the applied voltage increases, the attachment rate decreases but the detachment and extraction rates increase as anions are accelerated by the sheath fields. Correspondingly smaller optogalvanic signals are observed at these times.

The sign of the initial, transient current peak induced by photodetachment (Figs. 7 and 8), indicates the direction of the sheath field. When the current is negative (positive), electrons are accelerated toward (away from) the electrode and the field points toward (away from) the plasma body. From Fig. 7, therefore, it is apparent that the field

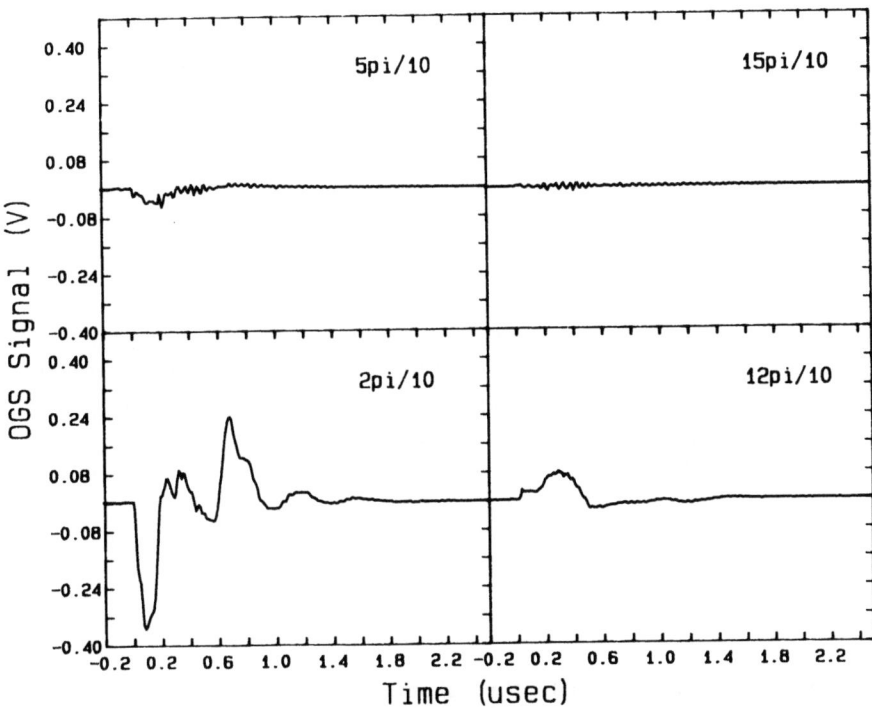

Fig. 7. Transient voltages produced across a 50-Ω resistor in series with the grounded electrode when a 10-nsec pulse from a N_2 laser is used to photodetach negative ions formed in a 50-kHz discharge through BCl_3 at 0.3 Torr and 0.1 W cm^{-3}. Optogalvanic signals at four different times during the rf cycle are shown.

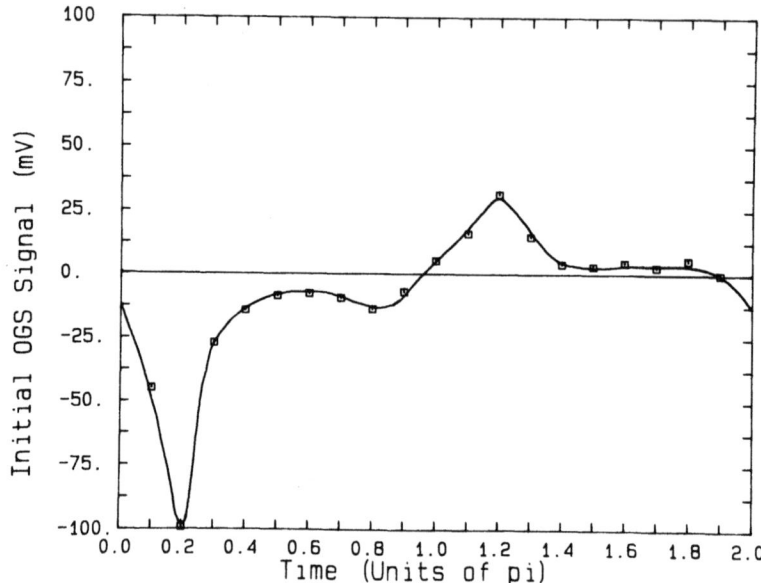

Fig. 8. Initial optogalvanic signal amplitude as a function of the applied voltage at 50 kHz for discharge through BCl_3 at 0.3 Torr and 0.1 W cm^{-3}. Time in units of pi refers to the phase of the applied voltage so that a value of 1.0 corresponds to 10 μsec at a frequency of 50 kHz.

points towards the plasma body throughout most of the anodic half-cycle, and the plasma potential always lies below the anode potential. This is contrary to the situation in most electropositive plasmas (Langmuir, 1929; Chen, 1974). A negative plasma potential relative to the anode potential is consistent with the electron density being orders of magnitude lower than the anion density. Otherwise, the large electron mobility would result in rectification of the anode sheath field as in electropositive discharges. Consequently, negative ions, as well as electrons, impact electrode surfaces during the anodic half-cycle of a low-frequency rf discharge through strongly attaching gases.

Low frequency discharges differ from both high frequency and dc glow discharges. In the latter discharges, the anion density is not strongly time-varying. In the dc case, the electron and ion energy distributions are not modulated; in the high frequency rf case, electron and ion energy modulation is incomplete since the applied voltage changes rapidly compared to electron energy relaxation (Flamm and Donnelly, 1985) and ion response times. Therefore, in dc and high frequency rf discharges the anion density reaches a steady-state level intermediate between the minimum and maximum levels attained during a single cycle of a low frequency rf discharge. This conclusion has been verified experimentally in two ways: (1) no optogalvanic signals resulting from anion detachment in the sheath have been observed by us at 750 kHz, which implies that the anion density in this region of the discharge is at least an order of magnitude smaller than at 50 kHz (Gottscho and Gaebe, 1985); and (2) the anomalously large anodic sheath fields, a result of high anion densities, are absent above 750 kHz (see ELECTRIC FIELD MEASUREMENTS below and Fig. 13).

Neutral Intermediates

Neutral species are of obvious importance in understanding discharge chemistry. Because they are highly reactive, their concentrations may be low. However, measurement of these concentrations can provide stringent tests for kinetic models. Their high reactivity with respect to surfaces often leads to large concentration gradients in the sheaths. Measurement of these gradients, in turn, can be a useful means for determining the reactivity of a particular species with respect to a surface (Selwyn et al., 1984).

Two techniques used widely in detection of neutrals are laser-induced fluorescence (Gottscho et al., 1983; Gottscho et al., 1984; Donnelly et al., 1982; Hargis and Kushner, 1981; Pang and Brueck, 1983; Gottscho and Mandich, 1984; Gottscho and Mandich, 1985; Gottscho, et al., 1982; Walkup et al., 1984; Di Mauro, 1984; Selwyn et al., 1984; Walkup et al., 1985) and emission actinometry (Coburn and Chen, 1980; d'Agostino et al., 1984; Tiller et al., 1981; Cramarossa et al., 1981; d'Agostino et al., 1981, d'Agostino et al., 1983; d'Agostino et al., 1982; Gottscho and Donnelly, 1984; Ibbotson, et al., 1983; Donnelly et al., 1984). Both techniques suffer from the disadvantage that they require the species of interest to be a strong emitter.

Emission actinometry suffers from the additional disadvantage of being an indirect technique. The unknown radical concentration is estimated by normalizing its emission intensity to that from an inert gas, such as Ar, having a similar electronic excitation function. In this fashion, the relative density of a species can be monitored while a plasma parameter, such as feedstock composition, is adjusted. Any changes in electron density are accounted for by normalization with respect to the actinometer. The major disadvantage of emission actinometry is that it relies upon the assumption that both actinometer and radical emission result from electron impact excitation and not from other processes, such as simultaneous excitation and dissociation of a molecular precursor, ion-impact excitation, surface recoil, sputtering, etc. Fortunately, these competing mechanisms generally exhibit line broadening, which can be used to distinguish them from direct electron-impact excitation (Gottscho and Donnelly, 1984; Benesch and Li, 1984; Cappelli et al., 1985). In the case of ion-impact excitation, broadening can result from energy transfer between accelerated ions and neutrals. In the case of molecular dissociation, Doppler broadening can result when chemical binding energy is transformed into translational energy. Examination of emission line shapes is a useful means for establishing the validity of an actinometer (Gottscho and Donnelly, 1984). Nonetheless, narrow linewidths are merely a necessary, not a sufficient, condition for excitation to be attributed to electron impact. In general, actinometry should be used when LIF is not a viable alternative. One example of the valid use of actinometry is in the detection of F atoms (Ar actinometer) in CF_4/O_2, SF_6/O_2, and NF_3/Ar discharges (Coburn and Chem, 1980; Gottscho and Donnelly, 1984, Donnelly et al., 1984; d'Agostino et al., 1981). From both downstream titration measurements and in situ measurements of F atom emission line widths, it has been concluded that emission from both F and Ar atoms results from electron impact on ground state atoms. At present, actinometry appears to be the best way to measure relative F atom densities in situ.

An example of the sort of information learned from LIF measurements of radical densities is shown in Fig. 9. Time-averaged LIF intensities of BCl (solid lines), formed in rf discharges through BCl_3 at three different frequencies, are shown as a function of position between the electrodes. Also shown are the corresponding PIE intensities (dashed lines). Differences between PIE and LIF profiles stem from differences in excited and ground state lifetimes, 19 ns and 0.3-2 ms, respectively

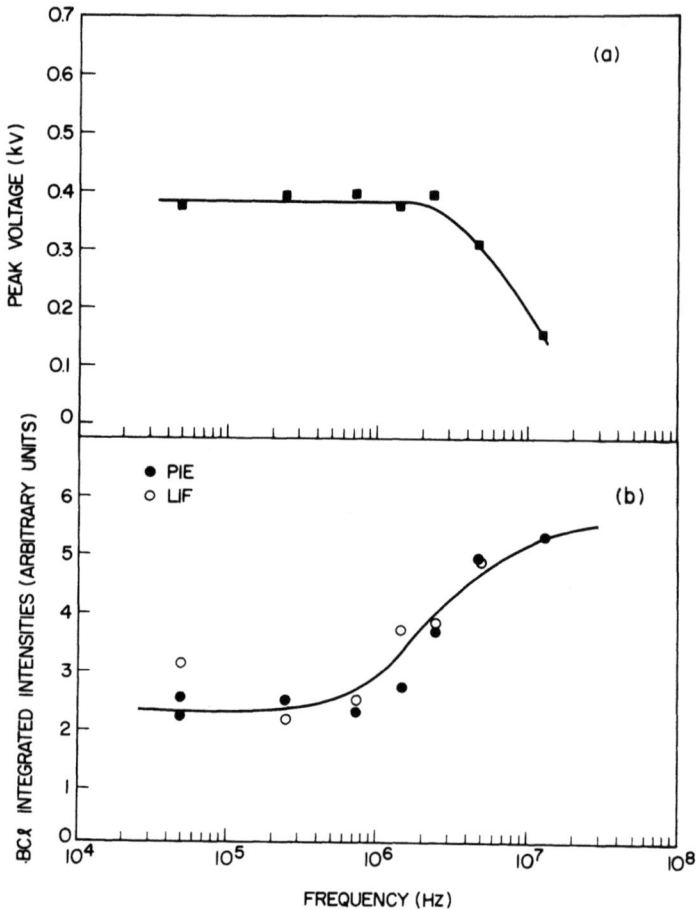

Fig. 9. Time-averaged spatial profiles of BCl radical excited (dashed lines) and ground (solid lines) state concentrations for three different operating frequencies of a BCl_3 discharge at 0.3 Torr and 0.1 W cm^{-3} (Gottscho and Mandich, 1985). The arrows indicate the positions of the stainless steel electrodes.

(Gottscho and Mandich, 1985). Emission profiles are determined by where the excited state is formed since radiation occurs before the excited state has time to diffuse or react. Ground state LIF profiles depend upon both where ground states have formed and where ground states have diffused. Similarities in the two profiles suggest that ground and excited state BCl radicals are formed by parallel mechanisms.

As the frequency is increased, at constant power, from 50 kHz to 13 MHz, both LIF and PIE intensities shift from the sheaths to the plasma body. This change in the distribution of intensity is indicative of a change in the mechanism by which power is dissipated. At low frequencies, ions respond to the instantaneous field in the sheath leading to power dissipation by ion conduction processes, i.e., heating of the gas and electrodes, sputtering, secondary emission, and etching. All of these processes are conducive to the formation of excited and ground sate radicals in the sheaths (Fig. 5) (Cappelli et al., 1985; Flamm and

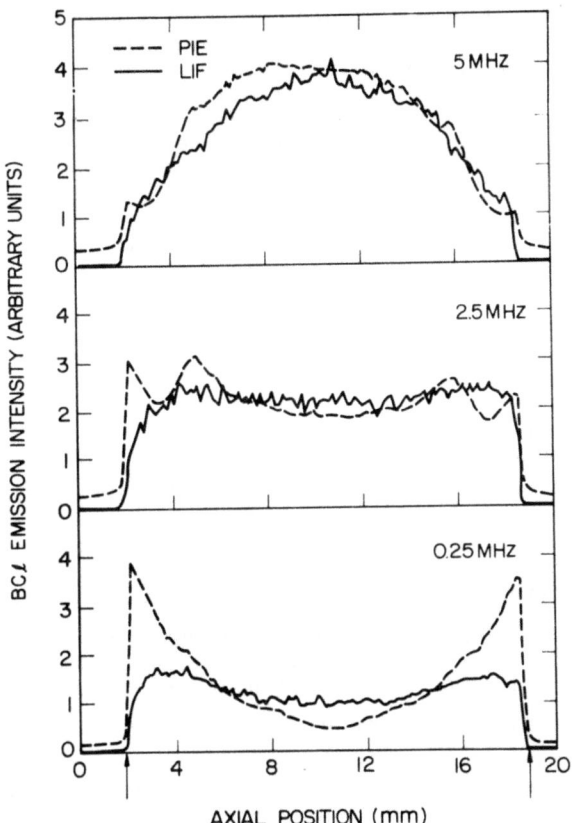

Fig. 10. (a) Peak voltage applied across discharge through 0.3 Torr of BCl_3 as a function of frequency. The power dissipated was maintained at a constant value of 10 W. (b) Spatially and temporally integrated BCl excited (closed circles) and ground (open circles) state radical densities as a function of frequency (Gottscho and Mandich, 1985).

Donnelly, 1985). Above the ion plasma frequency, ions do not respond to the instantaneous applied field (Bruce, 1981b), and the applied voltage and sheath field amplitude decrease for the same power density (Fig. 10a) (Gottscho and Mandich, 1985). Therefore, power dissipation shifts from ion processes in the sheath to electron processes in the plasma body (Fig. 9), i.e., dissociation, ionization, and excitation. Since less energy is dissipated by ion-surface impact processes, more energy is dissipated by electron-impact processes, leading to an increase in the excited and ground state formation rates. If the loss rates for excited and ground state radicals are frequency independent, as might be expected, an increase in the formation rates will manifest itself as an increase in the spatially integrated emission intensity and ground state density. This is consistent with observation (Figs. 9 and 10b).

A variation of the LIF technique, which has been applied recently to the detection of atomic species, is two-photon laser-induced fluorescence (DiMauro et al., 1984; Selwyn et al., 1984; Walkup et al, 1985). Because most light atom transitions of interest lie in the vacuum ultra-violet, one-photon, atomic LIF is a difficult technique to implement. Besides the problems associated with generating laser light at such short wavelengths, one must contend with the problems of detecting the light with a vacuum monochromator and propagating the light through the plasma, which will often be optically thick at such short wavelengths. By using two-photon LIF these problems are circumvented. Excitation is at twice the transition wavelength, which will usually be in the uv or visible, and detection is in the red or near infrared. Even without the use of a monochromator, the method has inherently high spatial resolution because the transition probability depends non-linearly upon the incident laser intensity and, therefore, is strongest at the laser focal point.

Successful application of this technique to the detection of O atoms in O_2 discharges has been achieved (DiMauro et al., 1984; Selwyn et al., 1984; Walkup et al., 1985). The two-photon LIF signal was calibrated in the original experiments by using electron paramagnetic resonance spectroscopy downstream to determine the absolute concentration of the O atom density (Di Mauro et al., 1984). In subsequent experiments, the LIF signal was calibrated using the known photodissociation yield of O_2 at 226 nm wavelength (Selwyn et al., 1984; Walkup et al., 1985). This technique was then used to assess the validity of actinometry for the determination of O atom density as a function of the concentration of CF_4 relative to O_2 in the feedstock, and to determine the concentration gradient of O atoms near a surface coated with an organic resist being etched by the plasma (Selwyn et al., 1984; Walkup et al., 1985).

ELECTRIC FIELD MEASUREMENTS

One of the most important diagnostics for understanding plasma chemistry and physics is measurement of the local electric field. In plasma modeling, the local field is obtained by self-consistent solution of the Boltzmann and field equations (see WHAT SHOULD BE MEASURED?). Thus, comparison of this calculated field with an in situ measurement provides a stringent test for theory. Beyond such tests, measurements of the local field may be useful in semi-empirical modeling, such as Monte Carlo simulations of electron and ion dynamics (Kushner, 1983; Kushner et al., 1985).

Two spectroscopic methods for measuring electric fields have recently been applied to plasma diagnosis. The first is an LIF technique and thus has all of the usual advantages of high spatial and temporal resolution (Moore et al., 1984; Mandich et al., 1985; Gottscho and Mandich, 1985; Sadeghi and Derouard, 1985). In the absence of an external electric field, a rotational level of well-defined parity can be prepared by laser excitation of a polar diatomic molecule. In the presence of an external electric field, however, a level of mixed parity is prepared and fluorescence which would ordinarily be forbidden (as a result of parity selection rules) becomes allowed. The forbidden line intensity is "borrowed" from the allowed line intensities. Thus, measurement of the forbidden-to-allowed line intensity ratio provides a means of determining the electric field amplitude. Such ratios are plotted in Fig. 11 as a function of the electric field applied between two parallel plates (Mandich et al., 1985). In accord with theory (Mandich et al., 1985; Alexander et al., 1984 in press; Moore et al., 1984), it is seen that the extent of parity mixing and, therefore, the forbidden-to-allowed line intensity ratio Q/R, is strongly dependent upon the excited state rotational quantum number. By tuning the laser to

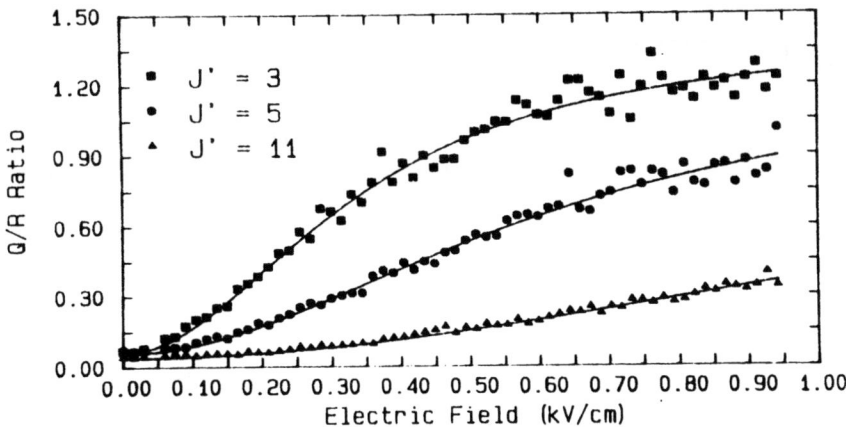

Fig. 11. Forbidden to allowed laser-induced fluorescence line intensity ratio Q/R from parity-mixed rotational energy levels in the BCl $A^1\pi - X^1\Sigma^+$ band system as a function of electric field. Three curves are shown corresponding to preparation of three different excited state rotational levels. The non-zero intercepts of the data at zero electric field are primarily due to underlying hot band transitions and collisions. Further details can be found in Mandich et al., 1985.

different rotational lines, a dynamic range of better than 10^4 can be achieved using this method.

The results of using this technique to measure the sheath fields in a parallel plate rf discharge at 50 kHz through 0.3 torr of BCl_3 are shown in Fig. 12. The sheath fields are measured as a function of position and time during a single rf cycle. Throughout most of the negative part of the cycle (right-hand side of Fig. 12), the fields are strong and decrease almost linearly throughout the sheath. This is consistent with expectations based on electropositive plasmas where ionization in the sheath is occurring (Warren, 1955). However, during the anodic half-cycle (left-hand side of Fig. 12), and at the beginning of the cathodic half-cycle, anomalous behavior is apparent. The strong field at the beginning of the anodic phase results from the conversion of electrons to anions when the applied voltage crosses zero (see Negative Ions). As a result, the negative and positive charge mobilities become comparable and there is no longer any mechanism by which the anodic sheath field can be rectified. As the applied voltage continues to increase, anions are partially reconverted to electrons and the field decreases in the anode sheath. It is also apparent from Fig. 12 that during this sequence of events, double layers are formed in both sheaths. The maxima in the fields are a direct consequence of there being layers of net negative and positive charge (see Eq. (2)).

Variation in the local field 1 mm above the powered electrode as a function of the phase of the applied voltage and for several different operating frequencies is shown in Fig. 13. As expected, the field is strongest during the negative part of the cycle for both low and high frequency. However, at high frequency the strong field is missing during the anodic part of the cycle near the voltage zero crossing. Investigation of the spatial dependence of the fields at 750 kHz shows no evidence for the double layers present at 50 kHz (Fig. 12). These differences are attributed to kinetic constraints imposed by the time-varying field on

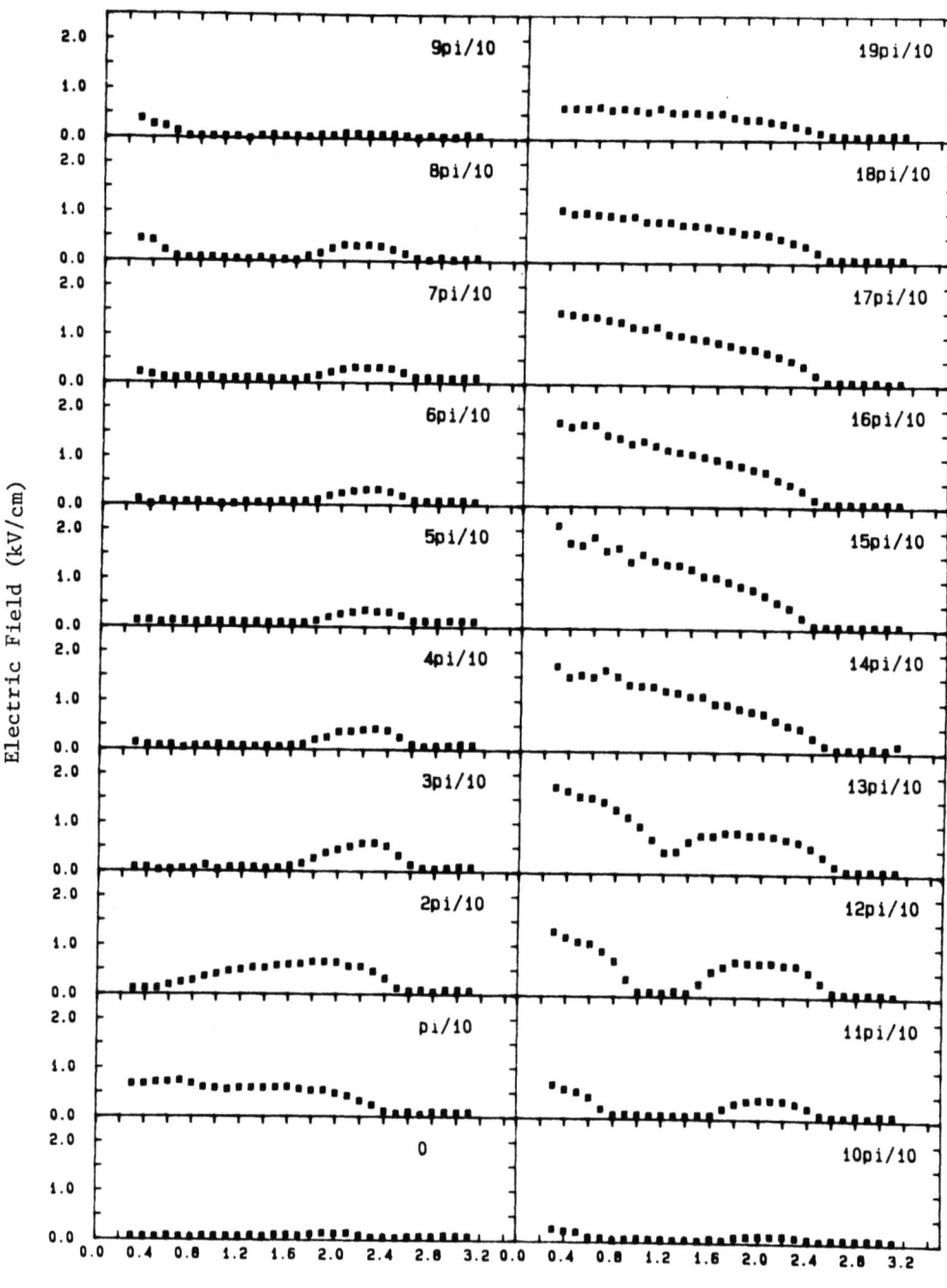

Fig. 12. Space-time profile of sheath fields in a 50 kHz discharge through BCl_3 at 0.3 Torr and 0.1 W cm^{-3}; one electrode is at 0 mm and the electrode gap is 16 mm. The times refer to the phase during a single rf cycle relative to sine wave excitation: for example, 5pi/10 (15pi/10) corresponds to the time when the applied voltage is a maximum (minimum).

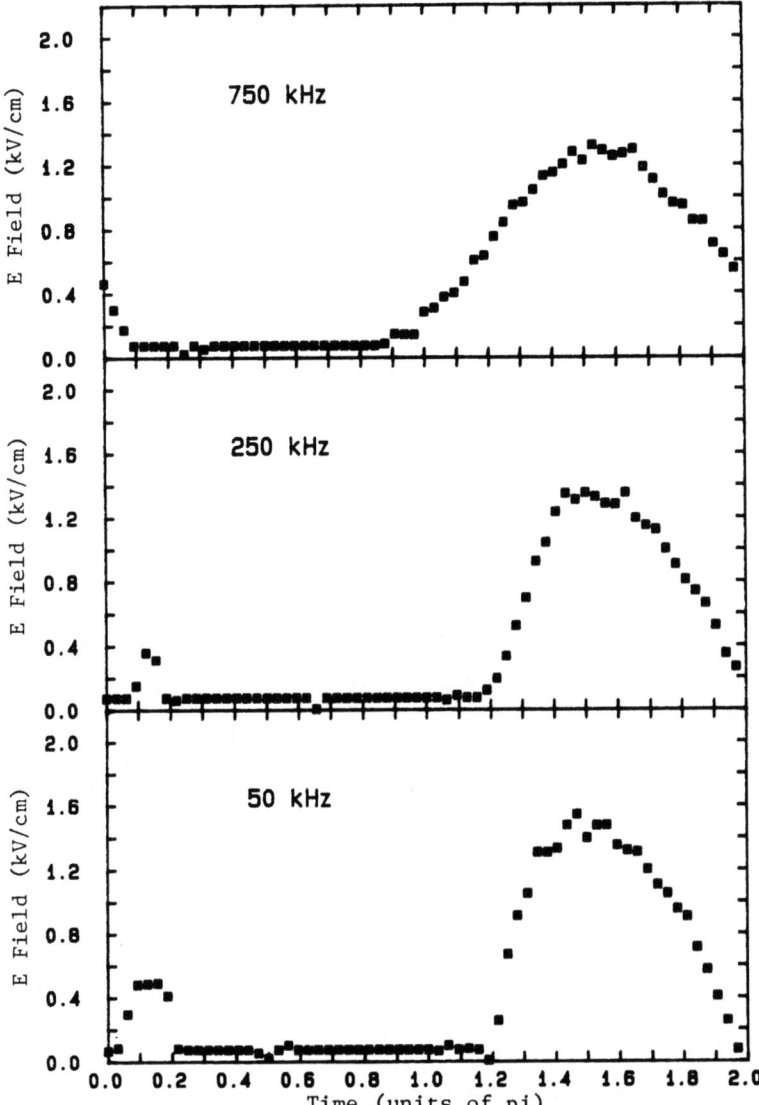

Fig. 13. Electric field amplitude as a function of time and frequency for discharges through 0.3 Torr of BCl_3. All other conditions are the same as in Figure 11. Note that the anion feature during the anodic part of the cycle disappears above 250 kHz because the electron energy distribution does not relax for a time sufficient for dissociative attachment to occur (see text).

negative ion formation: at higher frequency, the electron energy distribution cannot relax completely and the conversion of electrons to anions is incomplete.

The second method for measuring electric fields is more general than the method described above in that it relies upon excitation to high-lying Rydberg levels (Doughty et al., 1984; Doughty and Lawler, 1984;

Ganguly and Garscadden, in press). These levels correspond to quasi-one-electron atomic systems since the excited electron is in a very large orbital far from the ionic core. As a result, the orbital is very polarizable and easily perturbed by an external electric field. Large Stark shifts and broadening are readily observed by simply tuning a laser through the Stark-affected Rydberg spectrum and detecting transitions opto-galvanically. Rydberg atoms are easily ionized (either by collisions, absorption of black-body radiation, or tunneling to the continuum) thereby producing strong opto-galvanic signals. This technique is exceptionally sensitive and recent results indicate that a precision of 1 V/cm can be obtained (Ganguly and Garscadden, in press). This makes the method particularly well suited to the study of the local fields in the plasma body or the positive column of a dc glow. Since all atoms have Rydberg states, the method appears to be universal. Although this technique has been applied only to the study of dc glows, there appears to be no reason why the technique cannot be applied to rf discharges as well. Further discussion of this method is presented by Lawler and Garscadden, (this Proceedings).

ENERGY DISTRIBUTIONS

Few in situ measurements of energy distribution functions have been reported and this is one area where more information could be readily achieved. Rotational energy distributions have been measured using LIF, OGS, PIE, infrared, and microwave spectroscopy (Gottscho and Miller, 1984; Clark and DeLucia, 1981; Farrow and Richton, 1984; Knights et al., 1982; Oshima, 1978; Porter and Harshbarger, 1979; Donnelly et al., 1982; d'Agostino et al., 1981; Cramarosa, 1981). An example of a rotationally resolved LIF spectrum from which part of the rotational energy distribution of the CCl radical, formed in a 55-kHz discharge through 0.15 torr of CCl_4, was deduced is shown in Fig. 14 (Davis and Gottscho, 1983).

Generally, the distributions appear to be thermal because of rapid rotational-translational energy transfer. Using LIF, Schmidt et al., (1984) found the relaxation of the nascent SiH rotational distribution formed by electron impact in a low pressure SiH_4 plasma to occur in a few microseconds. Davis and Gottscho (1983) found the spatial dependence of the rotational temperature in CCl_4 and N_2 discharges to depend primarily on the electrode temperatures and the applied power density. Large thermal gradients were observed in the sheath regions when the surface temperature did not match the temperature in the plasma. Because the transfer of electronic translational energy to molecular rotational energy is relatively inefficient, emission spectroscopy is also a good way to measure ground-state rotational energy distributions when the excited state is created by electron impact (Davis and Gottscho, 1983).

Because vibrational relaxation is relatively slow, vibrational distributions are likely to be non-thermal with an excess of energy in higher lying vibrational levels. In accord with this expectation, all of the measurements made to date indicate that the average vibrational energy is higher than the average rotational or translational energies (Davis and Gottscho, 1983; Clark and DeLucia, 1981; d'Agostino, 1981; Cramarossa et al., 1981; Knights et al., 1982). It is likely that vibrationally hot molecules play an important role in plasma chemistry. For example, HF does not etch SiO_2 spontaneously and yet HF plasmas etch SiO_2 at a rate which is higher than what one might expect from the F atom concentration alone. It is possible that part of this etch rate enhancement is due to vibrationally excited HF in the plasma (Donnelly, 1985).

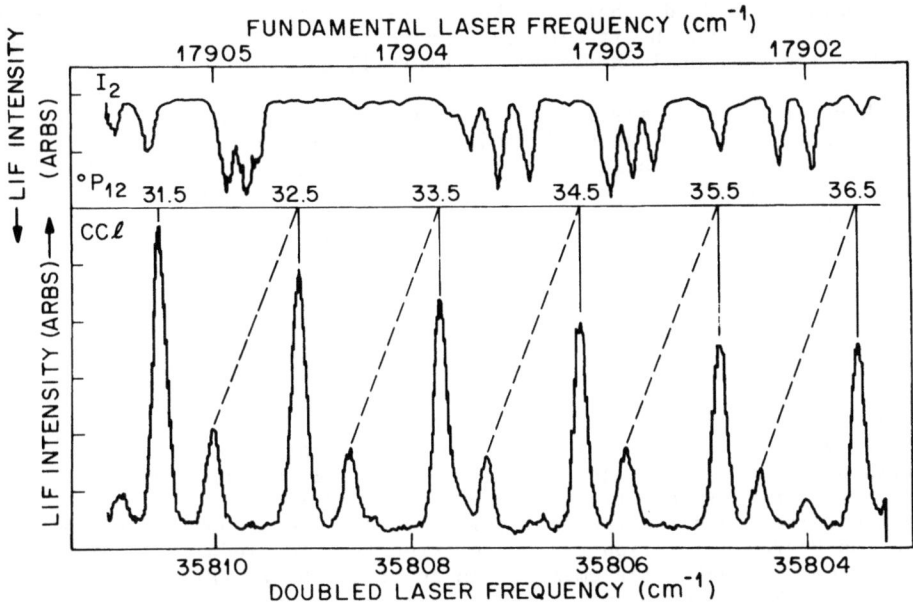

Fig. 14. High resolution (Δλ = 0.001 nm) excitation spectrum of the $^OP_{12}$ branch of the $A^2\Delta - X^2\Pi$ transition of the CCl radical, formed in a 55 kHz discharge through 0.15 Torr of CCl_4. The intensities of the individual rotational lines are used in determining the radical's rotational energy distribution, which in this case appears to be thermal with T = 600 K. The I_2 reference spectrum at the top is used for spectral calibration. Further detail can be found in Davis and Gottscho, 1983.

Translational energy distribution measurements are primarily of interest with respect to ions in the sheath region. The three-dimensional distribution function provides information on the acquisition of energy and momentum by the ions from the field and, therefore, about power dissipation and anisotropic ion transport. One way to obtain ground state ion energy distributions is to utilize the Doppler shift method. Due to the ion's motion the absorption frequency will be shifted proportionately to the ion velocity and the excitation frequency. Thus, by tuning the wavelength of a high resolution laser through the Doppler-limited line width and recording an optogalvanic or LIF signal, the ion velocity distribution along the laser propagation direction can be measured. This has been done in low pressure, magnetically confined plasmas by excitation of metastable Ne^+ and ground state Ba^+ ions (Hill et al., 1983; Stern, 1985). At higher pressures, the metastable ion density is likely to be too small, because of charge exchange reactions, for this to be a viable means for probing ion motion. Walkup et al. (1983) measured the translational energy distribution of N_2^+ in the sheath region of a dc glow discharge through N_2 using opto-galvanic spectroscopy. The average ion energy along the field direction was found to be small compared to the sheath voltage because of rapid charge exchange with cold neutrals. In a similar fashion, Doughty et al. (1985) found the translational temperature of Ne metastables to be significantly higher in the cathode fall region of a dc discharge. They attributed this heating to ion-atom charge exchange, which efficiently transforms ion energy gained from the field to translational energy in the neutralized atom.

Doppler-shifted emission and opto-galvanic measurements indicate that excited neutrals can be formed in the sheath regions of low-frequency and dc discharges as a result of charge exchange reactions, sputtering, and simultaneous recoil and neutralization of the ions when they impact the electrode surface (Benesch and Li, 1984; May et al., 1985; Cappelli et al., 1985). Such measurements permit an upper bound to be placed on the incident ion energy and, thus, the extent to which energy has been gained from the field.

KINETIC MEASUREMENTS

We have alluded to the rates of formation and relaxation in discussing many of the measurements above. These rates correspond to the collision terms on the right hand side of Eq. (1) averaged over the relevant energy distributions. Under some conditions, these rates can be measured directly using spectroscopic methods. For example, the rate of electron energy relaxation can be inferred from the modulation of short-lived atomic or molecular emission in an rf discharge (Flamm and Donnelly, 1982; Rosny et al., 1983; Flamm and Donnelly, 1985; Gottscho et al., 1984) (Figs. 2 and 5). Regardless of the excited state formation mechanism, the time-dependent emission intensity from a short-lived state provides a measure of the formation rate of that state. If the formation mechanism is electron-impact excitation, the time dependence provides further information about the electron density and energy distribution. Thus, if electron-impact induced emission is 100% modulated, electron energy relaxation must be fast compared to the modulation frequency. As the discharge frequency is increased, eventually the emission will be less than 100% modulated because of incomplete energy relaxation. However, this method is only useful for providing information about the relaxation of electrons above the excitation threshold.

A similar method using LIF has been employed to measure ground state ion and radical lifetimes (Gottscho et al., 1984; Gottscho and Mandich, 1985). Synchronous triggering of a laser and detection electronics with the applied voltage permits the concentrations of ions and radicals to be monitored as a function of time during a single rf cycle. The extent of modulation and the phase shift in the concentration waveform provide a measure of the ground state loss rate. An example is shown in Fig. 15. The ground state concentrations of N_2^+ (Fig. 15b) and Cl_2^+ (Fig. 15c) in pure N_2 and Cl_2 discharges, respectively, are only slightly modulated at an applied frequency of 55 kHz. The lifetimes are on the order of 30 to 40 µs and are governed primarily by diffusion. As Cl_2 is added to the N_2 feedstock, however, the extent of modulation in the N_2^+ ground state concentration increases while the phase shift decreases, indicating a shortening of the lifetime to ~ 1.5 µs (Fig. 15d). In contrast, the Cl_2^+ concentration modulation vanishes, indicating that an additional mechanism for the formation of Cl_2^+ has been created (Fig. 15e). Combined, these observations led to the conclusion that N_2^+ rapidly charge exchanges with Cl_2 to form Cl_2^+ (Gottscho et al., 1984).

An alternative means for measuring ground state lifetimes is to pulse the discharge and monitor the decay of the LIF signal after the discharge is turned off (Gottscho and Mandich, 1985). This has been done for BCl radicals formed in a pulsed rf discharge at 13.56 MHz through

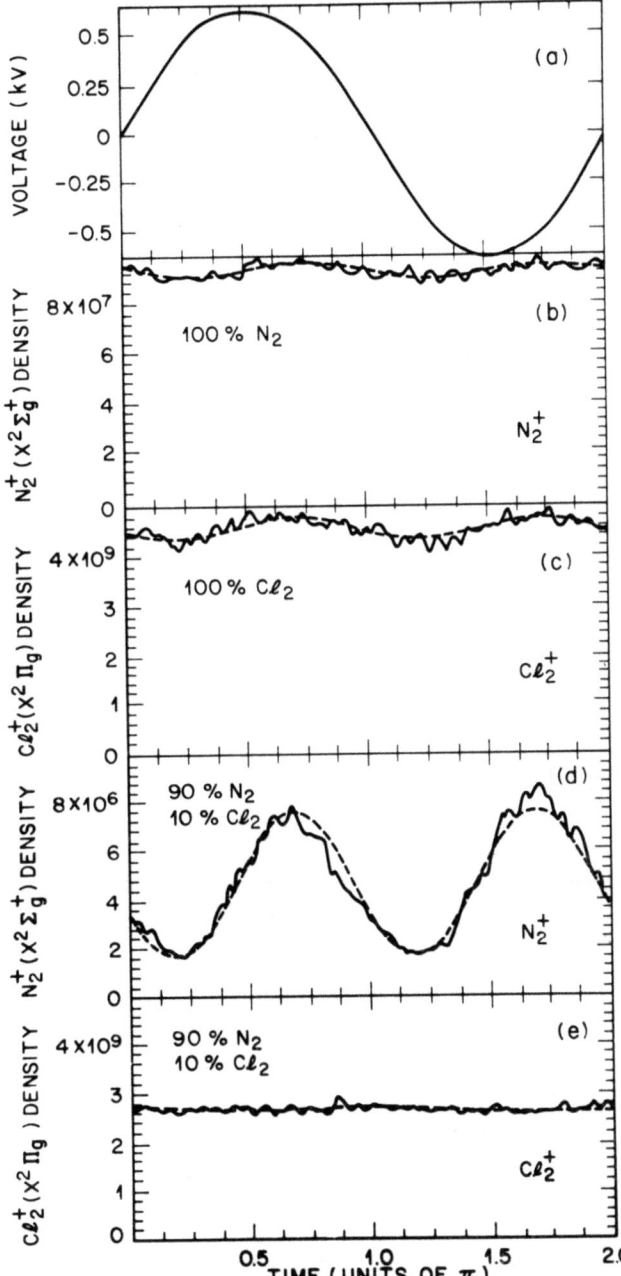

Fig. 15. Time dependent voltage and ground state ion densities in the center of discharges through N_2, Cl_2, and a mixture of N_2 and Cl_2 at a total pressure of 0.3 Torr, operating frequency of 55 kHz, and power density of 0.6 W cm^{-3}. (a) Voltage waveform from a discharge through pure N_2. (b) N_2^+ ground state density measured by LIF. (c) Cl_2^+ ground state density measured by LIF. (d) N_2^+ ground state density in a discharge through 10% Cl_2 in N_2. (e) Same as (d) except Cl_2^+ density.

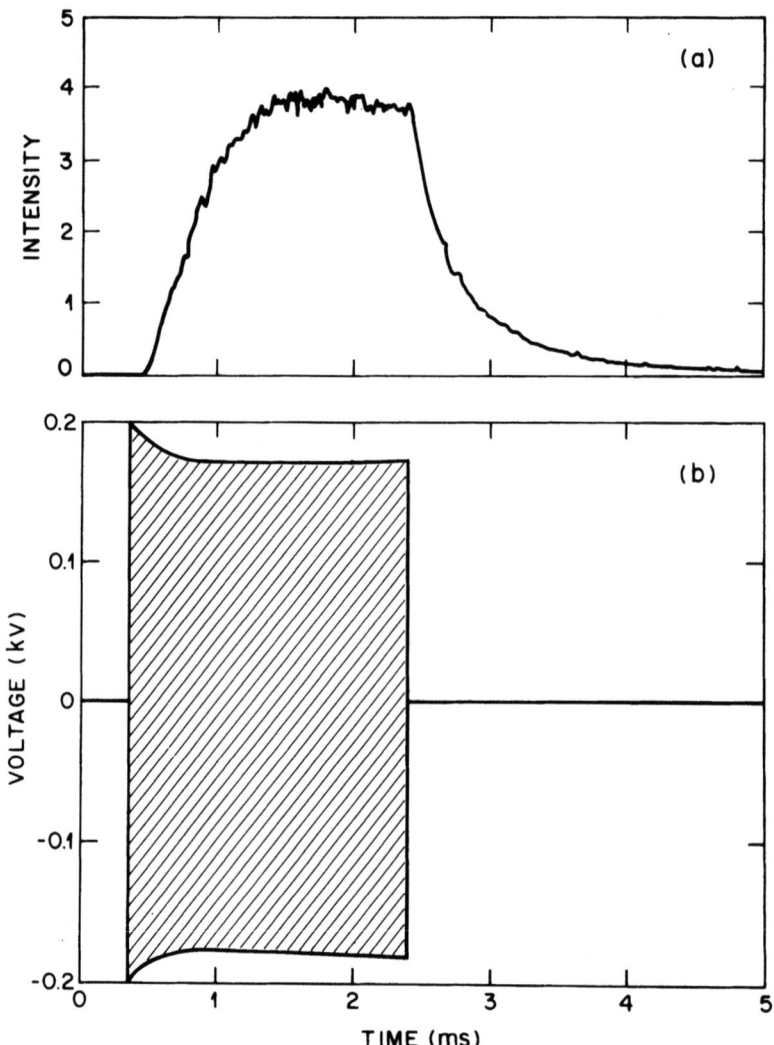

Fig. 16. (a) BCl ground state concentration, determined by LIF spectroscopy, and (b) rf voltage amplitude envelope as a function of time for a pulsed 13.56 MHz discharge through 0.3 Torr of BCl_3. The decay in the BCl density is used as a measure of the loss rates of the radical in the discharge (Gottscho and Mandich, 1985).

BCl_3 (Fig. 16). At a pressure of 0.3 torr, a biexponential decay, with time constants of 0.33 and 1.6 ms, was observed in the BCl concentration. This implies that the BCl decay mechanism is greater than first order. A similar experiment has been done recently by Greenberg and Verdeyen (1985), who used emission actinometry to monitor relative changes in F atom and N_2 molecule densities (Ne actinometer) in a pulsed dc discharge through NF_3. From the time-dependent emission intensities, they concluded that: (a) NF_3 is dissociated primarily by electron impact and not by dissociative attachment; (b) F atom emission arises from electron impact on ground state F atoms; and (c) N_2 is formed by the bimolecular reaction of NF radicals.

SUMMARY

We have reviewed recent progress in the development and application of spectroscopic methods to the study of rf discharge physics and chemistry with an emphasis on electronegative systems and the effects of anions on discharge properties. For the measurement of ion concentrations, radical concentrations, internal and translational energy distributions, local electric field amplitudes, and kinetic rates of formation and loss, the methods of laser-induced fluorescence, optogalvanic spectroscopy, and emission spectroscopy have all been found to be useful, particularly when used concurrently. It is anticipated that Raman and infrared spectroscopy will find greater application to the study of rf discharges in the near future.

ACKNOWLEDGMENTS

Much of the work described above was done over a number of years in collaboration with R. H. Burton, A. L. Cappelli, G. P. Davis, L. F. DiMauro, V. M. Donnelly, D. L. Flamm, M. L. Mandich, T. A. Miller, and C. A. Moore. We are most grateful for having benefitted from their insights and expertise.

REFERENCES

Alexander, M., J. Chem. Phys. (in press).
Bacal, M., Hamilton, G. W., Bruneteau, A. M., Doucet, H. J., and Taillet, J., 1979, J. de Physique, 40:C7-791.
Benesch, W., and Li, E., 1984, Optics Letters, 9:338.
Blades, M. W., and Horlick, G., 1980, Appl. Spectrosc, 34:696.
Bruce, R. H., 1981a, Solid State Technol., 24:64.
Bruce, R. H., 1981b, J. Appl. Phys., 53:7064.
Choi, B. S., and Kim, H., 1982, Appl. Spectrosc., 71:71.
Cappelli, A. L., Gottscho, R. A., and Miller, T. A., 1985, submitted to Plasma Chem. and Plasma Proc..
Chen, F. F., "Introduction to Plasma Physics", 1974, Plenum, New York.
Christophorou, L. G., 1980, Environ. Health Perspect., 36:3.
Clark III, W. W., and DeLucia, F. C., 1981, J. Chem. Phys., 74:3139.
Coburn, J. W., and Chen, M., 1980, J. Appl. Phys., 51:3134.
Davis, G. P., and Gottscho, R. A., 1983, J. APpl. Phys., 54:3080.
d'Agostino, R., Colaprico. V, and Cramarossa, F., 1981, Plasma Chem. Plasma Proc., 1:365.
d'Agostino, R., Cramarossa, F., DeBenedictis, S., and Ferraro, G., 1981, J. Appl. Phys., 52:1259.
d'Agostino, R., Cramarossa, F., and DeBenedictis, S., 1982, Plasma Chem. Plasma Proc., 2:213.
d'Agostino, R., Cramarossa, F., Colaprico, V., and d'Ettole, R., 1983, J. Appl. Phys., 54:1284.

d'Agostino, R., Cramarossa, F., DeBenedictis, S., and Fracassi, F., 1984, Plasma Chem. Plasma Proc., 4:163.
DiMauro, F. F., Gottscho, R. A., and Miller, T. A., 1984, J. Appl. Phys., 56:2007.
Donnelly, V. M., Flamm, D. L., and Collins, G., 1982, J. Vac. Sci. Technol., 21:817.
Donnelly, V. M., Flamm, D. L., Dautremont-Smith, W. C., and Werder, D. J., 1984, J. Appl. Phys., 55:242-52.
Donnelly, V. M., private communication.
Doughty, D. K., Den Hartog, E. A., and Lawler, J. E., 1985, Appl. Phys. Lett., 46:352.
Doughty, D. K., and Lawler, J. E., 1984, Appl. Phys. Lett., 45:611.
Doughty, D. K., Salih, S., and J. E. Lawler, J. E., 1984, Phys. Lett., 103A:41.
Emeleus, K. G., Woolsey, G. A., 1970, "Discharges in Electronegative Gases", Barnes and Noble, Inc., New York.
Farrow, L. A., and Richton, R. E., unpublished results.
Flamm, D. L., Gilliland, E. R., and Baddour, R. F., 1973, Ind. Eng. Chem. Fundam., 12:277.
Flamm, D. L., and Donnelly, V. M., 1982, Bull. Am. Phys. Soc., 27:97.
Flamm, D. L., and Donnelly, V. M., 1985 J. Appl. Phys. (submitted).
Gaebe, C. E., and Gottscho, R. A., 1985 in preparation.
Ganguly, B. N., and Garscadden, A., 1985 Appl. Phys. Lett. (in press).
Garscadden, A., in "Gaseous Electronics Vol. 1", 1978, N. Hirsh and H. Oskam, eds., Academic Press, New York, p. 65.
Gottscho, R. A., Burton, R. H., Flamm, D. L., Donnelly, V. M., and Davis, G. P., 1984, J. Appl. Phys., 55:2707.
Gottscho, R. A., Davis, G. P., and Burton, R. H., 1983, Plasma Chem. Plasma Proc., 3:193, 1983, J. Vac. Sci. Technol, 1:622.
Gottscho, R. A., and Donnelly, V. M., 1984, J. Appl. Phys., 56:245.
Gottscho, R. A., and Gaebe, C. E., 1985 (in preparation).
Gottscho, R. A., and Miller, T. A., 1984, Pure and Appl. Chem., 56:189.
Gottscho, R. A., and Mandich, M. L., 1985, J. Vac. Sci. Technol., A 3:617.
Gottscho, R. A., 1985; Proc. Mat. Res. Soc. Symp. Proc., 38:55.
Greenberg, K. E., Hebner, G. A., and Verdeyen, J. T., 1984, Appl. Phys. Lett., 44:299.
Greenberg, K. E., and Verdeyen, J. T., 1985, J. Appl. Phys., 57:1596.
Gottscho, R. A., Smolinsky, G., and Burton, R. H., 1982, J. Appl. Phys., 53:5908.
Gudeman, C. S., and Saykally, R. J., 1984, Annu. Rev. Phys. Chem., 35:387.
Hargis, P. J., Kushner, M. J., 1982, Appl. Phys. Lett., 40:779.
Hill, D. N., Fornaca, S., and Wickham, M. G., 1983, Rev. Sci. Instrum., 54:309.
Huq, M. S., Scott, D., White, N. R., Champion, R. L., and Doverspike, L. D., 1984, J. Chem. Phys., 80:3651.
Ibbotson, D. E., Flamm, D. L. and Donnelly, V. M., 1983, J. Appl. Phys. 54:5974.
Kinsey, J. L., 1977, Annu. Rev. Phys. Chem., 28:349.
Klein, R., McGinnis, R. P., and Leone, S. R., 1983, Chem. Phys. Lett., 100:475.
Knights, J. C., Schmidt, J. P. M., Perrin, J., and Guelachvili, G., 1982, J. Chem. Phys., 76:3414.
Kurepa, M. V., Belic, D. S., 1978 J. Physics B, 11:3719.
Kushner, M. J., 1983, J. Appl. Phys. 54:4958.
Kushner, M. J., Anderson, H. M., and Hargis, P. J., 1985, Proc. Mat. Res. Soc. Symp. Proc., 38:201.
Langmuir, I., 1929, Phys. Rev. 33:954.
Mandich, M. L., Gaebe, C. E., Gottscho, R. A., 1985 J. Chem. Phys., in press.
Massey, H. S. W., "Negative Ions," 1976, Cambridge University Press.
May, R. D., 1985, Appl. Phys. Lett. 46:938.

Miller, T. A., Bondybey, V. E., "Molecular Ions: Spectroscopy, Structure, Chemistry, 1983, North-Holland Amsterdam.
Moore, C. A., Davis, G. P., Gottscho, R. A., 1984, Phys. Rev. Lett. 52:538.
Oshima, M., 1978, Jpn. J. Appl. Phys. 17:1157.
Pang, S., Brueck, S. R., 1983, in: Laser Diagnostics and Photochemical Processing for Semiconductor Devices," edited by R. M. Osgood, S. R. J. Brueck, and H. R. Schlossberg, North Holland, New York.
Porter, R. A., and Harshbarger, W. R., 1979, J. Electrochem. Soc. 126:460.
Rosny, G., Mosburg Jr., E. R., Abelson, J. R., Devaud, G., and Kerns, R. C., 1983, J. Appl. Phys. 54:2272.
Sadeghi, N., and Derouard, J., 1985 unpublished results.
Schmitt, J. P. M., Gressier, P., Krishnan, M., De Rosny, G., and Perrin, J., 1984, Chem. Phys. 84:281.
Stockdale, J. A., Nelson, D. R., Davis, F. J., and Compton, R. N., 1972, J. Chem. Phys. 56:3336.
Selwyn, G. S., Saenger, K., and Walkup, R. E., 1984, Abstract F6.6, Materials Research Society Meeting, Boston, MA.
Stern, R., 1985 private communication.
Suzuki, T., 1981, Opt. Comm. 38:364.
An example of the latter is the recent report of metastable β-tungsten deposition from an rf discharge through WF_6. C. C. Tang, and D. W. Hess, 1984, Appl. Phys. Lett. 45:633.
Taillet, J., Legendre, R., 1969, C.R. Acad. Sc. Paris 269:52.
Tiller, H.-J., Berg, D., and Mohr, R., 1981, Plasma Chem. Plasma Proc. 1:247.
Thompson, J. B., 1959, Proc. Phys. Soc. 73:818.
Walkup, R., Avouris, Ph. Dreyfus, R. W., Jasinski, J. M., and Selwyn, G. S., 1984, Appl. Phys. Lett. 45:372.
Warren, R., 1955, Phys. Rev. 98:1658.
Walkup, R., Dreyfuss, R., and Avouris, Ph., 1983, Phys. Rev. Lett. 50:1856.
Webster, C. R., McDermid, I. S., and Rettner, C. t., 1983, J. Chem. Phys., 78:646.
Woods, R. C., Faraday Discuss. Chem. Soc. 71, 57 (1981).
Woods, R. C., 1983, in: "Molecular Ions: Spectroscopy, Structure, Chemistry, edited by T. A. Miller and V. E. Bondybey, North-Holland, Amsterdam.
Wormhoudt, J., Stanton, A. C., and Silver, J., 1983, in: "Proceedings of SPIE": 452 (October).
Walkup, R., Saenger, K., Selwyn, G. S., 1985, Mat. Res. Soc. Symp. Proc., 38:69.
Yamada, C., Nagai, K., and Hirota, E., 1981, J. Mol. Spectrosc. 85:416.
Zare, R. N., and Dagdigian, P. J., 1974, Science 185:739.

OPTOGALVANIC EFFECTS IN THE CATHODE FALL

J. E. Lawler, D. K. Doughty, E. A. Den Hartog, and S. Salih

Department of Physics
University of Wisconsin
Madison, WI, USA

INTRODUCTION

Optogalvanic effects are changes in the conductance of a gas discharge caused by illumination with radiation at a wavelength corresponding to an atomic or molecular transition. These phenomena were first discovered by Penning over fifty years ago, and were rediscovered a number of times in the intervening decades (Penning, 1928). Early work on optogalvanic effects was performed using incoherent light sources (Penning, 1928: Kenty, 1950: Meissner and Miller, 1953). The development of tunable dye lasers opened many new possibilities. Optogalvanic effects became widely used as a detection method in laser spectroscopy (Camus, 1983). Many unstable or difficult to produce species such as: free radicals, atoms of refractory elements, metastables, and atoms or ions in short lived levels are readily available in a discharge. Even high resolution Doppler-free spectroscopy is performed using optogalvanic detection (Lawler et al., 1979: Goldsmith et al., 1979).

A qualitative explanation of optogalvanic effects is fairly straightforward. The absorption of laser radiation in the discharge results in a change in the steady state population of bound atomic or molecular levels. Different levels will generally have unequal ionization rates or ionization probabilities. Hence the ionization balance of the discharge is perturbed. A perturbation to the ionization balance leads to a shift in the V-I characteristics of the discharge and a change in the current through the discharge or, equivalently, a change in the voltage across the discharge. Although a qualitative explanation is straightforward a detailed quantitative description of these effects is much more difficult. Quantitative treatments generally involve applying perturbation theory to a model of the unperturbed discharge (Doughty and Lawler, 1983a). The resulting model of an optogalvanic effect is only as good as the model of the unperturbed discharge.

Optogalvanic effects, besides serving as a detection method in laser spectroscopy, are enormously useful as a discharge diagnostic. Diagnostics based on optogalvanic effects are typically several orders of magnitude more sensitive than absorption spectroscopy. The effects make it possible to detect transitions which do not lead to detectable absorption or fluorescence. Optogalvanic diagnostics provide information in two ways: (1) the spectral features such as the Doppler width or Stark profile provide information on the discharge environment, and (2) the absolute magnitude of the effect can be compared to discharge models.

Examples of both types of optogalvanic experiments are discussed in this article.

Optogalvanic effects are particularly well suited to studying the cathode fall region of diffuse discharges. This is because the effects are amplified in the cathode fall (Doughty and Lawler, 1983b). An increase (or decrease) in free electron production at or near the cathode results in an avalanche (or the loss of an avalanche). Optogalvanic detection in the cathode fall is nearly five orders of magnitude more sensitive than absorption detection. There is a growing interest in developing a detailed theoretical understanding of the cathode fall region. This region is the most important, and yet least understood region of diffuse discharges. The cathode fall is critical in chemical vapor deposition and ion etching of semiconductors, in diffuse discharge switches, and in a variety of coherent and incoherent light sources. The reason it is so difficult to model the cathode fall is that standard discharge approximations based on hydrodynamic equilibrium fail in the cathode fall. The electron energy distribution function is not in hydrodynamic equilibrium with the local E/N (electric field divided by ground state atom density). The failure of the electron distribution function to be in hydrodynamic equilibrium is caused by the large and rapidly changing E/N and by the proximity of the cathode. Electron swarm data including drift velocities, Townsend coefficients, and excitation coefficients determined in hydrodynamic equilibrium are thus not directly applicable to the cathode fall. There is a persistent hope that it will be possible to parameterize the departure from hydrodynamic equilibrium in a fairly general fashion. Four years ago, at an earlier NATO ASI, Segur et al. (1983) reviewed their own work and the work of others on this problem (Segur et al., 1983). Optogalvanic diagnostics are the experimental tools needed to attack this important problem. Optogalvanic diagnostics are used to map various parameters in the cathode fall including: the electric field, the absolute gas temperature (density), the metastable density, the ion density, the fractions of discharge current carried by ions and electrons, and other parameters. A carefully chosen optogalvanic effect is being used to directly observe the size of the electron avalanche produced by releasing a calibrated quantity of electrons at some point in the cathode fall. These optogalvanic experiments in the cathode fall are described in this article. New theoretical approaches and optogalvanic diagnostics will ultimately lead to a more quantitative understanding of the cathode fall.

ELECTRIC FIELD MEASUREMENTS USING OPTOGALVANIC DETECTION OF RYDBERG ATOMS

The single most important parameter to be mapped in the cathode fall is the electric field. The electric field in the cathode fall is a non-uniform space charge field, and thus cannot be determined by bulk voltage measurements.

Our optogalvanic electric field measurements are based on linear Stark effects in Rydberg atoms. Rydberg atoms are highly excited atoms in a level near the continuum which has a large principal quantum number n. Every atom and molecule has Rydberg levels, thus the diagnostics are broadly applicable. The unperturbed energy of such a level is given by the Rydberg formula,

$$E_n = -R/(n-\delta_\ell)^2, \qquad (1)$$

where R is the Rydberg constant and δ_ℓ is a quantum defect. The quantum defect is independent of n but strongly dependent on ℓ, the orbital angular momentum. The quantum defect is a measure of the influence of the core electrons on the weakly bound valence electron. Electrons in

Rydberg levels with a large angular momentum do not penetrate to the core, thus the quantum defect vanishes for large . Rydberg atoms are extremely susceptible to the Stark effect because of their large size and small energy gap between levels of opposite parity. The "size" of a Rydberg atom scales as $n^2 a_o$, where a_o is the Bohr radius. The energy difference between levels of opposite parity scales as $2\delta_\ell R/n^3$. Very small electric fields can be observed by using a level with a large n.

Rydberg atoms are powerful gas discharge probes because they are easily perturbed and because the perturbations are straightforward to interpret. The calculation of a linear Stark effect in a Rydberg atom is essentially a Stark calculation for hydrogen with small corrections which are parameterized in terms of quantum defects. The calculation is performed independently for each value of "m", the angular momentum along the field axis. Electronic orbital angular momentum along the field axis is conserved, therefore m is a good quantum number for all field strengths. The total angular momentum ℓ is a good quantum number only for vanishingly small fields. The Stark effect in a Rydberg atom is calculated by diagonalizing an (n-m) by (n-m) Hamiltonian matrix (Zimmerman et al.,1979). The usual basis set is made of eigenfunctions of the unperturbed Hamiltonian. The diagonal matrix elements are the unperturbed energies of the levels as given by the Rydberg formula. Quantum defects for Rydberg series of many atoms can be deduced from C. E. Moore's tables of energy levels (Moore, 1971). The off diagonal matrix elements vanish except those connecting states differing in angular momentum "ℓ" by one unit. The off diagonal matrix elements are the product of the electric field and dipole matrix elements. The angular part of a dipole matrix element is an integral of spherical harmonics,

$$\int Y^*_{\ell,m} \cos\theta Y_{\ell-1,m} d\Omega = \sqrt{\frac{\ell^2-m^2}{(2\ell+1)(2\ell-1)}} . \tag{2}$$

The radial part is approximately determined from the radial wavefunctions of the hydrogen atom,

$$\int R_{n,\ell} er R_{n,\ell-1} r^2 dr = \frac{3}{2} a_o en \sqrt{n^2-\ell^2} . \tag{3}$$

A correction factor of order one has been tabulated as a function of quantum defects (Edmonds et al.,1979). The procedure described above is used to generate the Stark map of Fig. 1 for the n=11, m=0 singlet He states. Figure 1(a) is a plot of the energy of the various m=0 states as a function of electric field. The energies are the eigenvalues of the Hamiltonian matrix. The linear Stark effect is evident as a linear increase with field in the width of the Stark manifold. The eigenvectors of the Hamiltonian matrix also provide useful information. Suppose a transition from a low lying non-Rydberg s state to a Rydberg p state is driven. The amount of p state wavefunction mixed with each Stark component is proportional to the strength of the transition to that component. Relative oscillator strengths for selected components are shown in Fig. 1(b).

The visible and ultraviolet transitions connecting Rydberg levels to low lying levels are rather weak; the oscillator strengths scale as $1/n^3$. The Rydberg levels are excited from low lying levels using intense dye laser pulses. Rydberg atoms do not survive sufficiently long in a discharge to fluoresce in the visible or ultraviolet. They are collisionally ionized long before they radiate. Simple absorption spectroscopy is not usually sufficiently sensitive to detect the weak absorption features from transitions to Rydberg levels. Optogalvanic detection is uniquely suited to this task. Optogalvanic spectroscopy in a typical discharge is

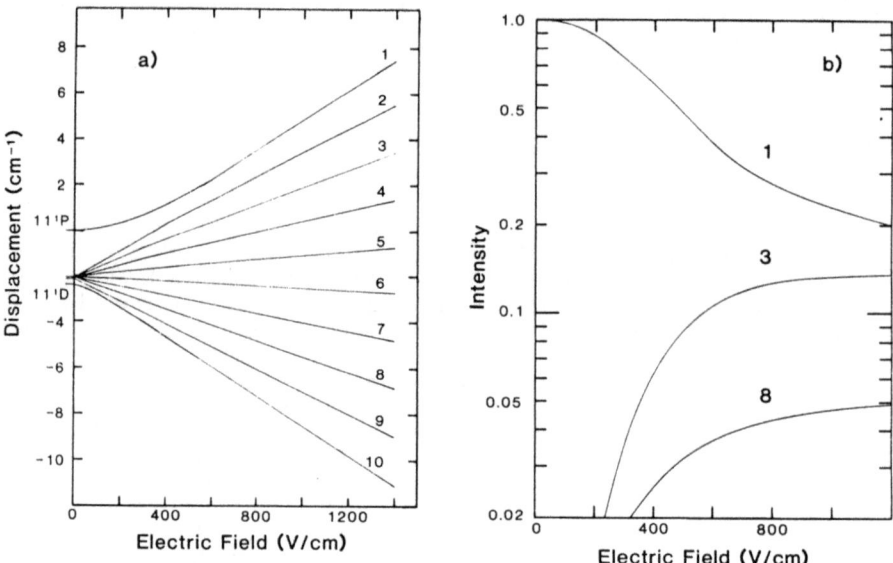

Fig. 1. (a) Theoretical Stark map for the n=11 singlet m=0 levels of He.
(b) Theoretical intensities of selected Stark components as a function of electric field.

two or three orders of magnitude more sensitive than simple absorption spectroscopy: optogalvanic spectroscopy in the cathode fall region is five orders of magnitude more sensitive than simple absorption spectroscopy.

Figure 2 is a schematic of the experiment with which we first demonstrated field measurements using optogalvanic detection of Rydberg atoms (Doughty and Lawler, 1984a). The discharge is formed between 1.5 cm diameter flat circular Al electrodes separated by 0.4 cm. The electrodes are enclosed in a much larger vacuum system which is constructed almost entirely of glass and metal. Most of the large seals are Varian Conflat flanges with Cu gaskets. The only exceptions are the high vacuum expoxy seals around the fused silica Brewster windows. The discharge system is connected to a diffusion pump station and ultra high purity He slowly flows through the system while the discharge is operated. The He pressure is 2.0 torr and the discharge current is 0.9 mA in this investigation. The discharge cell is on a precision translation stage so that spatial variations are observed without disturbing the laser alignment. A boxcar averager is used to detect the optogalvanic signals.

A N_2 laser pumped dye laser is used in our work. These dye laser systems are relatively inexpensive and widely available. The dye laser output is a 3 ns duration pulse with a peak power of 20 kW in a bandwidth of 0.2 cm^{-1}. A KDP (potassium dihydrogen phosphate) crystal frequency doubler is used to generate tunable ultraviolet. The dye laser is pressure tuned over ranges as wide as 30 cm^{-1}. The frequency scan is calibrated with a reference etalon. The dye laser second harmonic is used in this experiment, it is polarized perpendicular to the discharge electrodes. The polarization excites only the $\Delta m=0$ transitions. Only m=0 Rydberg states are excited because the lower level is an S state. The second harmonic is focused with a cylindrical lens to a strip 0.01 cm wide and 1 cm long parallel to the surface of the electrodes.

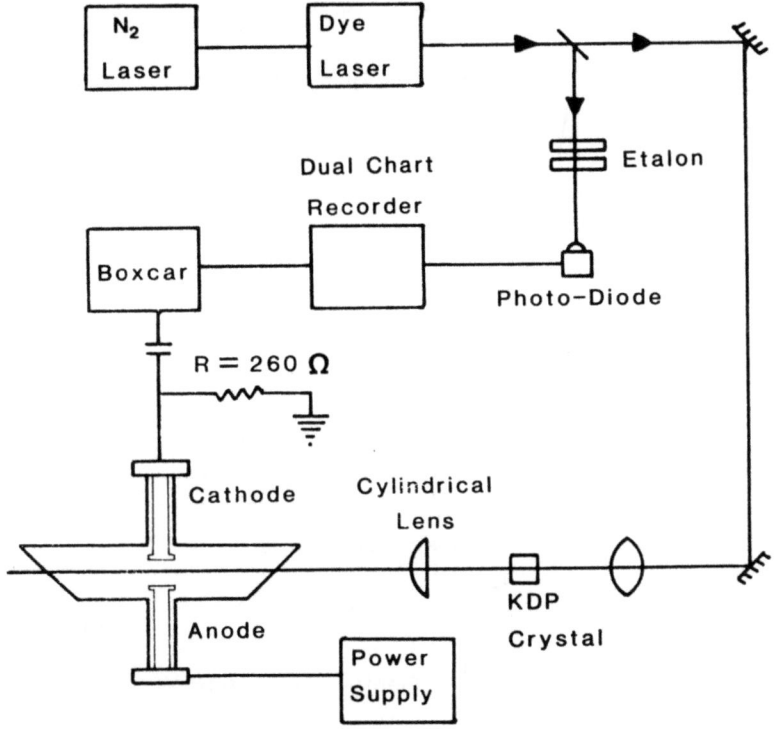

Fig. 2. Schematic of the electric field experiment in He.

The choice of transition is fairly important in the Rydberg atom experiments. A good signal-to-noise ratio is achieved by driving a transition from a heavily populated metastable level such as the 2^1S in He. The principal quantum number of the upper level is chosen in a trade off between oscillator strength (proportional to $1/n^3$) and the magnitude of the Stark effect (proportional to n^2). We found that the He $2^1S \to 11^1P$ transition at 321 nm was near optimum in our experiment. A lower principal quantum number is better for a higher pressure experiment because of the higher fields.

Stark spectra observed at two distances from the cathode are shown in Fig. 3. Ten Stark components are clearly visible in Fig. 3(a) which is recorded close to the cathode surface. The eleventh Stark component is the 11^1S state which has a large quantum defect, it is at -23 cm^{-1} and is not shown in the scan. The Stark spectrum of Fig. 3(b) is observed further from the cathode, and is contracted because the field is lower. The Stark component on the extreme left of the figures connects to the unperturbed 11^1P level at zero field. This component has the largest fraction of the available oscillator strength at small fields as shown in the calculation of Fig. 1(b).

The local electric field in the discharge is determined by comparing the width of the Stark pattern or the separation of individual components to the theoretical Stark map of Fig. 1(a). The field is also determined by comparing the relative linestrength of components or groups of components to the calculations shown in Fig.1(b). Two conditions must be

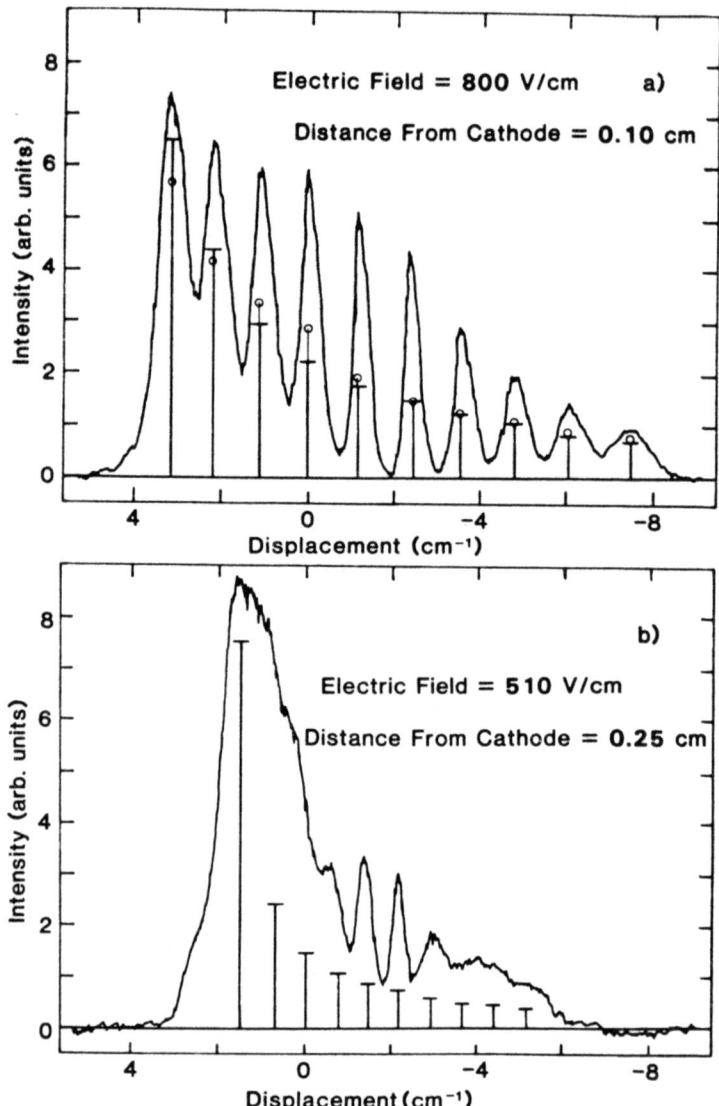

Fig. 3. (a) Stark spectrum observed 0.10 cm from the cathode in a He discharge. The calculated position of the ten Stark components at 800 V/cm are indicated by vertical lines. The height of the horizontal cross bar represents the calculated intensity. The height of the open circles is an experimental intensity.
(b) Stark spectrum observed 0.25 cm from the cathode.

satisfied in order to determine an electric field from a relative linestrength: (1) the transitions must not be saturated by the laser and the (2) optogalvanic effect must be equally efficient on each component. The first condition is experimentally verified by attenuating the laser. The second requirement is generally satisfied by rapid collisional mixing of Rydberg levels (Gallagher et al., 1977). The cross section for mixing among n=11 states in collisions with ground state He atoms in 1850Å^2. The mixing rate at 2.0 torr is 2×10^9 s^{-1} which is far larger than the radiative decay rate of the 11^1P level of 1.4×10^7 s^{-1} (Larsson et al., 1983). The comparison of linestrengths of Stark components is based on the integrated linestrength, not on the "height" of signal. It is essential to integrate the line profile because the linewidth varies across the Stark pattern. The variation is caused by sampling a finite volume containing an inhomogeneous field. The positions of the outer components are strongly dependent on field while those of the inner components are weakly dependent on field. The linewidth of an outer component provides a measure of the spatial derivative of the field. Electric field measurements in an obstructed glow discharge in He are shown in Fig. 4. The significance of the electric field maps is discussed later in this article.

The electric field diagnostic described in the preceeding paragraphs has many advantages over traditional techniques for measuring space charge fields. Traditional techniques are based on electron beam deflection (Warren, 1955). Optogalvanic techniques have better spatial and temporal resolution, they are applicable to higher pressure and higher current discharges, and they are simpler to use, especially in high purity discharges. One disadvantage of the optogalvanic technique described in the preceeding paragraphs is that the electric field measurement is an average along the laser beam path through the discharge. The disadvantage is overcome by using two-step optogalvanic effects produced with intersecting laser beams (Doughty et al., 1984b). An improved version of the electric field experiment is described in the following paragraphs.

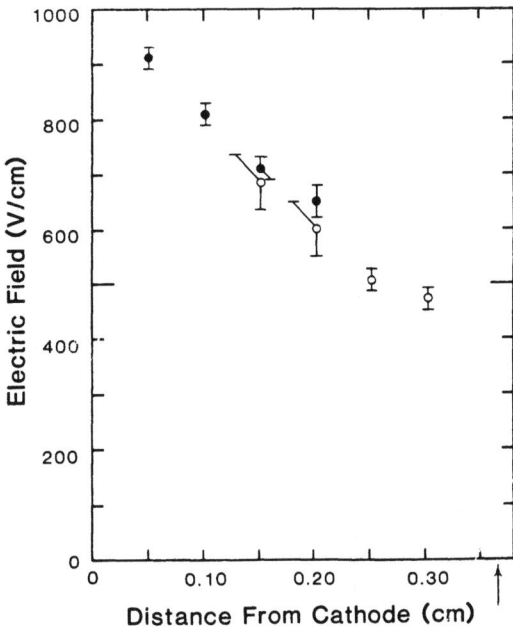

Fig. 4. Electric field as a function of distance from the cathode in a He discharge. Field determined from splittings are represented by solid dots. Fields determined from relative intensities of groups of components are represented by open circles. The arrow indicates the position of the anode.

A schematic of the improved experiment is shown in Fig. 5. It is similar to the schematic of Fig. 2 except that the frequency doubler is eliminated and a second dye laser is added. Both dye lasers are pumped by the same pulsed N_2 laser. Each dye laser beam passes through an aperture 0.5 cm x 0.05 cm. These apertures are then imaged in the discharge via the lenses with a magnification of 0.5. Both beams are normal to each other as well as to the discharge electric field. The resulting volume element containing both laser beams is 0.25 cm x 0.25 cm x 0.02 cm. The critical dimension of 0.02 cm corresponds to the edge of the focal volume which is normal to the surface of the cathode. At the intersection both laser beams have polarization parallel to the discharge electric field.

The discharge for the two-step experiment is in Ne because of the more convenient energy level structure. A good signal-to-noise achieved by exciting Ne Rydberg levels from the heavily populated $2p^53s\ ^3P_2$ level. Signal-to-noise considerations are especially important in this "pinpoint" diagnostic because only a small volume in the discharge is probed.

The first laser is tuned to 588.2 nm in order to excite Ne atoms from the $2p^53s\ ^3P_2$ level to the $2p^53p\ ^3P_1$ level. The second laser, which is delayed by a few ns, is tuned to 439.3 nm in order to excite Ne atoms from the $2p^53p\ ^3P_1$ to the $2p^5(^2P_{1/2})11d'$ Stark manifold. The primed notation refers to the Ne ion core coupling of $^2P_{1/2}$. The use of Ne Rydberg levels with $j = 1/2$ core insures fine structure splittings are small ($< cm^{-1}$). Figure 6 is a partial Grotrian diagram for Ne with the transitions used in the two-step excitation clearly identified.

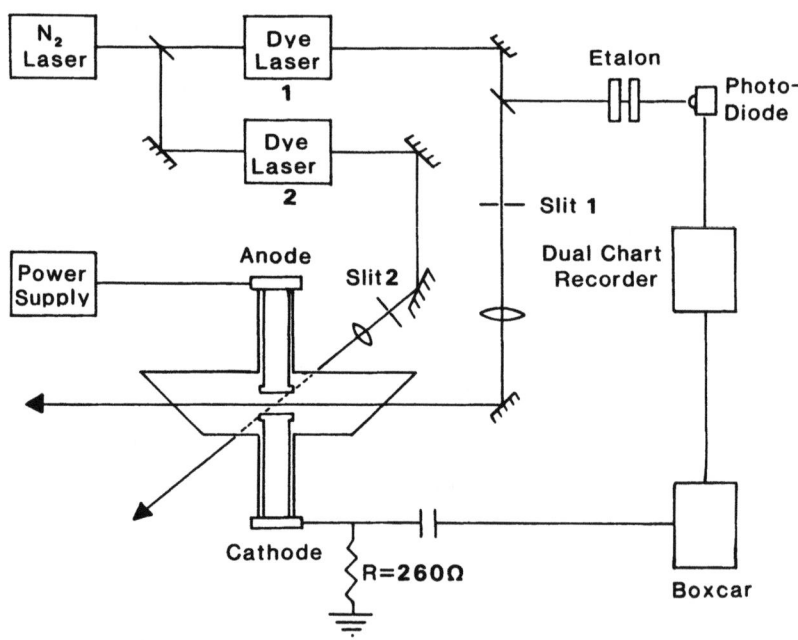

Fig. 5. Schematic of the electric field experiment in Ne. The use of intersecting laser beams provides a pinpoint measurement of the discharge electric field.

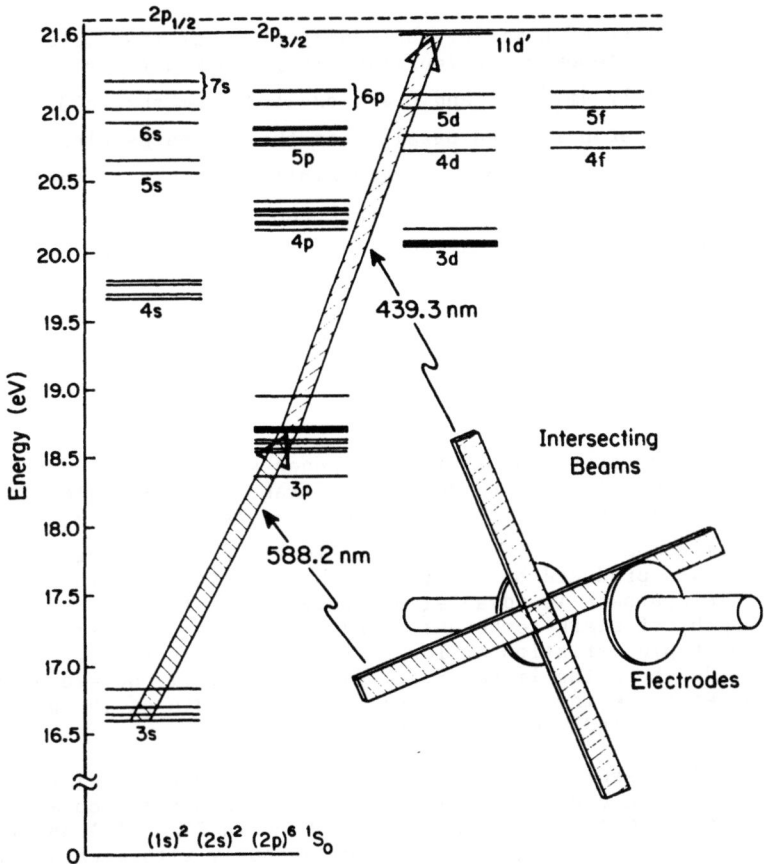

Fig. 6. Partial Grotrian diagram of Ne. The transitions used in the intersecting beam experiment are labeled.

It is desirable to use an intermediate level which will not produce a prompt or fast optogalvanic effect. The $2p^5 3p$ configuration is below the threshold for associative ionization in Ne, hence one key channel for fast effects is closed. Unfortunately there are large departures from Russell Saunders coupling in Ne and the $2p^5 3p\ ^3P_1$ level does not radiate to the $2p^5 3s\ ^1P_1$ resonance level. Thus a fast but weak optogalvanic effect is produced by enhancing the population of the short lived resonance level. The excitation of the Rydberg levels produces a much stronger fast optogalvanic effect. The Rydberg atoms are rapidly collisionally ionized, probably via associative ionization. The fast (~microsecond) effect decreases by more than an order of magnitude if either laser beam is interrupted. Similar observations of optical double resonance effects have been reported (Shuker et al., 1981). The excitation of Rydberg nd' levels in Ne is desirable because nd' levels have a small quantum defect of 0.018 (Moore, 1971). The small quantum defect leads to linear Stark effect at modest fields. The ns' and np' levels have quantum defects of 1.3 and 0.83 respectively (Moore, 1971). The calculation of a Stark map for Ne follows the procedure outlined previously for He. A Hamiltonian matrix with a small basis set including the 11d', 11f', 11g', ..., 11m', 11n', 12s', and 12p' states is diagonalized. We later expanded the basis set to test for n mixing at high field. The effects of n mixing were unimportant at the fields of this experiment. An independent calculation is performed for each value of $|m_\ell|$ excited. Only $|m_\ell|$ = 1 and 0 are excited from the $2p^5 3p$ configuration with the laser polarized parallel to the discharge electric field. The $|m_\ell|$ = 0 and 1 Stark maps are very similar because of the absence of nearby s' and p' levels. The nine Stark components are spread over 38 cm^{-1} at 3000 V/cm. The separation of $|m_\ell|$ = 1 and 0 components is <0.7 cm^{-1} at 3000 V/cm. We combine the two Stark maps according to the statistical weights in order to analyze our data. The Rydberg levels with a $^2P_{3/2}$ core at essentially the same energy are not a concern because of the very tight (780 cm^{-1}) spin orbit coupling of the ion core (Moore, 1971).

The relative strength of Stark components is used to make electric field measurements in He. We do not use transition strengths in Ne for two reasons: (1) the calculation was more complex because m_ℓ is not a good quantum number in the intermediate level, and (2) a better signal to noise ratio was produced by saturating the transitions with high laser power. This approach is advantageous only at low field where the relative strengths of stark components vary rapidly as a function of electric field. The rapid variation at low field is shown in Fig. 1(b). At higher fields where the linear Stark pattern is fully developed it is better to determine the field by measuring the width of the pattern or the separation of individual components.

Figure 7 presents Stark spectra obtained at various distances from the cathode. Nine clearly resolved Stark components are obtained in most of our spectra. (Not all components are shown in Fig. 7(a) and 7(b). Calibration fringes were generated simultaneously with each spectrum. The quality of these spectra is so high that ~1% field measurements are achieved.

Figure 8 is a plot of the observed field as a function of distance from the cathode. The Ne pressure is 2 torr, the discharge current is 7.8 mA and the discharge voltage is 725V. The uncertainty in the field measurement of Fig. 8 is approximately the size of the data points. These pinpoint electric field measurements make possible an important accuracy test. The integral $\int \bar{E} \cdot d\bar{z}$ from cathode is evaluated using the data of Fig. 8. The integral is 736V; it agrees (within 1.5%) with the discharge voltage of 725V determined using a digital voltmeter.

Optogalvanic detection of Rydberg atoms has other applications besides the experiments described in the preceeding paragraphs. Optogalvanic detection of Rydberg atoms in a positive column discharge is used to map radial fields with an accuracy ±1V/cm (Ganguly and Garscadden, 1985). Optogalvanic detection of Rydberg atoms should also be useful in measurements of ion density in the central regions of a discharge plasma. The width of transitions to Rydberg levels will be increased due to fluctuating (Holtsmark) fields from ion collisions. The essential advantages of optogalvanic detection of Rydberg atoms as a discharge probe are: (1) the sensitivity, (particularly in the cathode fall), (2) the large dynamic range provided by choice of principal quantum number and (3) the ease and accuracy with which Stark effects are interpreted.

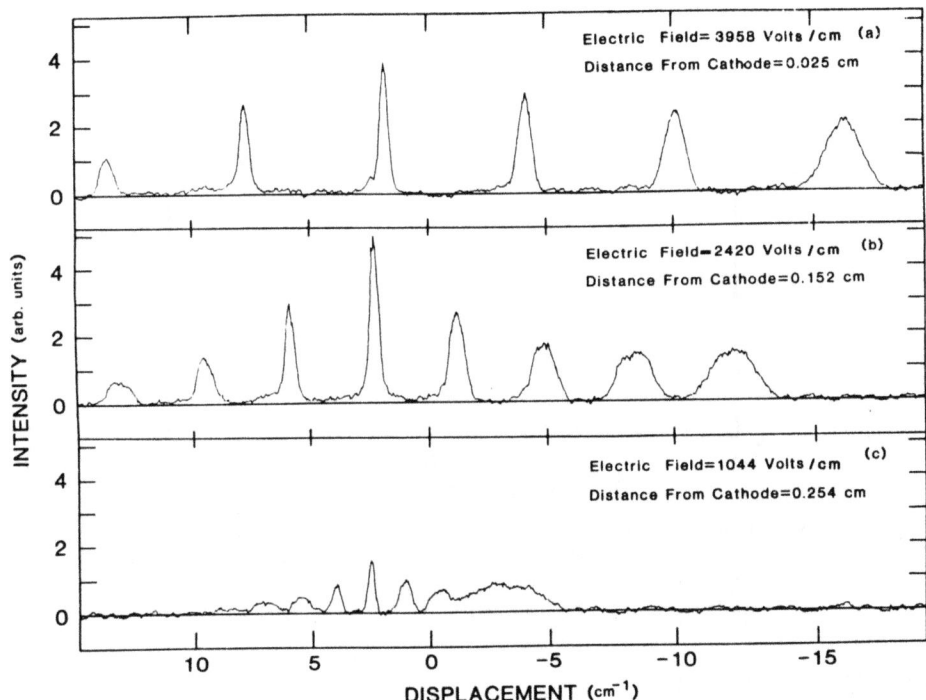

Fig. 7. Stark spectra observed at various distances from the cathode surface in a Ne discharge. The zero of the frequency scale corresponds to the unperturbed position of the $2p^5(^2P_{1/2})$ 11d' level.

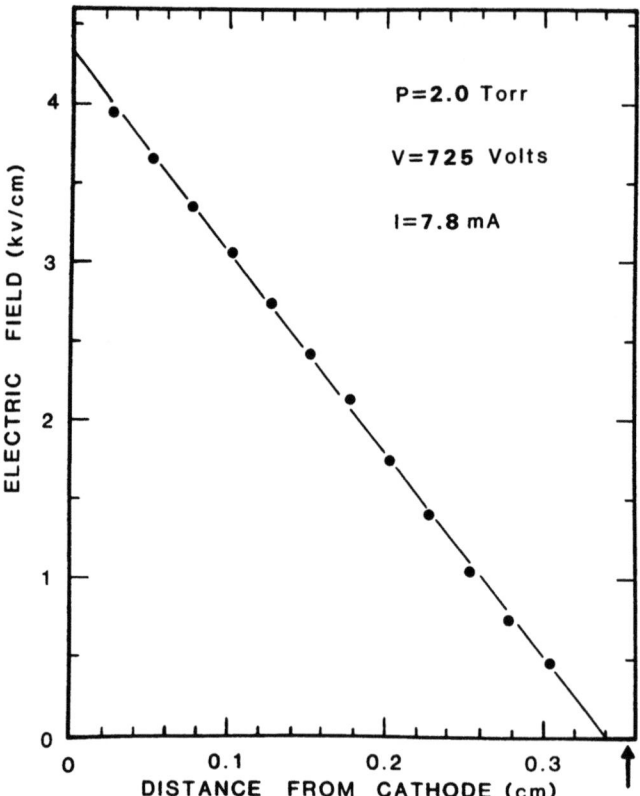

Fig. 8. Electric field as a function of distance from the cathode in a Ne discharge. The line is a linear square fit to the data. The arrow indicates the position of the anode.

TEMPERATURE MEASUREMENTS

We initially assumed that the local E/N in the cathode fall is determined from the field measurements and a measurement of the gas pressure in the discharge cell. The assumption is true only if the gas is at ambient temperature. We recently discovered that gas in the cathode fall is as much as a factor of 2 above ambient temperature even at modest mA/cm^2 current densities (Doughty et al., 1985). A factor of 2 increase in absolute temperature implies a factor of 2 decrease in N and a factor of 2 increase in E/N. The hot rarefied gas in the cathode fall has important implications for modeling the cathode fall, sputtering processes at the cathode, the glow to arc transition, and other phenomena associated with the cathode region.

The absolute gas temperature is determined from Doppler width measurements. A neon discharge was used in our initial studies of gas heating in the cathode fall. A thorough understanding of the line shape is essential in Doppler width temperature measurements. Large Stark effects are avoided by studying a transitions between low lying (non-Rydberg) levels. The 630.5 nm transition connecting the $2p^5 3s\ ^3P_1$ and $2p^5 3p\ ^1D_2$ levels of Ne ($1s_4 - 2p_6$ in Paschen notation) is used. The natural width of this line is 15.7 MHz (Wiese et al., 1966). Collisions with ground state atoms contribute an additional 10 MHz to the homogeneous width (Kuhn and Lewis, 1967). The Stark effect on this transition at a few kV/cm field is of order 1 MHz or less. It is a million times smaller than Stark effects observed on transitions to Ne Rydberg levels. The very small Stark effect is due to the absence of nearby levels of opposite parity, and to the comparatively small radial matrix elements. The nearest levels of opposite parity to the $2p^5 3p\ ^1D_2$ level are the levels of the $2p^5 4s$ configuration which are 9000 cm^{-1} higher in energy. The various broadening mechanisms are of order a few MHz to tens of MHz, they are negligible in comparison with the Doppler width of 1.3 GHz at 300K.

Figure 9 is a schematic of the experimental apparatus used in temperature measurements. The discharge cell is an improved version of the cell shown in Fig. 2. The anode is a 3.2 cm diameter Al disc. The cathode is segmented into an Al disc 1.6 cm in diameter, and a close fitting annulus with an outside diameter of 3.2 cm. The two parts of the cathode are maintained at the same potential during operation by adjusting the variable resistor. The segmented cathode enables us to measure the current distribution on the cathode and to estimate the importance of edge effects. The current distribution is uniform across the cathode, and thus the discharge is approximately one dimensional. A pressure of 1.5 torr with a 0.57 cm electrode separation is used in the experiment.

The laser used in this experiment is a single frequency, actively stabilized, continuous wave dye laser. Its very narrow bandwidth of 1 MHz is negligible in comparison to the Doppler width. Optogalvanic detection is used because it produces a better signal-to-noise ratio in the cathode region than either fluorescence or absorption methods. The 630.5 nm optogalvanic effect results primarily from a redistribution of excited atoms among resonant and metastable levels of the $2p^5 3s$ configurations (Doughty and Lawler, 1983a). A detailed quantitative model of the effect is not essential in linewidth measurements. The lineshape is modeled as the sum of two undistorted Gaussians (Doppler profiles), separated by the 1.68 GHz isotope shift, and weighted according to the natural abundances of Ne^{20} and Ne^{22} (Odintsov, 1965). Figure 10 is a comparison of experimental and model lineshapes.

Figure 11 is a plot of the absolute gas temperature and electric field versus distance from the cathode. Solid data points correspond to a current density of 1.39 mA/cm^2 and open data points correspond to a current density of 0.50 mA/cm^2. Throughout the cathode fall the temperature is fairly uniform and increases with increasing current density. The temperature drops off near the cathode fall negative glow boundary (where the electric field extrapolates to zero). A contraction of the cathode fall with increasing current density is clearly visible.

The gas heating in the cathode fall is due to fast neutrals produced by symmetric charge exchange. The large electric fields found in the cathode fall indicate that the power dissipation per unit volume, Ej, is large. Most of the discharge current near the cathode is carried by positive ions. The motion of atomic ions in their parent gas is limited by symmetric charge exchange.

$$Ne^+(fast) + Ne(slow) \rightarrow Ne(fast) + Ne^+(slow). \tag{4}$$

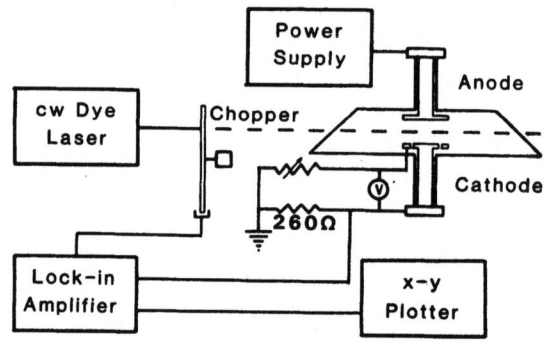

Fig. 9. Schematic of the gas temperature experiment.

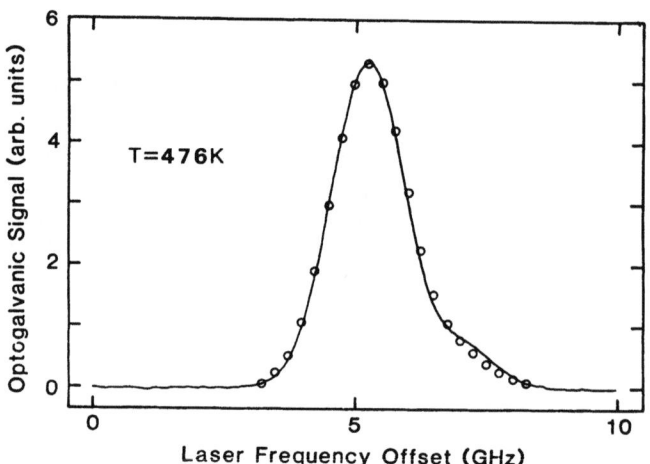

Fig. 10. Optogalvanic signal at 630.5 nm from the Ne discharge. The solid line is an experimental lineshape. The open circles are points on a theoretical lineshape for a temperature of 476K.

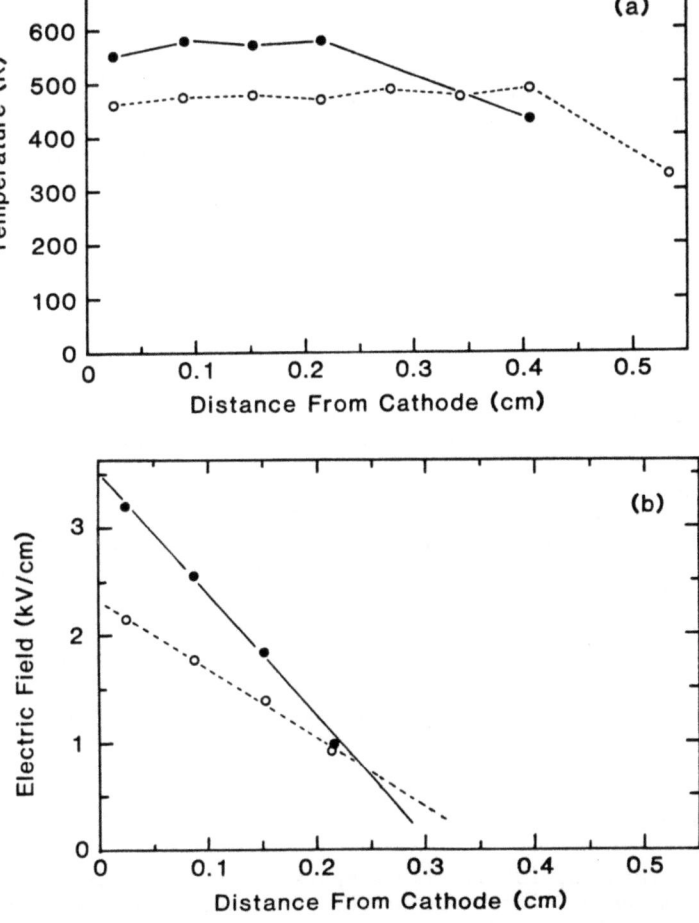

Fig. 11. (a) Temperature in the cathode fall as a function of distance from the cathode. The uncertainty is ±15K. The solid line corresponds to a current density of 1.39 mA/cm^2 and a discharge voltage of 520V. The dashed line corresponds to a current density of 0.50 mA/cm^2 and a discharge voltage of 425V.
(b) Electric field in the cathode fall as a function of distance from the cathode. The uncertainty corresponds to the size of the data points.

The fast neutrals play a role in sputtering at the cathode, but many are scattered before they reach the cathode. The cathode fall is a region of high power density in which much of the dissipated energy is converted directly to heavy particle translational motion.

MAPPING THE CATHODE FALL

Ionization Maps

Optogalvanic diagnostics are the experimental tools needed to develop a more quantitative understanding of the cathode fall. The most fundamental difficulty in modeling the cathode fall is due to the electron energy distribution function which is not in hydrodynamic equilibrium with the local E/N. The E/N maps which are produced using optogalvanic diagnostics provide information on the electron energy distribution function. The E/N maps are used to determine the position dependent ionization rate in the cathode fall. The spatial derivative of the electric field provides a space charge density from Poisson's equation,

$$\frac{dE}{dz} = \frac{e}{\varepsilon_o}(n_+ - n_e) \sim \frac{e}{\varepsilon_o} n_+. \tag{5}$$

The positive ion density dominates the space charge density. The local E/N and known ion mobilities determine the ion drift velocity, v_+, because the ions are in hydrodynamic equilibrium. Symmetric charge exchange produces a very short equilibration distance for singly charged atomic ions. The ion density and ion drift velocity are combined to determine the ion current density. The difference between the total discharge current density j and the ion current density is the electron current density,

$$j_e = j - en_+ v_+. \tag{6}$$

The spatial derivative of the electron current density provides the local ionization rate S from a continuity equation,

$$S = -\frac{1}{e}\frac{dj_e}{dz}. \tag{7}$$

An ionization rate map is directly comparable to ionization rate maps predicted from nonequilibrium solutions of the Boltzmann equation for electrons (Segur et al., 1983).

The feasibility of this technique for mapping the ionization rate is tested by using the data of Fig. 11 to compute the ion current density at the cathode surface. Nearly all of the discharge current should be carried by ions at the cathode surface. The spatial derivative of the electric field, $\frac{dE}{dz}$, is 6400 V/cm^2 at 0.50 mA/cm^2 current density. The electric field derivative corresponds to a positive charge density 5.67 x 10^{-10} Coulomb/cm^3 (an ion density of 3.54 x 10^9 cm^{-3}). The electric field at the cathode, 2300 V/cm, and the gas density, 3.1 x 10^{16} cm^{-3}, lead to an E/N of 7460 Td. It is essential to include the effect of gas heating in computing the E/N. The E/N and drift velocity data tabulated by Frost are combined to determine an ion drift velocity of 8.51 x 10^5 cm/s (Frost, 1956). The positive charge density and ion drift velocity correspond to a current density of 0.483 mA/cm^2 which is 97% of the measured current density 0.50 mA/cm^2. A similar computation for the 1.39 mA/cm^2 data of Fig. 11 gives 82% of the measured current density.

Avalanche Experiments

Optogalvanic effects are combined with laser induced fluorescence in an electron avalanche experiment. This experiment is an example in which

the magnitude of the optogalvanic effect is used to test a discharge
model. The essential idea in this experiment is to release a calibrated
quantity of electrons at some point in the cathode fall and to measure
the resulting avalanche as an optogalvanic effect. The best method of
releasing a calibrated quantity of electrons is to photoionize ground
state atoms. Unfortunately this requires vacuum ultraviolet radiation
for most atoms and molecules. An alternate method involves ionizing from
a metastable level. The interpretation of this type of optogalvanic
effect is not as simple as the interpretation of photoionization from the
ground state, because the former involves a depletion of metastables.
The loss of metastables results in a reduction of secondary electron
emission from the cathode, which partly cancels the effect of ionizing
the metastables. Fortunately the competing phenomena occur on different
time scales. The ionization of the metastables and subsequent electron
avalanche is a fast phenomena. It occurs on a time scale of ion drift
across the cathode fall. The reduction in metastable flux on the cathode
and subsequent loss of secondary emission is a slow phenomena. It occurs
on a time scale of metastable diffusion across the cathode fall.

The choice of method for ionizing the metastables is extremely
important. It is crucial that no atoms be driven from the metastable
level to a resonance level. A perturbation to a resonance level population
will result in a relatively fast optogalvanic effect due to secondary
emission from vacuum ultraviolet photons. Although the vacuum
ultraviolet photons are not as efficient as metastables in releasing
electrons from a cold cathode, the photons are faster. The time scale of
enhanced cathode emission due to an increase in resonance level population
is the trapped lifetime of the resonance level. Perturbations to
resonance level population must be avoided if the optogalvanic effect is
to be interpreted simply as an electron avalanche resulting from ionizing
a few metastables at some point in the cathode fall. It is also important
to choose a method for ionizing the metastables which is easily
calibrated, so that the number of excess electrons released at some point
in the cathode fall can be determined.

Helium is the ideal atom for the avalanche experiment. Russel-
Saunders coupling is extremely tight in low lying levels of He. Spin
triplet He atoms with $n \leq 3$ remain spin triplets during radiative decay
and during collisions with neutral He atoms. Electron collisions can, of
course, convert a triplet to a singlet but the electron density is extremely
low in the cathode fall. Helium atoms excited to the 3^3P, 3^3D,
or higher levels are associatively ionized.

$$He^* + He \rightarrow He_2^+ + e^- \tag{8}$$

The threshold for this reaction is just below the 3^3P level. Hence,
the 2^3S metastable is not associatively ionized. A dye laser at 388.9 nm
is used to selectively excite atoms from the 2^3S metastable level to the
3^3P level. Collisions with ground state He atoms transfer population to
the 3^3S and 3^3D levels. Thus, all three transitions from the $n = 3$
triplet levels fluoresce as shown in Fig. 12. Collisions with ground
state He atoms also associatively ionize some of the He 3^3P and 3^3D
atoms. The fraction of the excited He atoms which fluoresce and the
fraction which is associatively ionized is determined from well known
radiative decay rates, cross sections and the gas density (Wellenstein
and Robertson, 1972, Dubreuil and Catherinot, 1980). In summary, the He
atoms excited from the 2^3S metastable level to the 3^3P level suffer some
collisional redistribution to the 3^3S and 3^3D levels, but ultimately they

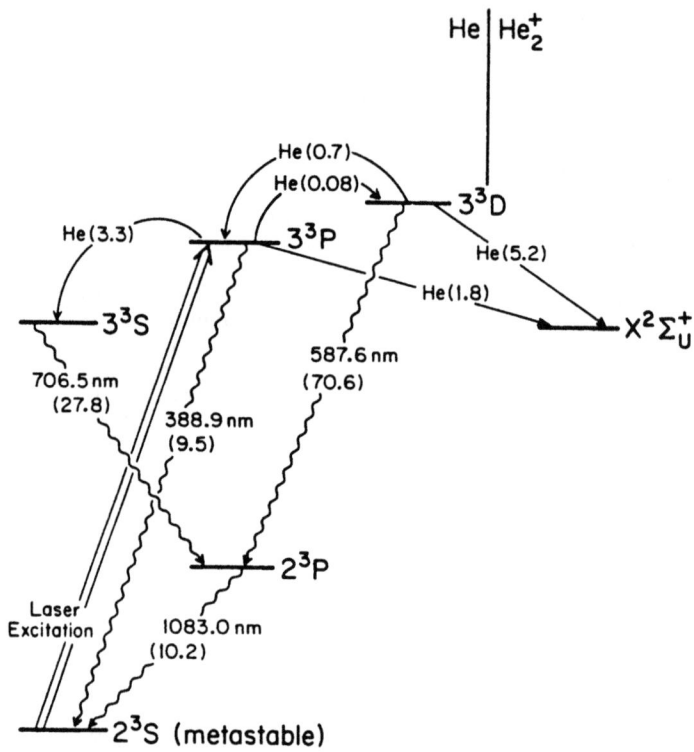

Fig. 12. Partial Grotrian Diagram shownig the He triplet levels involved in the 388.9 nm optogalvanic effect. Collisional and radiative decay rates are indicated in parentheses in units of $10^6 s^{-1}$. Collisional rates are for 2.0 Torr of He at 300K.

radiatively decay back to the 2^3S metastable level or they are associatively ionized. The fluorescence signal is proportional to the number of atoms excited by the laser pulse, and to the number of excess electrons released through associative ionization. The ratio of the optogalvanic signal to the fluorescence signal is the position dependent amplification of the optogalvanic effect. It is a measure of the size of the avalanche produced by releasing a calibrated quantity of electrons at some point in the cathode fall. Preliminary data from this experiment is shown in Fig. 13. The plot of "optogalvanic effect/fluorescence" increases with decreasing distance from the cathode as expected. Electrons released closer to the cathode produce a bigger avalanche than those produced further from the cathode. However, the plot of "optogalvanic effect/fluorescence" turns downward very close to the cathode. We did not expect this behavior. We have several tentative explanations but none are completely satisfactory.

The fluorescence signal of Fig. 13, besides providing a measure of the number of excess electrons released, also provides a spatial map of the metastable density. This map is a relative, not an absolute map, but it becomes an absolute map with a single absorption measurement. The metastable density map of Fig. 13 has several features in common with the 2^3S metastable production rate map calculated using a nonequilibrium solution to the Boltzmann equation for electrons (Segur et al., 1983).

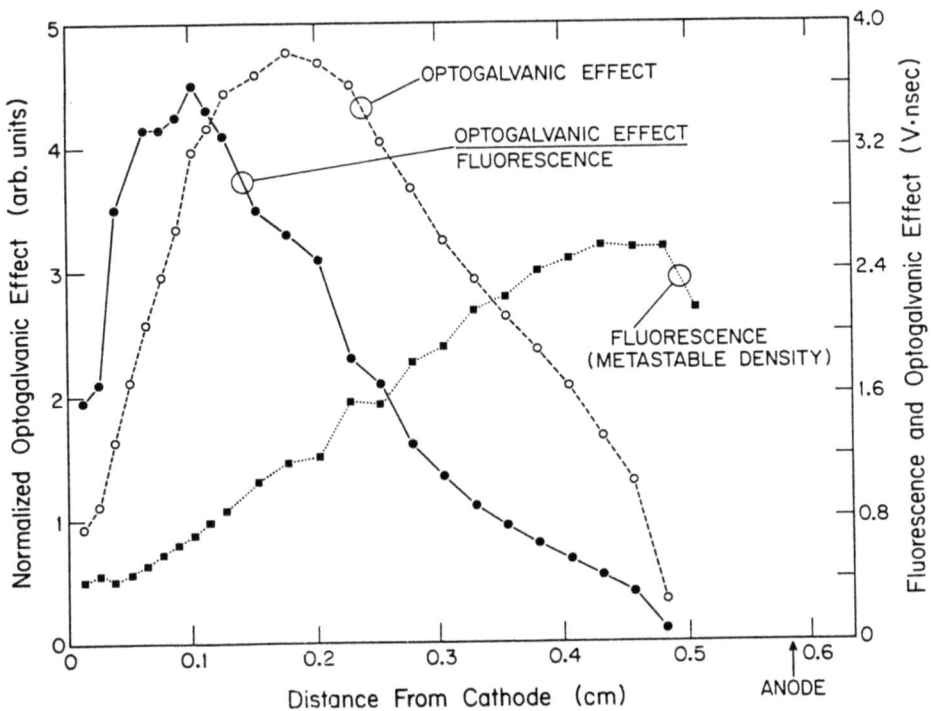

Fig. 13. Preliminary data from the electron avalanche experiment. The spatial dependence of the optogalvanic signal and fluorescence signal produced by laser excitation at 388.9 nm in a He discharge are represented by dashed and dotted lines, respectively. The ratio of optogalvanic signal over fluorescence signal is the position-dependent amplification of the optogalvanic effect in the cathode fall.

The density peaks near the cathode fall-negative glow boundary in agreement with the production rate. There is even a hint of a minor peak in the density near the cathode in agreement with the production rate. The minor peak in the production rate near the cathode is due to electrons from the cathode accelerating through an energy corresponding to the sharp peak in the triplet excitation cross section. The metastable density map is not an exact replica of the metastable production rate map because diffusion occurs between production and detection of metastables.

Figure 14 is a plot of time-resolved optogalvanic effects at 388.9 nm in the He cathode fall. The time-resolved measurements have some interesting features. The positive signal in Fig. 14 corresponds to an increase in discharge current. The duration of the signal increases with distance from the cathode. The signal starts to show temporal structure 0.4 cm from the cathode. This structure is probably due to the different time scales associated with the electron displacement current and the ion displacement current. The excess electrons and the resulting avalanche moves so quickly across the cathode fall that the electron displacement current peaks instantly. The associated cloud of excess ions produces a peak displacement current when it reaches the high field region near the cathode.

It is obvious that much effort will be required to model the spatial and temporal characteristics of the electron avalanche experiments in the

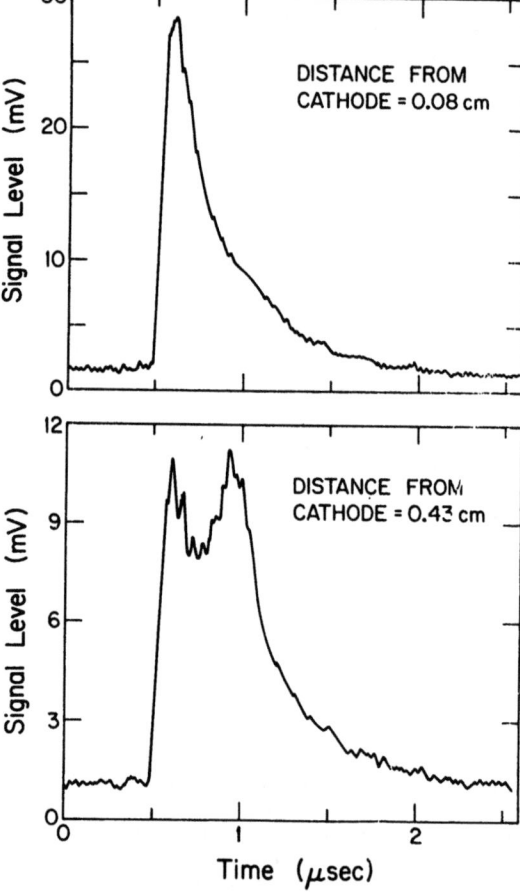

Fig. 14. Time resolved optogalvanic signals at 388.9 nm in a He discharge. A spatial variation of temporal features is observed.

cathode fall. The preliminary results reported here provide a clear indication that a great deal can be learned from such experiments.

SUMMARY

The cathode fall region is the least understood, and yet for many applications the most important region of diffuse discharges. The cathode fall is critical in diffuse discharge applications such as: semiconductor production processes, pulsed power switches, and coherent and incoherent light sources. The fundamental difficulty in modeling the cathode fall is the electron energy distribution function which is not in hydrodynamic equilibrium with the local E/N. The lack of hydrodynamic equilibrium is due to the large and rapidly changing E/N and to the proximity of the boundary. New theoretical approaches and new experimental tools are needed to develop a more quantitative understanding of this important region. It is the thesis of this paper that optogalvanic diagnostics are the experimental tools needed for this task. Diagnostics based on optogalvanic detection are extraordinarily sensitive in the cathode fall because of the intrinsic amplification of the effects in the cathode fall.

Electric fields in the cathode fall are mapped using optogalvanic detection of Rydberg atoms. Rydberg atoms exhibit dramatic linear Stark effects because of the large size of the atoms and the small energy gap between states of opposite parity. The large Stark effects are easy to measure and are straightforward to interpret. The Rydberg atoms are produced via laser excitation from metastable levels, and are detected using optogalvanic effects. Optogalvanic detection is essential because the fragile Rydberg atoms do not survive sufficiently long to fluoresce in a discharge. Electric field measurements to ~1% accuracy are achieved.

The gas temperature is mapped using the Doppler width of transitions to non-Rydberg levels which do not exhibit significant Stark effects. A factor of two increases in gas temperature observed at low (mA/cm^2) current densities. Production of fast neutrals through symmetric charge exchange of energetic positive ions is responsible for the gas heating. Accurate spatially resolved electric field and gas temperature (density) measurements are used to construct maps of other parameters in the cathode fall including: the ion density, the fractions of discharge current carried by ions and by electrons, and the local ionization rate.

The amplification of optogalvanic effects in the cathode fall is of intrinsic interest. A combination of optogalvanic and laser induced fluorescence measurements on a carefully chosen transition is used to measure the size of the avalanche produced by releasing a calibrated quantity of electrons at some well defined point in the cathode fall. These avalanche measurements are directly comparable to ionization rate maps based on non-equilibrium solutions to the Boltzmann equation for electrons. Optogalvanic diagnostics and new theoretical approaches will ultimately lead to a more quantitative understanding of the cathode fall.

ACKNOWLEDGMENT

Research supported by the U.S. Air Force Office of Scientific Research and the U.S. Army Research Office under Grant AFOSR 84-0328.

REFERENCES

Camus, P., 1983, ed. of "International Colloquium on Optogalvanic Spectroscopy and its Applications", J. de Physique Colloq., C7: 1-535.
Doughty, D. K., and Lawler, J. E., 1983a, Phys. Rev. A, 28:773.
Doughty, D. K., and Lawler, J. E., 1983b, Appl. Phys. Lett., 42:234.
Doughty, D. K., and Lawler, J. E., 1984a, Appl. Phys. Lett., 45:611.
Doughty, D. K., Salih, S., and Lawler, J. E., 1984b, Phys. Lett., 103A:41.
Doughty, D. K., DenHartog, E. A., and Lawler, J. E., 1985, Appl. Phys. Lett., 46:352.
Dubreuil, B., and Catherinot, A., 1980, Phys. Rev. A, 21:188.
Edmonds, A. R., Picart, J., Tran Minh, N., and Pullen R., 1979, J. Phys. B, 12:2781.
Frost, L. S., 1956, Phys. Rev., 105:354.
Gallagher, T. F., Edelstein, S. A., and Hill, R. M., 1977, Phys. Rev. A, 15:1945.
Ganguly, B. N., and Carscadden, A., 1985, Appl. Phys. Lett., 46:540.
Goldsmith, J. E. M., Ferguson, A. I., Lawler, J. E., and Schawlow, A. L., 1979, Opt. Lett., 4:230.
Kenty, C., 1950, Phys. Rev., 80:95.
Kuhn, H. G., and Lewis, E. L., 1967, Proc. Roy. Soc. A, 299:423.
Larsson, M., Mannfors, B., and Pendleton Jr., W. R., 1983, Phys. Rev. A, 28:3371.

Lawler, J. E., Ferguson, A. I., Goldsmith, J. E. M., Jackson, D. J., and Schawlow, A. L., 1979, Phys. Rev. Lett., 42:1046.
Meissner, K. W., and Miller, W. F., 1953, Phys. Rev., 92:896.
Moore, C. E., 1971, Atomic Energy Levels, Natl. Stand. Ref. Data Ser., Natl. Bur. Stand. (U.S.), 35, U.S. G.P.O., Washington, Vol. I, p. 5, 80.
Odintsov, V. I., 1965, Opt. Spectrosc., 18:205.
Penning, F. M., 1928, Physica, 8:137.
Segur, P., Yousfi, M., Boeuf, J. P., Marode, E., Davies, A. J., and Evans, J. G., 1983, in: "Electrical Breakdown and Discharges in Gases," edited by Kunhardt, E. E., and Luessen, L. H., Plenum, New York.
Shuker, R., Ben-Amar, A., and Erez, G., 1981, Opt. Comm., 39:51.
Warren, R., 1955, Phys. Rev., 98:1650.
Wellenstein, H. F., and Robertson, W. W., 1972, J. Chem. Phys., 56:1072 and 1077.
Wiese, W. L., Smith, M. W., and Glennon, B. M., 1966, Atomic Transition Probabilities, Natl. Stand. Ref. Data Ser., Natl. Bur. Stand. (U.S.) 4, U.S. G.P.O., Washington.
Zimmerman, M. L., Littman, M. G., Kash, M. M., and Kleppner, D., 1979, Phys. Rev. A., 20:2251.

RYDBERG STATES: PROPERTIES AND APPLICATIONS TO

ELECTRICAL DISCHARGE MEASUREMENTS

A. Garscadden

Air Force Wright Aeronautical Laboratories
Wright-Patterson AFB, OH, USA

INTRODUCTION

If an electron in an atom is excited to a level designated by a high principal quantum number n where n is typically greater than 10, the level is termed a Rydberg level (Stebbings and Dunning, 1983) and the atom is often called a Rydberg atom (Bayfield, 1983; Kleppner et al., 1983). These states have been observed from a wide variety of plasma sources in the laboratory and also from astrophysical sources. In interstellar space the Rydberg levels are formed by radiative recombination. Emission from Rydberg states of atomic hydrogen have been distinguished from the continuum radiation with n as high as 732. In laboratory plasmas, dissociative recombination, cumulative excitation and optical excitation are more likely excitation processes. The development of powerful tunable dye lasers into the ultraviolet permits the selective excitation of many atoms and molecules to high Rydberg states. Recently (Rinneberg et al., 1985) Rydberg states of barium with principal quantum numbers up to 290 have been observed under laboratory conditions. Figure 1 illustrates the photoexcitation method of the helium metastable states that we have used to investigate glow discharge conditions.

This paper will introduce Rydberg states and some of their rather unusual properties. We will summarize some of the elegant atomic physics experiments that have been performed recently on these states and then we describe some experiments that capitalize on the information thus gained and exploit Rydberg states for the measurement of electric fields and excited state densities in plasmas.

PROPERTIES OF RYDBERG ATOMS

The interaction of an atom with external fields or with other atoms is essentially determined by its electrons. These usually occupy a volume with characteristic length greater than 10^4 times the nuclear dimensions. When selected laser radiation is absorbed by the atom, an electron is excited to a higher level and the electron is further removed from the nucleus and the scale length becomes even larger. The electron orbital velocity is reduced and the orbital distance is increased so that properties such as the orbital period scale non-linearly with the principal quantum number (Table 1). There is also increased sensitivity to external fields as the Coulombic field is much reduced. The energy

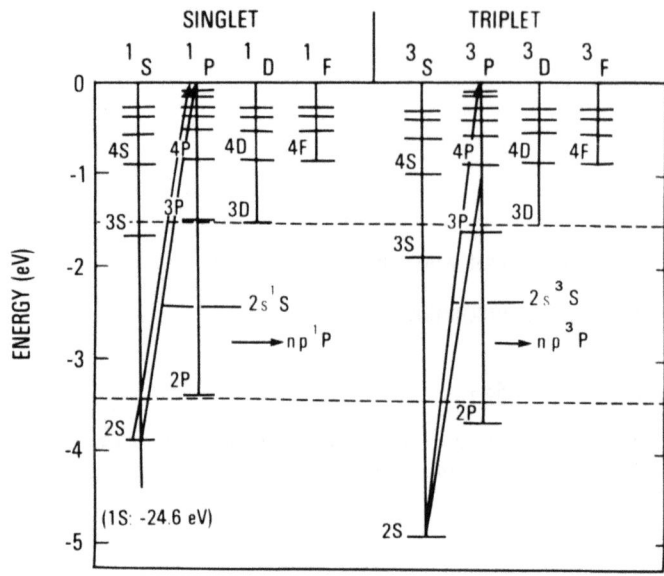

Fig. 1. Energy level diagram of helium. The energy is measured referenced to the ionization potential. The first excited levels of helium are at the comparatively high value of about 0.8 times the ionization energy. The lowest triplet term is strongly metastable owing to its parity being the same as the ground state. The absorption series lie in the far ultraviolet and the resonance line is at 584.4A.

Table 1. Properties of Rydberg States

PARAMETER	LEVEL SCALING	VALUES		
		$n = 1$	$n = 30$	$n = 100$
MEAN RADIUS (cm) # OF SUBLEVELS	n^2	5.3×10^{-9}	4.8×10^{-6}	5.3×10^{-5}
RMS ORBITAL VELOCITY (cm sec^{-1})	n^{-1}	2.2×10^8	7.3×10^6	2.2×10^6
BINDING ENERGY	n^{-2}	13.6 eV	15 mV	1.36×10^{-3} eV
ORBITAL FREQUENCY (Hz) ADJACENT LEVEL SPACING (eV) DECAY RATES EXCITATION RATES	n^{-3}	6.7×10^{15} 10	2.44×10^{11} 4×10^{-4}	6.7×10^9 10^{-5}

levels of the electron of the hydrogen atom are given by (Bethe and Salpeter, 1957)

$$E_{nj} = \frac{-13.6}{n^2} \times 1 + \left\{ \left(\frac{1}{137}\right)^2 \frac{1}{n} \left(\frac{1}{j + 1/2} - \frac{3}{4n}\right) \right\} \text{ (eV)} \qquad (1)$$

or approximately $E_{nj} \propto n^{-2}$. At larger n all atomic spectra become approximately hydrogenic. However, there remain small measurable differences due to the polarization of the core of the atom. The differences are related to the hydrogen levels by an equivalent Rydberg equation

$$E_n = -\frac{Rhc}{n^{*2}} = -\frac{Rhc}{(n-\delta)^2} \qquad (2)$$

$$E_n = E_H - \frac{2\delta Rhc}{n^3}$$

where R is the Rydberg constant = 109700 cm^{-1}, c is the velocity of light, h is Planck's constant, δ is termed the quantum defect, j is the electron angular momentum quantum number, and $n^* = (n - \delta)$ is the effective quantum number.

The observed quantum defects for helium levels (Wing et al., 1973) are given in Table 2. As will be noted the corrections are significant only for low principal quantum numbers. The frequency of a transition from a level in a neutral atom with principal number $n + \Delta n$ to the level n, where n is sufficiently large that the quantum defects are negligible, is then given by $\frac{dE}{dn} \propto n^{-3}$ or, more exactly, by

$$Rc \left\{ \frac{1}{n^2} - \frac{1}{(n+\Delta n)^2} \right\} \doteq 3.29 \times 10^{15} \left\{ \frac{1}{n^2} - \frac{1}{(n+\Delta n)^2} \right\} \text{ (Hz)} \qquad (3)$$

$$\doteq 6.58 \, \Delta n \left(\frac{100}{n}\right)^3 \text{ (GHz)}$$

The transitions for n = 40 have mm wavelength and those for n = 300 have meter wavelengths (and are therefore looked for in radio-astronomy). The radiative lifetime of a high Rydberg state (Wing et al., 1973) is given approximately by

$$t(n,\ell) = n^3(\ell + 1/2) \, 10^{-10} \text{s}. \qquad (4)$$

Therefore, for large n the levels become quasimetastable; n = 100 will result in a lifetime of $\sim 10^{-4}$ s. On the other hand, under discharge conditions, the collisional frequencies are very large and also background radiation redistributes the level populations so that experiments involving Rydberg states usually must be on a fast time-scale. The very close spacing of high Rydberg levels means that blackbody radiation even at 300 K appears as a rapidly varying field, whereas to a ground state atom it is a slowly varying field. (The Planckian curve for 300K has its peak at approximately 600 cm^{-1}.) The photon occupation number for $h\nu < kT$ is

$$\bar{n} = \frac{1}{[\exp(h\nu/kT)]-1} \doteq \frac{kT}{h\nu}. \qquad (5)$$

Table 2. Quantum Defects in Helium

Level	n^3S	n^1S	n^3P	n^1P	nD	nF
Quantum Defect	0.30	0.14	0.06	0.01	0.003	0.0003

This number is therefore inversely proportional to $h\nu$ and, thus, increases as n^3. The spontaneous emission rates are given by the vacuum fluctuation $\bar{n} = 1/2$. Therefore, at higher Rydberg levels when $\Delta E < 10$ cm^{-1} ($\bar{n} = 10$ for $T_{gas} = 300K$) the blackbody-induced transition rates become much larger than the spontaneous emission rates. This means that there will be a rapid redistribution of an initial laser-induced perturbation of the populations of Rydberg levels by the blackbody radiation of the surroundings.

Collisional Properties

A very important property of Rydberg states is their size scaling with n. The mean radius of the electron distribution is

$$r(n,\ell) = 2.6 \; [3n^2 - \ell(\ell+1)] \; 10^{-9} \text{cm} \qquad (6)$$

The dependence is dominated by the n^2 term. This means that for $n = 30$ the Rydberg atom is approximately 10^3 times larger than the unexcited state. This has important consequences on the collisional interactions of the atom. In the high Rydberg states, the orbital electron is moving relatively slowly (Table 1). Then a "fast" collision can occur when the colliding atom has a velocity much higher than that of the Rydberg electron. As it is the relative velocity of the colliding species that is important and the collision takes place far removed from the core of the atom, the situation reverts to the case of the free-electron model with the impulse approximation. It is found that these Rydberg electron-atom collision cross sections are very similar to the normal electron-atom impact cross sections and have the same structure and magnitudes. For high n, the cross sections are approximately independent of n (Butler and May, 1965).

The ion-Rydberg atom "fast" collisions have been analyzed by Olsen (1979) using a classical trajectory Monte-Carlo model and the free-electron model has been applied to collisions with hydrogen-like ions by Matsuzawa (1980). Very large cross sections are obtained. Thus, the cross section for n-changing collisions approaches 10^{-11} cm^2 and reaches a maxium when the relative velocity (v) is approximately twice the orbital velocity (v_n). The cross section for changes in the angular momentum of the Rydberg atom (ℓ-mixing collisions) by a charged particle is large as little energy is transferred. It is given by $Q_{\ell n} = n^4 \times 10^{-16}$ cm^2 (Butler and May, 1965). The ionization cross sections are also very large and when $(v/v_n) > 5$, the result is given by the classical equation for a fixed (v/v_n) ratio

$$Q_i = 6\pi a_o^2 n^4 q^2 \; (v/v_n)^{-2} \qquad (7)$$

or

$$Q_i = 6\pi a_o^2 n^2 q^2 v^{-2} \tag{8}$$

if v is measured in atomic units (2.19×10^8 cm s^{-1}) and q is the charge of the incident ion. These results, being derived classically, neglect ionization via tunneling and hence underestimate the cross section at higher $\frac{v}{v_n}$ by approximately 30%. At $\frac{v}{v_n} < 1$, electron capture by the ion is more probable by approximately an order of magnitude and this cross section tends to

$$Q_{cn} = 5.5 \pi n^4 q a_o^2 . \tag{9}$$

The Rydberg atom presents the total orbital diameter for charged particle collision because of the long range interaction. However, the atom is semi-transparent at high n for neutral particle collisions. Thus, a maximum in the Rydberg atom-atom collision cross section is reached around n = 10 and the cross section decreases at high n as n^{-4}.

The studies of Rydberg atom collisions with electrons fall into two energy categories. Measurements have been made by Schiavone et al. (1977) on helium states at energies E between 30 and 300eV. It was found that high ℓ states were populated by collisions of low ℓ Rydberg atoms with electrons. These authors obtained an ℓ-changing cross section for fast electrons:

$$Q\Delta\ell \text{ (cm}^2\text{)} = 5 \times 10^{-13} \left(\frac{n^4}{E} \right) \ln(100En^2) . \tag{10}$$

Collisions with slow electrons were measured by J-F Delpech et al. (1977) in a helium afterglow. Collisional coupling between the singlet and triplet systems is slow. However, the collisional coupling between the ℓ sublevels of a given n is extremely fast. Because of the larger statistical weight most of the Rydberg atom population will remain in high ℓ states. An analysis of the decay of laser perturbed levels gave the total electron collision depopulation rate coefficients for principal quantum numbers $8 \leq n \leq 17$, shown in Fig. 2. These rates increase very rapidly with n and attain enormous values, e.g., 10^{-4} cm^3 s^{-1} at n = 15. This means that in a plasma the collisional redistribution of laser excitation to a particular level will occur very fast even in a plasma afterglow.

Rydberg Atoms in an Electric Field

The potential for an electron in a Coulomb field with an imposed external field F along with the z-axis is (Bethe and Salpeter, 1957; Condon and Shortley, 1967; Kleppner et al., 1983)

$$V(r) = -\frac{1}{r} + Fz \tag{11}$$

where V has a maximum on the z-axis at $z = -\frac{1}{F^{1/2}}$. The cut along the z-axis of this saddle point is illustrated in Fig. 3. The saddle point potential is, neglecting angular momentum terms,

$$V_{sp} = -2F^{1/2}. \tag{12}$$

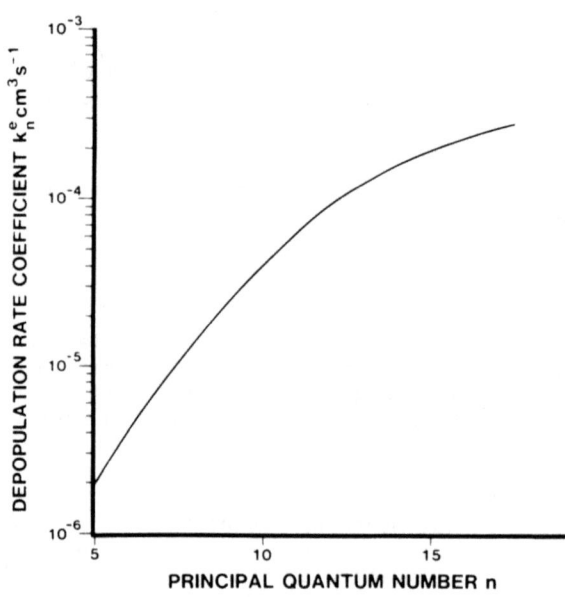

Fig. 2. Electron collision depopulation rates at low energies for principal quantum numbers 8 to 17.

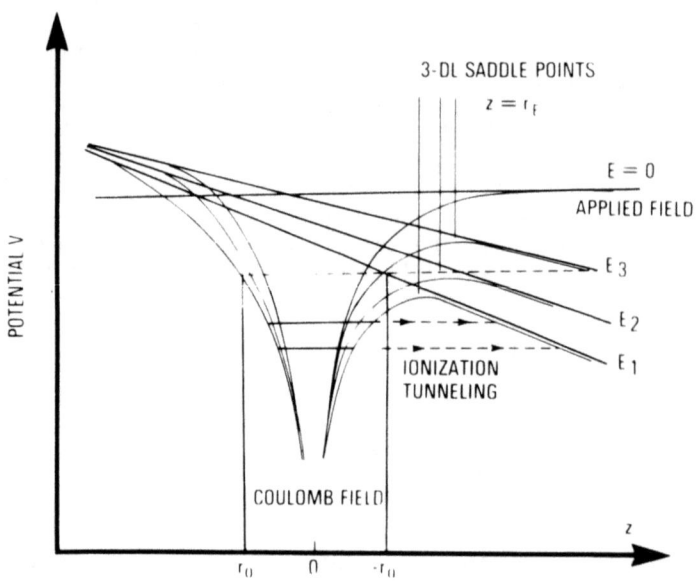

Fig. 3. Rydberg atom in an electric field, illustrating the addition of the Coulomb field and three different applied fields: $E_1 > E_2 > E_3$.

If a state is bound, then its energy lies below V_{sp}. Close to the top of the saddle point, ionization can occur by tunneling through the potential barrier. Above V_{sp} field ionization of the Rydberg state n can occur if the applied field is larger than the critical value F_c,

$$F_c = \frac{3.2 \times 10^8}{n^4} \text{ volts cm}^{-1} . \qquad (13)$$

(A more exact treatment is provided by Cooke and Gallagher (1978) to account for angular momentum.)

The Schroedinger equation describing the Coulomb potential is separable in both spherical and parabolic coordinates. However, if an applied field is superimposed it is then separable only in parabolic coordinates (Bethe and Salpeter, 1957). Hence, the Stark effect is usually described in terms of quantum numbers n, n_1, n_2 and m_ℓ where

$$n = n_1 + n_2 + |m_\ell| + 1 \qquad (14)$$

and n_1 and n_2 are the parabolic quantum numbers. The electric field breaks some of the energy degeneracy (n^2 neglecting spin). This effect is illustrated in Fig. 4 for the hydrogen levels n = 15 and 16 for m_ℓ = 0. At the low fields of great interest to discharge investigations, the level shifts are linear with applied electric field. The field ionization curve of the saddle point is in the upper right-hand corner of this diagram. Note that the states with $n_1 > n_2$ have their dipole moments aligned antiparallel to the applied field and increase in energy with the applied field. States with $n_1 < n_2$ have their dipole moments parallel to the applied field and decrease in energy with increasing applied field. At higher fields, applied to multi-electron atoms, the level shift behavior is considerably more complicated due to such phenomena as avoided crossings (Kleppner et al., 1983). The energy shift in the linear approximation is given by

$$E = \frac{3}{2} \frac{n}{Z} (n_1 - n_2) a_o eF \qquad (15)$$

where a_o is the Bohr radius for the n = 1 state (0.53 x 10^{-10}m), Z is the nuclear charge, or for hydrogen:

$$E \text{ (electron volts)} = 7.93 \times 10^{-9} n(n_1 - n_2) F(\text{volts/cm}) . \qquad (16)$$

It is noted that the Stark splittings from adjacent Rydberg levels overlap for electric fields higher than F_L which creates a practical limit to the effect as a diagnostic:

$$F_L \simeq 7.5 \times 10^6 (n^{-11/2}) \text{ kV cm}^{-1} . \qquad (17)$$

RYDBERG ATOMIC PHYSICS EXPERIMENTS

Rydberg Atom Detection and Resolution

Elegant experimental studies of Rydberg atoms in applied fields have been made in recent years using laser methods. The pulsed dye lasers are normally used to populate the high Rydberg states. As the oscillator

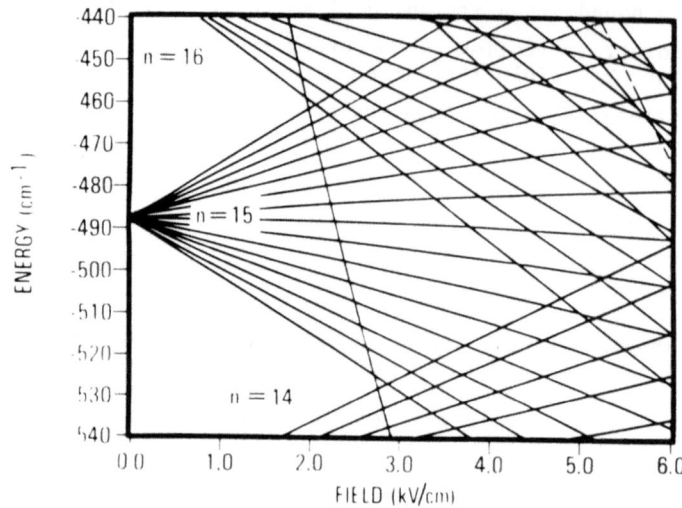

Fig. 4. Rydberg atom in an electric field, illustrating the splitting of the levels in a strong electric field. Dashed line is critical field (see text).

strength decreases rapidly with n (Table 1) very high power lasers are required to populate these levels. However, it should not be made too strong, otherwise, 'power-broadening' will distort the levels being investigated. Atomic beam methods have been favored because their low densities avoid unwanted collisional effects. Field ionization detection of resonance transitions to high n states has been applied with success by Kleppner et al., (1983). The method works in the following way: after a high Rydberg state has been populated by a tunable pulsed laser, a high electric field is rapidly applied and field ionization produces a free electron and an ion. A charged particle detector or multiplier is used to measure the signals. Different states will be field ionized at different field values as illustrated in Fig. 5. Depending on resolution, the selectivity is due to the binding energy being a function of n, ℓ and the magnetic sublevel m_ℓ.

It is possible to analyze Rydberg excitation by recording the fluorescence of the states. However, the fluorescence signal in the visible or UV regions is low and the method is limited to the study of relatively low n numbers.

A laser beam intersecting the narrow atomic beam at right angles will eliminate much of the Doppler effect. Resolutions of 20 MHz with a pulsed laser or 2 MHz with a cw laser can be obtained. To achieve further linewidth reduction counterpropagating beams can be used. Doppler-free pumping of Rydberg states using two-photon absorption has been reported. In these cases, transit time broadening begins to limit the resolution, especially if the atom beam originates from an ion beam by charge exchange.

Although the sensitive properties of Rydberg states are an advantage, the experimenter must be conscious of their sensitivity to perturbations other than the planned experiment. The diamagnetic Zeeman effect causes a level shift similar to the Stark effect. For most experimental situations the shift will be small. For example, a field of

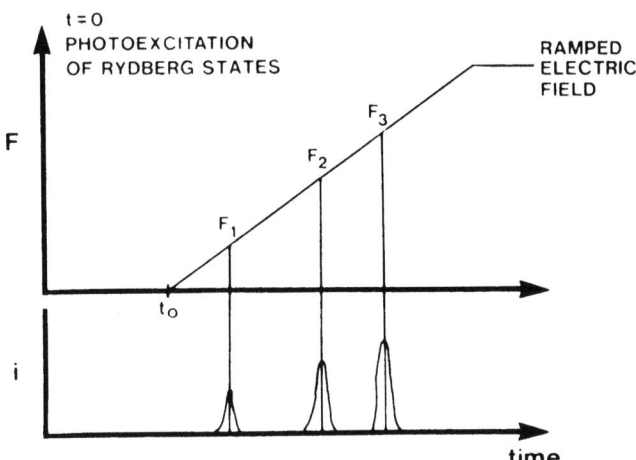

Fig. 5. Illustration of field ionization detection of Rydberg states.

one gauss shifts the n = 30 level by 200Hz. However, electrostatic perturbations are more dangerous. The static polarizabilities increases as n^7 and are also proportional to ℓ^5 for non-penetrating orbits. A stray electrostatic field of about 100 mV cm^{-1} induces a level broadening of about 5 MHz for n = 30.

While Rydberg states are relatively insensitive to visible or near-infrared fields this is not true for radio frequency or far infrared radiation. If the radiation is resonant the transition can be saturated with very low power levels. When the radiation is non-resonant the levels are shifted by the ac Stark effect. The blackbody radiation field of 300K background slightly shifts the Rydberg levels (approximately independent of n) by 24 kHz (Gallagher, 1983).

Many researchers have taken advantage of the unusual sensitivity and scaling of Rydberg state properties to determine their collision and reactions, interaction with radiation, and behavior in external fields. The early experiments used the alkali metals because the available lasers could excite high n levels of these atoms. Later studies used the advantages of two-photon spectroscopy to eliminate Doppler broadening and to derive polarizabilities for many levels. Recent work has used two-, three- and four-photon optical pumping to access the Rydberg states of many atoms. Some of the experimental methods to measure the high levels of hydrogen are summarized in Table 3.

Glab and Nayfeh (1985) recently made the first observations of the Stark induced resonances in the photoionization of hydrogen using the "straight forward" approach shown in Fig. 6. The n = 2 level of hydrogen was excited by two-photon absorption at 243 nm. A second beam tunable around 365 nm is used to excite Rydberg states near the continuum. [The generation of the pulsed laser beams at the appropriate wavelengths is non-trivial: a Nd:YAG laser is doubled to 532 nm, and a fraction is used to pump a dye laser at 630 nm which is then doubled to 315 nm and summed

Table 3. Experimental Methods to Access High Rydberg Levels

EXCITATION METHOD	REFERENCES
CHARGE EXCHANGE	KOCH & MARIANI (1981)
ATOMIC BEAM	W. SMITH (1981)
3-PHOTON ABSORPTION AT 273 nm	BJORKLUND et al. (1979)
TWO-STEP EXCITATION $\lambda_1 = 122$ nm, $\lambda_2 = 356$ nm	K. H. WELGE (1981)
2-PHOTON, 2-STEP $\lambda_1 = 243$ nm, $\lambda_2 = 365$ nm	GLAB & NAYFEH (1985)

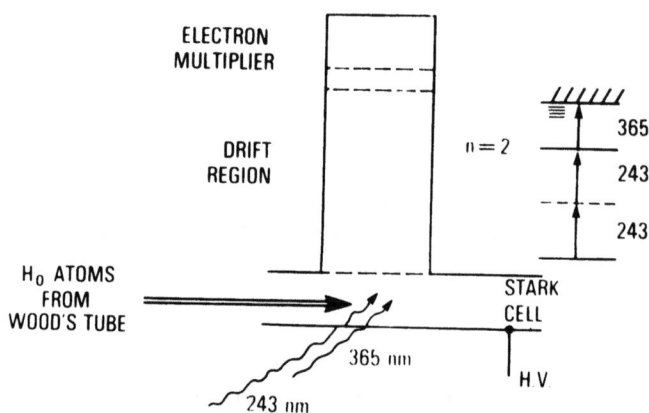

Fig. 6. Schematic of Glab and Nayfeh's experiment to observe Stark-induced resonances in the photoionization of hydrogen. Two-photon pumping at 243 nm is used to access the n = 2 state. A second beam is tunable around 365 nm to access high Rydberg states.

in a KDP crystal with the residual YAG fundamental to give 243 nm, bandwidth 1.5 cm^{-1}, pulse length 10 ns, energy 10 mJ. The remaining fraction of 532 nm pumps a second dye laser to give 555 nm which is summed with another part of the 1064 nm to produce 365 nm, bandwidth 0.6 cm^{-1}, pulse length 10 ns, energy 0.2 mJ.] The photoionization spectrum of hydrogen was obtained using π and σ polarizations. Modulations in the ionization signal were observed essentially only when both laser beams had π polarization with respect to the Stark field. The experimental Stark fields were several kV/cm. Below the critical energy (E_c) sharp symmetric peaks corresponding to the highly excited 'quasidiscrete' Stark states were observed. At short wavelengths, fairly broad resonances appeared and persisted above the critical energy. These broader resonances have asymmetry in the form of blue wings near E_c. They are interpreted as resulting from tunneling across the Stark-Coulomb barrier. (The detailed interpretation of these spectra is complicated by the fact that the n = 2 state is also split by the relatively large Stark fields; at 475 volts/cm the Stark effect and the Lamb shift are comparable and at 2.9 kV/cm the Stark effect and the fine structure splitting are comparable.) An equation derived by Luders (1951).

$$F_s = 10^5 \left(\frac{Z}{n}\right)^5 \text{ V/cm} \tag{18}$$

relates the field F_s to produce Stark shifts within a given n-manifold that are of the order of the fine-structure splitting. The extension of the Stark analysis to 'one' electron Rydberg atoms such as the alkali atoms has been made by Fano (1961), Harmin (1984), and by Cohen-Tannoudji and Avan (1977). The levels are divided into two categories due to the fact that the quantum defects of alkali atoms are negligibly small for states with angular momenta $\ell > 3$. This leads to the concept of an incomplete hydrogenic manifold and nonhydrogenic states that interact with each other. Experiments by Chardonnet et al. (1984) on cesium have given confirmation to these general concepts.

Argon ion (Ar III) laser excitation has been used to produce fast hydrogen and helium Rydberg states by intersecting a cw laser beam with (Doppler velocity tuned) fast atoms derived by charge transfer from an ion beam. Similar pumping can be achieved using CO_2 laser photons to excite from the n = 10 state. Using these methods (including double resonance) partial cross-sections for the production, rearrangement and ionization of Stark substates in various collisions and in electric and magnetic fields can be studied with great precision.

Field effects on the oscillator strength distribution near their ionization potentials of the heavier noble gases and helium have been investigated using absorption of synchrotron radiation. Although this method has some advantages in that it permits measurement on levels all the way from the ionization limit to low n-levels, the laser pumping of Rydberg states and field ionization detection permits higher energy resolution. The detailed agreement between theory and experiment leads to the conclusion by van de Water et al. (1984) that the field ionization of helium is "now understood in a detailed, quantitative and predictive sense."

Applications to Low Pressure Discharge Diagnostics

The Stark effect of atomic Rydberg states and Stark-induced mixing of molecular states recently has been used very successfully to measure electric fields in both dc and rf discharges. In addition, Doppler-free Stark broadening spectroscopy has been applied to arc discharges.

The Stark effect produces both a shift and a broadening of the line. The advantage of using Stark broadening as a method to measure electric field over Stark splitting is that the resolution in electric field (to 1 V/cm) is better. Also, the Rydberg transition measurements with $\Delta m = \pm 1$ polarization simultaneously determines the metastable densities and indirectly the electron density. Recent studies on helium and neon have emphasized opto-galvanic detection (OGS) advocated by Katayama et al. (1979). The schematic of a typical experimental setup used by Ganguly and Garscadden (1985) is shown in Fig. 7. A He-Ne laser type glow discharge tube (6 mm ID) equipped with side windows and Langmuir probes is operated at low milliampere currents at pressures around 1 torr. A Q-switched frequency doubled Nd:YAG laser pumps a DCM dye. The dye laser output (0.3 cm^{-1} FWHM linewidth) is doubled again in a KDP crystal which is continuously angle-tuned with an Inrad autotracker. This setup produces tunable UV output to excite, for example, selected atomic helium np^1P states from the $2s^1S$ metastable state. (Fig. 1 shows the levels

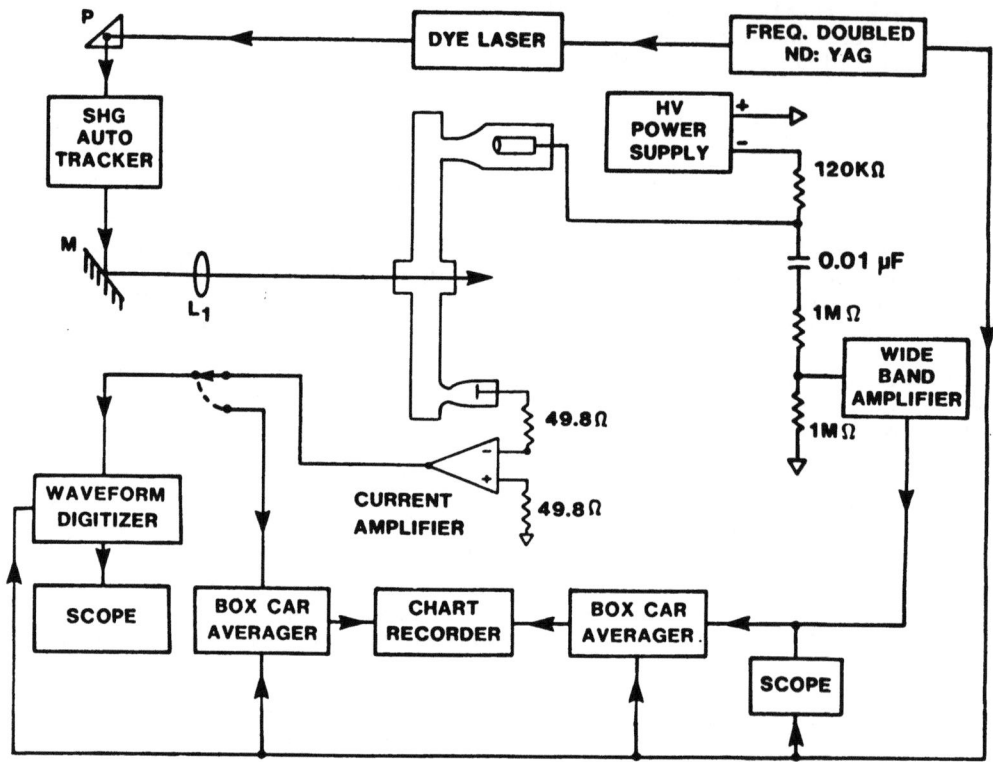

Fig. 7. Experimental arrangement to study Rydberg states in a helium low-pressure discharge by the optogalvanic effect.

that are being pumped from the metastable 2^3S and 2^1S levels). In this case the discharge performs the function of populating these levels and thus gives access to the Rydberg states of the rare gases with near-UV lasers. The laser polarization is variable but for the initial measurements it was perpendicular to the axis of the discharge. The dye laser pump beam and the UV beam are spatially separated using a Pellin-Broca prism. The dye laser output wavelength calibration is achieved by simultaneously measuring the opto-galvanic spectrum of a commercial hollow cathode neon discharge. Measurements are made with a wide bandwidth amplifier and a boxcar averager or a transient digitizer. To ensure constant interaction volume the positive column profile is measured by vertically translating the discharge tube. The laser beam is mildly focused to 0.3 mm x 2 mm spot size to avoid broadening at the modest laser energies (100-200 microjoules/pulse and pulse width 10 ns).

The atomic helium spectra for Rydberg levels n = 26 and higher are shown in Fig. 8 for five different radial positions in the positive column. There are several interrelated features of these spectra. First, note that there is a variation in the linewidth of each of the transitions, which increases radially. Secondly, the magnitude of the opto-galvanic signal decreases radially. Thirdly, the advance of the series goes to lower n the further off-axis. Fourthly, there is the appearance of satellite structure in the off-axis spectra. Not shown, if the discharge is run at higher currents, the structure of the higher transitions tends to disappear. The linear Stark line broadening, under the quasi-static approximation is given by:

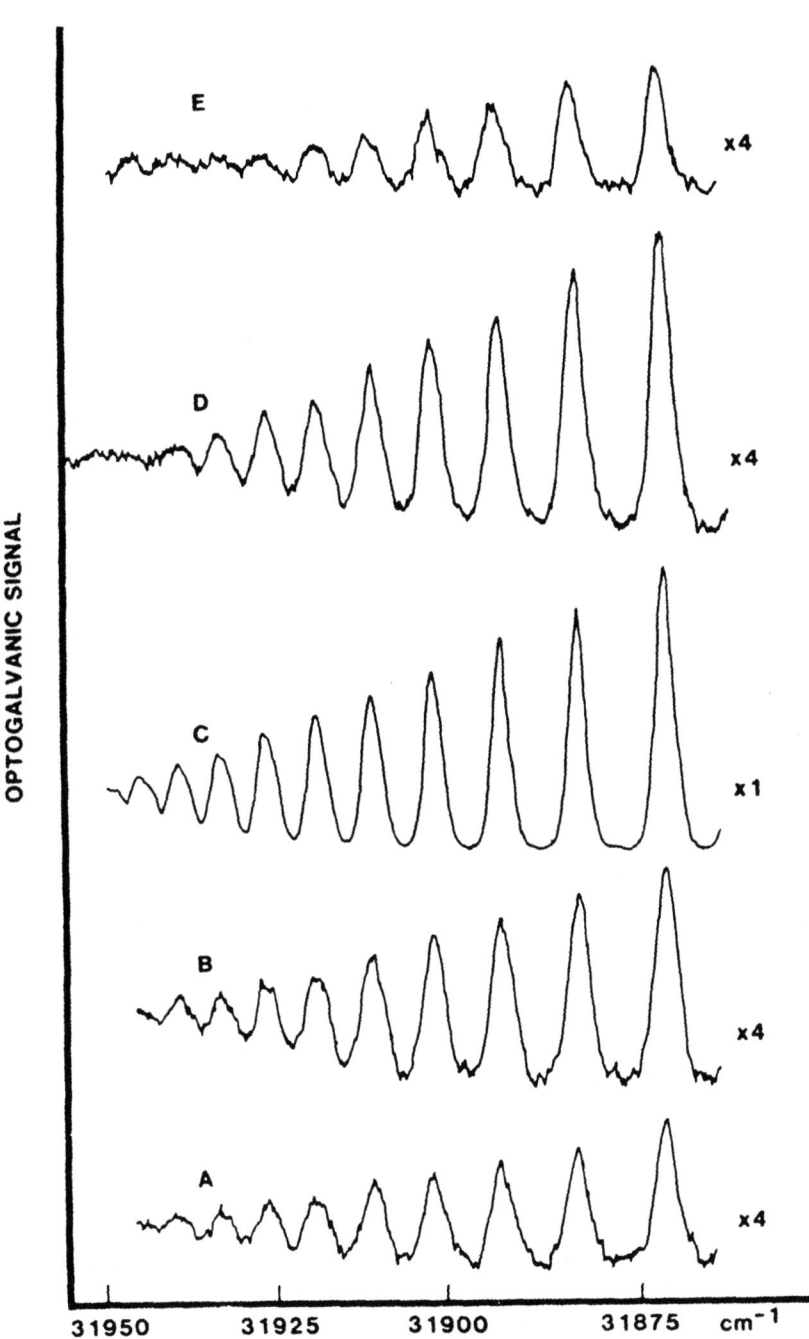

Fig. 8. Optogalvanic spectra for different radial positions in the positive column. The laser polarization was perpendicular to the discharge axis.

$$|\Delta\omega| = \frac{3\hbar}{2emZ}(n_i^{*2} - n_f^{*2}) F \qquad (19)$$

where $|\Delta\omega|$ is the HWHM of the transition. n_i^* and n_f^* are the initial and final state quantum numbers, and F is the total electric field in atomic units. Therefore, the net electric field at any given radial position can be obtained from the FWHM of the measured transitions. The electric field was calculated from the n = 26 to 33 lines of the Rydberg states. The measured radial profile of the net electric field in the positive column is shown in Fig. 9. The measured electric field on axis was 10 volts/cm ±1 volt/cm whereas the estimated axial field, based on the Schottky formalism, is 12-14 volts/cm. For these experimental conditions, it is difficult to estimate the axial field with a high degree of accuracy due to the marginal applicability of an ambipolar diffusion model and the perturbation of the column by the optical windows. Overall, the measured and estimated axial electric fields are in reasonable agreement. This experiment shows that there is no one $\frac{F}{N}$ parameter (N is total gas density) for Boltzmann transport calculations when discharges are strongly affected by boundaries. The limits on the resolution of the

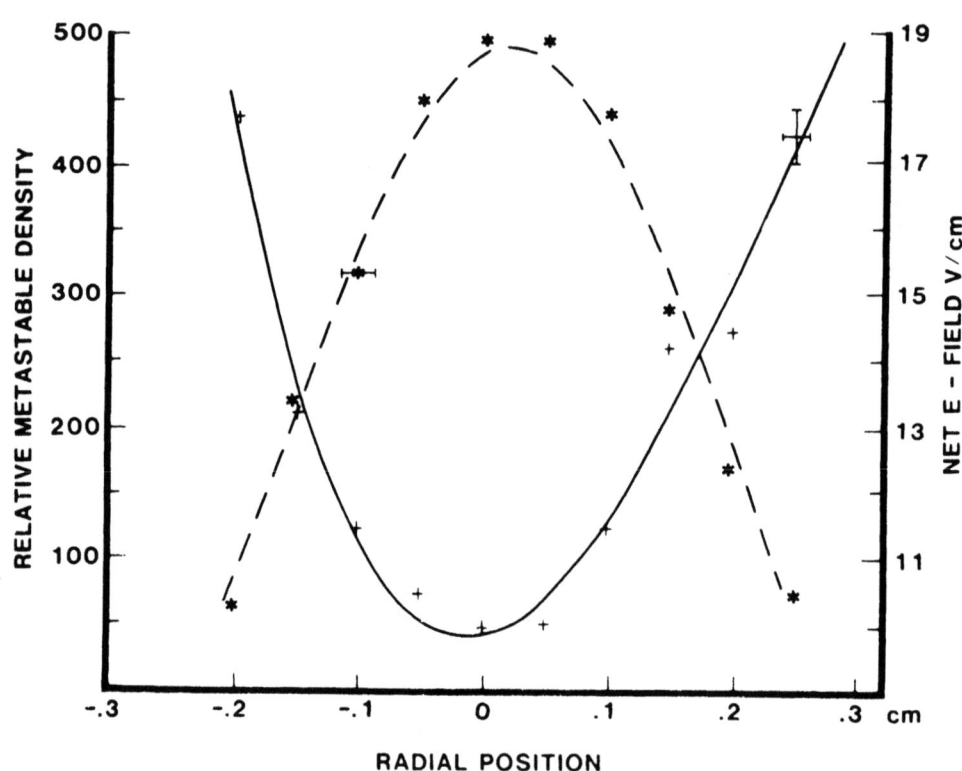

Fig. 9. Measured radial profile of the net electric field in the positive column of a helium discharge.

method at higher currents and higher charge densities are the effect and magnitude of the microfield. The Holtsmark field has been estimated for a range of ion densities and high electron temperatures. The results are shown in Fig. 10. The electron density for these discharges is estimated from the relation

$$I \text{ (tube current)} = 2 \, eW_D n_o \int_0^R J_0\left(\frac{2.4r}{R}\right) r \, dr \qquad (20)$$

where J_0 is the zeroth order Bessel function describing the electron density radial profile, n_o is the axial electron density, R is the tube radius and values of the drift velocity W_D were obtained from the tabulation of Gilardini (1972). At the low currents used the microfield contributes less than 1 volt/cm. However, it was observed that the linewidths are larger off-axis than on-axis, even though the electron density is decreasing. This is because the optical method measures the net electric field which includes the ambipolar radial field. Thus it is possible to obtain information about its magnitude also for axially symmetric conditions.

The progression of the series termination with the increase in radial field off-axis is consistent with the calculated magnitude of the net static Stark effect. Also, the appearance of the satellite structures is a consequence of the increasing radial field contribution to the net field. For the off-axis spectra, the laser polarization is no longer perpendicular to the E-field direction (i.e., the electric field of the discharge is no longer purely axial) and therefore $\Delta m_J = 0$ (m_ℓ in high fields) transitions are Stark-allowed and the "forbidden" ^1S states contribute to the spectra.

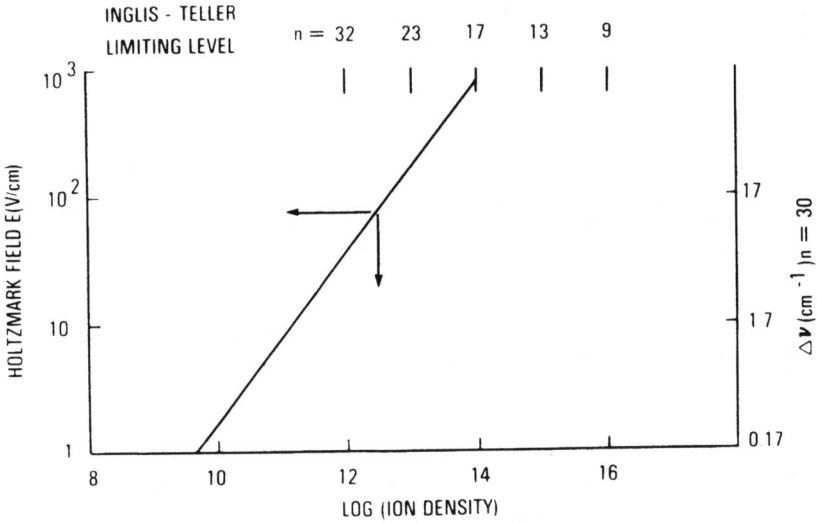

Fig. 10. Holtsmark field and maximum level dependencies on ion density.

It is possible to exploit this observation further to determine the electric field vector in the discharge. The discharge tube was rotated to an angle of 130° (chosen from previous experience) with the laser polarization. In this experiment we measured the Rydberg spectra of ^3S and ^3P states of atomic helium of n = 25 and above, photo-excited from the metastable state 2^3S. Figure 11 shows the ^3S and ^3P spectra for three different (representative) radial locations. The spectra labeled A and C are at radial locations 2 mm off-axis on either side (see inset). The spectrum B is at the center of the tube. As in the singlet experiments the metastable helium population is measured to be a maximum at the center of the tube. However, the linewidths are narrower there as the field is purely axial and the net field is lowest there. Again, note that the onset of the continuum occurs at relatively high principal quantum numbers (n ~ 35). Both the off-axis spectra show increased linewidths as expected due to the increased net field. From the linewidths it is possible to derive the magnitude of the net field. The FWHM of the $29p\,^3P$ transition in spectrum B is 1.85 cm^{-1}, corresponding to a field of 12 V/cm. In A and C the FWHM of the same transition is 2.8 cm^{-1} which gives a net electric field value of 17 V/cm. However, there are remarkable differences in the details of the spectra. In spectrum A, essentially only the ^3P transitions are observed and the forbidden ^3S transitions are not observed. Even though the net electric field at point A is higher than the axial field of point B, only the latter shows the ^3S transitions. At point A the electric field vector in the discharge is orthogonal to the laser polarization and, therefore, $\Delta m_J = 0$ transitions are not allowed. Consequently, the $2s\,^3S$ to $ns\,^3S$ transitions are forbidden even in Stark spectra. The relative magnitudes of the optogalvanic signals in the continuum overshoot region (above 38370 cm^{-1}) have the characteristic differences for $\Delta m_J = 0$ and $\Delta m_J = 1$ transitions observed in field ionization.

It is possible to test the validity of equation (19) by plotting the measured linewidths of different levels versus the parameter Γ where

$$\Gamma = (n_n^{*2} - n_i^{*2})/(n_{25}^{*2} - n_i^{*2}) \tag{21}$$

where n_n^* is the measured level, n_i^* is the initial level and n_{25}^* is the chosen reference level. Based on measurements of Ganguly and Preppernau, these calculations have been performed for both the triplet series (Fig. 12) and the singlet series (Fig. 13). The results are in the expected relationship.

Figures 14 and 15 show a comparison of the optogalvanic response for excitation in the helium discharge negative glow and in the positive column respectively. Whereas the series appears to terminate around n = 37 in the positive column, it extends out to as high as n = 50 in the negative glow. From the nearest neighbor approach to describe the Holtsmark field for ion broadening and the Inglis-Teller relationship for the maximum discrete level n_m, we obtain a relationship between the net electric field and the maximum level:

$$F_o \text{ (volts } cm^{-1}) = 1.2 \times 10^9 \, n_m^{-5} . \tag{22}$$

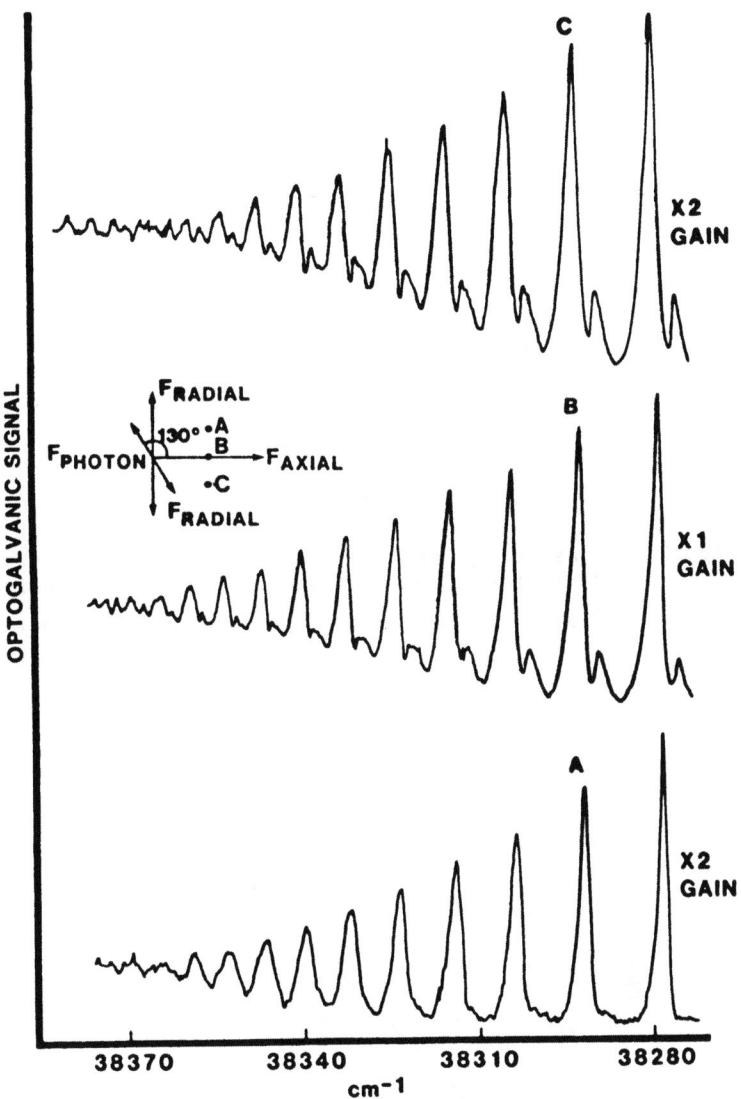

Fig. 11. Optogalvanic spectra for axial and radial positions in the positive column when the laser polarization is chosen to correlate with the net electric field vector at one radial position.

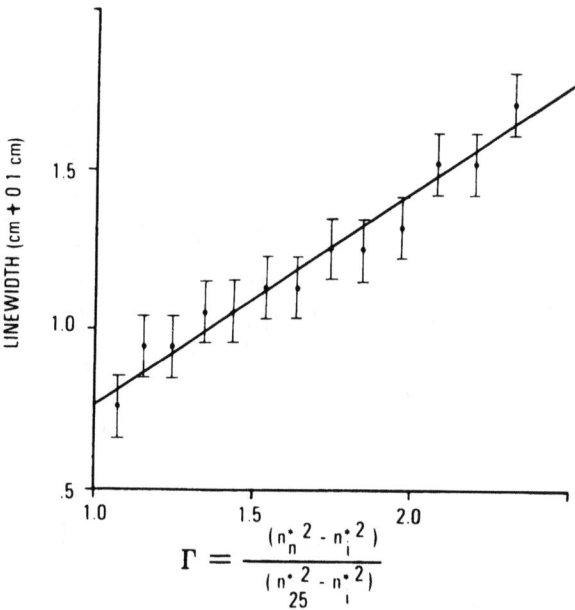

Fig. 12. Measured Rydberg state linewidths versus the parameters Γ for the triplet states. n_n varied from 27 to 33 (test made in positive column).

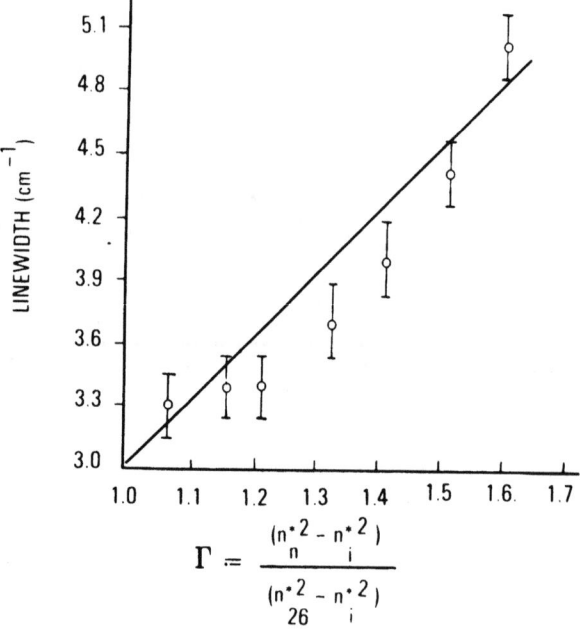

Fig. 13. Measured Rydberg state linewidths versus the parameter Γ for the singlet states. n_n varied from 26 to 38 (test made in negative glow).

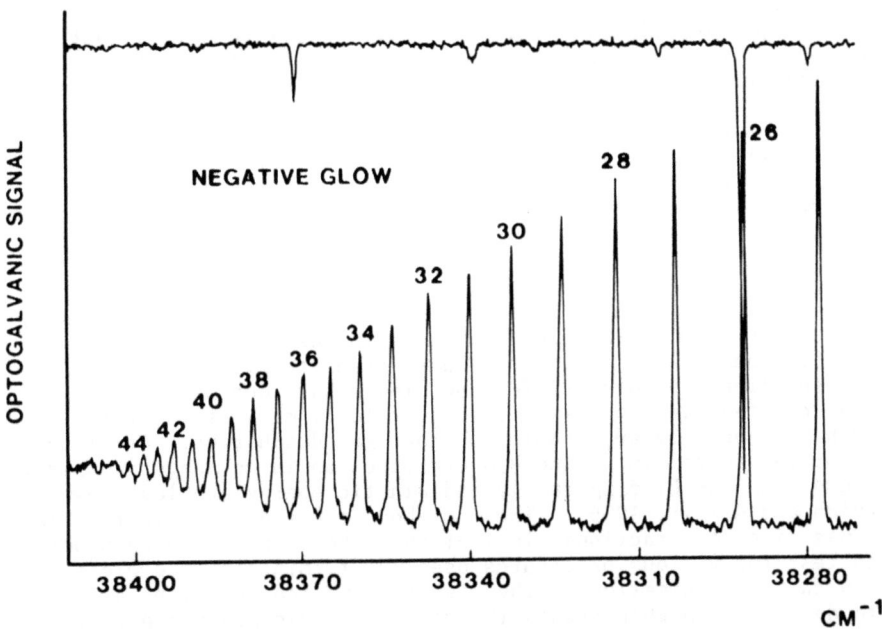

Fig. 14. Optogalvanic response in the helium negative glow.

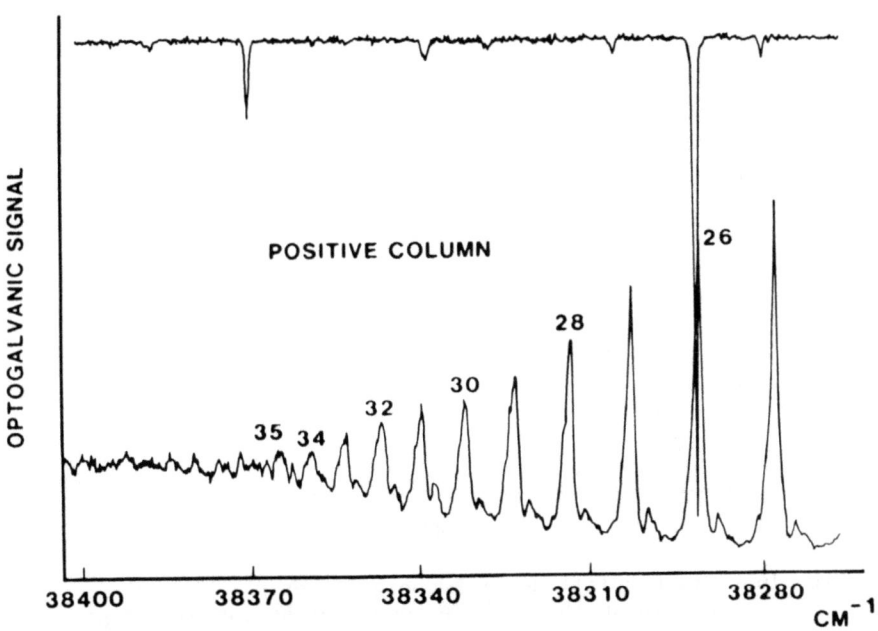

Fig. 15. Optogalvanic response in the helium positive column.

It is therefore estimated that the electric field in the negative glow must be less than 3 volts cm^{-1}.

In the cathode sheath the fields are much higher. With the resolution of the experimental system, it is therefore possible to illustrate the Stark splitting of the helium lines, e.g., at n = 14 and 15. (Lower lines exhibit less splitting, higher lines are more smeared for this estimated cathode sheath field of 600 volts/cm.) The results are shown in Fig. 16 for a measurement 0.25 mm from the cathode. They can be regarded as a vertical section of Fig. 4. Calculations of the effect of electric fields on helium spectra were performed by Foster (1927) as early as 1927. The estimate of the field was obtained using his methods. A full scan of the cathode fall is provided in another part of this Proceeding dealing with laser diagnostics.

The optogalvanic effect has been used very successfully by Doughty et al. (1984) to examine the electric field in the cathode fall. An additional refinement was to use two laser beams to give a pin-point spatial resolution. Bridges (1978) and others have demonstrated the optogalvanic effect in many gases and, therefore, many of the techniques outlined here should be applicable to other gas discharges. At higher electron or gas temperatures the Rydberg electron is no longer considered to be bound when its binding energy and kinetic energy are less than kT (Zel'dovich and Raizer, 1966). These methods therefore are most suitable for plasmas with low fractional ionization. There appear to be opportunities to test the various procedures used for terminating the electronic partition function summation if the electron and ion densities are allowed to decay. At high currents the electric field measurements will require discrimination against the isotropic microfield. The polarization effects described allow measurement of anisotropic fields to less than the microfield value.

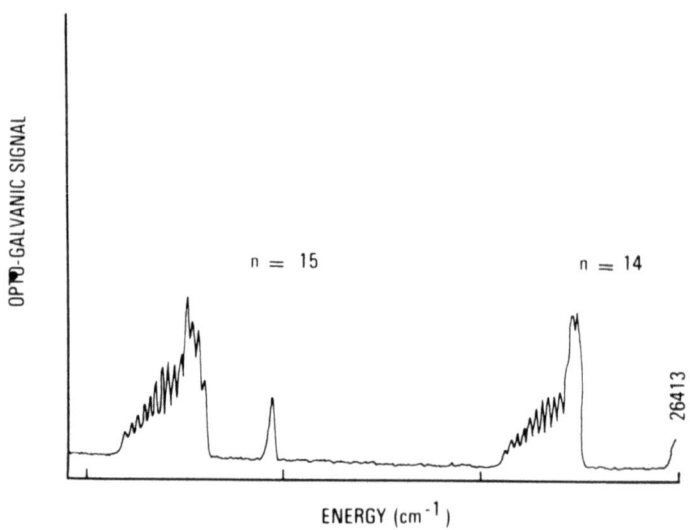

Fig. 16. Stark splitting of Rydberg states n = 15 and n = 16 in the cathode sheath of helium discharge. Scale: 38.7 cm^{-1}/division.

The combination of high Rydberg states, tunable lasers and sensitive detection methods such as optogalvanic spectroscopy offer great promise for further advances in discharge diagnostics (Dunning, 1985).

ACKNOWLEDGMENT

It is a pleasure to acknowledge the assistance and elegant experiments of my colleague Dr. B. N. Ganguly. These studies were supported by the AFOSR/AFWAL in-house basic research program.

REFERENCES

Bayfield, J. E., 1983, American Scientist, 71:375.
Bethe, H. A., and Salpeter, E. E., 1957, "Quantum Mechanics of One- and Two-Electron Atoms," Springer-Verlag, Berlin.
Bjorklund, G. C., Freeman, R. R., and Storz, R. H., 1979, Opt. Comm., 31:47.
Bridges, W. B., 1978, J. Opt. Soc. Am., 68:352.
Butler, S. T., and May, R. A., 1965, Phys. Rev., 137A:10.
Chardonnet, C., Delande, D., and Gary, J. C., 1984, Optics Comm., 51:249.
Cohen-Tannoudji, C. and Avan, P., 1977, in: "Colloque International," CNRS No. 273, Paris.
Condon, E. U., and Shortley, G. H., 1967, "Theory of Atomic Spectra," Ch 17, Cambridge University Press, Cambridge.
Cooke, W., and Gallagher, T. F., 1978, Phys. Rev., A17:1226.
Delpech, J. F., Boulmer, J., and Devos, F., 1977, Phys. Rev. Lett., 39:1400.
Doughty, D. K., Salih, S., and Lawler, J. E., 1984, Phys. Lett. 103A:41.
Dunning, F. B., 1985, Am. J. Phys., 53:944 and references therein.
Fano, U., 1961, Phys. Rev., 124:1866.
Foster, J. S., 1927, Proc. Roy. Soc., A114:47.
Gallagher, T. F., 1983, in: "Rydberg States of Atoms and Molecules," R. F. Stebbings and F. B. Dunning, eds., Cambridge University Press, Cambridge.
Ganguly, B. N., and Garscadden, A., 1985, Appl. Phys. Lett., 46:540.
Gilardini, A., 1972, "Low Energy Electron Collisions in Gases," John Wiley and Sons, New York.
Glab, W. L., and Nayfeh, M. N., 1985, Phys. Rev., A31:530.
Harmin, D A., 1984, Phys. Rev., A30:2413.
Katayama, D. H., Cook, J. M., Bondybey, V. E., and Miller, T. E., 1979, Chem. Phys. Lett., 62:542.
Kleppner, D., Littman, M. G., and Zimmerman, M. L., 1983, in: "Rydberg States of Atoms and Molecules," R. F. Stebbings and F. B. Dunning, eds., Cambridge University Press, Cambridge.
Koch, P. M., and Mariani, D. R., 1981, Phys. Rev. Lett., 46:1275.
Luders, G., 1951, Ann. Phys (Leipzig) Ser. 6, 8:301.
Mutsuzawa, M., 1980, J. Phys. B, 13:3201.
Olson, R. E., 1979, J. Phys. B, 12:L109.
Rinneberg, H., Neukammer, J., Jonsson, G., Hieronymus, H., Konig, A., and Vietzke, K., 1985, Phy. Rev. Lett., 55:382.
Schiavone, J. A., Donohue, D. E., Herrick, D. R., and Freund, R. S., 1977, Phys. Rev., A16:48.
Smith, W., 1981, in: "Symposium on H° in Strong Fields," Los Alamos, New Mexico, (unpublished).
Stebbings, R. F., and Dunnin, F. B., eds., 1983, "Rydberg States of Atoms and Molecules," Cambridge University Press, Cambridge.
van de Water, W., Mariani, R., and Koch, P. M., 1984, Phys. Rev., A30:2399.
Welge, K. H., 1981, in: "Symposium on Photoionization of Excited Atoms and Molecules," JILA, Boulder, Colorado (unpublished).

Wing, W. H., Lea, K. R., and Lamb, W. E., Jr., 1973, in: "Atomic Physics 3," S. J. Smith and G. K. Walters, eds., Plenum Press, New York.
Zel'dovich, Ya. B., and Raizer, Yu. P., 1966, "Physics of Shock Waves and High Temperature Hydrodynamic Phenomena," Academic Press, New York.

RESONANCE IONIZATION SPECTROSCOPY

C. Grey Morgan

Department of Physics
University College of Swansea
Swansea, Wales

INTRODUCTION

Resonance Ionization Spectroscopy (RIS) is one of the most powerful laser-based analytical techniques available to scientists. It is a very versatile method capable of, in the limit, detecting a single stable thermal-energy atom in a specific quantum state in the presence of an overwhelmingly larger number of other atoms of a different species. It is highly selective and has very good spatial and temporal resolution. It may be used with pulsed or cw lasers.

Since the conceptual basis of RIS occurred to Dr. G. S. Hurst of the Oak Ridge National Laboratory, Tennessee, in 1974 it has been developed into a practical laboratory tool which is having profound consequences in physics and chemistry and undoubtedly will have far reaching effects in many other disciplines.

RIS is essentially a highly selective multi- or multiple-photon photoionization process which takes advantage of the availability of moderately powerful tunable lasers whose photon energies can be adjusted to resonate with allowed energy levels in the selected atoms. Rapid multiple photon absorption causes ionization and either one or both of the electron-ion pair thus created can be detected and used to indicate the selected atom's presence. The atomic species discrimination and selectivity is ensured by the resonant multiple photon absorption into the bound states and by the fine tunability of the laser. With a large enough photon flux all the atoms of a given element and none of any other element present in the laser beam will be ionized and selective saturated ionization takes place. This saturation is required to ensure single atom detection. Even in a relatively large volume (several milliliters) one atom of a selected element can be ionized by a pulsed laser flash but not any of the extremely large number of atoms of any other element which may be present. Much smaller volumes are saturated by currently available cw tunable lasers on account of their lower powers. The free electron liberated from a single atom can readily be increased in number by accelerating it to cause avalanche ionization growth by collisions in the gas in a proportional or Geiger-Mueller counter, or alternately, the ion can be detected by a suitably modified electron multiplier such as a channeltron if a vacuum environment is used. The ion may also be mass analyzed to provide isotopic information.

Electrons and ions are generally easier to detect than photons so RIS has a number of distinct advantages over laser-induced fluorescence as a method for trace element detection with, of course, its now well-established ultimate limit of detectability.

The first experiment (Hurst et al., 1975) employing the RIS principle was the photoionization of metastable He (2^1S) by two-photon absorption via the intermediate He (3^1P) state to produce He$^+$ and an electron which was detected in an ionization chamber. The population of singlet helium atoms was created in the helium gas in the chamber by protons from a 2 MeV Van de Graaff generator. Only if the laser radiation was tuned to 501.5 nm, i.e., into resonance with the energy gap between 2^1S and 3^1P levels were these singlet metastables ionized. At other wavelengths no such ionization could take place due to the lack of resonant energy transfer from the laser beam to the singlet metastable atoms. This experiment established the RIS concept in practical form.

Hurst and his colleagues have perfected techniques, which, for the first time, can detect and identify a single free atom of any element with the current exception of ground state helium and neon. With the future attainment of shorter wavelength tunable lasers than are at present available these too will be detectable by RIS. In certain cases the selected atom can be stored for later retrieval and re-examination (Chen et al., 1980; Chen, 1984). This extended RIS technique has the propensity for isotope separation.

While it has been possible to detect single electrons and unstable, i.e., radioactive, atoms since the invention of the proportional counter there was no assured and reliable technique for single stable thermal-energy atoms until 1977 when Hurst and his colleagues (Hurst et al., 1977) unambiguously demonstrated the detection of single atoms of cesium. They achieved this notable advance by using a finely-tuned dye laser to photo-ionize a cesium atom by two-photon absorption and then used amplification of the electron so liberated by Townsend avalanches in a proportional counter to indicate its presence, and, by inference, that of the Cs ion and hence atom. In this crucial experiment to demonstrate the extraordinary selectivity and sensitivity of the RIS technique they showed that it was possible to detect a single Cs atom even when there were 10^{19} atoms of argon and 10^{18} molecules of methane in the same laser beam. These gases formed the counter gas.

A further spectacular demonstration of the power of RIS for single atom detection which again involved Cs was the application to the spontaneous fission decay of ^{252}Cf in which a Cs atom is emitted. This experiment proved that a single neutral daughter atom could be counted in temporal coincidence with the nuclear decay of its parent atom (Kramer et al., 1978.)

Figure 1 illustrates the RIS principle in its simplest form for the case of two-photon absorption leading to ionization and simultaneously occurring competitive processes of spontaneous decay and collisions resulting in depopulating the intermediate state. The selected atom in its ground state (1) is excited to level (2) lying more than half way up to the continuum by resonant photon absorption from a flux of photons having energy $h\nu$ equal to the energy difference between these states. If $h\nu$ has some other value the absorption cross section is negligible and no excitation takes place and the atom remains in its ground state. If a second resonance photon is absorbed by the excited atom before it is de-excited by spontaneous decay or by collisions with gas atoms, say at a combined rate D, it may be ionized and hence detected. With an appropriately large photon flux the ground and excited state quasi-equilibrate because the rates of absorption and stimulated emission greatly exceed the combined photoionization, spontaneous, and collision decay rates. Detailed

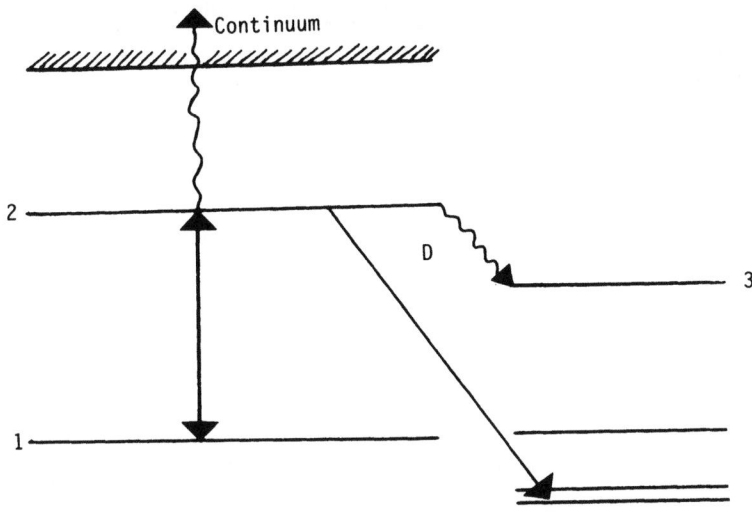

Fig. 1. Two-photon resonance ionization from ground state: (1) via intermediate state, (2) which is competitively depopulated by spontaneous decay to state (3) and collisions at a combined rate D sec^{-1}

consideration of the rate equations describing the populations in the levels as a function of the laser flux F, irradiation time T and photo-ionization cross-section σ of level (2) show that in order to achieve saturated ionization, i.e., to ionize all the atoms of the selected species in the irradiated volume, two conditions must be satisfied, viz,

σ F >> D, and

σ FT >> 1

With typical values for σ the corresponding laser requirement is FT>100 mJ cm^{-2} at the resonance wavelength. This is readily achievable with commercially available dye lasers.

For some other atoms it is necessary to use more complicated schemes which involve two or more synchronous tunable beams to excite the ground state atom to a first intermediate selected quantum state using a photon $h\nu_1$ and then to a second higher allowed level by a photon $h\nu_2$ from a second laser, and finally into the continuum by a photon from either. Some five multiphoton and multiple-photon schemes have been proposed by Hurst and his colleagues (Hurst et al., 1979) which are capable of resonantly ionizing all elements except, He and Ne in their ground state. These are shown in Fig. 2.

As the schemes become more complex the laser intensities required also increase and the wavelengths get shorter and considerable ingenuity has been shown in devising laser systems to achieve the appropriate wavelengths. An excellent example is the system developed for RIS studies of Ar, Kr and Xe (Kramer, 1984).

An enormous variety of investigations based on one or another of the above five schemes has recently been reported, ranging from searches for exotic atoms such as superheavy atoms and particles with fractional charges-quarks (Fairbank, 1984), the nature of the neutrino as deduced from double beta decay studies (Moe, 1984; Haxton, 1984), the solar neutrino anomaly (Cowan et al., 1984), to searches for magnetic monopoles (Hurst, 1984).

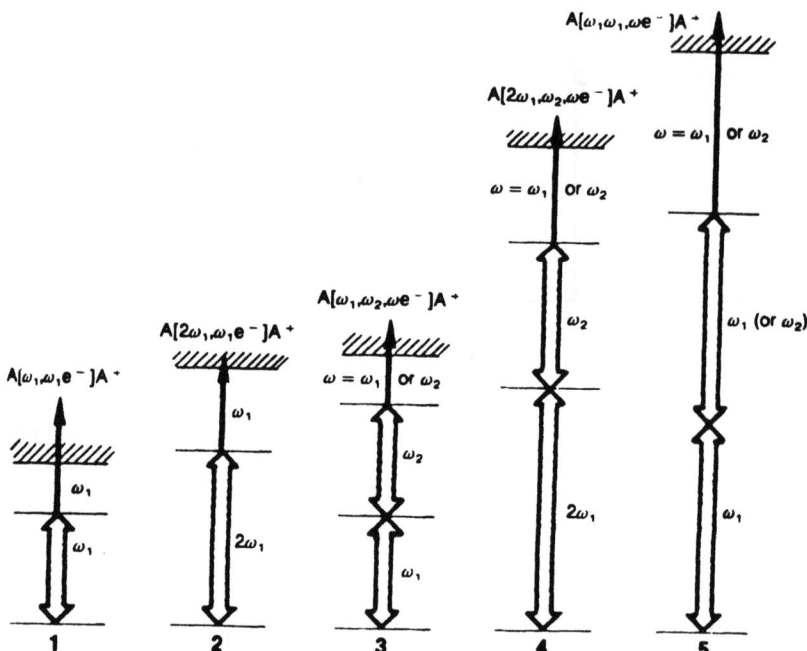

Fig. 2. Resonance ionization schemes for the elements.

RIS WITH MASS SPECTROMETERS

The coupling of a time-of-flight or magnetic sector mass spectrometer to the basic RIS apparatus is a logical step which provides a very powerful combination, generally designated RIMS (Resonance Ionization Mass Spectroscopy) in which a variety of techniques have been used to provide the gas phase needed for the RIS process. It should be noted here that although the gas phase is essential, this does not mean that RIS is limited to use only with gas samples. Any material which can be vaporized may be used and advantage taken of the various methods commonly used for atomization in mass spectrometer ion sources. These include ion bombardment of solids for the study of sputtering mechanisms (Robinson, 1984), surface structural and chemical analysis (Winograd et al., 1982, Winograd, 1984), glow discharges (Harrison et al., 1984), and the ubiquitous hot wire (Donohue et al., 1983, Nogar et al., 1983, 1984) to study isotopic ratio measurements of rare earths in the presence of isobaric interferences as well as for direct quantitative measurements of several elements (Fasset et al., 1983, Travis et al., 1984). An advantage of RIMS is that it almost automatically reduces the problem of isobaric interferences by virtue of its high selectivity.

SIRIS

Possibly of most direct interest to the present meeting is the development of the RIMS technique to the stage where it is a commercially available service for ultra-sensitive elemental analysis of a variety of materials. SIRIS (Sputter Initiated Resonance Ionization Spectroscopy), (Parks et al., 1983, 1984), employs a duoplasmatron ion source to provide energetic ions which are used to bombard the target to produce the gas phase needed for RIS as shown in Fig. 3. Argon is normally used in the source. The argon ions are passed through the analyzing magnet and are pulsed by deflection across a pair of chopping slits and focused on to the sample placed in the target chamber. Precise target positioning in

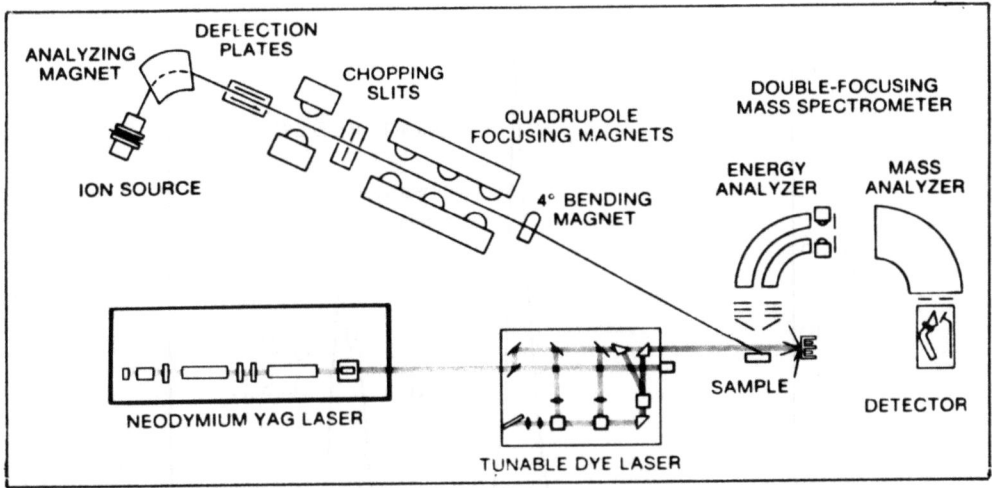

Fig. 3. Diagrammatic representation of SIRIS apparatus.

three dimensions by X-Y-Z rastering enables transverse and depth profiling to be carried out. A neodymium-Yag laser pumps a tunable dye laser to provide the resonant radiation to interact with the sputtered atoms. Its repetition rate is normally 30 Hz. The ions of the selected atoms so liberated are accelerated and analyzed with an electrostatic sector for energy determination and by a magnet sector for mass identification. Ion detection is achieved with an electron multiplier provided with a conversion dynode. UHV techniques are used to evacuate the system. The whole SIRIS system is micro-computer controlled and data acquisition and processing is also by computer. Both single ion counting and analog pulse averaging may be used for ion recording.

Figure 4 illustrates a typical SIRIS spectrum which shows two isotopes of gallium ^{69}Ga and ^{71}Ga for a target of silicon doped to a 0.5 ppm level with gallium. The data agree well with the naturally occurring abundance ratios of these isotopes.

Further experiments to establish the present sensitivity limit of SIRIS undertaken with gallium doped silicon again at the 0.5 ppm level suggest a sensitivity limit at 2 ppb with a five minute counting time at each mass. Anticipated improvements of the order of a factor of twenty should enhance this to 1 part in 10^{10}. These data illustrate the power and versatility of SIRIS as a readily applicable analytical tool.

In this brief talk it has been possible only to sketch in outline the RIS principles: for an up-to-date appraisal of the method, reference should be made to the forthcoming article by Sam Hurst and Marvin Payne in Reports on Progress in Physics, published by The Institute of Physics in London.

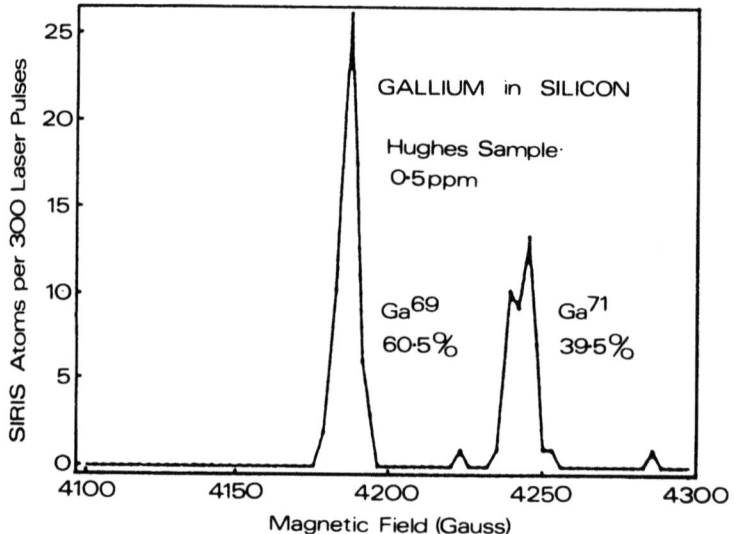

Fig. 4. SIRIS spectrum of gallium isotopes in silicon at a doping of 0.5 ppm.

ACKNOWLEDGMENTS

I wish to offer my best thanks to Dr. G. S. Hurst and his colleagues at Oak Ridge National Laboratory and to Dr. Harold W. Schmitt and Dr. J. E. Parks of Atom Sciences, for helpful discussions and permission to use their data and diagrams.

REFERENCES

References given as IOPCS71 (1984) appear in the Proceeding of the Second International Symposium on Resonance Ionization Spectroscopy and its Applications, Knoxville, 1984, edited by G. S. Hurst and M. G. Payne and published by the Institute of Physics, Bristol and Boston.

Chen, C. H., et al., 1980, Chem. Phys. Lett., 75:473.
Chen, C. H., 1984, IOPC71, p. 189.
Cowan, G. A., et al., 1984, IOPC71, p. 263.
Fairbank, W. M., Jr., 1984, IOPC71, p. 297.
Fasset, J. D., et al., 1983, in: "Analytical Chemistry Symposia," Vol. 19, edited by W. S. Lyon, Elsevier, New York.
Harrison, W. W., et al., 1984, IOPC71, p. 119.
Haxton, W. C., 1984, IOPC71, p. 269.
Hurst, G. S., 1984, IOPC71, p. 309.
Hurst, G. S., et al., 1975, Phys. Rev. Lett., 35:82.
Hurst, G. S., et al., 1977, Appl. Phys. Lett., 30:229.
Hurst, G. S., et al., 1979, Rev. Mod. Phys., 51:767.
Kramer, S. D., 1984, IOPC71, p. 205.
Kramer, S. D., et al., 1978, Optics Letters, 3:16.
Moe, M. K., 1984, IOPC71, p. 279.
Nogar, N. S., et al., 1983, in: "Analytical Chemistry Symposia," Vol. 19, edited by W. S. Lyon, Elsevier, New York.
Nogar, N. S., et al., 1984, IOPC71, p. 91.
Parks, J. E., et al., 1983, in: "Analytical Chemistry Symposia, Vol. 19, edited by W. S. Lyon, Elsevier, New York.
Parks, J. E. et al., 1984 IOPC71, p. 167.
Robinson, M. T., 1984, IOPC71, p. 151.
Travis, J. C., 1984, IOPC71, p. 97.
Winograd, N., et al., 1982, Chem. Phys. Lett. 88:581.
Winograd, N., 1984, IOPC71, p. 161.

APPENDIX A: POSTER PAPERS

Session A

A1	K. Akimoto	EM Radiation from Langmuir Turbulence
A2	C. Braun	Collisional-Radiation Coefficients from a Three-Level Atomic Model in Argon Plasmas
A3	M. Buchwald	Limiting Effects of Shock Waves on Spark Brightness
A4	J. Dakin	Radiation Transfer Calculations for the High-Pressure Sodium Lamp
A5	C. Deeney, C. Challis, P. Choi, A. Dangor, G. Nave	X-Ray Emission from a Gas Puff Z-Pinch
A6	T. Hammer	Transport of Precursor Radiation in Shock Waves
A7	A. Granier	Argon Surface Waves up to Atmospheric Pressure
A8	M. Hartney	Plasma Etching in Microlithography
A9	A. Kallenbach	Plasma Production by Resonant Laser Heating
A10	J. Kuno	Plasma Radiation Fundamentals
A11	R. Lagushenko	Isotope Effect in Resonance Radiation Transport
A12	G. Rogoff, A. Bellows, F. Feuersanger, H. Rothwell	High Pressure Lamp Arcs in Microgravity
A13	A. Policarpo, E. deLima, M. Salete, S. Leite, M. Alice Alves, R. Ferreira Marques	Photons and Self-Quenching Streamers
A14	N. Qi, H. Kilio, M. Krishnan	A Study of Resonant Photo-Excitation of CIII Ions in a Laser-Initiated Vacuum-Arc Discharge
A15	P. Reiser	Asymptotic Series Analysis of Self-Reversed Spectral Lines
A16	V. Roberts	Radiative Properties of Low-Pressure Hg/Ar Discharges in a Magnetic Field
A17	K. Schoenbach	Striations in High Pressure Diffuse Discharges
A18	M. Touzeau, A. DeSouxa, G. Gousset	Energy Transfer Between $N_2(A)$ Metastables and O or O_2

Session B

B1	F. Bayrakceken	LIF Lifetime Studies of Dibenzo (a,d) Cyclohepten 5-1
B3	W. Byszewski	Radiation From Transient Discharges
B4	J. Coutts, C. Webb, R. Hollins, D. Jordan	Time-Resolved Balmer Beta Measurements in High Pressure Discharges
B6	D. Erwin, M. Gundersen	Plasma Diagnostics Using LIF
B7	H. Gecim	Mixing of Nickel and Silicon by Incoherent Light Pulse Annealing
B8	D. Karabourniotis, S. Couris, J. Damelincourt	Determination of the Transition Probability of the 655-nm Line of Thallium
B9	Y. Lee	An Ionization Balance Model for Non-Equilibrium Plasmas
B10	K. McKnight, R. Stewart	Spatially Resolved Optogalvanic Studies of the Neon Positive Column
B11	E. McLean, J. Stamper, C. Manka, H. Griem, A. Ali, B. Ripin	Nitrogen Spectra in Interstreaming Plasmas
B12	M. Miller, A. Lesage	Ionic Line Stark Width Trends
B13	A. Oztarhan	Measurements of Low-Pressure Hydrogen Arc Discharge Properties
B14	R. Prasad, M. Krishman	Measurements of Temperature and Rotation in Magnetized Columns of Vacuum-Arc Discharges
B15	N. Sadeghi, J. Derouard	Laser Probing of E-Fields in NaK Plasma
B16	F. Williams	Laser-Induced Fluorescence
B17	M. von Dadelszen	UV-Sustained Glow-Discharge Opening Switch Experiments

APPENDIX B: ORGANIZING COMMITTEE, LECTURERS, AND PARTICIPANTS

ORGANIZING COMMITTEE

Lawrence H. Luessen
Naval Surface Weapons Center
Code F12
Dahlgren, VA   24448 USA

Dr. Arthur H. Guenther
Air Force Weapons Laboratory
AFWL/CA
Kirtland AFB, NM   87117 USA

Dr. Anthony Hyder, Jr.
Auburn University
202 Samford Hall
Auburn, AL   36849 USA

Professor Erich E. Kunhardt
Weber Research Institute
Polytechnic Institute
Rt 110
Farmingdale, NY   11735

Dr. John Waymouth
GTE Lighting Center
100 Endicott St.
Danvers, MA   01923 USA

Dr. Joseph M. Proud
GTE Laboratories
40 Sylvan Road
Waltham, MA   02254

Dr. Nicol J. Peacock
Culham Laboratory
Division B UKAEA
Abbingdon, Oxfordshire
OS 14 3DB England

Dr. Arthur V. Phelps
JILA
University of Colorado
Campus Box 440
Boulder, CO   80309 USA

Professor Timm H. Teich
FG Hochspannung
Federal Inst. of Technology
ETHZ
Physikstrasse 3
CH 8006 Zurich, Switzerland

LECTURERS

Dr. M. A. Cayless
Thorn EMI Lighting Ltd.
Research & Engineering Div.
Melton Road
Leicester LE4 7PD, England

Professor Robin Devonshire
Department of Chemistry
Sheffield University
Sheffield S3 7HF
England

Professor Carlos Ferreira
Centro de Electrodinamica de
  Universidade Tecnica de Lisboa
Instituto Superior
1096 Lisboa Codex, Portugal

Dr. Alan Garscadden
AF Wright Aeronautical Lab
POOC/3
Wright-Patterson AFB, OH 45433
USA

Dr. Richard A. Gottscho
Bell Labs
Murray Hill, NJ 07974 USA

Professor Hans Griem
Department of Physics
University of Maryland
College Park, MD 20740 USA

Mr. John Ingold
GE Lighting Business Group
Cleveland, OH 44112 USA

Dr. A. G. Jack
Nederlandse Philips Bedrijven
Lighting Div., EDW-5
Postbox 218
5600 MD Eindhoven
The Netherlands

Dr. Dimitros Karabourniotis
Department of Physics
University of Crete
Iraklion, Crete
Greece

Professor H. -J. Kunze
Inst. fur Experimentalphysik
Ruhr-Universitat
4630 Bochum
FR Germany

Professor James Lawler
University of Wisconsin
Department of Physics
Madison, WI 53706 USA

Professor E. Marode
Laboratoire de Physique des
  Décharges
Ecole Supérieure d'Electricite
Plateau du Moulon
91190 Gif-sur-Yvette, France

Professor Michel Moisan
Department de Physique
Universite de Montreal
CP 6128, Succ. A.
Montreal, H3C 3J7, Quebec
Canada

Professor Colyn Grey Morgan
University College of Swansea
Department of Physics
Singleton Park, Swansea
W. Glam, Wales

Dr. Nicol J. Peacock
Culham Laboratory
Division B UKAEA
Abbingdon, Oxfordshire
OS 14 3DB
England

Dr. P. van de Weijer
Philips Research Laboratories
P.O. Box 80000
5600 JA Eindhoven
The Netherlands

Dr. John Waymouth
GTE Lighting Center
100 Endicott St.
Danvers, MA 01923 USA

Dr. David Wharmby
Thorn EMI Lighting Ltd.
Research & Engineering Div.
Melton Road
Leicester LE4 7PD, England

PARTICIPANTS

Mr. Kazuhiro Akimoto
University. of Maryland
Dept. of Physics-Astronomy
College Park, MD 20742
USA

Professor G. Akovali
Dept. of Chemistry
Middle East Technical Univ.
Ankara
Turkey

Dr. M. Aubes
Centre de Physique Atomique
Universite Paul Sabatier
118 Route de Narbonne
31062 Toulouse Cedex
France

Dr. Fuat Bayrakceken
Physical Chem. Section, Chem. Dept.
O.D.T.U.
Ankara
Turkey

Mr. Christopher Braun
Univ. of So. California
Dept. of Elec. Eng.
SSC-409, MC-0484
Los Angeles, CA 90089-0484
USA

Professor H. Bruhns
Institut fur Angewandte Physik II
der Universitat Heidelberg
Albert-Uberlel-Str. 3/5
6900 Heidelberg I
W. Germany

Dr. Melvin I. Buchwald
Los Alamos Nat'l. Lab.
M.S. D-466
Group ESS-7
Los Alamos, NM 87545
USA

Dr. Wojciech W. Byszewski
GTE Laboratories, Inc.
40 Sylvan Road
Waltham, MA 02254
USA

Dr. Gaetan Chevalier
Ecole Polytechnique de Montreal
Campus de l'Univ. de Montreal
C.P. 6079 - Succ. A.
Montreal, Quebec H3C 3A7
Canada

Mr. James J. Childs, Jr.
Naval Surface Weapons Center
Code F12
Dahlgren, VA 22448
USA

Dr. L. Cifuentes
Thorn-EMI Lighting
Research & Eng. Lab.
Melton Road
Leicester LE4 7PD
England

Professor Euguene J. Clothiaux
Physics Department
Auburn University
Auburn, AL 36849
USA

Mr. J. T. Coutts
Clarendon Laboratory
Oxford Univ.
Parks Road
Oxford OX1 3PU
England

Dr. James T. Dakin
GE, Corporate R&D
P.O. Box 8
Schenectady, NY 12301
USA

Mr. Christopher Deeney
Imperial College
Plasma Physics Group
Blackett Laboratory, I.C.
Prince Consort Road
London SW7 2BZ
England

Dr. Robert N. DeWitt
Naval Surface Weapons Center
Code F-12
Dahlgren, VA 22448
USA

PARTICIPANTS (Cont.)

Mr. Daniel Erwin
Univ. of So. California
SSC-409
Dept. of Elect. Eng.
Los Angeles, CA 90089-0484
USA

Professor Kasra Etemadi
Electrical & Computer Eng.
SUNY at Buffalo
218 Bonner Hall
Amherst, NY 14260
USA

Dr. David B. Fenneman
Naval Surface Weapons Center
Code F12
Dahlgren, VA 22448
USA

Dr. H. Selcuk Gecim
Hacettepe University
Electrical Eng. Dept.
Beytepe - Ankara
Turkey

Dr. Myer Geller
Naval Ocean Systems Center
Code 843
Catalina Blvd.
San Diego, CA 92152
USA

Dr. Rudolf Germer
Fritz-Haber-Institute der
  Max-Planck-Gesellschaft
Faradayweg 4-6
1000 berlin 33/Dahlem
W. Germany

Mr. J. G. Gielen
Eindhoven Univ. of Tech.
Physics Department
P.O. Box 513
5600 MB Eindhoven
The Netherlands

Mrs. Agnes Cranier
Lab. de Phys. des Gas et Plasmas
Bat 212
Universite Paris-Sud Orsay
91405 Orsay, Cedex
France

Mr. Ulrich Greb
Dept. of Pure & Applied Physics
UMIST
P.O. Box 88
Sackville St.
Manchester M60 1QD
England

Dr. Martin Gundersen
Department of Electrical
  Engineering
MC 0484
Univ. of So. California
University Park
Los Angeles, CA 90089-0484
USA

Mr. Thomas Hammer
University of Hannover
Callinstrasse 38
D-3000 Hannover
West Germany

Mr. Mark A. Hartney
AT&T Bell Labs
1A-249
600 Mountain Ave.
Murray Hill, NJ 07974
USA

Dr. John Huennekens
Department of Physics
Lehigh University
Building 16
Bethlehem, PA 18015
USA

Mr. Arne Kallenbach
University of Hannover
Callinstrasse 38
D-3000 Hannover
West Germany

Professor Joseph A. Kunc
Dept. of Physics
Univ. of So. California
SHS-274
University Park
Los Angeles, CA 90089-1341
USA

Dr. Radomir Lagushenko
GTE Lighting Products
R&D Laboratory
100 Endicott St.
Danvers, MA 01923
USA

Dr. Walter P. Lapatovich
GTE Laboratories, Inc.
40 Sylvan Road
Waltham, MA 02254
USA

PARTICIPANTS (Cont.)

Dr. Yim T. Lee
Lawrence Livermore Nat'l Lab.
L-298
P.O. Box 808
Livermore, CA 94550
USA

Mr. Peter Leismann
Institute of Experimental Physics
Ruhr University
4630 Bochum
West Germany

Dr. M. Lerminiaux
Universite Paris Nord
Laboratoire de Physique des Lasers
Avenue J.B. Clement
93430 Villetaneus
France

Mr. Homero S. Maciel
Dept. of Eng. Science
Oxford University
Parks Road
Oxford - OX1 3PJ
England

Mr. Kenneth W. McKnight
Strathclyde University
Physics Dept.
John Anderson Building
107, Rottenrow
Glasgow, G4 ONG
Scotland

Mr. Edgar A. McLean
Naval Research Lab
Code 4732
Washington, DC 20375
USA

Dr. Myron H. Miller
State of Maryland & USNA
1133 Hampton Road, Rt. 4
Annapolis, MD 21401
USA

Mr. C. Nieswand
Physical Institute 2
Univ. of Duesseldorf
4000 Dusseldorf
Universitatsstrasse 1
West Germany

Dr. Ahmet Oztarhan
Dokuz Eylw Universites
Muhendislik Mimarlik Fakultesi
Makina Bollumu
Bornova - Izmir
Turkey

Dr. Leanne C. Pitchford
GTE Laboratories, Inc.
40 Sylvan Road
Waltham, MA 02254
USA

Prof. A. J. Policarpo
Physics Dept.
Universidade de Coimbra
3000 Coimbra
Portugal

Mr. Rahul R. Prasad
Yale Univ.
Mason Lab.
Dept. of Appl. Phys.
P.O. Box 2159
Yale Station
New Haven, CT 06520
USA

Mr. Niansheng Qi
Yale Univ.
Dept. of Physics
New Haven, CT 06520
USA

Mr. Paul Reiser
GTE Lighting Products
R&D Laboratory
100 Endicott St.
Danvers, MA 01923
USA

Mr. Christian Reventlow
Institute of Experimental Physics
Ruhr University
4630 Bochum
West Germany

Dr. Victor Roberts
GE Research & Development Ctr.
Bldg. K1 - Rm. 4C15
P.O. Box 8
Schenectady, NY 12301
USA

Dr. Michael Romheld
Siemens AG
ZFE TPH 34
Postfach 3240
8520 Erlangen
West Germany

Dr. N. Sadeghi
Laboratoire de Spectrometrie
   Physique BP 87
38402 St. Martin D'Heres Cedex
France

PARTICIPANTS (Cont.)

Professor Karl H. Schoenbach
Texax Tech Univ.
Depart. of Elect. Eng.
P.O. Box 4439
Lubbock, TX 79409
USA

Dr. R. S. Stewart
Strathclyde University
Physics Dept.
John Anderson Building
107, Rottenrow
Glasgow, G4 ONG
Scotland

Mr. M. Ribau Teixeira
Lab, Nac, Tecn. End.
Estrada Nacional 10
2685 Sacavem
Portugal

Dr. Michel Touzeau
Laboratoire de Physique des Gas
  et des Plasmas
Batiment 212
Universite Paris-Sud
91405 Orsay Cedex
France

Dr. E. M. Van Veldhuizen
Eindhoven Univ. of Tech.
Electrical Department
P.O. Box 513
5600 MB Eindhoven
The Netherlands

Dr. Michael von Dadelszen
Tetra Corp.
4905 Hawkins St, N.E.
Albuquerque, N.M. 87109
USA

Mr. M. Von Salisch
Physical Institute 2
Univ. of Duesseldorf
4000 Dusseldorf
Universitatsstrasse 1
West Germany

Dr. Bernhard Weber
Philips GmbH
Forschungslaboratorium Aachen
Post fach 1980
Weisshausstrasse
5100 Aachen
West Germany

Dr. Norbert F. Weigart
Brown Boveri Research Center
CH-5405 Baden-Dattwil
Switzerland

Professor Frazer Williams
Dept. of Elect. Eng.
Univ. of Nebraska
Lincoln, NE 68588-0511
USA

INDEX

Abel inversion, 56, 194, 288, 297,
Absorption coefficient, 160-162,
    177-180, 220, 283, 333,
    389
  effective, 14, 22
Absorption measurements, 481-491
  tunable laser diode (TLD)
      spectroscopy, 483-487
  frequency modulated absorption
      spectroscopy, 487
  laser source hook
      interferometry, 488-491
  TRISP, 487-488
Absorption oscillator strength,
    14, 177, 185
Active zone, 387, 399
Airy diffraction pattern, 365
Ambipolar diffusion, 331, 333,
    386, 389, 395
  coefficient, 355
  time constant, 313
Amplified spontaneous emission
    (ASE), 88
Arc temperature measurement,
    341-472
Arc tube
  HID lamps, 281, 282
Attenuation coefficient
  for traveling-wave discharge,
    400, 401
Auto-ionization (see Ionization)
Automatic collinear reflection,
    374
Avalanches (streamers), 120
Axial convection, 359

Bartels' method of temperature
    measurement, 341
Bartels' theory of radiation,
    200-219
Biberman-Holstein equations
  spectral forms of, 260
Black-body
  radiation law, 279
  spectral radiation intensity,
    269
Bohm-Gross dispersion relation, 44

Bohr's condition, 8
Boltzmann distribution
  over bound and free states, 159
Boltzmann equation, 173, 496
Boltzmann factors, 15

Cathode fall
  mapping the, 539-544
Charged particle balance
  condition for, 389
  in a steady-state discharge, 388
Coherent Anti-Stokes Raman
    Scattering (CARS), 474-478
Collision broadening, 209-210
Collision frequency
  for electron-neutrals, 114
  for momentum transfer, 369,
    383-386
Collisional ionization, 376-377
  rates, 18-19
Collisional processes, 15-21
  excitation and de-excitation,
    16-18
  ionization and three-body
    recombination, 18-19
  radiative and dielectronic
    recombination, 19-21
Continuity equation, 388
  electron concentration, 370
Continuous optical discharges
    (COD), 370-371
Continuous plasmas, 375-377
Continuum radiation, 12-13, 63
Cowan and Dieke's Theory, 219-244

De-excitation
  and excitation (see Collisional
      processes)
Dielectronic recombination (see
    Collisional processes)
Diffusion coefficient
  free electron and ambipolar,
    386, 388
Discharge
  surface-wave sustained, 455-464
Doppler broadening, 180
Doppler profile, 186

Doppler free spectroscopy
  two-photon, 479
Doppler shifted frequency, 41
Doppler shifts, 10, (see also Line broadening)
Doppler width, 10, 536-537
Drawin criterion, 174, 176, (see also LTE)
Dynamic form factor, 43

Effective maintenance field
  in a gaseous discharge, 398
  intensity, 395
Effective thermal conduction flux, 338
Efficiency of ionization, 384
Einstein coefficients, 8, 162, 177
Electric field measurements, 512-516, 526,
  using optogalvanic detection of Rydberg atoms, 526-534
Electron density
  in a steady-state discharge, 389-394
  in surface-wave-produced discharges, 401-412
  measurements with interferometers, 468-471
Electron energy distribution function (EEDF), 394-395, 417-418
  defined, 384
Electron plasma frequency, 468
Electron mean-free-path, 395
Elenbaas-Heller equation, 285, 296
Emission coefficient, 177-180
  continuum, 13
  line, 8, 177
Emission spectra, 162-163
Emission spectroscopy, 55, 483
Energy balance 23, 329-333
  calculations, 337-340
  equation for radial temperature profile, 356
  in a steady-state discharge, 389-394
Energy distributions, 516-518
Equation of radiative transfer, 190
Escape factor, 252, 273 (see also Imprisonment factor)
Excitation and de-excitation (see Collisional processes)
Excitation function, 223
Extinction coefficient (opacity), 253
Extreme wing broadening (see Line broadening)

Far wing broadening (see Line broadening)
Fermi's golden rule, 16
Filamentary discharge, 104-117

Fluorescent lamps, 65-68, 303-304, 309-314
  compact, 314-319
Fluorescent powder 313-314
Flux density, 264-268
Forbidden transitions
  in line emission, 9
Four-wave mixing, 474 (see also CARS)
  degenerate, 374
Franck-Condon principle, 149-152
  classical Franck-Condon Principle (CFCP), 149, 151
Frequency-modulated absorption spectroscopy, 487

Gaunt factors, 13-14, 17-18
Glows
  transient behavior of, 104-117

High frequency (HF) discharges, 381-382
  between parallel plates, 450-455
  diffusion-controlled, 441-450
High intensity discharge (HID) lamps, 281-299
  high-pressure-mercury, 288-299
  high-pressure-sodium (HPS), 297-299
  metal halide (MH), 290-297
High-pressure-sodium (HPS) arcs, 327-344
  arc operation, 328-329
  energy balance, 329
Holtsmark theory, 12
Holtzmark field, 126, 561-562

Impact approximation, 180-181, 185
Impact broadening, 180
Imprisonment factor, 262
Induced emission, 14, 22
Inhomogenity parameter, 224-227, 235, 241
Interferometry, 468-471
  hook, 488-491
Inverse bremsstrahlung absorption, 369-370
Ion detection, 491-493
Ionization
  auto-, 20-21
Ionization frequency
  in cascade ionization, 369
Ionization rate maps
  of cathode fall, 540
Isophotes, 365-366

Kinetic measurements, 518-520
Kirchoff's law (strict form), 269-270
Klein-Rosseland principle, 174

Laser-induced breakdown spectroscopy (LIBS), 371

590

Laser-induced fluorescence (LIF), 46, 497-500
  for measurement of neutral species, 471-473
Laser-produced plasmas, 363-377
Laser source hook interferometry, 488-491
Lindholm-Foley collision theory of line broadening, 181
Line broadening, 9-11, 164-167, 180-190, 236
  atom-impact, 183
  electron-impact, 181
  extreme wing, 167, 336
  far wing, 167
  radial variation of, 236-240
Line emission, 8-9
Line intensities, 61-63
Line profile, 9-12, 57-61, 180-190
  formation of the, 172-190
  function, 333
  Lorentzian, 186, 223
  position-independent, 226-228
  position-dependent, 228-231
Line radiation, 260-263
Local thermodynamic equilibrium (LTE), 29-30, 96, 173-177, 354-356
Long-lived states, 415-419
Low-pressure discharges, 348
  diagnostics, 557
  mercury-rare-gas discharge, 309-311
  sodium lamp, 321-323
  sodium-rare-gas discharge, 319-321

Maintenance field intensity, 390-391
Metal halide (MH) lamps, 290-297
Metal halide sources, 347-362
  characterization, 349-353
  physical principles, 353-360
Molecular intensities, 156-163
Molecular spectra
  intensities, 148-152
Molecular transition probabilities, 164
Multiphoton absorption, 367-368
Multiphoton transitions, 50-52

Negative ions, 506-508
  measurements, 491-492
Net emission coefficient, 270-273, 339
Neutral intermediates, 509-512
Non-equilibrium
  features of, 96-104

Optical depth, 14
Optical thickness, 202-203, 207,
Optically thin plasmas
  inhomogeneous, 56

Optogalvanic effect (OGE), 79-91
  in the cathode fall, 525
Optogalvanic spectroscopy (OGS), 500

Phase conjugation, 372-375
Phase modulation interferometer (PMI), 469-470
Planck function 269-270, 350
Plasma column
  inhomogeneous, 202-203
Plasma-induced emission (PIE), 500-502
Poisson's equation, 496
Positive column
  modeling of, 310-311, 321
Power absorption coefficient, 388, 398
Power balance, 311-314
Poynting's theorem, 364

Quasistatic approximation
  of line profile, 183-185

Radial demixing, 357, 359
Radiation transfer, 311
Radiation transport equation, 250-256
  in LTE, 333
Radiative processes, 8-15, 333-337
Radiative transfer, 190-194
Radiative transport, 13-15
Radiation transport equation, 250-256
Raman scattering, 48-50, (see also CARS)
Rayleigh scattering, 48-50
Recombination, 13
  by auto-ionization, 20-21
Resonance broadening, 185-186,
  (see also Line broadening)
Resonance ionization mass spectroscopy (RIMS), 572-573
Resonance ionization spectroscopy (RIS), 569-572
Resonance radiation
  lamps, 299-307
Resonantly produced laser plasmas
  continuous plasmas, 375-377
RF discharges
  laser diagnostics of, 495
RF and microwave discharges
  charged particle balance, 388-389
  classification of, 397-398
  energy balance, 389-394
  physical characteristics, 382-399
Rydberg atoms, 526-535
  detection and resolution, 553-557
  properties of, 547-553

Rydberg states
  properties of, 547–553
Rydberg levels, 547

Saha equation, 286
Scattering
  by plasma electrons, 40–45
  differential cross section for, 40
  incoherent, 42, 44
  resonant, by atoms and ions, 45–48
Self-absorption
  theory of line, 219–244
Self-reversal, 171
Similarity of discharges, 394–396
Single-color emitters, 291
Source function, 14–15, 268–270
Spectral exitance
  of blackbodies, 279
Spectral intensity, 333
Spectral line contour, 190–200, 214–215
Spectral radiance, 283–286
Spectral line radiation intensity, 250–252
Stark effect, 478–479
  in Rydberg atoms, 526–527
Steady-state discharges, 383
  physical processes in, 383–386
  charged-particle balance in, 388–389
  energy balance in, 389–394
Surface waves, 381–382
Surface-wave concept, 432
  discharges, 381–382, 399–401
  launchers, 403
  sustained plasmas, 401–409
    applications, 419–428
    radiative states, 412–415
    long-lived states, 415–519
  LMP applicator, 426–427
Surfatron, 419–422, 426–427

Temperature measurements, 536–539
Temperature profile
  of line contour, 224–225
  radial, 285
Thomson scattering
  by plasma electrons, 40–41
Three-body recombination, 18–19
Time-resolved infrared absorption spectroscopy (TRISP), 487–488
Time reversal
  in laser-induced plasmas, 373–375
Transient emission features
  in discharges, 117–143
Transitions
  multi-photon, 50–52
Two-photon absorption, 50
  fluorescence excited by, 50–52

Tunable laser diode spectrometry (TLDS), 483–487

Velocity modulated IR laser spectroscopy, 491
Vibrational excitation effects
  in discharges, 130–135
Voigt profile, 189–190

Weisskopf radius, 181, 184
White-light emitters, 291–293
Wien's function, 193, 203, 208

MIX
Papier aus verantwortungsvollen Quellen
Paper from responsible sources
FSC® C105338

If you have any concerns about our products,
you can contact us on
ProductSafety@springernature.com

In case Publisher is established outside the EU,
the EU authorized representative is:
**Springer Nature Customer Service Center GmbH
Europaplatz 3, 69115 Heidelberg, Germany**

Printed by Libri Plureos GmbH
in Hamburg, Germany